Ecological Responses to the 1980 Eruption of Mount St. Helens

Virginia H. Dale
Frederick J. Swanson
Charles M. Crisafulli
Editors

Ecological Responses to the 1980 Eruption of Mount St. Helens

With 115 Illustrations

With a Foreword by Jerry F. Franklin

Virginia H. Dale
Environmental Sciences Division
Oak Ridge National Laboratory
Oak Ridge, TN 37831
USA
dalevh@ornl.gov

Frederick J. Swanson
USDA Forest Service
PNW Research Station
Forestry Sciences Laboratory
Corvallis, OR 97331
USA
fswanson@fs.fed.us

Charles M. Crisafulli
USDA Forest Service
PNW Research Station
Olympia Forestry Sciences Laboratory
Mount St. Helens NVM
Amboy, WA 98601
USA
ccrisafulli@fs.fed.us

Library of Congress Cataloging-in-Publication Data

Ecological responses to the 1980 eruption of Mount St. Helens / Virginia H. Dale, Frederick J. Swanson, Charles M. Crisafulli, [editors].
p. cm.
Includes bibliographical references (p.).
ISBN 0-387-23868-9
1. Ecological succession—Washington (State)—Saint Helens, Mount. 2. Saint Helens, Mount (Wash.)—Eruption, 1980—Environmental aspects. I. Dale, Virginia H. II. Swanson, Frederick J. (Frederick John), 1943- III. Crisafulli, Charles M.

QH105.W2E265 2005
577.3′18′0979784—dc22 2004061449

ISBN 10: 0-387-23868-9 (hardcover) 10: 0-387-23850-6 (softcover) Printed on acid-free paper.
ISBN 13: 978-0387-23868-5 (hardcover) 13: 978-0387-23850-0 (softcover)

Printed in the United States of America. (TB/MVY)

9 8 7 6 5 4 3 2 1 SPIN 10941393 (hardcover) 10941409 (softcover)

springeronline.com

Foreword

Reconfiguring Disturbance, Succession, and Forest Management: The Science of Mount St. Helens

When Mount St. Helens erupted on May 18, 1980, it did more than just reconfigure a large piece of Cascadian landscape. It also led to dramatic revisions in our perspectives on disturbances, secondary succession, and forestry practices. The Mount St. Helens landscape turned out to be a far more complex place than the "moonscape" that it initially appeared to be. Granted, a large area was literally scoured and sterilized, and that vast expanse of newly formed rock, mudflows, and avalanche debris up and down the mountain made the Mount St. Helens landscape unique. But I still remember my surprise when, as I stepped out of the helicopter on first landing within the extensive "devastated zone," I saw hundreds of plants pushing their way up through the mantel of tephra.

Surviving organisms were stunning in their diversity, abundance, and the mechanisms by which they survived. They persisted as whole organisms living below ground, encased within late-persisting snowbanks, and buried in lake and stream sediments. They survived as rhizomes transported along with the massive landslide that accompanied the eruption and as stems that suffered the abrasion of mudflows. Mudflows floated nurse logs covered with tree seedlings and then redeposited them on the floor of a forested river terrace. Millions, perhaps billions, of plants survived as rootstocks and rhizomes that pushed their way up through the tephra, and others survived on the bases of uprooted trees.

Wood (snags, entire boles, and other woody debris) was evident many places in the devastated zone, beginning the fulfillment of its incredibly diverse functional role as habitat, protective cover, sediment trap, energy and nutrient source, and debris jam, among other aspects.

The multiplicity of disturbances, both individual and combined, resulted in an immense diversity of posteruptive conditions and potential study sites, including both terrestrial and aquatic sites ranging from minimally impacted to heavily disturbed. Almost every locale within the devastated zone was affected by two or three contrasting disturbances. Some impacts extended far beyond the devastated zone, particularly the aerial deposits of volcanic materials (tephra) that extended over hundreds of thousands of square kilometers.

The concept of *biological legacies* emerged from this very diverse collection of surviving biota and organically derived structures. Biological legacies encompass the array of organisms and organically derived structures and patterns that persisted from the predisturbance landscape to populate and influence recovery. Great insight was unnecessary: the noses of the ecologists investigating the devastated zone were daily rubbed in the concept of legacies and their importance.

Mount St. Helens was the catalyst for the biological legacy concept and many other important additions and modifications to our basic understanding of ecological recovery following disturbance, or "secondary succession," as old hands might have described it. Why should this have been so when the concepts are so obvious in retrospect? Perhaps it is because of the magnitude of the event, which led to the early and very incorrect perception of a moonscape at all locations. Or, it may be because many early studies of secondary succession focused on old fields with their limited legacies of organisms and near absence of structural legacies. Handy as the old fields may have been to early centers of ecological science, they were not representative of conditions and processes following natural disturbances.

Whatever the explanation, the importance of biological legacies, which include the "residuals" of Frederic Clements, is evident to disturbance ecologists everywhere, whether their focus is fundamental or applied. Indeed, *one might propose that the most important variables influencing postdisturbance recovery processes are the types and levels of biological legacies that are present*, not the type, intensity, size, or any other attribute of the disturbance. The legacies are at least among the most important predictors of what will happen during the initial recovery process.

The concept of biological legacies has had a similar effect on applied ecology, specifically forestry practices or silviculture. The dominant forest-harvest technique of clear-cutting proves to have little similarity to such natural disturbances as fire, wind, and even volcanic eruption. Clear-cutting leaves little in the way of biological legacies. Natural disturbances provide the appropriate models for silviculture, where maintenance of biological diversity and natural ecosystem functions is a primary or collateral goal of forest management, such as on U.S. federal timberlands. From research at Mount St. Helens has emerged the concept of variable-retention harvesting, a silvicultural system in which varying types, amounts, and patterns of living trees, snags, and logs as well as small forest islands are left to "lifeboat organisms" and structurally enrich the regenerating forest.

Mount St. Helens is providing us with other important conceptual or "big-picture" perspectives on large disturbances, such as patterns of recolonization and the role of large, slowly regenerating disturbed areas. Authors in this volume describe many spatial aspects of the recolonization process, including "hotspots" or focal points of community recovery that resulted from both survivors and from particularly favorable environmental conditions for organisms to establish, such as margins of ponds and wetlands. They also talk about "coldspots," such as sites where posteruption erosional deposits buried surviving plants and slowed recolonization.

Recolonization of the Mount St. Helens devastated zone has emerged from multiple centers and not primarily by incremental advances of invading organisms from the margins, as many predicted. The recovery process began with innumerable hotspots of surviving organisms, some as small as individuals and others as large patches of intact communities, such as those surviving in snowbeds. I have applied the term "metastasizing" to describe the recovery process within the devastated zone; according to Webster, metastasis is a pathological term, but it does convey the sense of multiple centers of colonization ("infection") that grow, spread, and eventually converge. This pattern of colonization provides an alternative to the "wave-front" model often favored in successional studies.

Mount St. Helens is also informing us about the important role that large, slowly reforesting disturbed areas may contribute to the maintenance of regional biodiversity. Large and diverse populations of major faunistic groups (such as songbirds, amphibians, and mesopredators) characterize the naturally developing (and unplanted) portions of the devastated zone. Who would have predicted western meadowlarks colonizing an area on the western slopes of the Cascade Range? Hardwood trees and shrubs, not conifers, dominate significant portions of this landscape.

We can expect that this structurally and environmentally diverse landscape is going to dominate the unsalvaged, unplanted portions of the Mount St. Helens devastated zone for many decades to come. The reestablishment of extensive tracts of closed coniferous forest so characteristic of this region is going to be a long time in coming!

The biological richness of the Mount St. Helens landscape has direct relevance to debates that are currently emerging regarding appropriate restoration policies following major wildfires and other disturbances. Timber salvage and rapid reforestation with conifers have been accepted policies that are aggressively pursued. As a consequence, naturally disturbed, unsalvaged, and unplanted early-successional habitat is the scarcest of the natural forestland states in the Pacific Northwest, much rarer than that of old-growth forest.

Research at Mount St. Helens is demonstrating the potential value of such naturally regenerating, early-successional landscapes as regional hotspots of biodiversity as well as demonstrating that such habitat is not to be confused with planted clear-cuts or even salvaged burns.

In this book, we have the first significant summary of this paradigm-busting science based on 25 years of research on one of the largest and most complex disturbance events accessible for intense study. The chapters provide both the details and idiosyncrasies of individual organisms as well as broad general lessons regarding the physical and biological processes associated with recovery.

The chapters of this book show the relevance of the Mount St. Helens eruption to fundamental ecological theory and not the "special case" that some scientific critiques once suggested. Ecologists should study it carefully because general ecological theory of disturbances and recovery processes must encompass the lessons from Mount St. Helens.

Similarly, applied ecologists (the foresters, wildlife managers, fisheries biologists, and other resource managers) will find much that they can incorporate into management regimes that are more closely based on natural disturbances and provide better for maintenance of biodiversity and ecosystem processes.

Finally, stakeholders and policy makers will find much in the Mount St. Helens science that should cause them to reflect on the role of natural, early-successional habitat as a part of our regional forest landscapes. When one reassesses resource management in the 20th century, the commodity-based perspective of "timber salvage and reforest" should be a major part of that reflection, and we can hope that it will be informed by a continuing flow of knowledge from the Mount St. Helens landscape.

This extraordinary volume provides an opening to the future. We owe major thanks to the hundreds of scientists, students, and technicians who have participated in this research and to the authors of these chapters for providing us with this stimulating synthesis. But the opportunities for further study are infinite and important. We hope that, among you readers, will be some who will assume the challenge of carrying the research on Mount St. Helens forward for the next 25 years.

JERRY F. FRANKLIN

Preface

The May 18, 1980 eruption of Mount St. Helens abruptly altered the geological and ecological systems of southwestern Washington State. The eruption was so well documented by the media that it was viewed around the world and it changed people's perception of volcanoes. The eruption created new landscapes that were subsequently studied by dozens of ecologists. This book integrates and analyzes much of the information learned from those studies and adds recent insights and findings by the contributors and their colleagues.

Many of the authors of this book have been studying ecological responses to the 1980 eruption since the early days. Several of us were on the first team of ecologists to enter the volcanic disturbance zones shortly after May 18. We were awed at the dramatic changes to the landscape and have returned for field studies in subsequent years. Others have joined the team over the ensuing years, and the loose-knit research group has met as a whole several times. Researchers working on the ecological recovery at Mount St. Helens gathered during the summer of 2000 when the USDA Forest Service's Pacific Northwest Research Station sponsored a week-long field camp, termed a "pulse." They visited each other's field sites and collected data on the 20-year status of ecosystems. The idea for this volume grew out of that pulse.

Over time, the physical and biological environment at Mount St. Helens has changed dramatically, yet the compelling character of the landscape remains. The eruption destroyed and buried much of the system of logging roads that had laced the landscape outside the remote, foot-access-only areas of Mount St. Helens and the Mount Margaret backcountry to the north. Thus, access was extremely limited in the first months and even years. Helicopters proved essential for many studies. As salvage logging proceeded outside the designated National Volcanic Monument and visitor access developed from 1981 to 1986, some of the preeruption road system was reestablished, and new roads were constructed, providing access to areas peripheral to the core of the volcanically disturbed area. With completion of salvage logging and closure of many roads by design and storm damage, access again became restricted in many areas. Yet scientists continued to return to find a fascinating, changing landscape.

Funding for ecological studies at Mount St. Helens has had a varied history. The Forest Service and National Science Foundation funded initial access and two 2-week-long field pulses in the summers of 1980 and 1981, which greatly facilitated cross-disciplinary interactions. Several National Science Foundation grants and Forest Service funding supported a series of studies from the 1980s to the present. Individual projects were funded by small grants from the National Geographic Society, Earthwatch, Washington Department of Fish and Wildlife, and several foundations. A great deal of work has been accomplished by personal initiative and by building upon related projects. The Forest Service has provided continuous support for work by Crisafulli, Swanson, and others at Mount St. Helens and for collecting, documenting, and archiving datasets from long-term ecological studies in the area.

This book is the direct result of the contributions of many people in addition to the authors. Frederick O'Hara did an excellent job as technical editor for the book. A special thanks is owed to the numerous scientists who reviewed drafts of the chapters. For this important work, we wish to thank Steve Acker, Wendy M. Adams, Joe Ammirati, Matt Ayers, Lee Benda, Edmund Brodie, Tom Christ, Warren Cohen, Kermit Cromack, Dan Druckenbrod, John S. Edwards, Roland Emetaz, Jerry F. Franklin, Scott Gende, Peter Groffman, Charlie Halpern, Miles Hemstrom, Jan Henderson, Sherri Johnson, R. Kaufmann, Jon Lichter, James A. MacMahon, Jon J. Major, Frank Messina, Randy Molina, Aaron Peacock, Daniel Schindler, Dave Skelly, Don Swanson, Lars Walker, Peter White, Amy Wolfe, Jingle Wu, and Wayne Wurtzbaugh. Theresa Valentine and Kathryn Ronnenberg (USDA Forest Service, Pacific Northwest Research Station) helped greatly with the preparation of maps and figures. Suzanne Remillard (USDA Forest Service, Pacific Northwest Research Station) assisted with information management. Jordon Smith assisted with editorial and compilation tasks. We also thank many colleagues at the U.S. Geological Survey, Cascades Volcano Observatory, for providing information and interpreting the events that occurred during the 1980 and other eruptions, particularly Jon J. Major, Dan Miller, Don Swanson, Richard Waitt, and Ed Wolf.

The editors' institutional homes provided essential support for their work at Mount St. Helens, including the writing and editing this book. Charlie and Fred gratefully acknowledge support of the Pacific Northwest Research Station and especially John Laurence, Peter A. Bisson, Tami Lowry, and Debby McKee. Virginia appreciates the support from the Environmental Sciences Division at Oak Ridge National Laboratory and specifically Linda Armstrong and Anne Wallace. The editors thank the Gifford Pinchot National Forest and Mount St. Helens National Volcanic Monument and their staffs for logistic support and access to records, maps, and research sites.

On a personal note, during the past 24 years we have spent much time in the volcanic landscape learning a great deal about disturbance ecology and Cascadian natural history and becoming quite familiar with the area. Perhaps most important have been the friends, colleagues, and family members with whom we have interacted and shared this fascinating landscape. Virginia especially thanks her family, who enjoyed assisting in the fieldwork and relinquished weekends and early mornings of her time. Fred gratefully acknowledges his family's tolerance of his Mount St. Helens fixation and the support of David Foster for the opportunity to work on the book while in residence at Harvard Forest. Charlie thanks James A. MacMahon, mentor and friend, for introducing him to Mount St. Helens and Charles P. Hawkins, Robert R. Parmenter, and Michael F. Allen for years of collaboration. Charlie thanks Hans Purdom, Josh Kling, Eric Lund, Aimee McIntyre, and Louise S. Trippe for their unwavering interest and collaboration at the volcano. Finally, Charlie thanks his daughters Erica and Teal Crisafulli, for their youthful wonder, and his parents, Helen and Carmelo Crisafulli, for tolerating his childish habitats of catching frogs and salamanders into adulthood. Collectively, the editors and authors owe special gratitude to Jerry F. Franklin, James A. MacMahon, and Jim Sedell for their personal commitments to science at Mount St. Helens and their colleagues who work there.

After 18 years of quiescence, Mount St. Helens broke her silence and entered an eruptive state on September 23, 2004. As we go to press, the volcano has been erupting for 18 continuous weeks; primarily building a new dome in the 1980 crater. Numerous small tephra falls have also been deposited near the mountain, and a few small mudflows have emanated from the crater and traveled down streams. Although it is not known how long this current eruption will last or if it will increase its activity, it is a testimony to the dynamic nature of Mount St. Helens.

As we reach the quarter-century anniversary of the major eruption, it is also timely for scientists who worked in the first posteruption period to begin passing the science baton to the next generation of scientists who will work at Mount St. Helens. This book describes observations, interpretations, and speculations from the first 25 years of ecosystem response and complements our efforts to leave well-documented, publicly accessible descriptions of long-term field plots and associated data. We hope to continue our research for years into the future but recognize the need and appreciate the opportunity to collect our thoughts and data at this juncture. Our greatest hope is that ecologists will continue to study and learn from the fascinating and complex interaction between organisms and their environment at Mount St. Helens.

VIRGINIA H. DALE
FREDERICK J. SWANSON
CHARLES M. CRISAFULLI
February 2005

Contents

Part V Lessons Learned

Note: A Web site has been established at http://www.fsl.orst.edu/msh/ containing background details (pictures, data details, graphs, etc.) to supplement the information included in this book.

Contributors

Wendy M. Adams
University of Michigan, School of Natural Resources and the Environment, Ann Arbor, MI 48104, USA. wmadams@umich.edu

Michael F. Allen
Center for Conservation Biology, University of California at Riverside, Riverside, CA 92521-0334, USA. michael.allen@ucr.edu

Joseph A. Antos
Department of Biology, University of Victoria, Victoria, BC V8W 3N5, Canada. jantos@uvic.ca

John G. Bishop
School of Biological Sciences, Washington State University at Vancouver, Vancouver, WA 98668, USA. bishop@vancouver.wsu.edu

Peter A. Bisson
USDA Forest Service, PNW Research Station, Olympia Forestry Sciences Laboratory, Olympia, WA 98512-9193, USA. pbisson@fs.fed.us

Daniel R. Campbell
North Coast-Cascades Network Exotic Plant Management Team, Olympic National Park, Port Angeles, WA 98362, USA. dan_campbell@nps.gov

Charles M. Crisafulli
USDA Forest Service, PNW Research Station, Olympia Forestry Sciences Laboratory, Mount St. Helens NVM, Amboy, WA 98601. ccrisafulli@fs.fed.us

Clifford N. Dahm
Department of Biology, University of New Mexico, Albuquerque, NM 87131, USA. cdahm@sevilleta.unm.edu

Virginia I. Dains
3371 Ayres Holmes Road, Auburn, CA 95603, USA. geobot@jps.net

Virginia H. Dale
Environmental Sciences Division, Oak Ridge National Laboratory, Oak Ridge, TN 37831, USA. dalevh@ornl.gov

Roger del Moral
Department of Botany, University of Washington, Seattle, WA 98195, USA. moral@u.washington.edu

John S. Edwards
Department of Zoology, University of Washington, Seattle, WA 98195, USA. hardsnow@u.washington.edu

Louise M. Egerton-Warburton
Chicago Botanic Garden, Glencoe, IL 60022, USA. warburton@chicagobotanic.org

William F. Fagan
Department of Biology, University of Maryland, College Park, MD 20742-4415, bfagan@glue.umd.edu

Jerry F. Franklin
College of Forest Resources, University of Washington, Seattle, WA 98195, USA. jff@u.washington.edu

Brian R. Fransen
Weyerhaeuser Company, Tacoma, WA 98063-9777, USA. brian.fransen@weyerhaeuser.com

Peter M. Frenzen
USDA Forest Service, Mount St. Helens NVM, Amboy, WA 98601, USA. pfrenzen@fs.fed.us

Keith S. Hadley
Department of Geography, Portland State University, Portland, OR 97207-0751, USA. hadleyk@geog.pdx.edu

Jonathan J. Halvorson
USDA Agricultural Research Service, Appalachian Farming Systems Research Center, Beaver, WV 25813-9423, USA. JHalvorson@afsrc.ars.usda.gov

Jasper H. Hardison III
PND Incorporated, Consulting Engineers, Anchorage, AK 99503, USA. j-hardison@pnd-anc.com

Charles P. Hawkins
Department of Aquatic, Watershed, and Earth Resources, Utah State University, Logan, UT 84322-5200, USA. hawkins@cc.usu.edu

Robert F. Holland
3371 Ayres Holmes Road, Auburn, CA 95603, USA. geobot@jps.net

Ann C. Kennedy
Washington State University, Pullman, WA 99164-6421, USA. akennedy@wsu.edu

Nicole C. Korbe
EDAW, Inc., Denver, CO 80202, USA. korben@edaw.com

Melissa J. Kreutzian
U.S. Fish and Wildlife Service, Albuquerque, NM 87113, USA. melissa_kreutzian@fws.gov

Douglas W. Larson
Department of Biology, Portland State University, Portland, OR 97207-0751, USA. franksooth@netzero.net

Rick Lawrence
Land Resources & Environmental Sciences Department, Montana State University, Bozeman, MT 59717-3490, USA. rickl@montana.edu

Robert E. Lucas
Fisheries Management Division, Washington Department of Wildlife, Vancouver, WA 98661, USA. blucas@scattercreek.com

James A. MacMahon
Department of Biology, Utah State University, Logan, UT 84322, USA. jam@cc.usu.edu

Jon J. Major
U.S. Geological Survey, Cascades Volcano Observatory, Vancouver, WA 98683-9589, USA. jjmajor@usgs.gov

Sherri J. Morris
Department of Biology, Bradley University, Peoria, IL 61625, USA. sjmorris@bradley.edu

Robert R. Parmenter
Valles Caldera National Preserve, Los Alamos, NM 87544, USA. bparmenter @vallescaldera.gov

Gary L. Parsons
Department of Entomology, Michigan State University, East Lansing, MI 48824, USA. parsonsg@msu.edu

Richard R. Petersen
Department of Biology, Portland State University, Portland, OR 97207-0751, USA. petersonr@pdx.edu

John D. Schade
Department of Biology, St. Olaf College, Northfield, MN 55057, USA. jschade@berkeley.edu

Jeffrey L. Smith
USDA Agricultural Research Service, Washington State University, Pullman, WA 99164-6421, USA. jlsmith@mail.wsu.edu

Sharon M. Stanton
Department of Geography, Portland State University, Portland, OR 97207-0751, USA. sstanton@pdx.edu

Patrick M. Sugg
Department of Zoology, University of Washington, Seattle, WA 98195, USA. rsugg@u.washington.edu

Frederick J. Swanson
USDA Forest Service, PNW Research Station, Forestry Sciences Laboratory, Corvallis, OR 97331, USA. fswanson@fs.fed.us

Jonathan H. Titus
Department of Biology, State University of New York at Fredonia, Fredonia, NY 14063, USA. titus@fredonia.edu

James M. Trappe
Department of Forest Science, Oregon State University, Corvallis, OR 97331-5752, USA. trappej@ucs.orst.edu

Louise S. Trippe
Department of Aquatic, Watershed, and Earth Resources, Utah State University, Logan UT 84322, USA. trippe@cc.usu.edu

Marc H. Weber
Department of Geography, Portland State University, Portland, OR 97207-0751, USA. weberm@geog.pdx.edu

Robert C. Wissmar
Center for Streamside Studies, University of Washington, Seattle, WA 98195, USA. wissmar@fish.washington.edu

David M. Wood
Department of Biological Sciences, California State University at Chico, Chico, CA 95929-0515, USA. dmwood@csuchico.edu

David K. Yamaguchi
5630 200 St. SW, Lynwood, WA 98036-6260, USA. trringzrus@aol.com

Donald B. Zobel
Department of Botany and Plant Pathology, Oregon State University, Corvallis, OR 97331-2902, USA. zobeld@science.oregonstate.edu

Part I
Introduction

1
Disturbance, Survival, and Succession: Understanding Ecological Responses to the 1980 Eruption of Mount St. Helens

Virginia H. Dale, Frederick J. Swanson, and Charles M. Crisafulli

1.1 Introduction

The ecological and geological responses following the May 18, 1980, eruption of Mount St. Helens are all about change: the abrupt changes instigated by geophysical disturbance processes and the rapid and gradual changes of ecological response. The explosive eruption involved an impressive variety of volcanic and hydrologic processes: a massive debris avalanche, a laterally directed blast, mudflows, pyroclastic flows, and extensive tephra deposition (Lipman and Mullineaux 1981; Swanson and Major, Chapter 3, this volume). Subsequent, minor eruptions triggered additional mudflows, pyroclastic flows, tephra-fall events, and growth of a lava dome in the newly formed volcanic crater. These geological processes profoundly affected forests, ranging from recent clear-cuts to well-established tree plantations to natural stands, as well as meadows, streams, and lakes. This book focuses on responses of these ecological systems to the cataclysmic eruption on May 18, 1980.

Initial ecological response to the 1980 eruption was dramatic both in the appearance of devastation (Figure 1.1) and in subsequent findings that life actually survived by several mechanisms in many locations (del Moral 1983; Halpern and Harmon 1983; Andersen and MacMahon 1985a and 1985b; Franklin et al. 1985; Crawford 1986; Adams et al. 1987; Zobel and Antos 1986, 1992). Ecological change occurred as a result of survival, immigration, growth of organisms, and community development. The pace of these biological responses ranged from slow to remarkably rapid. In addition, subsequent physical changes to the environment influenced biological response through weathering of substrates and by secondary disturbances, such as erosion, that either retarded or accelerated plant establishment and growth, depending on local circumstances. The net result of secondary physical disturbances was increased heterogeneity of developing biological communities and landscapes.

The sensational volcanic eruption of Mount St. Helens initially dwarfed the ecological story in the eyes of the public and the science community; but as the volcanic processes quieted, ecological change gained attention. The variety of disturbance effects and numerous interactions between ecological and geological processes make Mount St. Helens an extremely rich environment for learning about the ecology of volcanic areas and, more generally, about ecological and geophysical responses to major disturbances. More than two decades after the primary eruption, geophysical and ecological changes to the Mount St. Helens landscape have become so intertwined that understanding of one cannot be achieved without considering the other.

The 1980 eruption of Mount St. Helens and its ecological aftermath are the most studied case of volcanic impacts on ecological systems in history (Table 1.1). Ecological research at other volcanoes has often considered ecological responses based on observations made several years, decades, or even centuries after the eruption. In contrast to eruptions of some other volcanoes, lava surfaced only in the crater of Mount St. Helens; and most of the disturbance processes left deposits of fragmented volcanic rocks through which plants can easily root and animals can readily burrow. Furthermore, studies at other volcanoes typically investigated only one group of organisms (e.g., plants) and one type of volcanic process or deposit, which contrasts to the diversity of terrestrial and aquatic life and volcanic processes and deposits considered in this book.

Since the 1980 eruption of Mount St. Helens, analyses of ecological response to eruptions of other volcanoes and to ecological disturbance, in general, have made important advances. Ecological responses to other volcanic eruptions have been the subject of retrospective investigations of historic eruptions [e.g., Krakatau in Indonesia (Thornton 1996)] and analyses of responses to recent eruptive activity [e.g., Hudson volcano in Argentina (Inbar et al. 1995)]. More broadly, the field of disturbance ecology has blossomed through development of theory (Pickett and White 1985; White and Jentsch 2001; Franklin et al. 2002); intensive study of recent events, such as the Yellowstone fires of 1988 (Turner et al. 1998) and Hurricane Hugo (Covich and Crowl 1990; Covich et al. 1991; Covich and McDowell 1996); and consideration of effects of climate change on disturbance regimes (Dale et al. 2001). Lessons

FIGURE 1.1. Before and after photographs of the Mount St. Helens landscape: (a) view north from top of Mount St. Helens toward Mount Rainier across Spirit Lake before 1980; (b) same view in summer 1980. (*Source:* USDA Forest Service photos.)

about ecological response at Mount St. Helens are shaping understanding of succession (Turner et al. 1998; Walker and del Moral 2003), disturbance ecology (Turner et al. 1997; Turner and Dale 1998; del Moral and Grishin 1999), ecosystem management (Swanson and Franklin 1992; Dale et al. 1998, 2000; Franklin et al. 2002), evolution and the origin of life (Baross and Hoffman 1985), trophic interactions (Fagan and Bishop 2000; Bishop 2002), and landscape ecology (Foster et al. 1998; Lawrence and Ripple 2000).

In the context of this progress in disturbance ecology in general and ecological studies at Mount St. Helens more specifically, it is timely to synthesize knowledge of ecological response to the 1980 eruption of Mount St. Helens. In the first 7 years after the eruption, several compilations documented the numerous, intensive studies of ecological response (Keller 1982, 1986; Bilderback 1987); however, since 1987, the scientific community has not prepared a book-length synthesis of the scores of ecological studies under way in the area. Yet, more than half of the world's published studies on plant and animal responses to volcanic eruptions have taken place at Mount St. Helens (see Table 1.1) (Dale et al. 2005; Edwards 2005). The 25-year synthesis presented in this volume makes it possible to more thoroughly analyze the initial stages of response, to assess the validity of early interpretations, to examine the duration of early phenomena in a broader temporal context, and to consider landscape processes and patterns that were not evident in the early years. These studies provide an understanding of ecological change in a complex, continually changing environment. Hence, the Mount St. Helens volcano has come to hold a special place in the study of volcanic eruptions not only in the Pacific Northwest of the United States but also throughout the world.

TABLE 1.1. Summary of research on effects of volcanic activity on vegetation organized by types of physical impact.

Type of physical impact and volcano	Location	Dates of eruption	Reference
Lava			
Mount Wellington	New Zealand	9000 years before present (YBP)	Newnham and Lowe 1991
Mt. Fuji	Japan	1000	Hirose and Tateno 1984; Ohsawa 1984; Masuzawa 1985; Nakamura 1985
Rangitoto	New Zealand	1300, 1500, 1800	Clarkson 1990
Mt. Ngauruhoe and Mt. Tongariro	New Zealand	1550+	Clarkson 1990
Snake River Plains	Idaho, USA	~1720	Eggler 1971
Jorullo	Mexico	1759	Eggler 1959
Ksudach	Kamchatka, Russia	1907	Grishin et al. 1996
Waiowa	New Guinea	1943	Taylor 1957
Kilauea Iki and Mauna Loa	Hawaii, USA	1959	Fosberg 1959; Smathers and Mueller-Dombois 1974; Matson 1990; Kitayama et al. 1995; Aplet et al. 1998; Baruch and Goldstein 1999; Huebert et al. 1999
Surtsey	Iceland	1963	Fridriksson and Magnusson 1992; Fridriksson 1987
Isla Fernandina	Galapagos, Ecuador	1968	Hendrix 1981
Hudson	Argentina	1991	Inbar et al. 1995
Krakatau	Indonesia	1883, 1927	Whittaker et al. 1989, 1992, 1998, 1999; Partomihardjo et al. 1992; Thornton 1996
Pyroclastic flow			
Vesuvius	Italy	79	Mazzoleni and Ricciardi 1993
Kilauea Iki	Hawaii, USA	1750, 1840, 1955	Atkinson 1970
Miyake-Jima	Japan	1874, 1962, 1983	Kamijo et al. 2002
El Paracutin	Mexico	1943	Eggler 1948, 1959, 1963; Rejmanek et al. 1982
Mount St. Helens	Washington, USA	1980	Wood and del Moral 1988; Morris and Wood 1989; Wood and Morris 1990; Halvorson et al. 1991b, 1992; del Moral and Wood 1988a,b, 1993a,b; del Moral et al. 1995; Chapin 1995; Halvorson and Smith 1995; Tsuyuzaki and Titus 1996; Tsuyuzaki et al. 1997; Titus and del Moral 1998a,b,c; Bishop and Schemske 1998; Tu et al. 1998; del Moral 1998, 1999a; Fagan and Bishop 2000; Bishop 2002; del Moral and Jones 2002; Fuller and del Moral 2003
Avalanche			
Mt. Taranaki	New Zealand	1550	Clarkson 1990
Ksudach	Kamchatka, Russia	1907	Grishin 1994; Grishin et al. 1996
Mt. Katmai	Alaska, USA	1912	Griggs 1918a,b,c, 1919, 1933
Mount St. Helens	Washington, USA	1980	Russell 1986; Adams et al. 1987; Adams and Dale 1987; Dale 1989, 1991; Dale and Adams 2003
Ontake	Japan	1984	Nakashizuka et al. 1993
Mudflow			
Krakatau	Indonesia	1883	Tagawa et al. 1985
Mt. Lassen	California, USA	1914–1915	Heath 1967; Kroh et al. 2000
Mount Rainier	Washington, USA	1947	Frehner 1957; Frenzen et al. 1988
Mount Lamington	New Guinea	1951	Taylor 1957
Mount St. Helens	Washington, USA	1980	Halpern and Harmon 1983
Mount Pinatubo	Philippines	1991	Mizuno and Kimura 1996; Lucht et al. 2002; Gu et al. 2003
Tephra and ash deposition			
Auckland Isthmus	New Zealand	~9,500 YBP	Newnham and Lowe 1991
Krakatau	Indonesia	~1880	Whittaker et al. 1998
Laacher Volcano	Germany	~12,900 YBP	Schmincke et al. 1999
Laguna Miranda	Chile	~4,800 YBP	Haberle et al. 2000
Mount Usu	Japan	1977–1978	Tsuyuzaki 1991, 1995; Tsuyuzaki and del Moral 1995; Tsuyuzaki 1997; Tsuyuzaki and Haruki 1996; Haruki and Tsuyuzaki 2001
Lascar Volcano	Chile	1993	Risacher and Alonso 2001
Mount Mazama	Oregon, USA	~6,000 YBP	Horn 1968; Jackson and Faller 1973
Craters of the Moon	Idaho, USA	~2,200 YBP	Eggler 1941; Day and Wright 1989
Vesuvius	Italy	79	Dobran et al. 1994

(*continued*)

TABLE 1.1. (*continued*)

Type of physical impact and volcano	Location	Dates of eruption	Reference
Mt. Taranaki	New Zealand	1655	Clarkson 1990
Jorullo	Mexico	1759	Eggler 1959
Mt. Victory	New Guinea	1870	Taylor 1957
Krakatau	Indonesia	1883	Bush et al. 1992; Thornton 1996
Mt. Tarawera	New Zealand	1886	Clarkson and Clarkson 1983; Clarkson 1990; Clarkson et al. 2002; Walker et al. 2003
Soufriere	St. Vincent, BWI	1902	Beard 1976
Katmai	Alaska, USA	1912	Griggs 1917
Popocatapetl	Mexico	1920	Beaman 1962
Mount Lamington	New Guinea	1951	Taylor 1957
Kilauea Iki	Hawaii, USA	1959	Smathers and Mueller-Dombois 1974; Winner and Mooney 1980
Isla Fumandina	Galapagos, Ecuador	1968	Hendrix 1981
Usu	Japan	1977–1978	Riviere 1982, Tsuyuzaki 1987, 1989, 1991, 1994, 1995, 1996; Lamberti et al. 1992; Tsuyuzaki and del Moral 1994
Mount St. Helens	Washington, USA	1980	Mack 1981; Cook et al. 1981; Antos and Zobel 1982, 1984, 1985a,b,c, 1986; del Moral 1983, 1993; Seymour et al. 1983; Cochran et al. 1983; Hinckley et al. 1984; del Moral and Clampitt 1985; Frenzen and Franklin 1985; Zobel and Antos 1986, 1987a, 1991a, 1992, 1997; Adams et al. 1987; Harris et al. 1987; Wood and del Moral 1987; Pfitsch and Bliss 1988; Chapin and Bliss 1988, 1989; del Moral and Bliss 1993; Tsuyuzaki and del Moral 1995; Foster et al. 1998
El Chichòn	Mexico	1982	Burnham 1994
Hudson	Argentina	1991	Inbar et al. 1994
Mount Koma	Hokkaido, Japan	1929	Tsuyuzaki 2002; Titus and Tsuyuzaki 2003a,b; Nishi and Tsuyuzaki 2004
Santorini	Greece	~9,000 YBP	Bottema and Sarpaki 2003
Mijake-Jima	Japan	~9,000 YBP	Kamijo et al. 2002
Kula	Turkey	~9,000 YBP	Oner and Oflas 1977
Blowdown			
Mount Lamington	New Guinea	1951	Taylor 1957
Mount St. Helens	Washington, USA	1980	Franklin et al. 1985, 1988; Frenzen and Crisafulli 1990; Halpern et al. 1990

Source: Updated from Dale et al. (2005).

This chapter provides background on concepts of disturbance, succession, and the integration of ecological and geophysical perspectives that are explored further in this book. First, it defines disturbance, survival, and succession, and then briefly examines the major components of ecological response: survival, immigration, site amelioration, and community development. Next, the chapter addresses linkages among biotic and physical factors influencing succession. Finally, it considers the relation of events at Mount St. Helens to succession and disturbance ecology concepts. The chapter closes with an overview of what follows in subsequent chapters.

1.2 Ecological Change: Definitions and Descriptions of Disturbance, Survival, and Succession

1.2.1 Disturbance

Ecological disturbance has been defined as "any relatively discrete event in time that disrupts ecosystem, community, or population structure and changes resources, substrate availability, or the physical environment" (White and Pickett 1985). Rather than being catastrophic agents of destruction, many disturbances are normal, even integral, parts of long-term ecological dynamics. The composition, structure, and function of ecological systems are partially products of disturbances. In fact, some species and ecosystems are well adapted to frequent disturbances, so in some cases the absence of disturbance constitutes a disruption that can lead to changes in species, structures, or processes (White and Jentsch 2001; Dodds et al. 2004). To better understand any particular disturbance event, it should be considered in the context of the typical disturbance regime of the area.

Important characteristics of disturbance include their intensity (i.e., force exerted, such as heat release per unit length of fire front), severity (i.e., ecological effect, such as change in live plant cover), frequency, predictability, size, and spatial distribution (White and Pickett 1985). Severity and intensity are related but commonly differ because of differential species response to disturbance. Disturbance regimes span a broad range of frequency and predictability of occurrence. Disturbance size may be simply delineated when the area affected is uniform or may be quite complex where disturbance-impacted areas are patchy or the disturbance is variable in intensity and severity. Small, more frequent disturbances include individual tree falls; small fires; and small, patchy insect outbreaks. Large, infrequent disturbances include volcanic eruptions, crown fires, and hurricanes. Areas affected by large disturbance events commonly encompass complex patterns of disturbance intensity and severity, reflecting heterogeneity in the predisturbance landscape as well as complexity of the disturbance process itself (Turner et al. 1997). Timing of a disturbance can influence its effect on an ecological system. For example, ice storms can have more severe consequences in a deciduous forest if they are late enough in the spring that trees have leafed out (Irland 1998).

It is useful to distinguish between disturbance type and mechanism. Disturbance type refers to the geophysical or ecological phenomenon that has a disturbance effect, such as windstorm, fire, glacier advance or retreat, volcanic eruption, flood, wave action, insect or pathogen outbreak, or human activity. Mechanisms of disturbance are the specific stressors sensed by organisms, such as heat, impact force, and erosion or deposition. Both volcanic and nonvolcanic disturbance processes involve combinations of disturbance mechanisms. Intense forest fires, for example, can include high temperature and strong wind; and mudflows of volcanic or nonvolcanic origin involve impact force, scour, and deposition. Different disturbance types may have similar mechanisms of disturbance, such as occurs with both wildfire and volcanic processes that involve mechanisms of heating. Initial biological response to disturbance is reaction to the *mechanism* rather than the *type* of disturbance. Hence, understanding both the mechanisms involved in a particular disturbance process and the biotic response to individual mechanisms is critical to interpreting and predicting disturbance effects. There are several implications of this perspective. First, if the mechanism and intensity of disturbance by two different processes are similar, similar biological response would be expected, despite the difference in disturbance type (e.g., spores of some fungal species germinate when exposed to heat of wildfire or of volcanic eruption). Thus, some species may be adapted to mechanisms imposed by rare disturbance types (e.g., volcanic blast) because of adaptations to a more common disturbance type (e.g., fire). Second, where several mechanisms are involved in a particular disturbance type, the mechanism with the greatest severity overrides effects of the others.

1.2.2 Survival

Survival is a critical ecological process involving the interaction of organisms and disturbance processes, and survivors potentially play important roles in succession. Ecological effects of disturbances are determined, in part, by both living entities and nonliving biological and physical structures from the predisturbance system that remain after the disturbance (North and Franklin 1990; Foster et al. 1998; White and Jentsch 2001). The potential importance of residual plants and animals was noted by Clements (1916) and Griggs (1918a) but did not gain much prominence in early work on succession because of a focus on primary succession and old-field succession, where agricultural practices had erased any vestiges of previous forest. More recently, the term *biological legacy* has been defined as the types, quantities, and patterns of biotic structures that persist from the predisturbance ecological system. Living legacies can include surviving individuals, vegetative tissue that can regenerate, seeds, organisms in resting stages (particularly important for zooplankton), and spores. Dead biological legacies include standing dead trees, wood on the ground, litter, and animal carcasses. *Physical legacies* can strongly influence plant and animal survival, colonization, and growth. Important physical legacies after many disturbances are remnant soil, talus, rock outcrops, and aquatic habitats (e.g., seeps and springs).

1.2.3 Succession

The interplay of disturbance and response is an essential part of ecological change in landscapes. The process of gradual ecological change after disturbance, termed *succession* by Thoreau (1993), refers to changes that occur over time in biological and physical conditions after a site has been disturbed (Figure 1.2). Succession is the suite of progressive changes that occur to an ecological system and not the regular, seasonal, or interannual change in biological systems. In forests, succession can proceed over decades or centuries; whereas in microbial systems, succession occurs over days or months. Succession has intrigued ecologists since the first studies of ecology (McIntosh 1999). Early views of gradual ecological change were drawn, in part, from observations of sets of sites thought to represent different stages along a sequence of biotic development following some common initiating event, such as abandonment of farm fields, sand-dune formation, or deposits left by retreating glaciers (Cowles 1899; Clements 1916; Gleason 1917; Olson 1958). After a long period of debate about the processes and consequences of succession (Whittaker 1953; Odum 1969; Drury and Nisbet 1973; Bazzaz 1979; Odum 1983; McIntosh 1999), in recent decades ecologists have increasingly turned their attention to the study of disturbances and their ecological effects (Pickett and White 1985; White and Jentsch 2001).

Historically, ecologists distinguished *primary succession*, which follows formation of entirely new substrates and areas cleansed of biota, from *secondary succession*, which follows disturbances that leave substantial legacies of earlier ecological systems. Primary succession was thought to take place on entirely denuded sites, such as in the aftermath of a lava flow or glacier retreat, and in newly created habitats, such as lakes and streams on fresh landslide deposits. Primary succession is now commonly considered an endpoint along a continuum of abundance of residual organisms and biological structures left by a disturbance.

Following disturbance, succession does not follow an orderly path to a single endpoint. Instead, succession is commonly complex, having different beginning points, stages

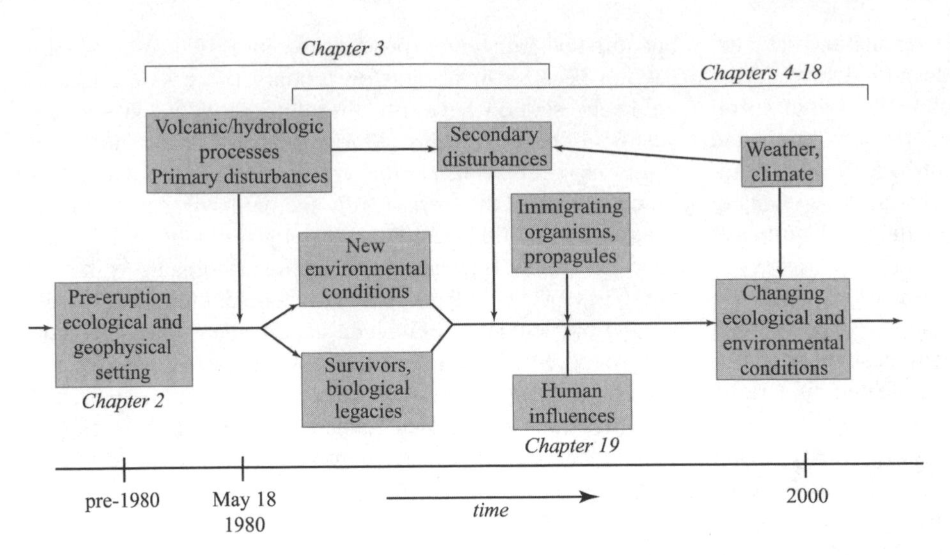

FIGURE 1.2. Sequential interactions of disturbance and succession processes over time and the relation of these topics to chapters in this book.

with different mixes of species and dominance patterns, and interruptions of successional trajectories by subsequent disturbances or other factors. Consequently, multiple pathways of succession may occur (Baker and Walford 1995). Ecosystems may undergo succession toward prior ecological conditions if the prevailing climate, species pools, and substrates have not been altered significantly. Yet, when one or more of these or other factors change, such as preemption of a site by a particular species or a profound change in soil conditions, a new stable state may be achieved (Paine et al. 1998).

1.3 Response to Disturbance: Processes of Change

Succession includes influences of biological and physical legacies, if any are present; immigration of organisms; establishment of some of these migrants; accrual of species and biomass; replacement of some species by others; and amelioration of site physical conditions. The replacement concept can be extended to include (1) replacement of one kind of community by another and (2) progressive changes in microbes, fungi, plants, and animal life, which may culminate in a community that changes little until the next disturbance.

1.3.1 Processes Affecting Survival

Survival depends on interactions between properties of the disturbance events and traits of organisms that allow them to avoid, resist, or respond neutrally or positively to disturbance impacts. Organisms may withstand disturbance by being in a protected location within the disturbed area (e.g., subterranean or under cover of lake ice) (Andersen and MacMahon 1985a) or being well adapted to withstand disturbance (Gignoux et al. 1997). Organism size can also foster survival; small macroalgae, for example, can better withstand the thrashing of intense wave action than can large algae (Blanchette 1997).

Some species may persist in a disturbed landscape by having either all or part of their populations away at the time of disturbance. For example, migratory birds and anadromous fish may be away from sites during disturbance and, upon their return, reoccupy the area if suitable habitat and food are present.

Despite apparently tight coupling of species–disturbance interactions, survival of individual organisms and populations is frequently a matter of chance. Nuances of site conditions, organism vigor, disturbance intensity at the site, timing of disturbance relative to the life history of the organism, and other factors can tip the balance of life versus death in ways that are difficult to anticipate.

1.3.2 Immigration, Establishment, and Site Amelioration

The early stages of succession are strongly influenced by persistence and growth of survivors; immigration, establishment, and growth of colonists; and interactions among these colonists. Mobility of organisms and propagules, as well as the conditions of the environment through which they move, affect dispersal patterns. Highly vagile organisms, such as those capable of flight and passive movement by wind, typically are first to reach disturbed areas distant from source populations. In contrast, low-mobility organisms, such as seed plants that lack structures for wind dispersal and animals which travel through soil, would be expected to slowly reach distant, disturbed sites. Many species with poor dispersal mechanisms can be transported great distances by hitchhiking in or on animals, moving in flowing water, or capitalizing on the influence of gravity. Dispersal is commonly thought to be determined by the distance to source populations, but complexities of disturbance processes and patterns and the state of the affected ecological system make such simplistic, distance-related interpretations unrealistic.

Once an organism disperses to a new location, its ability to establish, grow, and reproduce is determined by prevailing climate, site conditions, previously established organisms, and the organism's own requirements and tolerances. For animals, the requirements for successful establishment are often expressed as adequate cover and food. Cover provides protection from physical stresses as well as a place to hide from potential predators. Plant establishment requires appropriate light, moisture, and nutrient levels for germination and growth.

Site amelioration is an important process that can involve changes to soil conditions, microclimate, and microtopography. Soil development is often an essential factor in succession, especially in the case of primary succession, where soil is initially of poor quality. Soil formation involves physical and chemical weathering of rocks and minerals, accumulation and decay of biotic material, establishment of a microbial fauna, and marshalling of any legacies of earlier soil on the site. The death or stress of biota in response to the disturbance may deliver a pulse of litter to the soil surface or within the soil via root death. As a site ameliorates, plants establish and spread, species interact, and a community develops. Animals are tightly coupled to plant composition or physiognomy, so their colonization frequently tracks the development of vegetation.

Humans can profoundly influence the course of succession in many ways, both intentionally and unintentionally. A common influence is the introduction of invasive, nonnative species, which can have far-reaching ecological and management repercussions. Disturbance commonly favors establishment of invasive species, but predicting the vulnerability of a system to invasion is still a challenge (Mack et al. 2000). Planting of native trees and stocking with native fish can profoundly alter community structure and function.

1.3.3 Concepts of Change in Ecological and Environmental Factors During Succession

Changes in ecological and environmental conditions are both consequences and determinants of the path of succession. Today, concepts of succession and disturbance ecology have reached the point where they are examined and modified through experimental and modeling approaches as well as by studies of ecological change imposed by major disturbance events.

1.3.3.1 Community Development Through Succession

Species richness, biomass, and structural complexity of communities increase during succession. Various types of interactions among species drive community development, and these processes may change in their relative importance over the course of succession. In some cases, one species is replaced by another over time in what is called the process of relay succession. In these cases, the change in species is as abrupt as the handing over of a baton from one runner to another (building on the concepts of Egler 1954). Yet, such predictable and unidirectional transitions do not always occur.

In an attempt to advance understanding of succession, Connell and Slatyer (1977) proposed three models of mechanisms of succession, termed: facilitation, tolerance, and inhibition. These models describe the way in which species interact with their environment and with later-arriving species to either promote, hinder, or have minimal effects on the establishment and/or growth of some species and thus to shorten or lengthen the time to dominance by another species (Connell and Slatyer 1977). However, succession is highly variable; and in most cases, these three mechanisms, plus others, occur simultaneously during a successional sequence (McIntosh 1999). In addition to these models of succession, numerous species–species interactions, such as mutualism, predation, parasitism, and herbivory, help shape the pace and direction of succession.

Facilitation was first interpreted as the process of early successional species altering conditions or the availability of resources in a habitat in a way that benefits later successional species (Clements 1916; Connell and Slatyer 1977). For example, the first species to become established create shade, alter soil moisture, and ameliorate soil texture and nutrient conditions via decomposition of their parts and other processes. Commonly, nitrogen is a limiting factor in early successional stages, and the presence of plants with the ability to infuse the soil with nitrogen through association with nitrogen-fixing bacteria enhances soil development. Facilitation is now more broadly interpreted as positive interactions between species (Bruno et al. 2003) and as processes that improve a site's physical conditions (e.g., soil development) (Pugnaire et al. 2004). These beneficial interactions appear to be common under stressful environmental conditions (Callaway and Walker 1997).

In contrast to facilitation, the process of *inhibition* may slow or temporally arrest successional development (Grime 1977; Connell and Slatyer 1977). This process occurs when a resource, such as space, water, or nutrients, is so intensely used by one or more species that it is not available in life-sustaining quantities to other species. For example, following a mudslide along Kautz Creek on the flanks of Mount Rainier in Washington State, the depositional area was quickly colonized by an almost continuous mat of mosses and lichens. Germinating tree seedlings could not penetrate the mat and reach mineral soil, and thus tree establishment was inhibited for decades (Frehner 1957; Frenzen et al. 1988).

Tolerance refers to the situation where organisms best able to tolerate prevailing conditions are favored, but recognizes that prevailing conditions change with time. Under this model, later successional species are unable to become established without site amelioration by pioneer species that do not inhibit the later colonists (Connell and Slatyer 1977). A primary premise of the tolerance model is that later successional species can grow with lower resource levels than can earlier species and are better at exploiting limited resources. As later successional species grow and produce progeny, they replace the earlier, less-tolerant species and become dominant. Thus, life-history characteristics are critical in determining the sequence of species replacements.

1.3.3.2 Biotic and Geophysical Forces of Succession

Drivers of disturbance and succession can be viewed as falling on a continuum of relative influence of geophysical forces (*allogeneic* succession) versus biological factors (*autogenic* succession) (White and Pickett 1985). Where allogeneic succession dominates, physical forces (such as chronic, secondary geophysical disturbances) override biological causes of succession. Autogenic succession is driven by intrinsic properties of a community and the ability of organisms to affect their environment, such as when certain species preempt sites, create shade, and alter soil structure and chemistry.

Patterns of water runoff, sediment transport, and other geophysical processes can change dramatically after severe landscape disturbance. Some processes alter site conditions in ways that prepare a site to experience other processes. Analyses of drainage basin evolution (Koss et al. 1994) and sediment routing following wildfire and forest cutting (Swanson 1981; Swanson et al. 1982b; Benda and Dunne 1997), for example, reveal sequential interactions among geomorphic processes in ways that are akin to facilitation in biotic succession.

Often, biotic and geophysical patterns of succession occur in parallel following severe disturbance and involve both positive and negative feedbacks:

- Episodic disturbances, such as landslides, can erase a decade or more of ecological response following the primary disturbance event.
- Development of vegetation and its associated litter layers and root systems can suppress erosion processes.
- In some instances, erosion of new deposits exposes buried plant parts in the predisturbance soil, thus favoring plant and animal survival and development of ecological interactions.

Recognition of the succession of ecological and geophysical processes in severely disturbed landscapes can be useful in interpreting the direction, rate, and cause of ecological responses to disturbance. Ecological response to severe disturbance is, in part, a function of the pace at which the landscape stabilizes geophysically to a point where biological response can proceed with vigor. For example, fish reproduction may not occur in a disturbed site because of physical instability that degrades spawning habitat or conditions required for egg development.

Secondary disturbance processes (i.e., those that are influenced by a primary disturbance) often play important roles in ecological change. Examples of secondary disturbances include the increased pace of lateral channel migration as a result of increased sediment load and precipitation runoff. This chronic disturbance repeatedly removes developing riparian vegetation.

1.4 Linking General Concepts and the Mount St. Helens Experience

The wealth of knowledge about disturbance, survival, and succession briefly summarized above and elsewhere (Pickett and White 1985; McIntosh 1999; White and Jentsch 2001; Walker and del Moral 2003) provides useful concepts for examining the initial effects and subsequent ecological and geophysical change at Mount St. Helens during and following the 1980 eruption. These science concepts are in continuing states of development and searches for generality (McIntosh 1999; White and Jentsch 2001). No single concept or theory is adequate to structure the scientific analysis or the telling of the highly multifaceted Mount St. Helens story. On the other hand, lessons from studies at Mount St. Helens have influenced the development of these topics.

The Mount St. Helens landscape and the lessons drawn from research conducted there have changed substantially during the quarter of a century since the 1980 eruption. Initial observations emphasized the nearly desolate character of the landscape, the importance of surviving organisms, factors influencing patterns of species dispersal and colonization, and community development. Some of these initial ecological responses have had lasting effects, but others of them proved to be transient. After 25 years, much of the landscape has filled with plants, and the once stark gray area has been transformed to mostly green. Extensive tracts of the most severely disturbed areas remain in early seral stages dominated by herbs and shrubs and will require several more decades before becoming closed-canopy forest, if they ever do. Numerous conifer saplings are present in all disturbance zones, and the development of forest cover is accelerating in many locations. By 2005, the ash-choked lakes and streams of 1980 glisten with clear, cold, well-oxygenated water and support biota typical of the region. The growth and spread of surviving and colonizing species during the first 25 years after the 1980 eruption have provided many new opportunities to address questions about succession, patterns of landscape response, and consequences of secondary geophysical processes. Even so, many questions regarding ecological responses to the 1980 eruption remain unanswered. Continuing change of the Mount St. Helens landscape may bring new answers and certainly will bring new questions about ecological responses to major disturbances.

1.5 Overview of Book

This book presents much of the existing research that explores succession, disturbance ecology, and the interface between geophysical and ecological systems at Mount St. Helens (see Figure 1.2). Chapters 2 and 3 review the geological and ecological setting before the 1980 eruption and the geophysical environments created by the May 18, 1980, eruption. Chapters 4 to 8 focus on the survival and establishment of plant communities across diverse volcanic disturbance zones. Chapters 9 to 14 consider responses of animal communities, in particular, arthropods, fish, amphibians, and small mammals. Chapters 15 to 18 discuss responses of four sets of ecosystem processes: the symbiotic relationship between mycorrhizal fungi and plants in soils, animal decomposition in terrestrial environments, effects of a nitrogen-fixing plant on soil quality and function, and the complex biophysical processes of lake responses. Chapters 19 and 20 synthesize changes that have occurred across land-management issues, species, ecological systems, and disturbance zones during the first quarter century after the 1980 eruption.

Together, these chapters provide an in-depth analysis of ecological patterns of response after the 1980 eruption of Mount St. Helens. Conventional terminology is used throughout the book (see the Glossary at the end of the volume), and throughout the book locations of the various research studies are shown on a common reference map. A single bibliography for all chapters is at the end of the book. The major taxonomic source for species mentioned in the book is the Integrated Taxonomic Information System (*http://www.itis.usda.gov*). Additional information about the area and the research results is available at *http://www.fsl.orst.edu/msh/*.

Acknowledgments. We appreciate the reviews of an earlier version of the chapter by Dan Druckenbrod and Peter White. Oak Ridge National Laboratory is managed by UT-Battelle, LLC, for the U.S. Department of Energy under contract DE-AC05-00OR22725. The USDA Forest Service and its Pacific Northwest Research Station have supported many aspects of science at Mount St. Helens before and since the 1980 eruption.

2
Geological and Ecological Settings of Mount St. Helens Before May 18, 1980

Frederick J. Swanson, Charles M. Crisafulli, and David K. Yamaguchi

2.1 Introduction

Volcanoes and volcanic eruptions are dramatic players on the global stage. They are prominent landscape features and powerful forces of landform, ecological, and social change. Vesuvius, Krakatau, Pompeii, and, in recent decades, Mount St. Helens hold an important place in our perceptions of how the Earth works and the incredible, destructive effects of violent eruptions. Perhaps less appreciated is the great diversity of interactions between volcanoes and the ecological systems in their proximity.

Volcanic activity and ecological change at Mount St. Helens have been particularly dynamic and instructive. Frequent eruptions of diverse types have interacted with terrestrial and aquatic ecological systems to display a broad range of responses (Franklin and Dyrness 1973; Mullineaux and Crandell 1981; Foxworthy and Hill 1982). Leading up to the 1980 eruption of Mount St. Helens, Cascade Range volcanoes of the Pacific Northwest of the United States were the subject of a good deal of study for objectives that were both academic and applied, such as assessing volcanic hazards and prospecting for geothermal resources. The fauna and flora of forests, meadows, lakes, and streams of the region were generally well known and described. The 1980 eruption put a spotlight on Mount St. Helens, as the world watched volcanic and ecological events unfold in real time. These events also stimulated an interest to better understand the volcanic and ecological conditions that existed before 1980. The geological, ecological, and historical settings provide context for interpreting the physical and ecological responses following the 1980 eruption. [Here we use the term history in the broad sense to include geological time as well as recorded human history.]

Study of any ecological system should start with consideration of its context in space and time and in geographical, geological, and ecological dimensions. From a geographical perspective, the position of Mount St. Helens in a north–south chain of volcanoes along a continental margin sets up strong east–west geophysical and biotic gradients between the sea and mountain top and along a north–south climate gradient (Figure 2.1). Understanding of these broad gradients is useful in interpreting similarities and differences among different parts of a region. These gradients also organize fluxes of materials, organisms, and energy across broad areas. Marine air masses, for example, deliver water to the continental edge, and this abundant moisture flows back to the sea, forming a regional hydrologic cycling system. A well-connected marine–freshwater system fostered development of numerous stocks of anadromous fish. Similarly, the north–south climatic gradient and topographic features of mountain ranges and chains of coastal and inland wetlands form travel corridors for migratory birds. Movement of such wide-ranging terrestrial and aquatic species results in a flow of nutrients, propagules, genes, and organisms in and out of local landscapes within the region and even more widely.

Past activity of a volcano influences its surroundings and affects biophysical responses to new disturbance events. Legacies of earlier eruptive activity may be expressed in landforms, soils, lakes, streams, animal communities, and vegetation patterns. This pattern is especially true at Mount St. Helens, which has erupted about 20 times in the past 4000 years (Table 2.1 on page 16). Vestiges of both the preeruption ecological systems and recent eruptive activity can strongly influence the posteruption landscape and patterns of change in ecological systems. Across the region and over evolutionary time scales, climate and biota interact with disturbance regimes of fire, wind, floods, volcanism, and other agents. Thus, the ecological history of the local area and its regional context determine the pool of species available to colonize a disturbed area, the capabilities of those species to respond to disturbance, and the array of types and configurations of habitats available for postdisturbance ecological development.

Given the importance of spatial and temporal context, this chapter begins the analysis of ecological responses to the 1980 eruption of Mount St. Helens by describing the area before 1980. Our objective in this chapter is to set the stage for subsequent chapters, which detail the geological events and ecological responses unfolding on May 18, 1980, and during the subsequent quarter century. We characterize the Mount

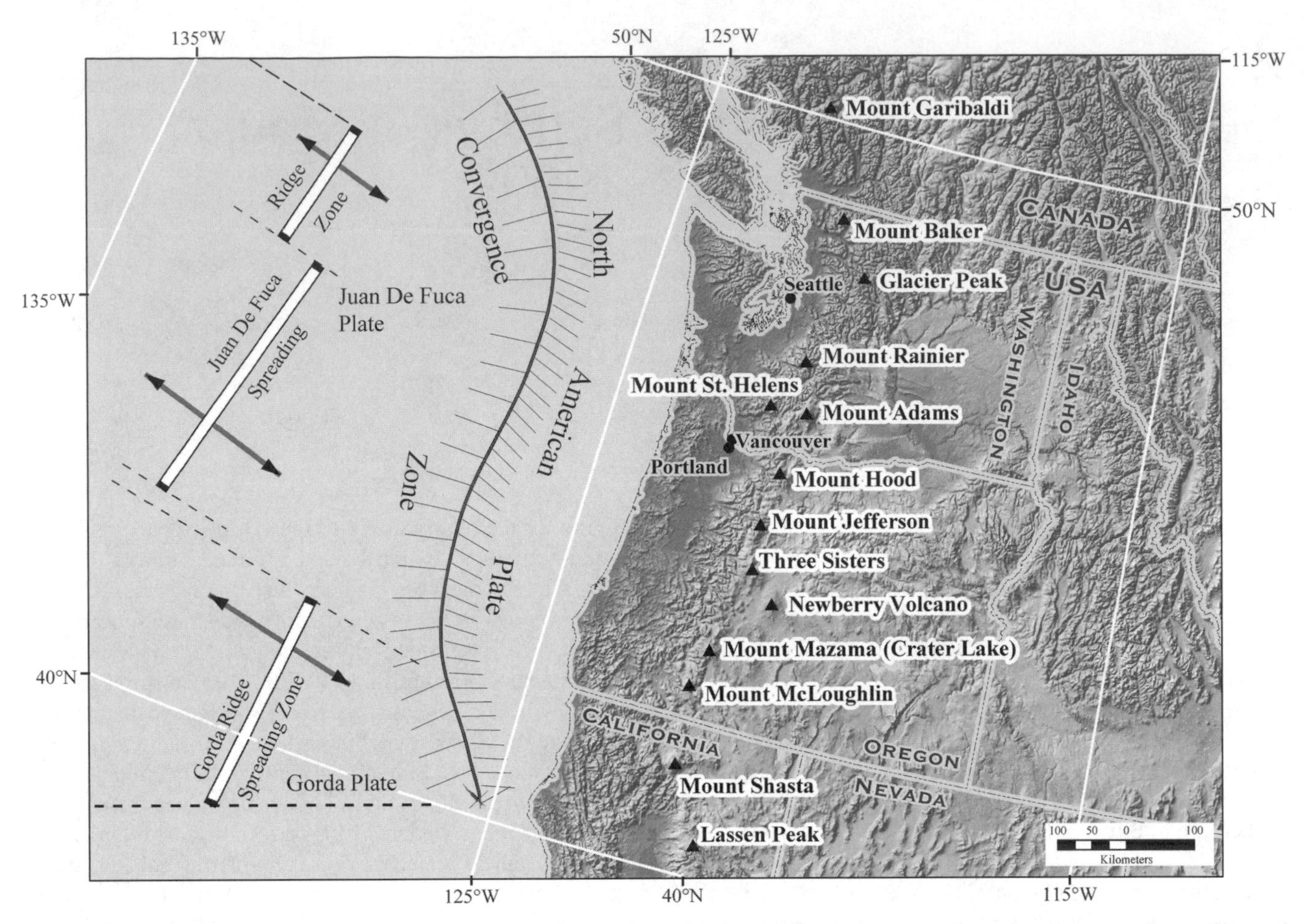

FIGURE 2.1. Regional context of Mount St. Helens, including major volcanic peaks and plate-tectonic setting in terms of spreading and convergence zones. [Adapted from Foxworthy and Hill (1982).]

St. Helens area in terms of its physiography, climate, geology, geomorphology, plant and animal assemblages, and ecological processes and its broader setting. Our geographical focus is the area affected by the 1980 event, generally within 30 km of the cone (Figure 2.2; see also Swanson and Major, Chapter 3, this volume).

Terminology in these discussions is summarized in the Glossary section of this book, generally following Lipman and Mullineaux (1981), Foxworthy and Hill (1982), and Fisher and Schmincke (1984). Plant-association and species nomenclature follows Franklin and Dyrness (1973) and the Integrated Taxonomic Information system (http://www.itis.usda.gov).

2.2 Geophysical Setting

2.2.1 Geological, Physiographic, and Geomorphic Setting

Mount St. Helens is part of the Cascade Range of volcanoes that extends from Canada to northern California (see Figure 2.1). The present and earlier alignments of Cascade volcanoes result from pieces of Pacific oceanic crust plunging beneath the North American continental plate (Figure 2.1). This geological setting has persisted for millions of years, thus shaping the broad outline of the region's physiography and the geophysical dynamics of chronic and catastrophic volcano growth and decay. These conditions are broadly representative of the circum-Pacific "ring of fire," where chains of volcanoes grow in response to geological forces operating within the Earth's mantle and crustal plates.

The structure of Mount St. Helens, as viewed before the 1980 eruption, had formed over the preceding 40,000 years on a geological foundation composed of volcanic rocks of Oligocene to early Miocene age (ca. 28 to 23 million years old). However, leading up to 1980, the entire visible cone had been constructed within only the preceding 2,500 years as an accumulation of volcanic domes, lava flows, and volcanic debris emplaced by other processes (Crandell and Mullineaux 1978; Mullineaux and Crandell 1981; Crandell 1987; Yamaguchi and Hoblitt 1995; Mullineaux 1996). The history of the volcano was read from deposits on its surface; from the types and ages of material it shed onto the surrounding countryside (subsequently exposed in the walls of deeply incised stream channels); and, after the 1980 eruption, in the volcano's internal anatomy exposed in the walls of the new crater. Deposits

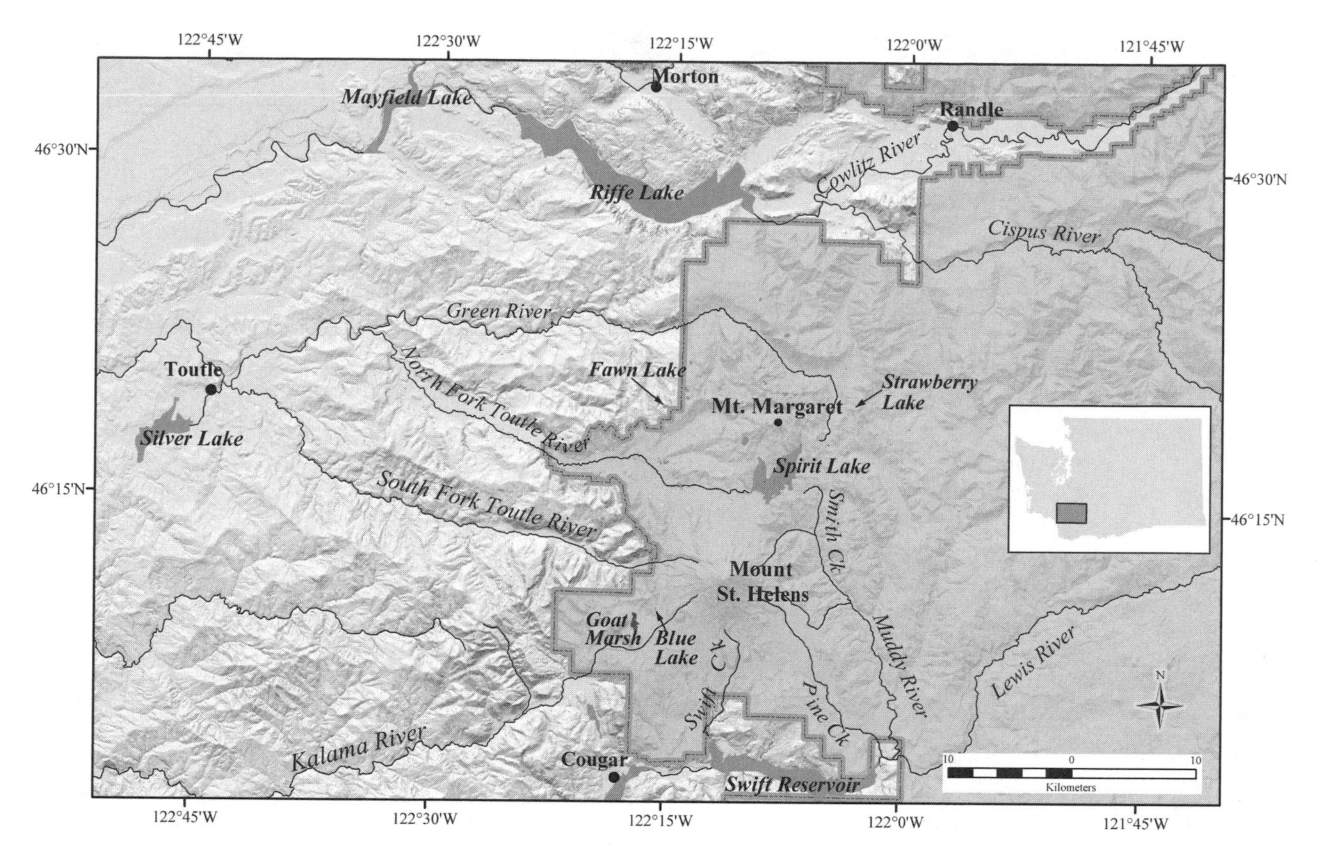

FIGURE 2.2. The Mount St. Helens area.

and events have been dated by analysis of tree rings, which give a record for much of the past millennium, and by radiometric dating of rock and organic material, which can extend much further into the past. The known eruptive history of Mount St. Helens spans periods of dormancy interspersed with periods of activity, which have been grouped into nine eruptive periods (see Table 2.1). Over the past seven eruptive periods, the length of dormant periods ranged from 50 to 600 years and averaged about 330 years.

Eruptive periods involved various combinations of a diverse suite of volcanic processes, which merit some definition. The term *tephra* refers to ejecta blown through the air by explosive volcanic eruptions. *Tephra fall* occurs when explosively ejected fine ash to gravel-sized rock debris falls to Earth and forms a deposit on vegetation, soil, or other surfaces. Eruption columns may extend kilometers into the air, and prevailing winds may cause tephra-fall deposits to accumulate in a particular quadrant around a volcano, generally the northeast quadrants of volcanoes in the Pacific Northwest. In contrast, hot (~800°C), pumice-rich eruption columns may collapse, forming *pyroclastic flows*, which move rapidly (tens of meters per second) down a volcano's flanks and onto the gentler surrounding terrain, accumulating in lobe-shaped deposits up to 10 m or more thick. Toward the other extreme of flow velocity, slow (e.g., millimeters per hour to meters per hour) extrusions of very viscous lava [e.g., with high silica (SiO_2) content] form *lava domes* with a circular or elliptical outline. Less-viscous lava may flow from vents and cool in *lava-flow* deposits, forming elongated lobes. Various interactions of water and the weak rocks (e.g., clay-rich or highly fractured) composing volcanoes can result in massive landslides, often termed *debris avalanches*. Volcanic debris avalanches may exceed a cubic kilometer in volume, enveloping a volcano summit and flank and spreading over tens of square kilometers at the base of the volcano. Volcanic *mudflows*, also termed lahars, may be triggered by many mechanisms, including drainage of debris avalanches, collapse of dams blocking lakes, and the movement of hot, volcanic debris over snow and ice. Mudflows have higher water content than do debris avalanches and, therefore, can flow at higher velocities and over greater distances (tens of kilometers) away from their sources. Less common volcanic processes are *lateral blasts*, which occur when superheated groundwater develops within a volcano by interaction of magma and infiltrating precipitation and then flashes to steam, producing an explosion. Such steam-driven blasts project large volumes of fragmented mountain-top rock laterally across a landscape. The resulting blast cloud, which can be hundreds of meters thick, topples and entrains vegetation along its path. Lateral blasts leave a blanket of deposits composed of angular sand, gravel, and fragments of organic material.

Some of these processes, such as dome growth and lava flows, contribute to volcanic-cone construction, while other processes contribute to the breakdown of volcanoes and the filling of surrounding valleys with volcanic debris. The Pine

TABLE 2.1. Summary of the Mount St. Helens eruptive history.

		Processes					
Eruptive period[a]	Approximate age (years)[b]	Tephra fall	Pyroclastic flow	Lava flow	Dome growth	Mudflow	Lateral blast
Current period	AD 1980–2005	X	X		X	X	X
Dormant interval of 123 years:							
Goat Rocks	AD 1800–1857	X		X	X	X	
Dormant interval of about 50 years:							
Kalama	AD 1480–mid-1700s	X	X	X	X		
Dormant interval of about 600 years:							
Sugar Bowl	1,080–1,060	X	X		X	X	X
Dormant interval of about 600 years:							
Castle Creek	Greater than 2,200–1,700	X	X		X	X	
Dormant interval of about 300 years:							
Pine Creek	3,000–2,500	X	X		X	X	
Dormant interval of about 300 years:							
Smith Creek	4,000–3,300	X	X			X	
Dormant interval of about 4,000 years:							
Swift Creek	13,000–8,000		X		X		
Dormant interval of about 5,000 years:							
Cougar	20,000–18,000		X	X	X	X	
Dormant interval of about 15,000 years:							
Ape Canyon	~40,000(?)–35,000		X			X	

The only lateral blasts interpreted within this record occurred during the Sugar Bowl period and on May 18, 1980.

[a] Dormant intervals are periods during which no unequivocal eruptive products from the volcano have been recognized.

[a] Ages of Goat Rocks–Kalama eruptive periods are in calendar years; ages of Sugar Bowl to Swift Creek periods, determined by radiocarbon dating, are expressed in years before AD 1950, following the calibrations of Stuiver and Pearson (1993). Ages of older periods are expressed less precisely in uncalibrated radiocarbon years.

Source: Adapted from Mullineaux and Crandell (1981), Mullineaux (1996), and Yamaguchi and Hoblitt (1995).

Creek, Castle Creek, and Kalama eruptive periods of Mount St. Helens (Table 2.1; Figure 2.3) were particularly voluminous, inundating neighboring areas north, southwest, and southeast of the volcano with pyroclastic-flow, mudflow, and lava-flow deposits. Lateral blasts were rare in the pre-1980 eruptive history of Mount St. Helens; only one has been noted in the geological record, and that was in the Sugar Bowl eruptive period (see Table 2.1). The numerous flowage deposits from Mount St. Helens significantly modified parts of all rivers draining the volcano. The deposits filled valleys, smoothing preexisting topography around the cone and disrupting earlier drainage patterns. The buildup of the Pine Creek assemblage diverted the Muddy River, which once followed the valley of Pine Creek, to the valley of Smith Creek. Similarly, accumulation of a broad fan on the north flank of the volcano intermittently dammed the head of the North Fork Toutle River, forming Spirit Lake. Periodically, this dam was partially breached, triggering massive mudflows down the Toutle River, several of which blocked Outlet Creek, forming Silver Lake, 45 km west-northwest of the summit of the volcano. Some streams draining the volcano subsequently cut deep canyons through these deposits, particularly on the south side of the volcano (Crandell and Mullineaux 1978).

Numerous eruptions spewed tephra on various trajectories to the east and northeast of the volcano (Table 2.1; Figure 2.4). The resulting deposits of fine ash to gravel-sized pumice and fragmented lava spread over many thousands of square kilometers, strongly affecting soil properties where their depth exceeded a few centimeters. These deposits have been dated with various tree-ring, radiocarbon, and other techniques, so they can be used as time markers to interpret landscape and vegetation conditions at times in the past (Mullineaux 1996). In some areas, such as 20 km northeast of the cone, tephra deposits of the past 3500 years exceed 5 m in thickness and contain several buried soils, including some trees buried in upright growth position (Franklin 1966; Yamaguchi 1993).

Lava flows during several eruptive periods covered parts of the southern and northern flanks of the volcano and flowed more than 10 km down the Kalama River and south-southeast to the Lewis River (Crandell 1987). Lava flows have been very resistant to erosion and therefore have tended to stabilize the land surfaces and deposits they cover. Hydrology is also strongly affected by lava flows, such as where massive volumes of water flow rapidly through lava tubes and beneath lava-flow deposits before discharging as large springs and streams with stable flow regimes.

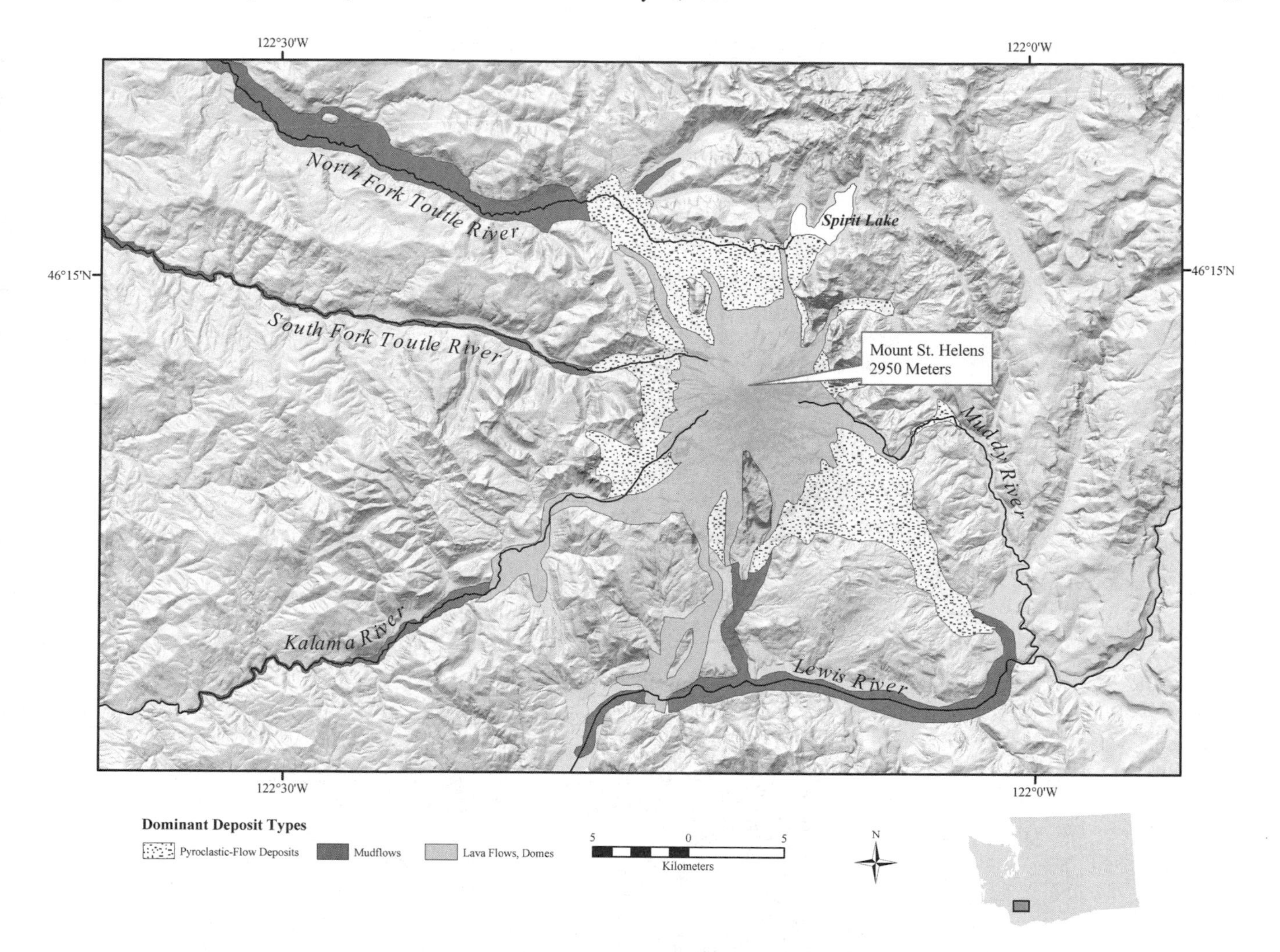

FIGURE 2.3. Approximate boundaries distinguishing Mount St. Helens and landforms composed of its products from older geological terrane of the southern Washington Cascade landscape. [Adapted from Crandell and Mullineaux (1978)].

Similar to many volcanic landscapes, landforms in the vicinity of Mount St. Helens can be broadly grouped into these categories:

- Extensive, older geological terrain with a long history of erosion, including glaciation, resulting in steep, rugged topography, and
- More gently sloping terrain, where younger volcanic flow deposits have accumulated during recent millennia, forming broad fans (see Figure 2.3).

The Mount St. Helens landscape has been sculpted during a long history of erosion and deposition by river, landslide, glacial, and other geomorphic processes both with and without the influence of volcanic activity. The steep, soil-mantled hill slopes in the Cascade Range landscape experience a variety of erosion processes, including subtle movement of the soil surface, transport of dissolved material, and diverse types of mass soil movement, most conspicuously shallow, rapid debris slides down hill slopes, and debris flows down narrow stream channels (Swanson et al. 1982a). Volcanic activity can greatly increase or decrease the rates of geomorphic processes that occur independently of volcanic influence.

During the Pleistocene (10,000 to 1,600,000 years before present), glaciers sculpted upper-elevation landforms in the vicinity of Mount St. Helens, creating very steep cliffs and cirques on north-facing slopes and broad, U-shaped valleys draining areas of extensive ice cover, such as the Mount Margaret high country about 15 km north-northeast of Mount St. Helens. Throughout the Holocene (the past 10,000 years), glaciers remained prominent features of most large Cascade Range volcanoes. Before the 1980 eruption, 11 named glaciers covered 5 km^2 of Mount St. Helens and extended from the 2949-m summit down to an elevation of about 1500 m (Brugman and Post 1981). However, the cone itself was so young that it had not been deeply dissected by glacial erosion.

This complex of processes, deposits, and landforms created numerous and varied terrestrial, lake, stream, and other types of habitats in a rather confined area around Mount St. Helens,

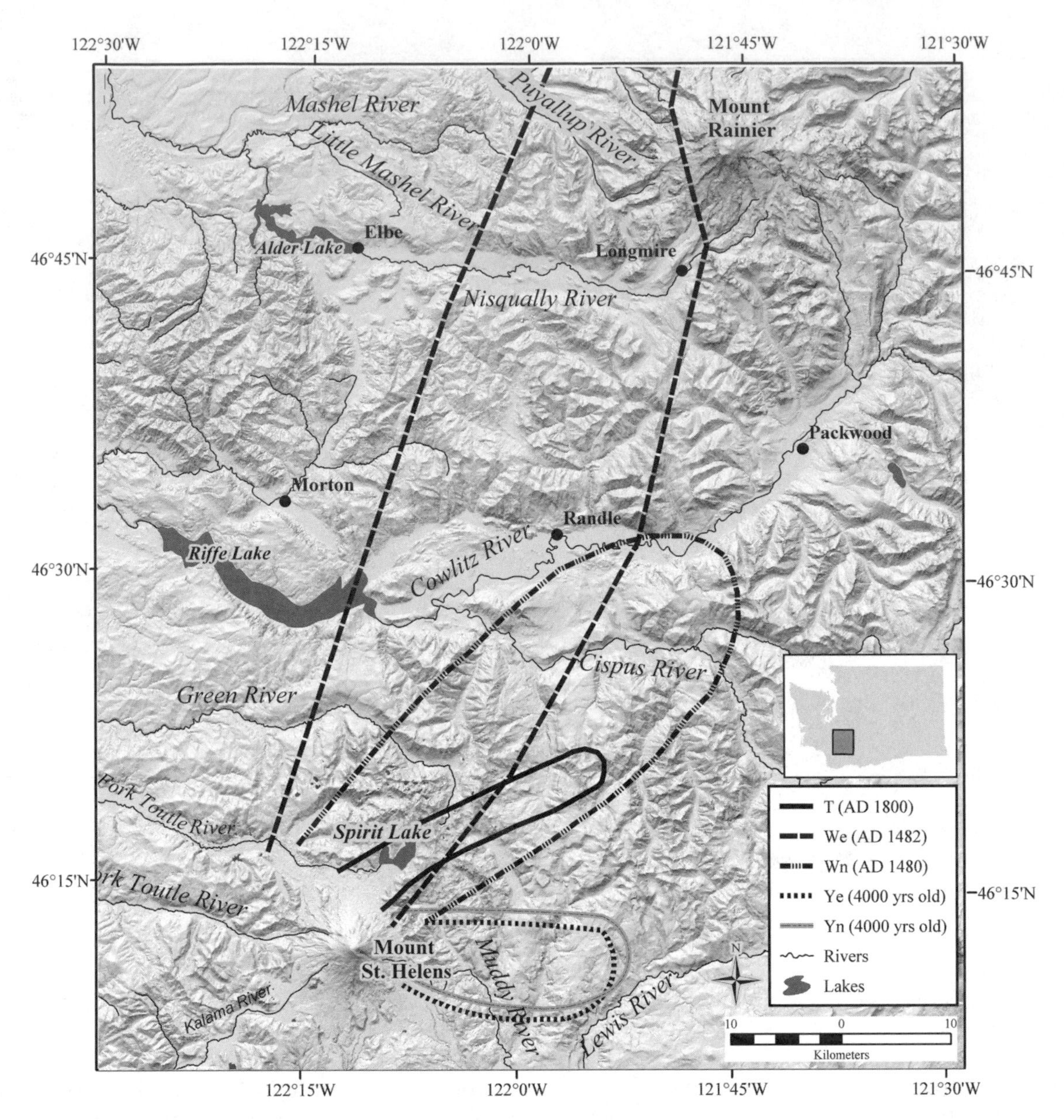

FIGURE 2.4. Extent of major pre-1980 tephra falls from Mount St. Helens. Mapped boundaries of deposits denote 20-cm isopachs. The letters T, We, Wn, Ye, and Yn refer to formal designations of particular pumice layer deposits. Date of each deposition is in parenthesis. [Adapted from Crandell and Mullineaux (1978) and Crandell and Hoblitt (1986).]

as it existed before the 1980 eruption. Lakes of various sizes and depths occur at a broad range of elevation as a result of their formation by both glacial excavation of bedrock basins and blockage of tributaries to the main channels draining the volcano's flanks. Stream and river habitats are similarly diverse, ranging from small, clear, headwater streams to large rivers. In addition to these extensive, common habitats, the geological complexity of the landscape created many localized, special habitats, such as cliffs, seeps, wetlands, lava caves, and dry meadows. Hydrothermal environments were limited to a few, small fumaroles before the 1980 eruption (Philips 1941), but these special types of habitats became important following the eruption.

2.2.2 Climate Setting

A large volcano and its associated large-scale disturbance events interact with the regional climatic regime in many ways. The regional climatic context sets the broad outline of moisture and temperature conditions, but local effects of a mountain edifice can substantially modify the regional climatic signal. A volcano of the stature of Mount St. Helens, and especially the more massive Mount Rainier 80 km to the north-northeast, captures moisture from the atmosphere, storing it in snowfields and glaciers high on the cone, and creates a local rain shadow in its lee. Thus, the climate of the Mount St. Helens area strongly reflects the effects of marine air masses moving eastward from

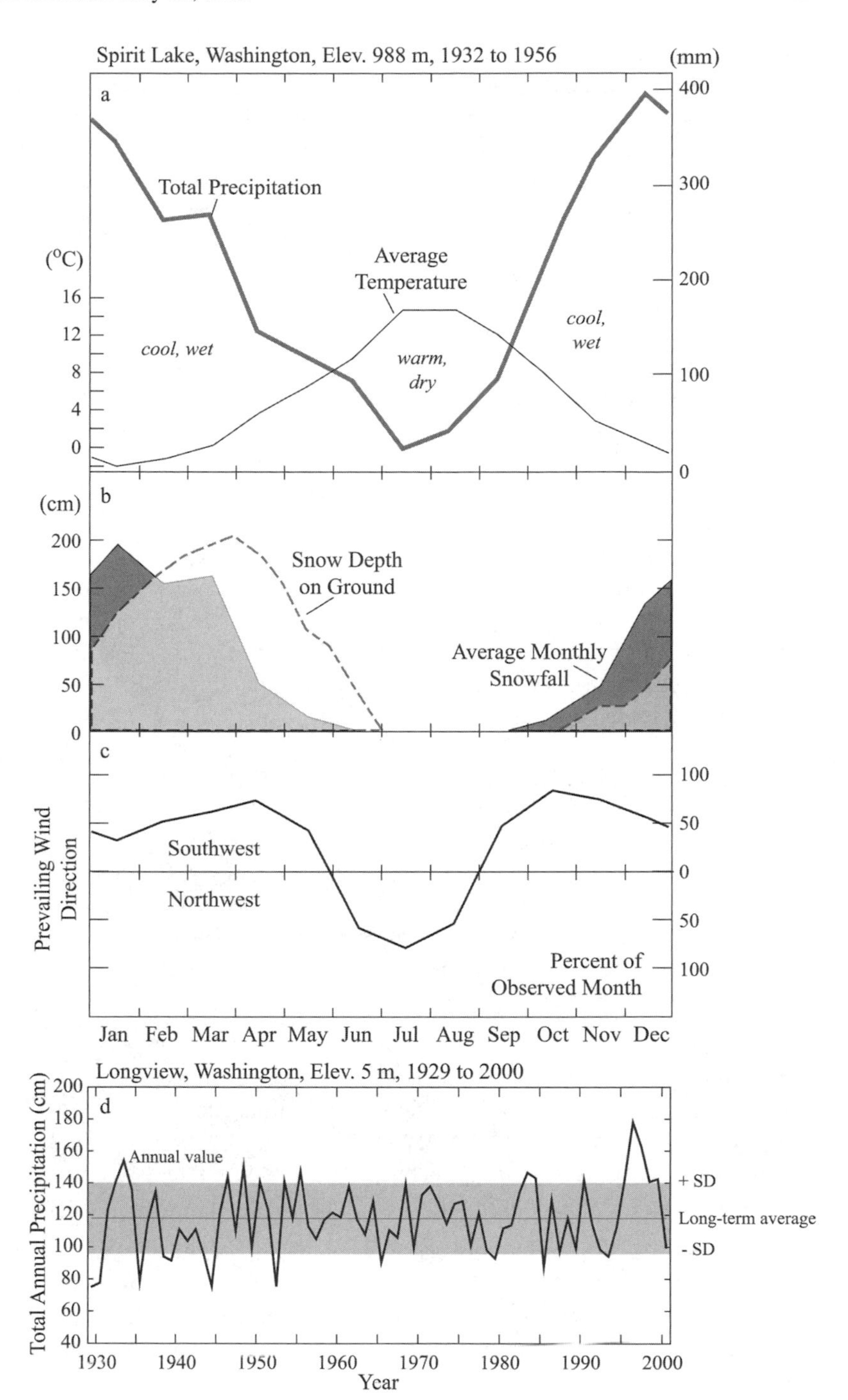

FIGURE 2.5. Climate of the Mount St. Helens area. (a) Climograph from Spirit Lake [after Walter (1973)]. (b) Average monthly snowfall and snow depth on ground. (c) Prevailing monthly wind directions and percent of observed months on record having prevailing directions. Mean number of monthly wind observations = 12. (d) Precipitation record for 1929 to 2000 at Longview, Washington, located 60 km west of Mount St. Helens. [(a), (b), and (c) are based on U.S. Weather Bureau (1932–1946, 1964) data collected at Spirit Lake (Figure 2.1; Easterling et al. 1996); (d) is based on U.S. Weather Bureau (1929–2000) data from Longview, Washington (Easterling et al. 1996) and updated by the Carbon Dioxide Information and Analysis Center, Oak Ridge National Laboratory.]

the Pacific Ocean and encountering the north-south Cascade Range. Broadly speaking, moisture decreases from west to east but increases with elevation. Temperature decreases from low to high elevation, and annual maxima increase from west to east as minima decrease.

The Pacific Northwest region experiences mild, wet winters and warm, dry summers. Mean annual precipitation at Spirit Lake (elevation, 988 m) was 2373 mm (from 1932 to 1962), of which only 162 mm (6.8%) fell between June and August (Easterling et al. 1996; Figure 2.5). Below about 600 m in elevation, precipitation falls mainly as rain. A seasonal snow pack occurs above an elevation of approximately 1000 m, and above an elevation of 1200 m, snowpacks of more than 3 m are common and may persist into July. An extensive area of intermediate elevation (approximately 300 to 1000 m) is within the transient snow zone, where rain-on-snow events can occur several times a year and can trigger major flooding (Harr 1981). Mean maximum and minimum temperatures in July are 22.3° and 7.3°C and in January are 0.4° and −4.4°C (Easterling et al. 1996). High topographic relief, steep slopes, and complex vegetation and wind patterns create very irregular patterns of snowpack across the landscape, especially in the spring when thick packs can persist on cold, north-facing, heavily forested sites while sunnier nearby sites at the same elevation can be snowless. These patchy, late-spring snowpack conditions existed at the time of the 1980 eruption, influencing the patterns of survival in both aquatic and terrestrial systems across several hundred square kilometers north of the mountain.

Precipitation in the vicinity of Mount St. Helens has varied substantially over the middle and late 20th century when records are available (see Figure 2.5d). Several decade-long periods of wetter- and dryer-than-average conditions may have affected terrestrial and aquatic ecological systems.

Prevailing winds are an important factor in the Mount St. Helens environment. They determine the pattern of tephra fall during eruptions, the spread of wildfire, and dispersal of wind-transported organisms and propagules that recolonize disturbed sites. Prevailing winds are from the southwest, except in summer, when they blow from the northwest. East winds during late summer and fall can be important drivers of wildfire, and during other seasons, east winds can cause windthrow and damaging ice accumulations on vegetation.

2.3 Ecological Setting

The Mount St. Helens area before 1980 was in many ways typical of Cascade Range fauna, flora, and ecological processes (Franklin 1966; Franklin and Dyrness 1973; Ruggiero et al. 1991; Matin 2001). Ecological conditions of large volcanic peaks within the Cascade Range can be considered in terms of three geographical strata:

- The volcano itself;
- Neighboring areas with little influence of past eruptive events (e.g., older geological terrain in the upwind direction, generally west and south of the volcano); and
- Neighboring areas with a strong, persistent legacy of tephra fall in soil profiles (e.g., northeast of the volcano).

These distinctions are particularly useful when examining effects of large-scale disturbance events, such as the 1980 eruption of Mount St. Helens, because conditions within each of these areas can strongly affect and are influenced by large-scale disturbance processes.

2.3.1 Soil

Soils in much of the southern Washington Cascades share common parent material broadly categorized as volcanic rocks, but differ in age and texture and, hence, in degree of development (Haagen 1990; D. Lammers, USDA Forest Service, Pacific Northwest Research Station, Corvallis, OR, personal communication). Despite this common parentage, shortly before the 1980 eruption, soil properties varied greatly in the vicinity of Mount St. Helens and were related to prevailing wind patterns and distance from the volcano. Soils differ substantially among the three geographical strata of the volcanic cone, areas with tephra deposits downwind from the volcano, and relatively tephra-free areas in other quadrants around the volcano.

The volcanic cone was steep and covered with young volcanic materials exposed to a cold-winter–dry-summer climate, so soils were shallow and undeveloped, termed Entisols by soil scientists. In vegetated areas on the lower flanks, a thin surface A horizon had formed over unweathered material of the C horizon.

Soils in extensive areas of southwestern Washington are formed in tephra deposits from prehistoric eruptions of Mount St. Helens and other volcanoes, and some sites have tephra from more than one volcano. In areas of young, thick tephra deposits, soils are poorly developed Entisols, and older deposits have had time to develop Andisols and Spodosols. Buried soils commonly occur within the tephra deposits because intervals between deposition events often were long enough for accumulation of organic material (e.g., forest litter) and for development of weathering zones in the fine-grained ash deposits that commonly cap tephra deposits (Franklin 1966). Trees can be observed rooted in older tephra units and apparently survived deposition of up to several tens of centimeters of tephra. Trees growing on one soil surface commonly extended their root systems downward through coarse-grained, "popcorn" pumice tephra and into more nutrient-rich, buried soils and finer-grained deposits, producing layered root systems (Yamaguchi 1993).

Upwind of the volcano, the land surface had received little tephra from recent eruptions, so soils had been in place for sufficient time to develop a B horizon by weathering of volcanic glass and formation of the amorphous material allophone. These soils are classified as Andisols and Spodosols.

In addition to volcanic influences, soil development in the vicinity of Mount St. Helens has also been influenced by glacial, alluvial, and soil displacement from hill slopes.

2.3.2 Terrestrial Vegetation

Terrestrial vegetation of the Pacific Northwest is dominated by a few coniferous tree species that have long life spans (e.g., 300 to 1000 or more years) and attain massive size of individual trees and huge biomass of forest stands (Franklin and Dyrness 1973; Waring and Franklin 1979; Franklin and Hemstrom 1981). Forest vegetation reflects environmental forces, including the wet, cool, seasonal climate and the influences of diverse disturbance processes, especially fire and wind (Waring and Franklin 1979). This environmental variation also results in some nonforest vegetation types in Pacific Northwest mountain landscapes (e.g., wet and dry meadows, frequently disturbed shrub communities in riparian zones, and varied low-elevation wetlands).

Before the 1980 eruption of Mount St. Helens, plant life within about 30 km of the volcano was diverse with regard to the number of plant species, assemblages, and physiognomy. The volcanic cone was dominated by alpine and subalpine communities, and its lower flanks and surrounding area were extensively forested. The biotic diversity of this area developed in response to the broad range of elevation (500 to 3010 m), complex topography, wet and cool climate, past disturbances (both natural and anthropogenic), and variation in snow conditions, among other factors, which created a complex template influencing species distributions, plant growth, and community development. Of the floristic surveys covering different parts of the Mount St. Helens area conducted before the 1980 eruption (Lawrence 1954; Franklin 1966, 1972; St. John 1976; Franklin and Wiberg 1979; Hemstrom and Emmingham 1987; Kruckeberg 1987), the most comprehensive was St. John's 1925 survey of the north and south sides of the volcano. St. John (1976) lists approximately 315 species, including 17 fern, 13 conifer, 68 monocot, and 217 dicot taxa. The total number of plant species found in the larger area, including lowland, wetland, and riparian habitats not sampled by St. John, was likely significantly greater.

Forests dominated the preeruption landscape. The area supported large expanses of coniferous forests that were punctuated by plantations of young forest created following clear-cut logging. Deciduous trees played a minor role and were often associated with areas of disturbance, notably riparian zones. Streamside areas supported deciduous shrubs and trees, especially red alder (*Alnus rubra*), black cottonwood (*Populus tricocarpa*), and willow (*Salix* spp.). The dominant tree species varied with elevation and time since past disturbance. Three general forest zones are included in this area: the western hemlock (*Tsuga heterophylla*), Pacific silver fir (*Abies amabilis*), and mountain hemlock (*T. mertensiana*) zones of Franklin and Dyrness (1973). The western hemlock zone extends up to about 900 m in the vicinity of Mount St. Helens and commonly includes Douglas-fir (*Pseudotsuga menziesii*) and western red cedar (*Thuja plicata*) along with western hemlock. The Pacific silver fir zone, which occurs from about 900 to 1300 m, also contains noble fir (*Abies procera*), Douglas-fir, western hemlock, western red cedar, and western white pine (*Pinus monicola*). The mountain hemlock zone, ranging from 1250 to 1600 m, includes Pacific silver fir and Alaska yellow cedar (*Chamaecyparis nookatensis*) as well as mountain hemlock.

The composition, abundance, and cover of forest understory communities were highly variable. Elevation and pre-1980 tephra-fall deposits played a major role in shaping understory development. In general, understory vegetation was composed of a mix of mosses, herbs, shrubs, and slow-growing, shade-tolerant tree species. Ericaceous shrubs were conspicuous components of the understory. Important herb species belonged to the Ericaceae, Saxifragaceae, Liliaceae, and Scrophulariaceae families. Ferns were widespread and frequently abundant. Mosses were common on downed wood and the forest floor, and lichens hung from the tree canopies and clung to tree boles. Beargrass (*Xerophyllum tenax*), a liliaceous species conspicuous in most Cascade Range subalpine landscapes, was relatively rare, perhaps because of repeated tephra falls and associated unfavorable soil conditions.

The vegetation at and above tree line at Mount St. Helens differed from that found on nearby volcanic peaks, such as Mount Rainier, Mount Adams, and Mount Hood, in several notable ways. First, the irregular tree line on Mount St. Helens was at an elevation of approximately 1340 m, compared to 1840 to 1980 m for the adjacent volcanoes, and was still advancing up the north slope of the volcano in response to volcanic disturbance in AD 1800 (Lawrence 1954; Kruckeberg 1987). This advance suggests that the tree line before the 1980 eruption was below the contemporary climatic limit for this latitude. Lawrence (1954) referred to this phenomenon as "trees on the march." Second, tree species composition at tree line consisted of an unusual mix of conifers and hardwoods (e.g., cottonwood, Douglas-fir, noble fire, western white pine, and lodgepole pine). Third, the meadows were limited in area, depauperate in species, and dry. At tree line, subalpine fir (*Abies lasiocarpa*) and mountain hemlock grew in dense, widely spaced patches, creating a parkland in the dry-meadow vegetation.

Similar to the well-documented Mount Rainier landscape (Hemstrom and Franklin 1982), the patterns of conifer forest age classes in the area surrounding Mount St. Helens reflected a millennium of wildfire and a half century of forest cutting. At Mount St. Helens, however, recent, vigorous volcanic activity gave volcanism a greater role in shaping the forests of the area (Yamaguchi 1993) than occurred in the vicinity of other volcanoes of the region, where most forests postdate significant volcanism. Forest age classes at Mount St. Helens had great diversity, such as very young stands in recently clear-cut sites, stands dating from fires in the late 19th and early 20th centuries in the Mount Margaret area, and older stands, including individual trees and groves dating to at least the 13th century (Yamaguchi 1993).

Nonforest vegetation types in the Mount St. Helens landscape were meadows, wetlands, cliffs, seeps, and avalanche paths. Collectively, these landscape features comprised only about 5% of the area. Meadow types included those located above tree line on the volcano that were composed of grasses and herbs and some subalpine heather species capable of growing on well-drained, steep scree slopes and more luxuriant meadows in upper-elevation areas, such as in the Mount Margaret area. Wetland communities occurred in scattered natural topographic depressions, near-shore environments of some lakes, streamside channels, and beaver ponds. Willows, sedges (*Carex* spp.), rushes (*Juncus* spp.), and bulrushes (*Scirpus* spp.) were common wetland plants. Snow-avalanche channels typified by steep slopes and deep snowpack supported dense growth of shrubs, especially sitka alder (*Alnus viridis*).

2.3.3 Terrestrial Animals

Before the 1980 eruption, the Mount St. Helens landscape provided habitat for a diverse fauna characteristic of montane, alpine, riparian, and aquatic habitats in western Washington (Aubry and Hall 1991; Manuwal 1991; Thomas and West

1991; West 1991; Martin 2001). Numerous invertebrate taxa, from nematodes to crawfish, and all five classes of vertebrates (amphibians, reptiles, fishes, birds, and mammals) were represented in the preeruption landscape. The presence, total coverage, and juxtaposition of specific habitat types in the landscape strongly influenced the distribution of species; and food availability, cover, parasites, predators, and weather determined their population sizes. Most forest-dwelling species were probably broadly distributed, whereas species associated with water, meadows, and rock (cliff or scree) were patchily distributed.

2.3.3.1 Vertebrates

Given the diversity of forest, meadow, and aquatic habitats that were present in the preeruption landscape, it is likely that most, perhaps all, of the vertebrate species indigenous to the southern Washington Cascade Range existed in the large area severely impacted by the 1980 eruption (see Swanson and Major, Chapter 3, this volume). The preeruption fauna was composed entirely of native species, with the notable exception of introduced fish (discussed below). Each of the five vertebrate classes is briefly described next.

Amphibians: Fifteen amphibian species, 5 frog and toad and 10 salamander and newt species, have ranges that extend into the Mount St. Helens area (Nussbaum et al. 1983; Crisafulli et al., Chapter 13, this volume). Of these, 3 are found only in forests; 2 live in streams as larvae and in forests as adults; 7 live in lakes as larvae and in forests, riparian zones, or meadows as adults; 2 are seep-dwellers; and 1 is a denizen of streams. Collectively, they are a taxonomically unique and diverse assemblage with several members having very specific habitat requirements. Because amphibians are thought to be highly sensitive to environmental change, have variable life-history strategies, and use several habitat types, they would presumably be profoundly impacted by volcanic-disturbance processes. Furthermore, 12 of the 15 species use streams or lakes for some part of their life cycles, so changes in the amount, distribution, and quality of water could have serious consequences for amphibians.

Reptiles: With a cool, wet environment, the Mount St. Helens area supported only four species of reptiles: two garter snakes (*Thamnophis sirtalis* and *T. ordinoides*), the rubber boa (*Charina bottae*), and the northern alligator lizard (*Elgaria coerulea*) (Nussbaum et al. 1983). Disturbances that remove forest canopy and increase temperature and dryness of localized habitats may improve conditions for all four of these reptile species. Thus, reptiles could respond favorably in the aftermath of past and future eruptions.

Birds: Approximately 70 to 80 permanent-resident and summer-breeding bird species reside in the southern Washington Cascade Range. In addition to these species, numerous other bird species pass through the area during fall and spring migrations. About 36 nonraptor species are associated with montane forests, including a core group of 10 ubiquitous species (Weins 1978; Manuwal 1991), about 6 species primarily associated with alpine habitats, 12 species that use lakes or streams, and about 16 raptors.

Perhaps more than any other vertebrate group, birds are ostensibly tightly coupled with habitat structure. Thus, alteration of avian habitat would be expected to lead to dramatic and predictable changes in bird species composition and abundance patterns across a disturbance-modified landscape. The pace of bird community response to disturbance would be expected to follow the pace of development of suitable habitat, which could be very slow, such as in the case of old-growth forest habitat.

Mammals: Likely 55 species of mammals, representing 7 orders and 20 families, were present on and adjacent to Mount St. Helens before the 1980 eruption (Dalquest 1948; Ingles 1965; West 1991; Wilson and Ruff 1999). These mammals include a diverse group of species that range in mass from the tiny and highly energetic Trowbridge's shrew (*Sorex trowbridgii*), weighing a mere 5 g, to majestic Roosevelt's elk (*Cervus elaphus roosevelti*), weighing nearly 500 kg. Other large mammals in the area included black-tailed deer (*Odocoileus hemionus columbianus*), mountain goat (*Oreamnos americanus*), American black bear (*Ursus americanus*), and mountain lion (*Puma concolor*). Many of the most abundant mammals (e.g., bats, mice, voles, and shrews) were inconspicuous because of their nocturnal schedule or cryptic habits. Most of the species associated with coniferous forest habitats were probably broadly distributed, whereas species found in meadows, such as the northern pocket gopher (*Thomomys talpoides*), or those tightly associated with riparian areas were likely spottily distributed across the landscape.

Mammals play important roles in ecosystem processes:

- By mixing soil
- Through trophic pathways as herbivores and secondary and tertiary consumers
- As scavengers
- As dispersers of seeds and fungal spores
- As prey for birds and reptiles

(See Crisafulli et al., Chapter 14, this volume, and references therein.) Gophers, for example, mix fresh, nutrient-deficient tephra with older soil as they forage for belowground plant parts. Bats and shrews consume enormous quantities of insects while foraging. Chipmunks (*Tamias* spp.) and deer mice (*Peromyscus* spp.) gather and cache seeds of trees, shrubs, herbs, and grasses, and many of these caches are not reclaimed, leading to the establishment of new plant populations. Similarly, ungulates consume seeds that are later egested at distant locations and germinate. Beavers (*Castor canadensis*), voles, and elk can strongly influence the cover, amount, form, and species of plants through herbivory.

We expect mammals to have quite varied responses to volcanic eruptions. Many species have high vagility, so they can flee some volcanic events and rapidly disperse back into disturbed areas. More-cryptic species with underground habits could survive some types of eruptions but not extremely severe ones; also, longer-term survival is not assured if food and other resources are inadequate for continued life.

2.3.3.2 *Invertebrates*

The invertebrate fauna of the Mount St. Helens landscape was poorly known before the 1980 eruption, but the inventory of an Oregon Cascade Range forest landscape by Parsons et al. (1991) provides a relevant frame of reference. This work at the H.J. Andrews Experimental Forest about 200 km to the south identified more than 4000 species of terrestrial and aquatic insects and other arthropods, including a rich spider fauna, occupying terrestrial habitats as diverse as deep within the soil, in the tops of 70-m-tall trees, and within the soggy interior of rotting logs. Several groups or habitats may not have been exhaustively sampled during this inventory, so the actual invertebrate richness in the preeruption Mount St. Helens landscape may have been substantially greater.

Arthropods are expected to have strong negative responses to volcanic disturbance because of the loss of habitat and food resources, such as foliage consumed by phytophagous insects; physiological effects of abrasive, desiccating tephra; and burial of the forest floor with tephra that would alter and seal off the belowground organisms from the surface. The reduction or elimination of insects has important implications during biological reassembly because of the roles these species play in pollination and as prey for secondary consumers, such as birds, mammals, and amphibians. On the other hand, many taxa are highly mobile and quickly reinvade disturbed landscapes.

2.3.4 Lakes

Before the 1980 eruption, approximately 39 lakes were located within 30 km of Mount St. Helens. Of these lake basins, 32 were formed by Pleistocene glacial activity, and the remaining 7 were formed where volcanic debris or lava flows blocked streams (see Figures 2.2 and 12.1). Glaciers carved cirques into bedrock on north- and northeast-facing slopes in upper-elevation areas of the highest parts of the landscape. Many of these cirques contain small lakes (surface areas ranging from 1.5 to 32 ha) at elevations ranging from 1000 to 1500 m. These lakes, extending in a band from Fawn Lake on the west to Strawberry Lake at the east, have persisted since their formation more than 10,000 years ago. Lakes above 1000 m were generally ice covered for 5 to 7 months of the year.

In contrast, lakes formed by volcanic-flow deposits from Mount St. Helens were generally restricted to a zone 8 to 12 km around the volcano, where slopes are gentle enough to cause volcanic flows from the cone to stop and form deposits thick enough to block stream drainages. In at least one instance, mudflows from Mount St. Helens traveled more than 45 km down the Toutle River to block a stream and create Silver Lake. These blockage lakes occur at lower elevations (760 to 1040 m) than do the cirque lakes and range in size from less than 1 ha to the 510-ha extent of Spirit Lake. The small, low-elevation lakes have shorter lifetimes than cirque lakes because they may fill with sediment or their outlet streams may cut through the weak deposits forming the blockage. Filled and partially breached lakes commonly became wetlands, such as Goat Marsh southwest of the volcano.

Before the 1980 eruption, ecological study of lakes in the vicinity of Mount St. Helens was scant, but some information from the area and other Cascade Range lakes gives a useful picture of pre-1980 lakes (Wolcott 1973; Bortleson et al. 1976; Crawford 1986; Dethier et al. 1980; Wissmar et al. 1982a,b; Bisson et al., Chapter 12, this volume; Dahm et al., Chapter 18, this volume). Lakes in this subalpine, forested, mountain landscape generally had clear, cold, well-oxygenated water with low levels of nitrogen, phosphorus, and other nutrients that reflect:

- The low-nutrient water delivered to the lakes by streams, groundwater, and precipitation;
- The young age of the lakes;
- The low solubility of geological parent material; and
- The cool, wet climate.

These water-quality conditions supported only sparse phytoplankton and zooplankton communities and only several tens to perhaps a hundred aquatic insect species. Amphibians, particularly the aquatic forms of the northwestern salamander (*Ambystoma gracile*), were probably widespread, occurred at high densities, and served as a top predator in most lakes without fish or before fish introductions. Lakes were important habitat for waterfowl and shorebirds, for semiaquatic mammals (such as beavers), and for riparian-associated species (such as the northern water shrew and water vole).

Fish are a critical part of the ecological story of Mount St. Helens lakes. Before active management by European settlers, most lakes were barren of fish, but widespread stocking for several decades resulted in the presence of introduced trout in many of the lakes. However, Spirit Lake was connected to a river without barriers, permitting the presence of coastal cutthroat trout (*Oncorhynchus clarki clarki*), coastal rainbow trout (winter steelhead, *O. mykiss irideus*), and coho salmon (*O. kisutch*). Beginning in the early 1900s and continuing through 1979 (the year before the eruption), fish were repeatedly planted in most of these lakes.

The decades-long presence of fish in lakes that were naturally without fish probably dramatically altered lake conditions, resulting in a biota that was far from pristine. Planted fish likely exerted a top-down effect through predation on amphibians, crawfish, large mobile macroinvertebrates, and zooplankton (Pilliod and Peterson 2001; Schindler et al. 2001). In turn, predation on these species likely rippled through the aquatic food web (Parker et al. 2001).

Volcanic and other disturbance processes can have both transient and persistent effects on lake ecosystems. Tephra deposits from pre-1980 eruption accumulated in pumice beaches and deltas along the shores of some lakes. These deposits had important biological effects, such as influences on development of emergent plant communities and breeding, rearing, and foraging sites for insects, snakes, amphibians, birds, and mammals. Disturbances that caused major, short-term alteration in water quality, water quantity, habitat, or biotic structure could have profound impacts on the biological diversity of both lakes and the landscape as a whole.

2.3.5 Streams and Rivers

The area within 30 km of Mount St. Helens is steep and highly dissected by a complex, high-density drainage network with origins in headwater seeps, springs, lakes, or glaciers (see Figure 2.2). Streams and rivers draining the older geological terrain of the Mount St. Helens landscape (see Figure 2.3) were cold (8°–15°C), clear, and fast-flowing over irregular beds of boulders, bedrock, and fallen trees in forested areas neighboring the volcano. Muddy River and other streams fed by glaciers on Mount St. Helens, on the other hand, ran turbid with silt in the summer. Streams cutting through recent volcanic deposits commonly had mobile streambeds and banks. Stream discharge through steep, lower-elevation channels was highly seasonal, driven by the wet-winter, dry-summer climate and by spring melting of the seasonal snowpack in upper-elevation areas.

Specific descriptions of the preeruption biota and ecological processes in these streams are unavailable in most cases (however, see general descriptions on amphibians, birds, mammals, and fish above), so we draw on findings from studies of similar stream systems elsewhere in the Cascade Range.

The vast majority of streams are small (1 to 3 m wide) and shallow, with high-gradient channels composed of steep cascades and small pools with boulder, cobble, and bedrock substrate. Large amounts of downed wood, including whole, massive trees, strongly influence the physical and biological characteristics of small Cascade streams (Triska et al. 1982). Dense coniferous forest canopies intercepted sunlight and limited primary productivity within these headwater systems. Thus, food webs were largely driven by organic matter entering streams from adjacent forests. The steep, straight tributary streams plunged rapidly down slope and entered larger, lower-gradient streams that flowed through broad, U-shaped valleys, such as Clearwater Creek and Green River valleys. Localized floodplains along these larger streams provided room for some channel meandering and formation of secondary channels. Streambeds were composed of cobble and gravel substrates and scattered log jams. These streams were wide enough to allow sunlight to reach channels and promote growth of diatoms, algae, and mosses on the streambed and banks. These streams also had steep reaches, including impressive waterfalls with deep plunge pools. Finally, these midsized streams flowed into the larger rivers draining the Mount St. Helens landscape. These large rivers had more extensive floodplains, gravel bars, and riffles, but large downed wood was less common than in smaller channels because of the transport capacity of high flows in wide channels. Ample sunlight reached the stream to promote food webs based on the stream's primary production (e.g., by diatoms, filamentous algae, and macrophytes) as well as on material transported from upstream and from tributaries. Productivity of the streams increased from small headwater streams to large rivers.

Fish are an important component of the four river systems draining the Mount St. Helens area, the Toutle, Kalama, Lewis, and Cispus (see Figure 2.2). These rivers eventually flow into the Columbia River, which, in turn, flows to the Pacific Ocean. Geological processes and landforms have modified these river networks over time, affecting their function as corridors for dispersal and creating natural barriers, such as waterfalls. These habitat features determined the fish species distributions within these watersheds before the 1980 eruption. Eight families, including about 25 species, represent the preeruption fish fauna (see Bisson et al., Chapter 12, this volume). The native fauna was a mix of anadromous and resident species, including lampreys (*Lampetra* spp.), salmon (*Onchorhynchus* spp.), trout (e.g., *Salvelinus confluentus*), and sculpins (*Cottus* spp.) (Reimers and Bond 1967; Wydoski and Whitney 1979; Behnke 2002). In addition to the native fishes, several trout species (e.g., brook, rainbow, and brown trout) had been stocked in the streams and rivers draining Mount St. Helens.

The specific distribution of most of these species was poorly documented as of 1980, but substantial information existed for a handful of salmonid species that had commercial or sport value. Most notable were the spectacular runs of steelhead (*O. mykiss irideus*) in the Toutle River and Kalama River systems, coho salmon (*O. kisutch*) runs in the Toutle River, and Chinook salmon (*O. tshawytscha*) in the Lewis River. Most of the nonsalmonids were likely confined to the lower reaches of these river systems, where the water was warmer, slower, and deeper.

The salmonids and some other fish species (e.g., sculpin) in the Mount St. Helens landscape required cold, clear water with gravel streambeds for foraging and spawning and would be expected to suffer deleterious effects of a major disturbance of these habitat features. Clearly, salmonids had to contend with major habitat disruption in the past but have several attributes that enabled them to persist or even flourish in this dynamic volcanic landscape. Chief among these characteristics are their tendency to stray from one watershed to another; their life-history trait of having one to several cohorts of their population at sea for several years; and their high mobility, which enables them to recolonize areas once conditions have improved.

2.4 Ecological Disturbance

Steep, forested volcanic landscapes in wet climates, such as the Pacific Northwest, are subject to a great variety of natural and human-imposed ecological disturbance processes, ranging from strictly geophysical processes, such as those associated

TABLE 2.2. Chronology of known, natural, stand-replacing disturbance events affecting forests near Mount St. Helens during the past 700 years.

Year or interval (AD)	Type of disturbance	Sector	Reference
May 18, 1980	Forests killed by lateral blast and associated events	N	Lipman and Mullineaux (1981)
1885	Forests on floor of upper South Fork Toutle River valley inundated by mudflow	W	
1844	Forests on north floor of South Fork Toutle River valley inundated by mudflow	W	Yamaguchi and Hoblitt (1995)
1800	Forests on flank of volcano inundated by lava flow; trees along flow edges killed or injured by heat	NW	Lawrence (1941); Yamaguchi and Hoblitt (1995)
1800	Forests killed by tephra fall (layer T)	NE	Lawrence (1939); Lawrence (1954); Yamaguchi (1993)
Ca. 1780	Hot mudflow kills forests on Pine/Muddy fan	SE	Yamaguchi and Hoblitt (1995)
1722	Mudflows down Muddy River briefly dam Smith Creek; forests at Muddy/Smith confluence buried	SE	
1647–mid-1700s	Mudflows down lower Muddy River kill at least some valley-bottom stands	SE	
1647	Pyroclastic flow inundates forests in upper valley of South Fork Toutle River, and, possibly, Castle Creek	W	Hoblitt et al. (1981); Yamaguchi and Hoblitt (1995)
1482	Forests killed by tephra fall (layer We); area, 30 km^2	E	
1480–1570	Forests on floor of Kalama River inundated by pyroclastic flows	SW	Yamaguchi and Hoblitt (1995)
1480	Forests killed by tephra fall (layer Wn); area, 65 km^2	NE	
Ca. 1300	Forests killed by extensive fire(s); area greater than 100 km^2	NE, E, also possibly W	Yamaguchi (1993)

Note that fire history was investigated only in the northeast quadrant (Yamaguchi 1993).
Source: Data from Yamaguchi and Hoblitt (1995) unless otherwise noted.

with volcanic eruptions, to strictly biological processes, such as insect and disease outbreaks (Lawrence 1939, 1941; Hemstrom and Franklin 1982; Yamaguchi 1993; Yamaguchi and Hoblitt 1995). Many disturbance processes involve interaction of geophysical and biological components, such as landform effects on fire and wind toppling of trees (Swanson et al. 1988; Sinton et al. 2000).

The pace and complexity of forest-disturbance history in the vicinity of Mount St. Helens are represented by a chronology of known stand-replacing disturbances during the 700 years leading up to the 1980 eruption (Table 2.2). The occurrence of disturbance events and the state of our knowledge about them are not uniform around the volcano. Tephra-fall events, well documented by the resulting deposits, are concentrated in the northeast quadrant from the volcano. On the other hand, in the past 700 years, wildfires probably affected nearly all forests surrounding the volcano with a mean recurrence interval of several centuries (Hemstrom and Franklin 1982; Agee 1993; Yamaguchi 1993). During the past 700 years, forests, lakes, and streams northeast of the volcano experienced several tephra-fall events and one extensive and several less-extensive wildfires. Many streams draining the volcano as well as many plant and animal communities were disturbed periodically by hot pyroclastic flows and cold mudflows from the upper slopes of the volcano. During the past millennium, large floods occurred without the influence of volcanic processes. These floods triggered small landslides on hill slopes, debris flows down stream channels, and lateral channel migration along larger rivers in the area, as observed for such storms elsewhere in the Cascade Range (Swanson et al. 1998).

Humans, first indigenous people and then European settlers, had many and varied effects on the landscape. Most conspicuously, forestry operations in the vicinity of Mount St. Helens in the several decades leading up to 1980 profoundly altered the stature, composition, and spatial pattern of forest cover. Clear-cutting opened the forest canopy in scattered 20-ha patches on USDA Forest Service lands east and south of the volcano. Much larger clear-cut patches were created on state and forest industry lands to the south, west, and northwest. These clear-cut areas were typically planted to monocultures of Douglas-fir below an elevation of 1000 m and to noble fir at higher elevation. These plantations developed into dense, young stands within one to two decades. A network of forest roads provided access for forestry operations, recreation, mining, and other uses. The rugged, sparsely forested Mount Margaret country north of the volcano experienced little logging and was designated mainly for high-country recreational use without roads. Fire suppression throughout the area tended to reduce fire extent at the same time that other human uses of the landscape increased fire ignitions.

Major geological and ecological changes have occurred at a similar pace in the vicinity of Mount St. Helens. The multicentury life spans of the dominant tree species in the area (Waring and Franklin 1979) and the frequency of nonvolcanic processes, such as major wildfires, are comparable to the mean quiet interval between eruptive periods. Close to Mount St. Helens, disturbance by volcanic processes may preclude some nonvolcanic disturbances, such as the effect of tephra-fall deposits that reduce the amount and continuity of forest fuels, thereby limiting spread of wildfire. Farther from the peak, volcanic disturbances may simply add to the disturbance regime of nonvolcanic processes.

2.5 Conclusions

Volcanoes are distinctive types of mountains. They repeatedly send rock debris into the surrounding landscape, reminding people of their presence. As a result of frequent eruptions during the past 40,000 years, Mount St. Helens has become an island of young, volcanically constructed deposits and topography set in a far older, deeply eroded landscape. Records and biological legacies of the many volcanic events are contained in deposits, landforms, and ecosystems around the volcano. The long temporal reach of volcanic influences prompts consideration of history in assessing the effects of new events, such as the 1980 eruption.

As discussed in Chapter 3, the 1980 events at Mount St. Helens follow the pattern to which this volcano has been accustomed. An eruptive period lasting less than a year to many decades is followed by several centuries of quiescence. A new eruptive period involves various volcanic and associated hydrologic processes occurring over a period of months to a few years. Both primary and secondary erosion processes alter the form of the main cone and the surrounding landscape, creating and modifying a variety of terrestrial and aquatic habitats.

Ecological features of the Mount St. Helens landscape are broadly representative of the Cascade Range in many specific respects and of more extensive temperate ecosystems in some general respects. These similarities include the following:

- A diverse flora is dominated by a few coniferous tree species that attain massive size and biomass. Also, many nonforest vegetation types comprise an important, but small, proportion of the landscape.
- The terrestrial fauna includes diverse vertebrate assemblages, composed of several regional endemic species, and thousands of invertebrate species.
- Stream and lake ecosystems are varied and, although limited in productivity by the clear, cold, low-nutrient waters, they provide habitat for numerous species that contribute to the biodiversity of the landscape.
- These ecological systems respond at varying paces to disturbances, ranging from the rapid chemical transformations occurring in lakes to the slow growth of long-lived trees to the multimillennial time scale of soil development.

Despite the history of frequent, severe disturbance in the vicinity of Mount St. Helens, these species and ecological systems have been very persistent for a variety of reasons. The complex terrain and diversity of microhabitats leaves many refuges after severe disturbance by fire, volcanic, or other processes. The wetness of the landscape, including areas with persistent snow cover, can buffer ecological systems from some types of disturbance. Another key factor is that the deposits originating from Mount St. Helens, with a few notable exceptions, are sufficiently unconsolidated that plants can root and animals (e.g., pocket gophers and ants) can excavate and burrow. The importance of this characteristic is clear when comparisons of ecological development are made to basalt flows. A variety of life-history strategies contributes to the resistance and resilience of many taxa to disturbance. For example:

- Anadromous fish spend long periods at sea away from the influences of terrestrial disturbances.
- Ballooning spiders and many other taxa have the capacity to disperse over distances much greater than the size of individual disturbance patches.
- The landscapes surrounding even the largest disturbance patches in the Cascade Range harbor an intact pool of native species available for recolonizing disturbed areas.

Mount St. Helens is an exceptional setting, where scientists can address many aspects of this interplay of ecology and volcanoes, particularly for the temperate, forested landscapes common around the Pacific Rim. The Mount St. Helens landscape continues to change at a rapid pace, and prospects for future eruptive activity seem high, so its role as a special place for learning about succession, disturbance, and interactions of humans with volcanoes will persist.

Acknowledgments. This work was funded by the USDA Forest Service's Pacific Northwest Research Station and Gifford Pinchot National Forest, the National Science Foundation, and the National Geographic Society. Yamaguchi's work was supported in part by a National Research Council–USGS research associateship at the U.S. Geological Survey, Cascades Volcano Observatory, Vancouver, Washington. We have benefited from the knowledge of many people and give special thanks to Jerry Franklin, Jim MacMahon, Chuck Hawkins, Virginia Dale, and an anonymous reviewer for helpful reviews and to Duane Lammers for information about soils.

3 Physical Events, Environments, and Geological–Ecological Interactions at Mount St. Helens: March 1980–2004

Frederick J. Swanson and Jon J. Major

3.1 Introduction

The diversity and intensity of volcanic processes during the 1980 eruption of Mount St. Helens affected a variety of ecosystems over a broad area and created an exceptional opportunity to study interactions of geophysical and ecological processes in dynamic landscapes. Within a few hours on the morning of May 18, 1980, a major explosive eruption of Mount St. Helens affected thousands of square kilometers by releasing a massive debris avalanche, a laterally directed volcanic blast, mudflows, pyroclastic flows, and widespread tephra fall (see Figure 1.1; Figures 3.1, 3.2; Table 3.1). These primary physical events killed organisms, removed or buried organic material and soil, and created new terrestrial and aquatic habitats. Despite these profound environmental changes, important legacies of predisturbance ecosystems, including live organisms, propagules, and organic and physical structures, persisted across much of the affected landscape. The physical characteristics of the volcanic processes (elevated temperature, impact force, abrasion, and depth of erosion and burial) in part determined the extent of mortality and the types and significance of biotic legacies in the posteruption landscape.

The primary volcanic events of May 18, 1980 triggered a succession of interacting biological, geological, hydrologic, and anthropogenic changes in the Mount St. Helens area (see Figure 1.2). The initial events altered landforms, watershed hydrology, sediment availability and delivery, and the roles of vegetation and animals in regulating physical and biological processes. Concurrently, hydrologic and geomorphic processes altered the paths and rates of ecological responses by persistently modifying habitats, thereby favoring some species and ecological processes while impeding others. Understanding the interactions between geological and ecological processes is critical for interpreting ecological change in the posteruption landscape.

To set the stage for examining the responses of plants, animals, fungi, microbes, and ecological processes to the major 1980 eruption, we summarize volcanic events immediately preceding the eruption, those associated with that major eruption and subsequent eruptions, and characteristics of the posteruption landscape. We also highlight ecological conditions in the weeks leading up to the major eruption because some conditions at this proximal time scale strongly influenced posteruption ecological responses. In addition, we discuss new environments created by the eruption, variations in climate and hydrology that affected both secondary disturbances and the potential for ecological responses, and the nature and pace of change imposed by secondary disturbances on ecological processes through the first quarter century after the eruption. We explore concepts linking ecological succession to the succession of geomorphic processes that occur in response to large-scale, severe disturbances. Broad geographical and historical contexts in geological and ecological terms leading up to 1980 are described by Swanson et al. (Chapter 2, this volume). Here, we summarize the recent eruptive activity at Mount St. Helens in terms of its relevance to ecological responses. More comprehensive technical discussions of the 1980 eruptive activity can be found elsewhere (Lipman and Mullineaux 1981; Foxworthy and Hill 1982; Major et al. 2005).

3.2 Events of March Through May 18, 1980

Intrusion of magma high into the edifice of Mount St. Helens culminated in the geophysical events that so dramatically affected the landscape on May 18, 1980. The first indication of volcanic unrest occurred only 2 months before the major eruption. After 123 years of quiescence, Mount St. Helens awoke, on March 15, 1980, with a series of small earthquakes (Endo et al. 1981). Earthquakes increased markedly in magnitude (to 4.2) and frequency (30 to 40 per day) on March 20, heralding 2 months of continuous earthquake activity, episodic steam-driven explosions from the summit, and progressive outward growth of the volcano's north flank as magma intruded into the cone (Table 3.2). The injected magma fractured, steepened, and weakened the north flank of the volcano.

FIGURE 3.1. Photographs of the Mount St. Helens landscape showing conditions produced by the May 18, 1980, eruption: (a) debris-avalanche deposit (North Fork Toutle River valley looking east); (b) blowdown zone with extensive toppled trees (upper Green River valley looking south); (c) scorch zone with standing dead forest (upper Clearwater Creek valley); (d) tephra-fall zone; (e) pyroclastic-flow deposits on the Pumice Plain (light-colored features that flowed from left to right); (f) mudflow zone [Muddy River looking downstream; note helicopter (in *circle*) at right of photo for scale].

TABLE 3.1. Characteristics of volcanic events at Mount St. Helens, May 18, 1980.

Event	Volume of uncompacted deposit (km^3)	Area affected (km^2)	Deposit thickness (m)	Temperature (°C)	Organic matter
Debris avalanche	2.5	60	10–195	70–100	Rare
Blast	0.2	570			
Blowdown zone		370	0.01–1.0	100–300	Common
Scorch zone		110	0.01–0.1	50–250	Common
Mudflows		50	0.1–10	30	Common
Tephra fall	1.1	1000	>0.05	<50	Rare
Pyroclastic flows	0.3	15	0.25–40	300–850	None

Note that the total area of the lateral blast includes zones mapped as tree-removal, blowdown, scorch, debris-avalanche, and pyroclastic-flow zones in plate 1 of Lipman and Mullineaux (1981). The pyroclastic-flow zone occupies a part of the debris-avalanche zone. The area shown for the blowdown zone is for the area mapped as such by Lipman and Mullineaux (1981) and is included in the blowdown zone shown in Figure 3.2. We do not list characteristics of the tree-removal zone because it is a composite of multiple processes, so thickness, temperature, and organic matter are quite varied and difficult to characterize. The tephra-fall zone is considered only for the area outside the blast-affected area, where thickness exceded 5cm.
Source: Various sources in Lipman and Mullineaux (1981).

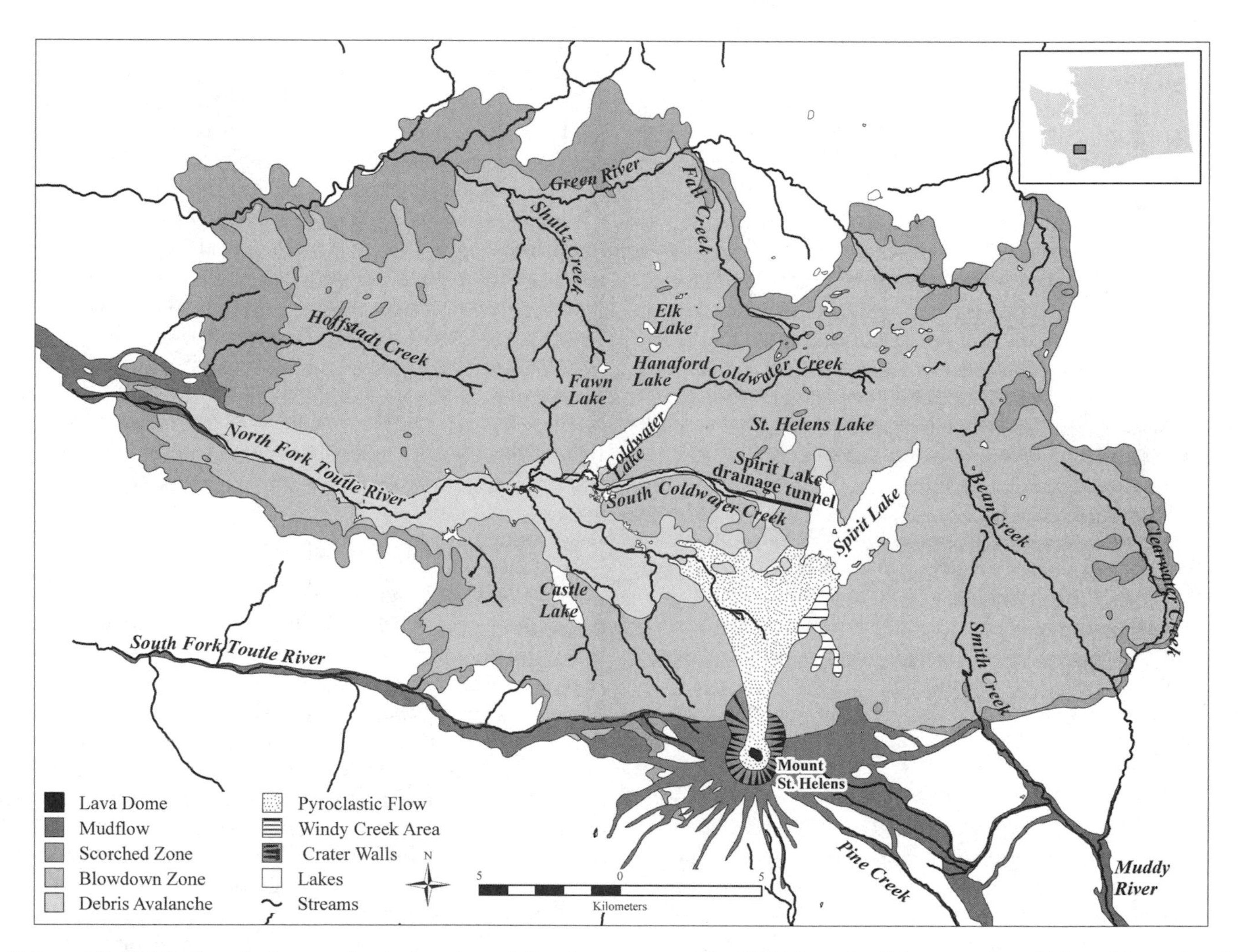

FIGURE 3.2. Distribution of primary volcanic deposits and disturbance zones of the 1980 Mount St. Helens eruptions. [Adapted from plate 1 in Lipman and Mullineaux (1981).]

TABLE 3.2. Chronology and characteristics of eruptive events at Mount St. Helens, 1980 to 2004.

Date	Events
March 27–May 18, 1980	Steam-blast explosions; deformation of north flank by magma injection; very little tephra fall beyond flanks of cone. $V = 0.0006$ km^3.
May 18, 1980	See Table 3.1.
May 25, 1980	Tephra fall greater than 1 cm thick over about 300 km^2 west and northwest of vent. $V = 0.031$ km^3.
	Pyroclastic flows reach Pumice Plain. $V = 0.001$ km^3.
June 12, 1980	Tephra fall greater than 1 cm thick over about 200 km^2 south. $V = 0.027$ km^3.
	Pyroclastic flows reach Pumice Plain. $V = 0.01$ km^3.
July to December 1980	Periodic growth of lava dome in crater. Some removal of dome during eruption of July 22.
July 22, 1980	Tephra fall less than 1 cm thick to east and northeast of vent. $V = \sim 0.004$ km^3.
	Pyroclastic flows reach Pumice Plain. $V = 0.006$ km^3. Partial removal of lava dome.
August 7, 1980	Tephra fall less than 1 cm thick to north and northeast. $V = 0.0008$ km^3.
	Pyroclastic flows travel more than 5.5 km from vent. $V = 0.004$ km^3.
October 16–18, 1980	Tephra fall less than 1 cm thick to southwest and southeast. $V = 0.0005$ km^3.
	Pyroclastic flows reach Pumice Plain. $V = 0.001$ km^3.
December 27, 1980 to October 21, 1986	Sixteen dome-growth events, two minor explosions, and three mudflows.
October 22, 1986 to September 2004	No magmatic eruptions; several minor phreatic explosions from dome.
September 2004	Lava dome growth; steam and ash eruptions

V, uncompacted volume of deposits.
Source: From Christiansen and Peterson (1981); Sarna-Wojcicki et al. (1981); Rowley et al. (1981); Swanson et al. (1983a); Brantley and Myers (2000).

Ecological and hydrologic conditions leading up to the morning of May 18 profoundly affected ecological responses to the eruption. For example, low- to mid-elevation (less than 1000 m above sea level) forests and meadows surrounding the volcano contained discontinuous patches of late-lying snow; but above about 1000 m, snow generally covered the terrain. That snow cover provided some protection from the eruption for terrestrial and aquatic organisms.

3.2.1 Debris Avalanche

The May 18, 1980, eruption commenced with a magnitude 5.1 earthquake and an associated collapse of the volcano's north flank at 8:32 A.M. (Voight et al. 1981, 1983). The resulting 2.5-km^3 debris avalanche, the largest landslide in recorded history, rushed northward in multiple pulses that broadly split into three lobes (Lipman and Mullineaux 1981; Voight et al. 1981; Glicken 1998) (see Figures 3.1a, 3.2). One lobe entered and passed through Spirit Lake, where it generated a seiche (an oscillating wave) that extended as high as 260 m above the pre-1980 lake level and raised the level of the lake by 60 m. A second lobe traveled northward 7 km at a velocity of 50 to 70 m s^{-1} (Voight 1981) and overtopped a 300- to 380-m-high ridge (later named Johnston Ridge). The bulk of the avalanche, however, moved 23 km westward down the North Fork Toutle River valley in about 10 minutes.

The resulting debris-avalanche deposit radically modified the upper North Fork Toutle River valley. It buried about 60 km^2 of the valley with hummocky, poorly sorted sand and gravel to a mean depth of 45 m (Voight et al. 1981; Glicken 1998; see Figures 3.1a, 3.2; see Table 3.1) and disrupted the drainage pattern (Lehre et al. 1983; Janda et al. 1984). The original locations of rock units and glaciers on the volcano and the flow paths of the various debris-avalanche pulses determined the distribution of rocks, soil, and ice in the heterogeneous avalanche deposit (Glicken 1998). The avalanche pulses that surged down the North Fork Toutle River valley smeared avalanche debris that contained little organic matter, a few toppled trees, and blocks of soil along the valley margin. The front of the debris avalanche scoured a forest from the valley floor and walls and left some of it as a tangled mass of vegetation at the downstream end of the debris-avalanche deposit. The margins of the upper half of the deposit chiefly comprise material from the first slide block that engulfed the outer flank of the volcano, which contained glacier ice and organic matter. That debris was emplaced at ambient temperature. In contrast, the central part, as well as the lower half, of the avalanche deposit comprises material that came mostly from deeper within the core of the volcano. That material contained negligible organic matter and a greater abundance of hot rock from within or near the intruded magma body. Temperatures were 70° to 100°C at depths of 1 to 1.5 m on the central part of the avalanche deposit 10 to 12 days after emplacement (Banks and Hoblitt 1981).

3.2.2 Directed Blast

Collapse of the volcano's north flank suddenly decompressed the shallow magma body within the volcano and the superheated groundwater circulating near it (similar to opening the vent on a hot pressure cooker) and triggered a devastating volcanic blast. The sudden decompression initiated both phreatic (steam-driven) and magmatic (exsolving gas-driven) explosions that propelled fragmented debris outward from the volcano until gravity caused the turbulent cloud of rock, ash, and gas to flow across the landscape (Hoblitt et al. 1981; Moore and Sisson 1981; Waitt 1981; Waitt and Dzurisin 1981). The blast followed the debris avalanche off the volcano but rapidly outran it. Along the shores of Spirit Lake and in the upper North Fork Toutle River valley, the blast raced ahead of the avalanche and toppled mature forest. The dense debris avalanche hugged the valley floor; but the lower-density, more-energetic blast flowed over the rugged topography it encountered and spread across a wide area.

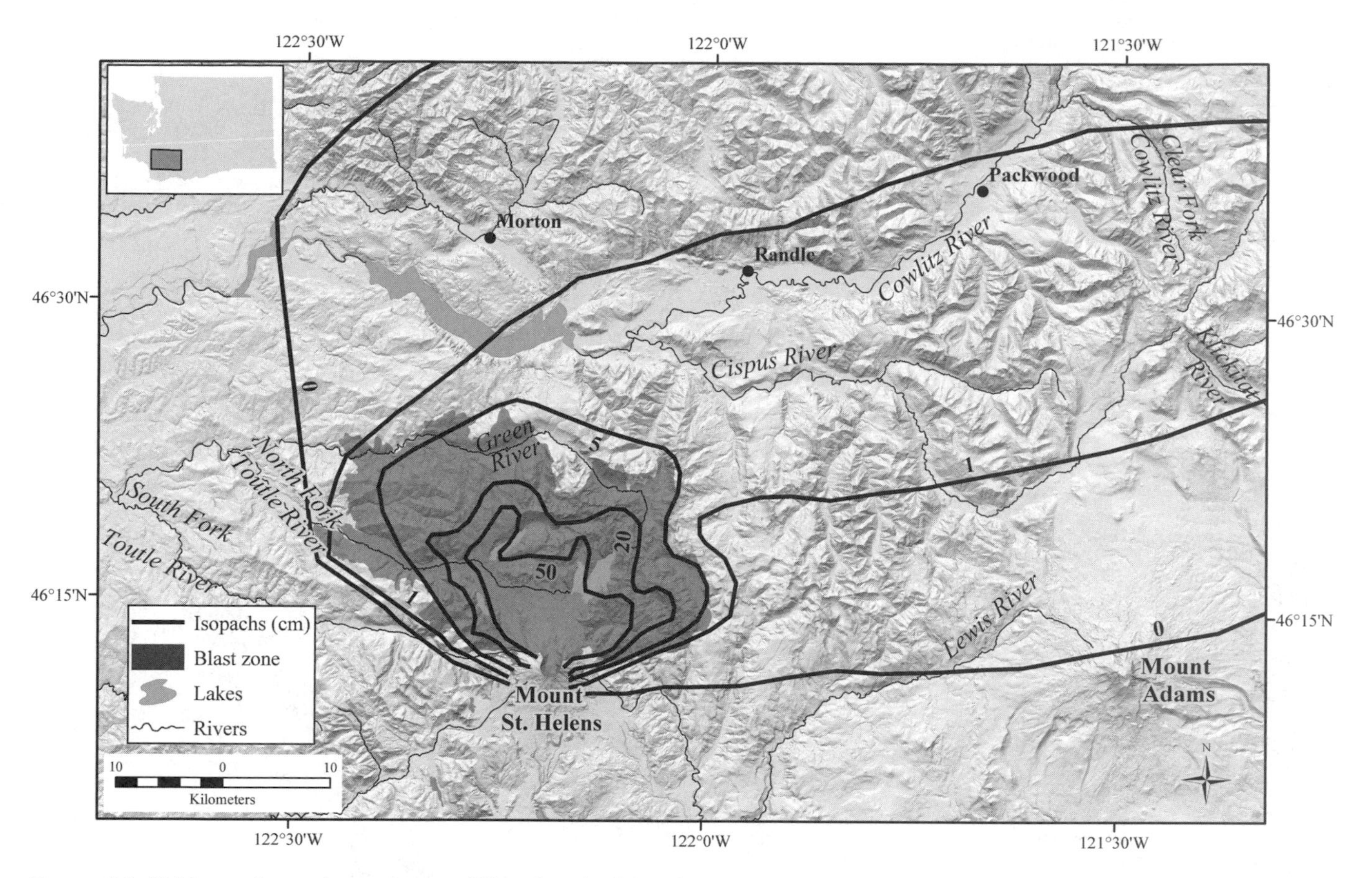

FIGURE 3.3. Thickness (in centimeters) map of blast deposit. (Isopachs are contours of equal deposit thickness.) Some of the blast deposit extends beyond the boundary of the blast area because fine material lofted high into the atmosphere drifted east with the wind. [Adapted from Figure 261 in Waitt (1981).]

The blast produced a hot cloud, charged with 0.2 km^3 of rock debris (see Table 3.1), that removed, toppled, or scorched most aboveground vegetation over an area of about 570 km^2 in a 180° arc north of the mountain (see Figure 3.2) (Moore and Sisson 1981; Waitt 1981). The area affected by the blast can be subdivided into three primary zones (see Table 3.1). Close to the volcano, the blast, the debris avalanche, and the avalanche-triggered seiche in Spirit Lake stripped most trees from the landscape. This area is called the tree-removal zone (Lipman and Mullineaux 1981, plate 1). The impact force of the blast diminished with increasing distance from the volcano, so it left progressively more of the trunk and limb structure of trees intact toward the margin of the zone. Beyond the tree-removal zone, the blast created a 370-km^2 blowdown zone of the blast-toppled forest (see Figures 3.1b, 3.2). Yet farther from the vent, the blast no longer had the force to topple trees, but instead left a 0.3- to 3.0-km-wide scorch zone, a 110-km^2 swath of standing dead forest killed by the heat of the blast cloud (see Figures 3.1c, 3.2). Our terminology of these zones differs slightly from that used by Lipman and Mullineaux (1981, plate 1).

The blast blanketed the landscape with a deposit that was progressively thinner and finer textured with increasing distance from the volcano (Figure 3.3). These deposits blanketed the tree-removal zone with 0.2 to 1.5 m of pebbly to sandy gravel overlain by sand, covered the blowdown zone with 0.1 to 1.0 m of sandy gravel to sandy silt, and draped 0.01 to 0.1 m of coarse sand to sandy silt over the scorch zone (Hoblitt et al. 1981; Waitt 1981). Within the tree-removal and blowdown zones, blast debris deposited on slopes in excess of 35° was unstable. On such steep slopes, the deposit commonly remobilized and formed secondary "blast-pyroclastic flows" that slid down hillsides and collected in stream valleys close to the volcano. The resulting secondary deposits are up to 10 m thick, and locally contain abundant organic matter stripped from the hill slopes. Emplacement temperature of these deposits ranged from 100° to 300°C (Banks and Hobblitt 1981).

The hot, rock-laden blast abraded, burned, and singed vegetation. Initial temperature of the blast deposit varied with the abundance of fragments of the hot magma, but it generally ranged from about 100° to 300°C (Banks and Hoblitt 1981; Moore and Sisson 1981). The hottest parts of the blast, typically northeast of the volcano, entrained and charred wood fragments (Moore and Sisson 1981). In the scorch zone, the blast

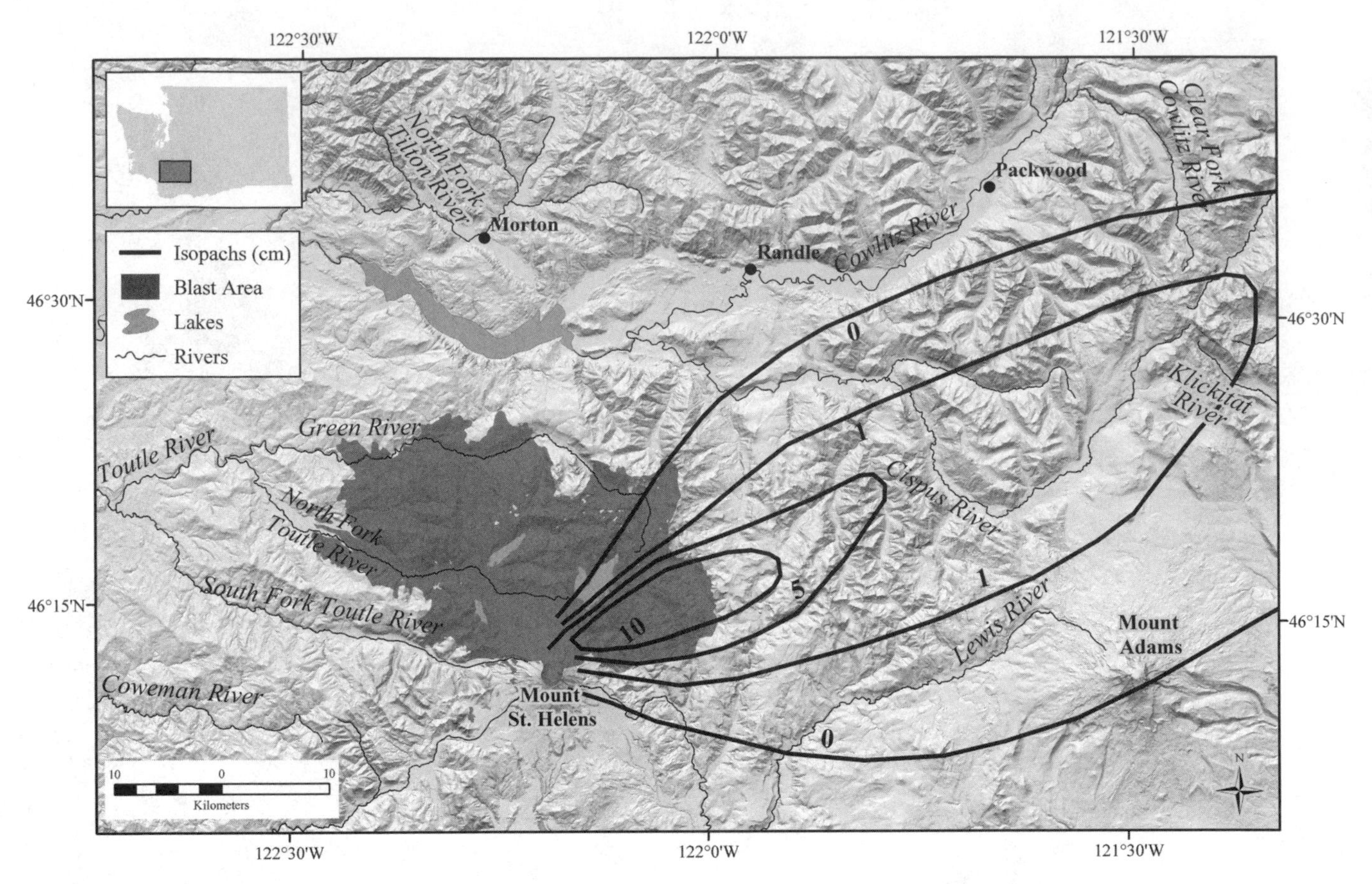

FIGURE 3.4. Thickness (in centimeters) map of May 18, 1980, tephra-fall deposit associated with vertical plume that developed after onset of eruption. This map does not include tephra falls associated with the blast or with post-May 18, 1980 eruptions. [Adapted from Waitt and Dzurisin (1981), Figures 358, 361.]

elevated air temperature to about 50°C (Winner and Casadevall 1981) and singed foliage, but it did not char the vegetation. Despite the high temperature of the blast, there is little evidence that it appreciably elevated the temperature of the buried preeruption soil.

3.2.3 Tephra Fall

Tephra fall caused the most widely distributed impact of the eruption. A billowing, vertical plume grew within minutes of the onset of the eruption, continued for about 9 hours, and ejected about 1.1 km^3 of tephra (calculated as uncompacted material) into the atmosphere (Sarna-Wojcicki et al. 1981) (see Figure 3.1d; see Table 3.1). Wind blew the tephra mainly to the east-northeast (Figure 3.4).

Tephra-fall texture and thickness vary greatly within the fallout zone (Figure 3.4). Pebble-sized pumice fell close to the mountain and formed a layer greater than 20 cm thick over an area of about 16 km^2, much of it within the area affected by the blast only hours earlier. Granule- to sand-sized tephra fell onto and through the forest canopy across hundreds of square kilometers northeast of the volcano to thicknesses of several centimeters. A silty layer as much as a few centimeters thick blanketed the entire area of tephra accumulation, including much of the blast area. Emplacement temperature of the tephra fall beyond the blast-affected area was apparently less than 50°C (Winner and Casadevall 1981). Vegetation beyond the limit of the blast area was covered by cool tephra fall.

3.2.4 Pyroclastic Flows

Subsequent to the debris avalanche and blast, pyroclastic flows inundated the upper North Fork Toutle River valley and Spirit Lake basin and locally spilled onto the flanks of the volcano. Pumice-rich pyroclastic flows spewed from the eruption vent for 5 hours, beginning about noon on May 18 (Rowley et al. 1981; Criswell 1987). These flows covered about 15 km^2 of the surface of the debris-avalanche deposit immediately north of the volcano and deposited approximately 0.3 km^3 of hot, loose, pumiceous sediment (see Figures 3.1e, 3.2; see Table 3.1). The inundated area is named the Pumice Plain to denote the composition and flat surface of these deposits. Individual pyroclastic-flow deposits ranged from 0.25 to 10 m thick; the total accumulation is as much as 40 m thick (Criswell 1987). Cobbles and boulders of pumice, and fragments of denser rock, cover the surfaces of these deposits, but their cores comprise

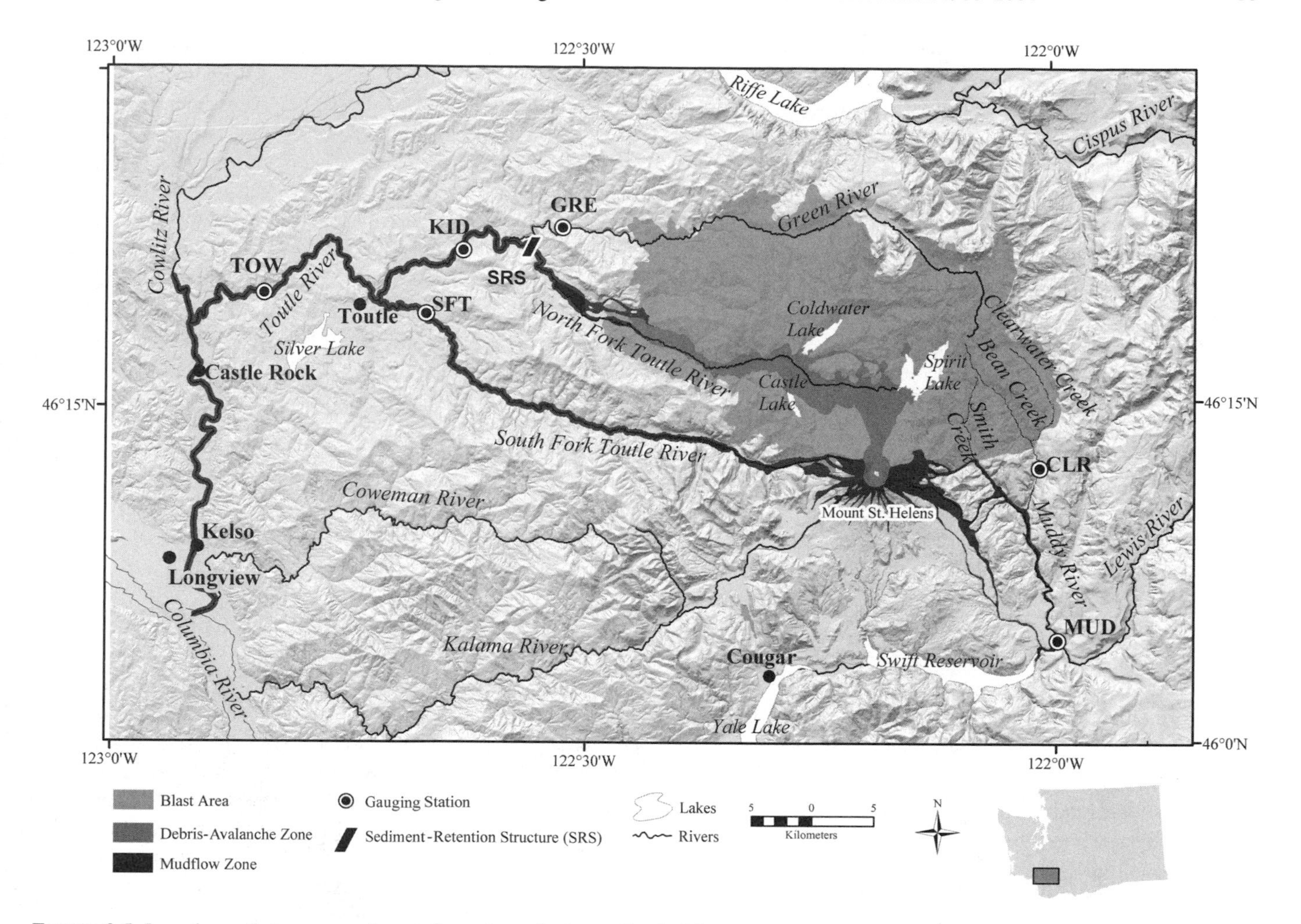

FIGURE 3.5. Locations of stream gauging stations along the lower Toutle River (*TOW*), North Fork Toutle River (*KID*), South Fork Toutle River (*SFT*), Muddy River (*MUD*), Green River (*GRE*), and Clearwater Creek (*CLR*). *SRS*, sediment-retention structure on the North Fork Toutle River.

gravelly sand (Kuntz et al. 1981). Emplacement temperatures of pyroclastic flows on the Pumice Plain ranged from about 300° to 730°C but were as much as 850°C close to the vent (Banks and Hoblitt 1981).

3.2.5 Mudflows

Eruption-triggered mudflows (commonly termed lahars) inundated the major channels that drained the volcano and flowed long distances downstream (see Figures 3.1f, 3.2; Figure 3.5). The debris-avalanche, blast, and pyroclastic flows generated the mudflows in several ways (Janda et al. 1981; Pierson 1985; Major and Voight 1986; Fairchild 1987; Scott 1988; Waitt 1989). The largest and most destructive mudflow on May 18 emanated from the debris-avalanche deposit and flowed down the North Fork Toutle River valley. That mudflow resulted from liquefaction induced by ground shaking (Fairchild 1987) and from muddy slurries produced by consolidation and slumping of water-saturated parts of the avalanche deposit (Janda et al. 1981; Glicken 1998). The combination of the great volume of this mudflow (1.4×10^8 m^3; Fairchild and Wigmosta 1983), more than 10 times larger than any of the other mudflows of May 18 (Major et al. 2005), and the relative confinement of the Toutle River valley permitted it to flow 120 km to the Columbia River (see Figure 3.5). Along its path, this mudflow extensively inundated the floodplain and entrained riparian vegetation and piles of logs from logging camps, which added to its volume and destructive force.

Parts of the blast cloud eroded and melted snow and ice on the west, south, and east flanks of the volcano, producing large (to 10^7 m^3), rapidly moving mudflows. Blast-triggered mudflows traveled tens of kilometers along the channels of Smith Creek, Muddy River, Pine Creek, and the South Fork Toutle River (see Figures 3.2, 3.5). Some of those flows gradually transformed into more-dilute, sediment-laden floods as they moved down the valley (Pierson 1985; Scott 1988). The blast also initiated small, thin, unchannelized mudflows on the broad southern side of the volcano (Figure 3.2), which inundated alpine and subalpine areas, such as Butte Camp and the upper Pine Creek fan (Fink et al. 1981; Major and Voight 1986).

Relatively small pumiceous pyroclastic flows that spilled onto the east and west flanks of the volcano during the afternoon of May 18 melted ice and snow and produced additional moderate-sized mudflows in the headwaters of Smith Creek, Muddy River, and South Fork Toutle River (Janda et al. 1981; Pierson 1985; Scott 1988). These pumice-rich mudflows were smaller (up to 10^5 m^3) than the blast- and debris-avalanche-triggered mudflows initiated in the morning, and they traveled along channels that had been extensively modified by the earlier, larger flows.

Dynamic characteristics of mudflows, such as volume, velocity, peak discharge, and impact force, generally varied among mudflows of different origin and decreased as flows traveled downstream and interacted with channel structures and riparian vegetation. The mudflows were anywhere from a few meters to more than 10 m deep as they flowed away from the volcano (see Figure 3.1f), but they left deposits of gravelly sand that generally were less than 1 m thick along most valleys. Some mudflows had slightly elevated temperatures and were described as having textures similar to "warm concrete" (Cummans 1981; Rosenbaum and Waitt 1981). Mudflows generally toppled and severely abraded vegetation close to the channel (see Figure 3.1d), but more tranquil passage through riparian forest farther from a channel axis resulted chiefly in deposition of a thin veneer of gravelly sand that caused little immediate tree mortality (Janda et al. 1981; Pierson 1985; Scott 1988; Frenzen et al., Chapter 6, this volume).

3.3 Volcanic Events Since May 18, 1980

Between May 18, 1980, and 1991, Mount St. Helens erupted an additional 21 times (Table 3.2) (Swanson et al. 1983a; Brantley and Myers 2000). These eruptions generally had minimal impact on the landscape and the biological response to the major eruption, except in the area immediately north of the crater. Five eruptions between May 25 and October 16, 1980, produced pyroclastic flows (Rowley et al. 1981) and tephra falls. None of these eruptions produced pyroclastic flows exceeding 0.01 km^3 (5% of the volume of the pyroclastic flows erupted on May 18), and the resulting deposits generally blanketed those of the May 18 eruption (Rowley et al. 1981). Post-May 18 tephra-fall deposits are generally less than 2 cm thick (Sarna-Wojcicki et al. 1981) and had little additional effect on the landscape. The volcano was largely snow free during these eruptions, and no major mudflows developed.

Eruptions between October 1980 and October 1986 involved predominantly nonexplosive growth of a lava dome within the volcano's crater, which by 1987 was about 270 m tall and had a volume of about 7.5×10^7 m^3 (Brantley and Myers 2000). Explosions during periods of dome growth between 1982 and 1986 commonly coincided with thick snowpacks in the crater, and some explosions led to rapid snowmelt and generation of mudflows, the most notable occurring on March 19, 1982 (Waitt et al. 1983; Cameron and Pringle 1990; Pierson 1999; Pringle and Cameron 1999). All these mudflows passed along the North Fork Toutle River and were gradually diluted to sediment-laden streamflow, but they were far less extensive than the mudflows of May 18, 1980.

The magnitude and intensity of eruptions at Mount St. Helens waned rapidly following the May 18, 1980, eruption. Significant magmatic explosions were largely complete by October 1980, and this phase of dome growth ended by October 1986. Between 1986 and 1991, the most significant eruptive activity involved minor steam explosions driven chiefly by rainfall seeping into fractures on the dome and contacting hot rock (Mastin 1994). Until dome growth commenced in 2004 no eruptions or explosions occurred after 1992. As eruptive activity declined in the mid 1980s, a glacier accumulated in the crater south of the lava dome (Anderson and Vining 1999; Schilling et al. 2004). As of 2001, the crater glacier had an ice volume (about 80×10^6 m^3) equivalent to 40% to 50% of the volume of glacier ice on the preeruption volcano.

3.4 Hydrology: Precipitation and Runoff

Temporal variation in climate and water runoff influenced both biological processes directly and the environments of posteruption biological succession. Such variations also affected geomorphic processes driven by rainfall and snowmelt runoff. Processes such as sheet and rill erosion, small landslides (debris slides), and lateral and vertical channel erosion dominated posteruption physical changes to the landscape, and they greatly affected the path and pace of biological succession.

Histories of regional precipitation and runoff provide a context for interpreting the physical and biological changes that have occurred on hill slopes and in channels between 1980 and 2000. Interdecadal periods of wetter- and drier-than-average conditions (Mantua et al. 1997; Biondi et al. 2001) were punctuated by higher-frequency El Niño–Southern Oscillation climate variability (McCabe and Dettinger 1999). Relatively dry conditions characteristic of the time of the 1980 eruption persisted for about 15 years (Mantua et al. 1997; Biondi et al. 2001), followed in the mid- to late 1990s by wetter-than-average conditions (Major et al. 2000; Major 2004). From 1995 to 2000, mean annual streamflow from basins near Mount St. Helens was 40% to 50% greater than during the period 1980 to 1994 (Figure 3.6b). This climatic variation in streamflow had a marked impact on erosion and sediment transport from basins affected by the 1980 eruption (Major et al. 2000; Major 2004).

Posteruption streamflow from the Mount St. Helens landscape reflects great changes in precipitation–runoff relationships, structural modifications to drainage networks, and reduction and regrowth of vegetation. Storms and snowmelt of similar magnitudes before and after May 18, 1980 produced substantially higher runoff after the eruption, as detected through comparisons of pre- and posteruption streamflow (Major et al. 2001). By comparing magnitudes and frequencies

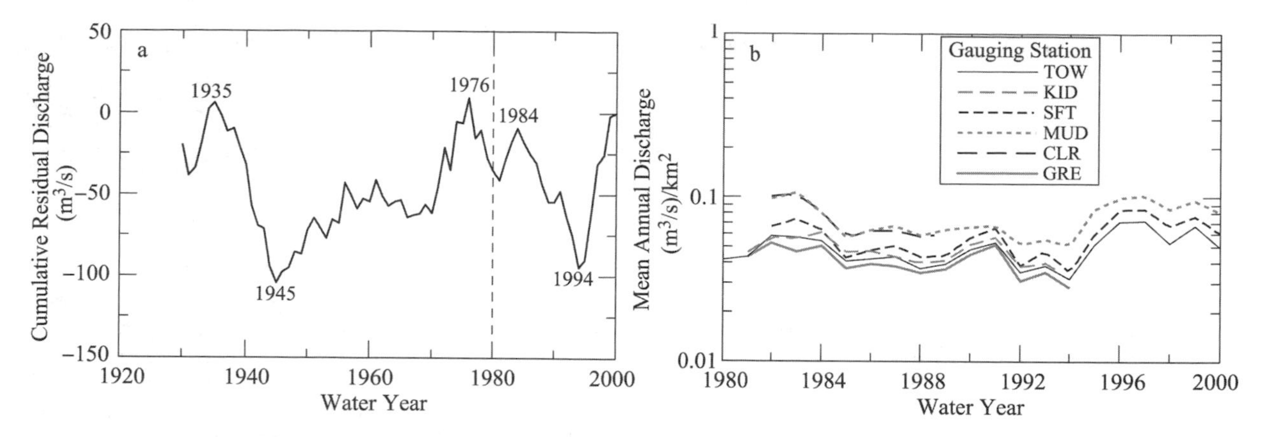

FIGURE 3.6. (a) Time series of the cumulative sum of annual deviations from long-term mean annual flow of Toutle River near TOW (see Figure 3.5 for station locations). A downward trend denotes periods of drier-than-average conditions; an upward trend denotes periods of wetter-than-average conditions. (b) Time series of posteruption mean annual unit area discharges for rivers at Mount St. Helens. [Adapted from Major (2004).]

of unit-area peak-flow discharges from basins affected by the 1980 eruptions with those of a nearby, relatively unaffected basin, Major et al. (2001) concluded:

- The major volcanic disturbances increased peak discharges over a broad range of magnitudes, on average by several tens of percent, for about 5 years;
- Streamflow from basins affected only by the blast or tephra fall (without mudflows) recovered to preeruption magnitudes within a few years, whereas basins having severe channel disturbances in addition to hill-slope disturbance recovered more slowly; and
- Streamflow from some basins that suffered severe channel disturbance displayed responses to the eruption even as late as 2000.

Dramatic increases in magnitude of peak flows immediately after the 1980 eruption resulted from changes to various conditions of the land. First, loss of canopy interception as a result of loss of the forest cover allowed more precipitation to reach the land surface and reduced evaporation. It also altered seasonal variation in soil moisture by reducing transpiration. Second, meager surface infiltration of the blast and tephra-fall deposits radically altered hill-slope hydrology and changed it from a regime that lacked overland flow to one dominated by it. Infiltration capacity of the ground surface decreased from about 75 to 100 mm h^{-1}, typical of forest soils in the region (Johnson and Beschta 1980), to less than 10 mm h^{-1} (Swanson et al. 1983b; Leavesley et al. 1989). In the blowdown zone about 20 km northwest of the volcano, the infiltration capacity of a silty crust developed on the blast deposit was as little as 2 to 5 mm h^{-1} in August 1980, and it only roughly doubled by the following summer (Leavesley et al. 1989). Salvage-logging operations, burrowing animals, trampling by large mammals, formation and melting of soil ice, and growth of vegetation partly mixed the tephra profile in the first few years after deposition (Collins and Dunne 1988). These processes, combined with local erosion, increased infiltration capacity (Swanson et al. 1983b; Collins and Dunne 1986, 1988). Parts of the blowdown zone deliberately disturbed by scarification and salvage logging had infiltration capacities as great as 28 mm h^{-1} by 1981 (Fiksdal 1981). As of 1998, measured infiltration rates of slopes in the blowdown zone northwest of Mount St. Helens that were not deliberately disturbed were locally as much as 25 mm h^{-1} (Major and Yamakoshi in press), suggesting that posteruption infiltration rates after nearly two decades have increased greatly without intervention but are still much less than likely preeruption rates. Consequently, common rainfall intensities (e.g., 10 mm h^{-1}, a 1-hour-duration storm having an approximately 2-year return interval) produce little surface runoff; but locally, lower-frequency, higher-magnitude storms can still induce appreciable surface runoff. Third, forest toppling by the blast altered the timing and rate of snow accumulation and melt during rain-on-snow events (Dunne and Leopold 1981; Orwig and Mathison 1981; Lettenmaier and Burges 1981; Janda et al. 1984; Simon 1999), similar to alterations observed following forest cutting (Harr 1981; Marks et al. 1998). Fourth, channel changes caused by the eruption altered the manner in which runoff moved downstream. Channel smoothing and straightening temporarily reduced flow resistance and allowed stormflow to travel faster and be attenuated less (Janda et al. 1984; Simon 1999). The alterations to both hill-slope and channel geomorphology, combined with enhanced sediment transport that increased flow volumes and damped turbulence (Janda et al. 1984), enhanced peak-flow discharges from many parts of the disturbed landscape.

In some parts of the landscape, the 1980 disturbances temporarily diminished water runoff. The Pumice Plain, for example, comprises highly permeable pyroclastic-flow deposits that produce virtually no surface runoff. In the upper North Fork Toutle River valley, the debris-avalanche deposit blocked several tributary channels and had an extensive, irregular surface

TABLE 3.3. Characteristics of river and riparian environments affected by the 1980 activity of Mount St. Helens.

Disturbance type[a] (example)	Initial disturbance of riparian vegetation	Initial change in channel complexity	Stability of channel location	Change in sediment load	
				Magnitude	Duration
Debris avalanche (North. Fork Toutle River)	Complete	Complete	Low	Very high	Many decades
Mudflow below debris avalanche (North Fork Toutle River)	Extensive; some sprouting of residual trees	Roughness reduced by scour and deposition	Low	Very high	Many decades
Mudflow (Pine Creek)	Extensive; some sprouting of residual trees	Roughness reduced by scour and deposition	Moderate	Moderate	Years
Blast area (Upper Green River; Bean Creek)	Extensive; some sprouting of residual shrubs and herbs	Increased by fallen trees; decreased by debris-flow scour and mass-movement sediment deposition	High	Moderate	Years
Downstream of blast area (Lower Clearwater Creek)	Minor	Minor	High	Moderate	Years
Tephra fall (Upper Clear Creek)	Minor	None	High	Low	Months
Downstream of tephra-fall zone (Lower Clear Creek)	Minor	None	High	Low	Months

Magnitude of "change in sediment load" ranges from low (less than 100% increase) to very high (more than an order of magnitude increase) (see Figure 3.8).
[a]Disturbance types are described in text.

composed of closed depressions that contained ponds. Lakes and ponds that formed adjacent to, and on the surface of, the debris-avalanche deposit trapped local runoff. Some lakes and ponds continue to trap runoff, whereas others filled and breached their impoundments and helped reconnect a drainage network across the deposit. It took more than 2 years before a drainage network reconnected headwater tributaries to the main channel of the upper North Fork Toutle River (Meyer 1995; Simon 1999). Obliteration of the stream network in the upper North Fork Toutle River valley by the debris-avalanche deposit partly counteracted landscape changes that enhanced surface runoff. Once the drainage network reintegrated sufficiently, posteruption peak-flow discharges of the North Fork Toutle River ranged from a few to several tens of percent larger than preeruption discharges of comparable frequency.

3.5 Environments Resulting from the 1980 Activity

The primary volcanic deposits created distinctive environments that posteruption processes modified at various rates (Collins and Dunne 1986; Smith and Swanson 1987; Meyer and Martinson 1989). Ecological responses to the 1980 eruptions depended, in part, on the character of these environments and on the type, intensity, and extent of modification of the primary deposits by secondary hydrologic and geomorphic events. The following discussion focuses on environments grouped as terrestrial, riparian, and riverine; lake and lakeshore; and hydrothermal.

3.5.1 Terrestrial, Riparian, and Riverine Environments

The degrees of primary and secondary landform modification and their impacts on biological legacies of the pre-1980 ecosystems varied among terrestrial, riparian, and riverine environments. Primary volcanic disturbances affected areas that are large relative to pathlengths of water and sediment transport down hill slopes, so physical processes in upland terrestrial environments respond almost exclusively to a specific primary volcanic disturbance. Rivers, on the other hand, transport water and sediment long distances and may flow through a variety of disturbance zones (see Figure 3.2). Thus, riverine and riparian environments respond to local disturbances as well as to primary and secondary disturbances farther upstream (Table 3.3).

3.5.1.1 Tephra-Fall Zone Outside the Blast-Affected Area

Biological legacies in the forms of residual organisms and organic structures, such as large logs on the ground, remained as significant factors influencing hydrology, erosion, and revegetation on tephra-mantled hill slopes beyond the blast area principally because of the limited thickness of tephra-fall deposits (see Figure 3.4), the low intensity of initial disturbance, and the modest intensity and short duration of secondary erosion. The thin tephra deposit did not obscure small-scale landforms, such as decomposed logs and root-throw pit-and-mound topography, but it damaged herbaceous and other ground cover while having little effect on trees and shrubs (see Antos and Zobel, Chapter 4, this volume). Minor sheet and rill erosion occurred on slopes steeper than approximately 30° (Swanson et al. 1983b), but the rate of erosion there was less than that observed within the blast area because litter derived from the still-living forest canopy mixed with tephra during deposition and also quickly covered the ground surface after deposition. This organic litter stabilized surfaces by reducing raindrop impact and increasing surface roughness, which fostered infiltration and reduced surface runoff. Rates of surface erosion within areas affected by tephra fall probably peaked and declined quickly over the first wet season, similar to rates of erosion documented in the blast-affected area (Swanson et al. 1983b; Collins and Dunne 1986; Antos and Zobel, Chapter 4, this volume).

Tephra fall had only minor effects on channel and riparian zones (see Table 3.3). Within a few years after the 1980 eruption, scattered patches of pumice gravel that had been deposited on bars and in the lee of obstructions provided the only obvious evidence for tephra fall within stream channels (Smith and Swanson 1987; Lisle 1995).

3.5.1.2 Blast Area: Blowdown and Scorch Zones

Posteruption environments in the blast area vary in the amount and condition of biological legacies, but many secondary erosion processes were common across the three zones affected by the blast. Many biological legacies survived in the blowdown and scorch zones because the blast deposit was relatively thin (less than 1 m). In the tree-removal zone, however, the blast removed most soil and vegetation.

The rugged topography of the preeruption landscape and the abundant trees toppled by the blast strongly influenced patterns of posteruption erosion in the blowdown zone. Sheet, rill, and gully erosion reworked the 1980 deposits in this zone (Swanson et al. 1983b; Collins and Dunne 1986; Smith and Swanson 1987). Unchannelized, shallow, linear depressions were gullied during the first significant posteruption rains in late 1980. Toppled trees created complex surface topography that obstructed runoff, encouraged infiltration, and reduced surface erosion. Posteruption vegetation development did not affect the rapid decline of hill-slope erosion that ensued within months of the eruption, even where artificial seeding was implemented (Stroh and Oyler 1981; Dale et al., Chapter 19, this volume). Surface erosion peaked and declined in response to physical factors before significant vegetation cover established on the blast deposit (Collins and Dunne 1986).

The blowdown and scorch zones experienced a common suite of secondary erosion processes whose rates diminished over time and differed between the two zones. In the blowdown zone, sheet and rill erosion were greater on steeper slopes and on slopes blanketed with deposits having fine-textured surface layers (Swanson et al. 1983b; Collins and Dunne 1986; Smith and Swanson 1987). Fresh tephra of blast and fall origins slid or was washed quickly from slopes steeper than about 35°. Collins and Dunne (1986) measured as much as 26 mm of erosion (computed as average landscape lowering) by 1981, but only an additional 1.8 mm occurred by 1982. The dramatic decline in sheet and rill erosion resulted from development of an armor layer of coarse particles, which followed disruption and removal of the surficial silty tephra. This rough armor layer increased water infiltration and reduced the magnitude and frequency of overland flow (Swanson et al. 1983b; Collins and Dunne 1986). Less than 15% of the tephra deposited in 1980 was removed before hill-slope erosion in the blowdown zone stabilized and geomorphic processes became more like those typical of forested areas (Collins and Dunne 1986). The rate of surface erosion in the scorch zone was generally less than 10% of that measured in the blowdown zone because litter from the scorched forest formed a protective layer and the thin tephra deposits retained forest-floor topography that disrupted surface erosion (Collins and Dunne 1986).

Temporal changes in a clear-cut area east of upper Smith Creek illustrate the general pattern of hill-slope erosion and vegetation development characteristic of natural recovery processes in the blowdown zone (Figure 3.7). Before the eruption, the site bore only stumps and a few logs left from cutting. After the eruption, the site was buried under 50 cm of blast and subsequent tephra-fall deposits. Between May 18 and early fall of 1980, minimal rainfall caused only minor erosion (Figure 3.7a). However, by January 1981 storm runoff had cut gullies into topographic depressions (Figure 3.7b). By then, however, gully erosion had ceased because the tephra surface had coarsened and reduced runoff, and the gullies had eroded down to the preeruption soil. Within 4 years, vegetation that included willow (*Salix* spp.) and perennial herbs colonized or resprouted from perennial rootstocks in the gully floors (Figure 3.7c); within 15 years, vegetation flourished, and Douglas-fir (*Pseudotsuga menziesii*) and several other species of trees and shrubs had colonized areas between gullies (Figure 3.7d).

In addition to sheet and rill erosion, small landslides also stripped tephra and soil from posteruption hill slopes. Numerous small, rapid landslides (100 to 100,000 m^3) occurred in steep areas of the blowdown and scorch zones, mainly within 5 years of the 1980 eruption and again in 1996. Field investigations and analyses of aerial photographs documented 278 slides that occurred between 1980 and 1984 over a 117-km^2 area within the watersheds of Smith Creek, Bean Creek, Clearwater Creek, and upper Green River (Swanson et al. 1983b; Swanson 1986). Landslide scars occupied less than 1% of the study area, but locally they covered up to 10% of the hill slopes and scoured as much as 30% of the channel lengths in these basins. Of the landslides inventoried in this area, 70% occurred in blast-toppled forest; the remainder occurred in areas roaded or clear-cut before 1980 or salvage logged after the eruption. An important cause of sliding may have been the loss of soil anchoring provided by tree roots, roots that had been pulled from the ground during the blast. Between 1980 and 1984, the frequency of slides in the blowdown zone was 35 times greater than that in undisturbed, forested areas and 9 times greater than that in areas clear-cut elsewhere in the Cascade Range (Swanson et al. 1981; Sidle et al. 1985). Many of these

FIGURE 3.7. Chronological sequence of photographs of gully and vegetation development at site east of Smith Creek, 9 km east-northeast of the crater. This preeruption clear-cut facing the volcano received about 50 cm of blast and tephra-fall deposits [Photographer: F.J. Swanson.]

early slides in the blowdown zone occurred despite the absence of major storms (e.g., recurrence intervals exceeding 10 years).

Intense regional storms in November 1995 and February 1996 triggered locally extensive sliding in parts of the blast area. In the basins of the upper Green River, Clearwater Creek, Bean Creek, and Smith Creek, numerous landslides mobilized (1) tephra deposited in 1980, (2) newly established forest, and (3) older tephra deposited by prehistoric eruptions of Mount St. Helens. Each of those basins is underlain by thick accumulations of prehistoric tephra deposits (Crandell and Mullineaux 1978; Mullineaux 1996; see Figure 2.4). Many of the landslides occurred at sites where planted 12- to 14-year-old Douglas-fir forests were well established. The landslide deposits spread onto valley floors and locally entered channels and riparian zones. Floods associated with these storms mobilized some of these landslide deposits and wood in the channels in the eastern part of the blowdown zone, causing severe damage to channel and riparian areas.

Channel impacts and persistence of riparian biological legacies within the blast-affected area varied greatly with distance from the volcano. Within 10 km of the volcano, but outside the upper North Fork Toutle River valley, the blast, mudflows, and secondary blast-pyroclastic flows locally buried valley floors with up to 15 m of hot debris (Hoblitt et al. 1981;Brantley and Waitt 1988), thick enough to obliterate preeruption channel form and to destroy streamside vegetation. Farther from the volcano, the blast deposit was thinner, and secondary blast-pyroclastic flows were less common, so that many manifestations of preeruption channel morphology and riparian vegetation survived. Channels within the blowdown zone unaffected by mudflows or secondary blast-pyroclastic flows (see Figure 3.2) remained relatively stable, in part because tree roots protected their banks. In general, channels within the blowdown zone widened by less than 10 m and incised less than 1 m (Meyer and Martinson 1989). Where the blast toppled mature forest into streams, channel complexity increased greatly, and the trees dissipated streamflow energy and reduced bed and bank scour. Such channel stability aided posteruption survival and establishment of riparian vegetation and development of aquatic habitat. Channel complexity decreased where wood was removed (Lisle 1995) and where debris flows scoured or sediment aggraded channel beds (e.g., lower Bean Creek and upper Clearwater Creek). Development

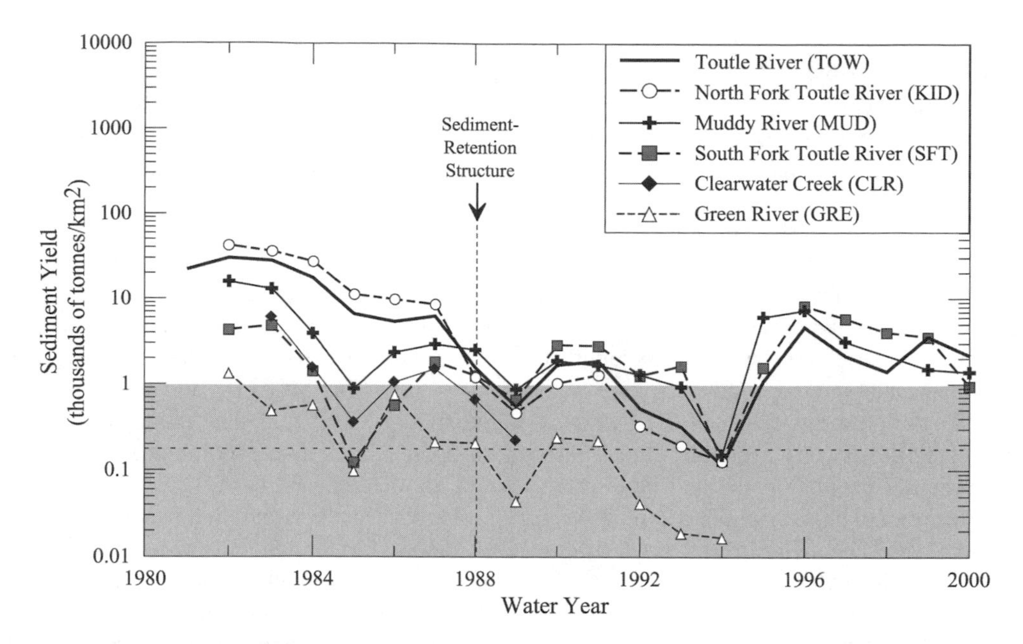

FIGURE 3.8. Annual suspended-sediment yields at gauging stations on rivers draining the Mount St. Helens area. (See Figure 3.2 for disturbance zones and Figure 3.5 for station locations.) *SRS* denotes date when the sediment-retention structure began to trap sediment, affecting sediment yield sampled at KID and TOW. *Shaded region* depicts the range of, and the *horizontal dashed line* depicts median value of, mean annual suspended-sediment yields from several rivers in the western Cascade Range. The yield value depicted by the horizontal line is comparable to average Cascade Range basin yields determined by Beschta (1978), Grant and Wolff (1991), and Ambers (2001). [Adapted from Major et al. (2000).]

of riparian and aquatic habitat along those channels was greatly delayed.

Sediment yield from watersheds where the blast severely disturbed hill slopes but minimally impacted channels was smaller and declined more rapidly than did the yield from basins that experienced severe channel disturbance. In the Green River basin, for example, sediment yield peaked as much as an order of magnitude above background level shortly after the eruption but returned to background level within 5 years (Figure 3.8; Major et al. 2000). This rapid decline was caused, in part, by the generally low streamflow during this period, but it occurred mainly because of the stabilization of rills and gullies that developed on hill slopes and the consequent reduction of sediment delivery to channels.

3.5.1.3 Debris-Avalanche Zone

The debris-avalanche deposit created a distinctive 60-km^2 landscape with complex, actively eroding terrain containing few biological legacies in the forms of soil blocks and vegetation fragments scattered along the deposit margin (see Figures 3.1a, 3.2). The topography of the central part of the deposit was a mosaic of steep-sided hummocks up to 30 m tall and closed depressions (Voight et al. 1981; Glicken 1998). Except for pond and wetland areas, which are common features of the deposit, this nutrient-deficient, erodible terrain presented an inhospitable environment for plant colonization until site conditions were ameliorated by weathering and other processes (Dale et al., Chapter 5, this volume).

Significant establishment of vegetation on the debris-avalanche deposit required development of a channel network across the deposit and a pervasive groundwater table within the deposit. Channel formation began at different times and progressed at varied rates in different areas (Rosenfeld and Beach 1983). On May 18, 1980, the North Fork Toutle River mudflow traveled across the lower half of the deposit and initiated channel formation (Janda et al. 1981, 1984; Meyer and Martinson 1989). Subsequent fill and spill of ponds on the avalanche-deposit surface and breaching of lakes along its margin triggered floods that carved new channels or modified existing ones. These breaches, combined with seasonal runoff and pumping of water from Spirit Lake across the deposit and into the North Fork Toutle River between 1982 and 1985 (Janda et al. 1984; Glicken et al. 1989), produced additional channel change (Rosenfeld and Beach 1983; Janda et al. 1984; Paine et al. 1987; Meyer and Martinson 1989).

Erosion and aggradation of channels across the debris-avalanche deposit created a complex mosaic of surfaces having varying ages, origins, and textures, and they frequently disturbed riparian vegetation. Within a decade of the 1980 eruption, about 35% of the avalanche-deposit surface had been reworked and replaced by braided channels and terraces (Meyer and Martinson 1989). In general, posteruption channels across the debris-avalanche deposit widened by hundreds of meters, incised tens of meters, and were locally aggraded by several meters (Meyer and Martinson 1989).

The groundwater system within the debris-avalanche deposit developed over several years after the eruption and strongly influenced the aquatic and riparian environments. The debris avalanche entrained a large proportion of hydrothermal groundwater that circulated within the volcano on May 18, 1980. That groundwater saturated parts of the debris-avalanche deposit and initially filled many of the depressions that formed on the deposit surface. Precipitation and streamflow contributed gradually to subsequent growth of the groundwater system. Groundwater feeds perennial and ephemeral ponds on the deposit, seep areas, and spring-fed channels. These hydrologic features, having persistently available water and no scouring flows, became hotspots of vegetation establishment and fostered development of plant and animal communities in pondshore, seep, and streamside environments (see Crisafulli et al., Chapters 13 and 20, this volume).

3.5.1.4 *Pyroclastic-Flow Zone*

The hot, thick, pyroclastic-flow deposits that blanketed 15 km^2 of the upper debris-avalanche deposit eliminated all components of previous ecosystems that may have survived the avalanche and blast and created a harsh environment for colonization. Erosion extensively modified the pyroclastic-flow deposits, carving deep, wide, steep-walled channels into the deposits (Meyer et al. 1986; Meyer and Dodge 1988), and several post-1980 mudflows enlarged the channels and locally coated the deposits with bouldery debris (Cameron and Pringle 1990; Pierson 1999; Pringle and Cameron 1999). In general, erosion dominated modification of the pyroclastic-flow deposits close to the volcano, and deposition predominated in more distant areas, especially where channels approach Spirit Lake. Dramatic channel incision into the pyroclastic-flow deposits followed a substantial increase in discharge of the North Fork Toutle River and an abrupt lowering of the local base level of the stream network in November 1982, when water was pumped out of Spirit Lake and into a channel that traversed the northern edge of the deposits (Janda et al. 1984; Paine et al. 1987; Glicken et al. 1989; Simon 1999; Dale et al., Chapter 19, this volume).

3.5.1.5 *Mudflow Zone*

Environmental modification by mudflows and effects of biological legacies of the pre-1980 floodplain ecosystem varied longitudinally with distance from the volcano and laterally with distance from a channel axis. On the flanks of the volcano, mudflows scoured substrates or destroyed vegetation and left deposits generally less than 20 cm thick across convex surfaces of alpine meadows and subalpine forest. Many biotic and abiotic remnants of the pre-1980 ecosystem survived in these environments, primarily along flow margins (Frenzen et al., Chapter 6, this volume). Farther from the volcano, channelized mudflows, mostly 3 to 10 m deep, swept river valleys and substantially modified landforms and aquatic and riparian ecosystems. Gross preeruption landforms generally remained intact, but the mudflows locally removed vegetation along river corridors and displaced, straightened, and smoothed river channels (Janda et al. 1981, 1984; Pierson 1985; Scott 1988). The mudflows moved large wood pieces and boulders onto floodplains and transformed rough, cobble- to boulder-bedded, forest-lined channels into smooth, sand-bedded channels having little riparian vegetation. Downstream of the blast area, mudflows completely removed riparian forest adjacent to channels, but commonly placed thin deposits on delicate understory plants and forest litter on floodplains (Frenzen et al., Chapter 6, this volume). Impact forces and abrasion by the wet-concrete-like fluid were greatest near the channel where the mudflows transported the coarsest sediment at the greatest velocities, but they diminished rapidly away from the channel (Scott 1988). Biological legacies persisting along channels included sprouts of standing but abraded hardwood trees and transported plant fragments (Frenzen et al., Chapter 6, this volume).

Posteruption sediment supply and transport affected the geomorphic response and stability of mudflow-affected channels. The greatest and most persistent changes to mudflow-affected channels occurred along the lower Toutle River and North Fork Toutle River below the debris-avalanche deposit (Meyer et al. 1986; Meyer and Dodge 1988). Initially, suspended-sediment yield along the North Fork Toutle River was as much as 500 times greater than yield typical of basins in the western Cascade Range (see Figures 3.8, 3.9; Major et al. 2000). Channel reaches downstream of the debris-avalanche deposit generally stabilized after sediment yield plummeted following closure of a large sediment-retention structure in 1988 (see Figures 3.5, 3.9). Secondary channel changes are evident, but less dramatic, along other mudflow-affected channels (Meyer et al. 1986; Meyer and Dodge 1988; Martinson et al. 1984, 1986) because of lower magnitude of sediment transport. The Muddy River and South Fork Toutle River, for example, had suspended-sediment yields that were initially about 20 to 100 times above typical background level (see Figure 3.8). Those high yields declined rapidly within 5 years but, even after 20 years, remained about 10 times greater than background (Figure 3.8; Major et al. 2000). Despite such high rates of sediment transport, the Muddy River and South Fork Toutle River each discharged only about one-tenth of the posteruption sediment load discharged by the lower Toutle River between 1980 and 1987 and about half the load discharged between 1988 and 2000 after closure of the sediment-retention structure (Major et al. 2000). Mudflow-affected channels, in turn, had suspended-sediment yields about 5 to 10 times greater than those from basins having headwaters located predominantly within the blowdown and scorch zones (e.g., Green River and Clearwater Creek).

Rates of stabilization varied greatly among reaches of mudflow-affected channels; but in general changes in channel geometry declined dramatically within a few years of the 1980 eruption (Meyer et al. 1986; Martinson et al. 1984, 1986; Meyer and Martinson 1989; Simon 1999; Hardison 2000). Channel locations and cross-sectional geometry changed most

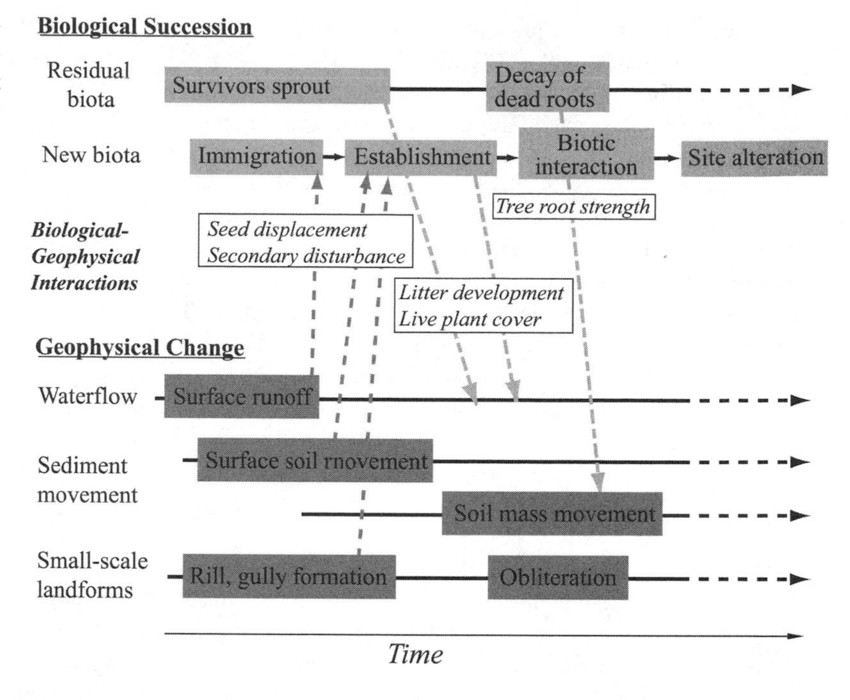

FIGURE 3.9. Schematic depiction of biotic succession, a sequence of geomorphic processes, and interactions of biotic and geomorphic processes at Mount St. Helens following severe landscape disturbance by the May 18, 1980 eruption.

dramatically through 1981 as rivers rapidly incised, widened, and transported the most easily eroded sediment. As channels widened and beds coarsened, rates of change slowed sharply, but some reaches exhibited progressive change through 2000. Mudflow-affected channels generally widened by tens of meters, incised up to 10 m, and were aggraded about 1 m (Meyer and Martinson 1989). Contrasting behaviors of stream reaches are broadly correlated to some extent with overall valley-floor gradient, floodplain width, and the sediment production capability of upstream and streamside areas (Hardison 2000). Although biological legacies persisted, riparian vegetation development along many reaches was delayed until channel geometry stabilized (Frenzen et al., Chapter 6, this volume). Even after channel stabilization, major flooding in 1996 severely disturbed riparian and aquatic systems along several channel reaches (Frenzen et al., Chapter 6, this volume).

3.5.2 Lakes and Lakeshores

The numerous lakes in the Mount St. Helens area after the major 1980 eruption included a mix of those that existed before the eruption (Swanson et al., Chapter 2, this volume) and newly created lakes. Ecological responses to these altered lake environments were similarly diverse (Bisson et al., Chapter 12, this volume; Dahm et al., Chapter 18, this volume).

Lakes existing within about a 30-km radius of the volcano before the eruption (see Figure 2.4; Swanson et al., Chapter 2, this volume) experienced impacts ranging from a light dusting of tephra to profound alteration of basin form and ecology. Of the existing lakes, Spirit Lake was modified most dramatically by the eruption. The debris avalanche raised its outlet, deposited 0.43 km^3 of sediment and woody debris in its basin, and raised its water level by 60 m (Voight et al. 1981; Glicken 1998). A large quantity of wood was also swept from hill slopes into the lake by the seiche generated when the avalanche slammed into the lake. Much of the wood in the lake formed a large, floating mat that has persisted to the present (2004). After 1980, the level of Spirit Lake rose gradually for 2 years until it was controlled by pumping and tunnel construction to form a stable outlet (Janda et al. 1984; Glicken et al. 1989; Meyer 1995). Lakeshore vegetation has recovered slowly because of the generally unstable nature of the lakeshore substrate and the (initially) fluctuating lake level. In contrast, the gross morphology of cirque lakes within the blast area and tephra-fall zone changed little because these lakes are large relative to the amount of sediment that they received. The amount of organic matter delivered to these lakes varied principally with the volcanic process that affected them. Twenty cirque-basin lakes within the blast-affected area received only a few tens of centimeters of blast deposit, including organic matter (Waitt and Dzurisin 1981). In the tephra-fall zone, little organic matter was delivered to lakes by the eruption. Vegetation generally recovered quickly along the shores of these cirque-basin lakes, especially where lake shores had been protected under snow.

Scores of new lakes formed during the May 18, 1980 eruption, and their character and longevity depended greatly upon their mode of formation (see Figure 18.1; Dahm et al., Chapter 18, this volume). Many of these lakes, however, were short lived; six of the nine largest lakes impounded along the margins of the avalanche deposit, for example, filled, overtopped, incised their impoundments, and emptied between 1980 and 1982 (Janda et al. 1984; Simon 1999). The U.S. Army Corps of Engineers constructed stable outlet channels at Spirit,

Castle, and Coldwater to prevent catastrophic breaching of their blockages (Janda et al. 1984; Simon 1999).

New ponds formed on the surface of the avalanche deposit when groundwater seepage and surface runoff filled depressions among the hummocks. These ponds initially contained little organic matter because the avalanche deposit was nearly free of organic material. Similar to several lakes formed at the deposit margins, many ponds on the deposit surface were short lived. Within the first few years of the eruption, several of these ponds breached their outlets and emptied. As of 2000, approximately 130 perennial and ephemeral ponds still existed on the avalanche-deposit surface, and about half of them dry out by the end of each summer (Crisafulli et al., Chapter 13, this volume).

3.5.3 Hydrothermal Environments

The broad spectrum of thermal conditions of the new crater and the 1980 deposits created diverse hydrothermal environments that differed in initial temperature, longevity, chemical composition, state (liquid or gas) of emitted water, and discharge (Keith et al. 1981). Locally, some hydrothermal systems created special habitats for posteruption biological activity; such habitats were absent or extremely rare in the preeruption landscape (Phillips 1941). The new hydrothermal environments ranged from hissing fumaroles on the lava dome within the crater to tepid seeps and springs that emerged on the debris-avalanche deposit, pyroclastic-flow deposits, and along the shoreline of Spirit Lake. Lava-dome and crater-floor fumaroles are "rooted" hydrothermal features; that is, they derive their heat from magma. The longevity (perhaps millennial duration) of magmatically supplied heat can potentially maintain long-lived fumaroles. "Rootless" hydrothermal features, for which volcanic deposits themselves provide the source of heat, developed in the pyroclastic-flow, blast, and debris-avalanche deposits. The pyroclastic-flow deposits, particularly in the western half of the Pumice Plain, produced several steam explosions, fumaroles, hot springs, and seeps (Moyer and Swanson 1987). Warm-water seeps occurred locally within secondary blast-pyroclastic-flow deposits and along channels cut into the debris-avalanche deposit. From 1980 to 1984, hydrothermal features formed in some valleys close to the volcano (e.g., Coldwater Creek, South Coldwater Creek, and upper Smith Creek), where secondary blast–pyroclastic-flow deposits accumulated. Unlike most of the other volcanic deposits in which hydrothermal environments developed, these secondary pyroclastic-flow deposits contained abundant organic matter stripped from adjacent hill slopes. Cooking of this organic matter produced gases and liquids having distinctive, short-lived organic chemical characteristics (Baross et al. 1982). Hydrothermal features formed on the debris-avalanche deposit only where channels were cut in the vicinity of isolated pockets of hot rock. Warm-water seeps also occurred along the south shore of Spirit Lake since 1980, but by 2000 these were barely warmer than adjacent lake water. Within only a few years after the 1980 eruption, all known rootless hydrothermal systems had depleted their heat sources.

3.6 Interactions Among Geophysical Processes: A Geomorphic Succession Perspective

Geomorphic responses to profound landscape disturbances follow trajectories somewhat analogous to those followed by ecological responses to volcanic, wildfire (Swanson 1981), and perhaps other disturbance types. Ecologists find it useful to consider biotic responses to major disturbance in terms of succession of biota, biotic processes, and their alteration of site properties by processes such as soil development. Geomorphic processes also interact in ways that resemble interactions among species and biotic processes during the course of ecological succession (see Figure 3.9). In both biotic and geomorphic cases, some processes change sites in ways that favor occurrence of other processes. Furthermore, trajectories of geomorphic and biotic successions interact with one another in important ways.

The primary disturbances caused by the eruption on May 18, 1980, altered landscape hydrology and sediment delivery, which greatly affected the nature and pace of secondary processes that responded to those disturbances. For example, fresh tephra that blanketed hill slopes initially had very low infiltration capacities. Hence, the first substantial rainfalls on tephra-covered slopes produced abundant overland flow that concentrated in hill-slope depressions and carved shallow gullies (see Figure 3.7). Overland flow, freeze-thaw cycles, bioturbation by plants and animals, and deliberate mechanical disturbance of posteruption surfaces (Collins and Dunne 1988) also removed or disrupted the surficial layer of silty tephra and exposed coarser tephra. Exposure of the coarser tephra increased the infiltration capacity of the ground surface and reduced runoff. In many areas affected by the blast, gullies stopped forming by early 1981 and began refilling as their walls collapsed (Collins and Dunne 1986). Increased infiltration through the tephra profile and gully floors also delivered more water to the preeruption soil, which facilitated the occurrence of many small landslides. Thus, successional sequences of processes affected the geomorphic responses of tephra-mantled hill slopes, and some processes had to occur before others could follow.

Another type of sequential interaction of geomorphic processes involves disturbance-triggered movement and storage of sediment on hill slopes and through channel networks. Propagating sediment pulses can cause transient or long-lasting sediment storage in localized depositional sites. Sediment storage in channels causes aggradation, which commonly produces bank erosion and consequent lateral channel migration. Bank erosion in turn feeds sediment to channels and increases stream sediment load, which induces local aggradation. Thus, interacting geomorphic processes can establish feedbacks that cause extensive secondary modification of disturbed landscapes. Interactions of this type were widespread on the debris-avalanche deposit and along mudflow-affected channels (Janda et al. 1984; Meyer et al. 1986; Meyer and Janda 1986; Meyer and Dodge 1988; Martinson et al. 1984, 1986; Hardison 2000). As a result, sequences of geomorphic processes and landforms developed over time and space and greatly affected biotic response.

The persistence of specific geomorphic processes following severe landscape disturbance is varied and depends on local hydrology, properties of soil and biota, and the timing of weather events that trigger episodic processes. Some geomorphic processes diminish within months, whereas others persist for years, decades, or possibly centuries. Such variability reflects the rates at which individual processes operate, the lingering effects of hydrologic perturbations following disturbance, and the timing of sediment transport through disturbed watersheds (Nicholas et al. 1995; Madej and Ozaki 1996; Major 2004). Rates of sheet and rill erosion, for example, were most vigorous in the first few storms after the eruptions but diminished quickly because of changes in the texture and hydrology of tephra-deposit surfaces, even before significant establishment of vegetation. Rates of change of surface infiltration (Leavesley et al. 1989; Major and Yamakoshi in press), the timing of intense storms, and gradual changes in the decay and cohesion of tree roots affected the timing of small-scale landslides from hill slopes in the blast area. Rates of posteruption channel change were most dramatic within the first few years but ebbed as channels widened, streambeds coarsened, and sediment loads diminished. Nevertheless, some reaches of channels that continue to move extraordinary amounts of sediment (Major et al. 2000) remain highly dynamic, especially along the North Fork Toutle River where it traverses the debris-avalanche deposit (Bart 1999).

Secondary geomorphic processes that act as disturbances interact with establishing biota and affect the trajectory of biotic succession (see Figure 3.9). In general, geomorphic systems must achieve a critical level of stability before vegetation can establish sufficiently to flourish and retard further erosion. The example of sheet and rill erosion in the blast area is most conspicuous. Rates of sheet and rill erosion subsided sharply within a year of the major eruption, and thereafter hill-slope erosion stabilized (Collins and Dunne 1986). In some areas, the first substantial vegetation developed along the floors of eroded gullies within a few years of the eruption. In those sites, removal of the sterile blast and tephra-fall deposits fostered natural revegetation processes. Elsewhere, grass and legume seeds applied to control surface erosion in the summer of 1980 washed off steep slopes before they germinated, or they failed to establish or reduce erosion until biological, physical, and chemical conditions of the soil improved and stabilized (Stroh and Oyler 1981; Franklin et al. 1988; Dale et al., Chapters 5 and 19, this volume). Frequent posteruption debris slides occurred in the blowdown and scorch zones before trees and their root systems reestablished and stabilized the new volcanic deposits. Yet, even in well-established, although young, planted Douglas-fir forest, extensive sliding occurred in 1996 when older tephra deposits below the rooting zone of the young forest mobilized. Lateral channel change along some river corridors, even after 20 years, persisted at a pace that suppressed establishment of extensive riparian vegetation (Frenzen et al., Chapter 6, this volume). Biotic stabilization along the banks of some unstable river reaches, especially on the debris-avalanche deposit, appears to be years in the future.

The unifying theme of this collection of observations of the Mount St. Helens landscape is that primary and secondary geomorphic processes change site conditions in ways that enhance or limit the occurrence of other geomorphic processes, much as biotic processes and species change site conditions and affect subsequent biotic processes and communities. Some degree of physical stabilization of a disturbed landscape must precede establishment of vegetation; as vegetation develops, the land surface is further stabilized, and soil development accelerates. In this way, geomorphic and biotic processes interact and affect the path and pace of geomorphic and biotic responses to disturbances.

3.7 Outlook

The combined effects of primary and secondary physical processes triggered by the May 18, 1980, eruption of Mount St. Helens have created an array of environments in which posteruption ecological systems are developing. Initial posteruption conditions ranged from entirely new environments (such as the debris-avalanche deposit, the Pumice Plain, lakes, ponds, and hot springs where conifer forest once stood) to forest, stream, and lake environments only subtly modified by a light dusting of tephra. Dramatic and continuing physical changes to some environments by various forms of hill-slope and channel erosion have regulated the paths and rates of ecological responses to the primary disturbances.

Ecological studies of severely disturbed landscapes can benefit by beginning with a broad stratification of the landscape with a general approach that includes the following:

- Distinguishing between disturbance mechanisms (e.g., erosion/deposition, impact force, and heat) and types of disturbance processes to give greater predictive power concerning ecological responses of both surviving and invading organisms;
- Considering successions of secondary hydrologic and geomorphic processes and geology–ecology interactions in an effort to anticipate how the landscape will change physically and how such change may affect biological legacies, ecological succession, and development of biotic landscape patterns; and
- Designing studies that examine the effects of landscape structure on processes, patterns, and rates of ecological responses.

The May 18, 1980, eruption dramatically altered the landscape, and secondary processes that responded to that disturbance continue to evolve. Future climate change, rare storms and floods, wildfires, and eruptive activity will adjust the course of geomorphic and ecological change. The stage set by the 1980 eruptions and subsequent geomorphic, hydrologic, and ecological responses may influence the geological and ecological responses to the next eruption. Although the importance of biological legacies of the 1980 eruptions to responses to future eruptions will diminish with time, the natural longevity of trees and the slow pace of some ecological processes (such as wood decomposition, soil formation, and development of complex forest structure) may extend those biological legacies for many centuries, a time frame that may encompass the next significant eruption of the volcano.

Acknowledgments. We acknowledge the contributions of many colleagues too numerous to mention. However, we give special thanks to Dick Janda (deceased), who worked hard to foster communication between ecological and geological science communities and to build broad understanding of dynamic landscapes. Charlie Crisafulli, Virginia Dale, Jerry Franklin, and Don Swanson provided very useful reviews.

Part II
Survival and Establishment of Plant Communities

4
Plant Responses in Forests of the Tephra-Fall Zone

Joseph A. Antos and Donald B. Zobel

4.1 Introduction

Tephra fall is the most widespread disturbance resulting from volcanic activity (del Moral and Grishin 1999), including the 1980 eruption of Mount St. Helens (Sarna-Wojcicki et al. 1981). Tephra is rock debris ejected from a volcano that is transported through the air some distance from the vent that produced it. Fine-textured tephra (less than 2 mm in diameter) is referred to as volcanic ash. Tephra may be transported far from a volcano and affect vegetation over thousands of square kilometers, well beyond the influence of other types of volcanic ejecta. Individual tephra deposits from volcanoes in the Cascade Range have been traced east into the Great Plains, and others cover much of the Pacific Northwest (Shipley and Sarna-Wojcicki 1983). Mount St. Helens has been the most frequent source of tephra in the Cascades for 40,000 years, producing dozens of tephra layers equal to or larger than the 1980 eruption, three experienced by trees alive in 1980 (1480, 1800, and 1980; Mullineaux 1996). The likely extent and magnitude of past volcanic eruptions are apparent in Cascade Range soils near or downwind from major volcanoes, soils that are largely formed from tephra (Franklin and Dyrness 1973), and in the large amounts of tephra in soils far east of the Cascade Range (Smith et al. 1968).

Tephra has had major effects on plants in many parts of the world (Table 1.1). Some trees may survive burial by tephra 2 m deep, but smaller plants are killed by much thinner deposits (Antos and Zobel 1987). Thin layers (a few millimeters thick) are likely to have little effect on plants. Between these extremes, various combinations of depth, texture, and frequency of deposition produce a wide range of plant responses (Antos and Zobel 1987). Effects of tephra on plants may include direct damage from impact, alteration of leaf gas and energy exchange by tephra adhering to foliage, modification of the soil environment, and burial of small plants and seed banks. Leachate from tephra may contain toxic elements that damage root systems. Conversely, tephra can be a source of plant nutrients, although it lacks nitrogen and most of its phosphorus is not easily leached (Hinkley 1987). Fine-textured tephra may harden after wetting to produce a surface crust with poor permeability to water. Such a crust is often dense and strong enough to restrict plant growth. For smaller plants, burial is the most important effect of tephra, and the ability to grow through the deposit is a key to survival (Griggs 1919, 1922; Antos and Zobel 1987).

From 1980 to 2000, we have studied the effects of Mount St. Helens tephra on understory plants in old-growth conifer forests with trees more than 500 years old, using two sites at each of two tephra depths, 4.5 and 15 cm, located 22 and 58 km northeast of the crater (Table 4.1). The sites are on flat topography at elevations between 1160 and 1290 m, in the *Abies amabilis* (Pacific silver fir) vegetation zone of Franklin and Dyrness (1973). Large conifer trees, including *Tsuga heterophylla* (western hemlock), *T. mertensiana* (mountain hemlock), Pacific silver fir, *Pseudotsuga menziesii* (Douglas-fir), and *Chamaecyparis nootkatensis* (Alaska cedar), dominate the sites, with a patchy distribution of smaller trees, primarily Pacific silver fir and *Tsuga* spp. (hemlocks). Before the eruption, understories contained ericaceous shrub layers 1 to 1.5 m tall with 17% to 45% cover, primarily *Vaccinium membranaceum* (big huckleberry) and *V. ovalifolium* (ovalleaf huckleberry). Herbaceous layers varied considerably among sites in cover (6% to 35%) and diversity (Table 4.1), but included a variety of growth forms. Bryophyte layers (9% to 36% cover) were dominated by *Dicranum* spp. (broom mosses) and *Rhytidiopsis robusta* (pipecleaner moss) (Zobel and Antos 1997). Wood more than 5 cm in diameter covered 3% to 11% of the preeruption surface. For 3 years after the eruption, from 1980 to 1983, we also sampled sites with 2- and 7.5-cm tephra at 550 and 880 m in elevation, respectively, in the western hemlock zone (Antos and Zobel 1985b, 1986). Our intent was to study long-term effects of tephra deposits on understory plants; thus, our intensive study sites have gentle slopes, although most Cascade Range topography in the tephra-fall zone is steep. On some very steep slopes, erosion was extensive, producing much greater understory cover. However, on many steep

TABLE 4.1. Study site characteristics of four intensively studied sites with permanent plots in the tephra-fall zone at Mount St. Helens.

Site code	Conditions (tephra depth/herb diversity)	Tephra depth (cm)	Number of herb species	Elevation (m)	Distance to crater (km)	Snow cover (%)
SP	Shallow/poor	<5	9	1245	58	29
SR	Shallow/rich	<5	26	1290	58	92
DP	Deep/poor	>12	12	1160	22	11
DR	Deep/rich	>12	32	1240	22	88

At each site, there were one hundred 1-m^2 plots on natural tephra and fifty 1-m^2 plots with tephra removed in 1980; at site DR, there were fifty 1-m^2 plots with tephra removed in 1982. Plots were read (cover and density by species) at all sites in 1980, 1981, 1982, 1983, and 2000; plots were read at DR and DP in 1984, 1987, and 1990 and at SR and SP in 1989.
Source: Details of the sampling procedures are given in Antos and Zobel (1985b) and Zobel and Antos (1997).

slopes erosion was limited to small channels. The residual tree canopy and resultant litterfall appear to have quickly stabilized the tephra surface. Conversely, few areas, including the flattest, completely lacked erosion channels. Our sites experienced minimal erosion. The conditions we studied represent the maximum likely impact from the deposition of 4.5 to 15.0 cm of tephra.

At each site, we used repeated sampling of permanent plots of two types (one with natural tephra and one with tephra removed soon after the eruption). Within each site, plots were arrayed in several transects, with plots at 2-m intervals, and transects several meters apart. The plots sampled at each site were located within an approximately 1-ha area. Each sample plot was 1 m square. At each site, 100 plots were established on undisturbed tephra. In 50 additional plots, we removed the tephra carefully during the summer of 1980. These cleared plots were used to estimate the preeruption vegetation of the site. Transects of cleared plots were interspersed among those with natural tephra. At the deep-tephra, herb-rich site (DR; see Table 4.1), 50 additional plots were cleared in late summer 1982 (2 years postdisturbance). These 50 plots were used to demonstrate vegetation change following delayed erosion of tephra.

Plant cover and density were measured for each species of vascular plant, and cover was measured for each bryophyte taxon. Seedling data were separated for preeruption and posteruption trees, and for first-year and older seedlings. All sites (see Table 4.1) were measured in 1980 to 1983, 1989 or 1990, and 2000. Sites with deep tephra were also assessed in 1984 and 1987. At the shallow-tephra, herb-rich site (SR; Table 4.1) we did not assess herb density in 1989. We consider 1981 values for cleared plots to be effective estimates of preeruption species importance for our sites (Zobel and Antos 1997).

We have used data from these permanent plots along with studies of plant growth through tephra (Antos and Zobel 1985a,c; Zobel and Antos 1987a,b), ability of plants to survive burial (Zobel and Antos 1986, 1992), and tree-seedling growth in tephra (Zobel and Antos 1991b) to evaluate the course and mechanisms of vegetation change following disturbance by tephra. Here we summarize and synthesize our work and that of others in the tephra-fall zone, present recent findings from sampling conducted in 2000, and evaluate possible factors controlling vegetation change, prospects for future change, and how vegetation change in the tephra-fall zone differs from that responding to other forms of disturbance.

4.2 Characteristics of the Tephra Disturbance

At elevations of 1160 to 1250 m near Mount St. Helens, snowpack is typically present during May. In May 1980, both of our two main study areas still had a patchy distribution of snowpack as well as areas without snow. At each area, a slightly higher elevation site with a dense herbaceous layer and moister soil was mostly snow covered, whereas a nearby site with a poorer herb layer was mostly snow free (Antos and Zobel 1982; see Table 4.1). No plants had developed new foliage at our main study sites. There was no evidence that the temperature of the tephra during deposition was high enough to damage plants.

Our sites received tephra depositions of about either 4.5 cm or 15 cm in depth, thicknesses that were first measured after several rains had compacted the new-fallen tephra. The compacted thickness of tephra is sometimes only half the noncompacted thickness displayed in geological reports, at least for fine tephra (Sarna-Wojcicki et al. 1981). We recognized three strata in the tephra deposits (Zobel and Antos 1991a): (1) debris from the lateral blast produced a dark, fine-textured, sticky basal layer, 4 to 7 mm thick in 1987; (2) a coarse, single-grained pumice layer 33 mm deep at 58 km from the crater and 114 to 135 mm deep at 22 km; and (3) a crust dominated by fine particles, 8 mm thick at 58 km and 16 mm thick at 22 km, which hardened after wetting and drying. Eruptions after May 18 contributed only scattered gravel and sand to the surface of the deposit. Texture was coarser near the volcano: particles less than 2 mm in diameter represented 82% of the mass of the crust at 22 km and more than 99% at 58 km; corresponding values for the single-grained layer were 70% and more than 99% for particles less than 2 mm in diameter, respectively (Zobel and Antos 1991a).

Tephra properties varied within stands. Depths ranged from 12 to 18 cm at the deep-tephra sites in areas without erosion or slumping of the deposit. Crust was thicker beneath the forest canopy than in openings (Zobel and Antos 1991a). At the base of large trees, the crust was thin, but the deposit of coarse pumice was thickest. On steep microsites and beneath logs, tephra was shallow or absent. Where tephra fell on snowpack, the crust cracked as the snow melted. The presence of cracks allowed us to map the snowpack in May 1980. Cracks filled during the first winter (Antos and Zobel 1982). Water ponded on the tephra crust for a few days after rain. With time, fine particles were eroded from the convex surfaces of the crust, which became thicker in concave spots and thinner elsewhere (Zobel and Antos 1991a). Deeper erosion of the tephra occurred primarily during the winter of 1980–1981, removing the entire tephra blanket from parts of some steep slopes. About 5% of two of our study sites had tephra removed to the old soil surface or received erosional deposits more than 2 cm deep.

Chemistry of the tephra and buried soil changed after 1980. By 1987, the tephra layers had lost 80% to 98% of their sulfate sulfur, 60% to 80% of their calcium, 20% to 70% of their magnesium, and 0% to 60% of their potassium (Zobel and Antos 1991a). During the same period, pH declined 1 to 2 units in deep tephra and 0.5 to 1 unit in shallow tephra, nitrogen increased to 7 to 100 times the original values, and phosphorus increased to 1.4 to 5.5 times the original values. These changes represent the net effect of leaching, plant uptake, and input from leaf litter and throughfall. Cations leached from tephra were captured in the buried soil profile. The thick, single-grained layer was less acid and had lower nutrient concentrations than finer-textured tephra layers. Pot tests using additions of fine Mount St. Helens tephra to Swiss chard and barley indicated no toxic effects from the tephra (Cochran et al. 1983).

4.3 Effects of the Tephra on Trees

Fine ash stuck to tree foliage and had a variety of effects, as indicated by several studies, all conducted away from our major sites. Although much of the ash washed off during the first winter, some remained for years. Amounts of ash retained on foliage varied considerably among species of conifers, but were particularly large on *Abies* spp. (fir). Near the volcano, needles embedded in tephra died from solar heating, because the heavily coated needles exchanged heat less effectively than did the needles alone (Seymour et al. 1983). This posteruption needle mortality is distinct from the effect of the heated air in the eruption cloud that killed foliage around the edge of the lateral blast (Winner and Casadevall 1983).

Tephra had various effects on tree growth. Needle loss led to reductions in tree diameter and elongation growth of fir during 1980; the deeper the tephra, the greater the growth reduction (Hinckley et al. 1984). Removal in 1980 of tephra from the foliage of fir saplings in a clear-cut increased twig elongation during 1981 and 1982 (Zobel and Antos 1985). Ash on tree foliage 113 to 480 km downwind produced minor, temporary physiological changes and increased surface fungal colonization, but there was no evidence of growth loss (Bilderback and Carlson 1987; Bilderback and Slone 1987). Abrasion from tephra during high winds damaged *Populus* (poplar) leaves, causing premature leaf abscission (Black and Mack 1984).

A delayed crown decline and mortality of mature and old-growth Pacific silver fir became obvious sporadically within the tephra-fall zone in 1986. This decline was most pronounced in areas where the thickness of the basal fine-textured layer was greatest, and its initiation was attributed to the effective, persistent retention of this type of tephra on foliage (Segura et al. 1994). Associated with substantial crown decline was a major loss of stem growth and development of a small, vigorous, apical crown segment with branches produced after 1980 (Segura et al. 1995a,b).

4.4 Initial Effects of Tephra on Forest Understory Plants

The initial effects of the tephra disturbance on the bryophyte and herb layers were primarily a function of deposit depth: 4.5 cm of tephra killed almost all mosses, whereas most herbs survived. Even 2 cm of tephra buried small bryophytes, but all except the smallest regained their preeruption coverage within 2 to 3 years (Antos and Zobel 1985b). With 4.5 cm or more of tephra, bryophytes were buried, except where tephra slumped off logs and tree bases, and recovery required new colonization of the surface from spores.

Although most herbaceous species could penetrate upward through tephra 4.5 cm deep or less, tephra 15 cm deep nearly eliminated the herb layer because few individuals could grow through deposits of this depth (Antos and Zobel 1985b). Herbs sometimes escaped burial in microsites that received little tephra or by growing through cracks in the crust caused by snow melting beneath the tephra. Survival of most herbs depended on growing through the tephra; however, most shoots that grew up into deep tephra failed to emerge through the crust.

The response of shrubs and small trees to tephra was complex. The spotty snowpack of May 18, 1980, produced patchy shrub survival. Where a snowpack was present at the time of the eruption, most shrubs and small trees were buried beneath tephra, even if the deposit was only 4.5 cm deep. Most shrubs and trees survived where snow was absent, even with 15 cm of tephra (Antos and Zobel 1982). Where snow was absent, tephra fell through the upright shrubs and small conifers, burying only those shorter than the tephra depth. However, the situation was much different where snow was present on May 18, 1980. Snowpack forms in fall, and new snow is added throughout winter and early spring. The snow continues to melt slowly within the pack and at the soil surface. Thus, woody plants that were covered by snow in fall 1979 eventually become prostrate in or beneath the snowpack. When the snow melted out from under the tephra, the heavy blanket of wet tephra covered the prostrate shrubs. By this mechanism, 4.5 cm of tephra removed most of the cover of a dense 1- to 1.5-m-tall shrub layer, and 15 cm obliterated shrubs almost completely (Antos and Zobel 1982). Some branches near the surface, or where the crust was disturbed, did produce emergent shoots. New shoots continued to emerge for several years after 1980.

The timing of the eruption was critical to the amount and characteristics of the damage caused by tephra. At higher elevations, plants that renew shoots each year were belowground during the May 18 eruption, and those with perennial shoots had not produced new leaves. Thus, tissue damage and loss of stored energy were minimal. In contrast, some shoots were damaged at lower elevations. For example, shoots of *Veratrum californicum* (California false hellebore) with expanded leaves collapsed beneath a thin tephra layer in Idaho (Mack 1981), but *Veratrum viride* (false hellebore) shoots that emerged after the eruption easily penetrated artificial tephra deposits twice the depth of deep natural tephra at our sites (Zobel and Antos 1987a). The situation was opposite for shrubs; while dormant and beneath snowpack, they were more subject to damage from tephra than during active growth because prostrate stems under snow became covered with a tephra blanket that was almost never penetrated, whereas tephra did little harm to upright stems without leaves (Antos and Zobel 1982).

4.5 How Plants Survived Burial

4.5.1 Growth of Buried Herbs Through Deep Tephra

For herbs with shoots that die annually, burial was inevitable except in rare microsites where tephra never accumulated (e.g., under some logs) or was quickly removed by erosion. Thus, for most herbs, survival was contingent on shoots growing through the tephra. Furthermore, prolonged survival of an individual requires either repeated penetration of the tephra by annual shoots, which could be a major energy drain, or placement of new perennating buds in or at the surface of the tephra, which would eliminate the necessity to penetrate the deposit annually. Perennating organs of most herbaceous species did colonize deep tephra (Antos and Zobel 1985a,c). Movement into tephra was accomplished by rhizome growth, development of buds on emergent stems, or both. Many combinations of tactics were employed. About 65% of species moved their perennating structures from the buried soil into the tephra; the same proportion produced numerous roots in the tephra (Antos and Zobel 1985a). Species that produced a new crown required one to three seasons to regain their usual depth beneath the new substrate surface. Perennating structures remained in buried soil for about 30% of species [e.g., *Erythronium montanum* (avalanche lily)], and shoots penetrated the entire depth of the tephra each year.

The responses required for survival varied with plant structure and phenology. Our model, which was proposed in 1981 and which integrates several factors required for survival of tephra burial (see figure 10.1 in Antos and Zobel 1987), appears accurate after 20 years of observations. However, for many species, emergence from deep tephra could not be predicted effectively from their usual growth form (Antos and Zobel 1985a; Zobel and Antos 1997). We anticipated that most herbs with short rhizomes or root crowns would only slowly, if at all, move the position of their perennating buds upward in the tephra. However, most of these species increased elongation of rhizomes or other stems and placed new perennating buds near the surface, an important change in morphology. This phenotypic plasticity allowed plants with short rhizomes, such as *Viola orbiculata* (round-leaf violet), to grow vertically a distance far greater than their usual horizontal spread, a response that we conclude was critical to their survival.

Although many species appropriately modified their growth form and occasionally grew through deep tephra (Antos and Zobel 1985a,c), almost all individuals failed where tephra was 15 cm deep. Plant survival was influenced by the composition of the deposits (ranging from loose pumice to hard, fine-textured ash) as well as by tephra depth. Other factors apparently affected survival, also:

- Many shoots reached the base of the tephra crust, but failed to penetrate it and grew in loops in the looser layer beneath that hard crust. During summer 1980, the crust in some locations was so firm that it was possible to drive a truck over the surface without breaking it.
- Abrasion from growing through pumice may have destroyed some shoot meristems.
- Energy reserves may have been inadequate.
- The soil environment (e.g., temperature or CO_2 or O_2 levels in the old soil) may have prevented growth.

In contrast to plants with plastic responses to burial, some growth forms have little capacity to escape tephra once they are covered completely. For example, a prostrate, evergreen, slow-growing subshrub, *Gaultheria humifusa* (alpine wintergreen), seldom survived tephra only 4.5 cm deep. *Xerophyllum tenax* (beargrass), a robust, evergreen, grasslike herb, has no capacity to break an intact tephra crust. Although this species is abundant and widespread in the Cascade Range, it is relatively sparse near Mount St. Helens, perhaps a response to the long history of major tephra deposits.

4.5.2 Adventitious Rooting in the Tephra by Woody Plants

Although tephra was not deep enough to bury most woody plants (except via the interaction with snow as already described), colonization of the new deposit by roots may be important to survival. Most woody plant species produced adventitious roots in deep tephra, and some shrubs produced long, new rhizomes (Zobel and Antos 1982; Antos and Zobel 1985a). Major exceptions were ericaceous shrubs, which dominated forest understories but produced few roots, and seedlings of Douglas-fir, which produced no adventitious roots. Eight other conifers, including Pacific silver fir up to at least 35 cm dbh (diameter at breast height), produced adventitious roots in deep tephra. Large adventitious roots in tephra from before 1980 occurred on old-growth trees near Mount St. Helens (Lawrence 1954), including many uprooted by the 1980 lateral blast.

4.5.3 Survival Subsequent to Prolonged Burial

Many plants survived burial for three growing seasons; some survived at least eight. Excavation of plots in late summer 2 and 7 years postdisturbance demonstrated a substantial capacity for survival beneath tephra by shrubs, herbs, and bryophytes (Zobel and Antos 1986, 1992). Burial for three seasons reduced cover for shrubs to 21%, for herbs to 13%, and for bryophytes to 18% of cover in plots excavated in 1980, but cover increased as plants grew after excavation (Zobel and Antos 1997). After eight seasons of burial, five species (two shrubs, one herb, and two mosses) survived and continued to live after release but did not grow or spread rapidly. However, many shoots of huckleberry emerged from tephra between 5 and 9 years after disturbance. Late-emerging shoots were connected to buried preeruption plants that had no other emergent shoots, so they were long-term survivors (Zobel and Antos 1992). Our results indicate that an initial lack of emergent shoots following the disturbance does not prove that there are no surviving buried plants.

4.6 Tephra Redistribution and Plant Survival

Processes that remove or thin the deposit increase plant survival and modify subsequent vegetation change. Erosion can produce conditions under which any species could survive. Places with thin tephra, produced by erosion or beneath large logs, serve as refugia that allow survival of some species, maintain stand-level species richness, and provide sources for future colonization of tephra. During summer in 1980 and 1981, erosion channels contained luxuriant communities of herbs and bryophytes embedded in an extensive matrix of barren tephra.

4.7 Twenty Years of Vegetation Change in the Tephra-Fall Zone

The forest canopy in the tephra-fall zone remained intact after the 1980 tephra deposit. Thus, tephra may seem to have minor or transient importance. However, the initial impact of 4.5 cm of tephra on bryophytes and shrubs was substantial. Where 15 cm of tephra fell on snowpack, most understory plants died. Twenty years after the deposition of 4.5 cm of tephra, the bryophyte layer remained sparse and of altered composition. With 15 cm of tephra, cover and diversity of most understory growth forms remained low after 20 years. The trajectory of vegetation change often did not approach preeruption conditions.

The substrate and forest floor changed greatly after 1980, as litter from the tree canopy accumulated. A new mor humus layer is developing on the tephra surface. In 2000, bare tephra was exposed only occasionally, usually on convex surfaces or where litter eroded. The forest floor looked much more like it did before 1980 than it did in the first few years after the eruption. The tephra deposit, however, remains largely intact below the new forest floor. Thus, the substrate for shallow-rooted understory plants differs from that before the eruption. Although roots and rhizomes were common in the tephra after 20 years, this new substrate, especially where deep, has produced long-lasting changes in the understory.

4.7.1 Bryophytes

Among major plant groups, bryophytes were the most affected in the tephra-fall zone both initially and after 20 years (Figure 4.1). In 2000 at all sites, tephra supported significantly less bryophyte cover than did cleared plots in either 1981 or 2000 (Table 4.2). However, increases from 1990 to 2000 were generally substantial (Figure 4.1), and the differences between tephra plots and cleared plots had decreased. Thus, total cover is likely to approach the predisturbance cover within a few decades. This change cannot be equated to complete recovery, however, because species composition may still be altered. The original dominant species (broom mosses and pipecleaner moss) still generally had less (often much less) than one-fifth as much cover in tephra as in cleared plots. At the herb-rich, deep-tephra site (site DR, Figure 4.1), total bryophyte cover in 2000 was slightly more than half that in cleared plots, but most of this cover (more than four-fifths of it) came from species that had colonized the tephra surface and were rare or absent before the eruption in 1980. These mosses include *Ceratodon purpureus* (fire moss, which is one of the most common bryophytes colonizing the Pumice Plain) and *Pohlia* spp. (a genus of mosses), which are common in disturbed habitats, especially bare soil. These mosses represent the only major case of invasion of the old forest by early successional species after the eruption.

4.7.2 Herbaceous Plants

Total herb cover was not substantially reduced by shallow tephra (Figure 4.2; see Table 4.2), because most herbs grew through the 4.5-cm deposit the first summer after the eruption. However, some changes in species composition occurred, and it is premature to conclude that the system will soon return to its pre-1980 state. Most notably, alpine wintergreen had only 0.1% cover on tephra in 2000, but 3.5% in cleared plots; reductions for density were even greater. In contrast, *Chimaphila*

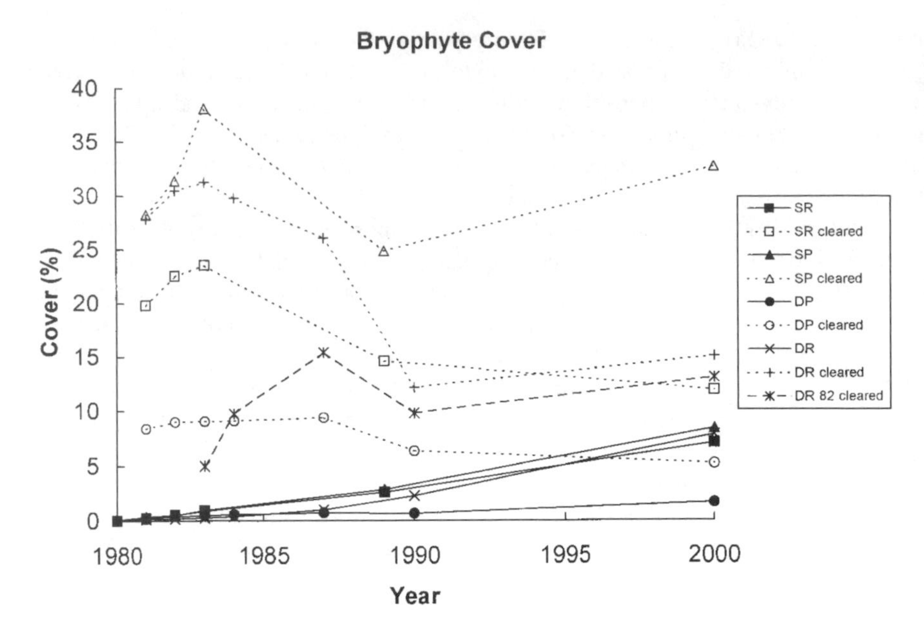

FIGURE 4.1. Change in total cover of bryophytes at four sites in the tephra-fall zone at Mount St. Helens during the 20 years following the 1980 eruption. At each site, values are given for plots on natural tephra and plots cleared during 1980 (these plots provide an indication of undisturbed conditions). At one site (*DR*), values are also given for plots cleared in 1980. Site codes are: SR, shallow tephra, herb-rich site; SP, shallow tephra, herb-poor site; DP, deep tephra, herb-poor site; DR, deep tephra, herb-rich site. (See Table 4.1 for site information.)

umbellata (prince's pine) was never reduced, but increased to more than three times its 1980 cover and density in both cleared and natural plots. Clearly, those two species responded differently to the disturbance and altered microenvironments.

With deep tephra, herbaceous cover was still low after 20 years (see Figure 4.2; Table 4.2). At the herb-poor site, cover and density were less than 5% of the values in cleared plots; all species were reduced. At the herb-rich site, cover was extremely low soon after the eruption but increased much more rapidly, reaching more than one-third of that in cleared plots by 2000. Rates of change varied among species. At one extreme was avalanche lily, the most abundant herb at the site before 1980, with almost half the cover in cleared plots; in 2000 it had less than 0.1% cover on tephra but 14.9% cover in cleared plots. Furthermore, cover on tephra changed little during the two decades, providing no evidence of recovery of this drastically reduced species. In contrast, other herbs had cover on tephra not statistically different from values in cleared plots in 2000, even though they were reduced to a few individuals with less than 0.1% cover on tephra after the eruption.

TABLE 4.2. Assessment of "recovery" of total cover and density of a growth form that employs two criteria: cover or density on tephra in 2000 was *greater than or equal to* the cover or density in (a) cleared plots in 2000 or (b) cleared plots in 1981.

(a) Criterion 1: Cover or density on tephra in 2000 was *greater than* the cover or density in the cleared plots in 2000

	SP		SR		DP		DR	
Growth form	Cover	Density	Cover	Density	Cover	Density	Cover	Density
Shrubs	=	=	–	=	–	–	–	–
Herbs	=	–	=	=	–	–	–	–
Bryophytes	–		–		–		–	

(b) Criterion 2: Cover or density on tephra in 2000 was *greater than* the cover or density in the cleared plots in 1981

	SP		SR		DP		DR	
Growth form	Cover	Density	Cover	Density	Cover	Density	Cover	Density
Shrubs	=	+[a]	+[b]	+[a]	=[b]	=[b]	–	–
Herbs	=	+[b]	+[a]	+[a]	–	–	–	–
Bryophytes	–		–		–		–	

Comparisons are based on the Mann–Whitney U test, p less than 0.05.

Codes: – indicates not recovered (less on tephra in 2000 than in cleared plots in 2000 or 1981); = indicates recovered (the same on tephra in 2000 as in cleared plots in 2000 or 1981); and + indicates more on tephra in 2000 than in cleared plots in 2000 or 1981. See Table 4.1 for site codes.

[a] Differs from criterion 1 without changing conclusion.

[b] Changes the answer to the question, "Has it recovered?"

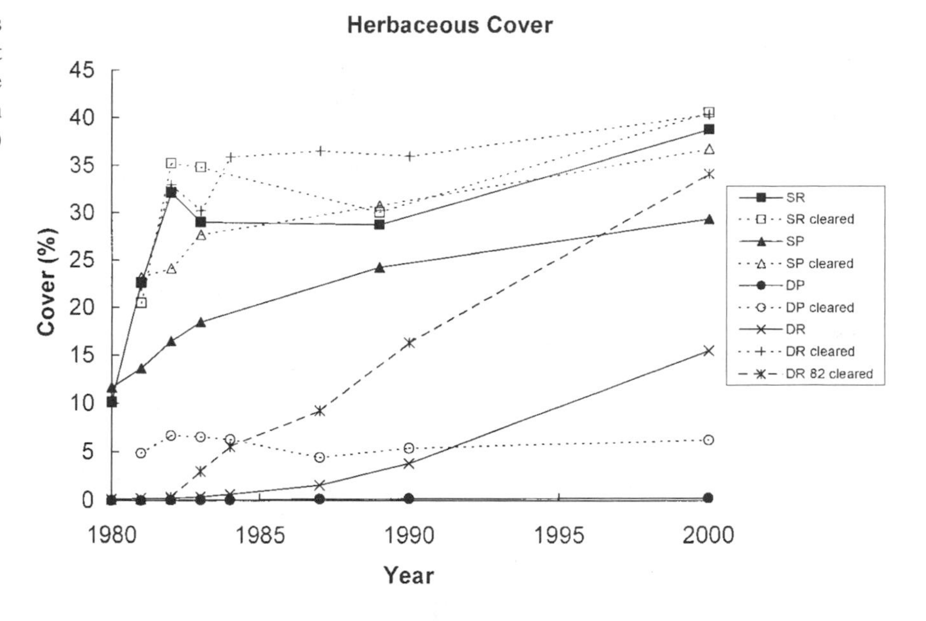

FIGURE 4.2. Change in total cover of herbs at four sites in the tephra-fall zone at Mount St. Helens during the 20 years following the 1980 eruption. (See Figure 4.1 for description of plot types and Table 4.1 for site information.)

These species include *Rubus lasiococcus* (dwarf bramble) and *Tiarella trifoliata* (coolwort foamflower), both of which had cover in 2000 that was greater than 100 times that in 1981. By 2000, they were the most abundant herbs in the stand with 9.0% and 2.3% cover, respectively. Thus, although total cover was increasing (see Figure 4.2), the species composition was changing dramatically. Some species were more abundant than before the disturbance while others were greatly reduced and showed little sign of recovery.

Much of the shift in plant composition in the tephra-fall zone is related to growth form. Clonal species with above-ground stolons [e.g., dwarf bramble, *R. pedatus* (five-leafed bramble), and *Linnaea borealis* (twinflower)] spread rapidly from their points of emergence from the tephra and covered areas several meters across within a few years. Stolon growth is a very efficient way of colonizing the tephra surface and allows a few surviving plants to have a major influence on the vegetation. Other successful species had little vegetative spread but flowered regularly and had many seedlings; the best example is coolwort foamflower (Antos and Zobel 1986). Avalanche lily, which failed to increase much beyond the few survivors, lacks vegetative spread, failed to move perennating structures into the tephra, and had few seedlings on the tephra, although survivors did produce seed.

Removing tephra within a few years of its deposit may compensate for the initial damage it caused. Plots cleared 2 years after disturbance showed a steady and rapid increase in herbaceous cover; by 20 years postdisturbance, they did not differ significantly in total herb cover from plots cleared in 1980 (see Figure 4.2). Furthermore, there were no significant differences between plots cleared in 1980 and in 1982 for any common species, including those that differed greatly in response to tephra. Thus, erosion within the first three growing seasons can almost eliminate the major effects of the disturbance within 20 years; without erosion, effects will persist for decades.

4.7.3 Shrubs

After 20 years, shrub cover was strongly related to the extent of snowpack present in the tephra-fall zone on May 18, 1980. In contrast, the increase in cover was related to tephra depth. At the shallow-tephra site with snow (SR), cover in 2000 had increased to 52%, although it was still significantly lower than in cleared plots (77%) (Figure 4.3). The assessment of change is complicated, however, because cover had also increased in the cleared plots to the extent that the cover in tephra plots in 2000 was significantly higher than the cover in cleared plots had been in 1981. Thus, the assessment of recovery in total shrub cover depends on the comparison used (see Table 4.2). This overall increase in shrub cover could relate to an especially favorable series of years but more likely reflects a positive effect of the tephra on shrub growth in these very old stands. This increase was species specific; cover of the two huckleberry species increased greatly, whereas that of *Menziesia ferruginea* (fool's huckleberry) decreased.

In contrast to the limited residual effects of shallow tephra on snow, shrub cover was still low on tephra compared to that on cleared plots (4% versus 38% cover) after 20 years where deep tephra fell on snow (see Figure 4.3; Table 4.2). This large difference remained even though shrubs in tephra plots increased by more than a factor of 20 in cover and by more than a factor of 50 in density from 1981 to 2000. Delayed shoot emergence was important for shrubs. Cover in the plots cleared 2 years after the disturbance was much higher than that in tephra plots and more than half of that in plots cleared in 1980. Thus, for shrubs, erosion that removed tephra within the first three growing seasons could mitigate much of the effect of the disturbance. As with shallow tephra, shrub cover increased in cleared plots at the deep-tephra sites (Figure 4.3). Shrub seedlings occurred occasionally on the tephra, but most grew slowly. Increases in cover resulted primarily from growth of vegetative sprouts.

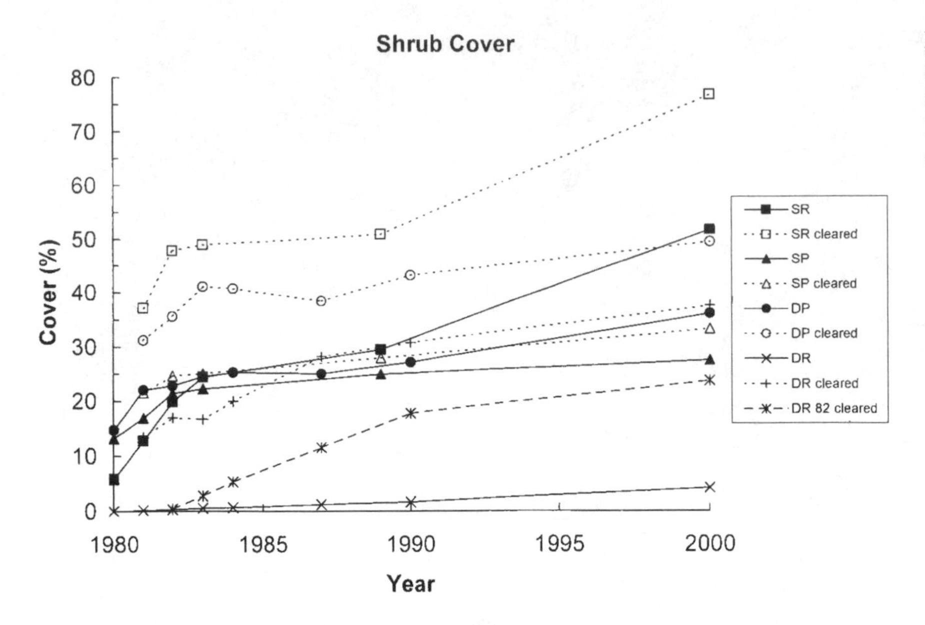

FIGURE 4.3. Change in total cover of shrubs at four sites in the tephra-fall zone at Mount St. Helens during the 20 years following the 1980 eruption. (See Figure 4.1 for description of plot types and Table 4.1 for site information.)

4.7.4 Survival and Growth of Small, Preeruption Trees

Snow at the time of the eruption and tephra depth had important effects on both the survival of small trees (less than 1.5 m tall) and their subsequent contribution to the understory community in the tephra-fall zone. Cover under most conditions remained similar or increased from 1981 to 2000 (Figure 4.4). Growth of individual trees was largely balanced by mortality of others as density of surviving trees decreased. Because of the effects of snow and tephra depth, cover of these trees in 2000 was significantly lower in tephra than in cleared plots at all sites except the shallow-tephra, snow-free site (Figure 4.4). At the deep-tephra, snow-covered site, more than 99% of these trees were lost to the disturbance. In contrast to other growth forms, small trees did not benefit from delayed erosion; most buried trees died within one season.

4.7.5 Establishment of a Tree-Seedling Layer

Perhaps the most dramatic development in the understories of forests in the tephra-fall zone since the eruption has been the establishment of a tree-seedling layer on the tephra. The new tephra supported more tree-seedling establishment and survival than did the original forest floor; seedling density was high soon after the eruption (Antos and Zobel 1986; Zobel and Antos 1991b, 1997). Although tree seedlings had been common in the stands before the eruption (cleared plots; see Figure 4.4), by 2000, the 26% cover of new tree seedlings at

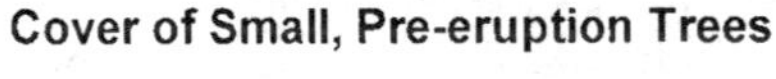

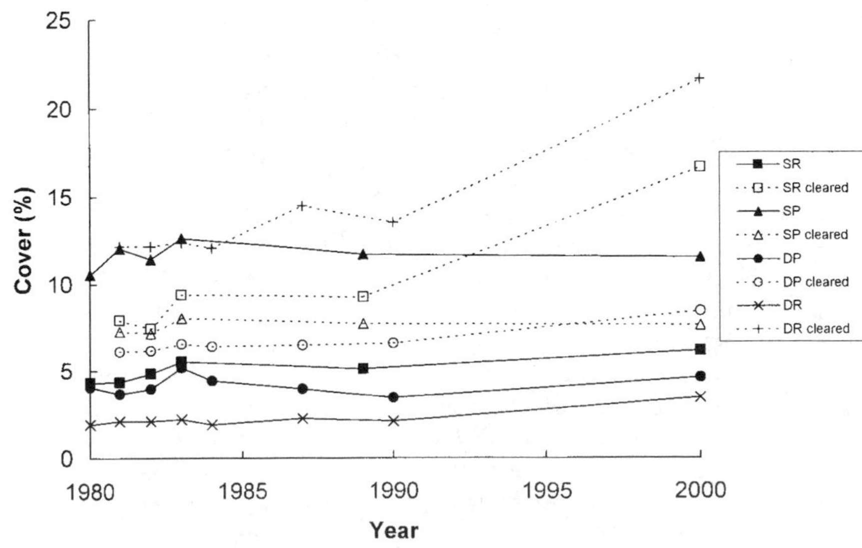

FIGURE 4.4. Change in total cover of small trees (less than 1.5 m tall in 1980) that established before 1980 at four sites in the tephra-fall zone at Mount St. Helens during the 20 years following the 1980 eruption. (See Figure 4.1 for description of plot types and Table 4.1 for site information.)

the deep-tephra snow site was more than double that in cleared plots at any site in 1981 (our best estimate of preeruption conditions). Increases for density were even larger than those for cover because these new trees were small. In deep tephra, the density of new tree seedlings was substantially higher than that of seedlings before the eruption (19.1 versus 3.6 m^{-2} at the no-snow site and 27.4 versus 10.1 m^{-2} at the snow site). Perhaps more importantly, the species composition of the seedlings had changed. Before the eruption, almost all seedlings had been Pacific silver fir, but the new seedlings contained a large component of hemlock (both western and mountain), which comprised 25% of the seedlings at the snow site and 66% at the no-snow site. Increases in the density of tree seedlings and the hemlock component were also apparent with shallow tephra, but the seedlings had grown slowly and still had little cover in 2000. At age 6, no new conifer seedling roots had yet reached the buried soil (Zobel and Antos 1991b).

4.7.6 Succession in the Forest Understory

The changes in the forest understory in the tephra-fall zone from 1980 to 2000 constitute an unusual form of succession. Although many plants died, nearly 100% in some situations, no species absent from the forest invaded to an important extent, except for bryophytes. Very few pioneer vascular plants entered the system, and these attained little cover (Zobel and Antos 1997). However, there were major compositional changes, which would be expected in a successional sequence. By 2000, however, these changes did not appear to be toward the predisturbance condition, as might be expected after 20 years. Although total herb cover is increasing toward the preeruption value at the deep-tephra snow site, the composition appears to be diverging because some species recover rapidly while others fail to increase. Furthermore, the changes in tree seedlings are a clear divergence from predisturbance conditions. The first 20 years of understory succession in the tephra-fall zone did not follow patterns for other types of disturbance, probably because of the combination of an intact tree canopy with a drastically altered substrate.

4.8 Factors Limiting Rates of Recovery

Why has understory vegetation not recovered more rapidly during the first 20 years? A variety of factors may have limited vegetation change and contributed to divergence in plant composition in the tephra-fall zone.

First, initial survival is critical (Zobel and Antos 1997), and few plants survived where tephra was deep.

Second, reproduction of many survivors was ineffective. Because few plants survived, seed production was limited, and foci for vegetative spread were few. Some survivors flowered rarely, if at all. Rates of vegetative spread are slow for many of these species (Antos and Zobel 1984, 1985c), and seed dispersal may also be very limited. Forest herbs can be slow at colonizing habitat from which they have been removed, even over short distances (Brunet et al. 2000; Donohue et al. 2000).

Third, the tephra is a very different substrate from the old forest floor, which had developed for 180 years after the previous major eruption. Although many forest herbs produced no seedlings, a few produced many. An altered substrate for seedling establishment probably is responsible for changes in the composition of seedlings. The new substrate, especially where deep, also provides very different conditions for vegetative growth. Spread by rhizomes is typical of many of these forest herbs (Antos and Zobel 1985a,c). Tephra may be less favorable for growth than was the old forest floor because of its abrasive qualities, low nutrient status (Zobel and Antos 1991a), and strength of its surface crust (Antos and Zobel 1985c).

Fourth, because the forest canopy is largely intact, light levels increased little after the eruption, in contrast to the effect of most disturbances. It could be argued that this lack of change would increase the likelihood and rate of return to predisturbance conditions, but low light could also limit both seedling establishment and growth.

Fifth, our sites are at elevations with prolonged snowpack, short growing seasons, and low temperatures, which limit rates of plant growth and succession. Following snowmelt, the growing season is often dry. Slow response should also be expected because the system is dominated by stress-tolerant plants with slow innate growth rates (Grime 1979). There is simply too much snow and too little summer in the upper-elevation forest understory to expect rapid responses.

4.9 Prospects for Future Changes in the Tephra-Fall Zone

The unusually dense layer of new tree seedlings is the most important change induced by the tephra, other than the direct killing of plants in the tephra-fall zone. Because these seedlings are potential canopy trees, changes to the seedling population may have long-lasting implications throughout the ecosystem. Few seedlings exceeded 50 cm in height even in the best conditions in 2000, and many will likely die. For example, in 2000 damage from felt blight (or brown snow mold, caused by fungi of the genus *Herpotrichia* on conifer foliage covered by snow; Hansen and Lewis 1997) was extensive at one site. However, many new seedlings were healthy, and slow growth is to be expected on these sites. We anticipate that a dense layer of saplings will develop atop the deep tephra, producing a major change in the forest structure. An unusually dense sapling layer had developed in some stands where the overstory trees were killed in 1980 and in some remaining stands near the blowdown zone northeast of Mount St. Helens, perhaps following the major tephra fall in 1800.

Many individuals of both western and mountain hemlock, which were rare in the understory before the eruption, are growing well, and some may reach the canopy. Thus, the eruption has provided a window of opportunity for hemlock to increase its importance in these forests.

A critical question for future vegetation change in the tephra-fall zone is how long the new tephra surface will alter establishment. The gradual buildup of litter should slowly transform the surface into a fully developed forest floor. Our deep-tephra sites have many layers of tephra (including thick pumice layers) below the current deposit. The 1980 tephra should eventually become very similar to the upper soil horizons present before the 1980 eruption. Thus, eventual convergence on the original vegetation would seem likely, but the timing is equivocal, and the system may diverge further in some properties before convergence occurs. It is possible that somewhat different compositional and abundance patterns will emerge as a result of the 1980 tephra fall. In either case, the establishment of the tree-seedling layer can be thought of as an alteration in the forest that will require centuries to disappear because of the potential longevity of the trees involved (600 or more years).

Even after 20 years, we can identify shrub-free locations in the tephra-fall zone that had snowpack at the time of the eruption. This effect is being erased as shrubs sprout and the new tree-seedling layer develops, but the fading of this effect does not necessarily indicate return to predisturbance conditions. At some sites, shrub cover has increased to levels higher than those before the eruption. This increase may be a transient phenomenon, but it is too early to tell. Increased shrub growth could be a result of reduced competition from other plants or accelerated nutrient release from the buried forest floor that results because the tephra increases decomposition by buffering temperatures and reducing drying (Edmonds and Erickson 1994). Another possibility is that initial nonlethal damage to canopy trees increased light levels. It also appears that more trees have died standing at our sites since the eruption than would be typical. In addition, at the deep-tephra sites, many pole-sized trees were snapped during the mid- to late 1990s, probably by unusual snow accumulations. This situation confounds interpretations of tephra effects. However, this snow damage was absent at the shallow-tephra sites, where the increases in shrub cover were most pronounced.

Thus, vegetation change in the tephra-fall zone is likely to be nonlinear and may include reversals. The processes responsible for such change will also differ with time and may include, for shrubs: (1) an initial reduction from tephra or snow-plus-tephra damage; (2) an increase in cover to a percentage that is higher than that before the eruption as extra resources become available because of canopy damage, reduced competition, and accelerated decomposition in the old soil; (3) a reduction in shrub cover as the new sapling layer develops and reduces light to low levels, perhaps for a couple hundred years; and (4) a final, gradual increase as this sapling layer thins and stand structure returns to that characteristic of old-growth forests.

The herb layer in the tephra-fall zone is also likely to undergo long-lasting and perhaps convoluted changes. Development of a dense sapling layer will reduce cover and modify species composition as the most shade-tolerant species become relatively more abundant. As a forest floor develops, initially successful species may lose their advantage. We anticipate ultimate convergence on predisturbance conditions as the alterations related to a developing sapling layer subside. In the meantime, major changes could occur in the herb layer, some opposite to convergence.

In contrast to herbs and shrubs, recovery of the bryophyte layer in the tephra-fall zone is complicated by the invasion of species previously absent or rare. However, these invasive species are largely restricted to tephra surface with little litter. They should decrease as a new forest floor develops. The originally dominant species were much reduced after 20 years, and if they are to increase to predisturbance cover, that might require many decades. Recovery of the bryophytes could be affected by changes in the microenvironment produced by a dense sapling layer, persistent substrate effects, or fluctuations in the shrub or herb cover.

Overall, we anticipate continuing change of the vegetation in the tephra-fall zone, not all of which can be described as recovery. Some attributes may diverge farther from the predisturbance condition, at least temporarily. Given climate change, a return to predisturbance conditions may never occur. A fundamental difficulty with assessing postdisturbance succession and rates or amounts of convergence on previous conditions is that even old forests are constantly changing. Because we know little about temporal variation in what we consider relatively stable old-growth forests, it is difficult to draw conclusions about some effects of tephra. For example, is the increase in shrub cover at the shallow-tephra sites a result of some favorable influence of the tephra or simply a fluctuation caused by a series of especially favorable years for shrub growth? Minor disturbances and variation in weather, herbivory, or plant disease can modify understory composition and abundance (Zobel and Antos 1997). A species can undergo major changes in abundance and population dynamics even in what appears to be a stable habitat (Bierzychudek 1999). Furthermore, even very old, apparently stable forests often undergo substantial and unidirectional change (Woods 2000), although other forests do appear fairly stable (Antos and Parish 2002). Although long-term, baseline data on temporal variation in forest understory plants are unavailable, it is clear that the tephra fall had a major impact on forest understories and has induced complex, long-lasting changes.

4.10 Comparisons to Other Habitats at Mount St. Helens

Large-scale disturbances vary greatly in both their intensity and the habitat affected (Turner et al. 1997, 1998). Nonforest sites in the tephra-fall zone at Mount St. Helens (i.e., meadows and recent clear-cuts) appear to undergo succession that is quite different from that in the forest understory. A fundamental difference is that, in a forest, trees provide litter and leachate to the new tephra. Seedling establishment is difficult on open tephra because of poor water-holding capacity, low nutrients, substrate instability, and the lack of shelter (Frenzen and Franklin 1985; Chapin and Bliss 1989; del Moral and Grishin 1999). In the open habitats east of the Cascade Range, even thin deposits of tephra had a major influence, killing the important cryptogam layer (Harris et al. 1987) and having various effects on vascular plants (Black and Mack 1986).

Closer to Mount St. Helens, where overstory trees were killed, successional changes differed greatly from the tephra-fall zone. Most studies examined habitats where few plants survived and primary succession predominated (Halpern et al. 1990; Dale 1991; del Moral and Bliss 1993; del Moral et al. 1995; Tsuyuzaki et al. 1997; Titus and del Moral 1998a; del Moral 1998). However, one generalization that can be made across disturbance types is that surviving plants were important to revegetation. This fact derives in part from difficulties in most habitats with the establishment of new plants. Survivors provided propagules and improved conditions for seedling establishment, both severe problems in the most disturbed habitats. Seeds often dispersed only short distances from established plants (Wood and del Moral 1987), and propagule availability also was limiting for many species at our forest sites. Although distances to surviving plants varied greatly among habitats, seed availability probably caused local clumping of seedlings in all habitats. For example, the few avalanche lily seedlings at our sites were almost always close to seed-producing plants. Thus, seed dispersal may structure spatial patterns of vegetation at all scales across this highly heterogeneous landscape.

In some situations, the volcanic disturbance had opposite effects on vegetation. An example is the effect of snowpack at the time of the eruption on the survival of small trees. Snow increased the damage to small woody plants in the tephra-fall zone, but in the blast zone, aboveground parts of plants were almost always killed except where protected under snow (Halpern et al. 1990). After 20 years, vigorous patches of conifers occurred in snow-protected areas in the blast area, where few other trees were yet present. Thus, the effect of initial conditions was contingent on site conditions at the time of the disturbance.

4.11 Tephra Fall at Other Volcanoes

On a global scale, the effects of tephra deposits on plants have been studied only occasionally (Table 1.1). Burial, a major mechanism by which tephra affects plants, has been studied in other situations, most notably sand dunes (Maun 1998; Kent et al. 2001), and some extrapolations to tephra can be made (Antos and Zobel 1987). Early studies by Griggs (1919, 1922) at Kodiak, Alaska, established that tephra depth was critical, that plants could succeed despite delayed emergence from tephra, and that survival was related to plant characteristics, such as the growth of adventitious roots. Direct comparison of our work with that from other volcanoes is tenuous because characteristics of both the eruptions and the vegetation differ. However, some generalizations can be drawn, especially from studies in temperate parts of Asia (Tsuyuzaki 1989, 1991, 1994; Grishin et al. 1996; Tsuyuzaki and Haruki 1996; del Moral and Grishin 1999). Although tephra exceeding 2 m deep is likely to kill almost all plants, survival at shallower depths is strongly related to the combination of plant and tephra characteristics, with the smallest plants often most strongly affected. Consistent with our results, erosion generally increases plant survival and facilitates vegetation recovery, although repeated erosion can have a negative effect on plants (Frenzen et al., Chapter 6, this volume). Erosion may also allow establishment from a seed bank in the old soil (Tsuyuzaki 1994; Tsuyuzaki and Goto 2001), a phenomenon we did not observe, perhaps because of limited erosion after the first winter. However, effects of seed banks were also not apparent in our experimentally cleared plots (Zobel and Antos 1986, 1992), suggesting that these old-growth forests had a limited seed bank. These contrasts reinforce the interpretation that characteristics of the initial vegetation are important in determining rates and mechanisms of postdisturbance succession. As anticipated, for the tephra-fall zone, recovery can be very slow (Grishin et al. 1996; del Moral and Grishin 1999).

4.12 Lessons from the Tephra-Fall Zone

Important lessons about the ecological effects of large disturbances have emerged from research at Mount St. Helens (Franklin and MacMahon 2000) and from comparisons with other large disturbances (Turner et al. 1997, 1998). Our studies of the least severe type of volcanic disturbance reinforce some generalizations and provide additional lessons related to low-intensity disturbance.

1. Disturbance effects were heterogeneous. Variations in preeruption flora, tephra depth, snowpack, and erosion produced substantial variation in posteruption vegetation within our sites.
2. The condition of the vegetation and environment (e.g., presence of snowpack) at the time of the eruption determined the extent and nature of disturbance effects. Tephra fall without snowpack present would affect woody plants shorter than the tephra depth, but tephra falling on a midwinter snowpack would devastate the shrub layer. These initial conditions have continuing effects; variation in plant cover and density after 10 years was more related to the initial damage than to differential recovery (Zobel and Antos 1997).
3. Processes occurring soon after disturbance were critical. Survival of herbs and mosses in the tephra-fall zone increased where tephra was thinned by erosion or by slumping from large logs. Some herbs penetrated tephra through cracks that filled during the first winter. Such processes provided refugia where plants could survive.

4. Prompt and thorough study of the details of the disturbance and initial plant survival was critical to observing important phenomena and to understanding disturbance effects (Zobel and Antos 1997). Examination of how individual plants initially respond to burial was essential to developing a mechanistic understanding of vegetation changes.
5. A strong surface crust on the tephra had a major influence on plant survival.
6. Plant survival was critical to postdisturbance succession in the tephra-fall zone. Many species of the old-growth forest showed great flexibility in growth form in response to burial and survived prolonged burial. Most woody species, including conifers, formed adventitious roots in the tephra, often in copious quantities. A diverse surviving flora including mosses, herbs, and shrubs emerged after burial for three growing seasons, and a few species survived burial for eight seasons.
7. Legacies are very important. The residual tree canopy contributed litter and seeds to a new forest floor and strongly modified microsite conditions, fostering vegetation change that was significantly different from that in the open. The buried forest floor, to which new tree seedlings were connected by copious mycorrhizal fungi, also contributed to the effect of the legacy (Zobel and Antos 1991b).
8. Even low-intensity disturbances can have severe effects on vegetation. With only 4.5 cm of tephra, the forest understory still differed from the preeruption condition after 20 years; some species remained depressed while others increased beyond 1980 levels, although total herb cover was little affected. Moss cover was still reduced but had increased greatly since immediately after the eruption. However, much of that increase was from species that were rare or absent before the disturbance, a pronounced contrast to vascular plants.
9. A major occurrence on the deep tephra was the establishment of a dense layer of new trees, which may dominate the understory for at least the next century.
10. Prediction of disturbance effects can be difficult, especially when no other disturbance provides a reasonable analogue (Antos and Zobel 1987). We now know that the effects of tephra burial are contingent on conditions including plant phenology, tephra depth and texture, erosion, and snowpack at the time of disturbance.
11. Vegetation change in the tephra-fall zone is not directly converging on the preeruption composition, and changes in that direction may be slow and convoluted. The composition and structure of these forests may best be viewed as continually changing in response to past disturbance as well as to current processes in this volcanic landscape.

Acknowledgments. We thank many field assistants for help over the years, and the USDA Forest Service for logistical support. Sampling in 2000 was made possible by a grant from Global Forest (GF-18-1999–45); earlier work was funded by the National Science Foundation, USDA Science and Education Administration, Natural Sciences and Engineering Research Council of Canada, and Oregon State University.

5
Plant Succession on the Mount St. Helens Debris-Avalanche Deposit

Virginia H. Dale, Daniel R. Campbell, Wendy M. Adams, Charles M. Crisafulli, Virginia I. Dains, Peter M. Frenzen, and Robert F. Holland

5.1 Introduction

Debris avalanches occasionally occur with the partial collapse of a volcano, and their ecological impacts have been studied worldwide. Examples include Mt. Taranaki in New Zealand (Clarkson 1990), Ksudach in Russia (Grishin et al. 1996), the Ontake volcano in Japan (Nakashizuka et al. 1993), and Mount Katmai in the state of Alaska in the United States (Griggs 1918a,b, 1919). Analyses have shown that as many as 18 previously undetected debris avalanches have flowed from the Hawaiian island volcanoes (Moore and Clague 1992). Following the debris avalanche at Mount Katmai in Alaska, Griggs (1918c) found that the deposit depth influenced plant survival. As a volcano collapses, glaciers, rocks, soil, vegetation, and other material are moved with great force down the mountain. Debris avalanches are typically cool and can bury surfaces with as much as 200 m of material. They tend to follow the original topography, have abrupt edges, and produce steep, undulating topography that can persist for many millennia.

The largest debris avalanche in recorded history occurred at Mount St. Helens on May 18, 1980. The debris avalanche flowed in three general locations north and west of the volcano. A portion slammed into and traveled through Spirit Lake, another portion overtopped a 300-m-high ridge about 7 km north of the volcano's summit and traveled down South Coldwater Creek, and the third and largest lobe traveled down the North Fork Toutle River valley (Voight et al. 1981). In this chapter, we focus solely on the portion of the debris avalanche that was confined to the North Fork Toutle River valley. That debris avalanche traveled 25 km, removing and burying nearly all components of the forest in its path. The debris-avalanche material was variable in temperature and included blocks of glacier ice and hot chunks of rock that originated from the volcano's cryptodome, the body of magma injected into the volcano leading up to the May 18, 1980, eruption. Near the mountain, the path of the deposit bisected areas of the blowdown zone; and toward its terminus, the deposit cut through forested areas that were largely unaffected by the 1980 volcanic events. The new deposit created environments in the valley floor that were largely free of living organisms, viable seeds, or organic matter as well as areas near the valley walls with clumps of organic material, soil, and surviving plants. The debris-avalanche deposit provided an opportunity to examine factors important under conditions closer to primary succession than the blowdown, tephra-fall, or mudflow zones created by the 1980 eruption of Mount St. Helens. Vegetation establishment has been affected by the nearly complete loss of plant life and seeds on the deposit, nature of the deposit, local climate conditions, herbivores, and surviving plant life and seeds in adjacent areas.

This chapter presents an overview of factors affecting plant establishment on the Mount St. Helens debris-avalanche deposit during the initial 20 years after the 1980 eruption. The chapter summarizes the initial physical and biological conditions on the debris-avalanche-deposit surface. Next, it describes a set of permanent plots, large-animal exclosures, and other measures used to document patterns, rates, and mechanisms of community development. It details the observations made and experiments performed to evaluate the role of these factors in determining patterns of plant establishment. The chapter concludes with a discussion of changes during the 25 years since the last eruption and implications for the future.

5.1.1 Formation of the Debris-Avalanche Deposit

The debris-avalanche deposit was created when the north side of Mount St. Helens collapsed during the May 18, 1980 eruption (Voight et al. 1981). Weakened by a bulging magma intrusion and jarred by earthquakes, 2.5 km^3 of material plunged down the side of Mount St. Helens and spread over a 60-km^2 area. The timing of the release and respective physical makeup of individual slide blocks greatly influenced the physical structure of the deposit (Glicken 1998). The magnitude 5.1 earthquake involved with initiation of the eruption triggered the release of the first slide block. Failure of the second slide block exposed the cryptodome, releasing a lateral blast that exploded outward through slide block two. The third slide block resulted from a series of failures caused by fragmentation from the exploding cryptodome (Glicken 1998). The successive timing of the debris avalanche and blast, coupled with the varying composition of individual blocks, produced a hummocky deposit composed of discrete blocks in a mixture of blast-homogenized material.

During the afternoon of May 18, 1980, earthquake-induced liquefaction of the debris deposit produced a massive mudflow that moved down the North Fork Toutle River to the Cowlitz River (Fairchild 1985). Just as a slurry of sand and water in a bucket separates when shaken, the many earthquakes on May 18 caused liquid to rise to the deposit surface, which eventually formed a mudflow that moved over the newly emplaced debris material. This mudflow eroded and deposited material on the debris-avalanche-deposit surface; erosion exceeded deposition by about 4 million m^3 (Fairchild 1985).

5.1.2 Initial Physical and Chemical Conditions

Deposits average 45 m deep and have a maximum thickness of 195 m (Voight et al. 1981). The landslide was hot and moist, both containing ice blocks and having an estimated emplacement temperature of 100°C in some locations (Voight et al. 1981). Probe measures of temperature varied from 68° to 98°C between 10 to 12 days after the May 18 eruption, depending on distance from the mountain (Banks and Hoblitt 1981).

The debris-avalanche deposit has heterogeneous topography. In areas where the mudflow moved across the debris deposit, the terrain is flat. These flatlands generally retain the thin (less than 1- to 2-cm) layer of air-fall tephra that was emitted by the volcano during subsequent eruptions. In other areas, mounds composed of lithic blocks derived from the interior of Mount St. Helens (Voight et al. 1981; Fairchild 1985; Glicken 1998) rise up to 50 m above the deposit surface. The extremely irregular topography consists of large mounds and pits typically measuring tens of meters in diameter. Pits resulted, in part, from subsidence caused by the melting of large blocks of glacial ice that were transported by the debris avalanche. Many new ponds, lakes, and wetlands were created on the surface as depressions filled with water.

Initial physical and chemical characteristics of the substrates were adequate, but not optimal, for plant establishment and growth (Adams et al. 1987). The debris-avalanche deposit consisted of poorly sorted material dominated by sand (63% by weight of the less than 2-mm fraction), but larger rocks constituted a variable component (36.4% ± 28.9% by weight). The sandy texture limited moisture retention. Nitrogen levels and conductivity were low (703 ppm NH_3 and 0.84 ± 0.71 mmho cm^{-1}, respectively). The low carbon-to-nitrogen ratio (1:0.23) reflected low levels of organic matter (0.31% weight loss on ignition). The soils were acidic (pH was 4.8 ± 0.5). Low moisture availability and high temperatures caused by solar radiation on the exposed surfaces were especially stressful during plant establishment. Growth conditions were also poor because of low fertility and moisture-holding capacity (saturation was 5.3% ± 0.9% per 15 atm). Jenny pot tests (Jenny et al. 1950), performed with the use of lettuce as a bioassay, showed a poor response to all nutrient additions to the debris-avalanche material but a positive response to the addition of a combination of nitrogen and phosphorus to mudflow material (Adams and Dale 1987). This lack of growth in the debris-avalanche substrate may have resulted either from soil-chemistry limitations or from structural limitations of the debris material.

Both erosion and deposition of material have occurred since the debris-avalanche deposit formed. Fluvial-erosion processes have been active on the debris-avalanche deposit primarily in developing new channels and in expanding and extending existing channels (Lehre et al. 1983; Swanson and Major, this volume Chapter 3). Channel and gully walls are typically steep, with a slope of 30° to 70°, a depth of 3 to 50 m, and a channel width of 3 to 120 m. Lehre et al. (1983) estimated that 42×10^6 m^3 of debris material (about 2% of the total volume of the avalanche deposit) eroded between June 1980 and May 1981. Deposition of material included up to 1 m of fragmental debris in the areas inundated during a small mudflow that occurred on the debris-avalanche deposit on March 19, 1982 (Waitt et al. 1983). Erosion remains a major factor, with annual suspended sediment yield from the debris-avalanche deposit being 100 times (10^4 Mg km^{-2}) above typical background levels ($\sim 10^2$ Mg km^{-2}) 20 years after the 1980 eruption (Major et al. 2000).

5.2 Methods of Monitoring Plant Establishment

During the first 20 years after the 1980 eruption, various sampling schemes have been used to characterize the establishment of the vegetation and the factors that influence observed patterns. First, extensive field reconnaissances were done to determine if any survival had occurred, and then a system of permanent plots was put in place to monitor emerging vegetation. A series of experimental treatments was subsequently established to assess the role of various factors on plant establishment. Our work focused on the central portion of the debris-avalanche deposit and not the marginal facies, where surviving plants were pushed to the edge of the river valley by the debris avalanche.

5.2.1 Initial Field Surveys for Vegetation

Field reconnaissance of the debris-avalanche deposit for vegetation was conducted numerous times during the summer of 1980. Several surviving plants were dug up to determine their mode of survival. About 1 m^3 of avalanche material was collected from several areas, hand sifted, spread into flats, and watered to determine if viable seeds were present.

5.2.2 System of Permanent Plots

A system of permanent plots for documenting vegetation development was established on the central portion of the

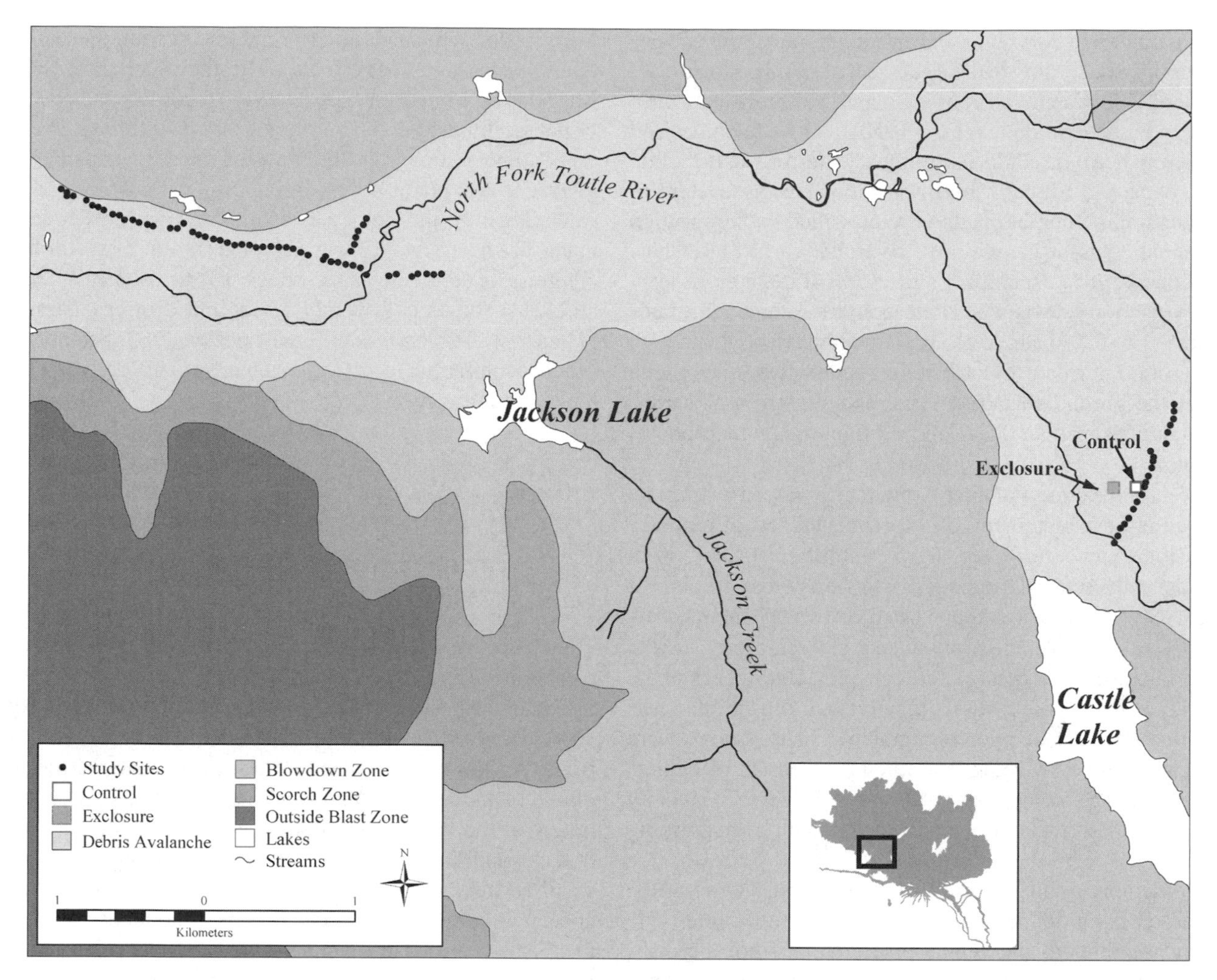

FIGURE 5.1. Location of study sites on the debris-avalanche deposit and of blowdown and scorched zones that held vegetation that was a source of seeds for the deposit.

debris-avalanche deposit during 1981 and 1982 and measured in 1981, 1982, 1983, 1984, 1989, 1994, and 2000. These circular plots are 250 m^2 and placed at 50-m intervals along transects between Castle and Coldwater lakes and down the length of the deposit (Figure 5.1). The initial 103 plots were placed to represent the variety of geological conditions on the debris-avalanche deposit and distances from surviving vegetation in the adjacent landscapes. However, during the ensuing two decades, many of these plots were lost to erosion and mudflows, and many could not be replaced because the steep terrain was impassable. By 1994, 98 plots remained; and by 2000, an additional 17 plots had been lost because of massive erosion and another 18 plots could not be reached because the bridge providing access to them was no longer passable. Thus, the 2000 estimate of plant cover is based on 63 plots.

In each sample year, the plots were monitored for the presence and cover of vascular plant species and for tree density (when individual trees could be distinguished). Plant density was recorded in 1981, 1982, and 1983. Plant cover was estimated by the line-intercept method (Mueller-Dombois and Ellenberg 2002) and then adjusted according to visual estimates because Bråkenhielm and Qinghong (1995) demonstrated that visual estimates provide the most accurate, sensitive, and precise measure of vegetation cover. Height and diameter of the largest tree and tree sexual maturity were recorded. In addition, topographic conditions of each plot were described in the field. Subsequently, each plot was assigned to one of six topographic categories: level, steeply sloped, deep channel cuts into level terrace, irregular mounds, high mound, and streambed.

Traps designed to catch wind-dispersed seeds were placed in the center of 31 plots between Castle and Coldwater lakes and 72 plots running the length of the deposit. These traps were monitored in 1981, 1982, 1983, and 1994. The traps were constructed of 0.25-m by 0.25-m squares of cloth made sticky and placed vertically, with their bases 0.25 m above the surface, for about 10 days. Seeds were counted and identified by comparing them to a reference collection of seeds collected from plants in the vicinity.

To characterize surviving and invading plants, the species were grouped according to the Raunkiaer (1934) life-form classification. This system of grouping plants had previously been related to climate gradients (Cain 1950), and we hypothesized that disturbance type could also influence life-form distribution of surviving and reestablishing vegetation. In this system of classification, perennial plants were classified by the location of perennating tissue in relation to the ground surface (Kershaw and Looney 1985). Raunkiaer's life form at the highest level of classification consists of chamaephytes (plants with buds that are 0.1 to 0.5 m above ground), cryptophytes (plants with belowground dormant tissue), hemicryptophytes (plants with buds at the ground surface), phanerophytes (trees or shrubs with buds greater than 0.5 m above ground), and therophytes (annuals).

To test for the effects of microtopography on seed resting site and seedling establishment, 60 experimental sets of mounds, depressions, and control sites were established at about 50-m intervals on the transect running east to west across the deposit (Dale 1989). Each test site consisted of two 25-cm^3 depressions, two mounds (made from the material removed from the depressions), and two controls sites per plot. One of each of the mounds and depressions was hand smoothed, and the other was left with rough surface microtopography. The treatments were conducted in June 1982 and monitored for seeds and seedlings in August and September of 1982 and in September 1983 by observing the surface of each treatment with a magnifying lens.

Because abnormally high precipitation during some of the 20-year observation period was likely to cause large-scale erosion, we refer to the annual precipitation from the nearest long-term weather station. Data have been collected since 1929 at Longview, Washington, which is about 60 km west of Mount St. Helens. Those data sets are available from the U.S. Historical Climatology Network (Easterling et al. 1996) and the Carbon Dioxide Information Analysis Center at Oak Ridge National Laboratory. The Longview station experiences less precipitation than the higher-elevation station at Spirit Lake reported in Figure 2.5d of this volume. The Longview precipitation record continued to be collected after the 1980 eruption.

5.2.3 Elk Exclosures

North American elk (*Cervus elaphus*) rapidly colonized the debris-avalanche deposit and attained large populations within the decade after the eruption. These large herbivores can have profound positive and negative influences on early plant succession (Hanley and Taber 1980; Hanley 1984; Hobbs 1996; Case and Kauffman 1997; Singer et al. 1998; Campbell 2001). The effects of elk on vegetation were quantified by constructing a 70.75- by 70.75-m exclosure on the vegetated area of the debris-avalanche deposit in 1992 (see Figure 5.1). An adjacent, companion site on the debris-avalanche deposit was identified to serve as an unfenced control. Within each site, one hundred twenty-five 1-m^2 microplots and fifteen 15-m line intercepts were randomly placed at least 5 m distant from the exclosure fence line to avoid edge effects. Thus, the effective sample area was 60 m by 60 m. Plant species were recorded, and cover was visually estimated in the microplots and measured to the nearest centimeter along the line transects in August of 1992 and 1999. Cover of plants not emerging from within microplots, but extending over microplot perimeters, was included in the estimates of cover. Species were divided into five growth forms: (1) grasses; (2) rushes and sedges; (3) forbs, ferns, and fern allies; (4) shrubs and woody vines; and (5) trees. Nonnative plants ($n = 22$) and modal forest species [e.g., species typical of local coniferous forests, *sensu* Curtis (1959); $n = 11$] were identified for analyses.

5.3 Results and Discussion

5.3.1 Vegetation Survival on the Debris-Avalanche Deposit

The coniferous forests and riparian vegetation that existed in the Toutle River valley floor before the eruption of Mount St. Helens were eliminated by the debris avalanche (Adams and Adams 1982; Fairchild 1985). No seedlings germinated from test flats of debris-avalanche-deposit material that were placed in a greenhouse and watered, and no seeds were found in any of the hand-sifted, debris-avalanche material samples (Adams and Dale 1987). Although these samples were very small relative to the size of the debris-avalanche deposit, it appeared that no viable seeds survived the landslide. Even if they had, their density would have been exceedingly low. In June 1980, the area was largely barren, and extensive searches revealed no seedlings and only a few vegetatively propagating plants (fewer than 1 km^{-2}). These surviving plants developed from rootstocks or stems that were transported in soil blocks which floated down valley in the debris-avalanche deposit and came to rest near the surface (Adams et al. 1987). Individual plants of at least 20 species survived on the debris deposit by this mechanism. The most common species were fireweed (*Chamerion angustifolium*), Canada thistle (*Cirsium arvense*), and broadleaf lupine (*Lupinus latifolius*). Lupine regeneration from root fragments was demonstrated by the regrowth of whole plants from root fragments that had been transported by the avalanche and subsequently placed in pots and watered (Dale 1986). No woody plants were found to have survived on the central portion of the debris-avalanche deposit.

Another feature of the debris-avalanche deposit was the abundance of vegetation and organic debris at the terminus of the deposit (Glicken 1998). During the movement of the avalanche, the vegetation in the Toutle River valley was scoured by the leading edge of the flow and pushed down the valley to the flow's terminus, much as a glacier pushes soil and organic material in its path. However, the Army Corps of Engineers removed this heap of vegetation during construction of a debris-retention dam, so it did not influence plant reestablishment.

5.3.2 Changes in Cover and Species Richness

Plant establishment and spread on the debris-avalanche deposit were slow during the first years after the eruption. Three years after the eruption, the average cover of the plots was less than 1%, although a few plots had as much as 30% cover. Between 1983 and 2000, plant cover dramatically increased to an average value of about 66% (Figure 5.2).

By year 14, the debris-avalanche-deposit plots averaged 38% cover, a value that was about double the cover estimated for the debris-avalanche deposit by Lawrence and Ripple (2000) for the same time period from remote-sensing imagery and classification and from regression-tree analysis. This difference between the estimate from ground-based plot measurements and that from the remotely sensed data indicates a need for considering the spatial arrangement of the study plots (as influenced by the river change that washed away many plots). Study plots remained only on the upland areas that appeared to support higher plant cover as suggested by our reconnaissance of the debris-avalanche deposit on foot and from low-flying helicopters. Vegetation cover was much reduced on the extensive riparian areas, partially because the meandering river removes many plants.

The average number of species per plot increased up to 1994 and then declined in 2000 (Figure 5.3a). The total number of plant species in the plots on the deposit also increased to 1994 and declined in 2000 (Figure 5.3b). Figure 5.3 shows only the data from those plots that survived throughout the two-decade sampling period to avoid the possibility that the decline in mean and total richness in 2000 related to the 37 plots that were not resampled. Those plots supported species typically found in moist woods [e.g., foamflower (*Tiarella trifoliata*) and youth-on-age (*Tolmiea menziesii*)] and in wetlands [e.g., horsetails (*Equisetum hyemale* and *E. palustre*)]. The loss of the 37 plots was caused by a large flood that was induced by heavy rains in 1996 and 1997 (see Figure 2.5d); that flood and the accompanying mudflow eliminated 17 of the plots and made another 18 inaccessible. Hence, ongoing disturbances need to be considered as an influence on species richness.

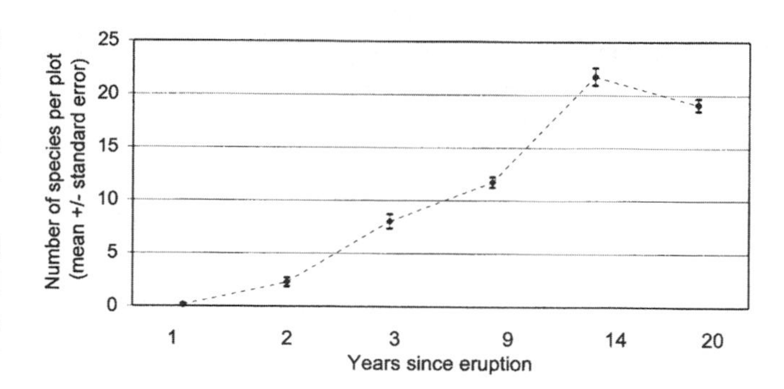

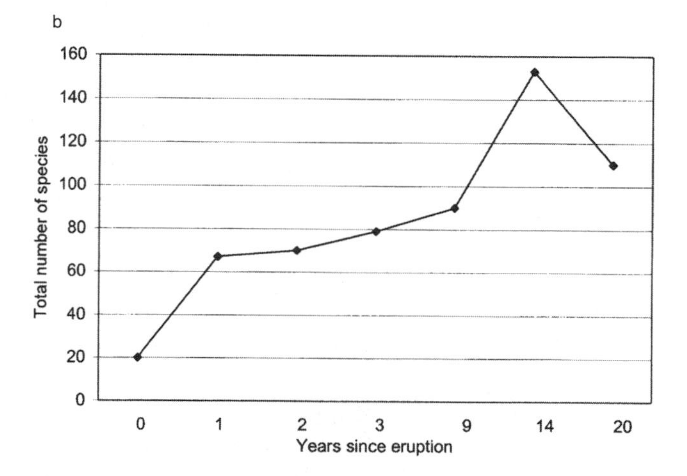

FIGURE 5.3. Changes over time in (a) the number of plant species per 250-m^2 plot on the debris-avalanche deposit for the plots that survived until 2000 and thus were sampled for all years and (b) the total number of plant species in all plots.

A shift in the life-form spectrum occurred during vegetation development on the deposit (Figure 5.4). All the plant species surviving on the debris-avalanche deposit were cryptophytes, species with dormant buds located below the surface. During the second year following the eruption, annual species invaded the debris-avalanche deposit, and in subsequent years they gradually became less abundant. All life forms were represented by 1981, although cryptophytes remained the most common life form 20 years posteruption. In 2000, hemicryptophytes, plants with buds at the ground surface, were the most different from their preeruption composition [as inferred from the species list for the slopes of Mount St. Helens complied by St. John (1976)], indicating the area was still in a reestablishment phase. There will undoubtedly be shifts in the life-form composition before the spectrum attains a composition similar to that before the eruption, as has occurred at other locations undergoing succession [e.g., Csecserits and Redei (2001) and Prach and Pysek (1999)]. Based on the preeruption life-form spectrum, it is anticipated that species with their buds below the ground will become less dominant and tall shrubs and vines

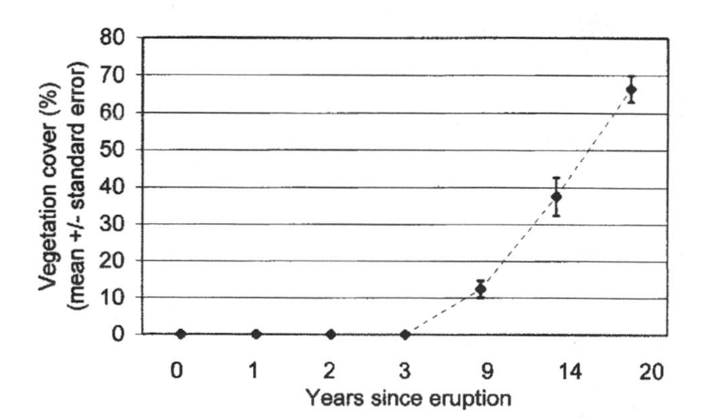

FIGURE 5.2. Changes over time in mean plant cover percent (with bars indicating the mean plus and minus the standard error) for the plots between Castle and Coldwater lakes and plots running the length of the debris-avalanche deposit.

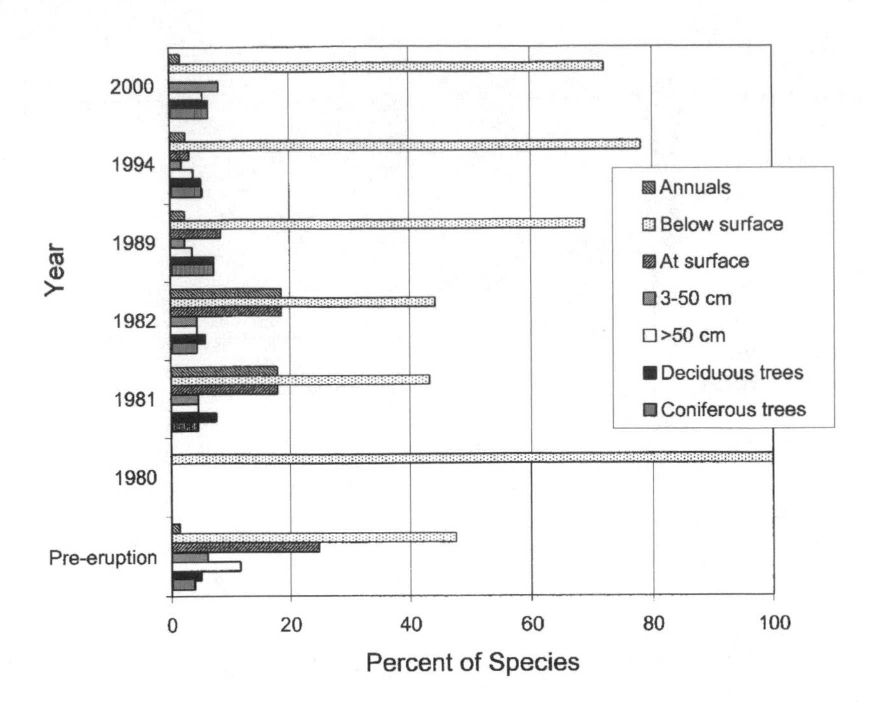

FIGURE 5.4. Changes in the life-form distribution over time by percent of species and mean number of plants based on the 250-m^2 plots between Castle and Coldwater lakes and plots running the length of the debris-avalanche deposit. The preeruption data are derived from the list in St. John (1976) for the entire mountain.

(species with their buds at the surface) will become more important.

The status of trees on the debris-avalanche deposit is of particular interest because a forest is expected to occupy the area eventually. No woody plants were found to have survived on the debris-avalanche deposit, and none was present in 1980 (Figure 5.5). Both deciduous and coniferous seedlings occurred on the debris-avalanche deposit by August 1981 (Dale 1986). A decrease in deciduous trees from 1981 to 1982 may have resulted from their elimination by the March 1982 mudflow because deciduous trees were more common than conifers close to the river channel and streambeds where the mudflow passed. The density of western hemlock (*Tsuga heterophylla*) seedlings increased greatly from 1982 to 1983 but declined in 1984 (Figure 5.5). Numerous noble fir (*Abies procera*) seedlings also died. Average tree-seedling density declined to 6 per 250-m^2 plot by 1984 from more than 16 per plot in 1983 (Dale 1986). The decrease in tree density in 1984 may have been related to the poor moisture-holding capacity of the soils and the fact that 1984 was such a dry year (see Figure 2.5d), or it may have been related to reduced seed rain coupled with the lack of suitable germination conditions. In 1984 there was a greater decrease of conifers than deciduous trees. Conifer seedlings are generally thought to be more tolerant of drought conditions than are deciduous trees; however, deciduous saplings were more tolerant of drought stress in (1) moisture-stress experiments that tested five conifers and red alder (*Alnus rubra*) growing on the debris-avalanche material (Adams et al. 1986b) and (2) field trials on the debris-avalanche deposit that used five conifers and four deciduous species (Russell 1986). Furthermore, deciduous trees tend to be rooted in seeps and along channel margins, where they experience drought impact to a lesser degree but are more susceptible to elimination by large floods and mudflows. By 20 years after the 1980 eruption, red alder was the tree species with the highest stem density on the debris-avalanche deposit.

Alnus became the most important tree genus on the debris-avalanche deposit. Up to 3 years after the eruption, red alder was rare on the debris-avalanche deposit, occupying only 15% of the plots (Table 5.1). By 9 years after the eruption, red alder occupied 37% of the plots, and those alder were vigorous and tall (up to 5 m). Predictions that red alder would become dominant (Dale 1986) have come to fruition. By 20 years posteruption, red alder contributed more to vegetation cover than did any other species (Table 5.1) and had the highest tree density; 26% of the 1066 red alder located in the plots were mature and producing seeds. The only other tree or shrub species found to be producing seeds on the avalanche deposit by 20 years posteruption were slide alder (*A. viridis* ssp. *sinuata*, of which 35% of the 46 shrubs in plots were mature in 2000) and black cottonwood (*Populus triochocarpa*, of which 2% of the 95 trees in the plots were mature in 2000). The tallest red alder in the plots was 10 m tall and had a diameter at breast height (dbh) of 21.1 cm by 20 years after the eruption. The largest trees on the debris-avalanche deposit were black cottonwoods (the largest in the plots having a height of 15 m and a dbh of 38 cm). However, black cottonwood was only 8% as common as red alder.

Even though alder often grows in more moist conditions, it likely became very abundant on the debris-avalanche deposit for four reasons. (1) Alder seeds germinated successfully on the debris material (23% germination rate; Adams and Dale 1987), and seedlings grew very quickly [e.g., 5 months after germination, the seedling-to-seed dry weight ratio was 3.6,

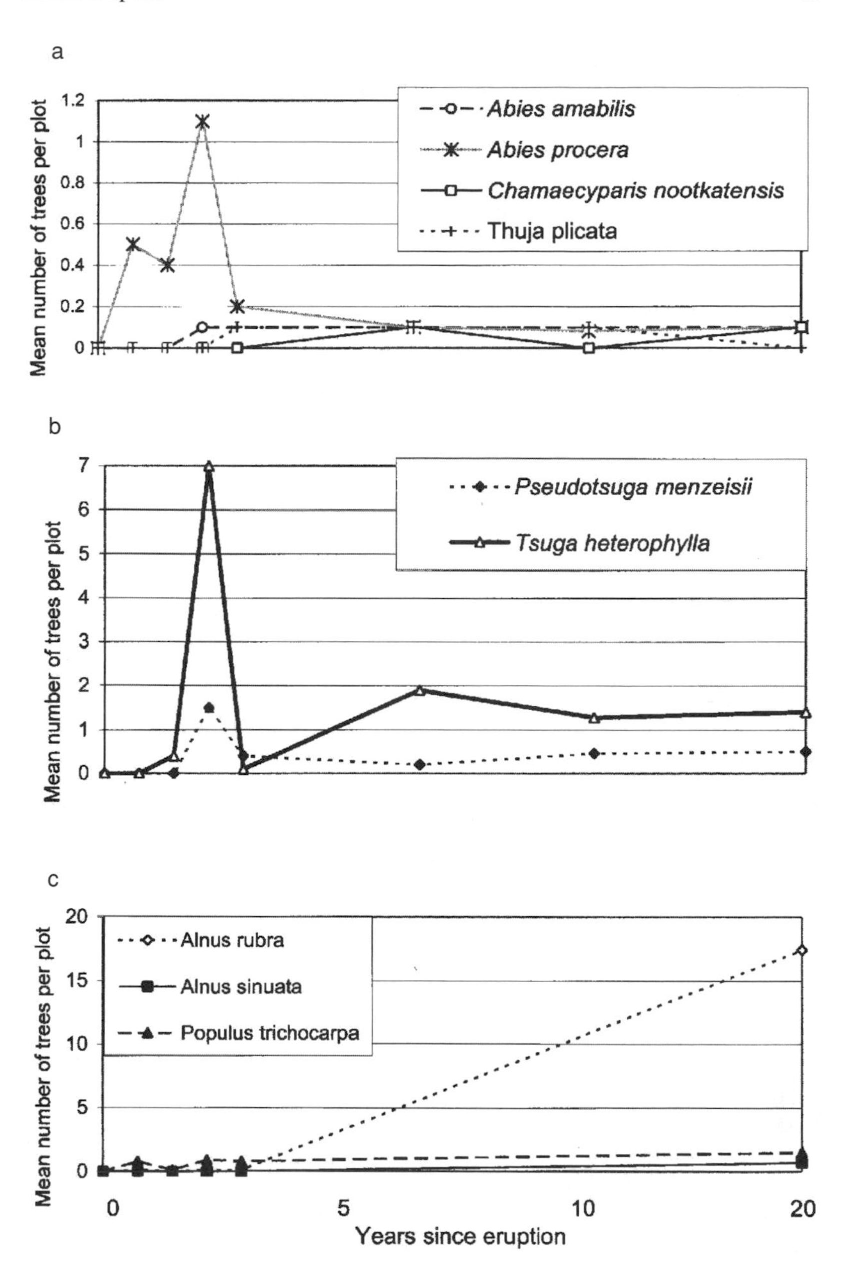

FIGURE 5.5. Number of trees per 250-m² plot over time on the debris-avalanche deposit for conifers with (a) lowest density and (b) highest density sampled in 1980, 1981, 1982, 1983, 1989, 1994, and 2000, and (c) for most dense deciduous trees sampled in 1980, 1981, 1982, 1983, and 2000 (except for *Salix* spp., because distinguishing individual willow plants was not possible).

more than three times higher than that for six conifers tested (Adams and Dale 1987)]. In field trials, planted alder saplings had a high survival rate (greater than 80%) and greater height development after 4 years than did eight other species (Russell 1986). In laboratory moisture-stress experiments with debris-avalanche soil, alder had the greatest height increment after 5 years for both very wet and very dry conditions compared with five conifer species (Adams et al. 1986b). (2) Alders fix nitrogen by means of association with symbiotic bacteria and, therefore, are better able to survive and grow in the nitrogen-poor environment. (3) Alders were less sensitive to browsing than other species (Russell 1986). Red alder was only rarely browsed, and slide alder was grazed but readily sprouts. (4) Alders produce abundant seed crops from plants as young as 3 years old (Schopmeyer 1974). Mature alders were an abundant seed source on the Toutle River mudflow directly downstream of the avalanche deposit (Russell 1986). By 1985, they were producing seed on the debris-avalanche deposit.

The most common conifer 20 years posteruption was western hemlock, but its high fluctuation in numbers over the two decades after the 1980 eruption suggested that future variation is likely (see Figure 5.5). The tallest conifer in the plots by the year 2000 was a 2.5-m Douglas-fir (*Pseudotsuga menziesii*), and the tallest western hemlock was 1.0 m in height. In 2004, a few Douglas-fir produced cones on the debris-avalanche deposit. The local production of seed on the deposit, rather than relying on long-distant transport, will likely

TABLE 5.1. Mean cover per 250-m^2 plot of selected species on the Mount St. Helens debris-avalanche deposit in six categories, with the frequency of occurrence by plots in parentheses.

Species	1983 (3 years since eruption; 97 plots)		1989 (9 years since eruption; 97 plots)		1994 (14 years since eruption; 97 plots)		2000 (20 years since eruption; 63 plots)	
Early-successional herbs with wind-dispersed seeds								
Pearly everlasting (*Anaphalis margaritacea*)	0.00701	(70)	0.28443	(74)	0.82619	(91)	0.0166	(98)
Canada thistle (*Cirsium arvense*)	0.00268	(27)	0.01196	(18)	0.07402	(23)	0.00144	(22)
Fireweed (*Chamerion angustifolium*)	0.00649	(65)	0.00536	(54)	0.04794	(69)	0.00299	(46)
Smooth willowherb (*Epilobium glaberrimum*)	0.00588	(59)	0.00372	(38)	0.30289	(51)	0.00289	(44)
Autumn willowherb (*Epilobium brachycarpum*)	0.0034	(34)	0.06423	(27)	0.08691	(47)	0.00175	(27)
Cat's ear (*Hypochaeris radicata*)	0.00598	(60)	0.34598	(69)	2.99505	(90)	0.28299	(100)
Wood groundsel (*Senecio sylvaticus*)	0.00588	(59)	0.01165	(14)	0.00103	(10)	0.05247	(16)
Sowthistle (*Sonchus arvensis*)	0.00082	(8)	0.00423	(42)	0.00082	(8)	0.00021	(3)
Grasses and upland sedges								
Thin bentgrass (*Agrostis diegoensis*)	0.00258	(45)	0.22113	(65)	0.00062	(6)	0.56041	(67)
Spike bentgrass (*Agrostis exarata*)	0.00052	(14)	0.41515	(38)	0.01577	(56)	0.00247	(38)
Rough bentgrass (*Agrostis scabra*)	0.00052	(5)	0.00165	(16)	0.21268	(72)	0.01206	(29)
Silver hairgrass (*Aira carophyllea*)	0.0001	(1)	0.03134	(6)	1.69402	(51)	0.07598	(65)
Slender hairgrass (*Deschampsia elongata*)	0.00041	(9)	0.00082	(11)	0.06515	(35)	0.05206	(11)
Rat-tail fescue (*Festuca myuros*)	0.00021	(2)	0.00021	(2)	0.98186	(33)	0.09464	(38)
Velvet-grass (*Holcus lanatus*)	0.00041	(11)	0.05381	(27)	5.22041	(68)	4.3633	(95)
Merten's sedge (*Carex mertensii*)	0.0001	(1)	0.00031	(3)	0.05402	(28)	0.00103	(16)
Showy sedge (*Carex spectabilis*)	0.00165	(16)	0.07474	(28)	0.2801	(21)	0.0132	(46)
Wetland species								
Water horsetail (*Equisetum fluviatile*)	0.00103	(10)	0.23784	(10)	0.66	(11)	0.02072	(5)
Sharp-fruit rush (*Juncus acuminatus*)	0.60398	(1)	1.12315	(5)	4.27459	(22)	0.93549	(3)
Toad rush (*Juncus bufonius*)	0.00062	(6)	0.08392	(16)	0.09351	(9)	0.0001	(2)
Common rush (*Juncus effusus*)	0.00052	(5)	0.21784	(15)	0.04216	(12)	0.01052	(5)
Swordleaf rush (*Juncus ensifolius*)	0.00072	(7)	0.02247	(20)	0.03247	(18)	0.0001	(2)
Field rush (*Juncus tenuis*)	0.0001	(1)	0.20701	(11)	0.45495	(21)	0.00052	(8)
Cattail (*Typha latifolia*)	0.00052	(5)	0.05227	(19)	0.05268	(13)	0.0001	(2)
Nitrogen-fixing legumes (introduced)								
Birdsfoot-trefoil (*Lotus purshiana*)	0.00041	(4)	5.24866	(29)	4.77526	(43)	0.7133	(54)
Creeping clover (*Trifolium repens*)	0	(0)	0.02134	(9)	1.28959	(20)	0.17987	(19)

TABLE 5.1. *(Continued)*

Species	1983 (3 years since eruption; 97 plots)		1989 (9 years since eruption; 97 plots)		1994 (14 years since eruption; 97 plots)		2000 (20 years since eruption; 63 plots)	
Nitrogen-fixing legumes (native)								
Broadleaf lupine (*Lupinus latifolius*)	0.00103	(10)	0.66186	(30)	0.98196	(46)	1.34175	(43)
Prairie lupine (*Lupinus lepidus*)	0.00031	(3)	1.28021	(26)	0.80784	(48)	1.55876	(54)
Trees								
Red alder (*Alnus rubra*)	0.00082	(15)	0.51835	(39)	12.52814	(68)	19.0933	(84)
Black cottonwood (*Populus trichocarpa*)	0.00309	(36)	0.12763	(47)	0.13	(67)	0.36381	(48)
Douglas-fir (*Pseudotsuga menziesii*)	0.00536	(56)	0.00134	(36)	0.0033	(33)	0.02278	(35)
Sitka willow (*Salix sitchensis*)	0.00289	(33)	0.11495	(26)	0.31237	(38)	1.99351	(87)
Western hemlock (*Tsuga heterophylla*)	0.00753	(82)	0.02464	(65)	0.02619	(48)	0.01289	(41)

change establishment dynamics, with Douglas-fir becoming much more abundant during the next few years. Within the coming decade, other conifers will likely mature and produce seed on the debris-avalanche deposit as well. For example, western hemlock can produce seeds on trees as young as 20 years. Subsequently, the density of conifers will rise, and as this next cohort of trees matures, vegetation cover will shift to being largely conifers. Until that time, deciduous trees will dominate vegetation cover on the debris-avalanche deposit. Indeed, red alder may preempt the debris-avalanche deposit for a few decades, thus limiting the development of coniferous forests. In the longer run, however, the presence of alder may benefit longer-lived, slower-establishing conifers by enriching nutrient conditions on the debris-avalanche deposit.

5.3.3 Factors Affecting Plant Establishment

Because residual plants were so rare, vegetation development of the debris-avalanche deposit depended on the ability of plants to colonize the deposit. Factors influencing plant establishment by seedlings were (1) distance from seed sources, (2) species-specific dispersal capabilities, (3) germination and growth characteristics of colonizing species, and (4) substrate conditions.

5.3.3.1 Seed Dispersal

Most of the seeds transported onto the debris-avalanche deposit were dispersed by wind (Dale 1989). Differences were observed among years 1, 2, and 14 posteruption in the number of seeds transported onto the debris-avalanche deposit, with the greatest number of the seeds occurring in the second year (1982) (Figure 5.6). This observation contradicts expectations that the number of seeds would increase over early-successional time. One explanation proposed for the increase in seeds trapped in August 1982 compared to August 1981 is that the source plants could have produced more seeds during the unusually wet summer of 1982 (see Figure 2.5d); however, 1983 was even wetter but had many fewer seeds than in August 1982. By 1994, the traps for the first time contained heavy lupine seeds, which probably came from the explosive release of seeds from the dried pods of nearby plants. The presence of lupine seeds in the traps was indicative of the importance of reproduction on the avalanche deposit itself.

The three major seed source areas were the slopes adjacent to the river valley in the blowdown zone, the downstream river banks and floodplain (which were inundated by the May 18, 1980 mudflow), and intact forests south and west of the volcano. The slopes above the North Fork Toutle River valley and the downstream areas had different patterns of plant

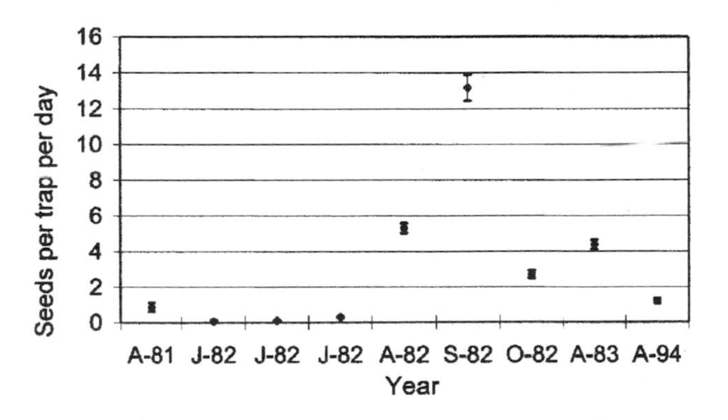

FIGURE 5.6. Changes over time in the number of seeds collected per 0.25-m by 0.25-m trap per day based on the 31 plots between Castle and Coldwater lakes and the 72 plots running the length of the avalanche deposit. Seeds were sampled for the entire growing season in 1982 [June (J), July (J), August (A), September (S), and October (O)] and in August of 1981, 1983, and 1994.

survival and vegetation development than did the valley where the avalanche deposit settled (Adams and Adams 1982; Adams et al. 1987). The slopes above the valley were a patchwork of forested and clear-cut land before to the eruption. Many early-successional annuals and perennials on the blowdown areas adjacent to the debris-avalanche deposit survived the eruption. Rapid revegetation of these tracts was partially caused by many sites having been previously logged; perennial herbs were already established on these cutover areas and quickly emerged from belowground buds (Adams et al. 1987; Stevens et al. 1987). Furthermore, the fine-grained composition of the west-side blast deposits contributed to rapid runoff and formation of gullies, which influenced survival and resprouting of plants on hill slopes (Lehre et al. 1983). These surviving plants provided abundant seed to the debris-avalanche deposit.

The second source of seeds was from areas downstream that had been inundated by the 1980 mudflow. That mudflow extended down the Toutle River valley and covered or washed away most of the herbaceous species. The main survivors in the direct path of the mudflow were trees that had branches above the surface of the flow (Adams et al. 1987). In contrast, herbaceous plants survived in marginal areas with minimal mudflow inundation. By 10 years after the eruption, the mudflow deposits were largely vegetated by red alder, willows, and other riparian vegetation. Thus, the abundance of alder and other deciduous trees and shrubs on the debris-avalanche deposit may have come from seeds that were dispersed up the river valley.

A third seed source for the debris-avalanche deposit was intact forests on the south-facing hill slopes above the South Fork Toutle River that were minimally affected by the eruption. It was likely that seeds of some tree species found on the debris deposit, such as the rare lodgepole pine (*Pinus contorta*), were dispersed from these distant, intact forests.

Both distance and wind patterns from a seed source affected the number of seeds arriving at this disturbed site. In 1982 and 1983, the number of seeds trapped on the avalanche deposit declined over a distance of less than 1.1 km from the edge but increased for the greater distances sampled (1.1 to 1.29 km from edge) (Dale 1989). Thus, more seeds were collected at the greatest distance from the vigorously establishing vegetation in the blast zone. The initial decline with distance probably relates to the ability of seeds to be transported by the wind from the adjacent blowdown zone. The increase in seed abundance for plots in the middle of the debris-avalanche deposit for 1983 (1.1 to 1.29 km from the edge) may be from seeds that were transported up the middle of the debris-avalanche deposit from plants which had colonized the mudflow deposit lower on the Toutle River. Although this is a longer distance than that from the adjacent blowdown areas, wind currents flow up the valley much of the time. Alternatively, the seed-dispersal pattern may result from seeds traveling over the blowdown-zone ridge lines and not descending precipitously over the valley wall. Instead, they may remain aloft and gradually descend to the center of the deposit. In either case, the observed pattern of wind-dispersed seed suggests that the general notion of declining numbers of seeds with increasing distance from parent plants should be modified to emphasize the variability at the tail end of the distribution.

The number of plants revegetating the debris-avalanche deposit decreased with distance from the edge of surviving vegetation and was related to seed abundance (Dale 1989). More plants were found on the debris-avalanche deposit in 1983 than in the previous year [an average of 494 plants per plot in 1983 compared to 66 plants per plot in 1982 (Dale 1989)]. This difference was likely caused by the high precipitation in 1983 (see Figure 2.5d). Time since the eruption and the abundance of seeds dispersed onto the avalanche deposit during the previous season also contributed to that increase.

As animals have become more abundant on the debris-avalanche deposit, they have likely begun to contribute to seed dispersal. Although initially quite rare, small mammals, birds, and elk were common by 20 years after the eruption. The mature forests that are expected to become established on the debris-avalanche deposit will likely support relatively few understory plants that depend on wind for seed dispersal, and the dominant trees will use both wind and animals to transport seeds.

5.3.3.2 Species Characteristics

The most dominant herbaceous species on the deposit can be classified into four groups: early-successional herbs, grasses and upland sedges, wetland species, and nitrogen-fixing legumes (see Table 5.1). The most common early colonists were early-successional species that produce many light, wind-dispersed seeds, grow rapidly, and mature early. The low cover of these species in the years just after the May 18 eruption belies their importance because total plant cover of the debris-avalanche deposit was also sparse. For example, pearly everlasting (*Anaphalis margaritacea*) increased in cover from 1981 to 1994 but thereafter declined (Table 5.1). The presence of these individuals probably facilitated establishment of other species by providing shade and organic material and by trapping seeds. Most of these early-successional plants established from seeds that were blown onto the debris deposit.

Grasses on the debris-avalanche deposit produce abundant light seeds and are able to survive droughts. Grass seeds are transported both by wind and by animals via their feces. During the 20 years since the eruption, of the 17 grasses found on the debris-avalanche deposit, 7 significantly increased in cover (Dale and Adams 2003; see Table 5.1). None of these were the grass species that were seeded onto the western debris-avalanche deposit in the summer of 1980 by the Soil Conservation Service to reduce erosion because those grasses failed to establish (Stroh and Oyler 1981). Velvet-grass (*Holcus lanatus*), a nonnative species, became one of the more abundant species on the debris deposit by 1994 and has since declined slightly in cover.

Wetland species quickly established in the numerous groundwater seeps and ponds that formed shortly after emplacement of the deposit. In the first years after the eruption, we would sink into the mud up to our knees as we sampled some of these plots, but by 10 years after the eruption, the ground was not saturated with water. Subsequently, most species have decreased in abundance (e.g., the horsetails, *Equisetum* spp.). Areas that were initially very wet, along with the riparian areas, now support some of the highest vegetation cover. Although wetlands (seeps, riparian zones, and ponds) were typically heavily vegetated, they were probably not important source areas for propagules for upland sites because many species growing in the wetlands are unable to establish on more xeric uplands. However, the presence of these species may have facilitated the establishment and growth of other plants. Species restricted to the wetlands include sedges (*Carex deweyanna* and *C. disperma*), three species of horsetail (*Equisetum fluviatile*, *E. hyemale*, and *E. palustre*), soft rush (*Juncus effusus*), and cattail (*Typha latifolia*).

Herbs associated with nitrogen-fixing bacteria have increased in cover during the period of record, probably because of their ability to grow in nitrogen-poor substrates. Broadleaf lupine increased in abundance on the debris-avalanche deposit. A few broadleaf lupine survived the eruption, and the species has expanded through seed production. The large seeds are primarily dispersed by gravity and water, so seedlings tend to occur near the parent plant. A few prairie lupine (*Lupinus lepidus*) also survived the eruption. By 1989, they provided more cover than broadleaf lupine did; and by 1994, they were more widely distributed across the debris-avalanche deposit (particularly close to the volcano). White clover (*Trifolium repens*), a common nonnative legume, also became widespread on the debris-avalanche deposit. Bird's-foot-trefoil (*Lotus unifoliolatus* var. *unifoliolatus*), a nonnative annual legume that was seeded to reduce erosion, is another nitrogen-fixing plant that has established high cover. More than 20 years after the 1980 eruption, this species dominated only those plots on the western part of the avalanche deposit. This dominance of the westernmost plots allowed us to document the role of nonnative species on plant reestablishment.

5.3.3.3 Effects of Nonnative Species Seeded onto the Debris-Avalanche Deposit

Nonnative species were introduced to the area through aerial seeding in hopes of reducing erosion. Because only the western portion of the avalanche deposit was seeded, the deposit offers the opportunity to compare successional processes with and without such introduced species. Long-term revegetation trends and effects of nonnative species on succession are important to understand because vegetation-recovery practices often rely on nonnative species for enhancing vegetation development of denuded sites along roadsides, strip mines, or other human-generated clearings.

Fifteen years after the eruption, plots invaded by nonnative species had greater vegetation cover and more native-plant richness than did plots that were not invaded (Dale and Adams 2003). These results suggest that nonnatives fostered the recruitment of native species, likely by trapping and nursing seeds. However, significantly greater mortality of conifers occurred in the plots dominated by introduced species shortly after the invasion of those species (Dale 1991), but no difference in conifer mortality occurred in the subsequent 5 years. The initial conifer mortality was likely caused by the rapid population increase of voles (*Microtus* spp.), which thrived with the abundance of seeds produced by the introduced grasses and herbs (Franklin et al. 1988). Under winter snows and with little other food, the voles apparently ate the living tissue around the conifer sapling stems and killed many of them (Franklin et al. 1988). The plots dominated by introduced species had fewer conifer trees 20 years after the eruption. Thus, the short-term pulse of conifer mortality after the invasion of introduced species may have long-term effects on the development of the forest vegetation. This pulse of conifer mortality demonstrates the importance of herbivory in early-successional situations, a theme that is revisited later in this chapter.

5.3.3.4 Substrate Conditions

Large variation in substrate conditions on the debris-avalanche deposit has produced a complex pattern of plant establishment. Differences between the upland areas and wetlands (e.g., riparian habitats, seeps, and streamside areas) are quite noticeable. Moist areas have greater species richness, higher plant density, and higher cover.

Groundwater seeps produced the greatest increase in plant cover. In these seeps, the surface of the deposit remains moist throughout the year but is not subjected to high flows that might disturb the plants. Twenty years after the 1980 eruption, vegetation in seeps was limited to comparatively few species, the most abundant being the common wetland horsetail, willow (*Salix scouleriana* and *S. sitchensis*), and willow-herb (*Epilobium ciliatum spp. watsonii*). Other species present include cattail, pearly everlasting, and fireweed.

The debris-avalanche surface is characterized by great variation in surface microtopography, which influences the distribution of seeds and seedling establishment. Pits tended to be moister than mounds, and plumed seeds were observed to fall into large depressions, probably because the plume diameter abruptly decreases with small increases in relative humidity (Burrows 1973). Experimental mounds and depressions on the deposit were tested for their ability to trap seeds and to support seedlings as an indication of the effect of microtopography on seed settlement and seedling establishment (Dale 1989). There were no significant differences between the seeds and seedlings in the depressions and on the mounds, and yet both had more seeds and seedlings than did level control sites. Mounds had a significantly greater density of seeds than the controls did because spider webs occurred on the mounds; these webs trapped

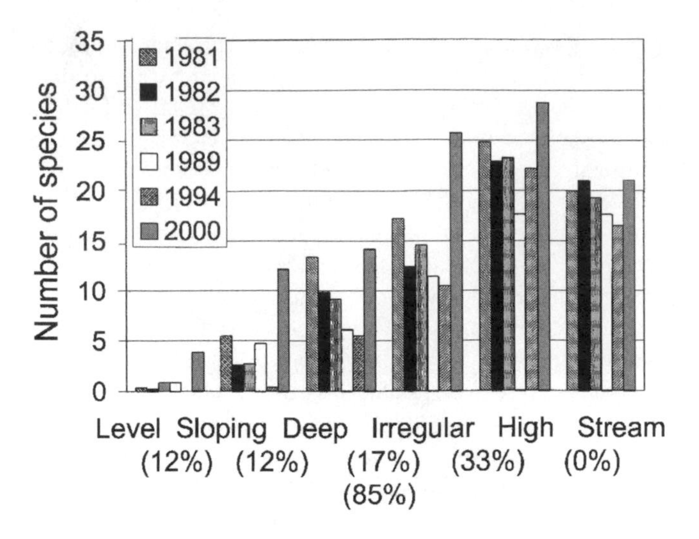

FIGURE 5.7. Number of species per 250-m² plot over time by topographic category: *Level*, *Sloping* (plots with a sloping side), *Deep* (plots with a deep channel between level terraces), *Irregular* (plots with irregular mounds), *High* (plots with a single, high mound), and *Stream* (plots containing a streambed).

windblown seeds, and moisture accumulated on the threads (Dale 1989). Thus, physical conditions were important in depressions in trapping seeds, and biological factors (the spider webs) were important on the mounds.

Time since the eruption was a major influence in development of species richness. Classifying the plots into topographic groups revealed patterns in revegetation over time (Figure 5.7). For example, in the year 2000, more species appeared in all topographic categories than in earlier years. The second growing season after the eruption (1981) supported a high number of species for all topographic groups except the level category. Yet, for all years except 2000, the level plots supported few species. The plots with streams running through them were disproportionately eliminated by the flood of 1997. The high mounds, however, consistently supported the greatest plant diversity because they were not affected by the mudflows.

5.3.3.5 Subsequent Physical Disturbances

The portion of the deposit that has consistently remained the most poorly vegetated is the area adjacent to the mainstream channels of the North Fork Toutle River. Deposits in these areas were subject to repeated disturbance as the channels meandered across and cut through the unconsolidated deposits. Most seedlings that became established in moist sediment along channel margins were swept away or buried by sediment the following winter.

Volcanic disturbance subsequent to the 1980 eruption has not had a major effect on plant reestablishment. Tephra fall has been minor. The largest single volcanic disturbance on the debris-avalanche deposit was an explosive eruption on March 19, 1982, that melted snow in the crater and produced a mudflow that deposited more than 1 m of sand and gravel in a few locations and lightly affected others (Waitt et al. 1983). The ground was cleared of plants in a few locations. Most vegetation reestablishment was not set back by this mudflow (Adams et al. 1987) because, just a few years after the flow, plants with underground buds that withstood the disturbance and annual species such as wood groundsel (*Senecio sylvaticus*) were more common on the areas covered by the mudflow than on areas not inundated. These disturbed areas evidently supported the survival of plants with underground buds and were good sites for colonists.

Heavy rain and snowmelt also produced floods and small mudflows down the North Fork Toutle River. Such disturbances caused erosion and several large-scale washouts of debris-avalanche material. More than 1.78 m of rain fell in 1996, the highest precipitation over the 71-year record (see Figure 2.5d), and that winter a large mudflow (Major et al. 2000) eliminated 17% of the study plots. In one case, a stretch of floodplain more than 200 m long was replaced by a channel with steep sides. Floods, mudflows, and streambank failures will continue to modify the debris-avalanche deposit.

5.3.3.6 Herbivory

Herbivores affected vegetation even in the first years after the eruption, when they were rare. For example, in the summer of 1982, the only bitter cherry trees (*Prunus emaginata*) observed were leafless and did not survive (apparently the leaves had been consumed by animals). Also, an outbreak of horned caterpillars (Sphingidae) eliminated a population of fireweed in the middle of the debris-avalanche deposit. The documented effect of herbivores on prairie lupine on the adjacent Pumice Plain (Fagan and Bishop 2000) may also be occurring on the debris-avalanche deposit. Experimental removal of insect herbivores from prairie lupine on the Pumice Plain increased the growth of lupine plants and the production of new plants (Fagan and Bishop 2000).

Elk also had an impact on vegetation. Deciduous trees (alder, willows, and black cottonwood) experimentally planted on the debris material near the North Fork Toutle River were more tolerant of elk browse than were the conifers (Russell 1986). Because of the seasonal (late winter and early spring) nature of their activity, elk preferentially browsed conifer species more than deciduous species because only coniferous trees were green at that time. Later in the season, the elk moved to higher elevations. In experimental plots, elk browse on deciduous trees caused minimal damage (except for trampling effects), whereas elk damage to conifers produced 82% mortality and reduced the height of surviving conifers by about one-half (Russell 1986).

The effects of elk herbivory were documented by an elk-exclosure study on the debris-avalanche deposit. Plant cover from 1992 to 1999 in both the elk exclosure and on an adjacent, unfenced, grazed control plot increased by almost

TABLE 5.2. Information from elk exclosures and controls on the Mount St. Helens debris avalanche (mean ± standard error, SE) for 1992 (12 years since eruption) and 1999 (19 years since eruption).

	1992		1999		Change from 1992 to 1999	
Type of data	Control grazed	Exclosure ungrazed	Control grazed	Exclosure ungrazed	Control grazed	Exclosure ungrazed
Species richness data						
Mean richness per plot	8.3 ± 0.3	10.3 ± 0.4	9.4 ± 0.3	7.7 ± 0.3	1.1	−2.6
Total richness	46	44	62	65	16	21
Shannon–Weiner diversity index	3.133	3.171	3.346	3.368	0.213	0.197
Sorensen's community similarity coefficient	0.89		0.8		−0.09	
Mean microplot cover						
All species	17.8 ± 2.3	29.8 ± 3.0	102.6 ± 4.6	116.7 ± 4.7	94.8	86.9
Exotic species	1.4 ± 0.2	1.3 ± 0.2	14.5 ± 1.9	4.3 ± 0.8	13.1	3
Modal forest species	0.3 ± 0.1	0.2 ± 0.1	0.2 ± 0.1	1.4 ± 0.4	−0.1	1.2
Cover by growth forms						
Grasses	0.4 ± 0.1	0.4 ± 0.1	1.8 ± 0.2	0.2 ± 0.1	1.4	−0.2
Sedges and rushes	1.6 ± 0.4	3.0 ± 0.4	1.9 ± 0.4	0.9 ± 0.2	0.3	−2.1
Forbs, ferns, and allies	6.3 ± 1.2	6.4 ± 0.8	26.5 ± 2.4	22.7 ± 1.9	20.2	16.3
Shrubs and woody vines	9.0 ± 1.4	18.2 ± 2.3	25.5 ± 2.6	33.3 ± 3.6	16.5	15.1
Trees	0.5 ± 0.1	1.8 ± 0.9	47.5 ± 4.0	59.6 ± 4.1	47	57.8

600% in the grazed plot and by nearly 400% in the ungrazed plot (Table 5.2). Plant species diversity (as measured by the Shannon–Weiner index) was similar in both years in the two plots. Species richness (the number of species per microplot) differed between sites and years. Mean species richness increased from 8.3 to 9.4 in the grazed plot but decreased from 10.3 to 7.7 in the ungrazed plot (Table 5.2). From 1992 to 1999, cover of nonnative species increased 935% in the grazed plot and 231% in the ungrazed plot. In contrast, mean microplot cover of modal forest species, such as sword fern (*Polystichum munitum*), increased by 1.2% in the ungrazed plot and decreased by 0.1% in the grazed plot. As expected, cover of grazing-tolerant graminoids increased in the grazed plot by 1.4% but declined by 0.2% in the ungrazed plot (Table 5.2). Again, as expected, shrub cover was greater in the ungrazed plot in 1999 (Table 5.2). However, the response of shrubs to ungulate herbivory differed from other ungulate-exclusion studies in that cover increased at a greater rate in the grazed site than in the ungrazed site (Case and Kauffman 1997; Table 5.2b). The rapid response in the grazed plot was indicative of vigorous compensatory growth of sitka willows, the dominate shrub species. The proliferation of nonnative species in the grazed site can be explained by elk herbivory and trampling. Moreover, the establishment and growth of modal forest species typical of local coniferous forests may have been facilitated by the greater amount of shade and litter provided by red alder and sitka willow in the ungrazed plot [in 1999, mean litter cover was 63.7% (SE = 3.4) for the ungrazed plot and 33.0% (SE = 3.5) for the grazed plot]. Although elk herbivory has influenced the composition of herbaceous species, the increase of woody cover in both sites suggests that elk herbivory was not intense enough to impede the development of a red alder/willow-dominated community on the debris-avalanche deposit. Herbivory was just one of many factors affecting the dynamic landscape as vegetation became established and plant–animal relationships developed.

5.3.3.7 Succession

In the first two decades after the 1980 eruption, plant species richness and vegetation cover increased. The characteristics of species that established on the deposit spanned a wide range, from short-lived, early-successional species (e.g., pearly everlasting) to long-lived, late-successional species (e.g., western hemlock). The simultaneous presence of both early and late seral species a few years after the May 18 eruption showed that establishment of late-successional species need not be facilitated by early-successional species [see discussion in Connell and Slatyer (1977)]. Thus, relay floristics (the replacement of one species by another over time), as strictly interpreted, has not occurred on the deposit. Instead, early-successional species were most abundant in the initial years of vegetation development. The late-successional species were present, but not common, in the early years. Some, such as western hemlock, have declined in density.

The extent to which early-successional species facilitated the development of the late-successional species by modifying the environment was variable. Most species played a role in facilitation only by their physical presence in trapping seed and providing shade and organic material, which increased soil moisture retention. However, red alder and lupines were exceptions. As these plants became established, they influenced ecological development by ameliorating the environment. By fixing nitrogen, alder and lupine can improve the soil fertility (Trappe et al. 1968). Moreover, the rapidly growing alder trees created shade and added organic material to the ground surface, thus facilitating seedling establishment of shade-tolerant species and inhibiting establishment of shade-intolerant species. Alder has also been an important factor in recolonization on other volcanic substrates in the Pacific Northwest (Frehner 1957; Frenzen et al. 1988).

Species with the highest cover on the debris-avalanche deposit in the first years after the 1980 eruption are characterized as ruderal because they can survive in areas with a high frequency of disturbances (Grime 1977). They are the early-successional species of Table 5.1. Ruderal species are of small stature, grow rapidly, produce abundant, widely dispersed seeds, and are typically annuals. Their dominance runs counter to Grime's (1979) proposition that plants have not yet evolved to survive in both high-disturbance and high-stress habitats. He defines stress as external constraints that limit plant production. However, the plants on the debris-avalanche deposit are abundant in high-disturbance areas (e.g., after fires or clear-cuts) and yet they occur on the highly stressed debris-avalanche deposit, where both water and nutrient deficiencies limit plant productivity.

The plasticity of ruderal species may be the primary feature allowing them to establish and reproduce in stressful habitats. Plasticity refers to the ability of an organism to adjust its morphology or life-history characteristics to accommodate prevailing environmental conditions. For example, mature wood groundsel is generally about 50 cm tall, yet it was 5 to 20 cm tall on the debris-avalanche deposit. As another example, in 1982 and 1983, several individual plants of the normally perennial prairie lupine, yellow monkey-flower (*Mimulus guttatus*), and miner's lettuce (*Montia siberica*) completed their life cycles in one season on the deposit. By putting all their energy into seeds at the end of the season and then dying, these typically perennial species altered their normal life cycle. As a result, the species were able to survive in the stressful habitat. Thus, plasticity of plants may allow them to reproduce in both highly disturbed *and* highly stressed habitats.

Tree growth and maturity are likely to be important forces of succession in both the short and long term. The conifers have such low densities and are so small that they are not likely to contribute to successional dynamics for some decades. Although these species can mature in 10 or 20 years, only a few Douglas-fir trees on the debris-avalanche deposit were producing seeds after 24 years. However, with red alder contributing the most to plant cover 20 years after the eruption, a forest is gradually being established in some locations. Where the canopy becomes closed, the understory is shaded, which reduces surface temperature and enhances soil moisture retention. Hence, in future years, the difference between sites dominated by closed forests and those open to the sun will likely become more pronounced.

5.4 Prospects for the Future

The long-term potential vegetation for the debris-avalanche deposit is a coniferous forest dominated by Douglas-fir, Pacific silver fir (*Abies amabilis*), and western hemlock on upland areas and by deciduous trees on riparian areas (see the discussion of preeruption vegetation in Chapter 2, this volume). We predict that, within the next decade, vegetative cover will exceed 75% on sites where red alder becomes dominant. Thus, plant cover is expected to increase and spread except on the sides of eroding hummocks and active floodplains, where the patchy nature of the vegetation will likely continue. Revegetation has been slow on most areas on the debris-avalanche deposit because (1) removal of the previous ecological system was so complete, (2) the seed sources are distant, (3) conditions on the debris-avalanche deposit are stressful, (4) disturbances continue to occur, and (5) herbivory has influenced the successional trajectory. Although vegetation cover was greater on areas with abundant nonnative species, those sites have also experienced higher conifer mortality and thus may be slower to develop into a coniferous forest.

Even though some of the anticipated climax species are already present on the deposit, we expect that the area will go through the typical stages of succession characterized by increases in plant cover, decreases in soil bulk density, and increases in litter (Odum 1969). The expected pattern of succession is similar to that observed after glacial retreat (Adams and Dale 1987), with early-successional species being dominant for the first 15 years, deciduous trees becoming dominant for years 15 to 80, and coniferous forest producing a transition in years 80 to 115 [e.g., Crocker and Major (1955)]. However, the timing of succession on the debris-avalanche deposit will likely be longer than in glacial succession because the source vegetation is distant (Adams and Dale 1987).

Future disturbances will influence pathways and rates of succession. Potential disturbances include drought stress; geomorphic processes, such as erosion and deposition; volcanic activity; mudflows associated with heavy rainfall; herbivory; fire; and plant breakage caused by snow, ice, or wind. These disturbances will not occur uniformly across the deposit but will be influenced by topography, distance from a stream, and status of the vegetation. Succession on the debris-avalanche deposit will likely be set back or redirected for those sites that experience such disturbances. The result will be a mosaic of successional states on the debris-avalanche deposit.

5.5 Summary

During the first 20 years after the 1980 eruption, plant colonization and the subsequent development of plant communities on the debris-avalanche deposit were influenced by the low numbers of survivors, proximity to source propagules, substrate moisture, nutrient conditions, and herbivory. We found:

1. The few surviving plants were herbaceous species that have their dormant buds below the surface. No seed survival was detected.
2. Invading plants were much more important to vegetation development than were the very few survivors.
3. The early colonists with the greatest contribution to plant cover were early-successional, wind-dispersed species that survived on the adjacent blowdown-zone hill slopes.
4. Average plant cover on the debris-avalanche deposit was less than 1% 5 years after the eruption but averaged greater than 65% by 20 years after the May 18, 1980 eruption.
5. The number of species increased linearly over time up to 14 years after the 1980 eruption and declined slightly thereafter.
6. The rate of vegetation development has been spatially variable and related to habitat heterogeneity. Wetlands are particularly important as areas of high plant diversity. Areas near streams are susceptible to mudflows and flooding events.
7. Trees established successfully on the debris-avalanche deposit, demonstrating long-distance dispersal. Red alder, a fast-growing, early-maturing, and nitrogen-fixing tree, will be particularly important in the vegetation-development process during the next several decades.
8. Herbivory has influenced successional patterns. Elk herbivory increased plant cover and caused the proliferation of nonnative species.

Eventual vegetation will be a conifer-dominated forest with alder, willow, and cottonwood abundant in riparian zones. In the next decade, vegetative cover is expected to reach more than 75% in sites where red alder is or becomes dominant, and patchy vegetation is expected to remain on steep slopes and floodplains. Ongoing disturbances will continue to shape this dynamic landscape.

Acknowledgments. Logistic support was provided by the USDA Forest Service and the Washington State Department of Natural Resources. Fieldwork was conducted with the assistance of A.B. Adams, D. Donohue, Margaret Evans, Howard Haemmerle, Bradley Hensley, Charlie Hensley, Bridgette Nyberg, Eric Smith, Mandy Tu, and John Wallace. Dale P. Kaiser of the Carbon Dioxide Information Analysis Center at Oak Ridge National Laboratory provided updates on precipitation records from Longview, Washington. Comments by Fred Swanson and two anonymous reviewers were helpful. The research was partially funded by the National Science Foundation; the National Geographic Society; EARTHWATCH; and The Center for Field Research of Belmont, Massachusetts. The University of Washington Herbarium was used to verify plant-species identifications. Oak Ridge National Laboratory is managed by UT-Battelle, LLC, for the U.S. Department of Energy under contract DE-AC05-00OR22725.

6 Geomorphic Change and Vegetation Development on the Muddy River Mudflow Deposit

Peter M. Frenzen, Keith S. Hadley, Jon J. Major, Marc H. Weber, Jerry F. Franklin, Jasper H. Hardison III, and Sharon M. Stanton

6.1 Introduction

6.1.1 Overview

Geomorphic disturbances are widely recognized as important processes that influence plant-community development and landscape-scale vegetation patterns [e.g., Veblen and Ashton (1978), Garwood et al. (1979), Swanson et al. (1988), and Malanson (1993)]. In volcanically active areas such as the Pacific Northwest, mudflows are locally important geomorphic disturbance events governing short- and long-term ecological conditions. Volcanic mudflows can scour and inundate river valleys with large volumes of debris (Janda et al. 1981; Pierson 1985; Vallance and Scott 1997; Scott 1988; Vallance 2000; Kovanen et al. 2001) and influence plant succession tens of kilometers downstream from their points of origin (Halpern and Harmon 1983; Adams and Dale 1987; Wood and del Moral 1987; Frenzen et al. 1988). In addition to altering plant succession, large volcanic mudflows can initiate a cascading chain of secondary disturbances that further modify the landscape and affect subsequent ecological responses (see Swanson and Major, Chapter 3, this volume).

The comparatively high disturbance intensity but spatially variable nature of volcanic mudflows provide unique opportunities to study complex interactions between geomorphic processes and ecological succession (Beardsley and Cannon 1930; del Moral 1998; Kroh et al. 2000). Nonetheless, few studies examined plant succession on mudflow deposits before the 1980 eruption of Mount St. Helens (Frehner 1957). Research subsequent to that eruption has shown that plant succession on mudflow deposits is highly variable in response to local substrates, plant reproductive strategies, distances to seed sources, and chance dispersal events (Halpern and Harmon 1983; del Moral 1998). "Biological legacies," such as floated logs, remnant snags, and shallowly buried residual plants, also play important roles in vegetation development on mudflow deposits (Frehner 1957; Franklin et al. 1985; Frenzen et al. 1988; Halpern and Harmon 1983; del Moral 1998; Kroh et al. 2000; Weber 2001).

Vegetation succession on mudflow deposits can follow an initial- or relay-floristics model (*sensu* Egler 1954) or some combination of the two (del Moral 1998) and can lead to the compositional convergence or divergence of neighboring communities (Franklin et al. 1985; Wood and del Moral 1987; Kroh et al. 2000). Although earlier studies of vegetation recovery on mudflow deposits provide important insights into the dynamics of herbaceous plant communities, few of these studies examined succession over decades after a disturbance (Frenzen et al. 1988; Kroh et al. 2000). In this chapter, we present a case study of geomorphic and vegetation responses at four sites along the Muddy River to large (up to 10^7 m^3) mudflows triggered by the May 18, 1980 eruption of Mount St. Helens. Our objective is to describe and qualitatively compare geomorphic changes and vegetation development along distinct reaches that represent a range of mudflow-induced disturbance intensities and environmental settings. We address three questions:

- How did the primary mudflow disturbance vary along the Muddy River valley?
- How have secondary disturbances varied among a comparatively stable but environmentally severe upper-elevation reach having incised channels (upper Muddy fan); a comparatively stable, low-elevation reach (middle Muddy fan); a chronically unstable, low-elevation floodplain (lower Muddy floodplain); and a stable, low-elevation river terrace (Cedar Flats terrace)?
- How have plant communities responded to the primary and secondary disturbances in each of these four locations?

6.1.2 Study Sites

The four study sites selected are within a 23-km-long reach of the Muddy River extending from the upper Muddy fan to the Cedar Flats Research Natural Area (Figure 6.1). The upper Muddy fan (900 to 1300 m in elevation) is an incised 3-km-long reach located approximately 4 to 7 km from the crater rim. The middle Muddy fan (550 m in elevation) is an incised 1-km-long reach located at the confluence of Muddy

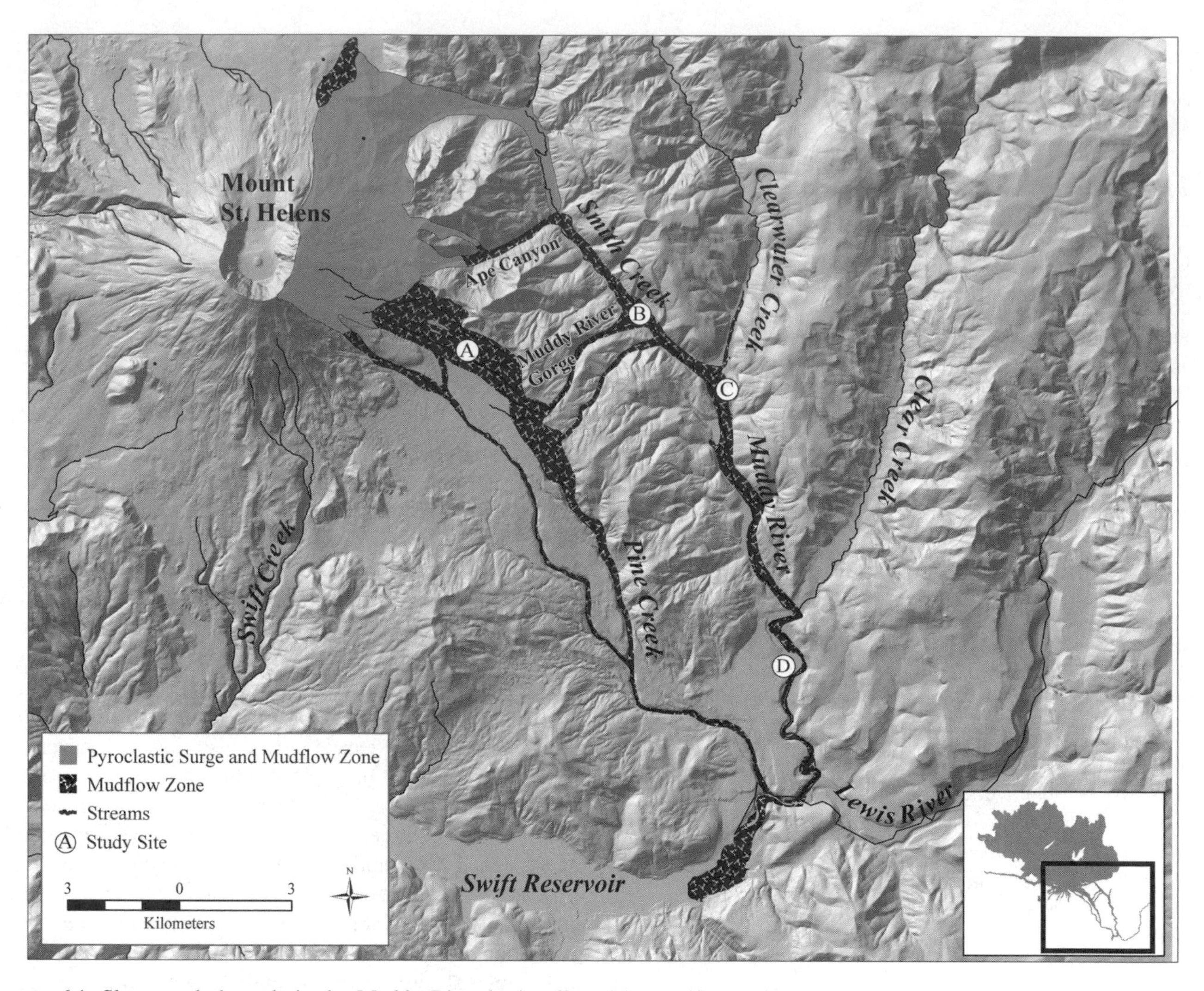

FIGURE 6.1. Slopes and channels in the Muddy River basin affected by mudflows triggered by the May 18, 1980, eruption of Mount St. Helens. Study sites of postmudflow geomorphic and vegetation development are *A*, upper Muddy fan; *B*, middle Muddy fan; *C*, lower Muddy floodplain; and *D*, Cedar Flats terrace. Two small hills that escaped mudflow inundation are located near *A* on the upper Muddy fan. [Modified from Pierson (1985).]

River and Smith Creek. The lower Muddy floodplain (360 to 520 m in elevation) is a 10-km-long reach located between the Smith Creek confluence and Cedar Flats terrace. Braided channels and an active floodplain characterize the lower Muddy River valley. The 650-m-long by 350-m-wide Cedar Flats terrace (365 m in elevation) is located 5 km upstream of the Lewis River confluence. The terrace rises approximately 3 m above river level and is bounded by a large meander of the Muddy River to the east and by steep valley walls to the west.

All the study sites except the lower Muddy floodplain are located in protected areas within the Mount St. Helens National Volcanic Monument or Cedar Flats Research Natural Area. In 1982 and 1983, the USDA Forest Service salvage-logged, piled, and burned woody material on the lower Muddy floodplain to reduce debris loading in Swift Reservoir. These activities may have subsequently contributed to local channel shifting, seasonal flooding, and the removal of large woody debris from 30-m-wide buffers initially placed around vegetation transects to protect them from disturbances related to the salvage logging.

6.1.3 Preeruption Vegetation

Preeruption vegetation on the upper Muddy fan consisted of old-growth stands of Pacific silver fir (*Abies amabilis*), western hemlock (*Tsuga heterophylla*), and Douglas-fir (*Pseudotsuga menziesii*) with scattered individual western white (*Pinus monticola*) and lodgepole pine (*P. contorta*). Common understory species included oval-leaf blueberry (*Vaccinium ovalifolium*) and beargrass (*Xerophyllum tenax*). Tree size and plant-community development varied across the fan, with larger trees and later successional stages occurring at lower elevations. Tree density decreased with increasing elevation, especially near channels and in the paths of small snow avalanches and outburst floods that flowed from Shoestring Glacier.

Preeruption vegetation development on the middle Muddy fan and lower Muddy floodplain varied with proximity to the river. Aerial photos show that islands of mature trees occurred on stable point bars as well as streamside terraces and that greatly variable community composition and productivity characterized the lower Muddy floodplain. In addition to species typical of the elevation, the floodplain also contained uncommon associates, such as subalpine fir (*Abies lasiocarpa*), mountain hemlock (*Tsuga mertensiana*), lodgepole pine, and Alaska yellow-cedar (*Chamaecyparis nootkatensis*).

Preeruption vegetation on the Cedar Flats terrace is inferred from the composition of residual trees and intact understory in areas of minor deposition. Large Douglas-fir (greater than 1 m in diameter) and a mixture of western red cedar (*Thuja plicata*), western hemlock, and Pacific silver fir were dominant species. Red alder (*Alnus rubra*), black cottonwood (*Populus trichocarpa*), big-leaf maple (*Acer macrophyllum*), and western red cedar dominated riparian areas near abandoned river channels toward the western edge of the terrace. Understory species on the terrace included vine maple (*Acer circinatum*), brambles (*Rubus* spp.), and ferns (family Polypodiaceae) (Weber 2001).

6.1.4 Effects of Prehistoric Eruptions on the Muddy River Drainage

The geomorphology of and vegetation in the Muddy River basin have been shaped by episodic eruptions of Mount St. Helens for more than 35,000 years (Crandell 1987). Eruptive styles have ranged from relatively quiescent eruptions that emitted basaltic lava flows to violently explosive eruptions that thickly mantled the landscape with volcanic ash and produced mudflows and fiery avalanches of pumice and hot gases (pyroclastic flows). Deposits from these eruptions built debris fans around the base of the volcano and partly filled valleys for many tens of kilometers downstream (Mullineaux and Crandell 1981; Crandell 1987; Scott 1988).

Pyroclastic flows and mudflows generated during the Swift Creek [about 13,000 to 8,000 years before the present (YBP)], Pine Creek (3000 to 2500 YBP), Castle Creek (2,200 to 1,700 YBP), and Kalama (AD 1480 to 1700) eruptive periods (Crandell 1987) partly filled the valleys of Pine Creek, Muddy River, and adjacent tributaries all the way to the Lewis River. The accumulated fill built a fan across the valley at Cedar Flats during the Swift Creek eruptive period (Crandell 1987). Flows generated during the Pine Creek and Castle Creek eruptive periods passed across the upper Muddy fan but moved mainly down the Pine Creek valley (see Figure 6.1). During the Kalama eruptive period, pyroclastic flows and mudflows buried the upper Muddy fan, and a series of mudflows built a thick fan of debris at the confluence with Smith Creek (Crandell 1987). Disturbance of the lower Muddy River by volcanic mudflows was largely confined to the Swift Creek (Crandell 1987), Kalama, and modern eruptive episodes. Following each eruptive episode, primary volcanic debris was periodically reworked by fluvial processes. Presumably, high sediment loads similar to those that followed the 1980 eruption (Major et al. 2000) triggered channel instability, bank erosion, and bed aggradation and transiently modified the characteristics of floods in the Muddy River system.

6.1.5 Preeruption Hydrology

Before the 1980 eruption, the Muddy River was hydrologically similar to comparably sized regional basins south and east of Mount St. Helens. Mean annual discharges per unit area and numbers of annual peak flow discharges above base flow of the Muddy River generally were similar to those of the East Fork Lewis River, a nearby river system. Gauged discharges from 1928 to 1934 and from 1955 to 1973 show that the greatest preeruption discharge of the Muddy River (496 $m^3\ s^{-1}$ in 1934) had a recurrence frequency of 25 to 50 years (Sumioka et al. 1998). The next highest annual peak discharge (312 $m^3\ s^{-1}$ in 1967) had a recurrence frequency of 5 to 10 years. All other annual peak discharges had magnitudes expected about every 5 years or less. Basin-level discharge measurements, however, do not necessarily characterize the fine-scale channel adjustments that influenced the composition of vegetation along the Muddy River valley before the 1980 eruption. Local reports of frequent road repair and persistent bank erosion along the lower Muddy floodplain confirm that Muddy River had an active channel system typical of rivers that drain active glaciers and sparsely vegetated uplands.

6.1.6 The 1980 Muddy River Mudflow

Within minutes, the May 18, 1980, eruption triggered mudflows on the volcano's western, southern, and eastern flanks. A fast-moving cloud of hot gases and fragmented rock (a pyroclastic surge) scoured snow, ice, and soil from the slopes of the volcano (Pierson 1985; see Figure 3.1 in Swanson and Major, Chapter 3, this volume) and triggered several energetic mudflow pulses that entered multiple tributaries of the Muddy River drainage. The multiple pulses coalesced into a single mudflow in the main stem of the channel below its confluence with Smith Creek (see Figure 6.1). That mudflow, together with another that flowed down Pine Creek, deposited more than 14 million cubic m^3 of sediment in Swift Reservoir (Janda et al. 1981; Pierson 1985).

One mudflow pulse swept across the upper Muddy fan as a broad sheet greater than 1 km wide and 3 to 5 m deep. It had an estimated peak velocity of 30 $m\ s^{-1}$ and a peak discharge of 190,000 $m^3\ s^{-1}$ (Pierson 1985). Along its flow path, almost all trees were sheared off at ground level; those remaining near the flow margin were severely battered and abraded (Figure 6.2; Table 6.1). The mudflow removed almost all vegetation along its central flow axis; roots and remnant plant parts survived mainly along flow margins and on two small hills (about 10 m

FIGURE 6.2. Photographs depicting early posteruption conditions on the Muddy River mudflow deposit. Study sites shown are: (a) upper Muddy fan (1980); (b) middle Muddy fan (1982); (c) lower Muddy flood plan (1980); and (d) Cedar Flats terrace (1981).

above mean surface level) that parted the mudflow and escaped inundation (see Figure 6.1). Across much of the upper Muddy fan, the mudflow deposited a veneer of poorly sorted, gravelly sand less than 1 m thick and scattered boulders up to 2 m in diameter.

Channelization and valley constrictions below the upper Muddy fan affected mudflow velocity, flow depth, and deposit thickness. Velocity and peak discharge became attenuated rapidly as the mudflow moved downstream (see Table 6.1). Along the Muddy River gorge, the mudflow reached a peak velocity of 23 m s^{-1} and a peak discharge of 20,000 m^3 s^{-1} (Pierson 1985). Near the confluence with Clearwater Creek, the flow had an average velocity of 14 m s^{-1} and a peak discharge of 22,000 m^3 s^{-1} after coalescing with a pulse of flow that transited Ape Canyon and Smith Creek. Near Cedar Flats, flow velocity fluctuated between 3 and 6 m s^{-1}, and peak discharge was less than 5,000 m^3 s^{-1} (Pierson 1985).

Within the Muddy River gorge, flow depth was as great as 20 m; along the main channel flow, depth was about 5 m. Despite the great flow depth, the resulting deposit was only 1 to 2 m thick on the valley floor and less than 0.5 m thick along the flow margin (Janda et al. 1981). As on the upper Muddy fan, deposits along the lower Muddy valley consisted primarily of poorly sorted, gravelly sand.

Vegetation damage along the lower Muddy River floodplain and middle Muddy fan was similar to, but less extensive than, that on the upper Muddy fan. Trees close to the central flow axis were sheared off at ground level or toppled and abraded (Janda et al. 1981). Trees along the flow margin were battered, abraded, and partly buried. The mudflow also deposited tangles of downed logs along the flow margin (Janda et al. 1981; Pierson 1985).

On the afternoon of the May 18, 1980 eruption, pyroclastic flows melted snow and ice on the volcano's flank and produced

TABLE 6.1. Characteristics of primary mudflow disturbance, initial biological impacts, and secondary disturbances at four sites in Muddy River valley.

	Upper Muddy fan	Middle Muddy fan	Lower Muddy floodplain	Cedar Flats terrace
Primary mudflow characteristics[a]				
Peak velocity	30 m s^{-1}	20 m s^{-1}	15 m s^{-1}	3 m s^{-1}
Maximum flow depth	5 m or less	10–20 m	5 m or less	2 m or less on terrace; 5 m in adjacent channel
Deposit thickness	1 m or less	1 m or less	0.5–2 m	0.2–1.5 m
Deposit width	300–1000 m	250–700 m	100–350 m	200–350 m
Biological characteristics				
Refugia	Deposit margins, local topographic highs	Deposit margins, rafted soil blocks	Deposit margins, logjams	Areas of shallow deposition, elevated substrates, rafted woody debris
Survivors	Perennial plants, roots	Perennial plants, roots	Perennial plants, roots	All types of overstory and understory species
Secondary disturbances				
Type and magnitude	Minor floods	Minor floods	Floods to more than 100-year event	None on surface
Geomorphic impact	Channel incision and widening, minor floodplain erosion; channels relatively stable from 1981 to 1996	Channel incision and widening, minor floodplain erosion; channels relatively stable from 1981 to 1996	Channel incision and widening, frequent channel shifting and lateral migration, extensive floodplain erosion	Little postmudflow change, surface stable; severe bank erosion by 1996 flood
Percent of area disturbed	Less than 20%	Less than 20%	Greater than 80%	Less than 10%
Frequency of disturbance	Channels reworked annually; floodplain rarely inundated	Channels reworked annually; floodplain rarely inundated	Channels reworked annually; floodplain inundated a few times per decade	Surface disturbance rare; bank erosion significant during rare flood (greater than 100-year event) in 1996

[a] Primary mudflow characteristics.
Source: After Pierson (1985).

a second mudflow that moved down Smith Creek and Muddy River. This mudflow overtopped terraces as far downstream as Cedar Flats and deposited more than 500,000 m^3 of debris into Swift Reservoir (Pierson 1985).

At Cedar Flats, the primary mudflow overtopped the 3-m-high terrace and deposited 0.2 to 1.5 m of gravelly to silty sand among standing trees. Flow velocity and deposit thickness were greatest where the mudflow initially overtopped the terrace and followed an abandoned river channel. The smaller, afternoon mudflow also overtopped the terrace and deposited up to 0.5 m of pebble-sized pumice, sand, and wood. Depositional patterns of both mudflows varied with premudflow topography and the presence of microtopographic features, such as downed logs, root collars, and tree-tip mounds.

6.2 Methods

6.2.1 Channel Surveys

Postmudflow changes in valley and channel morphology along the Muddy River were documented through repeated surveys of valley and channel cross sections at fixed locations (Martinson et al. 1984, 1986). Standard survey practices were used to measure cross-section morphology between monumented endpoints of 11 cross sections on the upper Muddy fan and 16 cross sections between the middle Muddy fan and Cedar Flats. Surveys were repeated at least annually through 1985 and less frequently between 1985 and 2003. Surveys in 1995 and 1996 bracket a major winter flood in 1996. Locations of cross-sectional and vegetation transects do not coincide precisely, and not all cross sections surveyed through 1985 were resurveyed subsequently. We selected channel cross sections having the longest time-series records within the upper, middle, and lower Muddy River study reaches for analysis of geomorphic changes that have influenced vegetation recovery on the Muddy River mudflow deposit.

6.2.2 Vegetation Surveys

Beginning in 1981, we monitored changes in vegetation composition along 16 cross-deposit transects between the top of the upper Muddy fan and the lower Muddy floodplain upstream of Cedar Flats terrace. Each transect consists of a series of 250-m^2 circular plots, placed systematically at 50-m intervals and marked at the center with a steel pin. The 50-m sampling interval ensured adequate coverage along each cross-deposit transect while maintaining a manageable total sample size for resurveying; the absence of vegetation on much of the mudflow deposit precluded the use of species–area curves in determining plot size. We placed 4 cross-deposit transects (59 plots) on the upper Muddy fan, 2 transects (15 plots) on the middle Muddy fan, and 10 transects (44 plots) on the lower Muddy floodplain. Plots located within active channels were not marked permanently, but surface conditions and presence or absence of vegetation resulting from surface erosion or deposition were recorded.

Vascular plant cover was estimated with Daubenmire's (1959) cover classes modified to include values less than 1%. Cover for all species was estimated on the upper Muddy fan from 1981 to 1983 and again in 1985 and 1991 and on the middle Muddy fan and lower Muddy floodplain annually from 1981 to 1984 and again in 1991. Taxonomic nomenclature follows the Integrated Taxonomic Information System (2003). In 1991, lack of resources limited remeasurements to 8 of 16 cross-deposit transects. We mapped surface features to scale on the upper Muddy fan, middle Muddy fan, and lower Muddy floodplain plots in 1982. In 1983 and 1984, we compared plot surfaces to the 1982 maps and updated the maps to reflect surface changes.

We initiated a separate study in 1981 to examine the mudflow-induced vegetation change in the largely intact forest on the terrace at Cedar Flats. We monitored overstory tree mortality and understory vegetation composition along two parallel, 50-m-wide transects oriented east to west across the terrace. These transects consist of 40 adjacent 25- × 25-m overstory plots having a total sample area of 2.5 ha. Beginning in 1981, trees 5 cm or more in diameter at breast height (dbh: 1.4 m above ground level) were marked with a numbered tag, measured, recorded by species, and mapped in each plot. We assessed mortality of tagged trees in 1982, 1984, 1991, 1995, and 1999. We remeasured all live trees 5 cm dbh or greater in 1991, 1995, and 1999, adding saplings that grew into the 5-cm minimum diameter or more class as ingrowth. In 1999, we increased the minimum-diameter class for red alder ingrowth to equal to or greater than 10 cm to facilitate measurement of a large number of new red alder stems.

We sampled understory vegetation at Cedar Flats terrace using forty 5- by 5-m vegetation plots placed systematically along the center of the two transects at 10- and 15-m intervals, respectively (total area = 0.1 ha). Understory cover was estimated annually on each vegetation plot from 1981 to 1984 and again in 1992, 1995, and 1999.

In 1999, we selected ten 25- by 25-m overstory plots (25% of the sample area) using a stratified random sampling technique (Mueller-Dombois and Ellenberg 1974) to examine the relationships among deposit thickness, overstory mortality, and postburial recovery. Within each of these plots, we cored all live conifers 5 cm or more and all live red alder 10 cm or more dbh to determine age-class distributions. Deposition in the 10 overstory plots was determined by measuring deposit thickness at each plot center and in four cardinal directions 5 m from the plot center.

We examined the severity of mudflow-induced disturbance by comparing overstory and understory vegetation responses among objectively determined plot groupings based upon detrended correspondence analysis (DCA) scores (Hill and Gauch 1980) and cluster analysis of tree mortality on each intensively sampled overstory plot (Weber 2001). Overstory plot groupings were as shown below:

	Description	Mean deposit thickness (cm)	Range (cm)	Number of plots
Minor mortality	Minimum deposition and little to no overstory mortality	29	23–32	3
Intermediate mortality	Intermediate deposition and partial overstory mortality	36	30–56	3
Heavy mortality	Maximum deposition and near total overstory mortality	59	48–120	4

Given the degree of local variation in deposit thickness within the 25- × 25-m overstory plots and the limited resolution of our overstory depth sample, we were unable to relate overstory mortality to deposit thickness except at a coarse level. For this chapter, we select three representative overstory plots, one from each of the three overstory plot groupings, to illustrate overstory responses across the mudflow deposit (Figures 6.3, 6.4).

Deposit thickness for understory plots was classified on the basis of natural groupings identified from a frequency curve of deposit-thickness values measured at 5-m intervals along each understory transect. Deposit-thickness groupings were classified as shallow (less than 53 cm; 9 plots), intermediate (53 to 73 cm; 24 plots), and deep (more than 73 cm; 7 plots).

6.2.3 Analytical Methods

Analytical methods were chosen to (1) document disturbance severity and vegetation response patterns at each site and (2) summarize field data collected at our study sites during the study period. To examine the relationship between type of disturbance and plot-specific vegetation response, we sorted plots into categories based upon annual surface changes documented through repeated mapping from 1982 to 1984. We categorized our 250-m^2 plots from the upper, middle, and lower Muddy sites into types of surface disturbance on the basis of the degree of surface change recorded on successive surface maps and annual survey notes that recorded locations of the active river channel and erosion of plot markers. Plots with missing sampling periods or surface documentation were removed before analysis.

We classified our plots on the upper Muddy fan over the 3-year surface-mapping period into one of three surface-disturbance types:

- Stable surfaces having no visible change (14 plots)
- Surfaces having annual changes confined to small erosion gullies (generally less than 20% of plot surface; 33 plots)

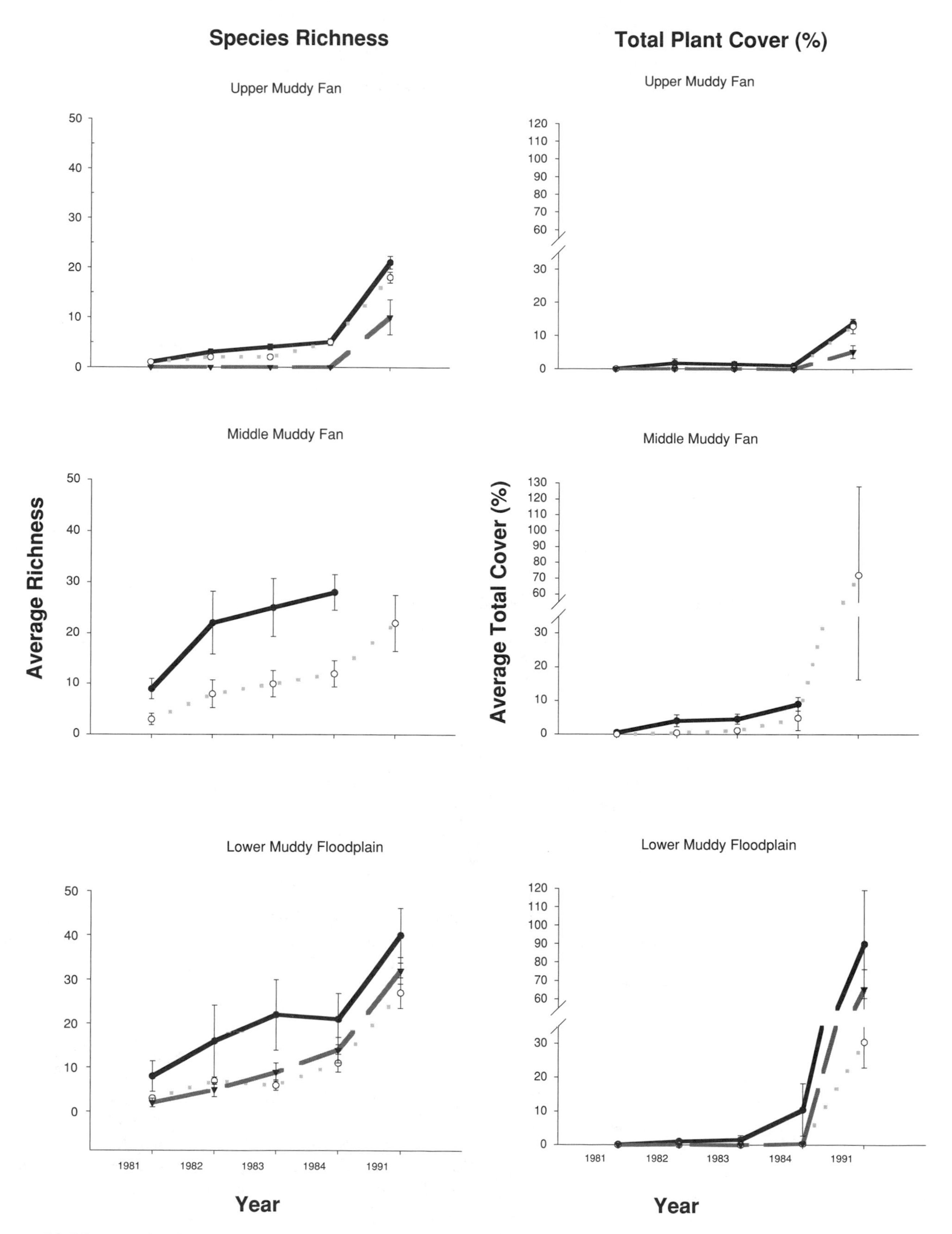

FIGURE 6.3. Mean species richness and total plant cover (%) for surface-disturbance types for eight transects at three study sites on the Muddy River mudflow deposit. *Solid line* represents stable surfaces; *coarse dashed line* represents moderately stable surfaces (less than 20% change annually); *fine dashed line* represents unstable channel margins (reworked annually). Unstable channel margins are not shown for middle Muddy fan because of insufficient sample size. For the lower Muddy floodplain, the *solid line* represents surfaces reworked once between 1980 and 1984, the *coarse dashed line* represents surfaces reworked two to three times, and the *fine dashed line* represents surfaces reworked annually. *Error bars* represent the 95% confidence interval.

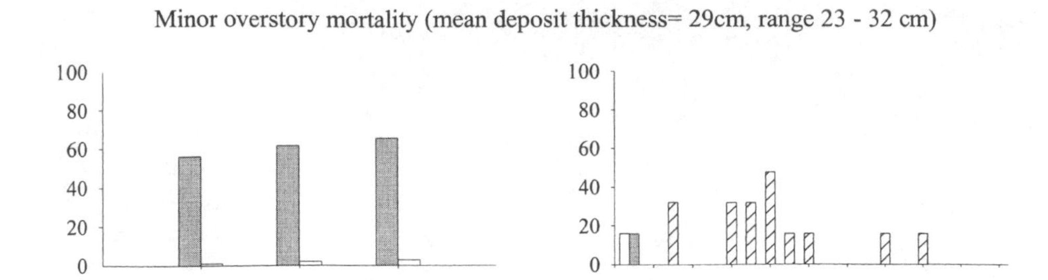

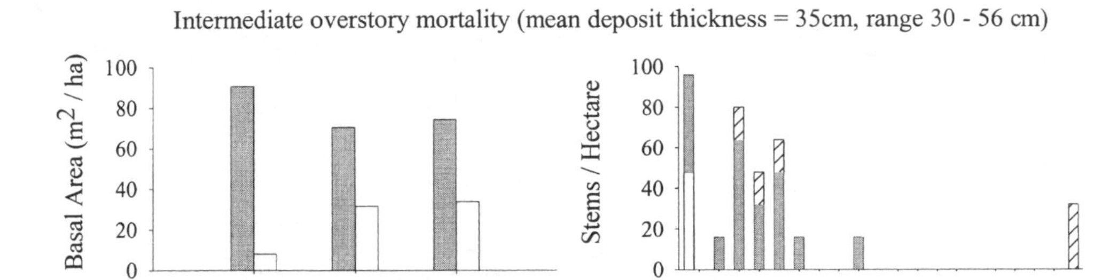

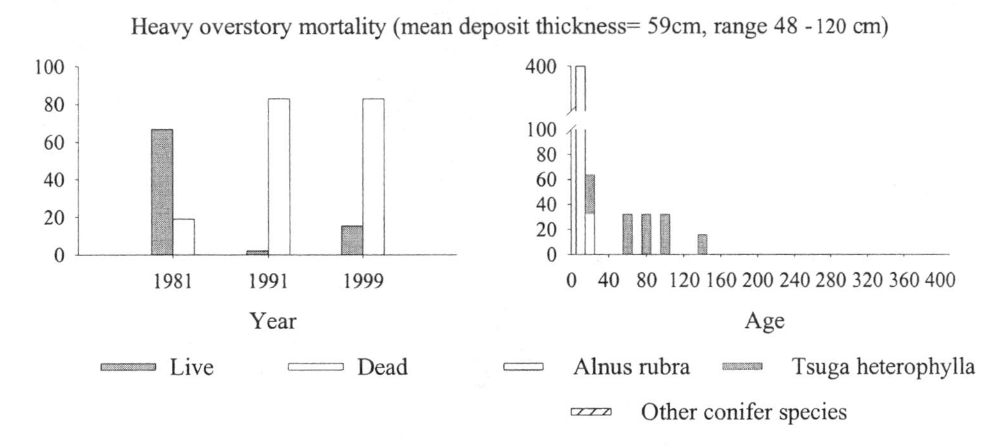

FIGURE 6.4. Mortality and recruitment of overstory trees (1981 to 1999) and age-class distributions by species (1999) in relation to mudflow-deposit thickness. Data are presented for three 25- × 25-m overstory plots, each representing one of three overstory mortality plot groupings (minor, intermediate, and heavy mortality) on the Cedar Flats terrace (7.5% of the total sample area). Deposit-thickness values and ranges are based upon five depth samples that were measured in each of the three selected overstory plots.

- Surfaces within incised channels where the entire plot was reworked annually (9 plots)

We classified plots as completely reworked when mapped surface features on more than 75% of the plot surface had changed between successive mapping periods. We used similar surface-disturbance types at our middle Muddy fan site:

- Stable surfaces (6 plots) and
- Annually reworked surfaces confined to small erosion gullies (9 plots).

Because of insufficient sample size, plots reworked completely on the middle Muddy fan were excluded from the analysis.

Our quantitative description of vegetation development on the middle Muddy fan excludes an undisturbed forest-edge plot or plots containing intact soil blocks deposited on the mudflow surface because of their small sample size ($n = 3$). Nonetheless, these plots illustrate the local influence of edge effect (shade, litter, and proximity of surviving plants) and rafted soil blocks on vegetation recovery along deposit margins.

We sorted plots on the lower Muddy floodplain into disturbance types on the basis of frequency of surface change: surfaces reworked once in 4 years (11 plots), surfaces reworked two to three times over 4 years (23 plots), and surfaces reworked annually (10 plots). This reach experienced frequent channel shifts, and all plots were subject to surface disturbance at least once during the 1981 to 1984 mapping period.

We mitigate the influence of varying sample sizes between sites and surface-disturbance types by comparing mean richness and total cover (%) values. Our temporal examination of surface stability and vegetation development represents (1) the 4-year period from 1981 to 1984, corresponding to annual vegetation composition and surface-stability sampling, and (2) vegetation change over an 11-year period from 1981 to 1991 for a subset of plots (8 of 16 plots; see Figure 6.3a–f; Table 6.2). We present our 1991 data using the 1981 to 1984 surface-disturbance groupings to ensure consistency of comparisons over time. We thought that the advantage of displaying comparable plot groupings for the entire 11-year study exceeded the risk of misclassifying the 1991 plots into outdated surface-disturbance categories. This assumption was supported by repeated observations of stream cross sections that documented a general stabilization of the deposit surface (Martinson et al. 1986) and slowly declining watershed sediment yield between 1985 and 1991 (Major et al. 2000).

We examined species distributions across surface-disturbance types for the upper, middle, and lower Muddy sites by using the occurrence frequency of common colonizing species and by counting the total number of species encountered by life form (see Table 6.2). We used Spearman rank correlations to examine the relationships between (1) the distance from the adjacent, undisturbed forest and (2) the species richness and total plant cover for each surface-disturbance type.

We examined overstory vegetation responses (for 1981 to 1999) on Cedar Flats terrace by comparing differences in species composition, tree diameter, and age-class distributions among overstory plots. Overstory changes between 1980 and 1999 were tracked by monitoring tree mortality and live- and dead-tree basal area. We used stand-age-structure diagrams and tree-growth rates to examine canopy-replacement patterns and recruitment rates.

Our analysis of understory vegetation at Cedar Flats examined shifts in species composition and abundance (from 1981 to 1995). These relationships were determined by calculating a Bray–Curtis similarity matrix (Bray and Curtis 1957), a semimetric distance measure derived from the number of species in each understory plot weighted by the percent-cover value for each species. The resulting value represents both relative richness and abundance. We plotted the Bray–Curtis measures of ecological distance using nonmetric, multidimensional scaling (NMDS) (Clarke 1993) for all 40 understory plots to illustrate patterns of community change between 1981 and 1995. NMDS reduces our multidimensional Bray–Curtis similarity matrix to a two-dimensional plot of similarity patterns that represent combined richness and abundance (percent-cover) values. Last, we created a list of indicator plants for each deposit-thickness class to identify which species are most representative of community change between 1981 and 1995 (Table 6.3).

6.3 Results

6.3.1 Channel Evolution

The May 18, 1980 mudflow obliterated a significant proportion of vegetation in the headwaters of the Muddy River basin and disturbed more than 20 km of river corridor between the volcano and Swift Reservoir (see Figure 6.1). Along their paths, mudflow pulses deposited a veneer of sediment ranging in thickness from a few centimeters on the upper Muddy fan to more than 2 m along the lower Muddy River valley. Overall, the main channel was displaced, straightened, and changed from a gravel-bedded to a sand-bedded system (Janda et al. 1984). These changes substantially altered runoff, erosion, and sediment transport in the basin (Major et al. 2000, 2001; Major 2003; see Swanson and Major, Chapter 3, this volume).

In the Muddy River system, channels rapidly incised and widened in all reaches. The most rapid changes occurred within 3 years after the eruption (Martinson et al. 1984; Hardison 2000) and were followed by more than a decade of relative channel stability when net channel geometry changed slowly or negligibly (Martinson et al. 1986). The period of relative stability abruptly ended with a large-magnitude flood in 1996 (more than 100-year event; Sumioka et al. 1998). Analysis of aerial photographs and channel resurveys in summer 1996 shows that the flood caused little net change in channel geometry on the upper and middle Muddy fans, but it triggered significant bank erosion and massive scour and overturn of the floodplain surface along the lower Muddy valley.

6.3.1.1 Upper and Middle Muddy Fans

On the upper Muddy fan, two parallel channels incised rapidly up to 4 m and widened up to 50 m between 1980 and 1981 (near A in Figure 6.1). Floodplains exhibited as much as 1.5 m of variable scour and fill in the same period (Martinson et al. 1984). From 1981 to 1985, the channel system on the upper Muddy fan stabilized, bed elevation fluctuated little, channel widening diminished, and the floodplain aggraded slightly. Over the following decade, the south channel widened as much as 25 m, and bed elevation remained stable, but the north channel aggraded as much as 2 m (Martinson et al. 1984, 1986). On the middle Muddy fan, a well-defined, narrowly incised channel rapidly aggraded as much as 5 m in late 1980. From 1981 to 1996, the channel and its floodplain slowly and broadly incised as much as 2.5 m.

6.3.1.2 Lower Muddy Floodplain

Below the confluence with Clearwater Creek, the Muddy River channel incised about 1 m and widened by 5 m from 1980 to 1981. From 1981 to 1982, valley-floor aggradation and channel shifting dominated channel adjustment (Martinson et al. 1984; Hardison 2000). Further channel shifting, incision of a new primary channel, and erosion of accumulated

TABLE 6.2. Number of species by life form and frequency of occurrence of common colonizing species for stable surfaces and active channel margins on the Muddy River mudflow in 1981 and 1991.

Study site		Upper Muddy fan				Middle Muddy fan				Lower Muddy floodplain			
Surface type		Stable surface		Active channel		Stable surface		Active channel		Stable surface		Active channel	
Year		1981	1991	1981	1991	1981	1991	1981	1991	1981	1991	1981	1991
Sample size (number of plots)		14	12	9	6	6	4	2	2	11	4	10	7
Number of species by life form	Life form												
Grasses	G	0	8	0	4	2	6	0	4	1	8	0	7
Sedges and rushes	SR	1	6	0	3	1	3	0	1	2	4	1	6
Forbs	F	8	21	1	13	11	20	0	13	15	58	4	48
Shrubs	S	3	7	0	2	5	7	0	2	4	8	0	5
Trees	T	3	9	1	5	5	8	0	4	6	8	6	7
Frequency of occurrence (%)													
Plots with no vegetation		8	0	88	0	0	0	100	0	18	0	38	0
Common colonizing species													
Natives													
Vine maple (*Acer circinatum*)	S	8	0			17	25			0	20		
Pearly everlasting (*Anaphalis margaritacea*)	F	0	42	0	50	17	100	0	+	9	100	0	100
Bearberry (*Arctostaphylos uva-ursi*)	S	0	8	0	17	50	25			0	20		
Fireweed (*Epilobium angustifolium*)	F	0	33	0	33	0	25	0	+	27	60	38	83
Fragrant bedstraw (*Galium triflorum*)	F											0	50
Salal (*Gaultheria shallon*)	S					17	25						
White hawkweed (*Hieracium albiflorum*)	F	8	42	0	50	0	100	0	+	0	80	0	83
Partridgefoot (*Luetkea pectinata*)	F	0	33	0	50								
Littleleaf miners lettuce (*Montia parvifolia*)	F							0	+	0	80	0	83
Cardwell's beardtongue (*Penstemon cardwellii*)	F	0	42	0	67	50	50						
Serrulate penstemon (*P. serrulatus*)	F	0	8							0	40	0	33
Arctic sweet coltsfoot (*Petasites frigidus*)	F									0	40	0	33
Wood groundsel (*Senecio sylvaticus*)	F	8	25							18	100	0	83
Blueberry (*Vaccinium* spp.)	S	0	42			17	50			0	40		
Willow (*Salix* spp.)	S	8	42	0	33	50	100	0	+	18	80	0	100
Nonnatives													
Thistle (*Cirsium* spp.)	F	8	8			50	0			18	40	0	100
Cat's ear (*Hypochaeris radicata*)	F	0	33	0	67	0	100	0	+	0	100	0	100
Tansey ragwort (*Senecio jacobaea*)	F					0	25			0	20	0	17
Nitrogen fixers													
Red alder (*Alnus rubra*)	T	8	0			17	75			36	80	38	100
Sitka alder (*Alnus viridis* ssp. *sinuata*)	S	0	33			0	50					0	17
Broadleaf lupine (*Lupinus latifolius*)	F	42	42	13	83	33	25	0	+	9	40	13	67
Prairie lupine (*Lupinus lepidus*)	F	50	25	0	83	67	100	0	+	9	60	13	83
Birdsfoot trefoil (*Lotus corniculatus*) (nonnative)	F	10	0							18	60	25	67
American birdsfoot trefoil (*L. unifoliolatus*)	F									0	20	0	67

TABLE 6.2. (Continued)

Study site		Upper Muddy fan				Middle Muddy fan				Lower Muddy floodplain			
Surface type		Stable surface		Active channel		Stable surface		Active channel		Stable surface		Active channel	
Year		1981	1991	1981	1991	1981	1991	1981	1991	1981	1991	1981	1991
Sample size (number of plots)		14	12	9	6	6	4	2	2	11	4	10	7
Trees													
Noble fir (*Abies procera*)	T	0	42	0	67	0	25						
Subalpine fir (*Abies lasiocarpa*)	T	0	17										
Bigleaf maple (*Acer macrophyllum*)	T					33	25	0	+	18	60	13	33
Lodgepole pine (*Pinus contorta*)	T	0	25	0	33								
Western white pine (*Pinus monticola*)	T	0	42	0	17					0	20		
Black cottonwood (*Populus trichocarpa*)	T									9	60	25	33
Douglas fir (*Pseudotsuga menziesii*)	T	75	42	13	33	100	100	0	+	64	40	50	67
Western red cedar (*Thuja plicata*)	T					0	25	0	+	45	40	25	33
Western hemlock (*Tsuga heterophylla*)	T	0	17			17	75	0	+	45	60	38	50
Mountain hemlock (*Tsuga mertensiana*)	T	0	17										

fill occurred in 1983. After 1983, the floodplain entered a period of comparative stability with little net change in channel geometry. Fluctuating aggradation and erosion of valley fill along main-stem channels persisted through 1994 (Hardison 2000).

Storm-induced runoff between 1994 and 1996 triggered local channel shifting and up to 2.5 m of channel incision but produced no general change in bed elevation. Near the confluence with Clear Creek, runoff triggered channel shifting and broad aggradation up to 1.5 m thick across the valley floor. Much of this change occurred as a result of the 1996 flood.

Near Cedar Flats, bedrock constrictions as narrow as 5 m pooled the mudflow, resulting in up to 2.5 m of deposition on the valley floor (Pierson 1985; Hardison 2000). The greatest postmudflow channel incision documented along the main stem of the Muddy River occurred in the middle of the Cedar Flats reach (see Figure 6.1). By 1981, the channel had incised up to 3 m despite having a gradient of less than 1% (Hardison 2000). Between 1981 and 2003, channel geometry changed negligibly.

6.3.2 Vegetation Responses

6.3.2.1 Upper Muddy Fan

Surface stability was the primary factor influencing vegetation development on the upper Muddy fan. During the first 2 years following mudflow emplacement, species-richness and cover values were greater on stable surfaces than on those associated with channel incision (see Figure 6.3a,b). Between 1982 and 1991, 41 new species had established on stable surfaces, and by 1991, plant cover remained low (less than 10%) and was dominated by species common to subalpine areas of the southern Washington Cascades: broadleaf lupine (*Lupinus latifolius*), prairie lupine (*L. lepidus*), partridgefoot (*Luetkea pectinata*),

TABLE 6.3. Understory indicator species by mudflow burial depth class on the Cedar Flats terrace in 1981 and 1995.

	Indicator species for each burial depth		
	Shallow (less than 53 cm)	Intermediate (53–73 cm)	Deep (more than 73 cm)
1981	Serviceberry		
	Salal		
	Red huckleberry		
	Red alder	Red alder	
	Pearly everlasting	Pearly everlasting	
	Vine maple	Vine maple	
		Western hemlock	Western hemlock
		Western red cedar	Western red cedar
1995	Salal		
	Douglas-fir		
	Western hemlock	Western hemlock	Douglas-fir
		Trailing blackberry	Trailing blackberry
		Sword fern	Sword fern
		Red alder	Red alder

Shaded background denotes greatest abundance for each species; shading of *Tsuga heterophylla* in both intermediate and deep burial sites in 1981 reflects its equal abundance.

Cardwell's penstemon (*Penstemon cardwellii*), and pussypaws (*Cistanthe umbellata*). These species occurred on 70% to 100% of plots. By 1991, colonizing tree species, including noble fir, Douglas-fir, and western white pine, were common but contributed little cover (see Table 6.2). In contrast, active channel margins experienced a net gain of 7 species following the loss of 3 early-colonizing species and the subsequent establishment of 10 new species during the period. Spearman rank correlations between total plant cover and distance to the adjacent, undisturbed forest were not significant on the upper Muddy fan throughout the measurement period.

6.3.2.2 Middle Muddy Fan

Species richness and cover were consistently greater on the middle Muddy fan than on the upper Muddy fan (see Figure 6.3c,d). On the middle Muddy fan, species richness was negatively correlated with distance to the adjacent, undisturbed forest in 1981 and 1984 ($p < 0.05$), and total plant cover negatively correlated with distance in all years.

Vegetation recovery along the forest edge was rapid, with 14 species present (1.4% total plant cover) by 1981, 32 species present (10.8% cover) by 1982, and 37 species present (114% cover) by 1991. Species composition and cover at the forest edge were evenly divided between colonizing species common to all Muddy River study sites and forest species that likely survived along the deposit margin. Forest species included salal (*Gaultheria shallon*), sword fern (*Polystichum munitum*), and trailing blackberry (*Rubus ursinus*). Bearberry (*Arctostaphylos uva-ursi*), a species common to pre-1980 mudflow deposits in openings and forest understory along the Muddy River, was also abundant.

Species richness in plots with rafted soil blocks was consistently higher than in the nonforested plots on the middle Muddy fan in 1981 (13 species) and 1991 (36 species). Total plant cover was substantially lower on the soil-block plots than on the adjacent forest edge (0.7% total cover in 1981 and 6.5% by 1991) but greater than that of other surfaces until 1984 when those sites were colonized by red alder.

6.3.2.3 Lower Muddy Floodplain

Between 1981 and 1983, the lower Muddy floodplain experienced frequent channel shifting, floodplain erosion, and aggradation that inhibited vegetation establishment near the active channel. During this period, the percentage of vegetation plots on stable surfaces (i.e., those surfaces exhibiting no visible change as reflected by comparison of successive surface maps) dropped from 65% in 1981 to 6% in 1983. Of plots surveyed in 1983, 55% were located on an actively changing floodplain surface, with 19% occurring on active channel margins. Species richness was greatest on stable surfaces, but total plant cover on all surface types remained below 0.2% in 1983 (see Figure 6.3e,f).

Between 1984 and 1995, the channel along the lower Muddy River valley experienced declining erosion, minor net changes in channel geometry (Hardison 2000), and dwindling suspended-sediment transport (Major et al. 2000). These conditions led to an increase in the proportion of vegetation plots on stable surfaces from 6% in 1983 to 16% in 1991. Species richness on the floodplain doubled, and total plant cover increased as much as a factor of 60 between 1985 and 1991 (see Figure 6.3e,f). Most of the increase in cover resulted from the establishment of red alder across the floodplain and along stabilized channel margins (with cover ranging from 3% to 85%). The marked increase in species richness corresponded to the establishment of five new species of grasses and four additional species of sedges (*Carex* spp.) and rushes (*Juncus* spp.). Richness also increased as herbaceous species common in the adjacent forest communities colonized moist and increasingly stable stream banks and the understory of newly developed red alder stands. New species in 1991 included coastal brookfoam (*Boykinia occidentalis*), broadleaf triflower (*Trientalis borealis*), Arctic sweet coltsfoot (*Petasites frigidus*), American speedwell (*Veronica americana*), and three species of ferns.

We found no significant statistical correlation between vegetation development on unstable surfaces and distance to the undisturbed forest along the lower Muddy floodplain. Species richness on stable surfaces (reworked only once in 4 years) was negatively correlated with distance from 1982 to 1984 (p less than 0.05).

6.3.2.4 Cedar Flats Terrace Overstory

Initial tree mortality on the Cedar Flats terrace was greatest along the deeply buried upstream edge of the terrace and least in areas of shallow deposition farther downstream (Weber 2001). The most prominent ecological consequence of mudflow deposition entailed replacement of the preeruption Douglas-fir and western red cedar overstory by colonizing red alder in thick-deposition sites (average depth = 59 cm). Most of the large-diameter overstory trees survived where the deposit was less than 30 cm thick (Weber 2001). By 1984 50% of the overstory trees had died where average deposit thickness exceeded 50 cm. This value can be compared to 36% mortality in areas of intermediate deposition (having an average thickness of 36 cm) and 15% in plots having an average thickness of 29 cm.

High mortality among deeply buried, large-diameter (greater than 1 m dbh) Douglas-fir resulted in large decreases in basal area and canopy cover of live trees (see Figure 6.4e) and subsequent establishment of dense stands of red alder (Figure 6.4f). Smaller patches of tree mortality occurred across the terrace, where local differences in deposit thickness, tree size, and species composition influenced survival. Small canopy gaps were most common in areas of intermediate deposition and were colonized mainly by western hemlock and some

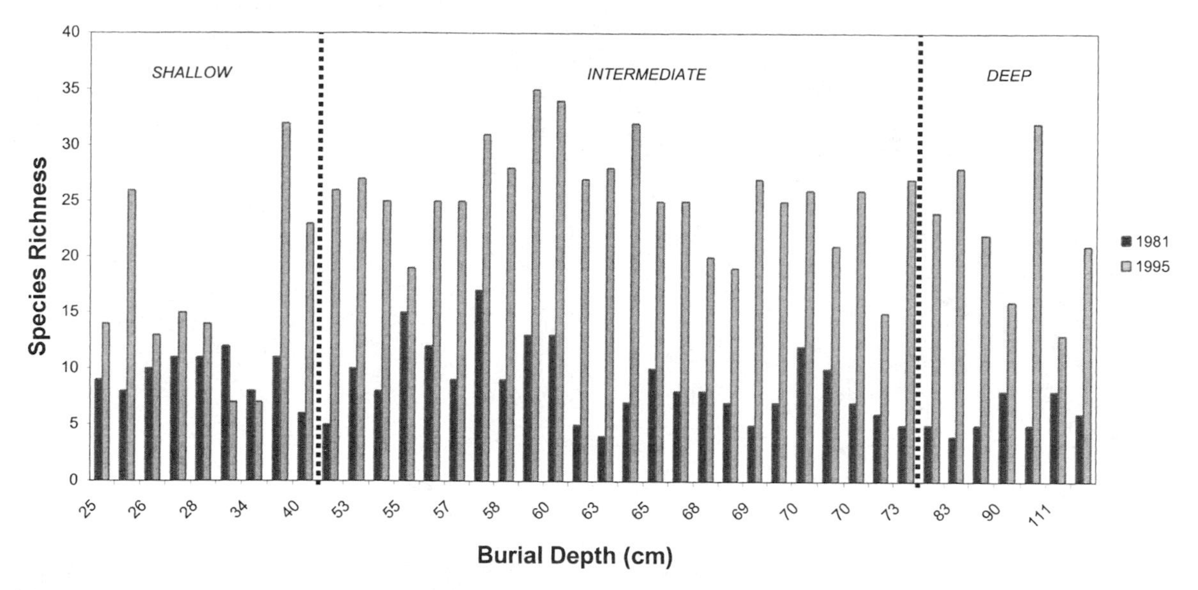

FIGURE 6.5. Understory species richness in 1981 and 1995 in relation to mudflow-deposit thickness on Cedar Flats terrace. *Dashed lines* indicate boundaries between understory burial-thickness classes (shallow, less than 53 cm; intermediate, 53 to 73 cm; deep, greater than 73 cm).

Douglas-fir (Figure 6.4d). Pacific silver fir and western hemlock saplings that survived on localized mounds and the root collars of dead, large-diameter Douglas-firs were also present on intermediate-deposition sites.

6.3.2.5 *Cedar Flats Terrace Understory*

In 1981, understory species richness (median, 5 species; range, 4 to 8; Figure 6.5) was least in sites of thick deposition (greater than 73 cm). At deeply buried sites, surviving overstory western red cedar and understory western hemlock rooted on elevated substrates and preeruption nurse logs were important postdisturbance species (see Table 6.3). In contrast, species richness (median, 8 and 10 species; range, 4 to 17 and 6 to 12, respectively; see Figure 6.5) was greatest on sites of intermediate (53 to 73 cm) and shallow deposition (less than 53 cm). Important postdisturbance species at sites of intermediate deposition included surviving vine maple, an upright shrub, western hemlock and western red cedar seedlings, and the early-colonizing herb pearly everlasting (Table 6.3). Surviving shrubs, including salal, western serviceberry (*Amelanchier alnifolia*), and red huckleberry (*Vaccinium parvifolium*), and colonizing red alder seedlings dominated understory species at sites of shallow deposition.

By 1995, the relationship between species richness and deposit thickness had reversed. Species richness was greatest at sites of intermediate (median, 26 species; range, 15 to 35) and deep deposit (median, 22 species; range, 13 to 32; see Figure 6.5). Surviving shrubs were the most abundant species in areas of shallow deposition. Sites of intermediate deposition supported a combination of colonizing forb, shrub, and tree species, whereas colonizing tree species dominated deeply buried sites (see Table 6.3).

Analysis with NMDS shows distinct groupings for the 1981 and 1995 species richness and abundance data and a high degree of dissimilarity within each of the three deposit-thickness classes in 1981 (Figure 6.6). By 1995, however, community composition in plots on intermediate and deeply buried sites had converged because of the shared dominance by four species: the colonizing species red alder and trailing blackberry, sword fern, and western hemlock (Figure 6.6; Table 6.3). Community composition in areas of shallow deposition remained dissimilar from 1981 to 1995, but the majority of those sites showed little change in community structure over time (Figure 6.6).

6.4 Discussion

6.4.1 Geomorphic Responses

Geomorphic and hydrologic responses to headwater (i.e., on the slopes of the volcano and the upper portion of the upper Muddy fan) disturbance and mudflow emplacement in the Muddy River system were dramatic and persistent. Channel adjustments occurred rapidly in all reaches within 3 years after the eruption (Martinson et al. 1984; Hardison 2000). Between 1983 and 2003, adjustments in channel geometry asymptotically approached an apparent equilibrium state, although severe flooding in 1996 disrupted the apparent equilibrium.

Runoff events from basins significantly affected by the 1980 eruption were disproportionately larger than those from unaffected basins for at least 5 years after the eruption (Major

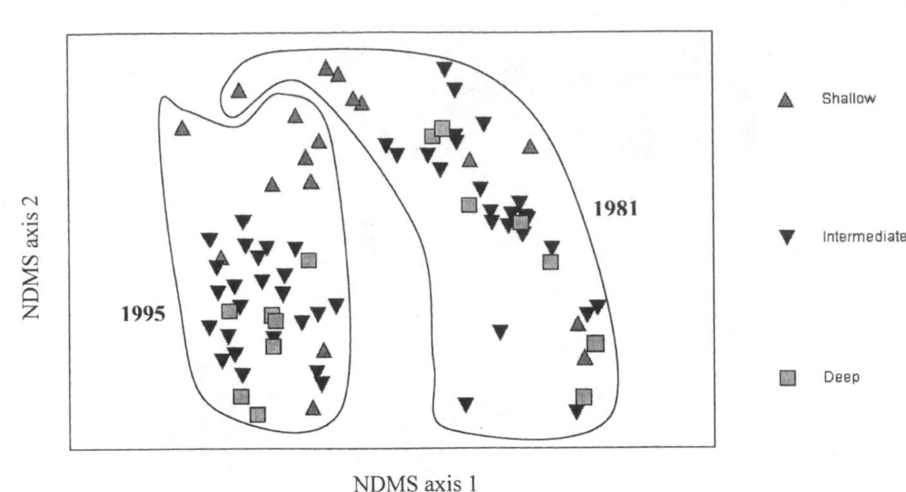

FIGURE 6.6. Bray–Curtis similarity patterns for Cedar Flats understory vegetation plots in 1981 and 1995. *Plotted values* include species richness weighted by species abundance. Increasing plot proximity indicates higher similarity values. *Links* identify similarity groupings for 1981 and 1995. *Symbols* designate mudflow deposit thickness classes for the respective plots: shallow (less than 53 cm), intermediate (53 to 73 cm), and deep (greater than 73 cm).

et al. 2001). In the Muddy River system, peak-flow discharges increased temporarily by several tens of percent for about 5 years. Through 1983, annual maximum discharges in the basin had apparent return intervals of 10 years or more, whereas nearby basins unaffected by the eruption had discharges smaller than flows expected about once every 5 years (cf. Swanson and Major, Chapter 3, this volume). Between 1984 and 2000, recurrence frequencies of most annual maximum discharges were similar to those of nearby unaffected basins. However, lingering effects of basin disturbance persisted. From 1995 to 2000, large peak-flow discharges had magnitudes ten times greater than comparable preeruption discharges (Major et al. 2001; see Swanson and Major, Chapter 3, this volume).

As a consequence of changes in posteruption runoff, straightening and smoothing of the river channel, and extensive deposition of volcanic debris, suspended-sediment yield in the Muddy River basin increased. Initially, yield increased by nearly a factor of 100 and after two decades remained nearly 10 times that of preeruption yield (Major et al. 2000). Channel incision, widening, and aggradation, driven chiefly by transiently enhanced runoff and extraordinary sediment transport, strongly influenced vegetation recovery, especially along the lower Muddy floodplain.

6.4.2 Vegetation Responses

6.4.2.1 Upper Muddy Fan

Disturbance severity was greatest on the upper Muddy fan, where the intensity of mudflow scouring was greatest. Plants survived mainly along deposit margins and on two small hills that escaped mudflow inundation. Postmudflow vegetation responses on the upper Muddy fan have been influenced chiefly by two factors: (1) local topography and its influence on surface stability and (2) a comparatively slow rate of seedling establishment on the high-elevation mudflow surface. Bank erosion hindered vegetation establishment along incised channel margins that comprise a comparatively small proportion of the upper Muddy fan. By comparison, stable substrates located outside of incised channels had significantly higher species richness and total plant cover values throughout the 11-year (1981 to 1991) measurement period (see Figure 6.3a,b). These results are similar to those found by Dale et al. (see Chapter 5, this volume), who document the impacts of secondary disturbance on vegetation succession on the debris-avalanche deposit along the North Fork Toutle River.

Seedling establishment on the upper Muddy fan proceeded slowly under comparatively harsh, postdisturbance growing conditions that existed at the upper-elevation site. Nonetheless, tree seedlings were common across the deposit by 1991. Wood and del Moral (1987) noted similarly slow rates of seedling establishment on deposits of small mudflows on the south side of Mount St. Helens in response to low soil-water potential, low soil nutrient availability, and high vapor-pressure deficits. Under such extreme environmental conditions, the trapping of windblown seeds and amelioration of environmental conditions by early colonizing plants on the upper Muddy fan were likely factors contributing to increased species richness between 1985 and 1991.

6.4.2.2 Middle Muddy Fan

Perennial plants rooted in rafted soil blocks and exposed soils along deposit margins accounted for the majority of plants that survived the mudflow on the middle Muddy fan (Halpern and Harmon 1983). The comparatively low elevation of the middle Muddy fan (550 m) made it more conducive to seedling establishment than the upper Muddy fan. The presence of an incised, fixed channel made the majority of the middle Muddy fan more stable than the lower Muddy floodplain. Consequently, for the first 4 years after disturbance, the middle Muddy fan exhibited higher species richness and total plant cover than either of its adjacent reaches (Figure 6.3c,d). The comparative suitability of the middle Muddy fan was further reflected in its accelerated colonization by several forest species, including broadleaf triflower, prince's pine (*Chimaphila umbellata*), false Solomon's seal (*Maianthemum stellatum*), and salal. After 1984, the establishment and rapid growth of red alder on the increasingly stable lower Muddy floodplain surpassed the rate of plant development on the middle Muddy fan.

6.4.2.3 Lower Muddy Floodplain

The 1980 mudflow removed vegetation along its central flow axis on the lower Muddy River floodplain, and it deposited uprooted trees and logjams along the valley margin (Janda et al. 1981). Plant roots survived in organic substrates, such as rootwads and stump bases, and individual plants survived in exposed, premudflow soils on stream banks along deposit margins (Halpern and Harmon 1983).

Channel adjustment occurred rapidly along the lower Muddy floodplain. By 1981, the channel incised as much as 1 m and widened up to 5 m (Hardison 2000). The frequency and magnitude of channel migration played a key role in the subsequent development and resetting of vegetation (see Figure 6.3e,f). The most stable parts of the floodplain, those reworked once between 1981 and 1984, had higher species richness and total plant cover than areas that were reworked two times or more during that 4-year period.

High species turnover rates between 1982 and 1984 (10 species lost and 26 new species recorded) reflect dynamic secondary disturbances rather than successional processes. The establishment of 68 new species on increasingly stable areas of the floodplain between 1984 and 1991 resulted largely from colonization rather than succession-related replacement. Consequently, the comparatively short-term duration of this study (combined with the high frequency and high severity of fluvial disturbance along the lower Muddy floodplain) precludes assessment of long-term successional pathways.

6.4.2.4 Cedar Flats Terrace

Postmudflow vegetation recovery on Cedar Flats terrace has been shaped strongly by biotic interactions and biological legacies. Competition, especially light interference, is common where deposits of intermediate thickness fostered accelerated canopy replacement by previously suppressed, subcanopy trees. Biological legacies in the form of rafted nurse logs and root mounds contributed to in situ survival of understory species and to the availability of elevated safe sites for seedling establishment. The presence of surviving and dead standing trees and rafted woody debris presumably contributed to seedling establishment by moderating surface temperatures, lowering vapor-pressure deficits, and reducing drought stress.

Postmudflow vegetation development on Cedar Flats terrace generally followed one of three developmental pathways related to deposit thickness (Weber 2001). Where a thick deposit killed overstory Douglas-fir and western red cedar, expansive canopy openings resulted in an early stage of primary succession dominated by red alder. In sites of intermediate deposit thickness, smaller canopy gaps resulted in accelerated regeneration and canopy recruitment of western hemlock. In areas of intermediate and shallow deposition, growth releases of formerly suppressed western hemlock and Pacific silver fir on localized safe sites, such as root mounds and preeruption nurse logs, contributed to accelerated replacement of the canopy of Douglas-fir and western red cedar killed by the mudflow.

Understory vegetation on Cedar Flats terrace shows a similar pattern of recovery in response to differences in deposit thickness. Initial shifts in understory communities, however, appear to be triggered by greater deposit-thickness thresholds than those that demarcate mortality thresholds for overstory species. For example, shifts in understory indicator species representative of sites of shallow deposition occur where the deposit is less than 53 cm thick, compared to a 30-cm mortality-depth threshold for tree species. These results suggest that highly localized safe sites corresponding to shallowly buried logs or other elevated substrates played an important role in the survival of understory species. In general, survival was the most important factor influencing vegetation recovery on shallow-deposition sites, whereas colonization was the predominant recovery process elsewhere (see Table 6.3). Regardless of deposit thickness, species composition is converging toward more similar communities across the entire terrace.

6.4.3 Interactions Between Geomorphic Processes and Vegetation Responses

For the first 5 years after the eruption, establishment of vegetation on floodplains and channel margins along the Muddy River was limited by frequent channel adjustments and unstable substrates. Between 1984 and 1991, the channel system stabilized temporarily and allowed vegetation to colonize previously active channel margins and floodplain surfaces. This change was most pronounced along the lower Muddy River, where stabilized surfaces on the floodplain were colonized rapidly by stands of red alder. On the upper Muddy and middle Muddy fans, where incised channels were relatively stable and occupied a small proportion of the deposit surface, the influence of secondary disturbance on vegetation development was comparatively minor. This was also the case at Cedar Flats, where the entire deposit is located above floodplain level.

Between 5 and 11 years after disturbance, the environment along the lower Muddy floodplain changed from one characterized by chronic channel instability to one characterized by increased stability and accelerated vegetation development.

This period of comparative stability continued until 1996 when a more than 100-year flood scoured large tracts of the floodplain and reset vegetation development.

Data on the relationship between the surface stability and vegetation development along the lower Muddy floodplain are limited to the 1981 to 1984 and the 1991 sample periods. From 1984 to 1991 and after 1991, information is derived from intermittent observations of species composition and cover in conjunction with data on channel adjustments obtained from repetitive surveys of stream cross sections. Our ability to draw conclusions about long-term relationships between geomorphic processes and vegetation development is hindered by our limited data. Existing data notably exclude relationships between geomorphic and vegetation responses to the 1996 flood, which removed much of the post-1984 vegetation in this reach. In active fluvial systems such as the lower Muddy River, infrequent, large-magnitude events exert substantial influence on the trajectories and rates of vegetation development and can reset decades of vegetation development following a previous catastrophic disturbance.

6.5 Summary

A voluminous mudflow triggered by the May 18, 1980 eruption of Mount St. Helens caused catastrophic, but not unprecedented, changes in the geomorphology and ecology of the Muddy River drainage. Results from four studies show that the geomorphic and ecological responses to the mudflow varied with disturbance severity longitudinally with distance from the volcano and laterally with distance from the channel center. Longitudinally, disturbance severity was related primarily to abrasion, impact force, and vegetation removal by the mudflow and not by burial depth. The severity of abrasion and vegetation removal decreased from the highest elevation site (upper Muddy fan) to the lowest elevation site (Cedar Flats terrace), but deposit thickness generally increased.

Premudflow topography, postmudflow channel adjustments, biotic interactions and legacies, and the occurrence of infrequent, large floods played important but variable roles in vegetation development at the four study sites. Premudflow topography affected mudflow emplacement and provided localized safe sites for plant survival in areas inundated by mudflow. For example, on the upper Muddy fan, roots and remnant plant parts survived on two small hills that escaped inundation (see Figure 6.1), and on Cedar Flats terrace, an abandoned channel affected mudflow emplacement and deposit thickness. Physical processes, however, largely controlled vegetation recovery and development during the first 4 years after mudflow emplacement. Channel adjustments, driven by transiently enhanced runoff and extraordinary sediment transport, influenced vegetation recovery in both the basin headwaters and along the main stem of the channel but especially along the lower Muddy floodplain, where lateral channel migration was most active.

Elevation, distances to seed sources, biotic interactions, and biological legacies further influenced trajectories and rates of postmudflow vegetation development. On the upper Muddy fan, seedling establishment was delayed by a short growing season and a drought-prone substrate. By comparison, more favorable environmental conditions existed at the three lower-elevation sites. On stable substrates, distance to seed source can be locally important, particularly on the comparatively wide middle Muddy fan. The general patchiness of vegetation establishment on the upper Muddy fan and the comparatively narrow width and frequent secondary disturbance of the lower Muddy floodplain mute any correlations between vegetation development and distances to seed sources at those sites. Biotic interactions, such as light competition at Cedar Flats and amelioration of conditions for seedling establishment by early-colonizing plants on the upper Muddy fan, illustrate the importance of community processes leading to postdisturbance recovery across a wide range of environmental conditions. Biological legacies were important catalysts that accelerated vegetation recovery on Cedar Flats terrace, where the primary disturbance was caused by burial rather than abrasion or removal of vegetation. Because of the biological legacies, vegetation recovery among the four sites has been most rapid at Cedar Flats.

In volcanically disturbed areas, riparian plant communities are composed of a variety of successional stages growing on surfaces that are periodically reshaped by flooding and other, infrequent, large-scale disturbances. Our study shows that the spatial responses of vegetation to mudflow disturbance in the Muddy River drainage basin are consistent with those noted for disturbed riparian systems in general (Tabacchi et al. 1990; Baker and Walford 1995; Bendix 1997). However, fully understanding the interplay between geomorphic and ecological processes following severe disturbance along tens of kilometers of river corridor requires a longer-term, more sustained, and more integrated approach than was incorporated into this study. With the advent of digital remote-sensing applications for channel and vegetation surveys and enhanced geographic information system (GIS) data-processing capability, such long-term, integrated geomorphic and vegetation studies may become more feasible in the future.

Acknowledgments. Little of our long-term vegetation research along the Muddy River would have been possible without the capable assistance of many dedicated volunteers. We are indebted to each of the following individuals for their able assistance: D. Fidel, L. Krakowiak, E. Sergienko, A. and S. Frenkel, C. Hessel, D. Dulken, M. Hyde, M. Huso, K. Lillquist, W. Petty, E. Edinger, H. Tobin, J. and M. VerHoef, Y. Borisch, C. Young, P. Fashing, J. Brown, J. Holmes, S. Heacock, W. Martin, S. Leombruno, S. Spon, S. Lundstrom, M. Kington, R. Jones,

J. Westman, D. Michola, M. Bailey, G. Busch, J. Hogan, N. Fortunato, J. Thompson, K. Halligan, C. Remmerde, S. Bondi, D. Jacobs, T. Loring, C. Antieau, J. Miesel, S. Franklet, K. Hibler, B. Owen, S. Campbell, J. Deyo, M. Lafrenz, and K. Pohl. We thank Charlie Crisafulli, Fred Swanson, Virginia Dale, and several anonymous reviewers for their insightful suggestions and editorial assistance. Support for this research was provided by National Science Foundation grants DEB 8109906 and BSR 8407213, the USDA Forest Service's Mount St. Helens National Volcanic Monument and Pacific Northwest Research Station, the U.S. Geological Survey, and Portland State University.

7
Proximity, Microsites, and Biotic Interactions During Early Succession

Roger del Moral, David M. Wood, and Jonathan H. Titus

7.1 Introduction

Our studies of succession on mudflows and pumice surfaces at Mount St. Helens support the view that plant succession is determined as much by chance and landscape context as by the characteristics of the site itself. Early primary succession is dominated by the probabilistic assembly of species, not by repeatable deterministic mechanisms. Before most plant immigrants can establish, some physical amelioration in the form of nutrient inputs or the creation of microsites may occur. As vegetation matures, there is a shift from amelioration to inhibition (Wilson 1999), but the magnitude of this shift varies in space and time. Species-establishment order is not preordained as stated by classic succession models (Clements 1916; Eriksson and Eriksson 1998). Life-history traits influence both arrival probability and establishment success, and the best dispersers are usually less adept at establishment. Therefore, interactions between site amelioration and proximity to colonists affect the arrival sequence and initial biodiversity. Unique disturbance events combine with usually low colonization probabilities to produce different species assemblages after each disturbance at a site. Early in primary succession, individuals just accumulate. However, over time, interactions begin that cause species to be replaced. Here we describe how a few struggling colonists slowly developed into pioneer communities (see Tsuyuzaki et al. 1997) and suggest how these communities may develop further.

7.1.1 Background

Until 1983, we focused on sites that had some survivors, for example, tephra-impacted and scoured sites at Butte Camp and Pine Creek (del Moral 1983, 1998). Descriptive efforts were gradually supplemented with experiments (Wood and del Moral 1987; Wood and Morris 1990; del Moral 1993; del Moral and Wood 1993a,b; Tsuyuzaki and Titus 1996; Titus and del Moral 1998b) as primary succession became our focus. Our first studies of primary succession on Mount St. Helens documented plant establishment on mudflows at Butte Camp. Subsequently, we focused on the Pumice Plain to explore surface heterogeneity (Wood 1987; Titus and del Moral 1998a; Tsuyuzaki et al. 1997), spatial patterns (Wood and del Moral 1988; del Moral 1993, 1998, 1999a; del Moral and Jones 2002), wetlands (Titus et al. 1999), and system predictability after disturbance (del Moral 1999b). This chapter provides an overview of early vegetation development on mudflow and pumice surfaces after 23 growing seasons at Mount St. Helens. These studies have modified and illuminated our understanding of primary succession. [See Walker and del Moral (2003) for a broad discussion of primary succession.]

Our view of primary succession is summarized in Figure 7.1. This perspective can be explained by considering vegetation life histories and strategies in isolated, barren habitats that were common immediately north of Mount St. Helens. Isolation from vegetation that survived the worst of the eruption's effects implies that most immigrating species were those with able wind dispersal. Mudflows that were near habitats with limited disturbance received many stress-tolerant species with poor dispersal in addition to the wind-dispersed species. Thus, the degree of isolation affected the types of species found in the first wave of colonists. The first successful immigrants established because of physical amelioration of the substrate and the presence of especially favorable microsites [*safe sites* (Harper 1977; del Moral and Wood 1988a)]. At first, there were few safe sites, but physical processes such as rill formation, rock fracturing, and freeze-thawing created more. Colonists eventually produced seeds, so local dispersal became possible. As more species established and populations became denser, biological effects created other types of safe sites, modified existing ones, or caused them to disappear entirely. Biological amelioration (facilitation) permits other species to invade the primary-successional landscapes, for example, in the shade or in litter. Established individuals can grow more robust and reproduce because of improved substrate conditions (fertility and water-holding capacity) or decreased exposure. In the future, we expect some species to fail because they cannot reproduce in the emerging environment, whereas others will be eliminated by competition (Aarssen and Epp 1990; del Moral and

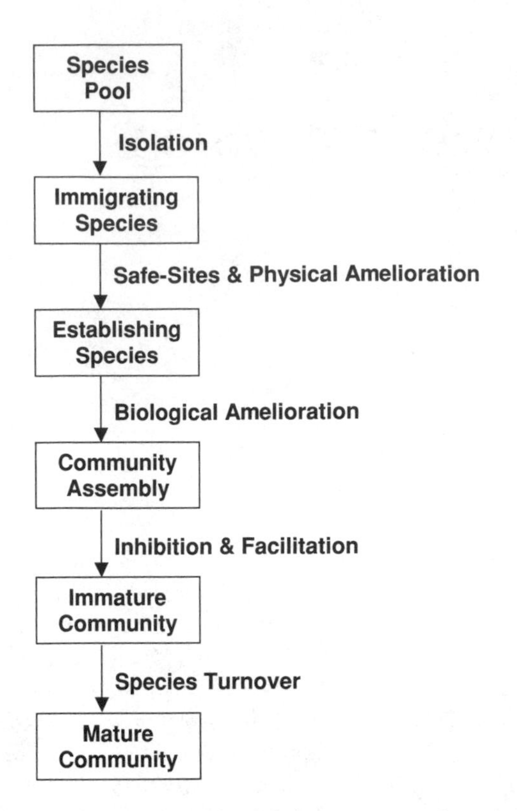

FIGURE 7.1. General model of primary succession at Mount St. Helens. Isolation is a sieve that permits only some species to reach a site. Of these immigrants, only those that find a particularly favorable safe site can establish; physical amelioration gradually improves the probability of seedling success, widens the spectrum of species able to establish, and improves the production of established plants; biological amelioration occurs as the biota modifies microhabitats; and a community of species gradually assembles. During assembly, individuals may inhibit others through competition or they may facilitate the success of other plants, thus leading to a dynamically changing community. Species turnover leads to a mature community that may have little in common with the initial vegetation on the site.

Grishin 1999). The net result is species *turnover*, one way to recognize succession. In our studies, we have observed little turnover, although shifts in the relative abundance of many species have occurred. Thus, as communities assemble, biotic interactions intensify, but only species well adapted to the new conditions thrive (Callaway and Walker 1997; Weiher and Keddy 1995). Notably, species that establish early by chance often persist even though they are not competitively superior. They can exclude seedlings of better-adapted species by control of the "space" resource [a *priority effect* (Drake 1991; Malanson and Butler 1991)]. For example, on wetland margins, where well-developed primary vegetation exists, upland species such as pearly everlasting (*Anaphalis margaritacea*) appear to exclude wetland species by virtue of the prior establishment. It will eventually be excluded, in all likelihood, only when tall shrubs dominate these margins. Thus, on mudflows and pumice surfaces at Mount St. Helens, as in many other primary-successional environments, vegetation heterogeneity was initiated by chance and may persist (Glenn-Lewin 1980; Mathews 1992; Savage et al. 2000).

7.1.2 Questions

Several questions sharpened the focus of our studies. Landscape ecologists suggest that the matrix within which a biota develops is crucial to early species accumulation (Kochy and Rydin 1997; Söderström et al. 2001). We first asked: How did isolation from propagule sources affect seed rain and seed availability and thereby the rate of vegetation development? Most early recruits did not flower, so further population growth (as distinguished from vegetative expansion) depended on continued long-distance seed dispersal from other populations. When did seedling recruitment switch from long-distance colonists to seedlings recruited from locally produced seeds?

Vegetation refugia on other volcanoes, such as *kipukas* (Hawaii) or *dagale* (Sicily) that are outcrops isolated by lava flows, can accelerate primary succession by providing adjacent propagule sources adapted to harsh environments. We were interested to determine if surviving vegetation on Mount St. Helens accelerated vegetation development and, if so, what were the mechanisms and extent of these effects?

We asked if the initial effects of chance colonization and of early arrival persist or if strong links between environmental factors and species composition were forged to create similar vegetation over space. We investigated changing statistical correlations between species composition and environmental factors in several habitats through time.

Most of the world is experiencing dramatic biological invasions, so recent disturbances have occurred in novel biological settings (Magnússon et al. 2001). Consequently, we asked if nonnative species could affect the trajectory of early primary succession to create species assemblages never previously observed.

7.1.3 Locations

Our main study sites focused on primary succession are on mudflows on the southwest and east flanks and on the pumice surfaces on the north side of Mount St. Helens (Table 7.1; Figure 7.2). Sites differed in their degree of isolation from potential sources of colonists. At Butte Camp, on the south side of the volcano, meadows and forests recovered quickly from thin tephra deposits (10 to 20 cm thick). However, several mudflows were deposited below the tree line when rapidly melting ice transported a jumble of rocks and mud that lacked any soil or seed bank. A large mudflow on the Muddy River was also studied. Mudflows are usually next to intact vegetation and, therefore, normally have a low degree of isolation. The north face of the cone collapsed spawning a directed blast and searing pyroclastic flows (see Chapter 3, this volume) and forming deep deposits of pulverized materials that have since

TABLE 7.1. Study sites used for studies of primary succession.

Site	Disturbance type	Elevation range (m)	Isolation	Type of study	Sampling dates
Mudflow 1	Fine Pumice	1380	Very low	Monitoring: plots	1980–
	Mudflow	1415–	Low	Monitoring: plots	2002
	Mudflow	1430	Low	Monitoring: grid	1982–2002
	Mudflow	1415–1430	Low	Dispersal: seed traps	1987–2001
		1425			1989–1990
Mudflow 2	Mudflow	1430–1460	Low	Monitoring: plots	1982–2002
	Mudflow	1430–1460	Low	Monitoring: grid	1987–2002
	Mudflow	1430–1460	Low	Dispersal: seed traps	1989–1990
Muddy River	Mudflow	790–1140	Low	Survey: convergence	1996
Pumice Plain	Pyroclastic flow	1100–1180	Moderate	Monitoring: grid	1986–1999
	Pyroclastic flow	950–1500	Moderate	Survey: habitats	1993
	Pyroclastic flow	1125	Moderate	Dispersal: seed traps	1982–1986
	Pyroclastic flow	1100	Moderate	Dispersal: seed traps	1989–1990
	Pyroclastic flow	1095	Moderate	Dispersal: seed traps	1989–1990
	Wetlands	950–1350	Moderate	Wetland surveys	1993 & 1999
Eastern Pumice Plain	Coarse pumice	1200	High	Monitoring: grid	1989–2002
				Monitoring: plots	1989–2002
	Coarse pumice	1200–1320	High	Mycorrhizae	1991–1995
	Refugia	1100–1525	High	Landscape effects: relicts	1997–1999
	Depressions	1280–1320	High	Monitoring: similarity	1992–1994; 1997–1998
Studebaker Ridge	Blast on lava: low	1050–1250	High	Monitoring: plots	1984–2002
	Blast on lava: high	1255–1450			1989–2002
Plains of Abraham	Blast, mudflow	1320–1360	Very high	Dispersal: seed traps	1989–1990
				Permanent grid	1988–2001
				Monitoring: plots	1995–2002

Location information is for the center of the study referenced. Sampling date ranges are annual. See Figure 7.2 for map of these locations.

been eroded (Wood and del Moral 1988). This area, termed the Pumice Plain, was substantially isolated from potential colonists. All plants were killed, except in a few refugia on steep terrain (concentrated in the eastern part of the north slope) that escaped pyroclastic flows. We continue to monitor pumice habitats north and northeast of the crater. On the eastern Pumice Plain, many sites are less exposed to physical stress and are closer to surviving vegetation found in refugia. Wetlands are also developing rapidly across the Pumice Plain. Typical sites were only moderately isolated from potential colonists, usually 1 km. Isolated from intact vegetation on the eastern Pumice Plains are depressions we call "potholes," which formed when over depressions on the debris-avalanche deposit thick pumice deposits to create a few hundred small self-contained depressions (del Moral 1999a). Wetlands and pyroclastic flows have been studied in a variety of ways since the mid-1980s (del Moral et al. 1995; Titus et al. 1999). Studebaker Ridge, on the northwest flank of the cone, received an intense blast during the early stages of the eruption that removed all plants and most soil to reveal old lava rocks. It is exposed and at a higher elevation than the Pumice Plain sites and, therefore, received a limited seed rain. East of the crater, the blast, a massive mudflow, and pumice deposits impacted the Plains of Abraham, and that area continues to be isolated from colonists by a ridge and the prevailing winds.

7.2 Methods

7.2.1 Permanent Plots

Permanent plots are located in four areas and provide the opportunity to nondestructively monitor vegetation through time (Table 7.1). Starting in 1980, these 250-m^2 circular plots (18 m in diameter) were sampled. The area of the vertical projection of the canopy of each species within a subplot is called percent cover. Percent cover was determined at the same 24 places each year with 0.25-m^2 subplots (del Moral 2000b). From these data, the total number of species (richness), mean percent cover of the plot, and other structural features were calculated (McCune and Mefford 1999). We compared changes in species richness in the same plot over time by employing repeated-measures analysis of variance with Bonferroni comparisons of the means (Analytical Software 2000).

We also sampled species richness, cover, evenness, diversity, and vegetation pattern in permanent grids formed of contiguous 10- by 10-m plots sampled with this cover-unit scale (Wood

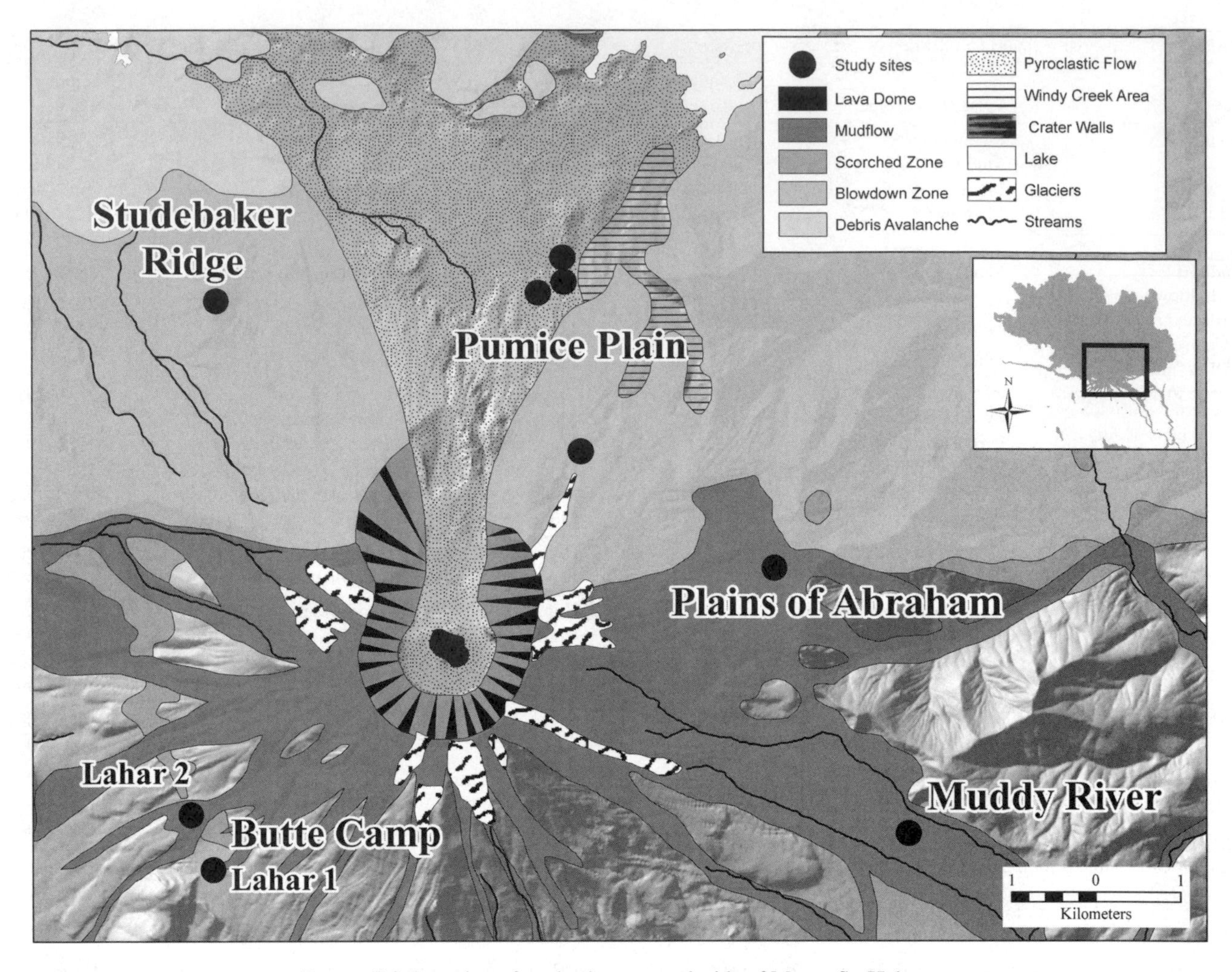

FIGURE 7.2. Location of study sites on north side of Mount St. Helens.

and del Moral 1988):

1. 1 to 5 plants
2. 6 to 20 plants
3. More than 20 plants *or* 0.25% to 0.5% cover
4. More than 0.5% to 1% cover
5. More than 1% to 2% cover
6. More than 2% to 4% cover
7. More than 4% to 8% cover
8. More than 8% to 16% cover
9. More than 16% to 32% cover
10. More than 32% cover

The index sacrifices precision for generality and provides reliable estimates of relative vegetation change. Grids document plant establishment and species expansion. The grids were established in 1986 in the pyroclastic zone (Pyroclastic, $n = 400$), on two mudflows at Butte Camp in 1987 (Mudflow 1, $n = 175$; Mudflow 2, $n = 317$), on pumice at the Plains of Abraham in 1988 (Plains of Abraham, $n = 400$), and on the eastern Pumice Plain in 1989 (Coarse Pumice, $n = 200$). Figure 7.2 shows their locations.

7.2.2 Colonization Patterns

We compared observed patterns of distribution on the Coarse Pumice Grid with the null hypothesis of random colonization using a simulation model. Input data were maps of each species distribution at 3-year intervals (with an empty grid used as the basis for predicting initial patterns) and N, the number of plots colonized between intervals. The model filled N quadrats randomly. The number of clusters (composed of contiguous plots containing the species) and the ratio of clusters to occupied plots were calculated. The simulation was repeated 100 times for each suitable species. The mean ratio and standard deviation of ratios were calculated and compared to the observed ratio with a t test (see del Moral and Jones 2002).

7.2.3 Relict Effects

The effects of relict sites, small patches of vegetation that survived the eruption within the eastern Pumice Plain region, were determined along a series of belt transects radiating from each of 37 refuges and from control plots located more than 100 m from any refuge (Fuller and del Moral 2003). Each relict site

TABLE 7.2. Community structure in permanent plots on Mount St. Helens after 20 years (1999).

Impact type	Richness [R] (species/plot)	Mean cover (%)	Evenness (H′/ ln R)
Recovered:			
Tephra ($n = 10$)	20.9	48.3	0.651
Primary succession:			
Mudflow ($n = 7$)	21.0	15.5	0.652
Blasted ridge ($n = 6$)	20.0	32.1	0.610
Coarse pumice[a] ($n = 11$)	19.3	7.1	0.777
Blast–mudflow[b] ($n = 10$)	17.0	5.0	0.796

[a] Eastern Pumice Plain.
[b] The Plains of Abraham.

was carefully searched to establish a complete species list. Then the percent cover of species found within the relicts in 1997 and 1998 were sampled by 1-m^2 quadrats until at least 90% of the species were encountered. The plant cover surrounding each relict was sampled along four transects consisting of 20 contiguous 1-m^2 quadrats each. Quadrats were oriented uphill, downhill, and along the contours in both directions from the relict.

7.3 Results

7.3.1 Patterns of Vegetation Development

Permanent plots and grids documented the development of species richness and cover after 20 years (Table 7.2). These plots include mildly impacted fine-tephra sites at Butte Camp for comparison and primary-succession sites on mudflows (Butte Camp), a blasted ridge (Studebaker Ridge), coarse tephra on the eastern Pumice Plain, and coarse tephra situated over the remains of a devastating mudflow on the Plains of Abraham. The mildly impacted tephra plots at Butte Camp returned to preeruption conditions of 48% cover within 5 years (del Moral 2000b). Richness fluctuated at about 20 species per plot since 1984 but has declined slightly since 2000. During this time, subalpine fir (*Abies lasiocarpa*), lodgepole pine (*Pinus contorta*), hawkweed (*Hieracium* spp.), orange agoseris (*Agoseris aurantiaca*), and mosses were among those accumulated. A few uncommon species disappeared, and other species were sporadic.

Richness of plots on the adjacent mudflows has approached that of tephra sites. However, species composition of mudflows differs from that of tephra sites, reflecting the difference between mature meadow vegetation and early primary succession. We expect richness to remain stable or to decline on mudflows. Pioneer species are beginning to be lost, and conifer density increases may exclude other species. Richness in all primary plots appears to be converging to a level of about 18 species per plot.

Species richness and percent cover in permanent plots (see Figure 7.2) are shown in Figure 7.3 for two mudflows near Butte Camp, two sites on (upper and lower) Studebaker Ridge, the eastern Pumice Plain, and the Plains of Abraham. The Butte Camp tephra plots are shown for comparison. Figure 7.3a shows species richness. All sites showed increased species richness during the monitoring period. Statistical analyses showed that even annual increments were often significant. The richness of the mudflow plots began to increase before that of the other plots, probably because of proximity to available seed sources (cf. Wood and del Moral 2000). Richness gradients extending from intact vegetation were pronounced for several years of monitoring. Mudflow 1 appears to be declining because of the exclusion of pioneer species by conifers, and Mudflow 2 also may be in decline. On Studebaker Ridge, richness continued to increase, but was reduced in the lower-elevation plots when prairie lupine (*Lupinus lepidus*) achieved strong dominance during the late 1980s. The upper plots lacked vegetation for 8 years, but after 20 growing seasons, they had achieved richness similar to that of the lower-ridge plots. Many of the Pumice Plain sites are windswept, which may contribute to their low mean richness. However, the less-stressful plots had relatively high richness values. The Plains of Abraham plots are more than 1 km from surviving vegetation, but achieved richness similar to that of the other sites after 17 growing season. These plots remain open, and richness continues to increase.

Mean plant percent cover (Figure 7.3b) contrasts with species richness. Vegetation cover on fine tephra fluctuated in response to summer precipitation (del Moral and Wood 1993a), a pattern similar to that of small-mammal abundances on Mount St. Helens (MacMahon et al. 1989; Crisafulli et al., Chapter 14, this volume). Cover development began significantly later on primary-succession sites than on other sites. The mudflows were the first primary sites to develop significant vegetation, and Mudflow 1 approached cover values found on tephra in 1983. Much of this cover was caused by conifers. Cover on the lower Studebaker Ridge fluctuated in response to variations in prairie lupine, but cover was comparable to that of tephra after 23 years. Cover of other species accumulated slowly. In 2001 and 2002, lupines exploded in cover on the eastern Pumice Plain to increase cover significantly. However, cover remained less than that on the mudflows and for lower-ridge vegetation.

The grids provide both species-composition and spatial data because we can determine where and when a species originated and how it expanded. Here we only address structure (Figure 7.4a,b). On the two mudflows at Butte Camp, richness increases were similar and had not increased appreciably during the last 5 years of the study. However, Mudflow 1 experienced a dense invasion of subalpine fir and lodgepole pine that sampling in 2002 suggested may eliminate pioneer species. Mudflow 2 had less tree invasion, but pioneer species such as fireweed (*Chamerion angustifolium*) and hairy catsear (*Hypochaeris radicata*) were declining. Field surveys in

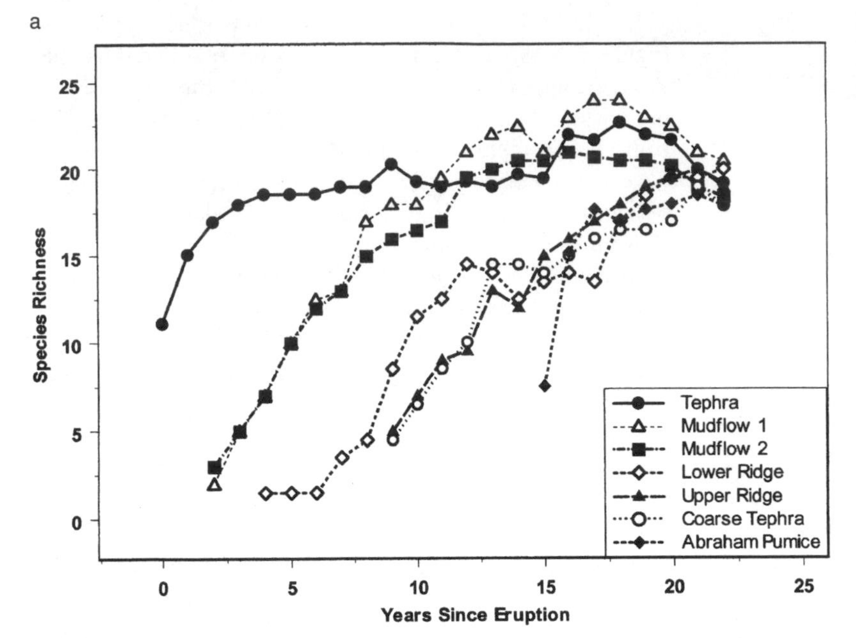

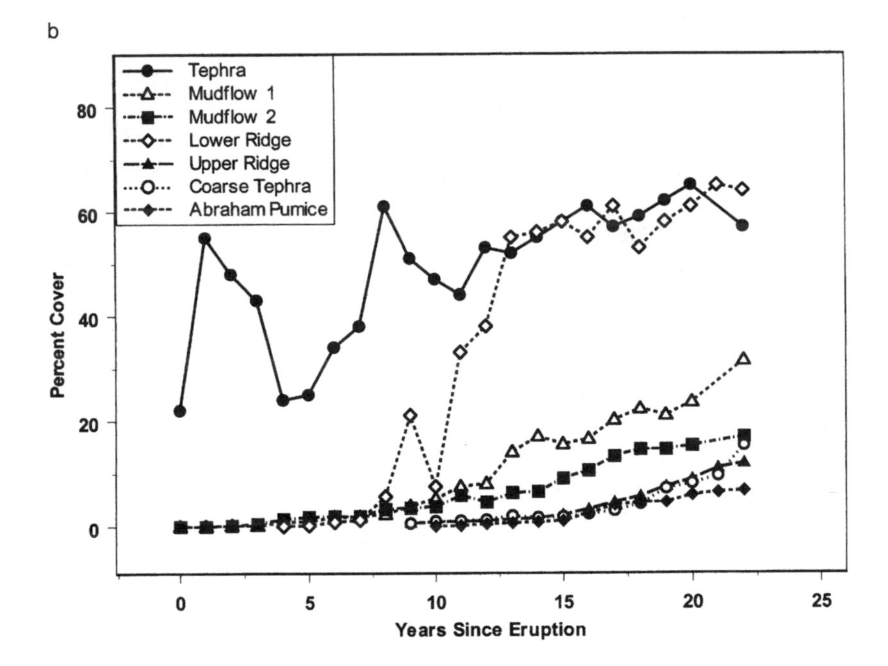

FIGURE 7.3. Structural changes in permanent plots on Mount St. Helens. (a) Richness; (b) total-cover percentage. Tephra ($n = 6$) represents recovered plots at Butte Camp and is shown for comparison to the other sites, all of which are primary-successional sites; Mudflow 1 ($n = 2$) and Mudflow 2 ($n = 5$) are near Butte Camp; Lower Ridge ($n = 4$) and Upper Ridge ($n = 4$) are on the blasted Studebaker Ridge on the northwest flank of the volcano; Coarse Pumice ($n = 11$) is on the eastern side of the Pumice Plain; and Abraham Pumice ($n = 10$) is on the eastern flank of the volcano.

1986 indicated that the eastern Coarse Pumice Grid lacked plants at that time, except in a gully. By 1989, the 10th growing season posteruption, vegetation was sufficiently developed to merit detailed sampling. The mean richness became similar to that of the mudflows, although richness may decline when prairie lupine becomes dominant in swales and other protected sites. In several more exposed parts of this grid, mosses formed mats that also may restrict seedling establishment. Mean plot richness increased substantially on the Pyroclastic Grid, although there was a slight reduction during the mid-1990s. Jumps in mean richness between other sample years were produced primarily by the expansion of existing species. Only on the Plains of Abraham do we expect further substantial increases in mean species richness. The site remains sparsely vegetated, and common genera [such as lupines (*Lupinus*), rush (*Juncus*), and pussypaws (*Cistanthe*)] are absent from many plots.

Total richness on each grid increased rapidly at first but then stabilized. No new species were encountered after 1999. Most species, and all that were dominant at the end of the century, had invaded by 1990. The Coarse Pumice Grid is more than 1 km from intact vegetation, as is the Plains of Abraham Grid (see Figure 7.2). After 10 years, these grids had received only 60% (30 of 50 species) and 70% (33 of 47), respectively, of their total after 22 years since disturbance. In contrast, after the same 10 years, Mudflow 1, within 100 m of intact vegetation, had received 84% (41 of 50) of its 20-year total; and Mudflow 2, which is 0.1 to 0.3 km from intact vegetation, had

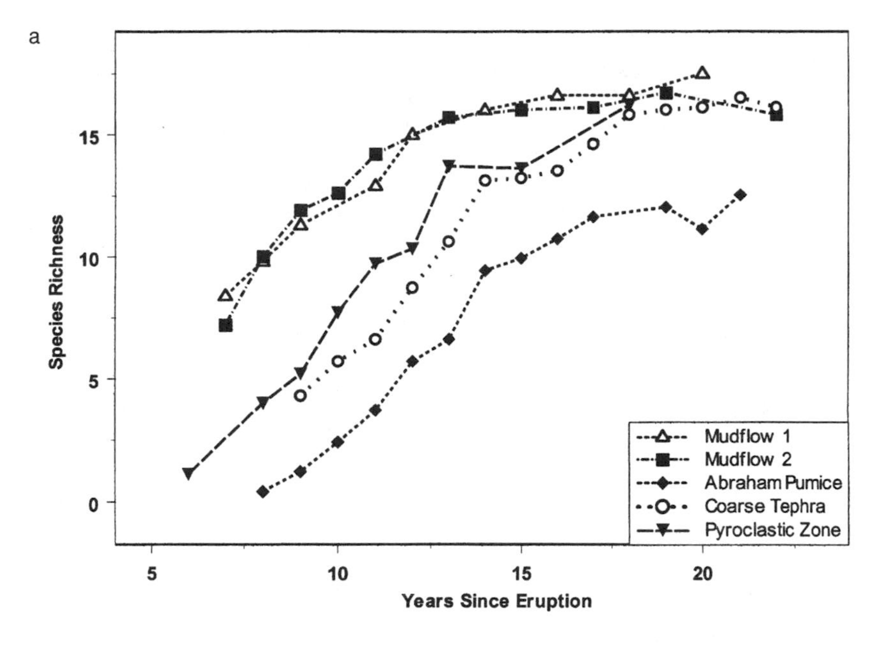

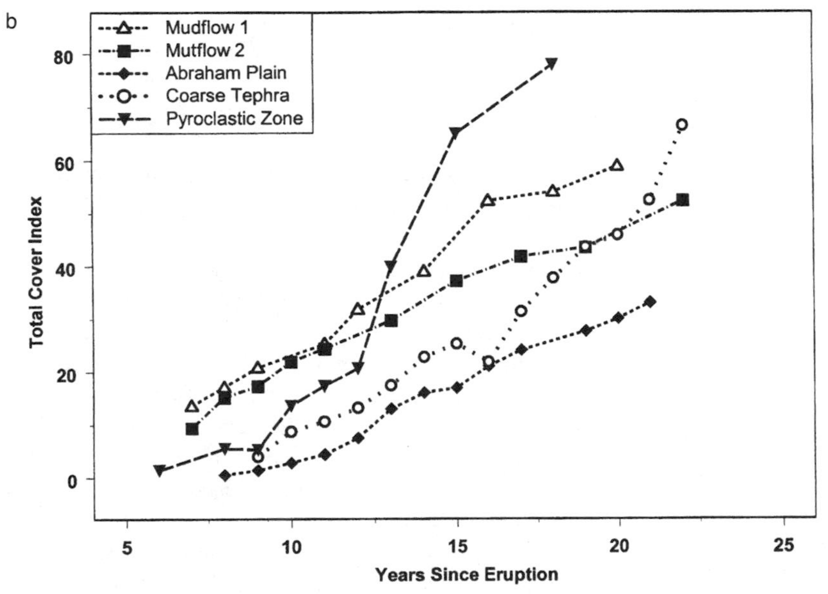

FIGURE 7.4. Structural changes on grids on Mount St. Helens. (a) Richness; (b) total-cover index. Mudflow 1 ($n = 175$ plots) and Mudflow 2 (317 plots) are near Butte Camp; Abraham Pumice ($n = 400$) is east of the mountain; Coarse Pumice ($n = 200$) is on the eastern edge of the Pumice Plain; and Pyroclastic ($n = 400$) is on the pyroclastic flow north of the crater.

76% (35 of 49) of its ultimate total. These data tend to indicate that the greater the isolation, the lower the percentage of species encountered within the first 10 years. The Pyroclastic Flow Grid is an exception in that it received 80% (58 of 72) of its species within 10 years despite being isolated to the same degree as the Plains of Abraham Grid. This difference may result from its more western location, where it more readily received input from logged sites. The abundance of wind-dispersed exotic species in this sample supports this suggestion. Thus, while dispersal was a major limiting factor with respect to the rate of primary succession (see Wood and del Moral 2000), most species entered the system within 12 years of the eruption and subsequently expanded from these centers of establishment.

Grid cover continued to increase in all cases through 2002 (see Figure 7.4b). Differences between successive samples greater than 4 cover units were significant. Because species richness was nearly constant in most cases since 1996, this change reflects the increase in cover of many taxa. On Mudflow 1, cover increases were dominated by firs (*Abies*) and pines (*Pinus*), while on Mudflow 2 alpine buckwheat (*Eriogonum pyrolifolium*) and lupines were major increasers. On the eastern Pumice Plain, lupines, rock mosses (*Racomitrium*), hair-cap mosses (*Polytrichum*), and sedges (*Carex* spp.) all increased significantly, while on the Pyroclastic Grid, cover increased exponentially during the mid-1990s because of the expansion of the lupine and willow populations. On the Plains of Abraham Grid, cover increased in most taxa, although early dominants (e.g., pearly everlasting and fireweed) declined during the last several years.

Species proportions changed dramatically during our studies. Most notable was a reduction in initial colonizers on the

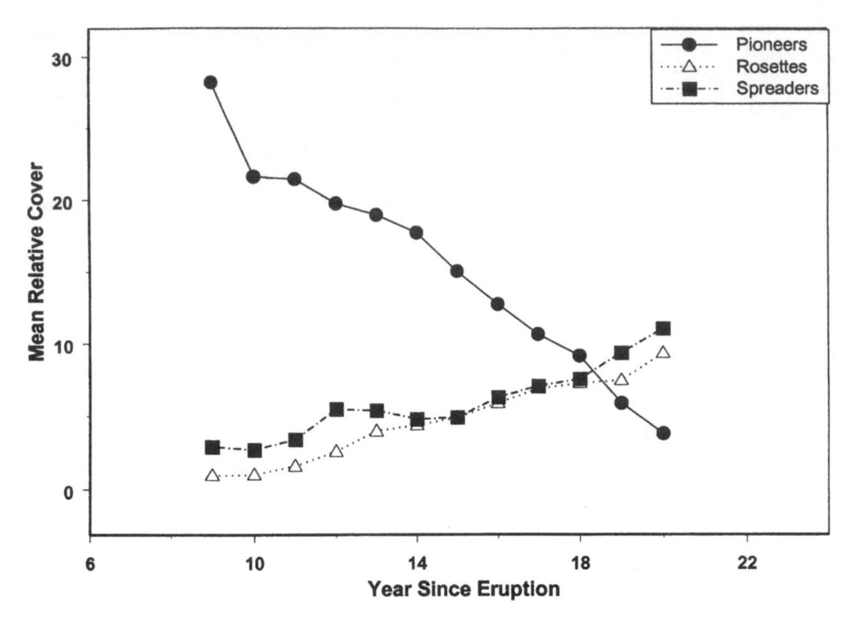

FIGURE 7.5. Relative cover of three types of species from the Plains of Abraham: the most common pioneers (pearly everlasting, fireweed, and hairy cats-ear); clumping rosette species that reproduce abundantly and become locally dominant (umbellate pussypaws, Parry's rush, and russethair saxifrage); and persistent, spreading species that were initially rare (seashore bentgrass, partridgefoot, and Cardwell's penstemon).

more-isolated grids, combined with an expansion of spreading and prolific species. On the Pyroclastic Grid, grasses expanded from trace in 1986 to 12.5% of the cover by 1999, whereas pearly everlasting declined from 22% to 4% by 2000. We suggest that grass expansion results because evapotranspiration is greater from the graminoid species because of their more expansive root systems, which results in a reduction of water for forb species. In addition, slide alder (*Alnus viridis* spp. *sinuata*) has expanded since 2000 in the pyroclastic zone, suggesting that it may dominate in less stressful habitats in this area. Figure 7.5 shows the shift from pioneers to dominance by later-colonizing species on the Plains of Abraham. Each curve is the mean relative cover of three species sampled between 1988 and 2000. (Relative cover is the proportion of the total represented by a species. Mean relative cover is the average relative cover of the three species in each category.) Pioneers (pearly everlasting, fireweed, and hairy cats-ear) were a declining proportion of the total, while less well dispersed rosette species [umbellate pussypaws (*Cistanthe umbellata*), Parry's rush (*Juncus parryi*), and russethair saxifrage (*Saxifraga ferruginea*)] and spreading species [seashore bentgrass (*Agrostis pallens*), partridgefoot (*Luetkea pectinata*), and Cardwell's penstemon (*Penstemon cardwellii*)] increased consistently. In 2000, the six rosette and spreading species accounted for more than 60% of the cover, while the three pioneers accounted for less than 12%.

7.3.2 Isolation

Isolation from sources of colonists can affect the species composition of a site because of differential seed rain. We noted repeatedly that similar sites have different species composition and that the species present differ in their dispersal mechanisms. Sites near intact or partially recovered vegetation are dominated by species with large seeds and poor dispersal, while those in isolated sites were initially dominated by species with good dispersal (see Fuller and del Moral 2003). Next, we examine isolation from several perspectives.

7.3.2.1 Seed Rain

Even distances as short as 100 m can restrict the species pool to wind-dispersed invaders soon after site creation. Therefore, isolated sites initially will be both depauperate and sparse. All our long-term data indicate that species richness increases more rapidly than cover. Eventually, less-adept species do arrive, and occasionally one may establish before wind-dispersed species do (e.g., prairie lupine on pyroclastic materials by 1981). Isolation also creates a stochastic effect that has been poorly appreciated. The chance that *any* seed would reach a particular, favorable, isolated microsite is very low; thus, the chance that two adjacent sites would receive the same seed rain is even lower. Therefore, species composition in two different isolated sites may initially be quite different and bear little relationship to the environment (del Moral 1993). Much unexplained variation on the landscape might have its origin in stochastic establishment in isolated habitats.

We sampled the seed rain when there was little onsite seed production (1982 to 1986) and when vegetation had recovered slightly (1989 to 1990; Table 7.3). Except where traps were located immediately adjacent to vegetation, seed densities were low, and variation among traps was high. The seed rain during the early years was dominated by "parachutists," species with excellent wind dispersal. Species with poor dispersal were rarely trapped, even if seeds were produced within 5 m (Wood and del Moral 2000).

7.3.2.2 Dispersal Ability

We investigated the nature of establishing species under different degrees of isolation. Grids were analyzed to determine the distribution of species grouped into five degrees of dispersal ability (from poor, with no obvious dispersal mechanism, to

TABLE 7.3. Seed rain in distinct habitats and years expressed as the rank of each species in the sample.

		Habitats						
		1982–1986			1989–1990			
Species	Dispersal	PZ 1	PZ 2	WS	MUD 1	MUD 2	PA 1	PA 2
Fireweed	Excellent	1	5	3	2	3	1	2
Pearly everlasting	Excellent	2	3	2			2	3
Fringed willowherb[a]	Excellent	3	1	1			3	5
Hairy cats-ear	Excellent	4	4	4	3	7	4	1
White flowered hawkweed	Moderate	5	6	5	6	6	5	4
Woodland groundsel[b]	Excellent	6				8		
Prairie lupine	Modest	7	2	6	1		6	
Canada thistle[c]	Excellent		7			9		
Umbellate pussypaws	Modest		8		4	2		
Newberry's knotweed	Moderate				5	1		
Cascade aster[d]	Moderate				8	5		
Slender hawkweed[e]	Moderate				9	4		
Few-fruited lomatium[f]	Poor				10			

Dispersal ability from del Moral (1998). Footnotes provide scientific names of species not mentioned in text.
PZ 1, pyroclastic zone; PZ 2, pyroclastic zone with high-density vegetation dominated by lupines; WS, pyroclastic-zone wetland area dominated by *Salix*; MUD, mudflow; PA, Plains of Abraham.
[a]*Epilobium ciliatum.*
[b]*Senecio sylvaticus.*
[c]*Cirsium arvense.*
[d]*Aster ledophyllus.*
[e]*Hieracium gracile.*
[f]*Lomatium martindalei.*
Source: Derived from Wood and del Moral (2000).

excellent wind dispersal). Each type was common at all sites, but there was a shift from poor to excellent dispersal with increasing site isolation (Table 7.4). This simple observation demonstrates the importance of the landscape context and emphasizes that early primary succession will be affected by the available colonist pool as well as by the physical characteristics of the site.

Wood and del Moral (1987) demonstrated that, although wind-dispersed species were most likely to reach a site, large-seeded, poorly dispersed species were more likely to establish

TABLE 7.4. Relative percent cover in each of five dispersal categories on four grids at Mount St. Helens.

Dispersal category	Adjacent mudflow	Isolated mudflow	Isolated pumice	Very isolated pumice
Poor	6.8	6.8	5.1	8.8
Modest	**58.5**	**27.2**	19.5	20.7
Moderate	7.4	18.0	14.8	6.7
Good	16.0	**32.1**	**23.9**	13.8
Excellent	11.5	16.6	**33.7**	**50.0**

Largest values in bold.
Source: From del Moral (1998); used by permission.

if they reached the site. Larger seeds also tend to produce more vegetatively spreading plants, so we predicted that these species would eventually come to dominate a site. We analyzed the Coarse Pumice Grid data from 1989 to 1999 (del Moral and Jones 2002) to test this prediction. Early dominants were those with good dispersal. In 1989, the 8 good dispersers accounted for more than 65% of the cover, while the 21 poor dispersers accounted for 31% and the 4 moderate dispersers accounted for 4%. Relative cover of poor dispersers increased to 68% by 1999 whereas that of good dispersers declined to 25%. Absolute cover of poor dispersers increased from a mean of 2 to a mean of 27 cover units; moderate dispersers increased from 0.2 to 3.3 units; and good dispersers increased from only 3.8 to 9.8 cover units.

7.3.2.3 Floristic Effects

Isolation affects which dispersal types reach a site, so that the vegetation of similar sites often differs floristically. Samples on mudflows at Butte Camp (adjacent to forests and meadows) and on Coarse Pumice, Studebaker Ridge, and the Plains of Abraham documented the dispersal processes. We compared species composition among grids and among permanent plots with the Spearman rank-order test. Mudflow grids were strongly correlated with each other ($r = 0.81$) but had low to negative correlations with the other grids. The Plains of Abraham Grid had low correlations with mudflow grids ($r = 0.37$; 0.22) and with the Coarse Pumice Grid ($r = 0.23$). The Coarse Pumice Grid was negatively correlated with the mudflow grids, indicating that the flora was drawn from different populations. Permanent plots on Studebaker Ridge were strongly correlated with plots on Coarse Pumice and moderately were correlated with those on mudflows.

Isolation also affects the rate of vegetation development. We demonstrated this effect using plant cover as a development index and comparing plots in comparable habitats across elevational gradients at several locations in 2000. At Studebaker Ridge, cover was negatively correlated with elevation, even though substrates, slopes, and aspects were similar. Cover at 1220 m was more than 70%. At 1285 m, cover dropped to 4%, whereas at 1340 m it was less than 2%. Above 1450 m, vegetation was sparse, and cover was less than 1%. Similarity also declined with distance between plots. Similarity of composition declined with distance between samples. Mean similarity of samples that are within 50 m, 200 m, 500 m, and 1 km declined from 54% to 42% to 31% to 26%, respectively. Thus, isolation can affect both species composition by filtering potential species and cover by reducing the frequency of colonization events by any species.

7.3.3 Seedling-Recruitment Patterns

For some species, we could determine when species recruitment shifted from dominance by seeds from long-distance dispersal to dominance by seeds from resident plants. This change is important because cover increases dramatically when seeds are produced locally, although the pace of species turnover may slow because recruitment is dominated by resident species that may inhibit invaders. The shift from donor-maintained species composition to locally controlled species composition was inferred by Wood and del Moral (1988) and elaborated by del Moral and Wood (1993a). del Moral and Jones (2002) modeled the invasion patterns and found that, when invasion is dominated by long-distance dispersal, the spatial pattern of invaders is random. If local plants produced seedlings, aggregation should occur. This simulation provided a conservative test of whether the observed pattern was consistent with a random invasion rather than expansion from local seed sources. The observed ratios of infrequent species were random, suggesting that rare species continue to invade from a distance.

Twenty-one species were analyzed on the Coarse Pumice Grid. Nearly all species with intermediate frequencies were clustered more than would be expected from long-distance dispersal. Species with random patterns included those for which only long-distance dispersal could provide seeds [e.g., firs and Douglas-fir (*Pseudotsuga menziesii*)]. However, willow species, which had not produced seeds on the site by 2001, were more clustered than random. Seashore bentgrass and Cardwell's penstemon, two common species of moderate dispersal ability, were not clustered more than would be expected of a random pattern, but species with good to moderate dispersal did demonstrate significant clustering. These species included hair bentgrass (*Agrostis scabra*), hawkweed, hairy cats-ear, Parry's rush, and russethair saxifrage. Three species of sedges, prairie lupine, small-flowered wood-rush (*Luzula parviflora*), and Sandberg's bluegrass (*Poa secunda* J. Presl.) also displayed clustering that suggested founder effects. We have observed similar patterns developing on the other grids, suggesting that the shift from donor to local control of species demographics is widespread, but occurs for each species at a unique rate. The process also occurs at different times for different species, depending on site isolation and the availability of safe sites for the species in question.

7.3.4 Microsites

7.3.4.1 Observations

On mudflows and pumice surfaces at Mount St. Helens, favorable microsites were crucial to early plant colonization in many cases (Wood and Morris 1990; del Moral and Bliss 1993; del Moral and Wood 1993a; Tsuyuzaki and Titus 1996). The phenomenon of initial establishment being localized in especially favorable sites is widespread (Oner and Oflas 1977; Tsuyuzaki 1989; Walker and del Moral 2001, 2003). Safe sites differ in environmental characteristics and provide relief from stress. For example, rills may be wetter than ridges because of longer snow retention, and near-rock microsites offer shade. The specific microclimate of a microsite within which a propagule is trapped may be critical for plant germination and growth. However, some microsites may permit dense colonization, leading to intense competition and low survival (see Lamont et al. 1993; Titus and del Moral 1998a), although greater biomass of a few individuals may result.

On a fine scale, initial colonization patterns were related to safe-site distributions (del Moral and Wood 1993a). Pioneers establish nonrandomly on pumice because favorable safe sites are strongly preferred by colonists. The stress of dry, hot, posteruption surfaces was emphasized by the observations that, although most seedlings were associated with safe sites, most safe sites lacked seedlings (del Moral 1993). These patterns have decayed on coarse tephra as plants expand from initial loci and as amelioration proceeds. Amelioration gradually improves all sites and blurs distinctions among microsites.

Eleven years after the eruption, we studied the distribution of seedlings with respect to rocks, rill edges, undulations, drainages, and flats on the Plains of Abraham. Pearly everlasting, fireweed, hawkweed, hairy cats-ear, seashore bentgrass, and pussypaws were all associated with rocks and negatively associated with drainages and flats. Rills and depressions supported the first three species disproportionately. These patterns became muted during the second decade as safe sites became more rare because of the breakdown of pumice and erosion within rills (del Moral 1999a) and were virtually nonexistent by 2001. Established species expand, preventing new seedlings from establishing. The habitat has ameliorated because of weathering, soil development, and continued inputs of organic matter from surrounding forests (Edwards and Sugg 1993; Sugg and Edwards 1998).

7.3.4.2 Experimental Studies

For several years, we manufactured safe sites on mudflows near Butte Camp, on the Plains of Abraham, on the eastern Pumice Plain, and in the pyroclastic zone to test the hypothesis that safe sites were indeed crucial to establishment success. Treatments were designed to mimic the effects of both abiotic and biotic amelioration. We showed that the addition of mulch, which lowers surface temperature, improves moisture, and traps seeds, produced at least 10 times more seedlings than when only rocks were provided. Creating shade, making rills, cultivating the surface, and adding nutrients all increased natural seedling recruitment (Wood and del Moral 1987).

Several aspects of plant colonization must be assessed when considering safe sites. We examined establishment and growth with native seeds sown in six microsites (flat, ridge, near rock, rill, dense vegetation, and dead lupines) on pumice (Titus and del Moral 1998a). We also studied colonization into constructed microsites. Maximum natural colonization did not occur in the same microsites as maximum establishment and growth from sown seeds. Colonization patterns also differed from year to year. This year-to-year shift in microsite colonization patterns illustrates the dynamic nature of the landscape and the important influences of climate, amelioration, and seed rain on plant establishment and community development.

Taxa found in seed traps were the major colonists on pumice (e.g., pearly everlasting, fireweed, and hairy cats-ear). These colonists also had greater biomass in dead lupine patches than in other sites, confirming that facilitation effects by lupines are delayed until most lupines have died (Morris and Wood 1989). Our studies also unexpectedly revealed that those microsites supporting the most seedlings were not always the most favorable for subsequent growth. Some sites (rills) trapped many seeds but did not support the best growth. As more vegetation developed, many seedlings colonized once-hostile sites because their seeds were trapped by previous colonists.

An overlooked aspect of sites lacking vegetation is that they are dynamic, with chronic erosion. Tsuyuzaki et al. (1997) examined erosion and tracked seedling survival on eroded pyroclastic sites. More seedlings established where eroded material accumulated and on coarse-textured surfaces, even though finer-grained surfaces had more organic matter and were moister. In addition, sites with higher cover of dead lupines, more rock and gravel substrate, and more rills had more seedlings.

Safe sites have been demonstrated to be crucial to the establishment of plants on exposed primary sites. The nature and frequency of safe sites has changed, and the competitive environment has undergone a profound shift. Future colonization will require species with different characteristics, and the next phase of succession will require species to compete effectively and to utilize biological facilitation by established individuals.

7.3.4.3 Mycorrhizae

Mycorrhizae, normally occurring as a root–fungus mutualism, are often important determinants of plant succession (Boerner et al. 1996). However, knowledge of their role during early primary succession is scant. During primary succession on volcanic substrates, it is unlikely that pioneer species would depend on mycorrhizae because nonmycotrophic and facultatively mycotrophic species could readily invade these sites. Species that require mycorrhizae cannot establish until a population of arbuscular mycorrhizal (AM) fungi is present (Allen 1991, chapter 14). Because mycorrhizal mutualism is a major investment for a plant, AM effects during early primary succession on infertile soils may be weak.

Mycorrhizal plants and AM fungal propagules (spores, hyphae, and AM-colonized roots) were common in sites with thick vegetation in the blowdown area but were extremely rare on pumice (Titus et al. 1998a). Only three AM fungal species were detected by Titus et al. (1998a). The vegetation of the Pumice Plain is composed primarily of facultatively mycotrophic species that remained nonmycorrhizal. On pumice, created microsites were inoculated with AM propagules, but these locations were no more favorable for the growth of six pioneer species than were uninoculated microsites (Titus and del Moral 1998b). There was, in fact, a trend for greater biomass in the nonmycorrhizal treatments, suggesting that AM were parasitic in this infertile environment (see Fitter 1986). Thus, the infertility of volcanic successional sites and the facultative nature of invading species preclude an early role for mycorrhizae. With substrate amelioration, increased plant density, and the invasion of species with greater AM dependency, AM may assume greater importance.

We conducted two greenhouse studies to examine the role of AM in pioneer species under three nutrient treatments and four competitive scenarios (Titus and del Moral 1998c). Nutrient treatments were either:

- Complete
- Complete but lacking phosphorus (−P)
- Tap water.

Phosphorus is the principal nutritive benefit plants receive from mycorrhizae. A negative effect from AM colonization was observed in tap water, perhaps because of a parasitic action of the AM fungi. A weak benefit from AM occurred in the −P treatment, where plants were similar in biomass to those in the complete nutrient treatment and where AM colonization levels were greater. AM did not significantly influence competitive outcomes between facultatively mycotrophic species. However, the performance of the facultatively mycotrophic hairy cats-ear in competition with the nonmycotrophic Merten's sedge (*Carex mertensii*) was significantly improved with AM compared to its growth in the absence of AM. Under field conditions at Mount St. Helens, it is unlikely that competitive dominance is affected solely by mycorrhizal associations. AM is only one of several interacting factors that, at best, only slightly alters plant-species composition.

Several conifer species, which are ectomycorrhizal, occur at low densities throughout the Pumice Plain, but rarely have they reproduced. Conifers may be limited by substrate infertility and elk browsing, but because ectomycorrhizal spores are well dispersed, the lack of mycorrhizae is not a likely source of population restriction. Conifers such as Douglas-fir, western white pine (*Pinus monticola*), and lodgepole pine grow very well and frequently produce cones on the slightly more fertile mudflows of Butte Camp and the Muddy River.

7.3.5 Relict Vegetation

Refugia (sites with surviving vegetation) are considered sources of colonists to disturbed landscapes (Cousins and Eriksson 2001). They contribute to ecosystem responses after disturbances in agricultural (Zanaboni and Lorenzoni 1989), industrial (Labus et al. 1999), and natural (Danin 1999) landscapes. They can contribute colonists after glaciation (Stehlik 2000) and lava flows.

In the eastern Pumice Plain, snow and northeast-facing slopes combined to permit survival of some species and soil. It was soon clear that refugia were vegetationally distinct from their surroundings (del Moral et al. 1995). Because refugia occur on steep slopes that are more shaded, cooler, and moister than coarse tephra, surviving species were not well adapted to colonizing coarse tephra. To assess the extent to which refugia

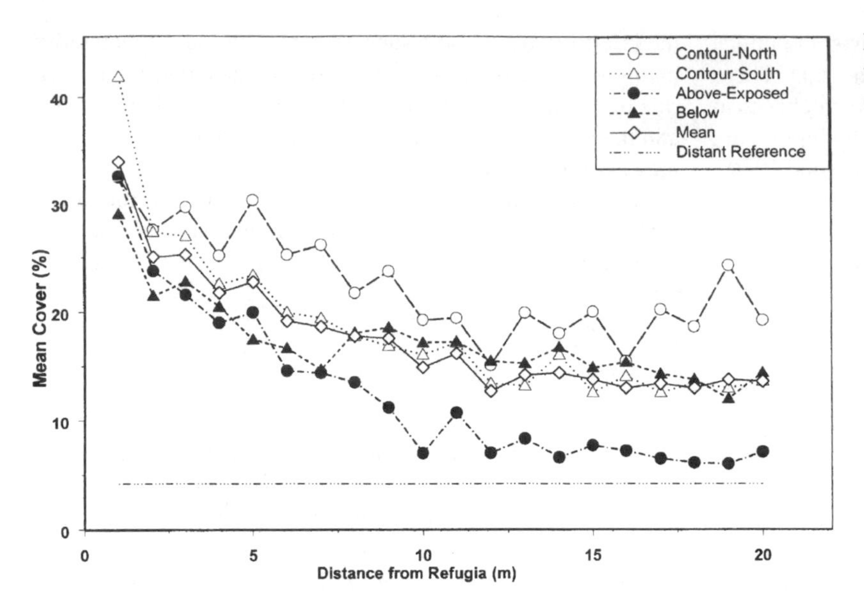

FIGURE 7.6. Changes in plant-cover percentage along transects from refugia. [From Fuller and del Moral (2003).]

and the coarse-pumice matrix are similar, we sampled each habitat type. There were 122 species (mean = 28.8 species) in refugia. Cover (mean = 82%) was dominated by 18 species, primarily shrubs and forest understory species. There were 25 species in the coarse pumice samples that were more than 100 m from any refuge. Of these, 23 reproduced abundantly in refugia, but few were refugia survivors. Nine species accounted for 90% of the coarse pumice cover (mean, 4.2%). These species included wind-dispersed taxa (e.g., pearly everlasting, fireweed, hawkweed, hairy cats-ear, and willow) and taxa that exhibit tumble dispersal (e.g., penstemon, bentgrass, and rush). Cover adjacent to a refuge was high but declined sharply with distance (Figure 7.6). At 20 m, cover above refugia was only slightly greater than the values for distant coarse tephra. Transects below refugia had the highest cover, and those above refugia had the lowest. Refuge effects will continue to expand as the habitat is ameliorated, but on coarse tephra, as of 2002, their present effects are constrained to no more than 50 m (Fuller and del Moral 2003).

Most species that survived in refugia have poor dispersal in the absence of animal vectors and a limited ability to establish in seasonally hot, dry, windswept sites. Refugia were invaded by the common wind-dispersed colonizing species, which found reduced competition and adequate soil fertility to reproduce abundantly. These species use refugia as "mega safe sites" and then contribute many more seeds to adjacent pumice than reach these sites from a long distance. These wind-dispersed species have three variants:

- *Parachute* species have plumes or other buoyancy mechanisms and routinely travel long distances.
- *Parasail* species have wings that permit some aerial movement.
- *Tumbler* species are adapted for movement along the ground.

Wind-dispersed species were disproportionately abundant in the Mount St. Helens refugia because large numbers of their seeds arrived. In 1989, they were also much more abundant on pumice, but by 1998 species with limited dispersal were more abundant than the wind-dispersed species (del Moral and Jones 2002) on pumice. The superior establishment ability and greater persistence of more poorly dispersed species (usually with larger seeds) compensates for their limited dispersal ability.

We investigated the effects of one refuge on its immediate surroundings (del Moral and Jones 2002). By analysis of the Coarse Pumice Grid, which includes a small refuge, we detected subtle effects of the refuge on local vegetation. The 100-m^2 plots were divided into those immediately adjacent to the refuge ($N = 17$) and those at least 10 m distant from the refuge ($N = 164$). The latter group was subdivided into plots that had less than 3% cover of prairie lupine ($N = 120$) because the cover of this species exploded and could swamp any relict effects. Richness, the cover index, and percent cover were calculated. Adjacent plots had significantly more species, 25% larger cover index, and 60% more cover than did distant plots with limited lupines.

Seedlings of bird-dispersed species such as black huckleberry (*Vaccinium membranaceum*) are increasingly becoming established downslope of refugia. Because refugia attract birds and rodents, vertebrates may be enhancing the rate of nutrient accumulation near refugia and therefore may accelerate the growth and dispersal of existing species. Ameliorated soils may become invasible for additional species, thus promoting succession. However, through 2002, refugia had rarely donated their own species to the surroundings. As coarse-tephra sites are ameliorated, for example, by the rapid expansion of prairie lupine, refugia should become more important.

7.3.6 Predictability and Determinism

Do environmental factors predict plant species patterns during succession? Most seres become more homogeneous through time, if only because the site fills in with vegetation. A more important question is whether the trajectory of succession (direction of change) leads to one community (convergence) or to multiple communities (divergence). The elapsed time has been insufficient to fully assess convergence on Mount St. Helens, but our work provides the groundwork for future studies. Next, we explore changes in vegetation heterogeneity to suggest how trajectories may develop.

7.3.6.1 Variation

Dlugosch and del Moral (1999) measured floristic heterogeneity along a short elevational transect on the Muddy River mudflow. The gradient mimicked a successional gradient because plant succession is slowed by the shorter growing season found at higher elevations. Species composition, measured in plots 100 m^2 in size, was similar over the gradient, but heterogeneity, measured by comparing variation within a plot, was lower at lower elevations because the vegetation sampled was more mature. This result suggests that species assembly in primary succession has a large element of chance. At higher elevations, chance effects are more prominent than in lower elevations because there have been fewer successful colonizations relative to the available area.

When first sampled, the pyroclastic-zone grids were nearly empty. Over more than 10 years, they filled in, and the variation among plots decreased. However, reduction in variation per se means neither that plots are becoming one homogeneous community nor that they are converging upon some mature vegetation type. Later dominance shifts may reverse the initial trend toward homogenization. These grids demonstrated a typical homogenization from their inception (1986) through 1993 as plot cover increased, and many species occurred in most plots. However, after 1993, some plots became dominated by grasses, others by lupines, and still others by other forbs or by low shrubs. As of 2000, heterogeneity was again substantial with cover ranging from less than 10% to nearly 100%. This result suggests that different portions of the pyroclastic zone are revealing alternative accumulation and expansion patterns that could lead to different successional trajectories when the vegetation matures. Even when (if) conifers dominate, significant compositional differences may result from events early in the process rather than from environmental differences.

The Coarse Pumice Grid provides an example of the complexity of vegetation development. Eleven samples each composed of 15 contiguous plots were formed. Detrended correspondence analysis (DCA) was used to analyze floristic changes at 3-year intervals between 1989 and 1998, plus 1999. Most variation occurred along Axis 1 as composition changed with time. The variation within each group, measured by mean similarity of the plots and by standard deviation of the DCA scores, declined significantly after 1989, indicating increasing homogeneity. However, while some groups converged as might be expected, others did not. Convergence occurred where the more-persistent or longer-lived taxa (lupines, penstemons, willow, and bentgrass species) increased at the expense of pioneer taxa (pearly everlasting, hawkweed species, fireweed, and hairy cats-ear). Two groups that were initially similar to each other developed in parallel, but each diverged from the other groups. Their divergence is based on increased dominance by roadside rock moss (*Racomitrium canescens*) and juniper hair-cap moss (*Polytrichum juniperinum*) where erosion has removed fine material.

7.3.6.2 Early Community Assembly

The shifts in plant species composition found in all our studies imply that several factors mold the vegetation populations. These factors could be environmental, such as moisture and nutrients, or they could be biological, such as competition and seed predation. We have explored the degree to which species patterns can be predicted by environmental variables under several circumstances.

In 1993, statistical correlations between measured environmental factors and species composition across the eastern Pumice Plain were very weak. The most important predictors of species composition were spatial, not moisture, fertility, or edaphic factors. This result indicated that landscape effects and dispersal limitations were more important than site factors (del Moral et al. 1995).

A unique situation occurred on the eastern Pumice Plain, where several hundred "potholes" formed soon after the first eruption. They provided an opportunity to study the relationships between species composition and the local environment. Potholes are environmentally similar to one another. They share a common internal slope, general aspect, and soil. They are of similar depth and were formed essentially simultaneously. However, their vegetation composition was very heterogeneous when first sampled, 13 years after the eruption (del Moral 1999a).

Between 1993 and 1999, percent similarity in species composition among potholes did not change significantly. We divided the potholes into seven spatial groups (to limit spatial effects) and found that, even though the vegetation cover increased substantially, species composition did not change appreciably. Richness increased by 3.2 species per plot and cover increased by a factor of 4. Each pothole appears to have developed along its own trajectory in which priority effects (see Section 7.4.2), the consequences of chance initial establishment, have dictated local succession.

The location (determined from a map based on field distance measures) and soil properties of these potholes were used to predict species composition and to determine if deterministic factors were coming into play. Using canonical correspondence analysis (ter Braak 1986), we tested for significant correlations. In 1993, the overall correlation between species patterns and environmental variables was not significant ($r = 0.62$ for the first axis). In 1998, the first axis was significant ($r = 0.72$, $p < 0.02$) with moisture (determined gravimetrically from three soil samples) and location variables being the leading predictors. By 2001, the relationship had

TABLE 7.5. Structure of community types.

Type	Dominants	Richness	Cover %	H?	E	% Hydrophytic
A	Willow/yellow willowherb[a]/mosses	10.0^{CD}	157.6^{A}	1.09^{AB}	0.484	97.9
B	Willow/goatsbeard[b]–Merten's sedge	13.7^{BCD}	83.7^{BC}	1.34^{AB}	0.508	80.9
C	Willow/fringed willowherb–bluejoint[c]/mosses	$13.0b^{CD}$	67.5^{CD}	1.42^{AB}	0.596	96.4
D	Willow/mixed herbs/mosses	20.5^{A}	28.9^{D}	1.68^{A}	0.562	91.5
E	Willow/Lewis's monkey flower[d]–fringed willowherb/swamp moss	19.6^{AB}	117.6^{AB}	1.61^{A}	0.540	92.6
F	Willow/rushes/golden short-capsuled moss	12.0^{CD}	85.5^{BC}	1.18^{AB}	0.481	99.9
G	Willow/rushes–fringed willowherb/golden short-capsuled moss	16.0^{ABC}	57.2^{CD}	1.70^{A}	0.613	99.1
H	Willow/field horsetail–rushes	14.2^{BCD}	72.9^{BC}	1.67^{A}	0.633	99.2
I	Willow/field horsetail	8.3^{D}	73.0^{C}	0.79^{B}	0.387	99.7
J	Cattail–toad rush[f]	12.8^{BCD}	52.6^{CD}	1.52^{AB}	0.603	97.2

Column values with different alphabetic (ABCD) superscripts are significantly different ($p < 0.05$, Bonferroni comparison). Richness is the mean number of species per sample; Cover % is the total percentage found in the sample, H? (diversity statistic) and E (evenness) are defined in text; and "% Hydrophytic" is the cover of species considered to be hydrophytes (e.g., obligate, facultative wetland, or facultative species). Footnotes a–f provide scientific names of species not mentioned in text.

[a]*Epilobium luteum.*

[b]*Aruncus dioicus.*

[c]*Calamagrostis canadensis.*

[d]*Mimulus lewisii.*

[e]*Brachythecium frigidum.*

[f]*Juncus bufonius.*

Source: From del Moral (1999b); used by permission.

strengthened slightly ($r = 0.74$, $p < 0.01$), with total nitrogen and location being significant factors. The increase in nitrogen as a predictor may have resulted in the large increase of lupines relative to other species. Thus, the relationship between the physical environment and species composition increased slightly. That the spatial dimensions were important implied that priority effects were important. Priority effects continue to influence development in these potholes, and trajectories have not converged.

To test if chance played a large role in determining the vegetation of these potholes, we developed a stochastic model to predict composition from initial species compositions (del Moral 1999a). This model accurately predicts mean richness, cover, frequency, and rank order of the potholes, a result that suggests that species assembly in such sites as these potholes is largely caused by chance. However, there was spatial homogeneity for several species, which implied that subsequent dispersal from adjacent potholes helps to structure vegetation.

7.3.7 Wetlands

Wetlands are a natural focus for studies of early primary succession because they develop rapidly, attract fauna, and may export materials to adjacent sites. New wetlands on primary surfaces north of the crater occur in depressions, on the new margins of Spirit Lake, along new springs, and along snow-fed streams. The first wave of colonists was dominated by wind-dispersed species (del Moral and Bliss 1993). Primary wetland vegetation remains variable, probably because of the combined effects of chance and the availability of several species that are able to dominate a site. For example, spike bentgrass (*Agrostis exarata*), field horsetail (*Equisetum arvense*), and toad rush (*Juncus bufonius*) can each dominate early in succession. As willows develop, a more consistent array of species adapted to shade may occur. Some upland species (e.g., pearly everlasting, fireweed, and hairy cats-ear) are common on wetland margins because of their broad ecological amplitudes in the absence of competition. As shrubs expand, the upland species should decline sharply.

Several community types described on the primary surfaces of Mount St. Helens have regional analogues (Titus et al. 1996, 1999; Table 7.5), although they differ because of their immaturity. These communities include a sitka willow (*Salix sitchensis*)/field horsetail (*Equisetum arvense*)–sweet coltsfoot (*Petasites frigidus*) association in northern Oregon similar to community type A and a lung liverwort (*Marchantia polymorpha*)–swamp moss (*Philonotis fontana*) association, similar to community type E, both described by Christy (2000).

Deterministic mechanisms that structure vegetation in wetlands on the Pumice Plain are increasing in strength. The principal structuring mechanisms on these wetlands are the moisture regime, which permits rapid development of plant biomass, and the competitive effects of willows. Developing wetlands demonstrated increasingly tight connection between vegetation and aspects of the environment. Titus et al. (1999) studied wetlands on the Pumice Plain after 14 growing seasons, and del Moral (1999b) sampled 78 new wetlands 6 years later to assess changes in vegetation and environmental relationships. Geographic, topographic, physiographic, moisture, and soil data were analyzed to determine which environmental features were correlated with vegetation patterns. During this interval, explained variation increased from 19% to 31%, when location, soil pH, and habitat type were the best predictors of species patterns. During the sampling interval, willow cover increased from 10% to 28%, resulting in reduced diversity and heterogeneity of understory vegetation. The understories of wetlands with more than 70% willow cover were significantly more similar to each other than were the understories of wetlands with less than 10% willow cover. These data suggest that the wetlands have begun to demonstrate deterministic effects because of greater competition and stronger coupling to moisture regimes.

We recognized 10 primary wetland vegetation types on Pumice Plain sites after 19 years (see Table 7.5; from del Moral 1999b). Richness, diversity, and evenness are inversely correlated with cover percentage, suggesting that, as willow dominance increases, fewer species will persist. However, common species that colonized wetlands by chance persist and seem to be resisting exclusion. If many species occur because of early stochastic events, then a strong, deterministic relationship between species patterns and the environment cannot develop. Eventually, these wetlands should continue to mature and develop tighter ties to the environment as competitive pressures from canopy dominants and from better-adapted understory species are exerted.

Willow thickets may eventually resemble communities observed elsewhere. However, herbaceous wetlands may change in less predictable ways. Clonal species, such as field horsetail and cattails, can persist indefinitely (Keddy 1989; Tsuyuzaki 1989; Prach and Pyšek 1994). It is likely that cattails (*Typha*) will continue to dominate some habitats while being excluded by willows in other, similar habitats (Tu et al. 1998). Physically unstable sites may continue to be dominated by horsetails or rushes. As of 2002, several primary wetland assemblages are not developing toward mature wetland communities.

7.3.8 Nonnative Species

Species that only appeared with Europeans are a part of the recolonization process. Most exotics are not adapted to higher elevations, having evolved in European cultivated land. Titus et al. (1998b) listed native and exotic species found around Mount St. Helens and found 341 vascular plant species in primary-successional habitats in 1995. Of these, 57 were exotic. The exotic flora is dominated by composites. On the Pumice Plain, 151 natives and 20 exotics occurred, but only 4 were common. On the Plains of Abraham, 65 natives and 4 exotics occurred. Primary wetlands had 110 natives and 11 exotics. The western Pumice Plain was positioned to receive many species from the clear-cuts to the west. As the vegetation of this area has developed, the number and dominance of exotic species [e.g., tansy ragwort (*Senecio jacobaea*), wild lettuce (*Lactuca serriola*), Canada thistle (*Cirsium arvense*), and velvetgrass (*Holcus lanatus*)] declined. The exception is hairy cats-ear, which appears to have become "naturalized" in many habitats. It is common where lupine has become abundant. We predict that hairy cats-ear will remain an integral part of the flora, although its importance will decline as shrubs and conifers come to dominate.

7.4 Implications and Conclusions

7.4.1 Dispersal Limitations

Our studies of succession on mudflows and coarse pumice at Mount St. Helens have offered valuable lessons and altered the traditional view of primary succession. Primary succession is usually slow (Walker and del Moral 2003), but its rate is strongly affected by proximity to sources of colonists as well as to resource availability. Mudflows at Butte Camp and on the Muddy River were invaded rapidly from adjacent, intact forests. However, these field studies, as well as the analyses of vegetation gradients surrounding refugia and remote sensing results (see Lawrence, Chapter 8, this volume), show that the effects of adjacent mature vegetation is limited. Beyond a surprisingly short distance, all sites are similarly isolated. Vegetation more than 100 m from forest margins or refugia is sparse and less diverse than vegetation within 20 m. The size and density of conifers also decline with distance, suggesting strong dispersal limitations. Exposed, sparsely vegetated sites have experienced little species turnover, the hallmark of succession. Species continue to assemble slowly, and only some pioneer species have shown relative declines. When a dense canopy of shrubs or conifers develops, species turnover can be expected. As the ground-layer vegetation becomes dense, as has occurred since 2000 in much of the eastern Pumice Plain as a result of lupine expansion, local extinction and colonization of different species may be expected. The development of forest vegetation on newly deposited pumice will take decades. In isolated wetlands formed on pyroclastic materials, species accumulation and growth has been rapid because seedling establishment was not greatly influenced by summer drought and because the dominant species have excellent wind dispersal. Biomass has developed quickly, dominated by tall shrubs that attract birds and mammals that can introduce additional species (see Crisafulli et al., Chapter 14, this volume). These results imply that dispersal, context, and vegetation structure are more important determinants of succession than is commonly thought and that rehabilitation projects should devote more effort to introducing species rather than depending upon natural dispersal.

7.4.2 Priority Effects

The term "succession," in the original, extreme sense, implied that a series of communities (or species) occupies a site sequentially and that there is a predictable trajectory toward a single stable community (but see the discussion in Dale et al., Chapter 1, this volume). The more recent term "assembly" describes a more stochastic case in which the course of species change is affected by the initial colonists, historical events, and landscape context (Belyea and Lancaster 1999). Many alternative communities that can persist indefinitely may result (see Young et al. 2001). The reality on successional surfaces that we have studied appears to be intermediate. Successional trajectories are not strongly deterministic. There appear to be a few, not many, communities during the first two decades of reestablishment. It will be many decades before the situation in profoundly disturbed, isolated sites on Mount St. Helens can be assessed fully. For now, we offer some preliminary comments about community convergence.

As succession unfolds, species fill in the landscape, and spatial variation in vegetation typically declines. Heterogeneity persists, although gradually becoming reduced, because there are only weak links among patches in a vegetation mosaic and local differences can persist because of inhibitory effects of the founding species. Local environmental conditions (such as the moisture regime) eventually exert their effects to reduce heterogeneity. In most cases, as seen in our wetland examples on the Pumice Plain, vegetation will tend to become more homogeneous, if only because few dominant species exist there. However, trajectories may not converge, and they may even diverge to form persisting novel assemblages. Mature communities will retain a residual of unexplainable variation linked to historical accidents (contingencies), stochastic invasion patterns, and landscape effects. Apparently, several alternative, equally "natural" communities can develop after an intense disturbance, and the one that ultimately results is initially poorly predictable. Restoration ecologists should acknowledge several potential alternatives rather than aiming for one sequence of species replacements. Restoration should be cast in more general plant functional types and conditions of the ecosystem (e.g., plant biomass, cover, and soil organic matter) and not specify particular target community types.

7.4.3 Phases of Primary Succession

Although recovery in response to disturbance forms a continuum from primary succession through secondary succession to mere damage repair, we have focused on several kinds of primary succession sites and compared them to secondary succession sites. Community development has three fundamental phases: assembly, interaction, and maturation. At some point after initial assembly, the processes cease to be unique to primary succession and occur also in secondary succession because of minor disturbances or senescence. In most primary habitats we have studied, the initial assembly phase appears to be nearly complete. Additional species, best adapted to forest understories, likely will wait until significant structural changes result from the maturation of tall woody species. Others, such as Newberry's knotweed (*Polygonum davisiae*), are common on some primary surfaces but remain absent from others. Eventually, they may colonize such habitats as the eastern Pumice Plain. Physical amelioration and biotic facilitation (e.g., nurse-plant effects) dominate, although negative interactions also occur. While species come and go on a microscale (van der Maarel and Sykes 1997), most survive in a larger study area, and other species invade.

The interaction phase has scarcely begun in most sites we have studied, although on mudflows near intact vegetation, in lupine patches, and in wetlands interactions have intensified. As a community shifts from assembly to interaction, the relative importance of facilitation and inhibition also shifts. Where woody species are gaining dominance, competition for light and other resources should exclude many pioneer species and should permit the invasion of species adapted to shade. Early in the interaction phase, diversity increases because pioneers persist in the gaps between woody plants. In our ongoing studies of invasions on mudflows, we note that virtually none of the invading species remained beneath alders or conifers, but they persisted in the gaps. The invasion of woody species appears to foretell reduced species diversity, but it is premature to determine whether subsequent invasions by forest herbs will replace lost diversity and lead to more or less heterogeneity. It may be possible to find general rules of establishment during this phase. The maturation phase involves strengthening dominance by large woody species. These species reduce environmental heterogeneity by casting more-uniform shade and by depositing litter that minimizes surface variations. Although diversity in nonvascular plants, arbuscular and ectomycorrhizal fungi, and saprophytes increases, the overall diversity of vascular herbs and low shrubs will decline, if adjacent forests are a guide.

7.4.4 Significance of These Studies

Long-term ecological studies are invaluable in fostering our understanding of how ecosystems assemble. Primary succession is stochastic and controlled initially by chance, contingency, and context. Prediction of trajectories is problematic and will be influenced both by the initial conditions (priority or founder effects) and by subsequent events, such as herbivore damage or minor disturbances. While our studies help to illuminate restoration guidelines being developed from studies of succession (see Walker and del Moral 2003), several implicit problems remain. Thoroughly degraded habitats, such as toxic mine wastes, will not recover without intense intervention, at least to alter the substrate and to introduce appropriate species. Appropriate intervention to rehabilitate human-disturbed habitats requires the scientific knowledge of basic succession mechanisms, appropriate trajectories, and management techniques. These attributes are all constrained by local social and fiscal factors. Studies on Mount St. Helens have done much to provide the scientific knowledge to effectively enhance restoration projects in this region. We have demonstrated that dispersal limitations alone often determine the course of early succession. Soil amelioration in the form of nitrogen fixation and inputs from outside the system accelerate the invasion process. The presence of resource oases, such as refugia and springs, clearly alters local succession, but they also may affect their surroundings.

The eruption of Mount St. Helens profoundly altered the lives of the present authors. We each followed professional trajectories remarkably different from what we could foresee in 1979. We hope that this chapter has conveyed some of the reasons why we were compelled to return consistently to document how recovery is unfolding on this unique volcano and to learn more about one of nature's most fascinating processes—primary succession.

Acknowledgments. We gratefully acknowledge funding from the National Science Foundation under grants DEB-80-21460 (L.C. Bliss, J.F. Franklin, R. del Moral, et al.), DEB-81-07042 (L.C. Bliss, J.F. Franklin, R. del Moral, et al.), BSR-84-07213 (L.C. Bliss, R. del Moral, et al.), BSR-88-11893 (L.C. Bliss, R. del Moral, et al.), BSR-89-06544 (R. del Moral and D.M. Wood), DEB-94-06987 (R. del Moral), and DEB-00-87040 (R. del Moral). Many people assisted us in this work. We are particularly indebted to P. Titus, S. Bard, A. Coogan, K. Dlugosch, A. Eckert, E. Ellis, T. Fletcher, R. Fuller, C. Jones, K. Pearl, R. Robham, L. Rozzell, S. Tsuyuzaki, M. Tu, M. Tweiten, and C. Wolfe.

8 Remote Sensing of Vegetation Responses During the First 20 Years Following the 1980 Eruption of Mount St. Helens: A Spatially and Temporally Stratified Analysis

Rick Lawrence

8.1 Introduction

The variety of disturbance mechanisms involved in the 1980 eruption of Mount St. Helens (e.g., heat, burial, and impact force) and the resulting diversity of vegetation responses have provided abundant opportunities for disturbance-zone-specific research (see other chapters in this volume). As evidenced by the research reported in this volume, tremendous amounts of knowledge can be acquired from studies that focus on vegetation responses within individual disturbance zones, such as the debris-avalanche deposit. As the responses to the eruption continue to develop, however, it becomes increasingly important to understand the larger context for specific study sites: What responses are common among disturbance zones? And what responses are distinctive to certain disturbance mechanisms and sites? Can the lessons from a local area be generalized throughout the disturbed area or even throughout a zone dominated by one disturbance type? These questions can be addressed in at least two ways:

- Studies at different sites can be compared for similar or divergent processes and states.
- Spatial-analysis tools can be used to evaluate response patterns across the disturbed area.

The first approach can be achieved through a comparative analysis of the disturbance-zone-specific studies reported in this volume and elsewhere. In this chapter, vegetation-response patterns present across disturbance zones are evaluated for, among other things, consistency with the ground-based observations and analyses available from other studies.

It is important in studying the first 20 years of vegetation response across the disturbed landscape at Mount St. Helens that the analysis be stratified, both spatially and temporally. Previous research indicated that early vegetation responses were largely dependent on the variable effects of disturbance mechanisms (Adams et al. 1987; Lawrence and Ripple 2000) as well as the substrates resulting from the volcanic events (del Moral and Clampitt 1985). Areas that had substantial biotic legacies (buried seeds, sprouting roots, or downed woody debris), for example, had much faster early vegetation development than areas dominated by primary successional processes, although posteruption management practices might be at least equally important (Figure 8.1). Within these broad categories, additional distinctions might be made based, for example, on the nature of the volcanic deposits (e.g., pumice deposits versus mudflow remnants). In addition, the area devastated by the 1980 eruption has experienced three distinctive posteruption management regimes (Franklin et al. 1988): (1) Mount St. Helens National Volcanic Monument (Monument), where natural processes of ecological recovery were allowed to dominate after its establishment in 1982 (although seeding with nonnative plants and salvage logging occurred in limited areas before that time); (2) Gifford Pinchot National Forest (GPNF) outside the Monument, which experienced salvage logging that was substantially completed by 1984 and then planted with commercial conifer species, mainly Douglas-fir [*Pseudotsuga menziesii* (Mirbel) Franco] seedlings; and (3) private forestlands, where portions were seeded with nonnative grasses and legumes, salvage logging was completed by 1982, and commercial conifer species were planted, also mainly Douglas-fir seedlings, although limited areas were left unplanted. Finally, spatial stratification by disturbance type and management regime enables important comparisons between analyses across disturbance zones and many of the analyses that provide more detail regarding individual disturbance zones.

Temporal stratification enables the analysis of changes in factors driving vegetation response over time. Early responses were likely most affected by factors relating to growth of survivors and success of colonizers, while later responses might have been affected more by factors relating to growth rates, both of individual plants and of established colonies through reproduction. Further, the timing of any such shifts in the relative influence of important factors could have depended on disturbance mechanisms or management practices.

The purpose of this study was to examine patterns of vegetation responses relative to selected ecological driving factors within each disturbance zone resulting from the 1980 eruption

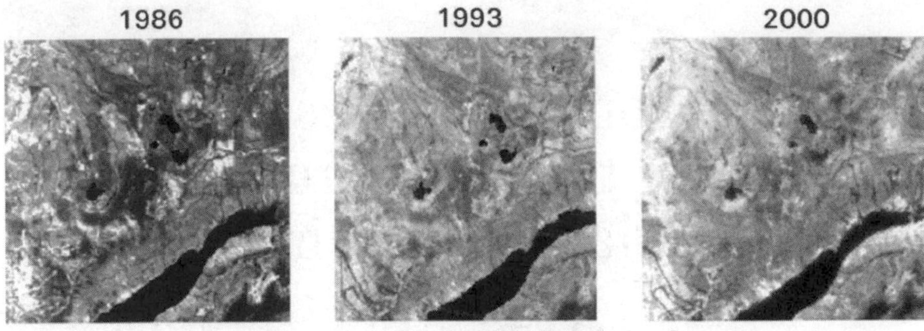

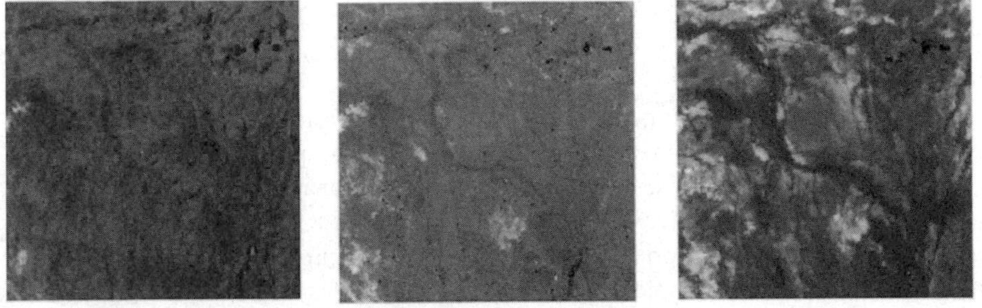

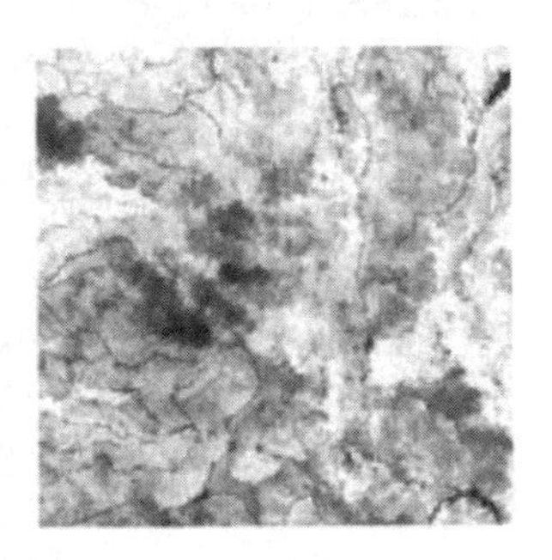

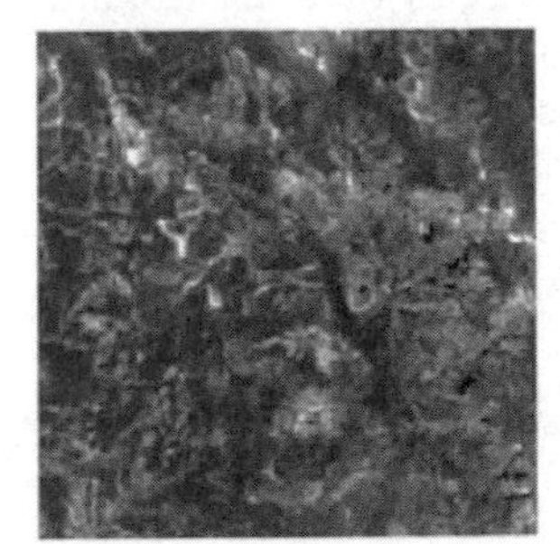
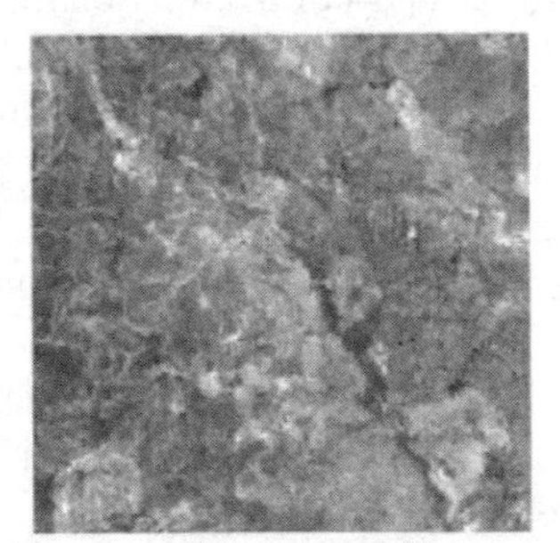

FIGURE 8.1. Patterns of recovery are visually strikingly different among different disturbance zones and different management regimes. These 12 images show normalized difference vegetation index (NDVI) for four locations at the end of each of the study periods. The secondary-succession-dominated location is primarily within the blowdown zone whereas the primary-succession-dominated location is primarily within the pyroclastic-flow deposits. Darker areas on the images had little vegetation and light areas had more vegetation. Individual images are approximately 4 km on each side.

of Mount St. Helens and the subsequent management zones and to compare the responses across zones. The first question examined, therefore, was why certain locations within disturbance and management zones were developing differently than were other locations within the same disturbance and management zones, while the second question addressed whether differences existed among zones. This analysis both provides a view of the importance of certain factors within the entirety of each zone, which can be viewed in conjunction with ground-based studies within each zone, and enables rigorous comparisons across zones because the same data and methods are applied to all zones.

The impacts of these factors were analyzed over three time periods for six broadly defined disturbance types within the portion of the study area allowed to respond in a largely unmanaged manner as well as within areas planted with trees by the USDA Forest Service and private industry. The study used a surrogate for vegetation response (change in a spectral index related to vegetation amount) rather than direct ecological measures, such as changes in absolute biomass, vegetation cover, or plant-community composition. That surrogate measurement was made at a spatial detail or resolution of approximately 25 m. Vegetation patterns can vary depending on how vegetation is measured, what vegetation characteristics are measured, and what resolution or scale is used. A patchy vegetation pattern at ground-plot scales, for example, might not be patchy at coarser resolutions, whereas a pattern that might show little variability across spatial extents measured on the ground might exhibit substantial variability when the entire zone is measured. The conclusions and inferences drawn from this study, therefore, as with all other observational studies, are limited to the nature of these measurements. A review of multiple studies at different scales and involving other vegetation measures provides a more complete picture of the developing vegetation patterns and their underlying processes.

8.2 Methods

The study area (Figure 8.2) included most of the area denoted by Lipman and Mullineaux (1981) as devastated by the 1980 eruption of Mount St. Helens. Excluded were talus deposits, lava domes, and new crater walls because they have not shown any detectable vegetation in the imagery acquired for the study. The central portion of the study area was part of the Monument where ecological response occurred with a minimum of human intervention (Franklin et al. 1988). Managed forests were present to the east and west.

The study area was divided into eight geographical strata, consisting of six disturbance zones within the Monument and two additional management strata outside the Monument (Table 8.1). The strata were not necessarily contiguous; areas of pyroclastic-flow deposits, for example, were scattered within the study area (see Figure 8.2). The six disturbance zones within the Monument were based on broad disturbance types mapped by the U.S. Geological Survey (USGS) (Lipman and Mullineaux 1981), including four zones dominated by primary successional processes (the debris-avalanche, mudflow, and pyroclastic-flow deposits and the tree-removal zone) and two zones that were dominated by secondary successional processes (blowdown zone and scorch zone) (Lawrence and Ripple 2000). It is important to note that these distinctions refer to the broad-scale dominance of certain types of processes and that, for example, isolated survivors within the mudflow and tree-removal zones resulted in some responses within those zones being of a secondary successional nature. The two strata of lands with intensive forest management included the eastern portion managed by GPNF and the western portion managed predominantly by private industry, each of which included a variety of disturbance types.

The study was stratified into three time periods, each encompassing seven growing seasons (the initial eruption occurred

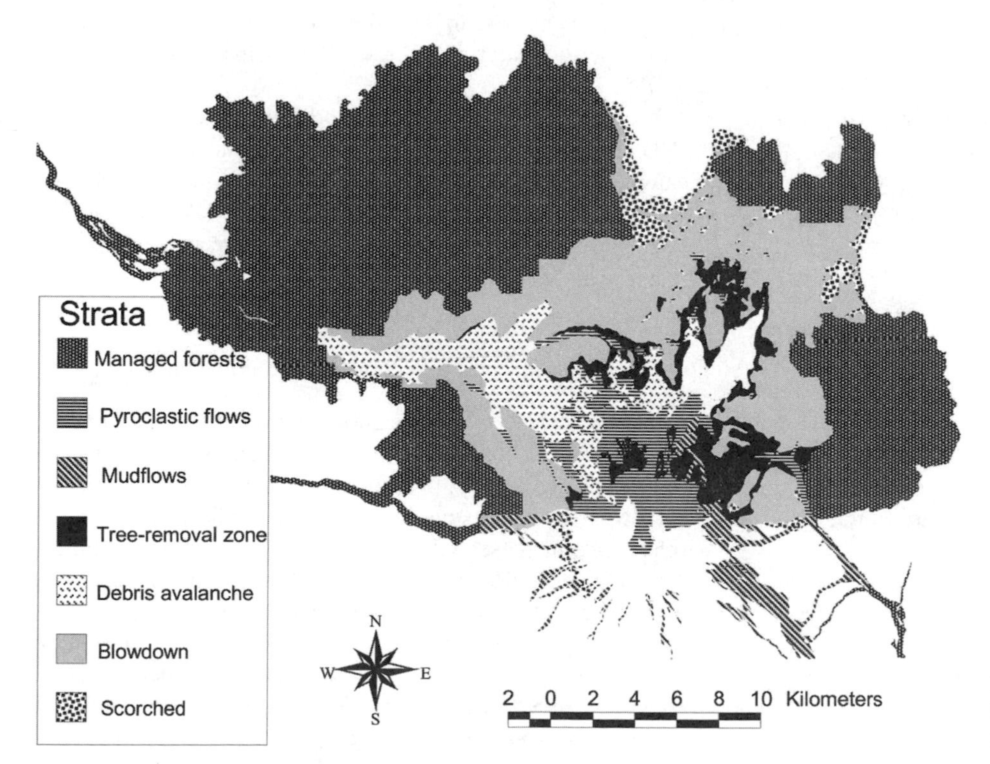

FIGURE 8.2. Map of the study area showing spatial strata based on major disturbance types and management regimes.

TABLE 8.1. Area of each stratum of the landscape studied.

Stratum or zone	Area (km^2)
Debris-avalanche deposits	36
Gifford Pinchot National Forest (NF) (outside national monument)	80
Mudflow deposits	17
Private industry land	247
Pyroclastic-flow deposits	30
Scorch zone	15
Blowdown zone	120
Tree-removal zone	32

Areas of strata other than the Gifford Pinchot National Forest and private industry lands include only those areas within the Mount St. Helens National Volcanic Monument.

before the 1980 growing season) and each beginning when the previous period ended: 1980–1986 (period 1); 1986–1993 (period 2); and 1993–2000 (period 3). Although these time periods were arbitrary, they coincided with available satellite imagery and enabled comparisons in intervals of equal length throughout the 20 years following the eruption.

Previous research suggested that three broad categories of factors affecting vegetation response should be examined (Lawrence and Ripple 2000):

- Direct effects of the eruption
- Posteruption physical forces
- Other habitat conditions

Although most factors could not be measured directly (as is often the case in observation-based ecological studies), it was worthwhile to examine variables related to these factors because, if a certain factor was important for vegetation development, then we would expect to see a correlation between a related variable and our measure of vegetation change. If temperature differences related to elevation were important, for example, we would expect to see a correlation between vegetation change and elevation even though we were unable to collect temperature readings for all locations consistently throughout the study period. Direct effects of the eruption examined included four factors:

1. More specific disturbance types were assigned to each geographical stratum originally containing multiple types, as mapped by the USGS, because the disturbance types affected the initial conditions for vegetation development (see Figure 8.1) in terms of both available substrates and potential for surviving vegetation and propagules. Seven zones within the debris-avalanche deposits were identified, for example; the tree-removal, blowdown, and scorch zones did not contain multiple disturbance types.
2. Direct exposure to the directed blast was assessed because topographic shielding from the blast could also affect initial conditions (Figure 8.3).
3. Distance from the crater was measured. That distance could be related to the intensity of the impacts and, among other things, the presence of surviving vegetation and propagules.
4. Initial air-fall tephra thickness from eruptions through July 22, 1980 was determined (Waitt et al. 1981) because it could affect the ability of survivors to establish and the ability of colonizers to reach organic soils.

Posteruption physical forces were also examined in terms of (1) slope gradient (Figure 8.4), with the assumption that steeper slopes experienced more erosion of tephra deposited in 1980, and (2) distance from surviving forests (Figure 8.5), which was potentially related to windborne and animal-borne seed dispersal (including in planted forests, where nonplanted vegetation, such as shrubs, forbs, and grasses, might have dispersed from surviving forests). Other habitat conditions examined included elevation and aspect (see Figure 8.3), which can affect growing conditions, such as temperature and incident solar radiation (Allen and Peet 1990). In addition, for period 2 and period 3, the effect of previous period vegetation on subsequent vegetation change was evaluated separately because the presence of established vegetation could result in higher growth rates and more-proximate seed sources.

Analysis was conducted with Landsat Thematic Mapper (TM) and Enhanced Thematic Mapper Plus (ETM+) imagery from August 26, 1986; August 29, 1993; and July 7, 2000. Although the 2000 image was from an earlier portion of the growing season than the other images and this difference might have affected the results, it was the closest date for which a clear image was available. Images were georeferenced to a Universal Transverse Mercator grid, resampled to a 25-m pixel size, and coregistered to less than 12.5-m root mean squared error. All image values were converted to exoatmospheric reflectance based on current calibration data, which accounted for differences in dynamic ranges of the TM and ETM+ sensors, sun angle and distance differences, and some atmospheric scattering (Clark 1986; Chavez 1996; NASA 2001).

For each image, the normalized difference vegetation index (NDVI) was calculated as a surrogate vegetation response because ground calibration data were not available to estimate vegetation parameters with the imagery. NDVI is the most widely used index of vegetation amount from remotely sensed imagery and is calculated as

$$\frac{(\text{near-infrared} - \text{red reflectance})}{(\text{near-infrared reflectance} + \text{red reflectance})}$$

(Rouse et al. 1973). NDVI values theoretically range from −1 to 1; but in practice, values near 0 (where near-infrared and red reflectance are roughly equal) represent little or no vegetation, whereas increasing positive values (where near-infrared reflectance increases relative to red reflectance) represent increasing vegetation amounts. NDVI has been found to be highly correlated with a wide variety of vegetation parameters, including leaf-area index, aboveground biomass, vegetation cover, and the fraction of photosynthetically active radiation (fPAR) absorbed by vegetation (Running et al. 1986; Anderson et al. 1993; Yoder and Waring 1994). NDVI is particularly

FIGURE 8.3. Three-dimensional rendering of a 2000 Landsat ETM+ panchromatic image illustrates several potential factors affecting vegetation responses. A careful examination shows that some slopes on the far side of ridges from the crater might have more vegetation, possibly as a result of topographic shielding from the directed blast. The analysis, however, generally did not show this to be statistically important. Possible effects of elevation and topographic aspect might also be seen.

useful, therefore, as a surrogate for vegetation amount when it is logistically impractical to collect ground reference data to calibrate remote-sensing data [for example, when spatial extents are large (e.g., Eastman and Fulk 1993) or when legal or safety restraints restrict sampling]. Some disadvantages to using NDVI as a surrogate are that it does not relate to any specific vegetation measure (such as biomass), its relationship to specific vegetation measures is not perfect, it can be

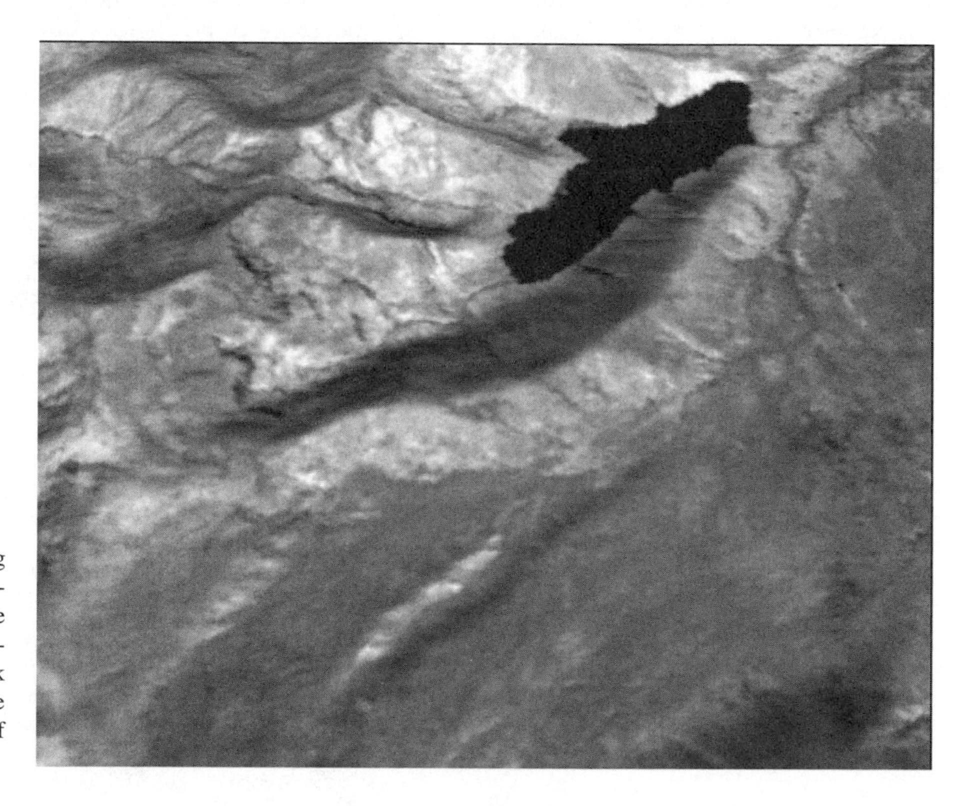

FIGURE 8.4. Three-dimensional rendering of a 2000 Landsat ETM+ panchromatic image shows effects possibly related to slope gradient. The steeper slopes show a different response than flatter locations. Black feature in upper part of image is Castle Lake, located in the upper left quadrant of Figure 8.3.

FIGURE 8.5. Three-dimensional rendering of a 2000 Landsat ETM+ panchromatic image shows areas of surviving forests next to disturbed zones, such as mudflow deposits. If seed dispersal from surviving forests was important in the varying responses across the mudflows, then we would expect a strong statistical relationship between distance to surviving forests and changes in NDVI. This statistical relationship was present during all periods but was especially strong during period 1. View is to the east. Mudflow and surviving forests are on the upper Muddy River fan (see Frenzen et al. Chapter 6, this volume).

sensitive to soil variations, and it loses sensitivity at high leaf-canopy densities. NDVI was found to perform better at Mount St. Helens than did other common vegetation indices, such as soil-adjusted indices and tasseled-cap greenness (Lawrence and Ripple 1998). NDVI was used, therefore, as a measure of relative vegetation response, although for this study the data were not available to statistically correlate it to any specific measure of vegetation amount.

Relative vegetation response for each period was calculated as NDVI at the end of the period minus NDVI at the beginning of the period (with the assumption of no vegetation immediately after the May 18, 1980 eruption detectable with 25-m-resolution satellite imagery) (Figures 8.6, 8.7, 8.8). This approach permitted an analysis of only vegetation growth during each period. Our previous study focused on vegetation trajectories during the first 15 years following the eruption without temporal stratification (Lawrence and Ripple 2000).

Statistical analyses were conducted with regression-tree analysis (RTA) (Breiman et al. 1984; Lawrence and Ripple 2000). Although multiple linear regression was examined, RTA consistently explained a larger portion of the variability in the data. Further, the spatially correlated nature of the data violated the assumption of independent observations for multiple linear regression. For each of the spatial strata, data from all pixels within the stratum were extracted for statistical analysis. The analysis was, therefore, a complete description of the relationships among the variables within each stratum because a complete census was analyzed rather than a sample. Analyses initially were conducted with all explanatory variables except previous-period NDVI to isolate the effects of previous vegetation amount from other factors, and then the effect of previous-period NDVI was added. For consistency of analysis, all regression trees were pruned to the 10 most important terminal nodes (response values), except when there were ties for the 10th node, in which event all tied nodes were evaluated. With eight strata and three periods, 24 regression trees were analyzed, plus an additional 16 to add the effects of previous-period vegetation for period 2 and period 3.

8.3 Results

The results of the RTA were examined for trends over time and for differences among strata. Importance of factors was determined primarily as a function of statistical deviance explained (i.e., the sum of the squared differences from the mean of the response variable) because this method provided a relative measure of the strength of the relationship between each explanatory variable and change in NDVI. Important factors considered were

- The total amount of deviance explained by the analysis;
- The amount of total deviance explained by each variable (the percent of total deviance);
- The relative amount of deviance explained by each variable (the relative deviance); and, most important,
- Changes in the relative importance of explanatory variables over time.

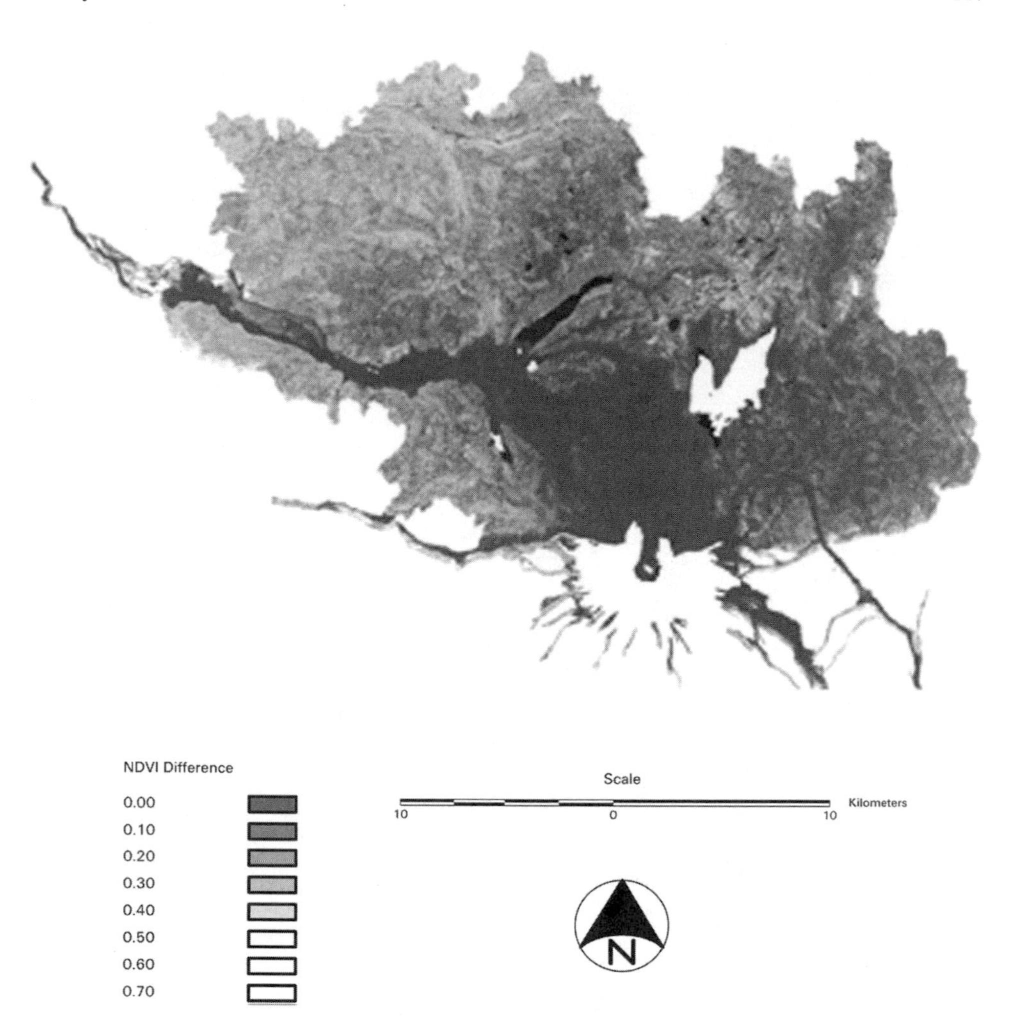

FIGURE 8.6. Differences in NDVI from 1980 to 1986. The study area was assumed to have no vegetation detectable by the Landsat imagery immediately after the May 18, 1980, eruption.

Many factors might affect vegetation responses, and no doubt many of these potential influences were not and could not be quantified for this study. Rather than attempt to explain all factors influencing vegetation responses, the variables that were included in this study reflected patterns and processes found to be potentially important in previous ground-based and remote-sensing studies (Lawrence and Ripple 2000).

RTA results are in the form of dichotomous trees that are often complex and reveal multiple interactions among predictor variables. An analysis of the resulting trees, however, revealed that the effect of important variables was as expected. Generally, as measured by predicted NDVI, when statistically important in the analysis, slower vegetation development was predicted by direct exposure to the directed blast, closer distances to the crater, thicker initial airfall-tephra layers, less-steep slope gradients, longer distances from surviving forests, and higher elevations.

8.3.1 Primary-Succession-Dominated Strata

During period 1 on the debris-avalanche deposits, the strongest factor related to vegetation response examined was distance to surviving forests, which accounted for more than half of relative deviance (Table 8.2). Avalanche units and slope each accounted for about 14% of relative deviance; aspect, distance from the crater, and tephra thickness were less important. On the mudflow deposits, distance to surviving forests was even more important, accounting for 78% of relative deviance. Tephra thickness, aspect, slope, and elevation were minor contributors. The pyroclastic-flow deposits exhibited a different pattern, with elevation and distance to the crater having the strongest relationship, collectively accounting for 65% of relative deviance. Distance to surviving forests and slope gradient had moderate relationships. Responses for the tree-removal zone were largely related to distance from the crater (79% of relative deviance), with moderate correlation to elevation.

During period 2 on the debris-avalanche deposits, distance to surviving forests explained somewhat less deviance; elevation was added as an explanatory variable, with the second largest impact; and tephra thickness, slope, and aspect ceased to be included in the model. The same pattern, but more pronounced, occurred on the mudflow deposits, where, compared to period 1, the importance of distance to surviving forests decreased by 68% and the importance of elevation increased by

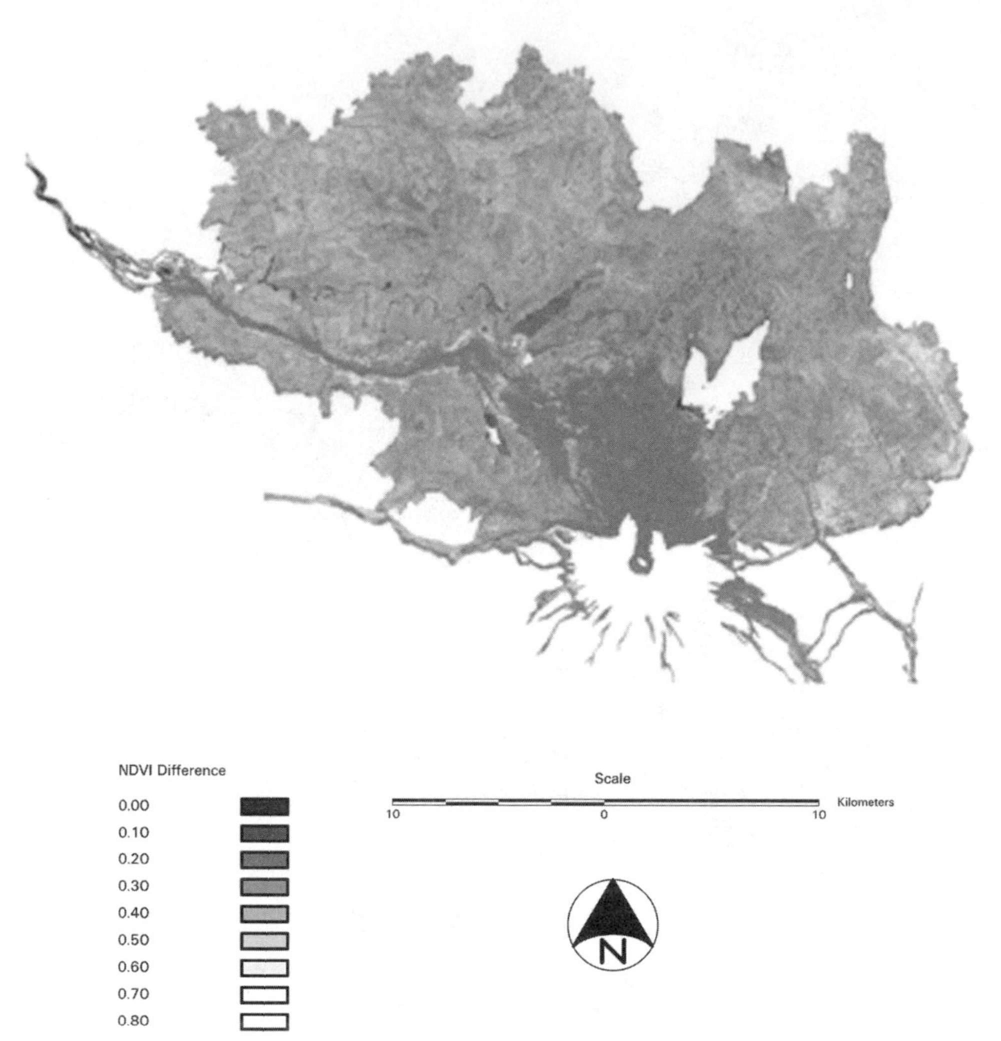

FIGURE 8.7. Differences in NDVI from 1986 to 1993.

70%. On the pyroclastic-flow deposits, elevation became the dominant factor, and some distinctions became evident among the various pyroclastic flows. An exception to this pattern was seen in the tree-removal zone, where elevation increased as a percent of relative deviance but was close to stable in terms of total deviance. There, distance to the crater was greatly reduced in importance relative to other factors.

During period 3, elevation became the dominant factor affecting vegetation change for the debris-avalanche and pyroclastic-flow deposits and especially for the tree-removal zone. On the mudflow deposits, distance from the crater increased greatly in importance whereas elevation decreased in importance. This trend might be misleading because, on the mudflow deposits more than any other stratum, elevation and distance from the crater tended to be correlated.

Including NDVI from previous periods had varying effects, depending on the stratum. In general, this measure of previous-period vegetation improved explanation of deviance by 5% ± 2%. The exceptions were for the mudflow deposits, where improvement was 9% in period 2 and 11% in period 3, and for the pyroclastic-flow deposits, where improvement was only 2% in period 3.

8.3.2 Secondary-Succession-Dominated Strata

For period 1 in the blowdown zone, most deviance was explained by slope gradient and tephra thickness, which together accounted for 70% of relative deviance (Table 8.3). Distance from the crater had a moderate relationship to vegetation change, and other factors were relatively minor. In the scorch zone, distance to surviving forests and aspect were most important, slope had a moderate relationship, and other factors were minor. The scorch zone in period 1 was the only instance in which direct exposure to the directed blast was included, accounting for 7% of relative deviance.

In period 2 for both the blowdown and scorch zones, the least amount of total deviance was explained for any of the analyses of unmanaged strata: 10% for the blowdown zone and 15% for the scorch zone. Of the deviance explained, the most important factors were aspect for the blowdown zone (61%) and distance from the crater and elevation for the scorch zone, collectively accounting for 70%.

During period 3 for both secondary-succession-dominated zones, elevation was the dominant factor. Also for both zones, aspect was a moderately important factor. Distance from the

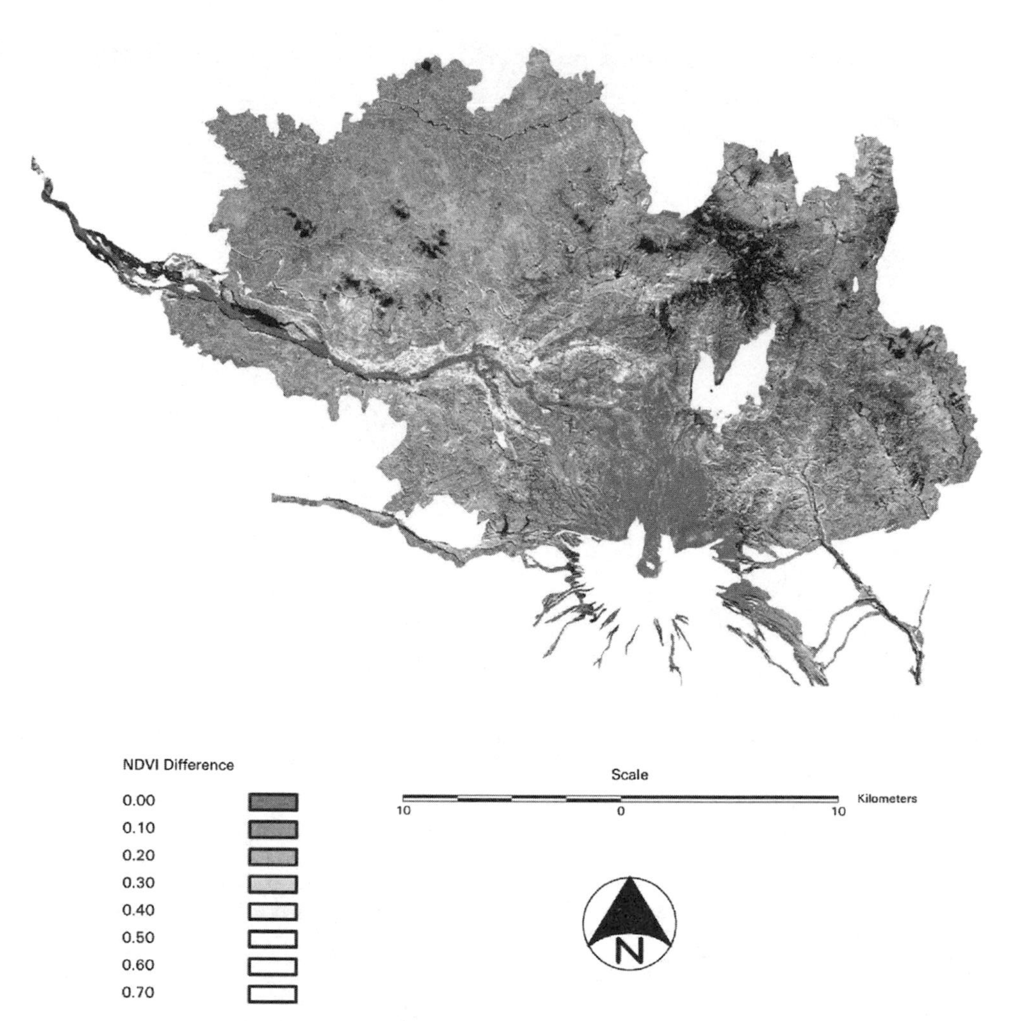

FIGURE 8.8. Differences in NDVI from 1993 to 2000.

crater was a moderately important factor for the scorch zone and a minor factor for the blowdown zone.

Inclusion of previous-period NDVI had a substantial positive effect on total deviance explained in secondary-succession-dominated strata during period 2, when total variability explained was otherwise low. Improvement was 6% in the blowdown zone and 23% in the scorch zone. For period 3, however, improvement was less than 3% in both zones.

8.3.3 Managed-Forest Strata

Vegetation response for the eastern managed forests was most related to distance to surviving forests during period 1 (Table 8.4). Distance from the crater and variation among disturbance zones were moderately important, whereas elevation and slope were minor factors. Responses for the western managed forests were strongly related to distance from the crater and disturbance type, with lesser impacts of tephra thickness, distance from surviving forests, and elevation.

For the eastern managed forests, disturbance zones, elevation, and slope increased in importance during period 2, whereas distance from the crater remained fairly constant in relative effect and distance from surviving forests decreased substantially in importance. For the western managed forests, elevation became the dominant factor, while distance from surviving forest had moderate impacts.

During period 3, elevation was the dominant factor for all the managed forests. Also, for all the managed forests, distance from the crater increased substantially in importance. For the eastern managed forests, disturbance zones continued to have moderate influence, while aspect also became moderately influential.

Adding previous-period NDVI to the analysis had a moderate impact on analysis of the eastern managed forests, similar to most other strata. Increases in total deviance explained were 8% for period 2 and 6% for period 3. The impact on the western managed-forest analyses was more dramatic, however, with increases in total deviance explained of 20% in period 2 and 11% in period 3.

8.4 Discussion

The most important factors for early (period 1) vegetation response were factors likely correlated with vegetation survival and establishment, consistent with some previous studies

TABLE 8.2. Importance in the regression-tree analysis of each explanatory variable for each period for the primary-succession-dominated strata.

	Period 1 (1980–1986)		Period 2 (1986–1993)		Period 3 (1993–2000)	
	Total deviance (%)	Relative deviance	Total deviance (%)	Relative deviance	Total deviance (%)	Relative deviance
Debris-avalanche deposits						
Disturbance	10.5	14.8	4.1	12.9	0.9	4.0
Exposure	—	—	—	—	—	—
Tephra depth	1.2	3.0	—	—	—	—
Crater distance	1.6	4.0	5.3	16.6	3.6	16.0
Forest distance	22.1	55.4	14.9	46.5	6.3	28.1
Elevation	—	—	7.6	23.8	11.6	52.1
Slope	5.5	13.9	—	—	—	—
Aspect	3.6	9.1	—	—	—	—
Total	39.8	100.0	32.1	100.0	22.3	100.0
Mudflow deposits						
Disturbance	—	—	0.6	1.9	—	—
Exposure	—	—	—	—	—	—
Tephra depth	2.6	6.0	0.9	2.6	—	—
Crater distance	—	—	1.3	3.8	10.6	42.1
Forest distance	34.1	77.6	3.2	9.2	6.6	25.9
Elevation	2.3	5.3	25.9	75.2	3.8	15.0
Slope	2.4	5.4	—	—	—	—
Aspect	2.5	5.7	2.5	7.3	4.3	16.8
Total	44.0	100.0	34.5	100.0	25.2	100.0
Pyroclastic-flow deposits						
Disturbance	—	—	5.0	12.9	3.7	15.4
Exposure	—	—	—	—	—	—
Tephra depth	3.3	7.0	0.6	1.6	—	—
Crater distance	12.6	26.3	2.4	6.3	2.7	11.4
Forest distance	6.0	12.4	1.9	4.9	2.8	11.6
Elevation	18.5	38.6	25.9	67.1	13.9	58.3
Slope	5.5	11.5	—	—	—	—
Aspect	2.0	4.2	2.7	7.1	0.8	3.2
Total	47.9	100.0	38.6	100.0	23.8	100.0
Tree-removal zone						
Disturbance	—	—	—	—	—	—
Exposure	—	—	—	—	—	—
Tephra depth	0.7	1.1	—	—	—	—
Crater distance	48.6	78.9	2.4	12.0	2.7	6.5
Forest distance	1.4	2.2	4.5	22.3	2.9	6.8
Elevation	6.9	11.1	7.5	37.5	33.9	80.2
Slope	3.1	5.0	4.3	21.6	—	—
Aspect	1.0	1.6	1.3	6.6	2.7	6.5
Total	61.6	100.0	20.1	100.0	42.3	100.0

Percent of total deviance is relative to total deviance in the response variable. Relative deviance explained is proportional to deviance explained by the first 10 terminal nodes of the regression-tree analysis.

(Halpern and Harmon 1983; Wood and del Moral 1987; Lawrence and Ripple 2000; but see Dale 1989). In primary-succession-dominated zones, where vegetation response was dependent almost exclusively on colonizers, the statistically most important factor was distance to surviving forests in most cases. This finding was consistent with some studies of early establishment that found that isolation appeared to be the primary limiting factor on primary successional sites. Although one study that examined this question did not find a statistically significant correlation between seedling density and distance from seed sources (Dale 1989), that study did show a general decline in density with distance up to 1.09 km. The lack of significance in the earlier study was attributable to the high variability in the seed and seedling locations and chance events, factors that were found to be important in several other primary-succession sites at Mount St. Helens (del Moral and Bliss 1993; del Moral et al., Chapter 7, this volume). High variability can interfere with the detection of statistically significant differences in studies based on sampling. The current study, however, is based on a complete census of the observational units (image pixels). The observed pattern was true for the debris-avalanche and mudflow deposits whereas distance to

TABLE 8.3. Importance in the regression-tree analysis of each explanatory variable for each period for the secondary-succession-dominated strata.

	Period 1 (1980–1986)		Period 2 (1986–1993)		Period 3 (1993–2000)	
	Total deviance (%)	Relative deviance	Total deviance (%)	Relative deviance	Total deviance (%)	Relative deviance
Blowdown zone						
Disturbance	—	—	—	—	—	—
Exposure	—	—	—	—	—	—
Tephra depth	10.9	32.1	—	—	—	—
Crater distance	4.3	12.5	1.7	17.3	0.7	1.9
Forest distance	2.2	6.6	0.3	3.4	—	—
Elevation	1.0	2.8	1.8	18.5	29.7	85.4
Slope	12.9	37.8	—	—	—	—
Aspect	2.8	8.2	6.0	61.0	4.4	12.7
Total	34.1	100.0	9.8	100.0	34.8	100.0
Scorch zone						
Disturbance	—	—	—	—	—	—
Exposure	1.4	6.6	—	—	—	—
Tephra depth	—	—	—	—	—	—
Crater distance	1.1	5.5	5.2	35.6	6.5	13.1
Forest distance	7.0	34.3	1.2	8.4	—	—
Elevation	1.4	7.0	5.0	33.9	32.0	64.2
Slope	2.3	11.4	0.9	5.8	—	—
Aspect	7.2	35.0	2.4	16.4	11.3	22.7
Total	20.5	100.0	14.6	100.0	49.8	100.0

Percent of total deviance is relative to total deviance in the response variable. Relative deviance explained is proportional to deviance explained by the first ten terminal nodes of the regression-tree analysis.

TABLE 8.4. Importance in the regression-tree analysis of each explanatory variable for each period for the actively managed strata.

	Period 1 (1980–1986)		Period 2 (1986–1993)		Period 3 (1993–2000)	
	Total deviance (%)	Relative deviance	Total deviance (%)	Relative deviance	Total deviance (%)	Relative deviance
Eastern managed forests						
Disturbance	7.8	17.9	9.8	33.6	1.4	13.3
Exposure	—	—	—	—	—	—
Tephra depth	—	—	—	—	—	—
Crater distance	5.8	13.3	3.6	12.5	3.0	29.2
Forest distance	23.5	53.8	2.9	9.9	0.3	3.1
Elevation	3.2	7.4	9.0	30.9	4.1	40.6
Slope	2.6	6.0	3.8	13.2	—	—
Aspect	—	—	—	—	1.4	13.6
Total	43.8	100.0	29.1	100.0	10.2	100.0
Western managed forests						
Disturbance	14.2	31.7	—	—	0.6	3.2
Exposure	—	—	—	—	0.6	3.3
Tephra depth	6.0	13.3	1.1	6.3	—	—
Crater distance	18.9	42.3	0.6	3.5	2.5	13.3
Forest distance	3.7	8.4	2.6	14.3	1.6	8.4
Elevation	1.9	4.3	13.7	75.8	13.8	71.8
Slope	—	—	—	—	—	—
Aspect	—	—	—	—	—	—
Total	44.8	100.0	18.1	100.0	19.2	100.0

Percent of total deviance is relative to total deviance in the response variable. Relative deviance explained is proportional to deviance explained by the first ten terminal nodes of the regression-tree analysis.

surviving forests was moderately important for the pyroclastic-flow deposits. At least one other study covering period 1 also found that seed dispersal was not an important factor on the pyroclastic-flow deposits (Wood and Morris 1990). The exception to this pattern was the tree-removal zone, which on average tended to be more distant from surviving forests (as were many, but not all, of the pyroclastic-flow deposits). For the tree-removal zone and the pyroclastic-flow deposits, the most important factor was distance from the crater, which might have been correlated with the intensity of the eruption's impacts.

In secondary-succession-dominated zones, where previous studies have shown biotic legacies potentially were more important than colonizers, at least initially (Franklin et al. 1985; Lawrence and Ripple 2000), a notably different set of factors was statistically most important in the early period. In the blowdown zone, which dominated secondary successional zones in area (Table 8.1; Franklin et al. 1988), the primary factors were tephra thickness and slope. This result was partially consistent with previous studies that found that factors favoring recovery of buried survivors should be paramount in this zone and that tephra depth was the most important site characteristic for this zone during period 1 (Halpern et al. 1990). Halpern et al. (1990) did not find slope gradient significant; however, this finding might have been the result of lack of statistical power ($n = 35$ for Halpern et al. 1990). Tephra thickness was the primary factor determining depth of burial, whereas steeper slopes resulted in greater erosion of tephra layers (Collins and Dunne 1986), thereby enhancing chances of establishment by buried survivors. In the scorch zone, distance from surviving forests was important (Lawrence and Ripple 2000), perhaps influenced by a combination of lessening of the force of the directed blast and the impedance of seed dispersal by upright, dead trees and shrubs.

In both primary- and secondary-succession-dominated zones, the most notable trend during period 2 and period 3 was a reduction in importance of the factors likely correlated with vegetation establishment and an increase in importance of factors more likely related to vegetation growth, including reproduction. Most important was the steady increase in importance of elevation as a factor. Increases in elevation are associated with decreased plant growth because of associated decreases in temperature and evapotranspiration and increases in snow depth and persistence (Allen and Peet 1990). In all primary-succession-dominated zones, elevation became at least one of the most important factors during period 2 and the most important factor by period 3, except for the mudflow deposits, where distance from the crater (which was highly correlated to elevation for that stratum) was most important. This finding was consistent with previous studies that found plant communities with reproducing individuals of multiple species beginning to develop in period 2 (del Moral and Wood 1993a). Slope, which was important during period 1, possibly because steeper slopes created relatively safe microsites through erosion, was not important during period 2 except in the tree-removal zone. Previous research found that, by 1993, persistent erosion in gullies resulted in reduced vegetation in some sites on the Pumice Plain (Tsuyuzaki and Titus 1996), which is consistent with this study. For the secondary-succession-dominated strata, very little deviance was explained during period 2, but elevation and, to a lesser extent, aspect were the dominant factors by period 3.

The importance of elevation and aspect was in marked contrast to our earlier study across disturbance zones, which found these factors to be relatively unimportant (Lawrence and Ripple 2000). That study, however, examined only response variables that incorporated, at least to some extent, early-period vegetation responses. When responses for each of the three periods were isolated, it was evident that an important shift over time had occurred throughout the study area. Factors probably related to vegetation reestablishment, whether through colonizers or survivors, were critical initially, but those processes might have largely run their courses or have been overwhelmed later by growth of reestablished vegetation. Recent vegetation responses appear to have been much more influenced by factors affecting growth of the early established vegetation.

Patterns observed in the managed forests were similar to some patterns in the unmanaged strata but also exhibited important differences. For the western managed forests, which were quickly salvaged and replanted (Franklin et al. 1988), the most important factors for early responses (period 1) were distance from the crater, disturbance zone types, and tephra thickness. These were all factors that likely affected the planting soil conditions and the ability of the planting crews to place roots in the relatively rich soils that existed before the 1980 eruption. In the eastern managed forests, which were not planted as quickly, early responses were most related to distance to surviving forests, with disturbance type and distance from the crater as moderate factors. The longer period to complete planting in the eastern managed forests might have resulted in a greater influence of colonizing and surviving vegetation relative to planted trees during the first 7 years in this area, thus explaining the importance of distance to surviving forests during this period.

As with the unmanaged strata, during period 2 and period 3, elevation became increasingly important for the managed forests and was the most important factor by the end of the study period. Compared to other strata, however, the impact of previous-period vegetation had a greater effect on the managed forests. This relationship was likely related to the growth patterns of the planted conifers. Because well-established trees often exhibit higher growth rates than poorly established trees and because tephra suppressed competing vegetation, areas where planted trees were thriving in one period might be expected to have higher growth rates in the next period.

The relation of NDVI to explanatory variables was consistent with the shifting of vegetation responses from a period dominated by establishment mechanisms to dominance of annual growth of established vegetation, including nucleation as previously noted (Franklin and MacMahon 2000). The spatial resolution of the satellite imagery used (25 m), as well as its spectral resolution, however, precluded a closer link between observed patterns and the processes of vegetation response. The Landsat imagery did not permit, for example, distinction of changes in plant-community composition because the spectral resolution did not enable the accurate identification of plant communities. Moreover, the spatial resolution was often larger than patch sizes and thus did not allow the analysis of vegetation patch patterns. The analysis *is* valuable, however, when evaluated in conjunction with ground-based studies conducted within each of the strata. Such multiscale analyses enable the evaluation of whether observed fine-scale patterns and processes are taking place throughout a stratum and across other types of disturbances and management regimes.

Part III
Survival and Establishment of Animal Communities

9 Arthropods as Pioneers in the Regeneration of Life on the Pyroclastic-Flow Deposits of Mount St. Helens

John S. Edwards and Patrick M. Sugg

Dedication

We dedicate this chapter to the memory of a great and intrepid volcano ecologist, Ian Thornton, 1928–2002.

9.1 Introduction

The eruption of Mount St. Helens, on May 18, 1980, affected an area of 600 km^2 within which communities of animals and plants sustained a wide range of impacts, depending on proximity to the volcano and local topography. The most extreme destruction occurred in the area immediately north of the crater, now known as the Pumice Plain (see map, Figure 9.1), where the eruption apparently destroyed the entire biota over tens of square kilometers. Our interest concerned the response of arthropods to the eruption and changed landscape, particularly in the most intensively disturbed area and one remote site. Early questions to address included the following:

- What was the pattern of survival across the area impacted by the blast and subsequent eruption?
- In particular, was the Pumice Plain truly devoid of eruption survivors and thus, with the biological calendar reset to zero, a classic landscape for the study of primary succession?
- What was the pattern of initial colonization?

Given the extraordinary (but largely underappreciated) capacity of arthropods for aerial dispersal and bearing in mind their ubiquity in the summer airstream, the eruption of Mount St. Helens gave us the perfect opportunity to test the hypothesis that, microorganisms aside, arthropods would be the true pioneers of the barren pyroclastic surfaces and the initiators of biological succession. Answers to these questions bear on the broader issue of spatiotemporal dynamics, recently described as the "final frontier for ecological theory" (Kareiva 1994).

The secret of the evolutionary success of the insects and spiders is, in large part, their capacity for widespread dispersal. Insects on their wings and spiders passively on silk threads become airborne on flights that may range from a few meters to hundreds of kilometers and on landing may become colonists in a landscape that is ever changing. On a summer's day, at least half the insect biomass may be airborne, a fact well known to swallows and swifts but little appreciated by earthbound humans.

One problem in studying the dispersal capacity of arthropods is the difficulty of differentiating immigrant dispersers from local residents. The eruption of Mount St. Helens provided an opportunity to study immigration because the Pumice Plain offered a large area devoid of local residents except those undertaking primary colonization. A wide variety of immigrant arthropods was expected at Mount St. Helens, but we surmised that few would be adapted for survival in barren and physiologically challenging sites like the Pumice Plain. Wide-ranging summer temperatures, abrasive mineral dust, which can abrade the superficial waxes on which water retention depends, and intermittent drought are hazards for all arthropods, especially small ones that have high surface-to-volume ratios. Those unable to survive in the radically changed landscape would nonetheless have ecological significance as an import of organic matter to the new pyroclastic surfaces. Hence, it was important to ask what the biomass of this arthropod fallout was and what it represented in terms of important nutrients, such as nitrogen and phosphorus.

Few studies have addressed the roles of arthropods in the context of primary succession. Primary succession is, by definition, initiated in an area devoid of plant life, but an absence of plants does not necessarily mean an absence of resident animal populations. Based on his studies following the catastrophic eruption of Krakatau in 1883, Dammerman (1948) showed that the primary successional sequence began there with scavengers feeding on organic material brought to the island by wind and ocean currents. But what held for a tropical island was not necessarily the case in the north temperate zone, because Lindroth et al. (1973) concluded from studies

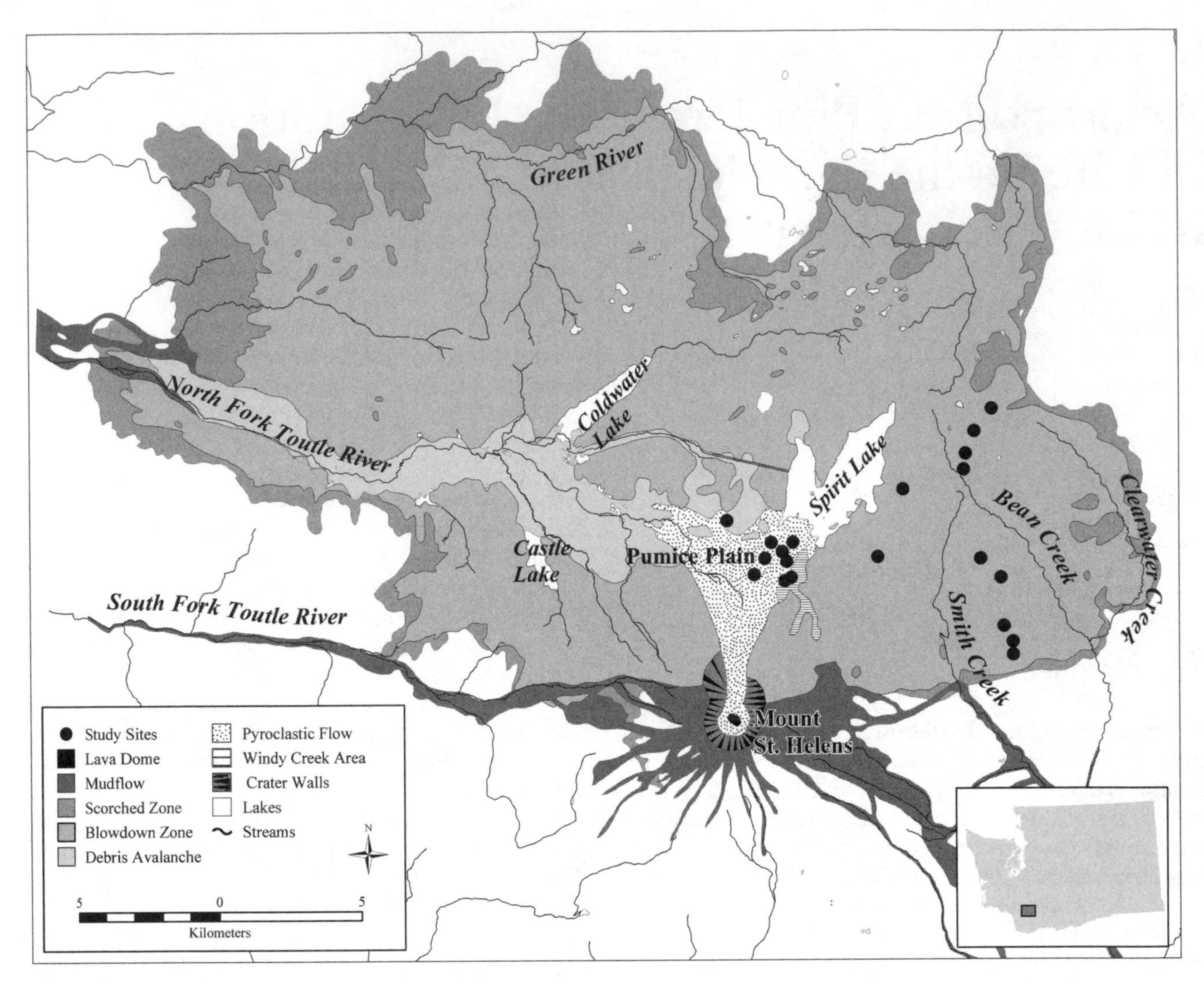

FIGURE 9.1. The area north of the crater, showing the distribution of pyroclastic-flow deposits, surrounding blowdown, and standing-dead zones. *Circles* denote arthropod sampling sites. [From Sugg and Edwards (1998); used with permission.]

of the volcanic island Surtsey, which rose above the surface of the Atlantic Ocean in 1963 about 30 km southeast of Iceland, that plants were needed as a primary resource before animal colonization could occur. This view of the primacy of plant establishment before colonization of viable animal populations was also supported by the observations of Delong (1966) from areas newly exposed by a receding Alaskan glacier. Yet even at Surtsey, Lindroth et al. (1973) reported transient populations of blowflies supported on allochthonous inputs, namely carcasses of seals and sea birds washed up on the shore. A more recent and detailed monograph on the ecology of recently deglaciated terrain, based largely on Scandinavian sites (Matthews 1992), makes no mention of arthropods as pioneers. It is, however, well established that areas with no significant vegetation, such as alpine zones, periglacial landscapes (Kaufmann 2001), and deserts (Koch 1960; Louw and Seely 1982) where little or no primary productivity occurs, can nonetheless support resident arthropod populations. Indeed, Swan (1963) coined the term *aeolian zone* for the habitat of such communities supported by allochthonous inputs of nutrients carried by winds; aeolian communities have been documented on several Cascade volcanoes (Mann et al. 1980; Edwards 1987). The importance of allochthonous inputs has long been appreciated for aquatic systems (Teal 1957), and studies of Polis and others (Polis et al. 1997; Polis and Hurd 1996) have more recently demonstrated its importance, as well, for terrestrial communities other than alpine zones.

Arthropod dispersal thus plays a dual role in the initial stages of succession, providing a source of primary colonists of new sites and nutrients in the bodies of immigrants that die upon reaching the site. Our working hypothesis for the study of primary colonization on Mount St. Helens was that an assemblage of arthropod predators and scavengers comparable to those found in other barren habitats would be the primary colonizers of the severely disturbed surfaces, where no residual biota remained.

9.2 Methods of Sampling in the Posteruption Landscape

In the area of the Pumice Plain, all biota was cooked, buried, blown away, or scoured clear. The inundation of the Pumice Plain by the debris avalanche, its subjection to the lateral blast, and then its coverage by extremely hot pyroclastic-flow deposits are described in Chapter 3 (this volume). Beyond this core area, the force of the lateral blast left a fan-shaped area of blown-down forest (blowdown zone; Figure 9.1) bordered by a narrow zone of standing trees singed and killed by the heat of the blast. Finally, the plineal phase of the eruption deposited tephra over a vast area immediately surrounding the volcano with the deepest deposits to the east-northeast.

We established sample sites in 1981 through 1986 at locations covering a range of disturbance intensities, from the pyroclastic-flow surfaces of the Pumice Plain to the tephra deposited in the plineal phase of the eruption in the eastern portion of the blowdown zone, where tephra depths at sample sites ranged from less than 10 cm to more than 0.5 m. Most of the blowdown sites were situated in preeruption clear-cuts because they provided a greater diversity of preeruption arthropod populations, especially ants, from which an index of destruction could be determined. Ant colonies are common in clear-cuts and aspects of their biology (e.g., being generalist predators and nesting below ground surface or in dead wood) enhanced their chances of survival. For comparison, sites were also established in clear-cuts and old-growth forest about 90 km north-northeast of the volcano. Clear-cuts in lightly dusted areas with less than 2 cm of tephra depth from the May 18 eruption were used as "control" sites, which ranged in age from 1 to 22 years, and samples from these sites provided a picture of secondary succession following clear-cutting. This pattern allowed an estimation of the community diversity and composition of clear-cuts near the volcano before the eruption.

We sampled arthropod activity using pitfall traps constructed from plastic tumblers set level with the ground surface in sleeves of polyvinyl chloride plastic pipe. The cups were partially filled with a 50% aqueous solution of ethylene glycol and were protected from rain and disturbance by a plywood cover supported about 2 cm above the soil surface by a tripod of nails. Pitfalls were sampled at about 2-week intervals, generally from May through October. Most results are based on linear arrays of 10 to 20 pitfall traps at each site, with pitfalls within a site placed at 10-m intervals. Comparisons between vegetated patches and the surrounding barren tephra were made at three sites with paired sets of five pitfalls each in vegetated and adjacent nonvegetated areas. Daily activity of colonizing arthropods, which were abundant in the Pumice Plain, was measured by captures in a grid of dry pitfalls in 1986. Forty-nine pitfalls were set at 10-m intervals in a 7×7 grid. The traps were sampled at 1- to 4-hour intervals for about 40 hours on three occasions. All captured individuals were identified and then released at least 2 m from the trap.

We sampled the flux of organic material from aerial fallout of arthropods and plant fragments at the ground surface in the Pumice Plain using linear arrays of 10 fallout collectors spaced 10 m apart. The collectors were composed of 0.1-m^2 wooden frames containing a monolayer of close-packed golf balls that simulated the desert pavement-like surface and provided dead space where material could accumulate (Edwards 1986b; Edwards and Sugg 1993). The frames were backed with fine nylon mesh to allow drainage but to retain the fallout. The accumulated contents of the collector, comprising fine mineral material and organic fragments, were removed at about 2-week intervals. Arthropod and other organic fragments were separated in the laboratory. Measures of the contribution of elemental nutrients over time by arthropod fallout were made by use of dried fruit fly samples set in situ as a fallout surrogate, as discussed later in this chapter.

9.3 Survival Patterns

One of the keys to the ubiquity of terrestrial arthropods is their capacity for dispersal. As an obligate phase of their life histories, the majority of insects and spiders exploit wind patterns either actively by flight or passively by ballooning on silk threads. It is a wasteful process to the extent that many dispersers end their lives in hostile places, but it ensures that some find their niche in diverse landscapes that are changing in space and time. Thus, the dispersal behavior patterns of arthropods ensured that transient populations of insects and arachnids were arriving on the Pumice Plain and other impacted areas shortly after the eruption. Indeed, search and rescue helicopter pilots reported seeing numerous insects on the tephra in the first days after the eruption. The problem lay in knowing whether individual arthropods present at a site were recent arrivals or from a surviving local population. Any species capable of aerial dispersal could conceivably have colonized a site since the eruption occurred. Only species limited to a pedestrian mode of dispersal (hereafter referred to as pedestrians), and hence limited to relatively small annual increments of population spread, could be used as unequivocal evidence of local survival.

The question of local survival in refugia in the posteruption landscape is a classic issue that has been argued since the eruption of Krakatau (Thornton 1996). Dammerman (1948) first raised the critical question: Given survival, what is the possibility of persisting when food resources are gone from the posteruption landscape? Ants at Mount St. Helens provide a good example of initial survival and subsequent death. Large numbers of carpenter ants (*Camponotus*) were found foraging on barren tephra at the border of the blast area 1 month after the eruption. These ants must have emerged from a nest in dead wood that was presumably sheltered in the lee of a ridge and covered by deep snow. Windborne insects and localized blooms of fungal growth around seepages appeared to be the only source of nutrients. The disappearance of ants from this location in subsequent years is presumed to have been caused by the absence of sufficient food. Kuwayama (1929) found a comparable pattern after the June 1929 eruption of Mount Komagatake in Japan, where ants were active 9 days after the eruption, but all colonies were dead 11 days later.

Because of their small size and cryptic habits, it was expected that, in contrast to the Pumice Plain, many arthropods initially survived the eruption in the blowdown zone. The possibility of survival was enhanced by the timing of the eruption in early spring, when many resident species were still dormant at the elevations immediately surrounding the volcano. Even further protection was provided in some locations by snowbanks in protected areas on north slopes or shaded by the forest canopy. Thus, there may have been widespread survival of the immediate impact of the eruption, but survivors in the blowdown zone would also find the local habitat drastically changed; long-term survival would depend on the ability of individuals to find adequate resources.

As expected, samples from the Pumice Plain indicated that local survival was nil. The extreme heat and great thickness of the deposits killed and buried any potential survivors. Virtually all arthropods taken in our sampling on the Pumice Plain from 1981 to 1983 were capable of aerial dispersal (the few individual exceptions are addressed below), whereas pedestrian species as well as aerial dispersers were common at all sites sampled in the blowdown zone (Table 9.1).

Arthropods in the blowdown zone had a range of food resources available. In clear-cuts especially, by midsummer 1980 there was emergent herbaceous vegetation dominated by fireweed (*Chamerion angustifolium*) that supported herbivores such as aphids, moths, and leaf beetles. Predators [e.g., carabid (Carabidae), staphylinid (Staphylinidae), and ladybird beetles (Coccinellidae)] had local herbivore populations available as well as the fallout of dispersing arthropods. Generalists, such as ants, were able to subsist on local arthropod survivors and new immigrants as well as on honeydew from aphids that colonized emergent vegetation. Dead wood provided habitat and food for a variety of arthropods [e.g., longhorn beetles (Cerambycidae), throscid beetles (Throscidae), and robberflies (Asilidae)]. Furthermore, wood-rotting fungi using the abundant resource of newly dead trees provided a resource for a variety of fungivores, such as the tenebrionid beetle *Iphthimus serratus*.

Although arthropod survival was apparent at all sites sampled across the blowdown zone, it was not uniform. Sampling of clear-cuts 90 km to the north-northeast of the volcano in 1983 revealed 9 to 13 ant species in all clear-cuts more than 5 years old, whereas sampling sites established in 12-year-old clear-cuts across the eastern portion of the blowdown zone revealed only 1 to 9 ant species. The number of surviving ant species in the blowdown was related to the depth of tephra deposited in the plineal phase of the eruption (Figure 9.2); where tephra depths were less than 20 cm, species number was similar to that of control sites.

Among pedestrians found at blowdown sites, and thus assumed to have survived the eruption in various refugia, were predators such as flightless carabids [e.g., *Pterostichus* (*Hypherpes*) spp. (including *P. neobrunneus*, *P. castaneus*, and *P. herculaneus*) and *Scaphinotus* spp. (*S. angusticollis* and *S. marginatus*)]. Other pedestrian predators included centipedes and nonballooning spiders, such as agelenids (*Cybaeus* sp.), hahniids (*Neoantistea* sp.), and antrodiaetids (*Antrodiaetus* sp.). Pedestrian scavenger/predators included

TABLE 9.1. Summary of presence or absence of select taxa from samples taken at 17 blowdown sites and 3 sites in the Pumice Plain from 1981 to 1983.

	Blowdown	Pumice Plain
Aerial dispersers:		
Formicidae (alates)	X	X
Carabidae	X	X
(e.g., *Bembidion, Harpalus, Amara*)		
Araneida	X	X
(e.g., Lyniphiidae, Lycosidae, Thomisidae)		
Orthoptera	X	X
(Acrididae)		
Pedestrian taxa:		
Formicidae (workers)	X	(x)[a]
Carabidae	X	(x)[a]
[e.g., *Scaphinotus* and *Pterostichus* (*Hypherpes*)]		
Araneida	X	
(e.g., Agelenidae, Antrodiaetidae)		
Phalangida	X	
Chilopoda	X	
Diplopoda	X	
Orthoptera	X	
(e.g., Gryllacrididae, Prophalongopsidae)		
Notoptera	X	
(Grylloblattidae)		

Pedestrian taxa are those incapable of aerial dispersal and so indicate probable local survival of populations following the eruption.
[a] Pedestrian taxa in the Pumice Plain are considered aberrant cases (see text).
Source: From Sugg and Edwards (1998); used with permission.

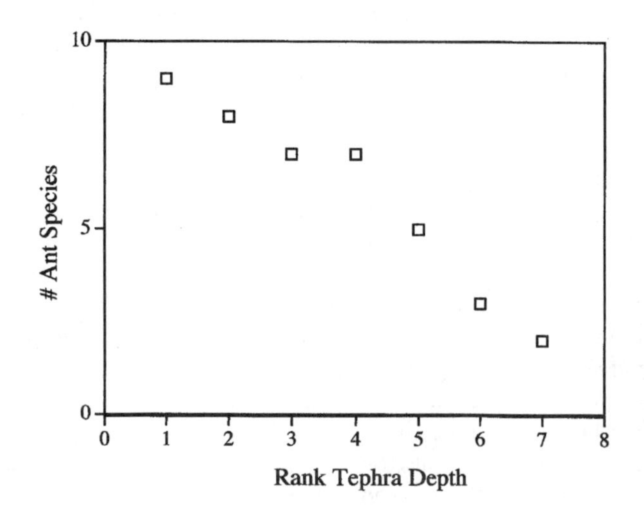

FIGURE 9.2. Numbers of ant species found at sites in the blowdown affected by different depths of tephra. Sites are rank ordered from least (1; 12 cm) to greatest (7; 52 cm) in tephra depth.

camel crickets (*Pristoceutophilus* spp.) and phalangids (e.g., *Leptobunus* sp.).

One dramatic die-off among pedestrian taxa was noted in the vicinity of Ryan Lake in September 1980, when innumerable dead millipedes (*Harpaphe* sp.) were found in the blowdown zone. These millipedes are important litter consumers and died with guts packed with ash ingested as they attempted to feed on ash-covered litter.

The diverse microarthropods that are characteristic members of the soil microfauna (e.g., mites and springtails) were present in low numbers in the Pumice Plain but were more abundant in the blowdown zone in the early years following the eruption. They are susceptible to wind transport because of their small size, and so their later presence in the Pumice Plain could not be taken as evidence of local survival. Sminthurid Collembola were the most numerous microarthropods taken in pitfall traps on the barren surface of the pyroclastic-flow deposits near Spirit Lake. These springtails live not only in soil but also on the foliage of conifers, where they are often abundant, and their presence in pitfall captures likely reflects their propensity for passive wind carriage from the distant forest canopy. The minimal presence of mites, springtails, and other soil microarthropods on the barren surfaces of the Pumice Plain is probably because of the dearth of organic material in the posteruption surface deposits.

9.4 Diversity of Arthropod Immigrants to the Pumice Plain

The capacity of arthropods for aerial dispersal is profound but little understood because of the difficulty of differentiating dispersers from local residents in most habitats. Mount St. Helens provided an opportunity to study this phenomenon. A wide variety of immigrants was expected to arrive on the denuded or new mineral surfaces, but few of those were adapted for survival in barren sites such as the Pumice Plain.

A sense of the diversity of arthropod immigrants to the Pumice Plain comes from analysis at the family level of pitfall samples, amounting to more than 20,000 specimens. From 1981 to 1983, when all materials taken in pitfall trap collections other than the few known pioneer residents (see following) were certainly of allochthonous origin, 150 families representing 18 orders were identified. Taxanomic identifications in this chapter are based on the Integrated Taxonomic Information System (http://www.itis.usda.gov).

Data at the species level for diversity of early arrivals are available for 25 of the 150 arthropod families sampled on the Pumice Plain. The total species count for this sample of 9 spider and 16 insect families is 250, giving a mean number of 10 species per family. Although the selection of families for identification to species level was made on the basis of available expertise and thus not in a strictly random fashion, it was not biased toward species-rich groups. An independent set of data concerning arthropod diversity in the Cascade Range comes from the detailed catalogue of arthropod species recorded at the H.J. Andrews Experimental Forest, a Long-Term Ecological Research site (Parsons et al. 1991). The average number of species per family in this tabulation source is 8.6, a figure close to the value of 10 for our Mount St. Helens material. On the basis of these data, we estimate that at least 1500 arthropod species reached the surface of the Pumice Plain during the first few years following the eruption.

Insect immigrants made up about 80% of the arthropod fallout. At the family level, flies (Diptera) dominated the fallout diversity with 42 families. The flies were followed by beetles (Coleoptera) with 30 families, bugs and close kin (Hemiptera/Homoptera) with 18, butterflies and moths (Lepidoptera) with 15, and wasps and relatives (Hymenoptera) with 10. Minor orders were represented by 1 to 6 families, and spiders contributed 9 families. Lacewings (Neuroptera) may be one example of a group represented in the fallout but unable to establish breeding populations on the barren mineral surfaces. On Mount St. Helens, 250 lacewing specimens from 23 species were collected in traps. The habitat preferences of the majority of the lacewing species captured indicate the prevalence of long-distance dispersal on winds blowing predominantly from forest and farmland southwest of the mountain (Sugg et al. 1994).

Spiders were well represented in the early posteruption fallout; from 1981 to 1986, 14,325 specimens comprising 125 species were taken in pitfalls and fallout collectors, making up 23% of the arthropods that reached the sampling sites. Overall, about half of them were wolf spiders (Lycosidae), with linyphiids contributing 34% of the catch (Crawford et al. 1995). The eruption of Mount St. Helens provided a unique opportunity to examine a little-studied aspect of spider dispersal on wind currents by means of ballooning on threads of silk. Although that behavior has been known for centuries, the issue of its ecological importance has been subject to debate, for there have been few opportunities to measure rates of landing at sites known to be otherwise devoid of spiders. This deficiency of data has led to underestimates of the importance of ballooning in dispersal (e.g., by Decae 1987 and by Wise 1993). During the 125-day field season in 1983, the average rate of spider arrival, as measured by occurrence in fallout collectors, amounted to 0.84 spiders m^{-2} day^{-1}. These observations bear on the issue of metapopulation dynamics (Hanski 1999), for which there are few data concerning spiders. Although standard pitfall sampling of spiders in the Pumice Plain during the first few years after the eruption revealed the presence of significant numbers, the apparent absence of reproduction indicates that these spiders represented sink populations of ballooners from distant sources, in some cases demonstrably more than 50 km distant (Crawford et al. 1995). They were unable to survive, presumably because of the rigors of the physical environment, for example, high summer temperatures, insolation, and desiccation. By 1986, six species of spider, comprising two lycosids and four linyphiids, had established reproducing populations on the Pumice Plain, but only

FIGURE 9.3. Primary colonist beetles from the Pumice Plain at Mount St. Helens. Carabidae: a, *Apristus constrictus*; b, *Bembidion planatum*; c, *B. improvidens*; d, *B. obscurellum*; e, *Nebria eshcholtzii*; f, *Opisthius richardsoni*. Trachypachidae: g, *Trachypachus holmbergii*. Agyrtidae: h, *Apteroloma caraboides*. Tenebrionidae: i, *Scaphidema pictum*.

at sites where sparse pioneer vegetation had already become established.

9.5 Primary Arthropod Colonists of the Pumice Plain

The main criterion for recognition of primary colonists on barren tephra surfaces in the Pumice Plain was firm evidence for successful breeding populations, such as juvenile stages (beetle larvae or hemipteran nymphs) or egg sacs (lycosid spiders). On this basis, we found four beetle families (Carabidae, Trachypachidae, Agyrtidae, and Tenebrionidae; Figure 9.3) and two families of true bugs (Lygaeidae and Saldidae) that established breeding populations on the Pumice Plain within 3 years of the eruption. Carabid species dominated in numbers of species and of individuals. Although it was not possible to match adults with larvae for all carabid and trachypachid species, we were able to assign larvae to five genera: *Apristus*, *Bembidion*, *Nebria*, and *Opithius* for carabids and *Trachypachus* (Thompson 1979). Larval tiger beetles (*Cicindela* spp.) are sedentary tunnel dwellers and thus not susceptible to pitfall capture but were observed in moist streamside tephra from 1984 on. The single species of agyrtid and tenebrionid beetles common in the samples had larvae that could be unequivocally recognized.

Our list of species assumed to have colonized in the first 5 years (Table 9.2) is based on presence of juvenile stages and, in most cases, clear evidence of increasing numbers of adults. This list is probably conservative. We have excluded taxa (e.g., the carabid *Trechus obtusus*) for which there was evidence of increasing numbers but no capture of larvae assignable to that genus. We also limited our account to those taxa apparently successful in breeding on the barren mineral surfaces. Small patches of vegetation that established in the Pumice Plain in the first few years could support a more standard food web of producers, herbivores, and predators, but these sites amounted to an insignificant fraction of the area of the Pumice Plain in the years immediately following the eruption. Nonetheless, vegetated patches had a significant, albeit localized, effect.

The first colonists of the Pumice Plain with apparent reproductive success were species of the carabid beetle *Bembidion,* a genus characteristic of disturbed habitats, such as the barren margins of braided rivers, gravel pits, open periglacial ground, and alpine snowfield fringes (Lindroth 1963; Anderson 1983; Mann et al. 1980). Three species, *Bembidion planatum*, *B. improvidens*, and *B. obscurellum*, were present in 1981, and these were the first to show a population increase during

TABLE 9.2. Species successfully colonizing barren pyroclastic-flow surfaces in the Pumice Plain at Mount St. Helens by 1985.

Aerial dispersers:
Coleoptera
- Carabidae
  - *Apristus constrictus*
  - *Bembidion breve (= incertum)*
  - *B. improvidens*
  - *B. obscurellum*
  - *B. planatum*
  - *B. quadrifoveolatum*
  - *B. recticolle*
  - *B. transversale*
  - *Cicindela oregona*
  - *Nebria escholtzii*
  - *N. mannerheimii*
  - *N. sahlbergii*
  - *Opisthius richardsoni*
- Trachypachidae
  - *Trachypachus holmbergi*
  - *T. slevini*
- Agyrtidae
  - *Apteroloma caraboides*
- Tenebrionidae
  - *Scaphidema pictum*

Hemiptera
- Saldidae
  - *Saldula* sp.
- Lygaeidae
  - *Geocoris* sp.

Pedestrian dispersers:
Orthoptera
- Gryllacrididae
  - *Pristoceutophilus* spp.

Notoptera
- Grylloblattidae
  - *Grylloblatta* sp.

subsequent years (Sugg and Edwards 1998). Two of these species, *B. planatum* and *B. improvidens*, are generally found in moist locations, such as the margin of streams or melting snowfields, whereas *B. obscurellum* is relatively independent of moist habitats (Lindroth 1963) and is found even in arid regions throughout the West. In the Pumice Plain, however, even reputedly hygrophilous species, such as *B. planatum*, colonized places distant from streams. It seems that the tephra deposits of the Pumice Plain had sufficient water-holding capacity to provide a suitable habitat for developing carabid larvae. For many *Bembidion* species, the nature of the substratum, especially grain size, may be more important than proximity to water (Anderson 1983). The *Trachypachus* spp. and *Apristus constrictus* are also associated with riparian habitats but are not restricted to the water's edge (Lindroth 1961, 1968). *Trachypachus holmbergi* was the most abundant caraboid at clear-cut sites in the blowdown and outside the impacted area, and these sites were probably the source of the immigrants to the Pumice Plain. Of all species, the most successful in terms of numbers of adults trapped was the carabid *Bembidion planatum*, which was abundant at all sites sampled. By 1985, the success of *B. planatum* was matched by the agyrtid *Apteroloma caraboides*, although it did not colonize as rapidly as did the *Bembidion* species.Before this study, little was known of the biology of *Apteroloma caraboides*. It was thought to feed on rotting vegetation (Van Dyke 1928), but from its success on the Pumice Plain, where there was no significant vegetation during its period of establishment, it is clear that *A. caraboides* is a generalist that can also subsist on organic fallout.

Carabid beetles were the most diverse of the primary colonists and the most successful, as evidenced by increasing numbers of adults and larvae. Carabids also manifested the greatest increases in species number and abundance from 1981 to 1985 (Sugg and Edwards 1998). The carabid genus *Bembidion*, in particular *B. planatum* and several of the *incertum* species group (*sensu* Lindroth 1963), mostly *B. improvidens*, showed the greatest reproductive success, with yearly increases in both adults and larvae in the years 1983 to 1985 (Figure 9.4). The greatest species diversity was found along stream margins, which is the characteristic habitat for most of the taxa showing evidence of reproductive success (Lindroth 1963).

In contrast to the carabids already discussed, the tenebrionid beetle *Scaphidema pictum* showed only limited population growth. A common inhabitant of river bars (Hatch 1965), the scavenging habit of this species was confirmed by our observation of several adults feeding within the body of a dead grasshopper on the Pumice Plain.

The two pioneer true bugs differed markedly in habitat preference. The saldid bug *Saldula* was found almost exclusively along stream edges, but the lygaeid *Geocoris* occurred far from open water. Lygaeids are usually thought to be seed feeders, but *Geocoris* is known to be a predator, as are other lygaeids that utilize arthropod fallout (Ashlock and Gagne 1983). While spiders were an abundant element of the immigrant fauna, they did not establish breeding populations until several years after the pioneer beetles and, for reasons that are not yet clear, only after the appearance of pioneer vegetation.

9.6 Colonization of the Pumice Plain by Pedestrians

The presence of two cricket-like insects, gryllacridids (*Pristoceutophilus* spp.) and grylloblattids (an undescribed species of *Grylloblatta*; R.L. Crawford, personal communication), at Willow Spring (WS) on the Pumice Plain late in October 1984 is notable as the first indication of colonization of the Pumice Plain by pedestrian species. Both occur in the surrounding forestland and were collected in the blowdown zone. Although these insects may seem out of place on open tephra surfaces, the presence of similar insects in other volcanic landscapes is well documented (Howarth 1979; Thornton 2000). A year later, in 1985, yet another pedestrian, the spider-relative harvestmen (phalangids), also reached the Pumice Plain on

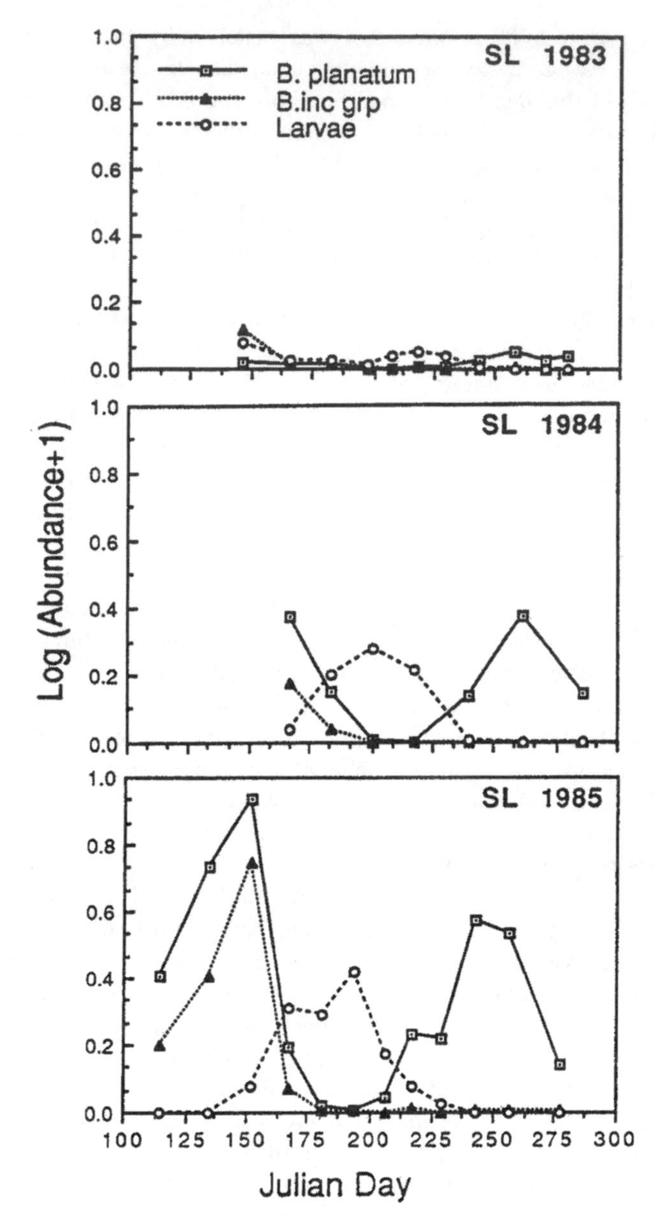

FIGURE 9.4. Relative abundance of *Bembidion planatum* and members of the *Bembidion incertum* species group, referred to as "B.inc grp" (*sensu* Lindroth 1963), mainly *B. improvidens*, and *Bembidion* larvae at sites near Spirit Lake from 1983 to 1985. Abundance is expressed as mean number of captures per pitfall day. [From Sugg and Edwards (1998); used with permission.]

foot. However, their arrival may not have been simply by walking, for harvestmen and grylloblattids are frequently encountered on snow surfaces during winter when the air temperature is above zero. Thus, they could have been blown for considerable distances across the snow surface.

Two exceptional pitfall captures of pedestrian insects warrant comment. The first is the capture of four worker ants (*Formica fusca*) on the Pumice Plain in 1982 at a site immediately below the ridge forming the northern boundary of the Pumice Plain. Ant colonies probably survived under snowpack on the lee side of this ridge, and, although wingless, workers may have been carried to the Pumice Plain by the frequently high winds that rake the area. The second case is the capture of a single specimen of the carabid beetle *Amerizus* (= *Bembidion*) *oblongulum* on the Pumice Plain near Spirit Lake in 1983. This remarkable species is subterranean in habit, with reduced eyes and nonfunctional wing vestiges (Lindroth 1963). Again, wind dispersal seems to be the most probable source of this single specimen because it was unlikely to have survived the eruption in situ or been transported by a mobile vector, such as an elk.

9.7 Temporal Resource Sharing by Pumice Plain Predators and Scavengers

It is at first sight seemingly paradoxical that so many species of predators and scavengers should be exploiting the same fallout resource base in the blast zone. This overlap can be explained, however, at least in part, by temporal differences in activity and life-history patterns. Specific patterns of activity are clear. Some, such as *Bembidion planatum* and *Apteroloma caraboides*, are nocturnal, whereas others, such as *Apristus constrictus* and lycosid spiders, are mainly day active (Figure 9.5). Similarly, there are differences in periods of adult reproductive activity and larval development. For example, the two pioneer species that were most abundant by 1985, the agyrtid beetle *Apteroloma caraboides* and the carabid *Bembidion planatum*, do not overlap in growth and development of larval cohorts. The carabid breeds in the spring and the agyrtid in the fall (Figure 9.6). The day-active species face high temperatures and desiccating conditions during mid- to late summer, but their capacity to thrive under these conditions is, at least in part, a result of their small size, enabling them to remain in the shade of pebbles and stones and to make only brief forays into direct sunlight to capture prey.

9.8 Arthropod Immigration as a Source of Nutrient Enrichment on the Pumice Plain

As already noted, the majority of winged insects enter a dispersive phase at some point during adult life. At that point, they enter the air column and actively migrate or are carried passively on wind currents, sometimes for great distances (Drake and Gatehouse 1995). They may then be deposited by local winds at inappropriate sites from which, because of fatigue, desiccation, or low temperature, they are unable to reenter the air column. These are the arthropods of the fallout fauna, the derelicts of dispersal (Edwards 1986b) that we have observed on the pyroclastic-flow deposits of the Pumice Plain.

The organic fallout not only provides a resource for the resident predators and scavengers but also proved to constitute a significant source of nutrient elements to the site. This

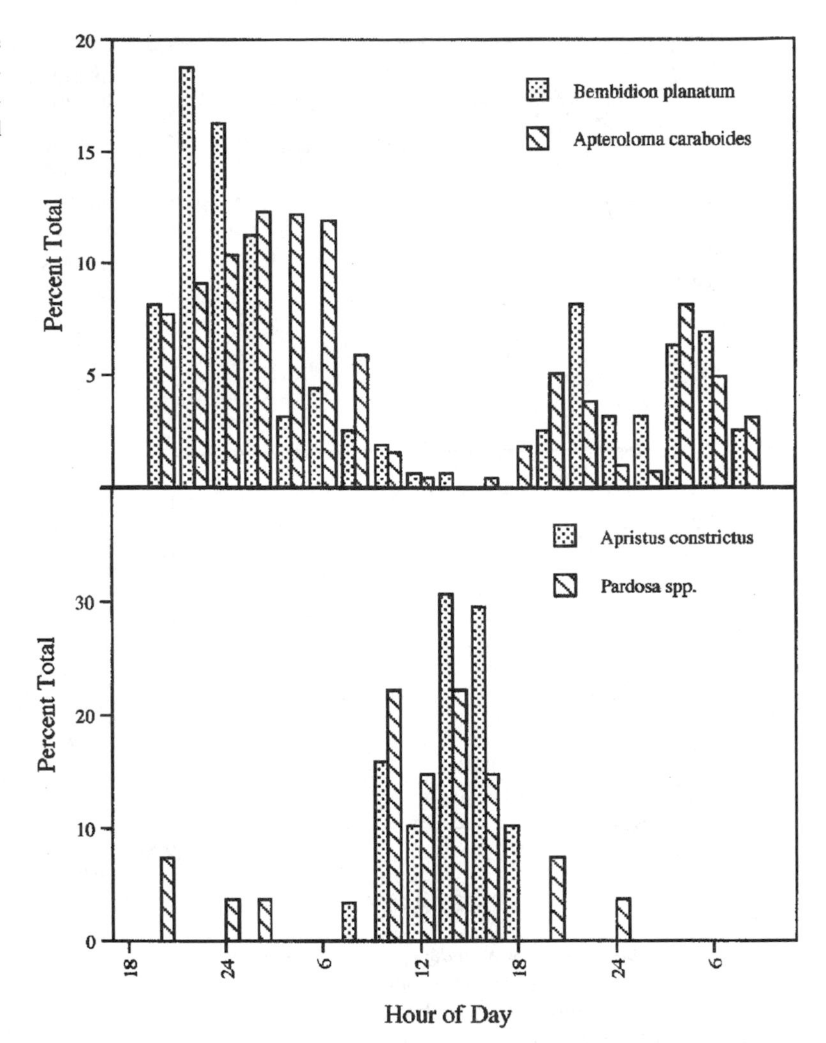

FIGURE 9.5. Diurnal activity patterns as reflected in live pitfall captures of three beetles, on the Pumice Plain. *Bembidion planatum*, *Apteroloma caraboides*, *Apristus constrictus*, and wolf spiders (Lycosidae). [From Sugg and Edwards (1998); used with permission.]

source was particularly notable for Mount St. Helens because the newly fallen tephra was nutrient poor, and the input of nutrients from arthropod fallout could play a significant part in generating soil fertility. Of course, the pioneer plants rapidly added organic material, carbon, and nitrogen to the soil, but phosphorus and other mineral nutrients in the siliceous tephra must have come from exogenous sources. Given the low phosphorus content of plant fragments in comparison with animal tissue, we chose to evaluate the contribution of arthropods alone. The quantity and composition of this organic fallout and its significance as a source of nutrients to the initially impoverished tephra was estimated for the early posteruption years. Estimates of fallout biomass were made with fallout collectors designed to simulate the surface of the pyroclastic-flow deposits and thus to give an index of the true flux of organic matter at the surface (Edwards and Sugg 1993). After the first year, known predator and scavenger species that may have been of local origin were removed from the sample before biomass determinations were made. Thus, the data underestimate the fallout flux to the extent that some of the scavengers and predators were new arrivals. Further, large insects, such as grasshoppers and butterflies that were not enclosed within the fallout collectors, were subject to bird predation and thus lost from our samples. Inputs as high as 18 mg m^{-2} day^{-1} were recorded for dried arthropod bodies and as high as 26 mg m^{-2} day^{-1} for nonarthropod material, consisting mainly of lichen and plant fragments. The overall average for organic fallout was in the range of 5 to 15 mg dry wt m^{-2} day^{-1} for 100 days encompassing the summer months. Of this material, 2 to 10 mg comprised bodies of immigrants, such as aphids, flies, and a broad range of other arthropods (Edwards and Sugg 1993). Comparable figures were found by Ashmole and Ashmole (1988) on the lava fields of Tenerife and by Heiniger (1989) in the Bernese Oberland. The main source of the windborne organic fallout carried on prevailing winds to Mount St. Helens is the fertile agricultural and forest lowlands to the west and southwest, while nearer the mountain are areas of blowdown, standing dead trees, mudflows, and riparian habitats with vegetation that supports a diverse arthropod fauna, also contributing to the input of fallout to the Pumice Plain.

Experiments were carried out to determine the rate of release of nutrients from insect bodies. Mesh bags containing weighed

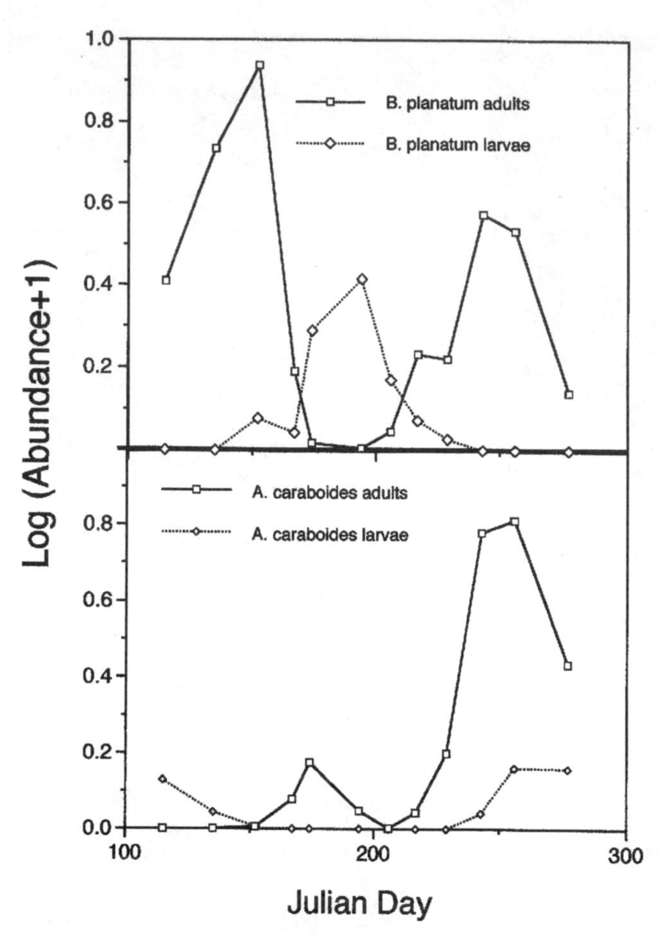

FIGURE 9.6. Annual pattern of abundance of the beetles *Bembidion planatum* and *Apteroloma caraboides*, adults and larvae, based on total numbers taken in pitfall traps on the Pumice Plain. [From Sugg and Edwards (1998); used with permission.]

amounts of dried adult fruit flies (*Drosophila melanogaster*) were placed on the pyroclastic-flow surface and lightly covered with tephra. The bags were subsequently sampled at intervals from 17 to 700 days. The nitrogen content of the insect material from bags from the field declined by 25% during the first 17 days and remained almost constant thereafter, probably because of the stability of the chitin component of the cuticle. The phosphorus content decreased by 73% during the first 17 days and thereafter by a further 10%, giving a total loss of phosphorus of 83%. Using these data, other estimates of the nitrogen and phosphorus content of arthropod fallout, and mean figures for the arthropod fallout on the pyroclastic-flow deposits at Mount St. Helens, we estimate that 80 mg fixed nitrogen and 5.5 mg phosphorus per square meter were added annually to the early posteruption surfaces. We know that these figures are minimal estimates, because data from our fallout collectors certainly underestimate net fallout. Plant and lichen fragments added their contribution, but they were not analyzed in our study.

The original pyroclastic-flow and tephra materials that formed the surface layers of the posteruption landscape contained very little carbon, nitrogen, and phosphorus. For example, the total organic carbon and total nitrogen content of 1980 samples of pyroclastic-flow materials taken near our arthropod sampling sites were reported as 0 (Engle 1983). By 1985, the levels were still relatively low, with 0.5 to 1 g kg^{-1} organic material, 10 to 90 mg kg^{-1} Kjeldahl-extractable nitrogen, and 0.3 to 0.4 g kg^{-1} phosphorus (Nuhn 1987). The measured increase must have been derived from imported material, of which arthropod fallout was a significant fraction, as revealed by our fallout collectors. The estimated rates of accumulation of total organic carbon (97.7 mg kg^{-1} $year^{-1}$) and nitrogen (2.8 mg kg^{-1} $year^{-1}$) in plant-free pyroclastic-flow material from Mount St. Helens (Halvorson et al. 1991a) are comparable to estimates for other volcanic areas at latitudes north and south of Mount St. Helens, for example, Mount Katmai (Griggs 1933) and Mount Shasta (Dickson and Crocker 1953). It should be borne in mind that the mean figures given here do not reflect the heterogeneity of the natural surfaces. Even on the superficially monotonous surface in the Pumice Plain area, there were many small cavities and declivities, sometimes covered by the webs of lyniphiid spiders, where we found aggregations of both arthropod fragments and seeds. Seeds in these sites were thus germinating in arthropod compost.

As already noted, the source of much of the arthropod fallout must have been distant (i.e., several to many kilometers), but there was also clear evidence of redistribution of organic material within the impacted area. For example, bodies of water, such as Spirit Lake, contained a rich organic soup derived from cooked vegetation. The soup provided the substrate for massive bacterial blooms during the first 2 years after the eruption. In Spirit Lake, those blooms reached the extraordinary figure of nearly a half billion cells per milliliter (Baross et al. 1982). This bloom provided the substrate, in turn, for mosquito larvae, which were able to tolerate the anoxic condition because of their air-breathing siphon that acts as a snorkel. The prolific productivity of mosquitoes in these waters showed up in high captures in 1981 pitfall traps situated several kilometers from the nearest open water and thus provided a striking example of nutrient redistribution at the landscape scale. Although the mosquito pulse was limited to the first year or two, other aquatic insects such as chironomid midges colonized the streams and lakes and maintained more local sources of fallout; seeps acted as nurseries, and the very open landscape permitted much greater wind dispersal than one normally sees in a forested landscape.

9.9 Role of Vegetated Patches

The newly established ecological systems also provided the opportunity to ask what effect plant establishment had on the primary community. Plants certainly are the base for most food webs, providing food for herbivores, which, in turn, are available for predators. Plants also provide physical habitat, with live and dead foliage giving shelter and holding moisture.

TABLE 9.3. Totals of select taxa in 1986 from sets of five pitfall traps at three sites on the Pumice Plain, PE, PL, WS, defined in the text below.

	PE		PL		WS	
	Vegetated	Barren	Vegetated	Barren	Vegetated	Barren
Bembidion planatum[a]	8	143*	87	216*	174	2460*
Bembidion incertum group[a]	—	—	6	28*	96	128
Nebria spp.[a]	145	427*	41	88*	76	415*
Apristus constrictus[a]	1	9*	23	73**	0	127
Apteroloma caraboides[a]	22	245*	120	259*	156	1068*
Scaphidema pictum[a]	7	12	88	77	24	66
Trachypachus spp.[a]	—	—	20**	9	22	39
Bembidion transversale[a]	—	—	—	—	3	33*
Bembidion dyschirinum[a]	—	—	—	—	12*	0
Trechus obtusus	—	—	—	—	50*	2
Amara sp.	12**	5	32*	4	43*	6
Pterostichus adstrictus	32*	5	19*	0	24*	2
Aphididae (nonwinged)	267*	18	2329	4413*	12	13
Other Homoptera	41*	2	35**	9	115*	4

Aphids and other homopterans are juveniles or nonwinged adults. Lack of an entry indicates that the total number of individuals from the paired transects for the taxon was less than 10. WS, Willow Spring; PE, PL.
[a] These taxa had established breeding populations on barren surfaces by 1985.
* $p < 0.05$, t test.
** $p < 0.1$, t test.

All in all, one might assume that vegetation would act as a magnet for predators attracted to the milder microclimate and populations of potential prey. Three vegetated sites were sampled with paired sets of pitfall traps, five traps placed at approximately 10-m intervals in the vegetated surface and another five in adjacent barren tephra set 10 m from the nearest plant. One site (WS) was a stream edge with the vegetation dominated by willows (*Salix* sp.) and pearly everlasting (*Anaphalis margaritacea*). A second site (PL, which was located on the Pumice Plain, 0.8 km south of Spirit Lake) was dominated by prairie lupine (*Lupinus lepidus*) established in well-drained pyroclastic-flow deposits. The third (PE, which was located on the Pumice Plain, 0.4 km south of Spirit Lake) was vegetated with a mix of grasses and forbs.

Vegetated areas supported populations of plant-feeding insects that were potential prey for the primary community of predatory arthropods. Notable were various homopterans, such as aphids and leafhoppers (Table 9.3). Yet, although this prey base was available in the vegetated patches, those taxa initially successful in the expanse of unvegetated tephra were generally still more abundant on the barren tephra surfaces. These are species that specialize in colonization of barren ground; for them, the presence of plants spoils the neighborhood.

Bembidion planatum and *A. caraboides* were most abundant, and both showed strong preference for unvegetated surfaces (see Table 9.3). Among the taxa that were reproductively successful on the barren tephra by 1985, a mixed response was seen only with *Trachypachus* spp. and the tenebrionid *S. pictum*, with higher numbers in the vegetation at the lupine patch site, PL (Table 9.3). Other carabids were more abundant on vegetated surfaces, namely *B. dyschirinum* and *T. obtusus* at the stream-edge site, WS, and *P. adstrictus* and *Amara* sp. at all sites (Table 9.3). The latter two taxa are of interest because they are characteristic of field habitats (Lindroth 1966, 1968) and were assumed to represent taxa that would become more dominant as vegetation increased, a prediction confirmed by later sampling (see Parmenter et al., Chapter 10, this volume).

The availability of prey by the secondary production of herbivores is not limited to the immediate area with vegetation cover. This pattern is made clear by results from site PL, where senescing lupines resulted in a wave of aphid pedestrian dispersal, moving in great numbers from the dying plants into the unvegetated surroundings and accounting for the greater number of aphids caught in the barren pitfall transect at the site (see Table 9.3). Thus, even if predatory arthropods are disinclined to move into vegetation, islands of plant establishment can provide an available food supply, supplementing arthropod fallout in the surrounding area.

The importance of arthropods in the larger context of primary succession is not limited to the brief time preceding significant vegetation establishment. The course of plant succession itself can be affected by insect populations (Fagan and Bishop 2000; Bishop 2002; Bishop et al., Chapter 11, this volume).

9.10 Conclusions: Primary Succession and Aeolian Communities

We conclude that the first stage of terrestrial primary succession, the colonization of the pyroclastic-flow zone, was initiated at Mount St. Helens by an assemblage of predatory and scavenging arthropods. The influx of unsuccessful immigrants, the "derelicts of dispersal," provided the resource base. The sustenance of communities by organic matter from external sources is well known for aquatic communities; leaf fall, for example, can be the major input to streams (Teal 1957). Aeolian communities are the terrestrial analogue and are widespread in alpine and desert areas (Edwards 1986b, 1987). The prevalence of arthropod dispersal and the magnitude of the biomass reflected in trap captures on the pyroclastic-flow deposits of Mount St. Helens (Edwards and Sugg 1993) imply that this pattern is a widespread and perhaps a general one for terrestrial primary successional habitats, such as retreating glaciers, landslide scars, flood-scoured river bars, and posteruptive volcanic surfaces. These environments dramatically exemplify the importance of spatial dynamics in ecology.

The continuing production of new mineral surfaces by crustal movement, volcanic activity, glacial retreat, isostatic rebound, and floods provides habitat for specialist arthropods, which, along with microorganisms, are the pioneer colonists, often preceding plant colonization by considerable periods. Except for very recent studies, for example, Kaufman (2001), the role of these pioneers has been largely neglected in studies of primary succession, where the emphasis has generally been on plants. We propose that comparable pioneer predatory and scavenging arthropods operate around the entire Pacific Ring of Fire and other volcanic areas, wherever volcanic activity produces new surfaces. Mount St. Helens has erupted at least 20 times in the past 4500 years (Crandall and Mullineux 1978), and in 1980 it was still recovering from the previous eruption 180 years earlier. With every eruptive cycle, ecosystems are destroyed, or altered and the cycle repeats, with pioneer arthropods playing their part in Act 1, Scene 1, of the succession play as consumers of the ubiquitous fallout of arthropod aerial plankton.

Acknowledgments. We thank the following for their assistance in identifying our samples: Rod Crawford (Arachnida), George Ball and David Kavanaugh (Carabidae), and Lita Greve (Neuroptera and Diptera). Paul Banko, Carrie Becker, Dan Mann, and Merrill Peterson assisted with collecting and processing samples. USDA Forest Service personnel from Randle and Mount St. Helens ranger districts, Gifford Pinchot National Forest, were helpful in the course of the fieldwork. This work was supported by NSF Grants DEB 80-21460, DEB 81-0742, and BSR 84-07213. Further support came from the University of Washington, Graduate Research Fund; Sigma Xi; and the Mazamas.

10
Posteruption Arthropod Succession on the Mount St. Helens Volcano: The Ground-Dwelling Beetle Fauna (Coleoptera)

Robert R. Parmenter, Charles M. Crisafulli, Nicole C. Korbe, Gary L. Parsons, Melissa J. Kreutzian, and James A. MacMahon

10.1 Introduction

Arthropods are important components of ecosystems because of the roles they play in pollination, herbivory, granivory, predator–prey interactions, decomposition and nutrient cycling, and soil disturbances. Many species are critical to the structure and functioning of their ecosystem, although some (particularly insects) are considered pests in farmlands and forests because of their detrimental effects from feeding on foliage and transferring pathogens to trees and crops. Arthropods also constitute a high-protein prey resource for vertebrate wildlife (especially small mammals, birds, reptiles, and amphibians), thus contributing to the existence and stability of these wildlife species. As such, studies of arthropod population dynamics and changes in species assemblages following natural disturbances are important for understanding ecosystem responses. In the case of the Mount St. Helens volcanic eruption, studies of arthropods not only can provide information on natural history and ecology of many different species but also are relevant for evaluating theories of disturbance ecology and postdisturbance successional processes.

The 1980 eruption of Mount St. Helens provided researchers with an opportunity to test a wide range of theories concerning the structure and functioning of ecosystems. In particular, the existence of a continuum of disturbance intensity across a large landscape made possible a suite of comparative studies that evaluated the influence of different levels of volcanic disturbance on the survival and initial recolonization patterns of plants and animals. For example, researchers to date have documented the survival and reestablishment of a number of plant species and described the patterns and rates of vegetation successional processes of the disturbed ecosystems (see Lawrence, Chapter 8, this volume; Antos and Zobel, Chapter 4, this volume; Dale et al., Chapter 5, this volume; del Moral et al., Chapter 7, this volume; and references therein). In addition, numerous faunal studies have quantified the eruption's impacts on survival and subsequent short-term responses of small mammals (Andersen 1982; Andersen and MacMahon 1985a,b; Adams et al. 1986a; Johnson 1986; MacMahon et al. 1989; Crisafulli et al., Chapter 14, this volume), birds (Andersen and MacMahon 1986), amphibians (Karlstrom 1986; Hawkins et al. 1988; Crisafulli and Hawkins 1998; Crisafulli et al., Chapter 13, this volume), and arthropods (Edwards et al. 1986; Sugg 1989; Edwards and Sugg 1993; Crawford et al. 1995; Sugg and Edwards 1998; Edwards and Sugg, Chapter 9, this volume).

Information collected on the fauna and flora of Mount St. Helens during the past 20 years facilitates the analysis of recolonization patterns in the context of two ecological theories: relay successional processes (MacMahon 1981) and the intermediate-disturbance hypothesis (Connell 1978). The term relay succession refers to the sequential replacement of species (plant and animal) in an ecosystem recovering from some form of disturbance. This process typically begins with species that either survived the disturbance or immigrated to the site shortly thereafter. Some of these species are well adapted to the disturbed conditions of the site and can greatly increase in abundance, whereas others are poorly adapted and become locally extinct. Biotic interactions (competition, predation, herbivory, and parasitic and disease infections), coupled with abiotic factors (extremes of temperature or moisture), often determine the success or failure of each species survival. Through time, as different species colonize the site, they alter the environment's characteristics (e.g., plant regrowth provides shade, cools soil surface temperatures, increases soil moisture and organic matter, and provides substrate for fungi and vegetation for herbivores). As the environmental conditions change, new opportunities are created for additional species to colonize and dominate, eventually replacing established species that have become competitively inferior in the altered environment; hence, the "relay" of species during postdisturbance succession. This process applies to both plant and animal species assemblages and inherently involves complex interactions among plants and animals (MacMahon 1981).

The second theory, the intermediate-disturbance hypothesis (Connell 1978), addresses the patterns of species richness across a gradient of disturbance. The theory predicts that both undisturbed sites and sites suffering severe, frequent, or large-scale disturbances should have fewer species than sites subjected to "intermediate" levels of disturbance intensity, frequency, or areal extent. The theory is based on the assumption that, in undisturbed sites, certain species are dominant because of their superior competitive abilities and exclude less-competitive species, thereby depressing overall species numbers. On severely disturbed sites, only a few species are capable of surviving in the disturbed environment; and the higher frequency, intensity, or spatial extent of the disturbance resets or arrests the successional process, thereby also depressing species numbers. Under intermediate levels of disturbance, species elements of both undisturbed and disturbed species assemblages would be present on a site, thereby exhibiting a greater composite number of resident species. Hence, over a range of disturbance intensities, frequencies, and/or sizes, the predicted pattern of species richness would be approximately bell shaped, with a peak in species numbers at the intermediate disturbance level (see Figure 1 in Connell 1978, p. 1303). Although intuitively attractive, the intermediate-disturbance hypothesis appears to apply in only limited situations (see Mackey and Currie 2001), suggesting that further testing of the theory is warranted.

The purpose of our study of the Mount St. Helens ground-dwelling beetle assemblages was to assess and analyze the posteruption successional processes with this taxonomically and ecologically diverse group of arthropods. Ground-dwelling beetles were selected for four reasons:

- By examining the *ground-dwelling* species, we can study species inhabiting a similar substrate (the ground's surface) common to all the sites impacted by a volcano's eruption. (Such is not the case with *plant-dwelling* arthropod species in forests and clear-cuts, because plants in some of these sites were totally removed by the eruption.)
- Beetles as a group are fairly well known taxonomically and can, in most cases, be readily and reliably identified to the species level.
- Ground-dwelling beetles are abundant and easily sampled with passively operating pitfall traps.
- Beetle assemblages not only comprise a taxonomically diverse group but also contain members of a wide variety of trophic groups (predators, herbivores, granivores, fungivores, scavengers, carrion feeders, dung feeders, parasitoids, etc.).

This last factor allows us to analyze the posteruption development of beetle trophic composition across the disturbance gradient of the Mount St. Helens region. Finally, previous research has shown that arthropods can have significant influences on successional processes on Mount St. Helens (Fagan and Bishop 2000; Bishop 2002), suggesting that further studies on the composition and temporal changes of the beetle assemblages would be of value for understanding the overall pattern of succession on the volcano.

Our specific goals for the study were to (1) describe the ground-dwelling beetle assemblages in the context of both taxonomic and trophic composition for sites previously supporting forested and clear-cut habitats on the Mount St. Helens volcano; (2) compare these beetle assemblages across the disturbance gradient from the highly disturbed Pumice Plain near Spirit Lake to the undisturbed sites well outside the eruption-affected area; (3) examine the temporal changes in species composition and abundances to ascertain the rate and extent of species turnover (relay succession) on different sites; (4) evaluate the disturbance-intensity component of the intermediate disturbance hypothesis with respect to the beetle assemblages; and (5) describe and compare the trophic composition of the beetle assemblages through time and across the disturbance gradient of the volcano. Our study ran from 1987 to 2000 and thus, when coupled with the extensive work on ground-dwelling arthropods from 1981 to 1985 by Edwards and Sugg (see Chapter 9, this volume), portrays a nearly complete record of change at Mount St. Helens.

10.2 Study Areas and Methods

The sampling sites for this study were located within the Pacific silver fir (*Abies amabilis*) zone, ranging in elevations from 1040 to 1175 m (Franklin and Dyrness 1973). The region had a long history of forest harvesting and had become a mosaic of clear-cuts, replanted forest stands of even-aged trees, and old-growth forest stands. As such, the study was designed to examine the succession of ground-dwelling beetle assemblages within areas that, just before the eruption, had been either standing old-growth forest or young-aged (~15-year-old) clear-cut plantations. Study sites were selected that were of comparable elevation, slope, aspect, and (for clear-cut sites) age since the most recent harvest and replanting. Seven study areas were chosen (Figure 10.1), representing two states of preeruption conditions (forested or clear-cut) and four levels of volcanic disturbance: (1) a single site within the pyroclastic-flow zone on the Pumice Plain located between the pediment slopes of the volcano and the newly formed shore of Spirit Lake (Figure 10.2a,b); (2) two sites within the tree blowdown zone: a clear-cut area in the Smith Creek watershed and a previously forested site near Norway Pass (Figure 10.2c,d); (3) two sites within the tephrafall zone (a clear-cut area and a forested site, known locally as the Hemlock Forest area; Figure 10.3a,b); and (4) two reference sites (a clear-cut and a forested site) unaffected by the volcanic eruption and located on Lonetree Mountain, about 40 km north-northeast of Mount St. Helens (Figure 10.3c). [See also Crisafulli et al. (Chapter 14, this volume, Table 14.1) for quantitative measures of vegetation and abiotic characteristics of these study sites.]

Within each study area, 10 arthropod pitfall traps were installed to sample the ground-dwelling beetle assemblages.

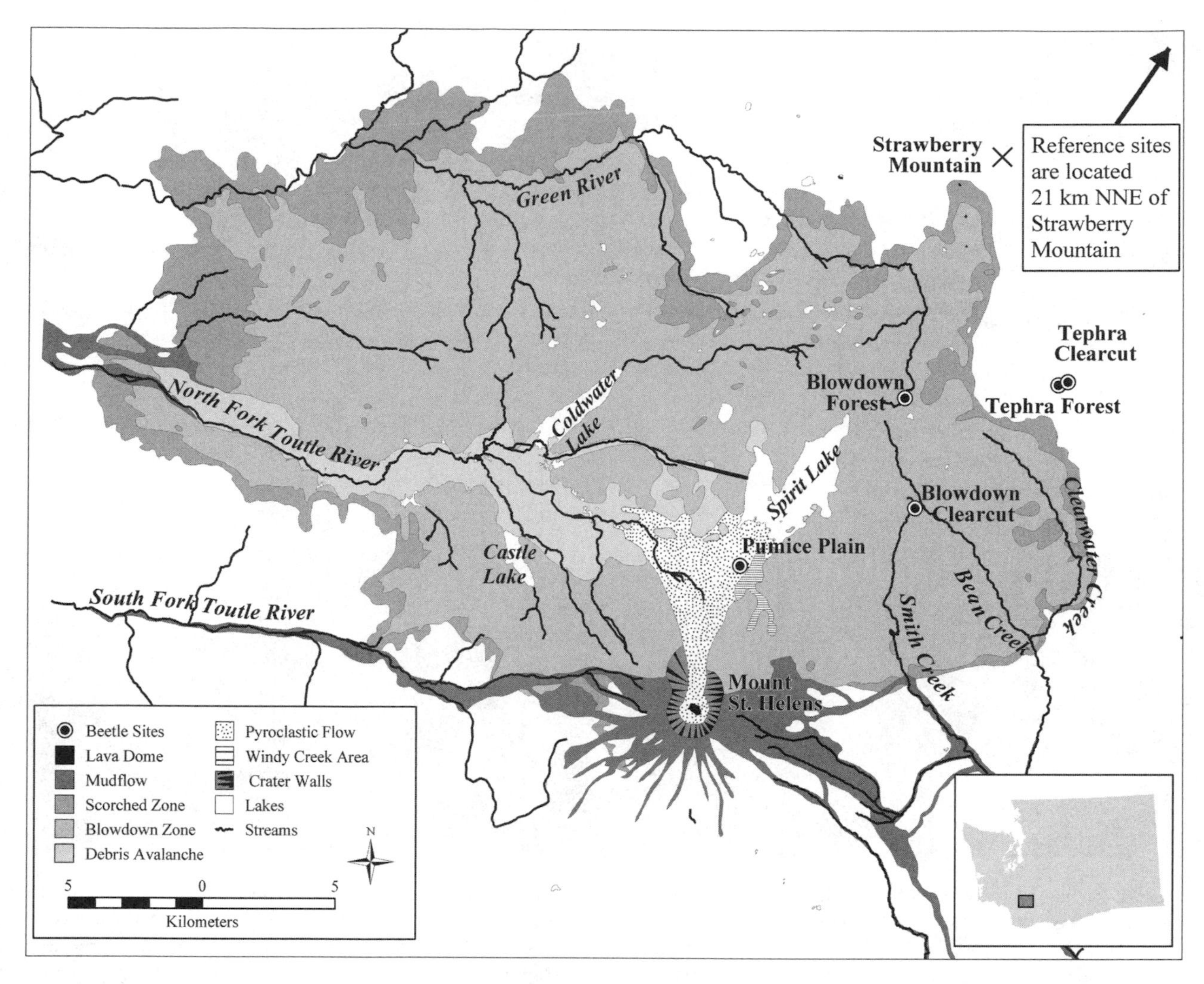

FIGURE 10.1. Locations of study sites within the Mount St. Helens eruption zone.

Traps consisted of metal cans or plastic cups inserted into the ground so that their tops were flush with the soil surface. A second plastic cup filled with propylene glycol was placed inside the trap; this fluid served to preserve captured arthropods, but also prevented freezing or evaporation. Traps were placed at 10- to 15-m intervals across each study area. Traps were opened in the spring (late May to mid-June, depending on time of snowmelt), left open continuously between sample collection times, and closed in autumn at the end of the growing season. Arthropods were collected from the traps in July, August, September, and October of 1987, 1990, 1995, and 2000. All arthropods were preserved in 70% ethanol, transported to the laboratory, sorted, identified, and counted. Species identifications were made by G. Parsons at Michigan State University, and series of voucher specimens are archived in the Division of Arthropods, Museum of Southwestern Biology, University of New Mexico, Albuquerque.

Once the beetle species from each site had been identified and counted, the beetle assemblages of the different study areas were compared with a similarity index that produced percentage similarities of species composition between pairs of study plots. The similarity index (S) was calculated as

$$S_{1,2} = [a/(a + b + c)] \times 100$$

where a = the number of species common to both sites 1 and 2, b = the number of species unique to site 1, and c = the number of species unique to site 2. These paired similarity indices were assembled into a resemblance matrix for each sample year (1987 to 2000) to evaluate patterns of similarity among the study areas and through time.

Patterns of beetle species relay succession on the study sites were evaluated by plotting the mean number of individuals per species collected on each site through time and comparing the temporal abundances of each species. If relay succession was occurring on the sites, then a series of species replacements would be expected in the disturbed study areas, and this phenomenon would be more pronounced in the more heavily disturbed areas of the volcano. For these analyses, we only used

FIGURE 10.2. (a) Northward view of the Pumice Plain study site in 1987, showing the bare tephra/pumice-covered substrate in the foreground and Spirit Lake in the background. (b) Southward view of the Pumice Plain in 2000, with Mount St. Helens in the background. Note the dense cover of lupines and other forbs and grasses; the *white squares* are the rain covers of the arthropod pitfall traps. (c) Norway Pass (blowdown forest) study site in 1986, illustrating the symmetrical patterns of logs from the old-growth forest leveled by the blast wave of the eruption. (d) Norway Pass study area in 2000. Note the surviving Pacific silver firs (*Abies amabilis*) as well as the settling of the logs into the soil's surface.

the most common species (i.e., those with large sample sizes) to reduce the chance of including apparent random extinctions and recruitments caused by small sample sizes [e.g., pseudo-turnover, as denoted by den Boer (1985)].

Finally, to evaluate the trophic structure of the beetle assemblages among the study areas and through time, the proportion of species belonging to five general trophic groups was determined. Those groups were predators, herbivores, omnivores, fungivores, and scavengers. These proportions were then compared across the disturbance gradient and over the time of the study (1987 to 2000). The predicted successional pattern was one of increasing importance of herbivores and fungivores through time as the disturbed sites recovered and as vegetation and decomposing litter (leaves and wood that provide substrate for fungi) became more abundant. Concomitantly, the more heavily disturbed sites would have relatively higher proportions of predators and scavengers in the early posteruption years when compared to the reference areas because of the lack of vegetation resources for herbivores and fungivores. Trophic-group assignments for each species were based on natural history information contained in Arnett and Thomas (2001) and Arnett et al. (2002) and references therein.

10.3 Results and Discussion

10.3.1 The Beetle Assemblages of Mount St. Helens

During this study, we collected 27,074 beetles representing 279 species and 39 families (Table 10.1). The most species rich families included the rove beetles (Staphylinidae, 75 species), ground beetles (Carabidae, 40 species), "click" beetles (Elateridae, 24 species), round fungus beetles (Leiodidae, 21 species), and weevils (Curculionidae, 18 species); the

FIGURE 10.3. (a) A forested area in the tephra-fall zone northeast of Mount St. Helens in 1980. The forest-floor litter layer and low understory vegetation have been buried by tephra. (b) The tephra-fall forest study site in 2000. The new litter layer has nearly covered the tephra surface, and many small shrubs and tree saplings have become established. (c) Old-growth reference-forest study site on Lonetree Mountain, approximately 40 km north-northeast of Mount St. Helens. Note the dense understory vegetation and the presence of decomposing stumps and logs.

178 species from these 5 families comprised 64% of all the species collected. In the sample, 16 families were represented by only a single species. It should be noted at the outset that pitfall traps would not be expected to capture all beetle species living on the study areas, particularly those species that typically live on upper layers of vegetation (e.g., shrubs and trees); hence, many of the rarer species in our data set were likely the result of accidental captures of individuals during short

TABLE 10.1. List of the 39 beetle families (with numbers of species, totaling 279) sampled on the study sites at Mount St. Helens from 1987 to 2000.[a]

Family	Number of species
Agyrtidae	2
Anthicidae	1
Bothrideridae	1
Byrrhidae	8
Cantharidae	2
Carabidae	40
Cerambycidae	6
Chrysomelidae	6
Coccinellidae	7
Corylophidae	1
Cryptophagidae	13
Cucujidae	1
Curculionidae	18
Dermestidae	1
Derodontidae	1
Elateridae	24
Endomychidae	2
Eucnemidae	1
Lampyridae	1
Latridiidae	5
Leiodidae	21
Lycidae	1
Melandryidae	2
Melyridae	1
Mordellidae	1
Nitidulidae	8
Oedemeridae	1
Ptiliidae	1
Salpingidae	1
Scarabaeidae	6
Scraptiidae	2
Scydmaenidae	6
Silphidae	1
Staphylinidae	75
Tenebrionidae	2
Throscidae	4
Trachypachidae	2
Trogossitidae	1
Zopheridae	2

[a] The complete data set for this study can be found on the Mount St. Helens National Volcanic Monument web page: http://www.fsl.orst.edu/msh/.

periods of activity on the ground surface. Nonetheless, most of the common families and species of beetles collected during the study were typical ground-dwelling taxa and would have been appropriately sampled with the pitfall traps.

A comparison of the total numbers of beetle species collected on each site during each year is shown in Table 10.2. Note that no samples were collected from the tephra-fall sites in 1987 because of extensive trap disturbances by elk. (*Cervus elaphus*) Several patterns are apparent in Table 10.2.

First, although the sampling effort was consistent among years, the 1987 sample yielded considerably fewer species (99) than the samples from later years (145 to 169 species); this difference even extended to the reference forest and clear-cut sites, indicating that some widespread phenomenon was responsible. The period from 1984 to 1987 was characterized by unusually dry summers (following the wet El Niño period of 1982 to 1983), and this dry period may have had a cumulatively negative effect on the beetle fauna. In addition, this drought may have had an influence on other faunal groups in the region; for example, MacMahon et al. (1989) reported a decline in small mammals during 1987 compared to the years from 1983 to 1986.

Second, in comparing the beetle species-richness values of the paired forest/clear-cut sites within each disturbance zone, we noted little consistent difference in the numbers of species; in some years, we found more species in the forests, while in other years we collected more species in the clear-cuts. Only the tephra-fall sites exhibited more species in forests than in clear-cuts in every year.

The third pattern that emerges in Table 10.2 is that the most disturbed site, the Pumice Plain (a former forest buried in the debris avalanche of the eruption), supported fewer species after 1990 than did most of the forested sites.

The fourth pattern is the apparent lack of species accrual between 1990 and 2000; none of the sites showed an increase in beetle species numbers during this decade. The numbers of beetle species per site collected during 1990 to 2000 varied from 26 to 84, but no obvious trend in the number of species was observed.

Finally, with respect to the intermediate-disturbance hypothesis, the Mount St. Helens eruptions of 1980 essentially provided a single event that created a disturbance-intensity gradient from the crater outward for tens of kilometers. As a result, we could examine the intermediate-disturbance hypothesis only in the context of this disturbance intensity because there was no range of disturbance frequencies (there was only one cluster of eruptions during the period) nor range of disturbance patch size (just one volcano with one large patch). Moreover, we did not conduct any sampling between 1980 and 1986. At that point, what we did was to test for disturbance intensity across the disturbance gradient. These constraints notwithstanding, in the early posteruption years, we expected to find more species on the sites subjected to intermediate disturbances (the tephra-fall and blowdown sites) than on the more-disturbed or less-disturbed sites (i.e., the Pumice Plain and reference sites). Such a pattern was observed in 1987 for the blowdown sites versus the other sites, but these numerical differences diminished in later years as the site vegetation recovered. The tephra-fall sites failed to exhibit consistent increases in species relative to any of the other sites. Thus, from these results, we conclude that, during the period from 1987 to 2000, there was only weak evidence in support of the intermediate-disturbance hypothesis. Given the limitations of this study, however, a more pronounced species pattern, perhaps more consistent with the intermediate-disturbance hypothesis, may have existed on Mount St. Helens shortly after the 1980 eruption, and it could have dissipated by the time we started conducting our study in 1987.

TABLE 10.2. Numbers of beetle species collected on the study sites at Mount St. Helens from 1987 to 2000.

Year	Study site							
	Reference forest	Reference clear-cut	Tephra-fall forest	Tephra-fall clear-cut	Blowdown forest	Blowdown clear-cut	Pumice Plain	Total species
1987	17	16	—	—	53	38	31	99
1990	62	50	50	36	60	84	52	169
1995	40	43	46	26	48	45	37	145
2000	60	36	45	29	51	62	41	169

10.3.2 Successional Patterns of the Beetle Assemblages

In examining the abundances of particular beetle species on the disturbed plots, we found considerable evidence of relay successional patterns through time. Beetle species that were abundant in 1987 on the Pumice Plain and on sites within the blowdown zone were often absent or greatly reduced in number by 2000. Comparative-abundance graphs for some of the common species on the Pumice Plain (Figure 10.4) and in

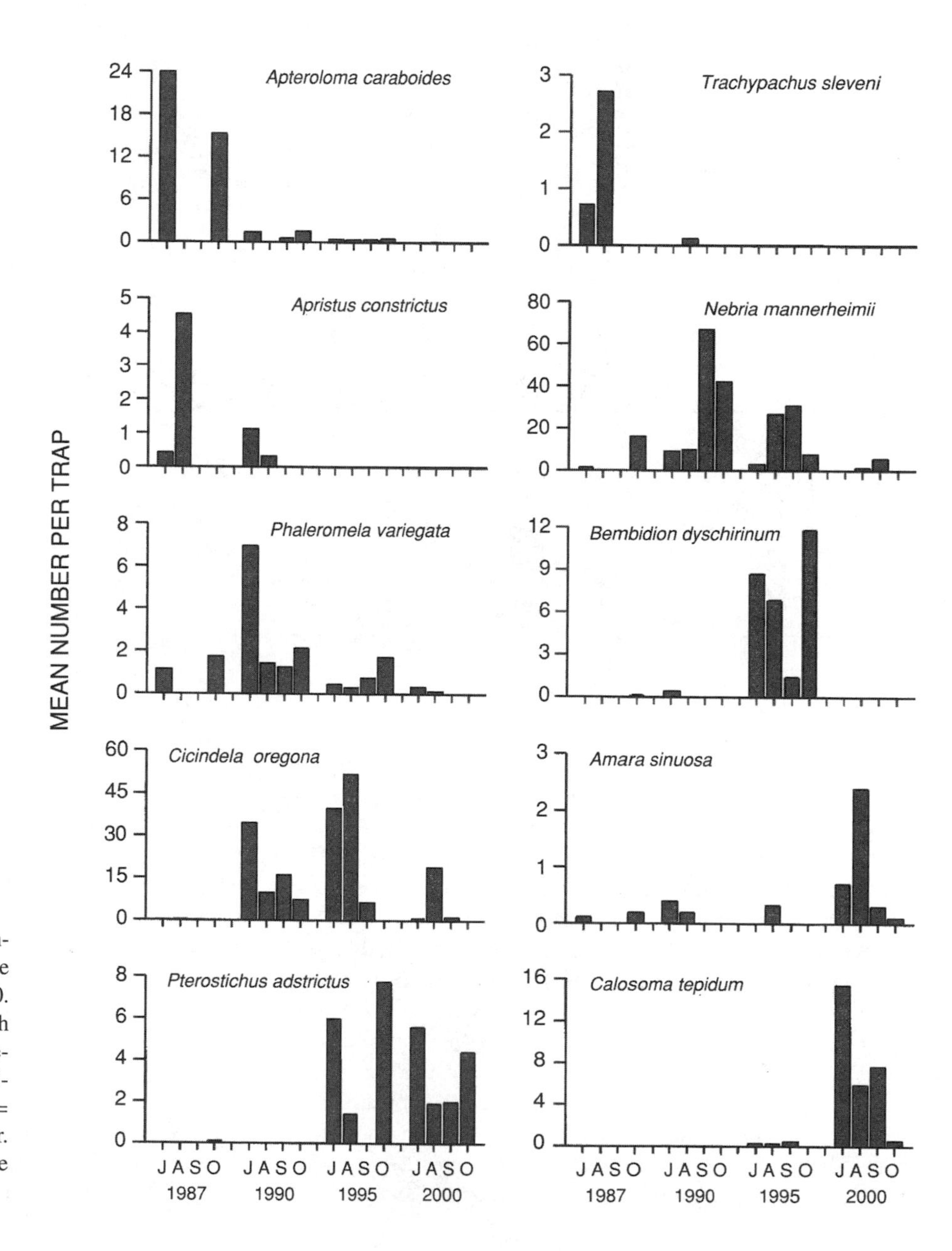

FIGURE 10.4. Temporal changes in the abundances of common beetle species on the Pumice Plain study site from 1987 to 2000. Species are sequentially replaced through time (relay succession). Sample dates reflect the beetles collected during the 4- to 5-week period before the date; J = July, A = August, S = September, and O = October. Note that the scale changes in the abundance axes.

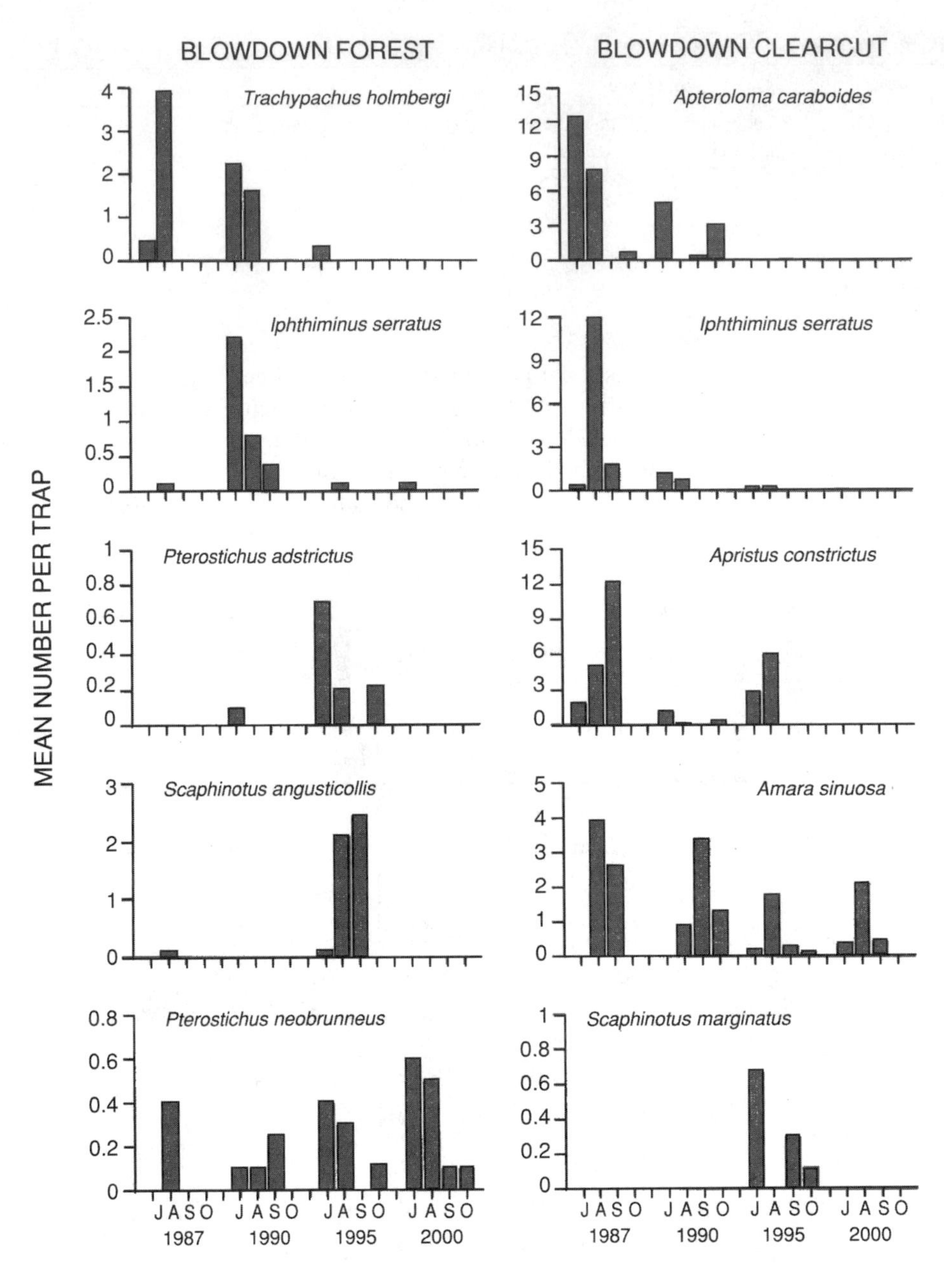

FIGURE 10.5. Temporal changes in the abundances of common beetle species on the two blowdown-zone study sites from 1987 to 2000. Species are sequentially replaced through time (relay succession). Sample dates reflect the beetles collected during the 4- to 5-week period before the date; J = July, A = August, S = September, and O = October. Note that the scale changes in the abundance axes.

the two blowdown-zone sites (Figure 10.5) illustrate this phenomenon. Edwards and Sugg (Chapter 9, this volume) documented 30 species of beetles on the Pumice Plain from 1981 to 1985, of which *Apteroloma caraboides* and *Apristus constrictus* were among the dominants (see Figure 9.5). These species were still common on our Pumice Plain study site in 1987, but became less abundant in 1990 and had nearly disappeared by 2000. Similarly, the various species of *Bembidion* observed by Edwards and Sugg (Chapter 9, this volume) on the Pumice Plain had been replaced by other *Bembidion* species, particularly *B. dyschirinum*. We also noted the appearance of new species at different times on the Pumice Plain; the carabid ground beetles *Nebria mannerheimii* and *Pterostichus adstrictus* appeared in large numbers in 1990 and 1995, respectively. Populations of the tiger beetle (*Cicindela oregona*) virtually exploded between 1990 and 1995, becoming one of the most abundant predatory beetles on the Pumice Plain. Similarly, the caterpillar hunter beetle, *Calosoma tepidum*, became highly abundant between 1995 and 2000, coincident with the large increase in herbaceous plant cover and the increase in moth caterpillar larvae (Fagan and Bishop 2000; Bishop 2002). Thus, the pattern of relay succession was well evidenced in the beetle species of the volcanically disturbed landscape.

In analyzing the patterns of the beetle assemblages among the study sites through time, we compared the percentage species similarities in all paired combinations within each sample year (Table 10.3). From these computations, the impacts of the different levels of volcanic disturbance become

TABLE 10.3. Similarity coefficients and numbers of shared beetle species among study sites across the Mount St. Helens disturbance gradient from 1987 to 2000.

	Reference forest	Reference clear-cut	Tephra-fall forest	Tephra-fall clear-cut	Blowdown forest	Blowdown clear-cut	Pumice Plain
				1987			
Reference forest	—	14			9	12	0
Reference clear-cut	4	—			15	17	2
Blowdown forest	6	9			—	34	17
Blowdown clear-cut	6	8			23	—	23
Pumice Plain	0	1			12	13	—
				1990			
Reference forest	—	32	56	24	23	20	11
Reference clear-cut	27	—	32	34	29	26	17
Tephra-fall forest	40	24	—	25	26	22	15
Tephra-fall clear-cut	19	22	17	—	23	22	23
Blowdown forest	23	25	23	18	—	41	30
Blowdown clear-cut	24	28	24	22	42	—	26
Pumice Plain	1	15	13	12	26	28	—
				1995			
Reference forest	—	22	32	22	22	18	8
Reference clear-cut	15	—	22	17	21	19	8
Tephra-fall forest	21	16	—	26	25	17	6
Tephra-fall clear-cut	12	10	15	—	16	16	7
Blowdown forest	16	16	19	10	—	33	15
Blowdown clear-cut	13	14	13	10	23	—	22
Pumice Plain	6	6	5	4	11	15	—
				2000			
Reference forest	—	37	38	24	28	16	4
Reference clear-cut	26	—	37	25	26	20	7
Tephra-fall forest	29	22	—	35	30	19	5
Tephra-fall clear-cut	17	13	19	—	23	15	4
Blowdown forest	24	18	22	15	—	19	8
Blowdown clear-cut	17	16	17	12	18	—	16
Pumice Plain	4	5	4	3	7	14	—

Upper-right portions of matrices show percentage species-similarity values; lower-left portions show numbers of shared species.

clear. Table 10.3 is arranged to show increasing levels of disturbance from left to right (reference forest through Pumice Plain). In all the sample years, percentage similarity in beetle species composition also declines from left to right, indicating lesser similarity in species with increasing disturbance level. For example, in 1987, the Pumice Plain had no beetle species in common with the reference forest, and the most similar beetle assemblage was found on the blowdown clear-cut site (23%). Similar patterns are seen in later years. As expected, the reference forest has the highest similarities with the tephra-fall forest (see Table 10.3). The sites in the blowdown zone have many species in common with both the reference sites and with the Pumice Plain, but, as described above (Table 10.2), the blowdown-zone sites did not consistently have an overall total species number in excess of the other sites.

The other pattern of interest in Table 10.3 is the inconsistent progression in similarity from 1987 to 2000 among the sites. If successional processes during this period had allowed the sites to substantially recover from the eruption disturbance, we would have expected to see a pattern of increasing similarity between the disturbed sites and the reference sites. In the most favorable example, the reference-forest and blowdown-forest comparisons yielded values of 9%, 23%, 22%, and 28% from 1987 to 2000. However, such was not usually the case; for example, similarity percentages between the reference forest and the tephra-fall forest in 1990, 1995, and 2000 were 56%, 32%, and 38%, respectively. Even the Pumice Plain showed little directional change in relation to the reference forest, fluctuating from 0% similarity in 1987 to 11%, 8%, and finally 4% in subsequent years. So, we conclude from these analyses that, although the sites were clearly changing through time in a relay successional fashion, their beetle assemblages after 20 years of posteruption recovery were still not comparable to their presumed preeruption species composition. This pattern was obviously produced by the differences in site characteristics (vegetation, soils, and microclimate) that have resulted from site-specific successional processes and by the fact that we would not expect convergence of all the beetle assemblages until all the sites have developed similar habitat attributes (i.e., old-growth coniferous forest, such as that existing on the reference sites).

10.3.3 Trophic Structure of the Beetle Assemblages

As described earlier, the ground-dwelling beetle fauna on each site consisted of a large number of species, and these assemblages were made up of a wide variety of species feeding on an even wider variety of food items. The success of any species in surviving and reproducing in their environment depends upon (among other things) finding sufficient food resources. In the harsh environment of the Pumice Plain immediately after the 1980 eruption, only scavenger and predator species could survive on the aeolian rain of dispersing arthropods (Edwards and Sugg 1993; Sugg and Edwards 1998) because other food resources (e.g., plants or fungi) were not present. However, as plants slowly established in the disturbed areas, herbivorous arthropods began recolonizing as well, followed by other predators, fungivores, and carrion and dung feeders (taking advantage of carcasses and dung of mammals and birds).

In an effort to view the succession of the beetle assemblages from a functional-role perspective, we analyzed our beetle data set with respect to the trophic groups represented within each site's assemblage. We considered only five broad categories: predators, herbivores, omnivores, fungivores, and scavengers. Although the dietary habits of some species were either speculative or unknown, the majority of the species has reasonably well-known food preferences and could thus be classified into one of the five categories (Arnett and Thomas 2001; Arnett et al. 2002).

In comparing the trophic structure of the ground-dwelling beetles in the forested sites (Figure 10.6), we found that, between 1990 and 2000, the trophic structure remained relatively stable in the reference forest. In this site, predators (mostly Carabidae and Staphylinidae) dominated the assemblage, with fungivores being the second most common trophic group (fungi were abundant on the decaying logs and litter of the old-growth forest floor). Herbivores, omnivores, and scavengers made up a relatively small proportion of the assemblage. The tephra-fall forest exhibited greater change in trophic structure, with increases in herbivorous and omnivorous beetle species and a decrease in predators, fungivores, and scavengers. The blowdown-forest site showed a relatively stable trophic structure through time. This site was characterized by large numbers of decomposing logs with abundant fungi (e.g., the coral tooth mushroom, *Hericium abietis*) and a variety of polypores, which presumably contributed to the large percentage of fungivorous beetles found on this site (see Figure 10.6).

In the clear-cut sites and on the Pumice Plain site (Figure 10.7), notable differences in beetle trophic structure existed between the levels of disturbance, and some proportional change occurred between 1987 and 2000 (species numbers also increased markedly; see Table 10.2). The reference clear-cut, much like the reference-forest site, was dominated by predators and fungivores; the blowdown clear-cut, having less decomposing woody debris than the blowdown forest (see Figure 10.2c,d), was dominated by beetle predators, herbivores, and omnivores instead of fungivores (compare Figures 10.6 and 10.7). The Pumice Plain beetle trophic structure was dominated by predators but did show an increase in herbivores (from 6 to 10 species) and omnivores (from 4 to 11 species) between 1987 and 2000; this observation corresponds to the increase in herbaceous vegetation during this same period (see Figure 10.2a,b). Hence, the beetle trophic structures of the study areas on and around Mount St. Helens corresponded well with the predicted patterns, based on the availability of food resources (plants, fungi, and other arthropod prey species).

It is interesting to note that the patterns observed in the beetle trophic groups on Mount St. Helens correspond to the posteruption observations made on another volcano, Krakatau, during the late 19th and early 20th centuries. A Dutch naturalist, K.W. Dammerman, studied the successional patterns of plants and animals after the Indonesian volcano exploded in 1883 and noted that, among the animals, the "first [to] come [are] those animals that subsist upon refuse—chiefly vegetable debris, but sometimes also animal waste [carrion and dung]; then come omnivorous species, equally undiscriminating as to their food; the true herbivorous species follow later, while the predaceous animals and the parasites are the last to settle" (Dammerman 1948, p. 201).

10.4 Conclusions and Lessons for the Future

In total, our study found that the patterns of ground-dwelling beetle recolonization and succession on the volcanically disturbed sites were consistent with predicted patterns from relay successional theory, with individual species being sequentially replaced through time as the environmental characteristics of each site changed. The beetle assemblages exhibited predictable patterns of similarity across the landscape, with the least similar assemblages being at opposite ends of the disturbance gradient. Trophic structures of the assemblages also conformed to predicted patterns based on the available food resources of the sites. The major pattern that did not wholly concur with predictions was that from the intermediate-disturbance hypothesis; we did not consistently observe a greater number of beetle species on sites subjected to intermediate volcanic disturbance when compared to undisturbed or severely disturbed sites. Finally, from our analyses of the first 20 years of posteruption recovery, it appears that the beetle fauna of the disturbed sites bears little resemblance to that of the undisturbed (reference) forested and clear-cut sites and that considerable development of the beetle assemblages will be occurring in future decades.

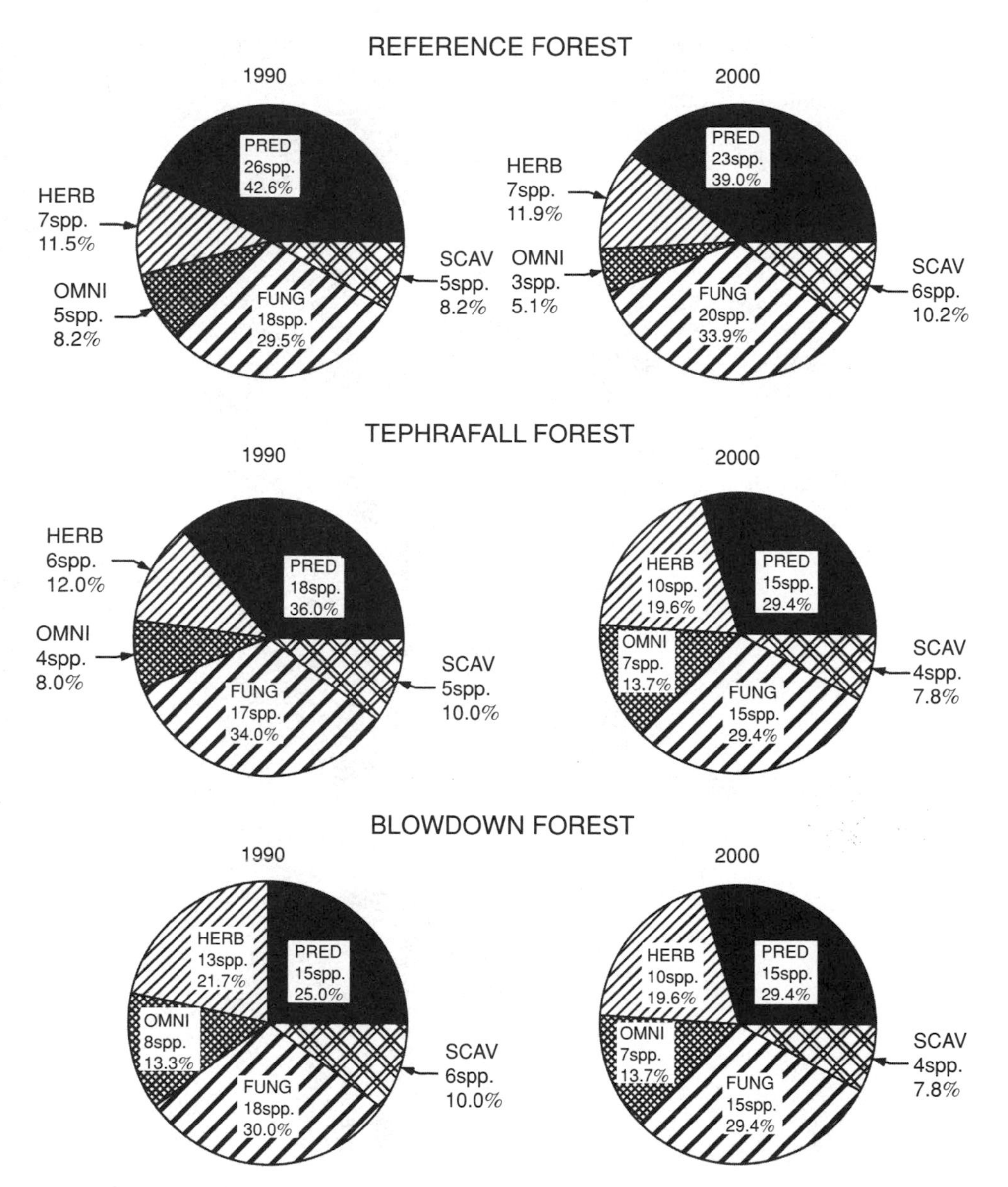

FIGURE 10.6. Comparisons of beetle-community trophic structure among forest sites in 1990 and 2000. The beetle community in the reference-forest and blowdown-forest sites remained relatively stable during this time, while the tephra-fall forest's beetle community exhibited a small shift toward herbivores and omnivores.

A final question is: What might future changes in the beetle assemblages look like? In the absence of further large-scale disturbances (more eruptions, fire, and anthropogenic development), the various sites on Mount St. Helens will likely continue to progress toward a forested ecosystem (presumably similar to the preeruption forests on the volcano). During this successional process, each site's biotic and abiotic characteristics (vegetation composition and structure, soil development, litter, and microclimate) will possibly converge in similarity, thus providing a more homogeneous forested landscape conducive to domination by forest beetle species. This successional process will proceed at different rates and time steps for the variously disturbed sites. Although the beetle assemblages of forests and clear-cuts in the tephrafall and blowdown zones might converge with the reference sites within several decades, the areas on the Pumice Plain may require centuries to attain a forest assemblage.

In addition to the need for appropriate habitat characteristics for successful establishment, different beetle species have varying capabilities of dispersal and will exhibit differences in their abilities to immigrate to areas of suitable habitat. Although many species are strong fliers and can easily move from distant source populations, others are poor fliers or are completely ambulatory, while still others are polymorphic for wingedness. These latter species may need to await the development of continuous suitable habitat corridors or archipelagos of habitat islands through which the beetles could safely disperse.

Clearly, beetles form a species-rich and trophically diverse group of animals that will play important roles in the successional processes on disturbed areas of Mount St. Helens. The 279 beetle species identified from the pitfall traps in this study represent only a fraction of the total number of beetles that inhabit the area. Many more undoubtedly live higher up

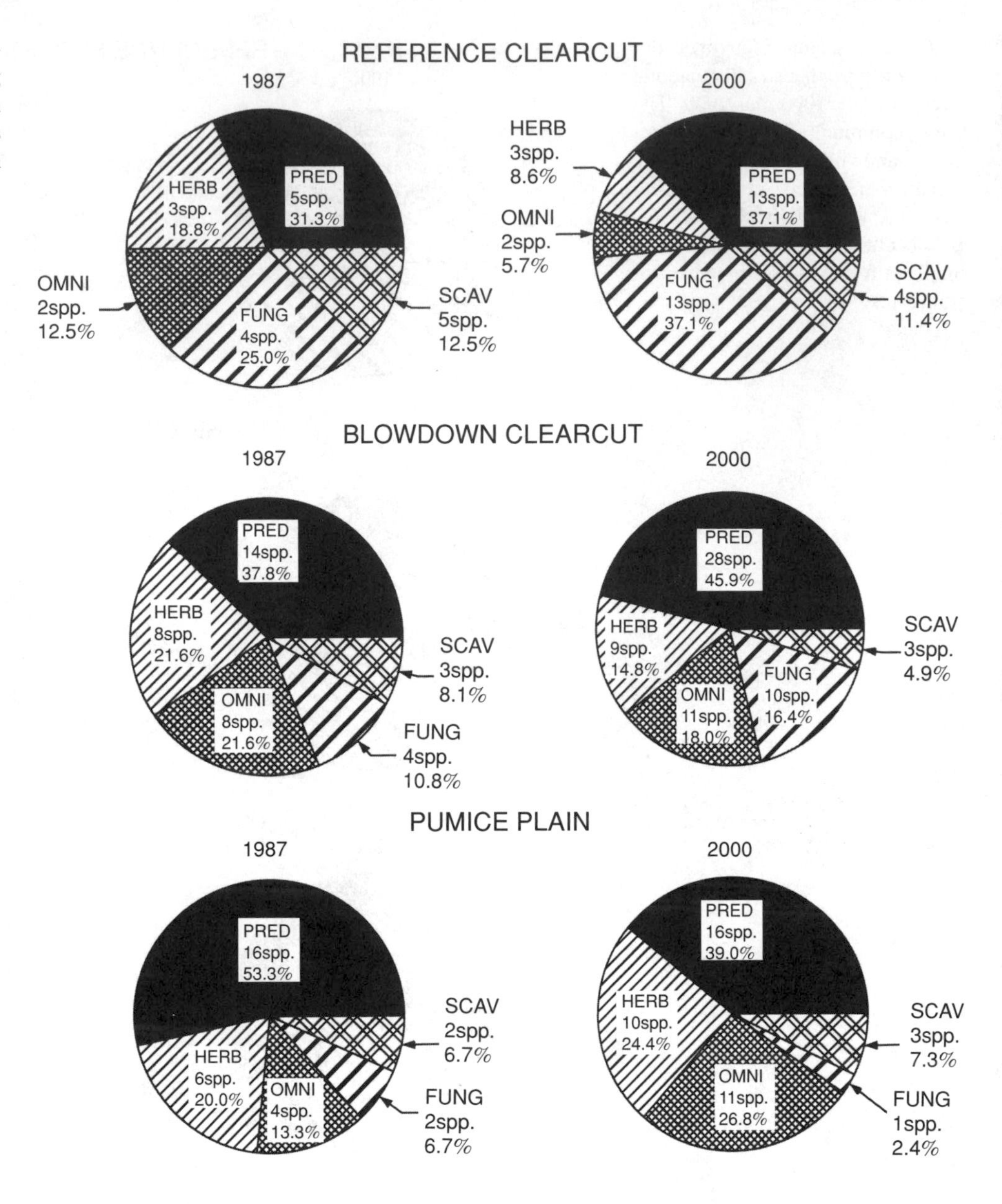

FIGURE 10.7. Comparisons of beetle-community trophic structure among the clear-cuts and the Pumice Plain site in 1987 and 2000. Herbivores and omnivores on the Pumice Plain site increased from 1987 to 2000.

in vegetation and in specialized microhabitats that we did not sample. These species will continue to influence ecological processes through their activities of herbivory, predation, nutrient cycling, pollination, soil disturbance, and dispersal of fungal spores. These activities, in turn, may prove critical in determining the rate of each site's succession, the pathways of successional trajectories, and perhaps even the species composition of other taxa (plants and animal) in the later successional stages. Future studies on these important and conspicuous beetle groups, and other arthropod taxa, will undoubtedly provide valuable insights into the ecosystem responses to the 1980 eruption of Mount St. Helens.

11 Causes and Consequences of Herbivory on Prairie Lupine (*Lupinus lepidus*) in Early Primary Succession

John G. Bishop, William F. Fagan, John D. Schade, and Charles M. Crisafulli

11.1 Introduction

Primary succession, the formation and change of ecological communities in locations initially lacking organisms or other biological materials, has been an important research focus for at least a century (Cowles 1899; Griggs 1933; Eggler 1941; Crocker and Major 1955; Eggler 1959; Miles and Walton 1993; Walker and del Moral 2003). At approximately 60 km^2, primary successional surfaces at Mount St. Helens occupy a minor proportion of the blast area, yet they are arguably the most compelling. The cataclysmic genesis of this landscape, its utter sterilization, and the drama of its reclamation by living organisms stimulate the imagination of scientists and nonscientists alike. These primary successional surfaces are the most intensively monitored areas at Mount St. Helens because of what they may teach us about the fundamental mechanisms governing the formation and function of biological communities. At a practical level, understanding successional processes provides a conceptual basis for the restoration of devastated landscapes (Bradshaw 1993; Franklin and MacMahon 2000; Walker and del Moral 2003).

Succession is a fundamentally multitrophic process. It involves not only plants but also herbivores, predators, and decomposers. Yet, the important effects of these other trophic levels are sometimes ignored. Study of trophic interactions (i.e., interactions between consumers and resources) in primary succession can provide new insights into mechanisms of primary succession and can inform the debates surrounding what controls the level of herbivory in terrestrial ecosystems. In this chapter, we describe often-devastating attacks by insect herbivores on the prairie lupine, *Lupinus lepidus* var. *lobbii* (Dougl.), and how they have affected the (spatial) spread of this little plant, generally considered the most important colonist during the first two decades of primary succession on the Mount St. Helens Pumice Plain. We also discuss a surprising spatial pattern in the intensity of lupine herbivory on the Pumice Plain and outline two hypotheses for this pattern. One hypothesis is based on gradients in plant quality; the other is based on gradients in the density and diversity of the herbivores' natural enemies. Both hypotheses involve processes that are inherent to primary succession and that are likely relevant to systems beyond Mount St. Helens.

11.2 Herbivory in Primary Succession

Ecologists have long debated whether animal and plant populations are regulated by their consumers (so-called "top-down" forces) or by access to resources ("bottom-up" forces). Primary successional systems on land have figured little in this debate, perhaps because primary consumers (e.g., herbivores) are rarely identified as a significant factor in terrestrial primary succession. Instead, in most studies of terrestrial primary succession, the initial composition of plant communities appears to be controlled by the availability of propagules, by the ability of the plants to establish despite extreme abiotic conditions and an absence of soil resources, and by various stochastic events (Miles and Walton 1993; Walker and del Moral 2003). Subsequent changes in community composition are influenced by diverse factors, but scientists have paid the most attention to competition and facilitation among plants, which mediate resource availability, a bottom-up factor. In their classic critique of successional research, Connell and Slatyer (1977) noted that the effect of herbivores on succession had been largely ignored, essentially tacked onto a list of factors that may influence the outcome of competition or facilitation. Subsequently, potentially strong impacts from herbivores, including invertebrates, have become better incorporated into ecologists' conceptions of secondary successional change. This recognition has occurred faster in marine and freshwater systems, where consumers are ubiquitous and the importance of multitrophic interactions in succession has been recognized for decades (Lubchenco 1983; Farrell 1991; Hixon and Brostoff 1996).

Discussion of herbivore effects is nearly absent from the literature on terrestrial primary succession (Walker and Chapin 1987; Glenn-Lewin et al. 1992; Miles and Walton 1993; McCook 1994; Walker and del Moral 2003). Yet consumers do accompany most cases of primary succession. For example, numerous vertebrate herbivores, both wild and domesticated, frequent glacier forelands (Matthews 1992, chapter 6.1), and invertebrate consumer faunas have been described in various early-successional volcanic habitats (Howarth 1987; Thornton et al. 1990; Ashmole et al. 1996; Ashmole and Ashmole 1997; Sugg and Edwards 1998; Parmenter et al., Chapter 10, this volume). At Mount St. Helens, primary and secondary arthropod consumers (i.e., herbivores and their predators) were the very first colonists of pyroclastic-flow substrates, ahead of even ruderal (i.e., highly colonizing) plants (Edwards 1986a; Sugg and Edwards 1998). The question arises as to whether consumers are truly unimportant in primary succession or whether their effects have simply been overlooked. Plant damage by herbivores, especially by invertebrate herbivores, is sometimes subtle. Damage may be readily apparent only during outbreak periods or narrow seasonal windows, and highly mobile, belowground, or endophytic herbivores may be difficult to observe. Relatively limited herbivory may have strong impacts if concentrated in portions of a patch that otherwise contribute the most to population growth and spread (Fagan and Bishop 2000). A final difficulty involves the practicalities of science itself. Elucidating the role of herbivory has taken decades, even in common and intensively studied secondary successional systems, such as old fields; large primary successional systems are infrequent and less studied.

A few studies of consumer effects in terrestrial primary succession do exist. Most exceptional is the work on floodplains of the Alaskan taiga, where primary succession on silt terraces has been intensively studied (Walker and Chapin 1986; Walker et al. 1986; Van Cleve et al. 1993). In that system, browsing by moose (*Alces alces*) accelerates replacement of willows by alders during early primary succession and affects biogeochemical processes such as nitrogen and carbon accumulation both directly through fecal deposition and indirectly through effects on nitrogen-fixing plants (Kielland and Bryant 1998). The snowshoe hare (*Lepus americanus*) also appears to accelerate change in those riparian systems from felt-leaf willow [*Salix alaxensis* (Anderss.) Coville] to alder (*Alnus* spp.) (Bryant 1987), and both types of herbivores cause a shift from nutrient-rich, early-successional shrub stages to nutrient-poor and better defended evergreens (Bryant and Chapin 1986). Fast-growing, early-successional shrubs, such as willow (*Salix* spp.) and alder, were found more palatable to herbivores than slow-growing plants, such as spruce (*Picea abies*) and labrador tea (*Ledum*, an evergreen shrub). This difference in susceptibility, attributed to the high allocation to carbon-based defenses (i.e., tannins and other phenolics) by slow-growing plants in resource-poor habitats, led in part to the landmark proposal that resource availability is a major determinant of a plant's allocation to defensive chemicals and the type produced (Bryant et al. 1983; Coley et al. 1985).

Vertebrate herbivores are common in other primary successional systems as well, but their effects are less studied (Faegri 1986; Wood and Anderson 1990; Matthews 1992). At Mount St. Helens, herds of elk (*Cervus elaphus*) are frequently observed on primary successional surfaces of the Pumice Plain and debris-avalanche deposit. Excluding elk from the latter site has led to dramatic increases in woody plants (see Dale et al., Chapter 5, this volume). In a separate study at Mount St. Helens, aster (*Aster ledophyllus*) seedlings increased by 76% in elk exclosures surrounding individual plants on the Pine Creek mudflow, suggesting that elk browsing may explain the slow rate of colonization of aster on that surface (Wood and Anderson 1990). Whether elk are as influential at Mount St. Helens as are moose in the taiga or ungulates in successional grassland ecosystems (McNaughton 1985; Ritchie and Tilman 1995; Olff and Ritchie 1998; Ritchie et al. 1998) is a critical question for understanding controls on succession at Mount St. Helens.

The only detailed work on insect herbivory during primary succession is Catherine Bach's demonstration (Bach 1990, 1994) that brief periods of intense insect herbivory on a single, common plant species can strongly affect the rate and direction of succession on Lake Huron sand dunes. Bach experimentally excluded a specialist chrysomelid beetle (*Altica subplicata*) feeding on willow (*Salix cordata*) during a 3-year outbreak. Herbivory slowed primary succession by indirectly facilitating growth of early-successional species (Bach 1990, 1994) and affected the community trajectory (e.g., causing a decrease in monocots) (Bach 2001a). Negative effects on willows themselves were exacerbated over time because herbivore-damaged plants were more susceptible to sand accretion (Bach 2001b). Other reports also hint at an important role for arthropod consumers. On glacier forelands, spruce bark beetle (*Dendroctonus rufipennis*) damages Sitka spruce (*Picea sitchensis*) at Glacier Bay, killing forest canopy (Eglitis 1984). At Mount St. Helens, Wood and Anderson (Wood and Anderson 1990) documented insect seed predators on aster, but found no short-term effects, and Dale (1986) noted devastation of fireweed (*Chamerion angustifolium*) patches by a sphingid caterpillar. Fire tree (*Myrica faya*), an invasive nitrogen-fixing tree that severely alters successional trajectories on Hawaiian lava flows, has recently had large portions of its canopy destroyed by the introduced homopteran (leafhopper) *Sophonia*. Removal of fire tree canopy leads to dominance of nonnative grasses that take advantage of the unusually high soil nitrogen produced by fire tree (Adler et al. 1998). Although not a native system, the *Myrica–Sophonia* interaction demonstrates the potential for important top-down effects that insect consumers can have on succession when acting on keystone species (i.e., species that disproportionately affect the organization of their communities).

In what follows, we discuss the effect of insect herbivores on prairie lupine (*Lupinus lepidus* var. *lobbii*) during primary succession on the Pumice Plain of Mount St. Helens (see map, Figure 11.1). The demographic impact of herbivores on lupine is of special interest because prairie lupine acts as a keystone species in early primary succession. Consequently, what controls herbivory on lupine is also of great interest. Herbivory in this system is strikingly absent from large, high-density lupine patches, a fact that allows us to examine how top-down and

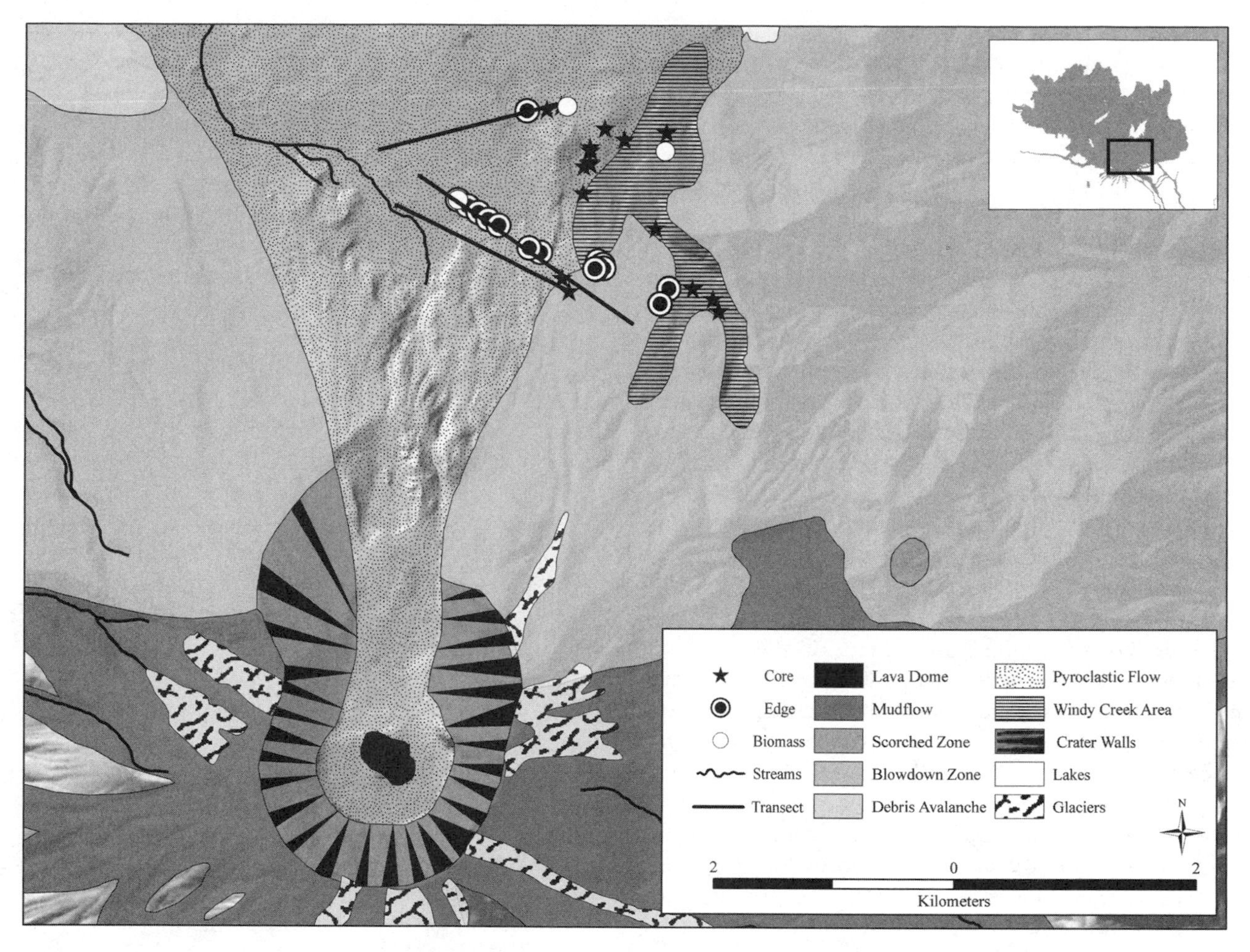

FIGURE 11.1. Map of Mount St. Helens showing the Pumice Plain, debris-avalanche deposits, and study areas.

bottom-up forces may control plant and herbivore populations during primary succession.

11.3 Lupine Effects on Succession

Prairie lupine is a nitrogen-fixing legume characteristic of alpine and subalpine pumice communities in the Washington and Oregon Cascades. Following the 1980 eruption, it was the first plant to successfully colonize pyroclastic-flow and volcanic-debris avalanche deposits on the north flank of Mount St. Helens. Isolated individuals that were discovered during reconnaissance surveys in 1981 and 1982 were of a size consistent with colonization by seed the year before. Colonization probably occurred via seeds, rather than root fragments, because root fragments might require transport by water. Initial colonists were found on uneroded pyroclastic flows several kilometers from the nearest possible source population, suggesting that water transport did not account for dispersal. The initial long-distance colonization events were greeted with surprise (del Moral and Grishin 1999) because prairie lupine seeds lack any obvious means of long-distance dispersal and are larger than those of weedy colonizers. However, prairie lupine seeds, although not abundant, were caught regularly in traps set for windblown seeds (Wood and del Moral 2000), indicating that long-distance dispersal of lupine seeds occurs frequently.

As nitrogen fixers and as the only successful primary producer established on pumice and other rock substrates, lupines were expected to drive the pace and pattern of early succession. Prairie lupine has met expectations in this regard. Initial substrate concentrations of organic matter and nitrogen (a nutrient that often limits plant growth) were zero (Engle 1983). Lupines accelerate soil development through direct nutrient and organic-matter input; trapping of windblown debris and propagules; attraction of insects that ultimately die in situ; and attraction of animal dispersal vectors, such as birds, elk, and mice, which transport seeds and microorganisms (Allen 1988). For example, one of us documented seven species of vascular plant (four graminoids and three herbs) and three species of fungi dispersed to a lupine patch in elk feces in 1998. Soils under lupines have much higher levels of total nitrogen, organic matter, and microbial activity than do adjacent bare areas (Halvorson et al. 1991b, 1992; Halvorson and Smith 1995;

Fagan et al. 2004). Experiments demonstrate a net positive effect of lupines on growth of such ruderal plant species as hairy cats-ear (*Hypochaeris radicata*), fireweed (*Chamerion angustifolium*), and pearly everlasting (*Anaphalis margaritacea*) (Morris and Wood 1989; Titus and del Moral 1998b), although these species may have to wait for lupines to die to take advantage of the site. In a recent survey, del Moral found that cover of other plant species is higher within lupine patches than outside them and that species composition differed substantially inside and outside patches. Some differences in plant composition are striking. For example, paintbrush (*Castilleja miniata*), which is hemiparasitic on lupine, has become abundant in many lupine patches but is rare outside them. Nutrient-responsive species, such as hairy cats-ear and bentgrass (*Agrostis pallens*), are also much more common in lupine patches. Our survey data also indicate that biomass and species richness of arthropods are much greater within large lupine patches than outside them. Later successional species, including woody plants such as willow (*Salix commutata*), Sitka alder (*Alnus viridis*), conifers, and animal-dispersed shrubs such as huckleberry (*Vaccinium* sp.), are also colonizing the Pumice Plain, and once they establish their vertical structure will dramatically alter plant and animal communities. Whether lupine facilitates establishment and growth of these species and thereby influences later stages of succession remains to be investigated. However, it is clear that any factor limiting the rate of spread of lupines, such as herbivory, may affect succession by delaying the early stages of soil formation and by changing the floristic trajectory.

Despite prairie lupine's impressive impacts on soil and community development, a number of early papers discounted lupine's facilitative effects because lupines occupied such a small portion of the primary successional landscape. This observation reflects the fact that prairie lupine failed to colonize the vast majority of available habitat during the first two decades of succession. Populations springing from the founding colonists exhibited rapid rates of growth and spread during the first few years after the 1980 eruption, but these rates were much reduced by the late 1980s (Fagan and Bishop 2000; Bishop 2002). For example, although the average rate of population increase in the patch founded in 1981 was 11.2 offspring per individual (geometric average from 1981 through 1985; Figure 11.2a), the rate of increase in similar patches founded in 1988 to 1990 averaged only 1.5 (1991 to 1995) (Fagan and Bishop 2000; Bishop 2002). Beginning from this core of initial patches, lupines rapidly colonized the 1.5-km expanse sloping down to Spirit Lake, but colonization up or across slopes (and newly created gullies) was far more circumscribed. For the sake of description, we will refer to the large, high-density patches formed during the early 1980s as "core" areas. We refer to the smaller, younger patches uphill of the core as the "edge" areas. We use these terms because the two areas loosely represent the core and edge of an invasion front. Limited dispersal ability can partially explain slow rates of spread from the core in the uphill direction. However, by 1990, a few plants located in small, isolated patches (i.e., edge

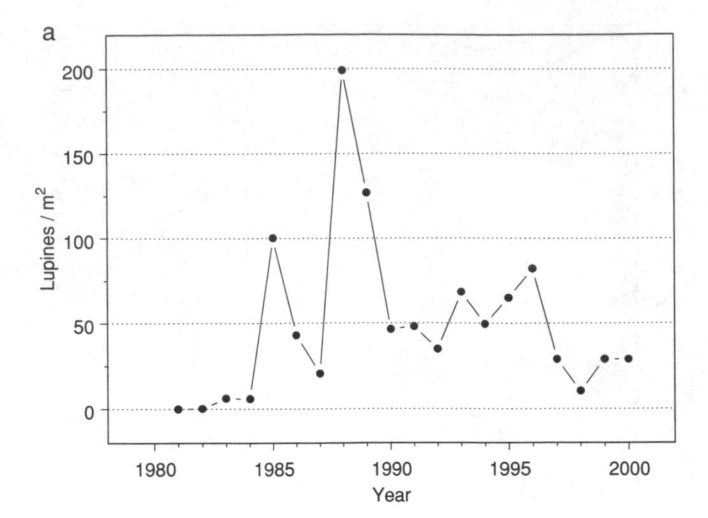

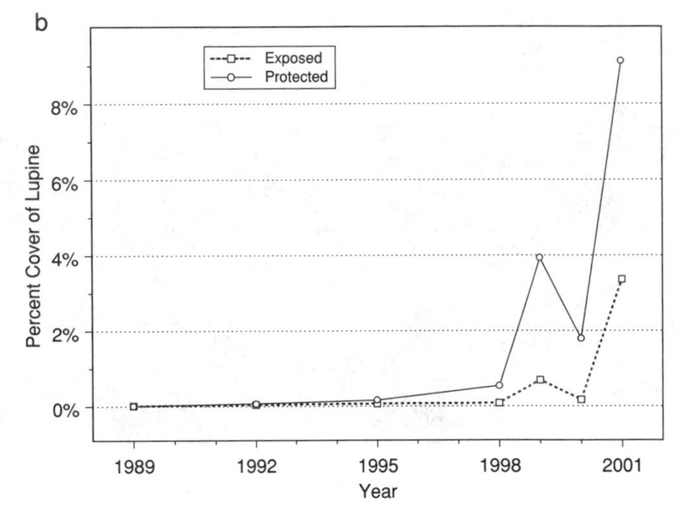

FIGURE 11.2. (a) Prairie lupine population growth in the founding core lupine patch, 1981 to 2000. Because plants outside the 168-m^2 sample grid were not counted, the data are presented as plants/m^2. (b) Growth in percent cover of prairie lupine, 1989 to 2001, on a 20,000-m^2 grid on the east Pumice Plain. Note the abrupt increase in cover from 1998 to 1999 (attributable to record snow accumulation), followed by a sharp decline in 2000 (caused by tortricid root borers and noctuid cutworms). Protected plots are on a north-facing slope. Data for 1989 to 1999 are from del Moral and Jones (2002) and are the average of $n = 87$ 10-m × 10-m plots in protected areas and $n = 97$ plots in exposed areas. (Data for 2000 and 2001 are courtesy of R. del Moral.)

patches) were present uphill or at the same elevation as initial patches (Bishop 2002). Occasionally, these new edge-area patches expanded rapidly. More often, however, they experienced little or no population growth, despite establishment in sites nearly identical to successful patches, suggesting that failure to spread was not caused by an exhaustion of suitable habitat (Bishop 2002). Even though lupine populations have dramatically increased since 1999, large portions of available habitat are still only sparsely occupied. In 2002, we sampled six 2.5-km transects on the Pumice Plain, stopping every 25 m to estimate lupine cover in 40-m^{2}circular plots (thus sampling

0.12% of the survey area). We found that 24% of the plots had no live lupine and 70% of plots had less than 10% cover of lupine. Although some of these sites were unsuitable for lupine, these data indicate that, 20 years after initial colonization, relatively little of the Pumice Plain is directly affected by prairie lupine.

11.4 Herbivory on Lupines at Mount St. Helens

By 1985, insect herbivores were observed in the founding lupine patch. Their effect is illustrated in Figure 11.2. In 1984, a large seedling cohort was produced, which then experienced high mortality attributable to intraspecific competition (see Figure 11.2a). However, the continued decline in lupine density from 1986 to 1987 was caused mostly by the mortality of adult plants attacked by the root-boring larvae of two tortricid moths, *Hystricophora* spp. (near *H. roessleri*), and *Grapholita spp* (near *H. lana*). A new, even larger seedling cohort was produced in 1987, again followed by high mortality attributed to a combination of intraspecific competition and tortricid root borers. An outbreak of lupine aphids (*Macrosiphum albifrons* Essig) in 1987 likely acted synergistically with the tortricid root borers. In fact, the mortality of adult prairie lupine, the cause of which was largely underground and therefore unseen, was so striking and ubiquitous that several ecologists suggested lupines possessed a deterministic life span of 4 to 5 years. Similar die-offs have also been noted in prairie lupine populations on Mount Rainier and the Olympic Mountains in Washington State (Ola Edwards, University of Washington, Seattle, Washington; personal communication). Whatever its cause, high mortality combined with a relatively even-aged population structure causes prairie lupine populations at Mount St. Helens to undergo extreme fluctuation in numbers (del Moral 2000a,b). However, it is unlikely that this mortality is caused by a deterministic life history. A survival analysis of about 9000 individual plants that were followed for 5 to 7 years beginning in 1991 found that, although prairie lupine is short lived with only a few individuals surviving 7 years, mortality rates were relatively constant except in years with high mortality from tortricid root borers (Bishop 2002). Now we can thoroughly document an alternative cause for massive lupine die-offs: tortricid root borers [along with a more minor root borer, *Walshia* spp. (Cosmopterygidae)] "girdle" the root by eating the vascular tissue, thereby killing plants (Bishop 2002). This kind of "stand-level" mortality also occurs in bush lupine (*Lupinus arboreus*), a dominant plant of coastal prairie in Northern California, where it is killed by root-boring larvae of the moth *Hepialus californicus* (Hepialidae) (Strong et al. 1995; Maron 1998). Interestingly, whole-patch die-offs of subalpine lupine (*L. latifolius*) in the blowdown area of Mount St. Helens were documented during 1987, and there too a stem-boring lepidopteran appeared to be the cause.

Here we offer three additional examples of caterpillars affecting prairie lupine. The first is drawn from Roger del Moral's 2-ha grid, divided into 200 squares and monitored since 1989, in what we designate the edge region (del Moral and Jones 2002). Percent cover of prairie lupine on the grid increased slowly during the first 10 years of monitoring; then cover growth accelerated suddenly and dramatically in 1999, followed by an equally dramatic die-off in 2000. This die-off reduced total lupine cover by 50%, with the median 100-m^2 plot declining by 72% (see Figure 11.2b). We carefully examined plants in 11 plots on the grid and found 60% mortality of nonseedlings. Excavating the roots of dead plants revealed unequivocal evidence for massive root-borer activity, in the form of frass and galleries, in 77% of dead plants but no root-borer evidence in a sample of living plants. The remaining dead plants were too decayed to examine and may also have been killed by insects. In addition, one-third of the surviving 40% of plants had been completely defoliated by a noctuid cutworm, *Euxoa extranea*. Recovery of defoliated plants (along with the growth of the 1999 seedling cohort) accounts for the rapid recovery in 2001 (Figure 11.2b).

Decimation by tortricid root borers and noctuid cutworms not only reduced cover in edge areas in 2000 but also caused spatial contraction of some core patches. Figure 11.3 summarizes data from transects running along radii from the center point of patches to the patch margins. The 2000 growing season witnessed complete mortality of lupines on the outer one-half to two-thirds of these patches, caused primarily by tortricid root borers. In addition, noctuid cutworms heavily defoliated the few surviving plants in this area and plants at the new patch margin. By 2001, cutworm-attacked plants were mostly dead. Seedlings were abundant at the new patch margin but scarce in the region attacked by tortricid root borers. This caterpillar-induced dieback represents a 90% reduction in spatial extent of these patches. However, because lupines were at low density at the fringes of these patches, the actual proportion of plants killed in the patch was modest. Nevertheless, these losses are important because dead plants were concentrated at the margin, where they would have contributed disproportionately to future spatial spread.

Many other insects also feed on prairie lupine (Halvorson et al. 1992; Bishop and Schemske 1998; Fagan and Bishop 2000; Bishop 2002). Mortality induced by tortricid moths has provided the most dramatic herbivory, but the leaf-mining gelechiid caterpillar *Filatima* attacks more consistently. Damage by that caterpillar is conspicuous because it binds leaves into a woven yellow or white mass (Bishop 2002). Demographic surveys in 1991 and 1992 encountered no leaf-miner damage, but damage by leaf-mining gelechiid caterpillars became common in 1993 (Bishop 2002) and has continued at high levels in many areas through 2002 (Figure 11.4). Such damage is almost entirely confined to small, low-density patches and to low-density margins of large patches. For the years in which we have good sample sizes (Figure 11.4), the proportion of photosynthetic area consumed by leaf-mining gelechiid caterpillars

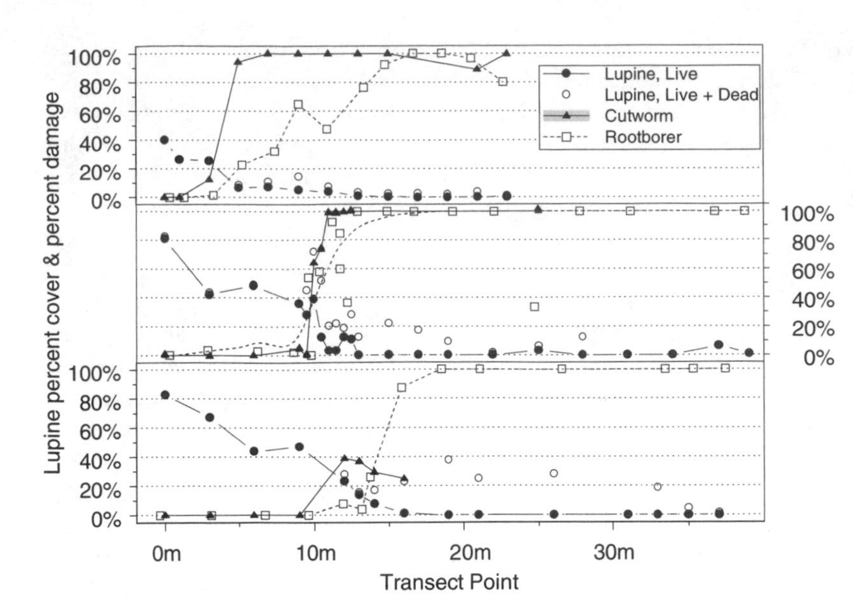

FIGURE 11.3. Herbivory along transects running from the centers to the margins of core patches. Each panel depicts prairie lupine surface area in August 2000, percent mortality caused by tortricid moth root borers from 1999 to 2000, and percent of lupine surface area attacked by noctuid cutworms in 2000 along a transect. *Open circles* depict surface area of live and dead lupines combined; *closed circles* are lupines alive at the time of the census. At each point on the transect, cover was estimated within a 0.25-m^2 frame. Sampling at a point was repeated along a line perpendicular to the transect until at least 200 cm^2 of lupine cover was recorded.

averaged 32% in edge patches but only 3% in core centers (i.e., areas with greater than 20% plant cover). Other important herbivores, such as noctuid cutworms, lupine aphids (*Macrosiphum albifrons*), and a guild of dipteran and lepidopteran predispersal seed predators, occasionally cause extreme damage in these patches, but they exhibit great spatial and year-to-year variation in their level of herbivory (Bishop 2002). As with outbreaks of tortricid root borers and leaf-mining gelechiid caterpillars, eruptions of noctuid cutworms, when they occur, also tend to be concentrated in areas of low-density lupine (see Figure 11.3).

Within the edge area, plants in the smallest patches were less likely to be attacked by tortricid root borers (Bishop 2002), suggesting that temporary escape from herbivory might be important for the establishment of new lupine patches. This kind of escape possibly underlies the explosive growth of lupine that began in 1999 (see Figure 11.2b). An extraordinarily deep snowpack in 1998–1999, with snow and cold weather lingering into July, delayed insect development. Although leafminer damage usually peaks in mid-August, repeated surveys in 1999 found that peak herbivory levels were postponed until September. Although lupine growth and flowering were also delayed, high soil moisture and the timing of herbivory permitted extremely high seed set.

Bishop (2002) provides extensive correlative evidence for strong demographic impacts of seed predators, leaf miners, and root borers from 1992 through 1995. These demographic impacts were largely confined to patches in the low-density edge region. In high-density core areas, seedling mortality caused by interspecific and intraspecific competition appeared to be the dominant population regulator (Bishop 2002). Experimental removal of leaf feeders and seed predators from edge patches in 1995 resulted in a 300% to 500% increase in population growth compared to control plots (Fagan and Bishop 2000). Herbivore removal in core areas resulted in higher seed production but did not lead to any increase in population growth rate (r), which remained near 0. A model of spatial spread predicted that herbivore effects on population growth would cause a 50% reduction in lupine's rate of advance across the Pumice Plain, a delay that could diminish the effect of lupines on primary succession (Fagan and Bishop 2000).

When we carefully examine small portions of the Pumice Plain, the predicted delay in lupine advance appears qualitatively correct. Herbivory clearly reduces population growth rate, causes local spatial collapse of some patches (cf. Figures 11.2b and 11.3), and causes a shifting mosaic of small patches and extreme fluctuation in percent cover (del Moral 2000b).

11.5 Why Is Herbivory Spatially Structured?

The demographic impacts of lepidopteran herbivores are impressive enough, but their spatial distribution and relationship to lupine density are truly remarkable. For 10 consecutive years, gelechiid leaf miners have been abundant but largely confined to low-density edge patches and core margins (see Figure 11.4). Although more sporadic, outbreaks of tortricid moth root borers and noctuid cutworms have also been far more frequent in low-density areas (see Figures 11.3, 11.4b). Apparently, above a certain density of lupine, herbivores are not sustained or are somehow repelled.

Negative relationships between host density and herbivory are known in other systems, but the underlying mechanisms in any particular case remain obscure (Thompson and Price 1977; Courtney and Courtney 1982; Courtney 1986; Kunin 1999). The simplest, if biologically unrealistic, hypothesis is that female moths disperse uniformly across the landscape and

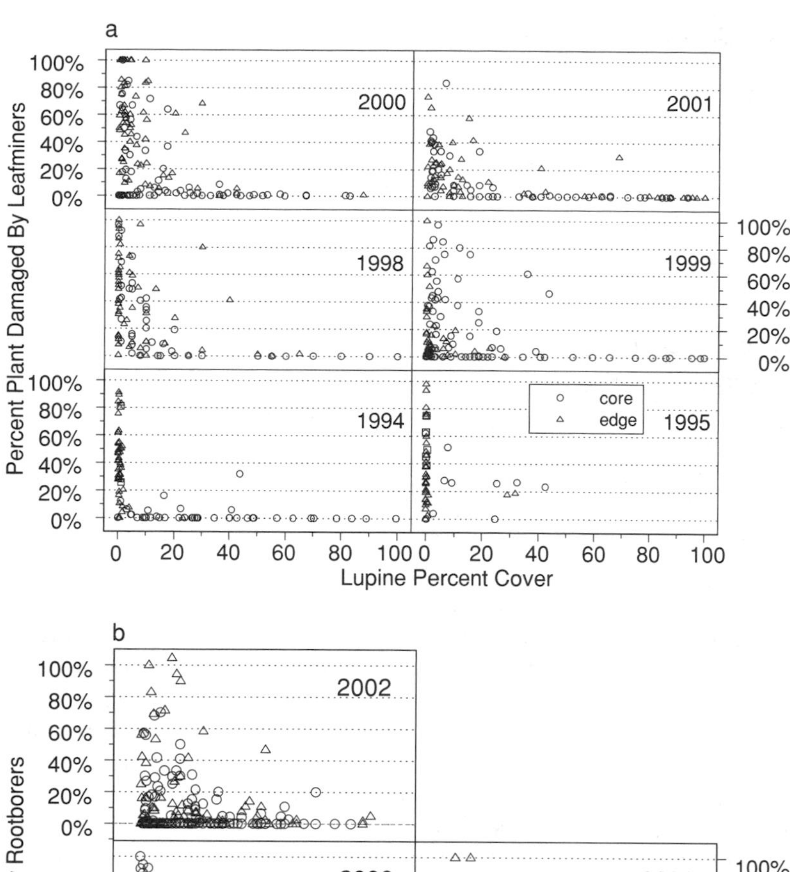

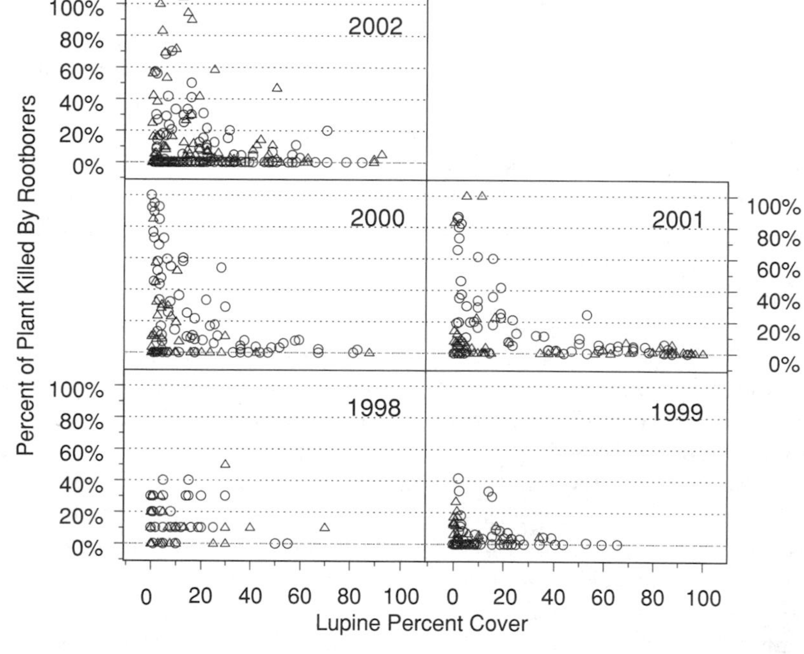

FIGURE 11.4. Relationship between lupine density and herbivore damage. (a) Leaf-miner (*Filatima* spp.) damage to prairie lupine, 1994 to 2001; (b) mortality caused by tortricid root borers (*Hystricophora* and *Grapholita* spp.), 1998 to 2002. Each point is an estimate of photosynthetic area consumed (leaf miner) or killed (root borer) in a sample. Sample area ranged from 0.25 m^2 to 315 m^2, depending on plant density. *Circles* denote core areas; *triangles* denote edge areas.

then oviposit on the nearest lupine. If the number of females per square meter is the same everywhere on the landscape, lupines in low-density areas would receive more eggs per plant than those in high-density areas because each low-density lupine draws moths from a greater area. We tested this null model for gelechiid leaf miners in 2001 by comparing caterpillars per square meter in high- and low-density areas. The data were obtained by harvesting all plants from four 0.64-m^2 plots in the core and two 100-m^2 plots in the edge, then carefully quantifying the number of caterpillars on each plant. Contrary to what the uniform distribution null model would predict, the number of caterpillars per square meter of land surface was four times higher in the core area than in the edge area (Table 11.1). Furthermore, we found 70 times more caterpillars per plant and 13 times more caterpillars per square meter of lupine surface in the edge plots than in the core plots, contrary to expectations based on the distribution of caterpillars per square meter in core and edge areas (Table 11.1). A second hypothesis is that greater diversity in core patches makes finding a host more difficult for caterpillars or adults. This concept also seems unlikely because prairie lupines are far more abundant in the core than in the edge. In core patches, cover of potential hosts is often greater than 60%, and host plants form a contiguous "canopy" (Fagan and Bishop 2000; Bishop 2002).

Succession itself suggests two additional explanations for the observed distribution of herbivory. First, by virtue of their

TABLE 11.1. Plant and caterpillar density in several core and edge quadrats during 2001.

	N	Surface area sampled (m^2)	Number of plants sampled	Leaf area sampled (m^2)	Plants/m^2	Percent cover	Caterpillars/m^2 surface area	Caterpillars/m^2 lupine	Caterpillars per plant
Core	4	0.64[a]	88.3	0.62	137.9[a]	96.3[a]	5.47[a]	9.5[a]	0.04[a]
Edge	2	100	60.0	0.95	0.6	0.95	1.23	128	2.77

[a]Denotes significant difference between core and edge ($p < 0.05$, t test).

longer existence, larger size, or other properties, core patches may have accumulated a greater abundance and diversity of arthropod and vertebrate predators (e.g., spiders, insects, and birds), which may, in turn, suppress or eradicate herbivores. Second, successional changes in soil development and fertility or increased intra- and interspecific competition may result in lower-quality lupine tissue in core areas, mediated by increased concentration of defensive chemicals, decreased nutrient concentration, or unfavorable nutrient ratios. These general explanations correspond to top-down and bottom-up controls, respectively, on herbivore populations.

11.5.1 Top-Down Explanations

Patch age alone might predict more predators in core areas because the probability that a population of predators will discover a patch increases with time. In addition, core plant communities are larger, denser, more diverse, and probably more productive. Accompanying these changes are increased habitat complexity and physical structure. The biomass of soil microbes is also greater in core plant communities (Halvorson and Smith 1995). Any of these conditions could lead to a larger and more diverse prey base that might sustain larger predator populations. For example, colonies of ants, such as *Formica lasioides* and *F. pacifica*, are commonly found in core areas but only rarely in edge areas, perhaps because minimum resource requirements for colonies are not met in edge areas. Throughout the summer, it is not uncommon to see these ants dragging noctuid cutworm caterpillars across the pumice.

In August 1995, we censused, by visual inspection, predators on and under lupine plants in edge and core areas. We found the density of predatory spiders, beetles, and true bugs to be four times higher per plant on core plants than on edge plants (Fagan and Bishop 2000). In 1999, we further characterized the distribution of arthropods using pitfall traps. We placed 53 traps in edge patches and along transects running from the centers to the margins of core patches. Traps were sampled every 10 days for 90 days from July to September (a total of 4770 trap-days). (The deep snowpack of 1999 prevented us from reaching the Pumice Plain until July 1, about 1 month later than usual.) Of the many arthropods caught in these traps, we discuss here the relative abundance of moths, several predatory beetles in the family Carabidae, and parasitoid flies (family Tachinidae, some of which attack larger caterpillars, such as noctuids). We focus on three beetle taxa: tiger beetles (*Cicindela oregona*), a large and highly mobile predator; *Calasoma* spp., one of the largest ground beetles whose common name, "caterpillar hunter," reflects active predatory behavior and a predilection for caterpillars; and *Nebria* spp., which are nocturnal predators. Many small moths could not be identified, so we have lumped them as "microlepidoptera." This category includes the tortricid moths *Hystricophora* and *Grapholita*, the gelechiid *Filatima*, and several other taxa.

The distribution of adult microlepidoptera was quite similar to the pattern of herbivory; they were more than three times as frequent in edge traps as in core traps (Table 11.2a). In contrast, the larger-bodied noctuid moths, the three predatory beetles, and parasitoid flies were all more common in core patches (Table 11.2a). The distribution of insects within core patches, along transects running from center to margin, is instructive. The frequency of beetles, tachinid flies, noctuids, and microlepidopterans along the transect all increased as total plant cover decreased (Figure 11.5). However, in many cases, the distribution of predator taxa was better predicted by the presence of food items than by plant cover (see Table 11.2b). For example, *Calasoma* beetles (the caterpillar hunters) are strongly predicted by the distribution of tachinid flies, noctuid cutworm moths, and microlepidoptera but not by plant cover (Table 11.2b). The tachinid *Peleteria* was closely associated with its putative host, noctuid moths. In fact, four of six predators and parasitoid taxa showed strong association with noctuid cutworms. In contrast, the catch density of the tiger beetle *Cicindela* is predicted strongly by low plant cover, by noctuid cutworms, and by *Nebria* but not by microlepidoptera or by tachinids. *Nebria* was predicted only by plant cover (not shown). Overall, the pitfall data confirm that core patches have a greater abundance and diversity of predators. However, many of these predators reside in the lower-density margins of core patches, where lepidopteran prey is most abundant.

The distributions of predators and both larval and adult lepidopterans in core patches suggest that these predators are not responsible for the absence of herbivores at core centers. Otherwise, both moths and predators would occur in our center traps at some point in the season. Other predators not sampled by these traps or not yet analyzed, such as birds, spiders, or ants, could be responsible for the lack of herbivores at core centers, but this hypothesis would require that these other predators discover moths, including night-flying noctuids, so efficiently that moths have no opportunity to die in traps. We have observed that some core patches in some years sustain

TABLE 11.2. Spatial distribution of selected insects caught in pitfall traps in 1999.

	Noctuidae*	Micro**	*Cicindela***	*Calasoma****	Nebria	Tachinidae*
Edge	1.3 (61)	10.7 (257)	3.3 (78)	0.0 (0)	5.4 (130)	4.1 (99)
Core	3.4 (215)	3.0 (86)	8.4 (245)	6.7 (194)	8.7 (252)	7.5 (218)

One-tailed Wilcoxon-signed rank test: * = $p < 0.05$; ** = $p < 0.01$; *** = $p < 0.001$.

a. Comparison of herbivore and predator abundance in core and edge areas: number of individuals/trap (total individuals) for 29 core traps and 24 edge traps.

	Plant cover	*Calasoma*	*Nebria*	*Arctophyto*	Tachinidae	Micro	*Euxoa*	*Schinia*	R^2
Calasoma					0.67***	0.67*	0.52*		0.54
Cicindela	−8.6*		0.22**				0.42**		0.66
Peleteria							0.20*	0.46***	0.54
Arctophyto		0.28***							0.40
Linnaemyia				0.26***			0.06**		0.79

b. Linear-regression results for five predator and parasitoid taxa in core traps. Dependent variables are in the left-most column. No transformation resulted in a significant coefficient for plant cover when insects were in the model, except in the case of *Cicindela* and *Nebria* (not shown). For *Calasoma*, the predictor variables microlepidoptera and *Euxoa* were collinear. Table entries for these two variables are from two separate regressions. *Peleteria* spp., *Arctophyto* spp., and *Linnaemyia comta* are tachinid taxa; Tachinidae denotes total tachinids, Micro denotes total microlepidoptera.

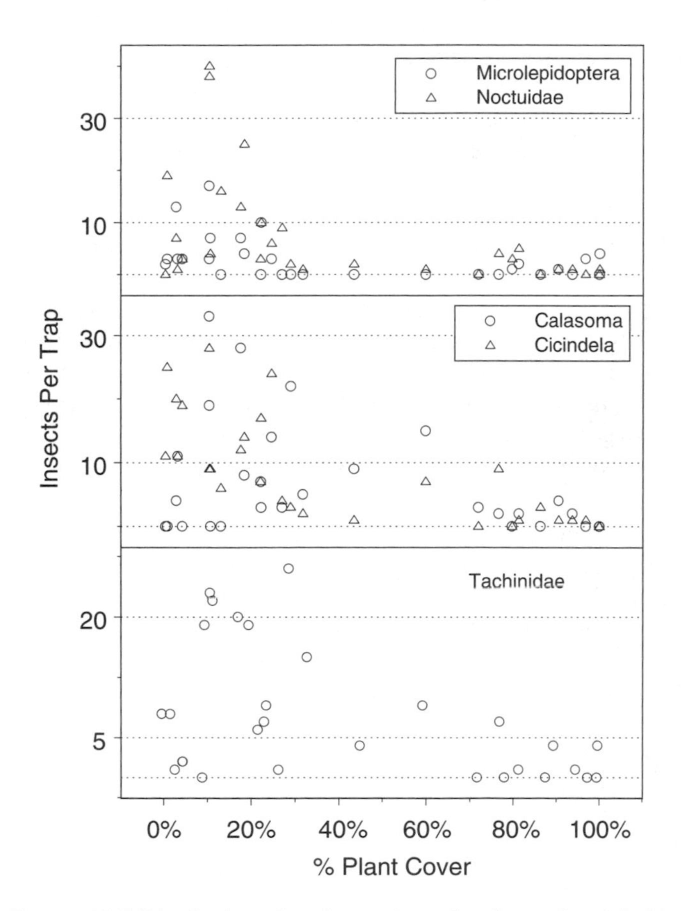

FIGURE 11.5. Distribution of moths, predatory beetles, and parasitoid flies in core patches in pitfall traps in 1999. Note that all predator taxa are more common at patch margins.

very little herbivore damage, even in low-density margins. In these cases, the pitfall data suggest that depletion of herbivores by predators is highly plausible.

11.5.2 Bottom-Up Explanations: Plant Quality

Specialist herbivores routinely identify and prefer their host taxa for purposes of oviposition as well as for feeding. Moreover, ovipositing females are known to discriminate among different-quality plants within a single host species (Bernays and Chapman 1994; Renwick and Chew 1994), suggesting that variation in lupine tissue quality could drive the negative correlation between plant density and herbivory. This discrimination might occur if core plants are either more effectively defended or are nutrient poor relative to edge plants. In fact, in preliminary experiments, we found that edge plants are indeed a higher-quality resource: caterpillars of gelechiid leaf miners exhibited a significantly reduced relative growth rate when fed leaves of core plants than when fed leaves of edge plants (Fagan et al. 2004).

Many studies conclude that nitrogen is a limiting nutrient for insect herbivores (McNeill and Southwood 1978; Mattson 1980; White 1993). On the other hand, recent comparative analyses of carbon, nitrogen, and phosphorus in plants and their insect herbivores conclude that phosphorus may be limiting more often than realized (Markow et al. 1999; Elser et al. 2000). These latter studies use elemental ratios (i.e., stoichiometry) to develop a common scale from which to gauge nutrient limitation. For example, most consumers are far more nutrient rich (i.e., lower C:N and C:P ratios) than their food. Likewise, if there is a cost to overconsuming one nutrient to obtain another, comparing N:P ratios between a consumer and its food can identify when phosphorus is more limiting than nitrogen.

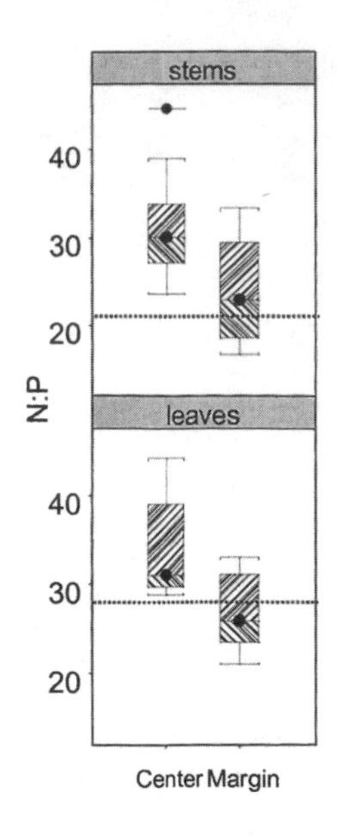

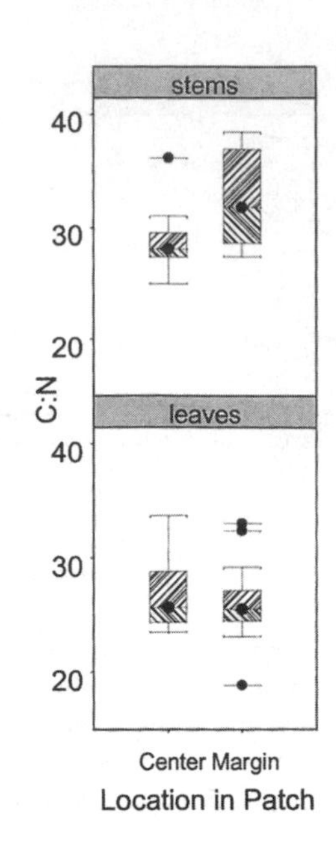

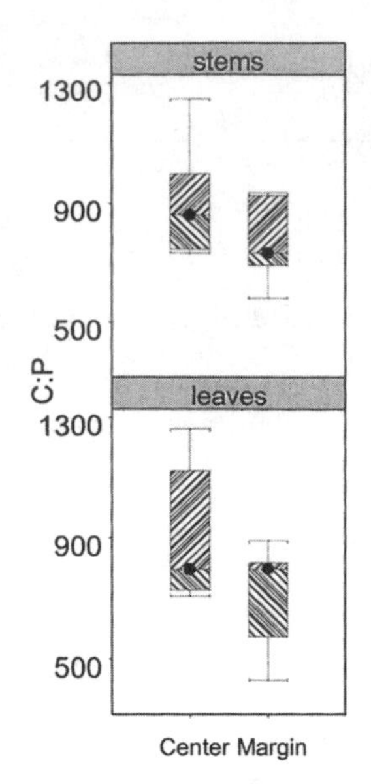

FIGURE 11.6. Elemental nutrient ratios from inner and outer portions of the transects depicted in Figure 11.3. *Shaded boxes* comprise the upper and lower quartiles around the median (*horizontal line*); *connected fences* and *unconnected dots with horizontal lines* indicate the extents of data and outliers, respectively. The *dotted horizontal line* shows N:P for *Hystricophora* root borers (stems/roots) and gelechiid leaf miners (leaves). Herbivore C:N and C:P ratios are lower than the graph minimum. Methods and the full data set are given in Fagan et al. (2004).

To examine whether nutrient concentration might explain herbivory patterns, we measured tissue concentration of carbon, nitrogen, and phosphorus in the year 2000 from undamaged or slightly damaged lupines at both ends of the transects shown in Figure 11.3 and in gelechiid leaf miners and tortricid root-borer larvae. We divided tissue into "root" parts fed upon by tortricid moths (mainly the upper root and caudex) and leaves (eaten by gelechiid leaf miners). Plants at the margin are somewhat more phosphorus rich (i.e., they have a lower C:P) for both tissue types than are plants at the core, and stems were nitrogen poor, but these differences were not statistically significant (Figure 11.6). Both herbivores were extremely nitrogen and phosphorus rich compared to the tissue they feed upon. However, plants from low-density margins were 23% more phosphorus rich per unit of nitrogen than were core plants. Whether this difference matters may depend on the N:P of the herbivores, illustrated in Figure 11.6 by dotted lines. In fact, herbivore N:P closely matches that of margin plants but is phosphorus rich compared to plants near the center of the core patches. Because herbivore C:N and C:P ratios do not change greatly in response to host-plant nutrient content (Elser et al. 2000), these results suggest that both herbivores could be phosphorus limited on center plants. Although reports of phosphorus limitation in insect herbivores are rare, Schade et al. (2003) recently showed that soil phosphorus availability affected abundance of a curculionid weevil by influencing the C:P ratio of its nitrogen-fixing host, mesquite (*Prosopis*). Such an effect at Mount St. Helens could be exacerbated by other factors. For example, because accumulated thermal units can limit larval growth rates, in montane environments (where thermal units may be marginal) more nutritious host plants may be required (Scriber and Lederhouse 1992; Hunter and McNeil 1997).

Nutrient availability may also be affected by differences in plant defensive chemistry. Like many plants, lupines produce an array of secondary chemicals that assist in the defense against herbivore attack (Wink et al. 1982; Johnson and Bentley 1988; Adler et al. 1998). Chief among these are quinolizidine alkaloids (QA), which may constitute as much as 5% of foliar nitrogen (Johnson et al. 1987, 1989) and which exhibit intraspecific variation (Carey and Wink 1994; Wink and Carey 1994; Wink et al. 1995). We have not yet assessed the content of QA or other defensive chemicals in prairie lupine. Whether center or margin plants are likely to be better defended is difficult to predict, as are the precise effects of QA. QA concentration is correlated with tissue nitrogen concentration in bigleaf lupine (*Lupinus succulentus*) and yellow bush lupine (*L. arboreus*), and QA content initially increases in response to herbivore damage (Johnson and Bentley 1988; Johnson et al. 1989; Bentley and Johnson 1994), both of which observations suggest that margin plants should produce more QA. QA and other defensive chemicals are often more effective against generalist herbivores, such as noctuid cutworms, than they are against specialists, such as gelechiid leaf miners and tortricid root borers (Smith 1966; Stermitz et al. 1989; Montllor et al. 1990; Bennett and Wallsgrove 1994), and even QA deterrence of generalists can be mitigated by increased tissue nitrogen (Johnson and Bentley 1988). Thus, it would be somewhat surprising if QA governed host use for all three herbivore species. Nevertheless, investigating lupine secondary chemistry is a critical component of future research.

11.6 Conclusions

Insect herbivores have strongly affected population dynamics of the prairie lupine on the Pumice Plain, but herbivores have not prevented the lupine population from growing to an enormous size and becoming the dominant plant on the landscape. Nevertheless, areas of intense herbivory still cause substantial local reductions in lupine patch size and spread rate, and a large proportion of the landscape has yet to be occupied. Prairie lupine would undoubtedly have achieved landscape-dominant status sooner and to a greater degree in the absence of herbivory. Remarkably, the activities of at least three different herbivore species are more or less confined to low-density areas of lupine. This distribution is counterintuitive, because the high-density areas offer much greater biomass of lupine. Although more plant species are present in these high-density patches, they often contain an interconnected carpet of lupine constituting a more continuous resource than in low-density areas. Our current research is focused on uncovering the reasons for this surprising spatial pattern. Our observations of herbivore and predator distributions indicate that suppression of herbivores by predators and parasitoids in core areas provides one plausible explanation. On the other hand, variation in plant quality is also consistent with observed patterns of herbivory. As is the case with many other successional processes, revealing the mechanisms that underlie observed patterns of herbivory requires experimental manipulation.

Although we have focused much of our efforts on understanding why herbivory is so strongly concentrated in edge areas, it is also important to recognize that the near absence of herbivory from core areas may be critically important to successional processes. Large mats of lupines trap seeds and detritus blowing across the landscape and change soil chemistry, facilitating subsequent colonists (Morris and Wood 1989; Titus and del Moral 1998b; del Moral et al., Chapter 7, this volume). Long-established core patches harbor numerous species that are rare or absent in nearby bare areas, including orchids (*Spiranthes* spp.), paintbrush (*Castilleja miniata*), several grasses, and numerous fungi (Allen et al., Chapter 15, this volume). As shrubs and trees continue to colonize and grow in lupine-prepared areas, core patches may provide habitat for birds, rodents, and other species that add to the diversity of these areas. Thus, areas in which herbivores are only weakly affecting lupines may become foci of colonization for other taxa.

Dichotomous views have dominated the ecological literature regarding the prevalence of population control by top-down processes (i.e., control by consumers) versus bottom-up processes (resource limitation). Recently, a more synthetic view has emerged that holds that both forces will interact to determine population size and density of primary consumers, with bottom-up forces providing a resource template upon which top-down forces can act (Hunter and Price 1992; Price 1992). In this view, environmental variation will determine plant quality and, hence, the bottom-up template. Important predictions to test include these:

- Do higher trophic levels increase in importance with increasing primary productivity (Oksanen 1990)?
- Does variation in plant quality caused by successional processes cascade up to affect higher trophic levels?
- And, most importantly, under what combinations of soil development, plant-community structure, and abiotic conditions do natural enemies dominate trophic interactions during succession?

Lupine patches on the Pumice Plain comprise a suitable system for examining the interaction of top-down and bottom-up factors because patchy colonization and accelerated soil development create sharp resource gradients on a relatively uniform environmental background. Spatial spread of nascent patches, both through wavelike expansion from initial patches and by creation of new foci through jump-dispersal, replicate and recreate earlier stages of the resource gradient. Superimposed on these resource gradients are temporal changes stemming from successional processes, especially the addition of higher trophic levels. However, while the recovering landscape is suffused with scientific opportunity, the complexity of patch types and the dynamic nature of succession also present challenges. Drawing generalizations can be difficult if only a few patches are studied in only a few years, and opportunities to test for successional mechanisms can be fleeting. Twenty-five years after the volcano's cataclysmic eruption, there remain exciting prospects for new studies and experiments into the effects of food web dynamics on the recovery of Mount St. Helens.

In summary, although population size and the area occupied by lupine have expanded dramatically in recent years, our 2002 survey showed that even after the recent explosive growth, prairie lupine is still absent, or present only at very low density, on most (70%) of the Pumice Plain. In light of our work, it seems likely that the continued absence of lupines, along with high rates of patch turnover, is explained in part by herbivory and in part by physical factors. Herbivory by several lepidopteran larvae, each with different modes of feeding, is strikingly elevated in low-density areas of lupine. Thus far, patterns of nitrogen and phosphorus availability appear to explain this pattern of herbivory better than suppression of herbivore populations by predators. Given the known effects of prairie lupine on soil and community development on the Pumice Plain, herbivory on prairie lupine has likely altered the pace and pattern of succession.

Acknowledgments. We thank Norman Woodley (USDA), Jerry Powell, Jennifer Apple, and especially Rick Sugg for insect identification. Ian Dews ran the trapping experiment. We thank Candan Soykan, Amy Novotny, Nicholas Murchison, and Pawel Drapala for field assistance. Roger del Moral kindly shared unpublished data. This work was supported by grants DEB-0089843 to J. G. Bishop and OCE-9973212 to W.F. Fagan.

12
Responses of Fish to the 1980 Eruption of Mount St. Helens

Peter A. Bisson, Charles M. Crisafulli, Brian R. Fransen, Robert E. Lucas, and Charles P. Hawkins

12.1 Introduction

Fish are important components of the Mount St. Helens aquatic system. Historically, no other region of Washington State supported as many native freshwater and anadromous species (anadromous fish mature in the ocean but spawn in freshwater) as did the region near Mount St. Helens (Table 12.1; McPhail 1967; McPhail and Lindsey 1986). Many of the anadromous species, including Pacific salmon (*Oncorhynchus* spp.) and eulachon (*Thaleichthys pacificus*), are keystone species that provide an important trophic link between aquatic and terrestrial ecological systems and are the foci of food webs that depend on marine-derived nutrients (Willson and Halupka 1995; Bilby et al. 1996; Levy 1997; Cederholm et al. 2001). In addition, fish are important consumers within rivers and lakes and can influence the species composition and structure of biological communities of these aquatic systems through herbivory, predation, and competition (Power 1990).

Rivers and lakes near Mount St. Helens have traditionally been managed as separate entities by the Washington Department of Fish and Wildlife and its predecessor agencies. The emphasis for lakes has been to manage primarily for recreational fisheries. The emphases in rivers have been (1) to manage salmon for commercial harvest and angling and steelhead (*Oncorhynchus mykiss irideus*) and sea-run coastal cutthroat trout (*O. clarkii clarkii*) for sport harvest and (2) to assure adequate reproduction of wild stocks. Some of this harvest has occurred at sea or in the Columbia River. Although fisheries-management strategies differed between streams and lakes, fish have often moved from one environment to the other. Because most lakes have tributaries with barriers to upstream migration, the majority of movement has probably been from lake to stream rather than the reverse. These barriers proved especially important in the aftermath of the 1980 eruption.

The streams and lakes surrounding the volcano are numerous and diverse with respect to physical habitat and biological communities. Streams range from small, steep, cascade-dominated mountain channels to large, floodplain rivers (e.g., the Cowlitz River and Lewis River). Lakes range from high-elevation, cool, subalpine lakes that are common north of the mountain to low-elevation, relatively warm systems [e.g., Silver Lake (which is discussed in Chapter 2 and Chapter 18 of this volume)]. They were subjected to an array of volcanic disturbances during the 1980 eruption that ranged from the relatively low-level impacts of a few centimeters of tephra fall to burial of an entire river drainage or lake beneath the enormous debris-avalanche deposit (see Swanson and Major, Chapter 3, this volume).

In this chapter, we address five questions.

1. Did fish survive the initial impacts of the 1980 eruption, and, if so, what factors aided their survival?
2. Were fish able to recolonize streams and lakes during the ensuing 20 years, and, if so, by what means?
3. What were the key factors influencing fish survival and growth in posteruption streams?
4. What role did management play in reestablishing fish populations? and
5. Are there lessons from fish responses to the 1980 eruption that could be applied to other managed landscapes?

These questions are addressed first for fish in streams and rivers and then for fish in lakes.

12.2 Fish in Streams and Rivers

12.2.1 Preeruption Conditions

The streams and rivers draining Mount St. Helens (Figure 12.1) were among the most productive for anadromous fish in southern Washington. They supported large commercial and recreational fisheries. Before the 1980 eruption, tens of thousands of salmon, steelhead, and sea-run cutthroat trout spawned and reared in several major tributary systems of the Columbia River below Bonneville Dam that drained the Mount St. Helens area. Chinook salmon (*Oncorhynchus tshawytscha*), coho salmon (*O. kisutch*), steelhead, and sea-run coastal (anadromous) cutthroat trout were historically the most abundant anadromous salmonids in Mount St. Helens river systems, although a few chum salmon (*O. keta*) and sockeye salmon

TABLE 12.1. Fish species present in the Lewis, Kalama, Toutle, and Cispus river drainages prior to the 1980 eruption of Mount St. Helens.

Family	Common name	Scientific name	Life history	Origin
Petromyzonitidae	Western brook lamprey	*Lampetra richardsoni*	Resident	Native
	Pacific lamprey	*Lampetra tridentatus*	Anadromous	Native
Salmonidae	Mountain whitefish	*Prosopium williamsoni*	Resident	Native
	Brown trout	*Salmo trutta*	Resident	Introduced
	Coastal cutthroat trout	*Oncorhynchus clarkii clarkii*	Anadromous	Native
	Westslope cutthroat trout	*Oncorhynchus clarkii lewisi*	Resident	Introduced
	Coastal rainbow trout	*Oncorhynchus mykiss irideus*	Anadromous/resident	Native
	Brook trout	*Salvelinus fontinalis*	Resident	Introduced
	Lake trout	*Salvelinus namaycush*	Resident	Introduced
	Bull trout	*Salvelinus confluentus*	Resident/anadromous	Native
	Coho salmon	*Oncorhynchus kisutch*	Anadromous	Native
	Chinook salmon	*Oncorhynchus tshawytscha*	Anadromous	Native
	Sockeye salmon	*Oncorhynchus nerka*	Resident	Introduced
Osmeridae	Eulachon	*Thaleichthys pacificus*	Anadromous	Native
Cyprinidae	Redside shiner	*Richardsonius balteatus*	Resident	Native
	Longnose dace	*Rhinichthys cataractae*	Resident	Native
	Speckled dace	*Rininchthys osculus*	Resident	Native
	Northern pikeminnow	*Ptychocheilus oregonensis*	Resident	Native
	Peamouth	*Mylocheilus caurinus*	Resident	Native
Catostomidae	Largescale sucker	*Catostomus macrocheilus*	Resident	Native
Gasterosteidae	Three-spine stickleback	*Gasterosteus aculeatus*	Resident	Native
Percopsidae	Sandroller	*Percopsis transmontana*	Resident	Native
Cottidae	Coast sculpin	*Cottus aleuticus*	Resident	Native
	Shorthead sculpin	*Cottus confusus*	Resident	Native
	Torrent sculpin	*Cottus rhotheus*	Resident	Native
	Riffle sculpin	*Cottus gulosus*	Resident	Native

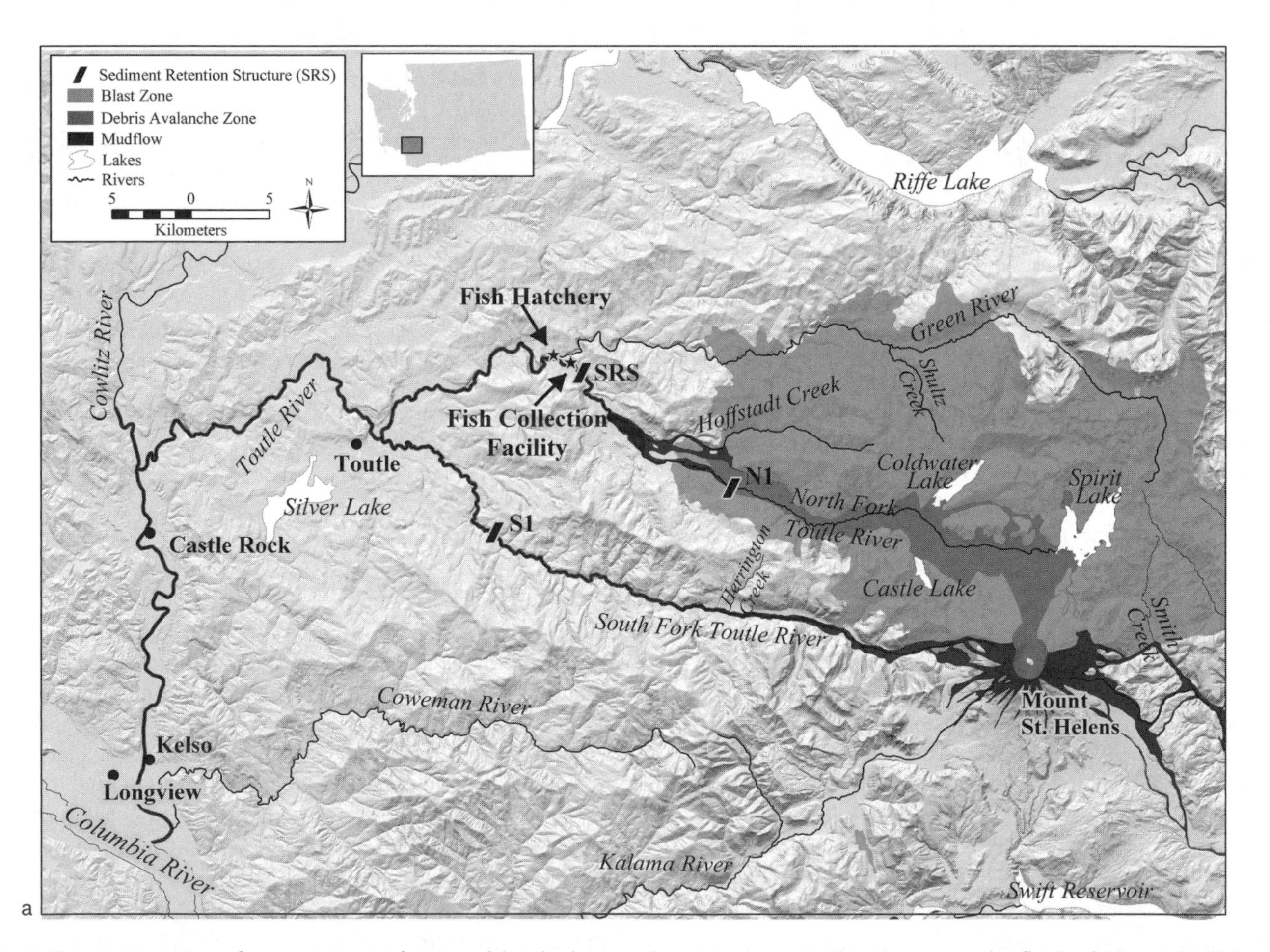

FIGURE 12.1. (a) Location of some streams, dams, and hatcheries mentioned in the text. The streams on the flank of Mount St. Helens, including Smith Creek, are part of the Lewis River system.

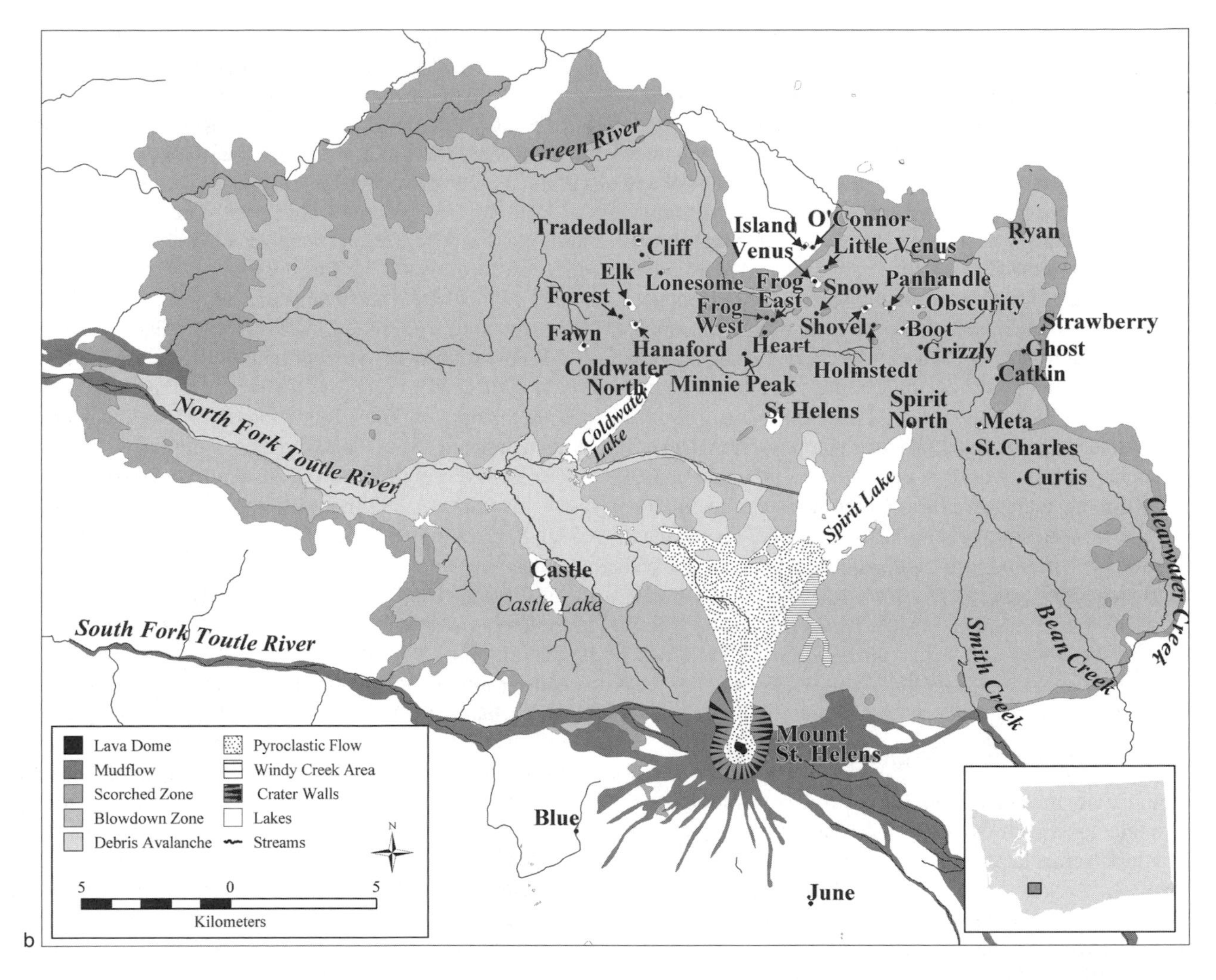

FIGURE 12.1. *(continued)* (b) Location of lakes mentioned in the text.

(*O. nerka*), possibly strays from other Columbia River populations, were occasionally observed. Other anadromous fish used some of the rivers draining the Mount St. Helens area and included eulachon, white sturgeon (*Acipenser transmontanus*), green sturgeon (*A. medirostris*), Pacific lamprey (*Lampetra tridentatus*), and an introduced species, the American shad (*Alosa sapidissima*). Some anadromous species (Chinook salmon, coho salmon, and eulachon) have supported valuable commercial fisheries in addition to recreational fisheries; others (steelhead and sea-run cutthroat trout) have supported popular recreational fisheries (Tacoma Power and Washington Department of Fish and Wildlife 2004).

Reimers and Bond (1967) reported eight families and 29 species of freshwater fish from tributaries of the lower Columbia River in the vicinity of Mount St. Helens. The most diverse taxa were sculpins (Cottidae), which were represented by 7 of the 12 species known from the Columbia River basin, and minnows (Cyprinidae), which included redside shiners (*Richardsonius balteatus*), dace (*Rhinichthys* spp.), peamouth (*Mylocheilus caurinus*), and northern pikeminnow (*Ptychocheilus oregonensis*). Mountain whitefish (*Prosopium williamsoni*), the only whitefish species known from the Columbia River west of the Cascade Mountains, and largescale suckers (*Catostomus macrocheilus*) were widespread and abundant in many of the streams. Distribution limits of fish upstream in the drainage networks were usually bounded by waterfalls, with most species absent from reaches above large falls unless colonization occurred (1) by stream capture (i.e., diversion of a stream from one drainage network to another, often through headward erosion or damming and diversion by natural geological processes), (2) before formation of the falls, or (3) via deliberate introduction by humans.

Because the Mount St. Helens area has a long history of large natural disturbances (including eruptions by nearby Cascade Mountain volcanoes as well as several periods of glaciation within the Quaternary Period), most species have undergone episodes of local extirpation followed by recolonization (Wydoski and Whitney 2003). The lower Columbia River provided important refugia during many of these disturbances and served as a source of new colonists following local catastrophic events (McPhail and Lindsey 1986).

Three major river systems drain the slopes of Mount St. Helens: the Cowlitz River, Kalama River, and Lewis River (see Figure 12.1). Of these, only the Kalama River has remained undammed; each of the others have several flood-control and hydroelectric dams that block the migration of anadromous fish and seasonal movements of resident fish. Some salmon and steelhead adults in the Cowlitz River are captured in traps and trucked above the dams. Additionally, each of the rivers possesses several fish hatcheries that propagate both anadromous and resident salmonids. A salmon hatchery on the Kalama River is one of the oldest in the Pacific Northwest, having been in operation since the 1890s. At its completion in 1968, a hatchery on the Cowlitz River produced more salmon than any other hatchery in the world. It continues to maintain annual releases of more than 10 million Chinook and coho salmon. The hatcheries were built to mitigate for losses of habitat upon completion of dams and to provide additional fish for harvest (Lichatowich 1999).

12.2.2 Posteruption Conditions in Streams and Rivers

12.2.2.1 New Conditions Created by Eruption

The immediate impact of the May 18, 1980 eruption was catastrophic to stream-dwelling fish populations inhabiting the Toutle River, a large tributary of the Cowlitz River (see Figure 12.1). An enormous debris avalanche and numerous mudflows caused the most damaging effects. The debris avalanche buried the uppermost 25 km of the main-stem North Fork Toutle River. Mudflows occurred in the north and south forks of the Toutle River, and smaller mudflows affected the upper Kalama River and tributaries of the Lewis River (see Swanson and Major, Chapter 3, this volume). The mudflows contained exceptionally high suspended-sediment levels (greater than 10,000 mg l^{-1}), with a peak suspended-sediment concentration of 1,770,000 mg l^{-1} being recorded in the Toutle River. In addition, some mudflows contained heated water (greater than 30°C) from pyroclastic flows and from water contained in hot deposits of the main debris avalanche (Dinehart et al. 1981). Fish mortality could not be documented, but it must have been nearly complete in the parts of rivers experiencing the debris avalanche or mudflows. During the summer and early fall of 1980, suspended sediment levels in the Toutle River remained at 300 to 1000 mg l^{-1}, a range that was found to be lethal for fish exposed to Mount St. Helens mudflow and tephra particles (Stober et al. 1981). Yet, some adult steelhead returned to the river. Mudflows continued in the Toutle system in the years that followed the 1980 eruption (Swanson and Major, Chapter 3, this volume), some of which led to large changes to channel morphology.

Before the 1980 eruption, the Toutle River system contained approximately 280 km of streams used by salmon, steelhead, and sea-run coastal cutthroat trout. Volcanic mudflows and the main debris avalanche in total inundated 169 km (58% of the length of streams available to anadromous salmonids), including the entire main stem of the Toutle River. The mudflows also blocked access to some tributaries, but the number of streams and the length of potential habitat lost were not known.

In addition, all streams within a 350-km^2 area in a wide arc north of Mount St. Helens were affected by the powerful lateral blast that leveled the forest, sending trees into and across stream channels. These streams and adjacent hillsides also received varying amounts of blast material, coarse pumice, and ash. Rainstorms after the eruption mobilized hillside deposits, which continued to change channel substrates and morphology for the first few years following the eruption. Nonetheless, many fish survived and persisted during the early 1980s in tributaries of severely impacted mainstem rivers. For example, in 1981, juvenile coho salmon were present in tributaries of the Green River, and steelhead and cutthroat trout were observed at high densities in several tributaries of the south and north forks of the Toutle River. These survivors played an important role in the recolonization of streams that were severely impacted by the eruption.

Adult salmon and steelhead straying was an important survival mechanism. One or more age cohorts of the anadromous species were at sea during the 1980 eruption. The first spawning adults of these cohorts returned to their natal rivers in summer and early autumn of 1980. For the next 2 years, many adult salmon and steelhead returning to Mount St. Helens streams had been born and reared under preeruption conditions. Physical changes to the rivers and to the chemical properties of the water were so great that olfactory cues guiding adults back to their natal streams were probably disrupted, and many fish strayed from their rivers of origin. Lieder (1989) found that large numbers of adult steelhead entered relatively unaffected tributaries of the Columbia River near the most heavily impacted rivers, the Cowlitz River and Toutle River. Estimates of adult steelhead straying from volcanically impacted rivers increased from 16% preeruption to 45% posteruption. Leider (1989) believed that winter-run steelhead strays originated from both the Cowlitz and Toutle rivers while summer-run steelhead strays were primarily from the Toutle River. Small numbers of adult salmon and steelhead, however, did navigate sediment-laden waters of main-stem rivers to return to small tributaries of the Cowlitz and Toutle rivers in 1980 (Martin et al. 1984; Lucas 1985), where they presumably spawned. Somewhat surprisingly, Leider (1989) found no evidence that the temporary infusion of Cowlitz River and Toutle River steelhead strays into the neighboring Kalama and Lewis Rivers produced significant increases in adult returns to these rivers during the mid-1980s.

Elevated concentrations of volcanic sediment in the Cowlitz and Toutle rivers were believed to be the primary cause of returning salmon and steelhead adults straying to other river systems (Whitman et al. 1982). Chronically elevated suspended sediment continued to affect juvenile and adult salmon wherever they encountered it for a number of years after the 1980 eruption. Redding and Schreck (1982) exposed juvenile steelhead to relatively low (500 mg l^{-1}) and high (2000 to 3000 mg l^{-1}) levels of Mount St. Helens ash. They found evidence of sublethal stress (i.e., elevated corticosteroid and hematocrit) when exposure to either concentration was continuous for 48 hours, but fish were able to tolerate this relatively short exposure to volcanic sediment without exhibiting prolonged physiological stress responses. Redding and Schreck (1982) believed that extended exposure to abrasive sediment would erode the mucous coating on gills and cause respiratory impairment.

Elevated volcanic sediment concentrations have continued in the North Fork Toutle River below the permanent sediment-retention structure (see Figure 12.1a). Olds (2002) reported suspended-sediment levels of 252 to 1970 mg l^{-1} during the salmon smolt (juvenile seaward migrant) emigration period of March through May from 2001 to 2003. These concentrations, combined with high water velocity, were linked to reduced smolt survival in the Toutle River on the basis of sediment tolerances of salmonids in laboratory studies (Olds 2002).

12.2.2.2 Sediment-Retention Structures

After the eruption, concerns about the threat of erosion and flooding of the heavily developed lower Cowlitz River and Columbia River floodplain led to construction of several sediment-retention structures (dams) during the 1980s. Two low structures were completed within several months of the May 1980 eruption, one on the north fork and one on the south fork of the Toutle River. The structures were less than 10 m high and were meant to contain sediment eroded during the initial posteruption years. The temporary North Fork Toutle River sediment structure blocked upstream salmon and steelhead migrations, but the South Fork Toutle River structure was equipped with a fish ladder that passed adult salmon and provided a means to assess numbers of returning fish. Floods breached the structure on the North Fork Toutle River the first year, and the U.S. Army Corps of Engineers, at the request of Washington Department of Fish and Wildlife and numerous sports groups, removed the South Fork Toutle River structure 2 years later.

The very large sediment-retention structure on the North Fork Toutle River was completed by the Corps in 1989. It was placed about 1.2 km upstream of the mouth of the Green River. This structure was designed to be permanent, and upon completion it extended approximately 50 m above the streambed and 600 m across the North Fork Toutle River valley. The structure is located 48.8 km upstream from the mouth of the Toutle River and is too high to include a fish ladder; thus, it is a barrier to salmon and steelhead using the upper North Fork Toutle River watershed. The species most heavily impacted by the structure are Chinook salmon, coho salmon, steelhead, and sea-run cutthroat trout, as well as other fish moving upstream, such as minnows and suckers.

To mitigate for upriver losses, a fish-trapping facility was built about 1 km downstream of the sediment-retention structure in 1989. Upstream-migrating salmon and trout adults trapped at this facility are trucked above the dam and released into tributaries to spawn. Hatchery-origin summer steelhead that are captured in the collection facility are moved back downstream so they will not interfere with the success of the native Toutle River winter-run steelhead, which are transported upstream. Initially, sand accumulated at the trap's intake, preventing fish from entering it. The problem became acute during the mid-1990s as the pool behind the dam filled with sediment and large quantities of sand began passing over the spillway, aggrading the river near the fish-trap intake. The Corps has repeatedly improved access to the fish trap by removing sediment. The dam also caused sedimentation in the lower reaches of Alder Creek and Hoffstadt Creek when sediment-laden water pooled behind the dam deposited sand near the mouths of the two streams. This sedimentation caused severe degradation to some of the highest quality coho salmon and steelhead spawning areas that existed before the eruption.

Downstream-migrating smolts pass over the dam's spillway, a concrete channel 670 m long with an average slope of 7%. On average, 22% of the coho salmon smolts passing over the spillway received external injury from the spillway in 2001 and 2002 (Olds 2002). Impacts were found to be greater when velocities over the spillway were elevated from heavy rains and melting snow, which resulted in higher suspended-sediment levels and greater risk of external injury to the fish from striking the spillway surface or trapped wood debris.

12.2.2.3 Salmon and Steelhead Returns

Following the 1980 eruption, many fishery managers predicted that recovery of salmon and steelhead populations would take decades because riverine habitats had been so extensively damaged. Returning adults were, in fact, scarce in the first 3 years after the eruption (Lieder 1989). Many adults strayed to nearby, unimpacted Columbia River tributaries, and others were unable to successfully swim through the warm, sediment-rich Toutle and Cowlitz rivers to reach spawning streams (Whitman et al. 1982). The recreational fisheries for salmon and wild steelhead returning to the Toutle River were closed immediately after the eruption and remained so until 1987 (Lucas and Pointer 1987). However, fishing for hatchery steelhead reopened on the mainstem Toutle River in 1983.

Most salmon and steelhead returning to the Toutle River during the late 1970s and in 1980 before the eruption were hatchery-produced fish. After the eruption, what few steelhead adults returned to the South Fork Toutle River were mostly naturally spawned, suggesting that the homing fidelity of wild steelhead was greater than that of hatchery fish (Lucas 1985; Quinn et al. 1991). The two major Toutle River tributaries (South Fork Toutle River and Green River) eroded through mudflow or tephra-fall deposits and returned to preeruption streambeds within a few years. In the absence of sport harvest, adult steelhead returns rebounded much more rapidly in these rivers than many managers and biologists had predicted. Numbers of steelhead redds (egg deposition sites) observed in the main stem of the South Fork Toutle River rose from 0 in 1980 to an average of 5.7 redds km^{-1} in 1984 and further to 21.5 redds km^{-1} in 1987 (Lucas and Pointer 1987). The remarkable recovery of the wild steelhead population in the South Fork Toutle River during the mid-1980s exceeded all expectations. Since that time, returns of naturally spawning salmon and steelhead declined somewhat, as did anadromous salmonids in virtually all Columbia River tributaries until 1999 when improved ocean conditions resulted in sharply increased runs.

Salmon and steelhead have been stocked into many Mount St. Helens streams to accelerate population recovery and to provide commercial and recreational harvest opportunities in the wake of the 1980 eruption. For example, in 2001, 150,000 spring Chinook salmon fry were stocked in the Lewis River at the confluence of Crab Creek, which is upstream of the three mainstem Lewis River dams, and 140,000 Chinook salmon and 220,000 coho salmon fry were stocked into the Muddy River, a tributary of the Lewis River that extends into the mudflow and blowdown zones created by the 1980 eruption. In addition to these fish stockings, 7,000 adult coho salmon and 54 adult Chinook salmon were stocked in Swift Reservoir on the Lewis River in 2001 to enrich the reservoir and its tributaries with the nutrients from their carcasses. Emphasis on supplementing Toutle River populations with juvenile hatchery salmon ended about a decade after the 1980 eruption; however, transport of adult salmon and steelhead from the North Fork Toutle River fish trap to Hoffstadt Creek and the Green River is continuing.

12.2.2.4 Refugia for Resident Fish

Colonization of streams containing resident (nonanadromous) fish after the 1980 eruption was strongly influenced by the presence of several different types of refugia. Headwater streams typically had waterfalls and other natural barriers that prevented fish from entering them from downstream. Other streams possessed anthropogenic barriers, such as dams and impassable culverts at road crossings, which likewise limited upstream movement. If fish survived the eruption in refugia, these barriers could have important implications for the repopulation of fish in disturbed drainages. Hawkins and Sedell (1990) examined the dispersal of aquatic flora and fauna into Mount St. Helens streams based on a long-term monitoring study of the Clearwater Creek drainage, a tributary of the Lewis River within the blowdown and tephra-fall zones. For brook trout (*Salvelinus fontinalis*) and cutthroat trout, the primary refuge sites were headwater lakes that were covered by ice and snow at the time of the eruption (Hawkins and Sedell 1990; Crisafulli and Hawkins 1998). Resident cutthroat trout may also have survived in the upper reaches of tributaries that were impacted only by tephra fall. Once suitable habitat was present, these fish or their progeny colonized downstream portions of the drainage network from which fish had been extirpated during the eruption. Sculpins appeared to have survived within tephra-fall streams and rebounded to relatively high abundance during the 1980s in response to favorable habitat (shallow, sandy reaches with dense algae and large midge populations). Within 3 years of the eruption, sculpins recolonized the main stem of Clearwater Creek, either from headwater tributaries impacted by tephra fall or springs located in the floodplain of the stream (Crisafulli and Hawkins 1998).

Efforts to reestablish steelhead may have resulted in the establishment of resident rainbow trout populations (steelhead are the anadromous form of rainbow trout). In the west fork of Schultz Creek, a Green River tributary, hatchery steelhead were stocked in the 1980s above a barrier falls and have since established a nonmigratory local breeding population. Our data show that these rainbow trout may have hybridized with native cutthroat trout, because trout with intermediate morphological features began to appear in the late 1980s. The frequency of putative rainbow and rainbow–cutthroat trout hybrids gradually increased through the 1990s, and by 2003, no visually distinct cutthroat trout were observed in this stream, suggesting that introduced rainbow trout had completely displaced the native cutthroat trout population. A stream draining Venus Lake (see Figure 12.1) also supported rainbow and cutthroat trout (and apparently their hybrids). These species were observed during electrofishing surveys conducted in 1994, 14 years after the eruption. These fish likely originated from fish stocked in Venus Lake because natural barriers prevented dispersal upstream into Venus Lake from the Green River.

In our electrofishing surveys of 19 second- and third-order streams within the Muddy River and Green River systems of the blowdown zone during 1994 and 1995, cutthroat trout were found in 6, brook trout in 4, rainbow trout in 1, and sculpins in 6 streams. Most of the streams were connected to lakes where fish were known to have survived the eruption. Even if the stream populations were extirpated, headwater lakes may have served as source populations for recolonization. Sculpins, capable of seeking refuge in interstitial spaces within the substrate (Bond 1963), were assumed to have survived in the stream network, particularly in systems with tributaries extending into the less-disturbed tephra-fall zone.

Resident fish in the blowdown zone relied on at least two forms of refugia: ice-covered lakes and connected tributaries that flowed through the less-impacted tephra-fall zone. The distribution of these refugia influenced the initial patterns of survival and subsequent recolonization events. The importance of refugia for resident fish appeared to have varied by species. Lakes served as important refugia for some species, whereas in-stream cover and microhabitats facilitated the survival of other stream fish.

12.2.2.5 Stream Habitat Development

Streams affected by the 1980 eruption experienced a variety of impacts and developed habitat conditions suitable for fish colonization or population growth at different rates during the next 25 years. Headwater streams in the blowdown zone received 30-cm to more than 100-cm tephra deposits. Their riparian zones and valley-wall forests were leveled by the lateral blast, leaving stream channels open to warming via direct solar radiation. During subsequent autumn and winter freshets, these streams carried very high levels of suspended sediment, and they received even more sediment from erosion of hill slopes, including landslides. Debris flows were common in some portions of the blowdown zone during the first few years after the 1980 eruption and again in 1996, resulting in many small streams having log jams at or near their mouths. The large jams may have inhibited salmon and trout from entering some Toutle River system tributaries.

Mudflows along the North Fork Toutle River and South Fork Toutle River (the Green River did not experience mudflows) created elevated sediment terraces that blocked confluences of the small streams flowing into them. Many tributaries created entirely new channels as they cut across these mudflow terraces before entering the main-stem stream (Figure 12.2). The new channels were almost completely devoid of riparian vegetation in the first year after the eruption. Their streambeds contained very little large wood and other potential habitat structures. Mudflow terrace reaches of tributary streams ranged in length from less than 100 m to several kilometers, and they were gradually colonized by aquatic invertebrates and fish from both the headwater streams and the rivers into which they flowed. Because these reaches were essentially new channels, they presented an extraordinary opportunity for studying early-successional patterns in low-gradient stream ecological systems.

Martin et al. (1982, 1986) studied nine streams in volcanically disturbed and undisturbed drainages near Mount St. Helens in 1981 and 1982. They found that juvenile coho salmon survived poorly in the volcanically disturbed streams in summer because they experienced lethal and highly variable water temperatures. They also found that apparent winter survival (i.e., accounting for actual death plus emigration) of juvenile coho salmon was low in volcanically disturbed streams, which they attributed to lack of in-channel habitat complexity and hiding cover. It is possible that survival of these largely hatchery-bred fish was lower than survival of naturally spawned coho (Nickelson et al. 1986) or that they responded differently to the altered habitat. Nevertheless, Martin et al. (1986) concluded that two factors would improve the recovery of fish habitat in Mount St. Helens streams: (1) reestablishment of riparian vegetation, which would moderate stream temperatures, and (2) recruitment of large woody debris, which would provide

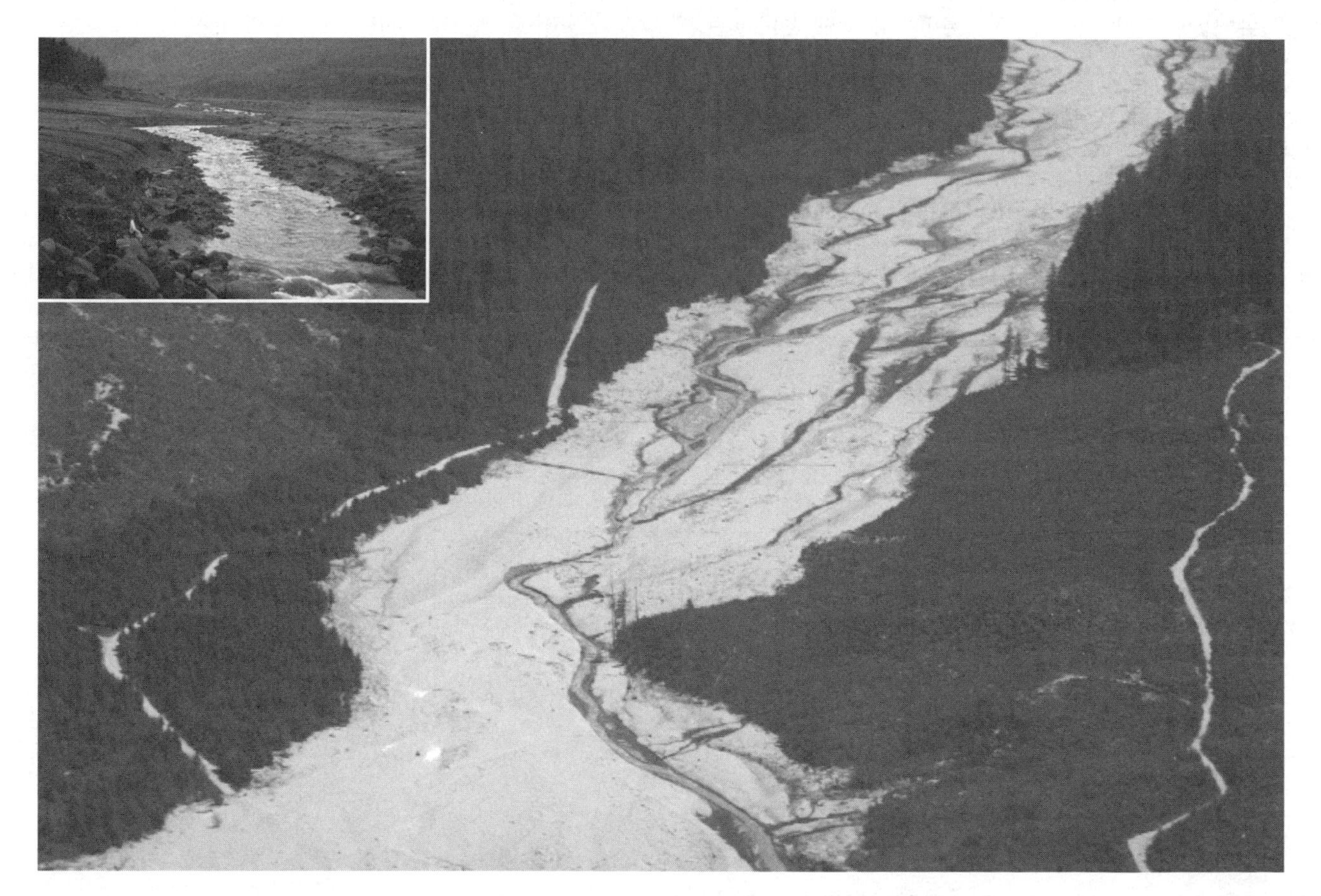

FIGURE 12.2. South Fork Toutle River in the Herrington Flats area in 1981. Photo *inset* shows the lower section of Herrington Creek, where the stream cut a new channel through the mudflow terrace. (Photo credits: P.A. Bisson.)

physical cover in winter. They estimated that 5 to 20 years would be needed for riparian vegetation to provide effective shading for productive fish habitat and that 50 to 75 years would be required for new wood to be recruited from riparian forests.

Bisson et al. (1988) studied coho salmon recovery in three Toutle River streams from 1983 to 1986, including one of the streams used by Martin et al. (1986) in their investigation from 1981 to 1982. Coho salmon stocked in the three study sites (one site in a mudflow terrace stream and two sites in blowdown-zone streams) exhibited increasing summer production during this period despite poor physical habitat (less than 30% of the stream area in pools) and high temperatures. By 1986, juvenile coho salmon production, expressed in milligrams of new tissue produced per square meter per day, was found to be twice as great as production rates of juvenile coho salmon stocked in nearby old-growth forested streams (Bilby and Bisson 1987). At the Mount St. Helens sites, these values ranged from 2.3 to 21.6 g m^{-2} over a 150-day warm-season period. The remarkably high productivity occurred during a period when summer temperatures in one of the streams reached 29.5°C, several degrees above the assumed lethal threshold for salmonids (Bjornn and Reiser 1991). Bisson et al. (1988) attributed this extraordinary productivity to an abundance of both aquatic and terrestrial food resources, caused in part by increased light levels and associated increases in net primary productivity and by a relative absence of predators and competitors.

Peak temperatures in Herrington and Hoffstadt creeks (Figure 12.3) frequently exceeded the 24°C assumed lethal threshold for salmonids during the 1980s, but episodes of these lethally high temperatures declined during the 1990s. In spite of potentially hazardous thermal conditions during the first 22 years after the eruption, all three streams continued to support salmon and trout, and one stream supported sculpins. Fish may have survived high temperatures by making temporary use of cool groundwater seeps and other thermal refugia (Bilby 1984), where these features were available. However, more research is needed on the actual mechanisms of survival in a thermally hostile environment.

Recovery of stream habitat and fish populations was followed in the three streams studied by Bisson et al. (1988) from the early 1980s to 2002. Maximum summer water temperature (see Figure 12.3) gradually declined at both the mudflow terrace stream (Herrington Creek) and the two blowdown-zone streams (Schultz Creek and Hoffstadt Creek). Peak temperature declines were related to recovery of riparian vegetation, particularly red alder (*Alnus rubra*), that formed dense riparian stands and shaded the stream channels (Figure 12.4). Riparian vegetation recovery appeared to be influenced by the proximity of seed sources. The riparian zone adjacent to Herrington Creek was rapidly colonized by alder; a nearby intact upland alder stand likely served as a seed source.

From 1983 to 2002, the relative abundance of pool habitat [the other factor believed by Martin et al. (1986) to potentially limit fish recovery in Mount St. Helens streams] has been measured annually by computing the surface area of pools relative to other habitat types. The percentage of pools (Figure 12.5) increased in both Herrington and Schultz creeks but not in Hoffstadt Creek. The rapid rise in pool habitat in Herrington Creek from 1981 to 1983 resulted from exposure of large boulders by channel incision through mudflow deposits. The subsequent increase in pools to 70% of the stream area from 1987 to 1989 was caused by beaver (*Castor canadensis*) activity. When the beavers abandoned lower Herrington Creek in 1990, the beaver dams were breached by high flows in winter storms, and the percent of the wetted channel composed of pools declined to about 50%. The increase in pools in Schultz Creek was related to the recruitment of boulders and wood during winter floods. In Hoffstadt Creek, which was scoured nearly to bedrock by posteruption debris flows, pool area has remained an almost constant 35% through 2002. The relative scarcity of pools in Hoffstadt Creek might favor riffle- and cascade-dwelling fish, such as sculpins and longnose dace (*Rhinichthys cataractae*), but such species had not yet successfully colonized this stream by 2002. In spite of the lack of pool habitat in Hoffstadt Creek, juvenile steelhead and coho stocked in the 1980s survived and grew at rates equal to or greater than those in relatively undisturbed streams in the region. Progeny of

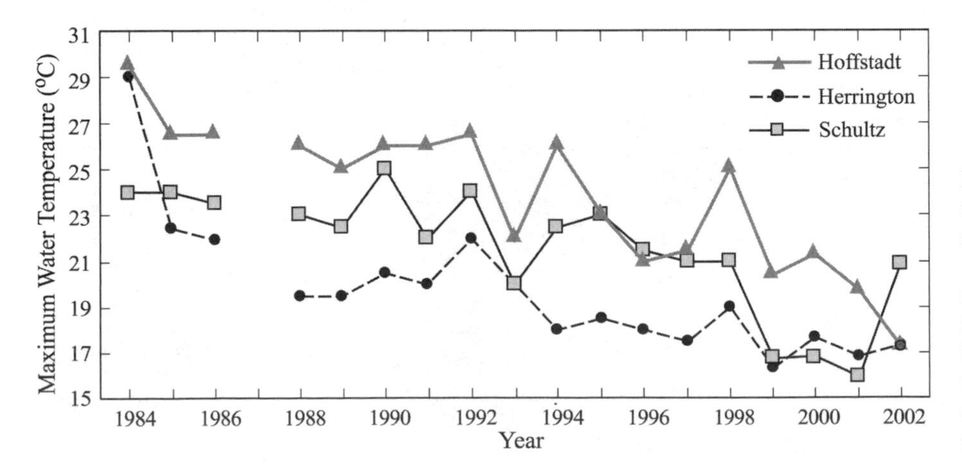

FIGURE 12.3. Maximum summer water temperatures measured in three tributary streams of the Toutle River from the early 1980s to 1999. Herrington Creek is a mudflow terrace stream (see Figure 18.3); Schultz Creek and Hoffstadt Creek are blast-zone streams. Missing years are denoted by *broken lines*. [Based on Martin et al. (1986), Bisson et al. (1988), and thermograph data collected from 1989 to 1999.]

FIGURE 12.4. Chronosequence of photographs taken from approximately the same location on lower Herrington Creek, a mudflow terrace tributary of the South Fork Toutle River. (Photo credits: P.A. Bisson and B.R. Fransen.)

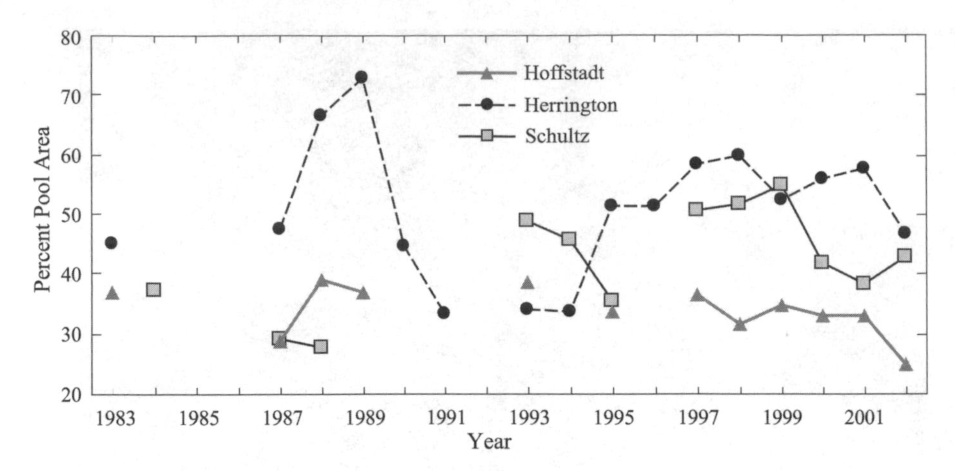

FIGURE 12.5. Trends in the percentage of pool habitat in three Mount St. Helens streams. Percentages are based on the fraction of the wetted channel occupied by pools during summer flow conditions. [Based on Martin et al. (1986), Bisson et al. (1988), and midsummer field surveys from 1989 to 1999.]

adult steelhead transported to Hoffstadt Creek have also survived well. In the upper reaches of the stream, cutthroat trout persisted in the absence of competing steelhead.

Excluding the first 2 posteruption years, the biomass of salmonids (Figure 12.6) has been both variable and high relative to many streams in the region. Cascade Mountain Range streams tend to be cold, oligotrophic ecological systems with total salmon and trout biomass typically ranging from 1 to 4 g m^{-2} (Bilby and Bisson 1987). In Herrington, Hoffstadt, and Schultz creeks, salmonid biomass has rarely dipped below 2 g m^{-2} and has occasionally exceeded 10 g m^{-2} between 1983 and 2002. In Herrington Creek, highly variable summer fish biomass during the 1980s resulted from differences in the timing and numbers of stocked coho salmon, but after hatchery releases were discontinued, the total biomass was relatively stable. Schultz Creek, one of the blowdown-zone streams, exhibited a peak in total salmonid biomass in the mid-1980s when coho salmon were stocked in it, but a second increase in biomass resulted instead from naturally spawning cutthroat and rainbow trout. In Hoffstadt Creek, relatively high biomass in the 1980s resulted, in part, from hatchery coho salmon stocking and in the 1990s from steelhead stocking. All three streams have received fish of hatchery origin, and those fish have been able to survive and grow in hostile conditions. The high growth rates suggest that the streams are relatively food rich, although we do not know with certainty whether the food resources are primarily of aquatic or terrestrial origin. Additionally, it is possible that the hatchery fish possessed some characteristics (such as tolerance for high densities or higher growth rates) that were well adapted to the posteruption environment.

Partial support for the speculated importance of food for fish survival and production was provided by a study of the physical and biological factors influencing the abundance of trout in Clearwater Creek (see Figure 12.1a; Baker 1989). Baker observed a significant positive association between cutthroat trout density and drifting invertebrates, although the association between brook trout abundance and estimated food availability was not statistically significant. Baker also found that pool quality (measured by size and presence of cover) was more important than simply the number of pools. Large, deep pools held proportionally more adult trout than did small pools. In terms of overall trout abundance, food availability slightly outweighed physical factors, such as substrate size and pool size, in influencing the abundance of trout in pool habitats. This finding was consistent with other monitoring studies of salmonid populations near Mount St. Helens (Bisson et al. 1988) and suggests that posteruption conditions favored an abundance of terrestrial and aquatic food organisms.

12.2.2.6 Long-Term Fish Productivity in Streams

Many fish populations at Mount St. Helens recovered more quickly than originally thought possible after the 1980 eruption. On the basis of the overall results of different monitoring studies, a generalized disturbance and recovery figure (Figure 12.7) illustrates some of the important processes and factors underlying the initial crash and subsequent rebound of fish in streams impacted by the eruption. These processes, described next, seem to have occurred regardless of the type of volcanic disturbance; that is, response patterns in debris-avalanche, mudflow, and blowdown-zone streams were similar.

The causes of initial declines were obvious. Many fish were killed outright during the May 18 eruption and associated mudflows by physical forces (i.e., abrasion and heat). Extreme posteruption sediment and temperature levels far surpassed tolerance thresholds for the cold-water fish species inhabiting streams near Mount St. Helens. Of long-term concern was the loss of cover, previously provided by large logs, in streams experiencing posteruption debris flows. However, many streams in the blowdown zone actually accumulated wood as the result of widespread forest blowdown. Within 3 years, fish population recovery was under way in many areas. Juvenile anadromous and resident fish surviving in the less-impacted tephra-fall and blowdown-zone streams were important sources of colonists throughout many of the drainage networks. The presence of these survivors, coupled with their dispersal ability and the connectivity of aquatic systems, led to a surprisingly rapid

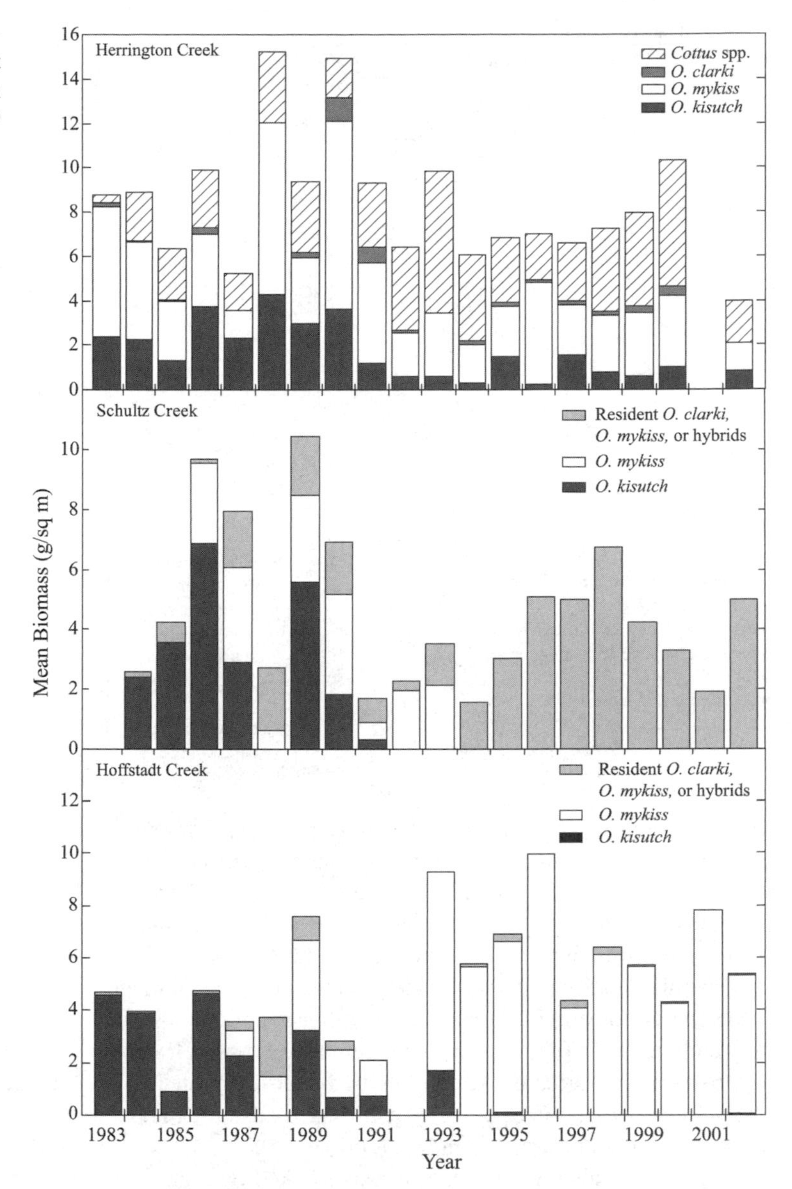

FIGURE 12.6. Total biomass of salmonid fishes (*Oncorhynchus kisutch*, *O. mykiss*, and *O. clarki*) in three Mount St. Helens streams. [Based on Martin et al. (1986), Bisson et al. (1988), and summer electrofishing surveys from 1989 to 1999.]

recovery of some populations. In addition, cohorts of salmon and steelhead that were at sea during the eruption returned to spawn; although many returnees initially strayed to nearby unimpacted rivers, some returned to their natal streams and refounded wild populations.

Fishery managers also played a hand in rebuilding some populations during the 1980s by imposing a temporary moratorium on sport harvest. This management decision coincided with rapid increases in both the number of successfully returning adults and total juveniles rearing in streams having anadromous species. The role that aggressive stocking programs played, using both native fish and stocks from other river basins, in salmonid recovery is less clear but probably did increase the number of returning fish. However, the planting of nonnative salmon in waters that still had surviving native stocks may have resulted in changes in the genetic composition and fitness of naturally spawning populations (Independent Scientific Advisory Board 2003).

The rapid posteruption rebound was also driven by a variety of ecological processes. In many channels, fine sediment from the eruption was flushed during high flows so that original gravel-cobble streambeds were recovered within 1 to 3 years. Dense riparian plant communities, including herbs and shrubs [salmonberry (*Rubus spectabilis*), buttercup (*Ranunculus* sp.), coltsfoot (*Petasites frigidus*), fireweed (*Chamerion angustifolium*), and pearly everlasting (*Anaphalis margaritacea*)] as well as vigorously growing deciduous trees (red alder and willow), began to provide shade and moderate stream temperature, reduce erosion, and provide habitat for terrestrial invertebrates specialized for living on herbaceous vegetation. However, many streams continued to receive abundant sunlight during the first 10 years following disturbance. The high solar

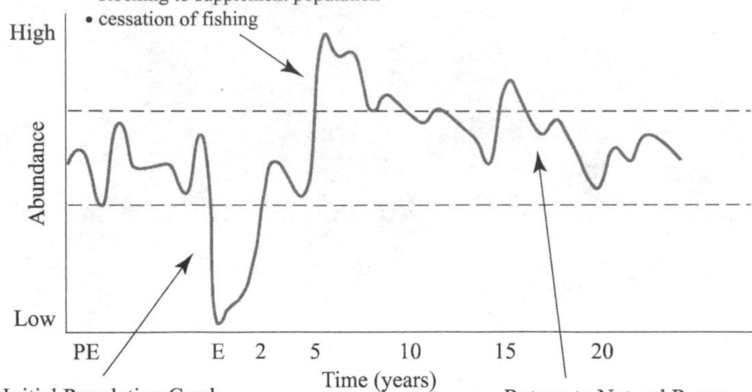

FIGURE 12.7. Hypothetical causes of posteruption fish-population crash, subsequent rebound, and gradual return to the range of abundance (bounded by *dashed lines*) typical of streams in the western Cascade Mountains. Time is shown for pre-eruptive conditions (PE), during the eruption (E), and in the years following the 1980 eruption.

radiation levels provided energy for elevated primary production by green algae and diatoms, which in turn enhanced grazer and collector-gatherer aquatic invertebrates, such as baetid mayflies (Baetidae) and orthocladiid midges (Chironomidae), which are common prey of juvenile salmonids (Mundie 1974). In general, fish tended to recolonize streams, often with human assistance, more rapidly than did potential aquatic competitors and predators (e.g., crayfish and amphibians), which were not restocked in streams where they had been extirpated. As early-successional top carnivores, fish were likely able to take advantage of temporarily abundant food.

The second posteruption decade was marked by a gradual return of stream-dwelling fish to abundances more typical of the western Cascade Mountains (see Figure 12.7). In some streams, pool habitat was scarce and may remain so until large trees grow and fall into the channel, a multidecadal process. However, summer water temperatures continued to decline as riparian trees provided more shade, and primary productivity presumably returned to levels typical of forested streams in southwestern Washington, with a corresponding reduction in mayflies and midges. The development of tree-dominated riparian zones has also reduced the amount of herbaceous vegetation, which may have lowered the number of terrestrial invertebrates falling into streams from herbs and shrubs. Over time, it is likely that a more complete assemblage of top aquatic carnivores will return to Mount St. Helens streams, leading to more competition and predation.

The pattern of initial fish response to the eruption and subsequent rebound observed in Mount St. Helens streams (see Figure 12.7) has been observed in fish exposed to other severe disturbance events, such as floods and wildfires (Minshall et al. 1989; Spencer et al. 2003). Waters (1983) found that a flood accompanied by a debris flow in a Minnesota stream caused a sharp drop in the abundance of fish, but after several years there was a rapid increase in trout populations as the stream habitat recovered. Minshall et al. (1989) and Spencer et al. (2003) documented significant changes in the food webs of streams impacted by large wildfires in the northern Rocky Mountains. They found that reduction of the forest canopy facilitated a shift in favor of productive autotrophic food webs (those based on algal production within the streams) and away from heterotrophic food webs (those based on terrestrial leaf litter and other allochthonous food sources). Minshall (2003) noted that, after fire, streams tended to converge on prefire conditions within 10 to 15 years. These case studies suggest that the pattern of decline and rebound observed at Mount St. Helens streams is consistent with disturbance-recovery patterns documented elsewhere.

Literature on fish responses to volcanism is relatively scant, but there are at least two examples from the North Pacific. Eicher and Rounsefell (1957) studied the effects of tephra deposits on sockeye salmon following the 1912 eruption of Mount Katmai, Alaska. Four thousand adult salmon died during the eruption because of suffocation in a sediment-choked stream that received 25 cm of tephra fall. Several months after the eruption, streams still contained abundant sediment, which was believed to have caused the death of several hundred additional sockeye salmon and the nearly complete loss of their food base. Two years after the eruption, intense storms added more sediment to the streams. Fewer than average salmon returned to spawn 4 to 8 years after the eruption, supporting the idea that the event had depressed the populations for at least a few generations. However, after this period, the salmon populations rebounded to preeruption levels. Eicher and Rounsefell (1957) suggested that, following an initial depression in productivity, the tephra fall may have promoted sockeye growth because smolt sizes from the volcanically impacted systems were substantially larger than those from adjacent undisturbed sites.

The second example is from the 1955 eruption of Mt. Besymjanny, Kamchatka, Russia, where Kurenkov (1966) reported a decline in the number of adult sockeye salmon returning to Lake Asabatchye beginning 4 years after the eruption. Adult returns remained depressed until 1964, the last year reported. The duration and amplitude of the population crash and subsequent responses were influenced by the extent and severity of the disturbance, and apparently the recovery of the lake system had not progressed enough to support a normal abundance of sockeye 10 years after the eruption.

Management actions can profoundly influence recovery of native fish assemblages in postdisturbance stream environments, and there is strong evidence that management caused some changes at Mount St. Helens that may be irreversible and undesirable in terms of sustaining native species and stocks. Minshall (2003) showed that altering riparian plant communities after large disturbances (e.g., by salvage logging) changed the food web of streams. The salvage logging after the Mount St. Helens eruption was one of the largest such efforts in history and removed many downed trees from streams and adjacent riparian areas. Waters (1983) found that an introduced species [brown trout (*Salmo trutta*)] displaced a native species (brook trout) when it was stocked in a flood-disturbed Minnesota stream. Given the widespread history of salmon, steelhead, and trout stocking, as well as nonnative fish stocking in low-elevation lakes near Mount St. Helens, long-term changes in fish communities resulting from fisheries management are a likely possibility. As a consequence, alteration of river and riparian recovery processes by forest management and sediment-control structures, combined with deliberate stocking of nonnative species, suggest that some streams and rivers near Mount St. Helens have been changed substantially and will likely remain so for several decades or longer.

Despite these concerns, one of the most surprising findings of research on streams and rivers at Mount St. Helens is the speed with which many have returned to near-preeruption conditions. Most experts held a pessimistic view of the prognosis for long-term recovery immediately after the eruption because changes in streams were so extreme. Twenty-two years after the eruption, many stream habitats and fish communities had returned to levels commensurate with the range of conditions found in Southwest Washington streams not affected by the volcano. Recovery has occurred more quickly than originally thought possible. Studies of stream-dwelling salmon and trout populations at Mount St. Helens have suggested that these fish may thrive in postdisturbance environments where food is abundant and predators and competitors are largely absent, even where habitats are distinctly suboptimal. The situation in lakes, however, has been different.

12.3 Lake-Dwelling Fish

12.3.1 Historical and Preeruption Conditions

Most of the 33 lakes (see Figure 12.1) that were studied in the vicinity of Mount St. Helens after the 1980 eruption were barren of fish before the arrival of Euro-American settlers (Crawford 1986). Waterfalls or other barriers on outlet streams prevented fish colonization from downstream sources. The only exception was Spirit Lake, which was connected to the Toutle River without barriers to dispersal, enabling access of sea-run coastal cutthroat trout, winter steelhead, and Coho Salmon to the lake. With the arrival of white settlers in the mid-1800s, the Mount St. Helens area was used for logging, mining, hunting, and exploration. With these activities, came the desire to provide recreational fishing in the backcountry lakes. The first recorded official fish stocking occurred in Spirit Lake in 1913, and most of the lakes were stocked by the 1950s. Few, if any, spawning surveys attempted to quantify the reproductive success of trout stocked in the lakes following either single or multiple stocking.

From 1913 through 1979, the year before the eruption, at least 1.8 million fish comprising four species and one subspecies had been stocked in the 24 lakes capable of supporting fish (Lucas and Weinheimer 2003). This preeruption stocking estimate should be viewed as the minimum number of fish stocked because it is based entirely on Washington Department of Fish and Wildlife records, which include stocking done by counties, sportsman groups, and individuals but do not include illegal or other stocking. Although the remaining 9 lakes were likely stocked, they could not support fish because of their shallow depths, producing high summer temperature and anoxia under the ice in winter.

The species stocked in Mount St. Helens lakes were brook trout, rainbow trout, westslope cutthroat trout (*O. clarki lewisi*), coastal cutthroat trout, and lake trout (*S. namaycush*). Brook, rainbow, and westslope cutthroat trout were commonly stocked in the lakes, whereas coastal cutthroat and lake trout were confined to 1 or 2 lakes and were seldom stocked. Following initial stocking, brook trout, which is a species capable of successfully spawning in lake-bottom substrates, maintained reproducing populations in the 14 lakes where they had been released. In contrast, rainbow and cutthroat trout require streams for spawning, and access to these habitats was very limited because of the presence of waterfalls on inlet and outlet streams. Consequently, reproductive success was limited, and several lakes required regular stocking to maintain their fisheries.

Lakes at Mount St. Helens were created through past volcanic eruptions, landslides, and glaciation and varied considerably in area, depth, elevation, and exposure to sunlight. As is typical of temperate montane lakes, these lakes had a fall and spring overturn period of vertical circulation and a summer thermal stratification. Lakes were ice covered for 3 to 8 months, depending on elevation. Information on the physical, chemical, and biological conditions of many of the lakes before the 1980 eruption is scant. Crawford (1986) reports very low densities of plankton from 1979 sampling of Spirit Lake. Several lakes studied by Bortleson et al. (1976) possessed low concentrations of nutrients for phytoplankton growth (phosphorus and nitrogen), high oxygen levels in the epilimnion, and water with high clarity (Secchi-depth measurements of 7 to 14 m). Mount St. Helens lakes were oligotrophic because of low nutrient levels, low phytoplankton, high light transmission, and cool temperature (Swanson et al., Chapter 2, this volume; Dahm et al., Chapter 18, this volume). Accordingly, the lakes supported relatively small standing crops of fish, and populations were often composed of stunted individuals (Crawford 1986). These fish likely foraged on the low densities of aquatic and terrestrial invertebrates typical of oligotrophic montane lakes in Washington State. Such prey consisted primarily of calanoid copepods, chironmid midge larvae and pupae, a variety of terrestrial insects, and freshwater amphipods (*Gammarus* spp.) (Crawford 1986). More than 50 years of stocking in most of these formerly fishless waters undoubtedly altered the structure of indigenous biotic communities (e.g., plankton, macroinvertebrates, and amphibians), and most lakes were no longer pristine (Knapp et al. 2001).

12.3.2 Posteruption Condition

Lakes in the Mount St. Helens landscape were disturbed by a complex set of physical mechanisms (heat, scour, and deposition) during the 1980 eruption (Swanson and Major, Chapter 3, this volume). The types, intensity, and severity of lake disturbances were related to distance and direction from the volcano and also to specific conditions at each lake at the time of the eruption (Dahm et al., Chapter 18, this volume). The immediate posteruption response of fish varied according to this array of volcanic disturbances and ranged from lakes with high survival to others with complete mortality (Crawford 1986).

Mount St. Helens lakes can be assigned to four categories based on the type of volcanic disturbance they received during the 1980 eruption. Four lakes (Blue, Island, June, and O'Connor) received only tephra fall; 28 lakes were influenced by blowdown or scorch disturbance (hereafter, blowdown); and Spirit Lake was influenced by multiple volcanic disturbances (debris avalanche, lateral blast, pyroclastic flows, and tephra fall). In addition to the 33 lakes present before the eruption, 2 large new lakes developed when debris-avalanche deposits blocked tributaries to the Toutle River, creating Coldwater and Castle lakes. For each of the lake categories, we describe species survival, subsequent natural and human-mediated (i.e., stocking) colonization, species persistence, and reproductive status of fish in these lakes from 1980 to 2002. Species are considered to have survived if they were observed in a lake within the first 5 years after the eruption and if that lake had not been stocked since the eruption (Crawford 1986).

12.3.2.1 Tephra-Fall Lakes

Blue, Island, June, and O'Connor lakes were stocked before the eruption, and each had fish populations that survived deposition of 5 to 15 cm of tephra. Additionally, three of the lakes maintained self-perpetuating populations of trout for 20 years posteruption in the absence of stocking. Several factors contributed to their success. At the time of the eruption, at least three, and perhaps all four, of the lakes were covered by snow and ice. Tephra entered them slowly during the spring as their ice cover melted. This slow process may have minimized the impacts of tephra on water quality and the biota. Wissmar et al. (1982b) reported that the chemical and physical properties of tephra-fall lakes were typical of undisturbed lakes of the region when measured on June 30, 1980, about 6 weeks after the eruption. The tephra that entered the water column settled out within a few weeks. Rapid settling of tephra suggests that phytoplankton were not severely impacted and that phytoplankton were sufficient to support zooplankton, an important trout food resource. However, some of the tephra was highly buoyant pumice that floated on the surface and was blown to the shoreline or became waterlogged and sank. The tephra and pumice that sank presumably covered benthic invertebrates, but the extent to which this affected trout food availability is not known. The survival and successful reproduction of fish in these lakes indicate low to moderate amounts (5 to 15 cm) of tephra fall had little or only a transient effect on the water quality, invertebrate abundance, and spawning habitat for trout. In the late 1990s, large hatches of caddisflies (Trichoptera) and mayflies (Ephemeroptera) were observed emerging from the lakes; thus, aquatic insects had become potential food resources for fish populations during the second posteruption decade.

12.3.2.2 Blowdown-Zone Lakes

Fish surveys with gill nets and angling were conducted in 28 lakes in the blowdown zone. Fish were assumed present in 19 of these lakes at the time of the 1980 eruption on the basis of stocking records for September 1979, preeruption creel censuses, and fish surviving the eruption (Crawford 1986). The remaining 9 lakes were thought to have been too shallow to support fish, lacked any stocking records, and had no other records that would have confirmed fish presence. Fish survived in 13 of the 19 lakes assumed to support fish before the eruption. The number of lakes occupied by the various trout species decreased following the eruption. Brook trout survived in 8 of 13 blowdown-zone lakes, rainbow trout survived in 1 of 5, westslope cutthroat trout survived in 3 of 18, resident coastal cutthroat trout survived in 1 of 1, and lake trout survived in 1 of 1 where they had existed previously.

Several factors contributed to fish survival and persistence in the blowdown-zone lakes, which were more heavily impacted

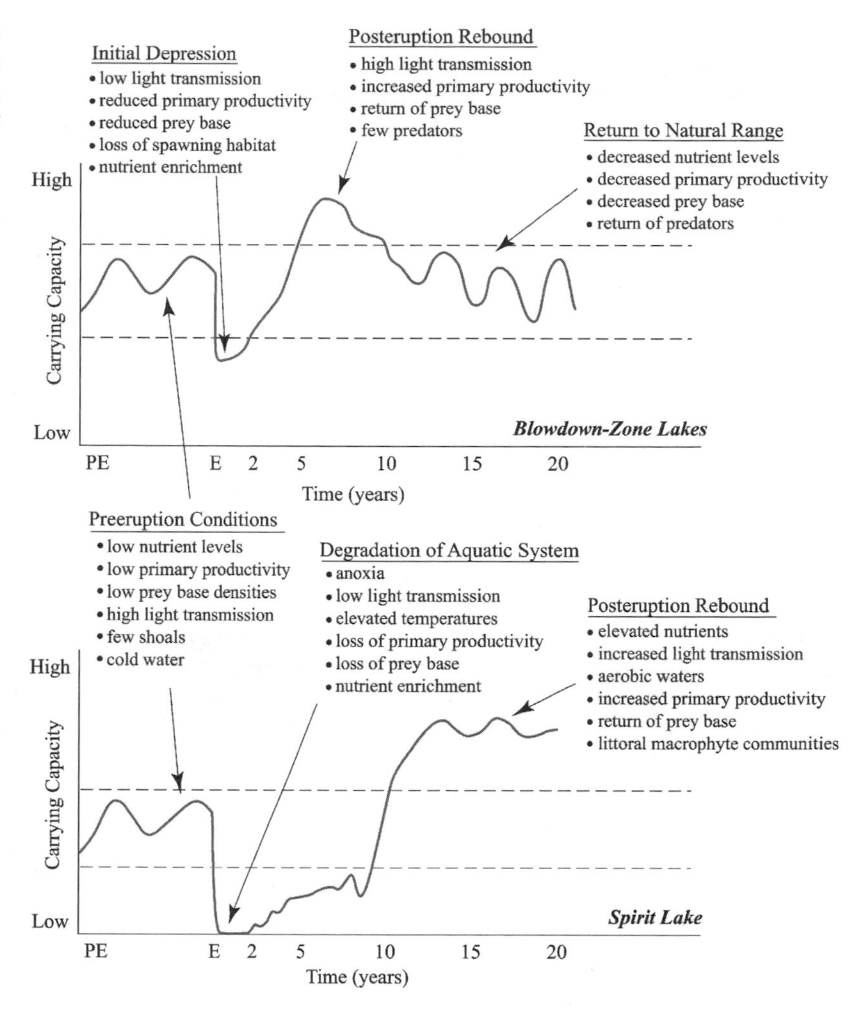

FIGURE 12.8. Hypothetical causes of fish-population changes in blowdown-zone lakes (*upper*) and in Spirit Lake (*lower*) for preeruptive conditions (PE), conditions during the eruption (E), and years after the 1980 eruption.

than the tephra-fall lakes. Of primary importance was the eruption's timing. Ice protected fish during the eruption, but ultimately survival depended on the degree to which the chemical, physical, and biological conditions of the lakes were changed. Of particular importance were water-quality characteristics and the type and abundance of food items.

A conceptual diagram of fish response in blowdown-zone lakes is presented in Figure 12.8 (upper). Blowdown-zone lakes received large amounts of volcanic blast material and tephra and organic constituents of the shattered and pyrolized forest. The latter material included nutrients that enriched lake waters. Wissmar et al. (1982b) found greatly reduced water clarity in blowdown-zone lakes during June 1980 because of suspended material in the water column. However, Crawford (1986) reported that clarity of these lakes had greatly improved by late summer 1980, and Carpenter (1995) found that by 1990 water transparency was typical of undisturbed lakes of the region.

Crawford (1986) suggested the eruption had short-term effects on the aquatic food web as a result of decreased light penetration. He surmised that lowered light penetration as a result of suspended particles reduced primary producers, which in turn severely reduced the abundance of their zooplankton consumers, an important prey for fish. With improved light transmission, zooplankton numbers rebounded, and limited data suggest this process did occur. In St. Helens Lake, zooplankton (*Daphnia* spp.) were present at a density of 4 m^{-3} in 1980, and by 1981 they had increased to 2260 m^{-3}; during this period, cyclopoid copepods increased from 3 m^{-3} to 305 m^{-3} (Crawford 1986). Scharnberg (1995) studied the zooplankton in several of the blowdown lakes from 1992 to 1994 and found a diverse fauna composed of more than 50 species, including rotifers, copepods, and cladocerans. He noted that the zooplankton community structure of these lakes was influenced by the presence of fish and that highly palatable species (e.g., *Daphnia* spp. and *Bosmina* spp.) were lower in abundance when fish were present. Scharnberg (1995) compared zooplankton communities from Mount Rainier and Mount Hood with those of Mount St. Helens and concluded that zooplankton communities had largely recovered from the 1980 eruption.

Benthic insects often comprise an important component of trout diet in lakes. Crawford (1986) reported that deposition of tephra in lakes severely reduced bottom-dwelling insects but did not eliminate them. A wide array of invertebrate prey was observed in fish stomach samples during the first few years following the eruption. From 1980 to 1981, the dominant food item taken by fish was midge larvae, but by 1983 prey also included beetles, black flies, mayflies, caddisflies, stoneflies, dragonflies, and terrestrial insects (Crawford 1986). By 2000, additional insects were present in fish stomach samples, particularly terrestrial species such as ants, termites, and grasshoppers that were in greater abundance because of the colonization and growth of terrestrial vegetation. Phantom midge larvae (*Chaoborus* sp.) are large, mobile invertebrates that are often preferred prey of trout. These insects were present at much lower densities in lakes with fish compared to fishless lakes (Scharnberg 1995). Other food items that appeared in fish diets since the eruption were the signal crayfish (*Pacifasticus leniusculus*), amphipods (*Gammarus* spp.), and the northwestern salamander (*Ambystoma gracile*).

Results from fish stomach analyses suggest that plankton and insect prey were initially depressed but recovered quickly following the settling of suspended volcanic material. It remains unclear to what extent residual plankton and insects played a role in lake recovery relative to the importance of new colonists. The eggs and resting stages of zooplankton can persist for prolonged periods (Vogel et al. 2000) and are readily dispersed by wind or animals (e.g., dragonflies and waterfowl) (Maguire 1963). Most aquatic insects have highly mobile terrestrial life stages that are adept at colonizing new habitats (Merritt and Cummins 1996). Regardless of the contribution of residuals and colonizers, immediately after the eruption, enough prey was available to support some fish. Within a few years of the eruption, blowdown lakes supported a diverse prey assemblage.

The presence of suitable spawning substrates was also critical if fish were to persist in the blowdown lakes following the eruption. Crawford (1986) reported that, by 1985, successful spawning had occurred for brook trout in 3 lakes (Meta, Obscurity, and Shovel) and for westslope cutthroat in 1 lake (Lower Venus). Resident coastal cutthroat trout were observed spawning at Ghost Lake in 1983, and additional spawning may have occurred at other lakes during the first 5 years after the eruption. By 2000, natural trout reproduction was confirmed at 11 lakes, predominantly by brook trout (Lucas and Weinheimer 2003).

Trout populations in six lakes (Boot, Forest, Grizzly, Holmstedt, Ryan, and Snow) apparently experienced complete mortality during the May 1980 eruption. The reason fish perished from those lakes is unknown, but lake depth may have played a key role, given that four of the lakes where fish mortality was complete were among the shallowest lakes (maximum depth, less than 6 m). Boot Lake is both deep (21 m) and fairly large (6.5 ha), so reasons other than depth and size were responsible for fish extirpation there. Crawford (1986) noted that Boot Lake remained very turbid because of a large inlet stream that carried high sediment loads into the lake for several months after the eruption and speculated that turbidity was responsible for fish mortality. Ryan Lake, a shallow body of water that was ice free at the time of the eruption, was more severely disturbed than other blowdown-zone lakes. Dahm et al. (1983) reported lethally low oxygen levels in the epilimnion of Ryan Lake during August 1980 and total anoxia in deeper water.

During the first 20 years following the eruption, breeding fish populations were observed in all 13 lakes in the blowdown zone where they had survived the initial eruption. Three apparent colonization events occurred between 1980 and 2001. Brook trout appeared in Forest Lake in 1993 and in Ryan Lake in 2001. It is very likely that anglers transplanted these fish. Rainbow trout were captured in Little Venus Lake in 1993. These fish probably originated from fish stocked in nearby (upstream) Venus Lake in 1989. The only documented species extirpation in the 20 years following the eruption has been the loss of westslope cutthroat trout (a nonnative subspecies) from Venus Lake. In addition to these lakes, 8 other blowdown-zone lakes were stocked by Washington Department of Fish and Wildlife since the eruption (Lucas and Weinheimer 2003).

By 2000, fish populations were present in 15 of 18 blowdown-zone lakes thought to be capable of supporting fish. Three lakes in which fish were absent, but that could probably support fish, were managed in a fishless condition in accordance with the Mount St. Helens National Volcanic Monument Comprehensive Management Plan (USDA Forest Service 1985). Brook trout populations in 10 lakes were present at high densities, and individuals typically grew slowly. Westslope cutthroat and rainbow trout were present in 4 and 3 lakes, respectively. These fish had limited spawning success and were probably not capable of sustaining themselves for more than 10 years in the absence of stocking. Resident coastal cutthroat trout occurred in only one location, Ghost Lake, where they maintained a viable population. Brown trout were found in a single lake (Elk), where they required stocking to persist; lake trout were self-sustaining only in St. Helens Lake.

12.3.2.3 Spirit Lake

Spirit Lake was the most severely impacted lake during the 1980 eruption. The debris avalanche slid into the lake, displaced the water, and dramatically altered the basin morphology. In the aftermath, the lake's surface elevation increased by approximately 60 m; its surface area increased by 80%; its depth was greatly reduced; and new, large shoal areas were created. Thousands of tree boles and other forest debris floated on the surface, forming a large log mat. The once cold, oligotrophic water became highly enriched.

Spirit Lake was sampled in 1983, and no fish were captured (Crawford 1986). During several dozen limnological surveys between 1983 and 1986, no fish were seen by several investigators. Similarly, no fish were observed along the southern shoreline during littoral-zone surveys from 1982 through 1988 or during a snorkel survey in 1989. It was clear from these survey efforts and from the dramatically altered water-quality data (Dahm et al. 1981; Larson and Glass 1987; Larson 1993; Dahm et al., Chapter 18, this volume) that fish perished during or shortly after the eruption.

TABLE 12.2. Spirit Lake rainbow trout sample size (*n*), snout-fork length, and mean weight for 2000–2002.

Year	Sample size	Fork length (mm)	Weight (g)
2000	62	519 (77)	1662 (548)
2001	16	544 (45)	2163 (643)
2002	73	565 (46)	2122 (429)

Standard deviations are in parentheses. Fish were captured by netting and angling.

In 1993, a single rainbow trout was captured in a gill net (Lucas and Weinheimer 2003). Return visits in 1994 and 1997 resulted in the capture of 1 rainbow trout each year. With the confirmed presence of fish, sampling efforts increased substantially, and 16 trips were made to the lake between 2000 and 2002. During these trips, 151 rainbow trout were captured by netting and angling (Table 12.2), 50.3% of which were males, 35.7% were females, and 14% were of undetermined sex. The age of 29 fish captured in 2000 was determined from otoliths and scales. Rainbow trout in the samples were present in three age classes: age 1+ (3.4%), age 2+ (13.8%), and age 3+ (82.8%), a very unusual age structure. The population appeared to be composed of few age classes and dominated by large fish (although the apparent age structure was likely influenced by sampling methods). Since 2000, a few age 4+ fish have been recorded.

Each of the major tributaries to Spirit Lake was either visually inspected for spawning adults or redds or electrofished for fry during the spring or summer of either 2000 or 2002. No evidence of spawning was found, and no fry were captured. Moreover, two snorkeling surveys, each several hours long, along the south end of the lake during 2002 failed to detect any small fish, although numerous large fish were observed. Reproduction has not been directly confirmed by observing redds or fry, but the presence of multiple age classes suggests successful reproduction has occurred or clandestine stocking has taken place.

Overall, there is a very limited amount of stream habitat available for spawning near Spirit Lake. The best spawning habitat is found in cobble- and gravel-dominated tributaries entering the lake from the north through the blowdown zone; but because of steep channel gradients (waterfalls or cascades), the length of streams available for spawning is small. Streams entering the lake from the south pass through the pyroclastic-flow deposits and carry high loads of fine sediment, yielding poor-quality spawning habitat. Fish may be spawning in the numerous springs on the lake bottom within the littoral zone. Gravel-sized pumice has sunk to the lake bottom and, in combination with cold, oxygenated springs, may provide suitable spawning sites. Numerous patches of bare pumice 0.5 to 1 m^2 in area observed in the densely vegetated littoral zone during 2002 may have been trout redds, but this suspicion has not been confirmed. The source of the Spirit Lake rainbow trout is unknown, but most likely they were illegally stocked.

Regardless of how trout entered the lake, they have recently found excellent conditions for growth, but major changes were necessary before the lake could support fish at all (see Figure 12.8, lower). In the aftermath of the 1980 eruption, the initial responses included greatly reduced light transmission because of suspended particles, elevated water temperatures, anoxia, and high levels of nutrients and reduced metals. The preeruption biota perished and was replaced by a prolific heterotrophic microbial community. Oxygen was depleted, and a rapid succession of biological, physical, and chemical transformations ensued (see Dahm et al., Chapter 18, this volume). By summer of 1982, heterotrophic bacteria had processed much of the fine organic matter, and inputs from rainwater and streams further diluted metal and nutrient concentrations. Reduction of particulates and solutes greatly increased water clarity, allowing light to penetrate several meters into the water column. Dissolved oxygen increased, and phytoplankton, zooplankton, and aquatic insects were starting to appear.

Spirit Lake's flora and fauna have changed considerably since the eruption. Immediately after the eruption, phytoplankton disappeared, but between 1983 and 1986, 138 species were present (Larson 1993). Zooplankton were observed in Spirit Lake in 1982, and high densities of rotifers were present by 1983 with fewer numbers of cladocerans and copepods (Dahm et al., Chapter 18, this volume). Zooplankton samples in 1986, 1989, and 1994 yielded rotifers, copepods, and cladocerans. *Daphnia pulex* and phantom midge larvae, both important fish prey, became very abundant. Aquatic insects, especially diving beetles (Dytiscidae), midges (Chironomidae), dragonflies and damselflies (Odonata), water boatmen (Corixidae), caddisflies (Trichoptera), stoneflies (Plecoptera), and mayflies (Ephemeroptera) all recolonized Spirit Lake from 1983 to 1986. These insects attained very high densities along Spirit Lake's southern shoreline, where the extensive shoal environment developed a species-rich and structurally complex macrophyte community dominated by pondweed (*Potamogeton* spp.). Several amphibians also recolonized Spirit Lake during the late 1980s, and the aquatic life stage of the northwestern salamander (*Ambystoma gracile*) was exceedingly abundant in the lake (Crisafulli et al., Chapter 13, this volume). The salamanders were readily preyed upon by large trout.

When fish colonized the lake in the early 1990s, a prey base was well established, and conditions promoted rapid growth. Water quality was favorable for salmonids, food was abundant, and predators were scarce; the rainbow trout population expanded quickly, and growth rates were high. The mean mass of age 3+ fish in 2000 was 2035 g ($n = 24$, SD = 226 g), a very large size for trout in their fourth year. Unfortunately, during the time of fish population growth, limnological measurements had largely ceased, and it is unclear how fish influenced food web dynamics. Fishing has remained closed in Spirit Lake, but natural predators have been observed in the area during the late 1990s, including the bald eagle (*Haliaeetus leucocephalus*), osprey (*Pandion haliaetus*), mink

(*Mustela vison*), and northern river otter (*Lontra canadensis*), and predation on trout has been observed. The fish population will likely continue to grow and be dominated by large fish into the early years of the 21st century, but eventually the lake's food web will likely return to preeruption conditions as the nutrient legacy from the eruption is diminished. At that point, the rainbow trout population will contain smaller individuals and will probably have a reduced standing crop.

12.3.2.4 New Lakes Formed by the Eruption

Coldwater and Castle lakes began to form when the May 18, 1980 debris-avalanche deposit dammed Coldwater and Castle creeks. Any fish in these new water bodies were recruits from tributary-stream populations or were intentionally stocked. Coldwater Lake was first surveyed in 1985, and no fish were captured or observed (Crawford 1986). During this same initial survey, a small segment of North Coldwater Creek was sampled to determine if fish had survived there, but no individuals were detected (Crawford 1986). Resident coastal cutthroat and rainbow trout as well as juvenile anadromous fish (steelhead, sea-run coastal cutthroat trout, and coho salmon) known to inhabit North Coldwater Creek in 1979 were all apparently killed by the eruption.

Coldwater Lake was assumed barren of fish until 1989, when about 30,000 under-yearling rainbow trout were stocked. Sampling conducted in the lake from 1990 through 2001 indicated a vigorous and healthy rainbow trout population had become established. Between 1997 and 2001, four or five age classes were present in the population. Most were 1- to 4-year-old fish and, in both 1998 and 2001, a few age 5+ individuals were also present (Lucas and Weinheimer 2003). During the 2001 survey, both westslope and resident coastal cutthroat trout were also captured (Lucas and Weinheimer 2003). It is possible that a few westslope cutthroat trout were inadvertently mixed with rainbow trout at the hatchery and stocked in the lake during 1989. The most plausible explanation for the presence of resident coastal cutthroat trout was that some individuals had survived the eruption in tributary streams and eventually colonized the lake. Lucas and Weinheimer (2003) report that the first evidence of trout reproduction was in 1992 when trout fry were observed in both Upper Coldwater Creek and South Coldwater Creek, and spawning trout were captured in the South Coldwater Creek during 1993.

Castle Lake was first sampled in 1985, and no fish were captured or observed there (Crawford 1986). Resident coastal cutthroat, rainbow trout, steelhead, sea-run cutthroat trout, and coho salmon that were known to be present in Castle Creek before the eruption were assumed to have perished during or shortly after the eruption. Rainbow trout were first seen in Castle Lake in 1991. These were apparently age 2+ hatchery fish (evidenced by their eroded dorsal fins) that emigrated from Coldwater Lake following the 1989 stocking and reached nearby Castle Lake through the Toutle River system. Subsequent sampling of Castle Lake in 1993 and from 1995 to 2001 revealed only rainbow trout. During this time, the population consisted of at least three and sometimes four age classes (Lucas and Weinheimer 2003). Growth appeared to be greatest during the first few years after introduction (1991 to 1993) and then declined somewhat from 1995 to 2001. The average fork length and weight of trout from 1997 to 2001 were nearly identical for Castle and Coldwater lakes, suggesting similar productivity levels within these new lake systems. Lucas and Weinheimer (2003) found numerous rainbow trout fry in Castle Creek during 1991 and 1993, confirming successful spawning.

Colonization of the new lakes by other native species required either the movement of fish from the Toutle River into the upper drainage network or deliberate human introductions. It is possible that a few resident coastal cutthroat trout survived in headwater streams and later founded the population in Coldwater Lake. Movement of anadromous fish into Coldwater and Castle lakes was limited by the sediment-retention structure located downstream on the North Fork Toutle River; however, migrating adult fish are captured and transported above the dam and may be able to reach Castle Lake. Coldwater Lake possesses a barrier to upstream fish movement at its outlet channel. In contrast, Castle Lake does not have impediments to fish colonization, and steelhead and coho salmon passed over the sediment-retention structure have access to the lake.

When Coldwater and Castle lakes were forming from 1980 to 1983, limnological conditions of these waters were not suitable for trout (Dahm et al. 1981; Wissmar et al. 1982b). By 1989, when rainbow trout were first stocked in Coldwater Lake, water quality had greatly improved. Secchi-depth readings of 7.0 to 8.0 m were recorded during 1989 and 1990 (Dahm et al., Chapter 18, this volume), indicating high light transmission into the water column. By 1989, oxygen levels within the epilimnion were also suitable for trout. Kelly (1992) found a diverse and abundant phytoplankton flora in both lakes in 1989. During that year, zooplankton reached high densities (up to 3088 m^{-3}), and the zooplankton community included 28 cladoceran, copepod, and rotifer species. *Daphnia* were the most common zooplankton. Additionally, by 1989, the aquatic insect community in Coldwater Lake was typical of lakes of the region. Crisafulli et al. (Chapter 13, this volume) found aquatic forms of the northwestern salamander in both Castle and Coldwater lakes.

Coldwater and Castle lakes developed water-quality conditions, aquatic invertebrate communities, and spawning habitats that supported self-sustaining trout populations within 10 years of the eruption, but the length and weight of fish has decreased during the decade since initial stocking. This decline may have been a response to reduced nutrient levels that were initially elevated following the eruption, or it may have been a compensatory response to the increasing number of individuals. In either case, the size of fish may be expected to decrease further in the coming years.

12.4 Summary

As mobile aquatic organisms, fish depend on an interconnected network of streams and lakes to fulfill many of their life-cycle needs. The impact of the Mount St. Helens eruption on aquatic systems was not uniform. Some streams were virtually devastated while others were barely affected. Existing lakes were altered by inputs of nutrients, tephra, and wood, and new lakes were formed. Human activities influenced fish recovery through stocking programs and creation of sediment-control structures through which fish movements were impeded or blocked. The extensive timber-salvage operation and planting on private and some public lands altered the composition of upland and riparian plant communities and changed posteruption forest succession, thus affecting the amount and types of organic matter available to some streams and lakes. All these activities made it difficult or impossible to conduct controlled field experiments on the recovery of fish populations after the eruption. Even so, there are common patterns in the ecological processes that have influenced fish at Mount St. Helens in the aftermath of 1980.

1. Two decades of monitoring have shown that changes in the trophic structure of streams and lakes can have a dominant influence on fish recovery in postdisturbance environments. The remarkable rebound of stream-dwelling fish populations, sometimes supplemented with salmon and steelhead of hatchery origin, in the food-rich conditions that prevailed in the late 1980s and 1990s demonstrated that even cold-water species such as salmon and trout could prosper in relatively poor habitats (high temperatures, scarce pools, and infrequent cover) if prey were abundant. Likewise, the proliferation of abundant, large trout in nutrient-rich lakes formed by the debris avalanche provides further support for the importance of food resources. The temporary abundance of certain types of food created by successional processes occurring in the first two decades posteruption has been largely responsible for the rapid recovery of fish at Mount St. Helens.
2. The type and intensity of volcanic disturbance have strongly influenced the rate of response toward ecological conditions typical of lakes and streams of the region. Streams and lakes in the tephra-fall zone and in the periphery of the blast area experienced little change, whereas those in the path of pyroclastic flows or inundated by volcanic mudflows were severely altered. The extent of damage to lake-dwelling fish populations was mediated by the amount of snow and ice cover at the time of the May 18, 1980 eruption. Had the eruption occurred at some other time of year or had the volcano not produced a large lateral blast, changes in streams and lakes, and their subsequent response rates, would have been quite different.
3. Finally, management activities have altered the response trajectory of native fish communities in many streams and lakes by changing riparian vegetation, sediment dynamics, and species distribution (through the introduction of non-native species and hatchery fish stocks). Unfortunately, the concurrent timing and spatial overlap of many of the management actions have made it impossible to evaluate the importance of each environmental factor separately. Some of the changes will probably be irreversible; however, the mosaic of managed and unmanaged watersheds, dammed and undammed rivers, and stocked and unstocked streams and lakes has created an unprecedented opportunity to compare long-term changes in fish across a landscape with a varied disturbance history, provided long-term monitoring programs remain in place.

Acknowledgments. We gratefully acknowledge the financial support of the USDA Forest Service, Pacific Northwest Research Station, the Weyerhaeuser Company, Washington Department of Fish and Wildlife, Utah State University, and the National Science Foundation (Grant BSR-84-16127 to C.P.H.) for field work, data management, and preparation of this chapter. We thank John Weinheimer, Ann Baker, Jason Walter, John Heffner, and James Ward for assisting with fieldwork. The manuscript benefited substantially from reviews by Frederick Swanson, Virginia Dale, and two anonymous referees.

13 Amphibian Responses to the 1980 Eruption of Mount St. Helens

Charles M. Crisafulli, Louise S. Trippe, Charles P. Hawkins, and James A. MacMahon

13.1 Introduction

Volcanism is a major agent of natural disturbance in the Pacific Northwest and other regions of the world. Volcanic eruptions alter surrounding landscapes and ecosystems and strongly influence the distribution and abundance of species. Although ecologists documented responses of vegetation (del Moral and Grishin 1999; see Chapters 4 to 8, this volume), mammals (MacMahon et al. 1989; Crisafulli et al., Chapter 14, this volume), and arthropods (Edwards and Sugg, Chapter 9, this volume; Parmenter et al., Chapter 10, this volume) to volcanic disturbance, no equivalent work exists for amphibians. The 1980 eruption of Mount St. Helens created an opportunity to examine the initial responses of an amphibian assemblage to a diverse array of volcanic disturbances and to describe patterns of species colonization in areas that were influenced by the eruption. In addition, amphibian responses to the 1980 eruption may provide insights into how amphibians respond to major environmental changes over large spatial scales. This information is important because of the apparent declines of amphibian species during the past two to three decades (Blaustein and Wake 1990, 1995; Pechmann et al. 1991; Pechmann and Wilbur 1994; Sarkar 1996; Green 1997; Corn 2000).

In this chapter, we describe amphibian responses to the 1980 eruption of Mount St. Helens based on 21 years of observation. We focus on the following questions:

- Did amphibians survive the eruption?
- If so, what factors aided survival?
- Given that amphibian habitat was greatly altered, did survivors persist in the "new" landscape?
- If areas were defaunated, what was the rate and pattern of recolonization?
- What biotic interactions developed between amphibians and other species, and what were some potential implications of these interactions for colonization and amphibian populations?

13.1.1 Preeruption Amphibian Assemblage

To discuss the effects of the Mount St. Helens eruption on amphibians, we compiled a list of the species in the preeruption fauna by using two approaches. First, we queried several major natural history museums in the United States for records of preeruption specimens. The first documented collections were obtained in 1904 from the Spirit Lake basin. Subsequent collecting in the study area from 1920 to 1979 resulted in 818 museum specimens, representing 11 species (Table 13.1; preeruption occurrence = "H"). Second, several lines of evidence imply that 4 additional species also occurred in the study area before the eruption (Table 13.1; preeruption occurrence = "A"). Literature sources (Slater 1955; MacMahon 1982; Nussbaum et al. 1983; Zalisko and Sites 1989), personal communications with herpetologists and naturalists who worked in the area before the eruption, and data we collected during surveys of the study area in 1980 and 1981 (the first 2 years after the eruption) indicated that long-toed salamander (*Ambystoma macrodactylum*) and Pacific tree frog (*Pseudacris regilla*) were present before the eruption. The ensatina (*Ensatina eschscholtzii*) and Larch Mountain salamander (*Plethodon larselli*) were found at several locations during surveys conducted from 1989 through 2003 at about 80 sites south and east of the volcano (areas that sustained only minor disturbance) and north of the severely impacted area. We therefore assume their presence in the preeruption fauna and include them on our list. These results indicate that, before 1980, the study area (Figure 13.1) supported an amphibian fauna representing 9 families, 10 genera, and 15 species (see Table 13.1).

13.1.2 Species Life History and Habitat Associations

The 15 species vary widely in their life histories. Species can be assigned to four groups based on their habitat associations

TABLE 13.1. Amphibian species documented ("historical occurrence") in the vicinity of Mount St. Helens (i.e., within the boundaries of the map, Figure 13.1) before the 1980 eruption and assumed to be present in the area at the time of the eruption.

Order	Family	Species	Common name	Occurrence preeruption	Primary habitat
Caudata	Ambystomatidae	*Ambystoma gracile*	Northwestern salamander	H	Lake
		Ambystoma macrodactylum	Long-toed salamander	A	Lake
	Dicamptodontidae	*Dicamptodon copei*	Cope's giant salamander	H	Stream
		Dicamptodon tenebrosus	Coastal giant salamander	H	Stream
	Rhyacotritonidae	*Rhyacotriton cascadae*	Cascade torrent salamander	H	Seep
	Salamandridae	*Taricha granulosa*	Rough-skinned newt	H	Lake
	Plethodontidae	*Plethodon larselli*	Larch Mountain salamander	A	Forest
		Plethodon vandykei	Van Dyke's salamander	H	Seep
		Plethodon vehiculum	Western red-backed salamander	H	Forest
		Ensatina eschscholtzii	Ensatina	A	Forest
Anura	Leiopelmatidae	*Ascaphus truei*	Coastal tailed frog	H	Stream
	Bufonidae	*Bufo boreas*	Western toad	H	Lake
	Hylidae	*Pseudacris regilla*	Pacific tree frog	A	Lake
	Ranidae	*Rana aurora*	Red-legged frog	H	Lake
		Rana cascadae	Cascades frog	H	Lake

Historical records are based on museum specimens collected from the area before the eruption. Species "assumed present" are based on literature reviews, personal communications, and surveys from the disturbed areas during the first year after the eruption. Also listed is the primary or breeding habitat of each species.
H, historical record; A, assumed present.

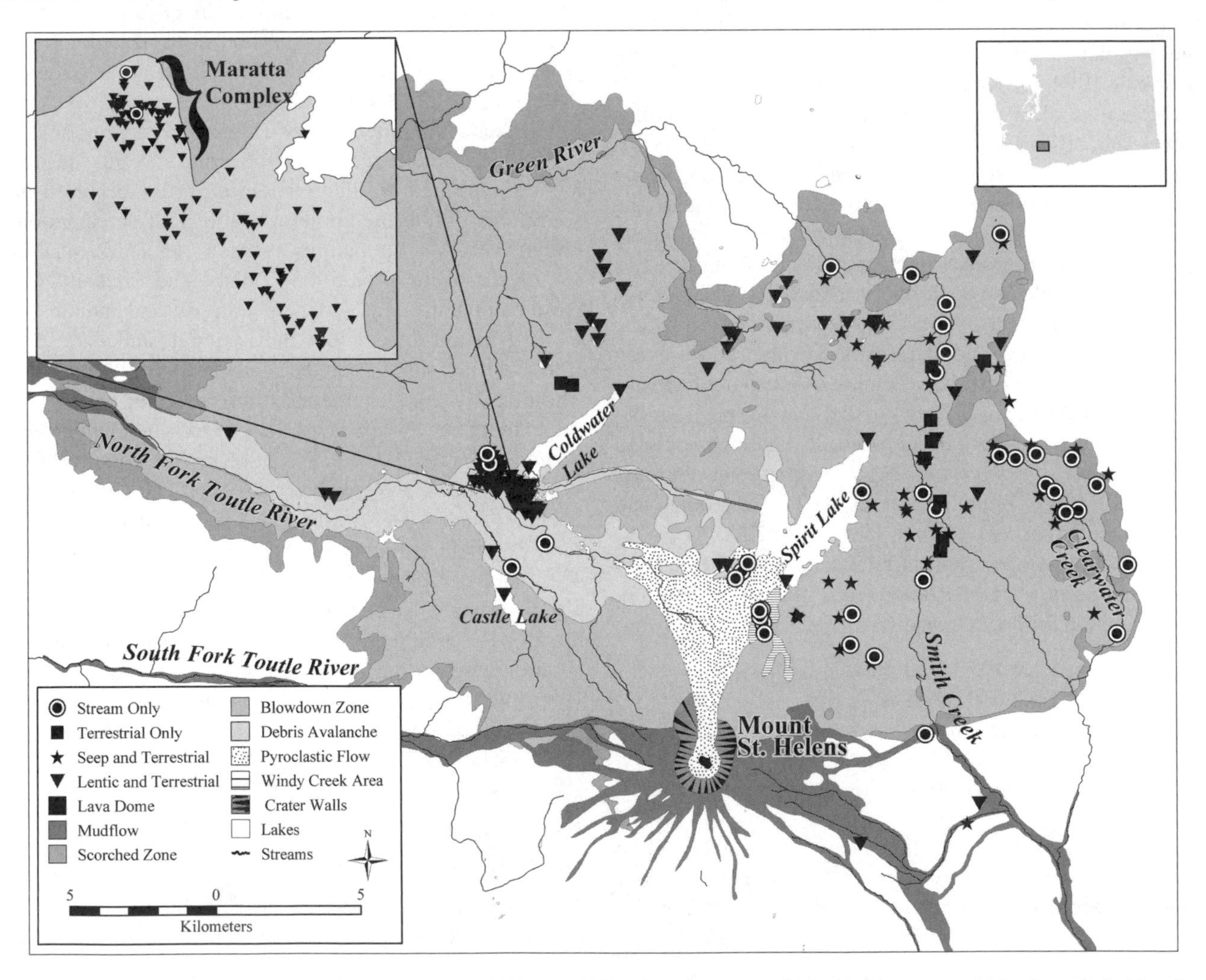

FIGURE 13.1. The Mount St. Helens region showing amphibian sampling locations according to habitat type within the scorch, blowdown, debris-avalanche, and pyroclastic flow disturbance zones and in the Windy Creek area. In this chapter, we pool the blowdown and scorch zones as BDSC zone and the debris avalanche and pyroclastic flow zones as DAPF zone.

and primary breeding areas: (1) lakes and ponds, (2) streams, (3) seeps, and (4) forests (see Table 13.1). Eleven species breed in lakes and ponds, streams, or seeps and have complex life histories characterized by gelatinous eggs that develop in water, an aquatic larval stage, and a primarily terrestrial adult stage. Three of these species, the northwestern salamander (*Ambystoma gracile*), coastal giant salamander (*Dicamptodon tenebrosus*), and Cope's giant salamander (*Dicamptodon copei*), may not undergo metamorphosis (Nussbaum et al. 1983). These neotenes or paedeomorphs retain larval characteristics at sexual maturity and remain in the water throughout their lives (Duellman and Trueb 1986). Sprules (1974) hypothesized that neoteny may be an adaptation for populations living in landscapes with harsh terrestrial environments. The 4 remaining species are terrestrial (3 species) or seep-dwelling (1 species) and lack an aquatic larval stage.

Seven species [four frogs and toads (hereafter anurans) and three salamanders] use lake and pond systems for breeding. Once breeding has ended, adults disperse from the breeding sites to various terrestrial and aquatic locations. Anuran tadpoles and the larvae of rough-skinned newt (*Taricha granulosa*) and long-toed salamander undergo metamorphosis within 2 to 4 months. The larvae of the northwestern salamander metamorphose in about 15 months, unless they are neotenic.

Three species [Cope's and coastal giant salamanders and the coastal tailed frog (*Ascaphus truei*)] are associated with streams. The two species of giant salamanders breed in both high-gradient and low-gradient streams. Coastal giant salamanders have an aquatic larval period of 18 to 24 months (Nussbaum et al. 1983). Neoteny appears to be the dominant life-history strategy in Cope's giant salamander, and it is common in the coastal giant salamander (Nussbaum 1976). Metamorphosed individuals are forest-dwellers. The coastal tailed frog is the only stream-breeding frog in the study area. Coastal tailed frogs are typically associated with steep, fast-flowing streams but also occur in larger, low-gradient streams (Nussbaum et al. 1983). Females deposit eggs on the underside of stream boulders and cobbles (Adams 1993). Larvae require 2 to 3 years to complete development in the study area (Hawkins et al. 1988). Adults are terrestrial and occur in forests, often distant from water.

The Cascade torrent (*Rhyacotriton cascadae*) and Van Dyke's (*Plethodon vandykei*) salamanders occur in valley and headwall seeps; areas of colluvial slumping along the stream valley walls; and splash zones located at the base of waterfalls, chutes, and cascades of streams (Nussbaum et al. 1983; Jones 1999). Larval Cascade torrent salamanders are aquatic and require 3 to 4 years to develop (Nussbaum and Tait 1977). Van Dyke's salamanders develop from eggs deposited on land (Blessing et al. 1999). Adults of these species are terrestrial but are generally found within a few meters of seeps or streams (McIntyre 2003).

Three species [western red-backed (*Plethodon vehiculum*), Larch Mountain, and ensatina salamanders] are associated with forests. Each belongs to the lungless salamander family Plethodontidae. These species are fully terrestrial and lack an aquatic larval stage. Eggs are deposited on land in cool, moist subterranean locations or in logs (Nussbaum et al. 1983). When the young hatch, they appear as miniature adults. Females typically have a biennial reproductive cycle and produce 4 to 19 eggs per clutch (Nussbaum et al. 1983). Sexual maturation takes 3 to 5 years (Nussbaum et al. 1983).

13.1.3 Predisturbance Habitat Conditions

Before the eruption, the area around Mount St. Helens consisted of stands of mature and old-growth coniferous forests with well-developed shrub, herb, and bryophyte layers that were interspersed with natural meadows and plantations resulting from more than three decades of clear-cut logging (Franklin and Dyrness 1973). Typical lakes in the area were oligotrophic and located within deep glacial cirques. Streams generally originate in headwall seeps high on the ridges. Seep water moving down slope incised the valley walls, creating a diverse network of small high-gradient streams that eventually flowed into larger, lower-gradient streams in alluvial valleys. In addition to these streams, others emanated from glaciers on the volcano. Channel substrates were dominated by gravels, cobbles, and boulders. Stream water was cool (generally below 16°C) and well oxygenated. Chapter 2 of this volume provides a more comprehensive description of preeruption conditions.

13.2 Study Locations

13.2.1 Description of Study Areas

The events of May 18, 1980 severely altered forest, lake, stream, and seep habitats in an area of about 600 km^2. (See Chapter 3 of this volume for a detailed description of the disturbance events.) Across the volcanically disturbed landscape, amphibian habitat was eliminated, reduced, or in some cases created. For example, 550 km^2 of forest was severely altered when trees were removed, blown down, or left standing dead and a variety of volcanic substrates were deposited on the ground (Lipman and Mullineaux 1981; Swanson and Major, Chapter 3, this volume).

Before the eruption, more than 30 lakes occurred in the study area, with a total surface area of 662 ha. After the eruption, when numerous lakes and ponds were created, this same area had more than 163 lakes and ponds (a 494% increase), with a cumulative surface area of 1679 ha (a 253% increase). These increases resulted from newly created lakes that developed behind blocked stream channels and ponds that were created on the debris-avalanche deposit in depressions among the hummocks or from pits created during phreatic explosions and where large blocks of glacier ice melted (Swanson and Major, Chapter 3, this volume).

During the main eruption on May 18, 1980, several disturbance types interacted with a highly variable preeruption landscape to create several broadly defined disturbance zones (see Figure 13.1). We surveyed amphibians within four volcanic disturbance zones: (1) scorch, (2) blowdown, (3) debris avalanche, and (4) pyroclastic flow. Some amphibian species survived in blowdown and scorch zones (about 480 km^2). It is likely that all amphibian species perished in pyroclastic-flow and debris-avalanche zones (about 75 km^2).

13.2.2 Sites Without Residual Amphibians

13.2.2.1 Pyroclastic-Flow Zone

Study sites in the pyroclastic-flow zone (15 km^2) were located on the Pumice Plain (see Figure 13.1). In the 1980 eruption, all forest cover was removed, and the area was buried beneath several different volcanic deposits, including super-hot (800°C) pyroclastic flows (Figure 13.2a). Streams and seeps were obliterated. The south shore of Spirit Lake was the only lake and pond habitat on the Pumice Plain. Spirit Lake experienced the full force of the eruption, and only microbes survived (see Dahm et al., Chapter 18, this volume).

13.2.2.2 Debris-Avalanche Deposit

In May 1980, an enormous rock-fall landslide (termed the debris avalanche) removed the forest and deposited volcanic debris (area, 60 km^2) up to several tens of meters deep (Swanson and Major, Chapter 3, this volume). Lake and pond habitat included the creation of about 130 new ponds and portions of the shorelines of two new lakes (Castle and Coldwater) in 1980. Ponds in the debris-avalanche zone are shallow (mean maximum depth = 1.1 m; SD = 0.96; n = 90) and have extensive littoral zones (see Figure 13.2b). Hydroperiod is variable across the ponds, and both permanent and ephemeral systems occur. Depending on the year, 40% to 50% of the ponds were dry by August and 55% to 65% by October. Streams and seeps were initially destroyed by the debris avalanche, but within several months to a few years new seeps and drainage networks were established. Streams were subjected to significant channel shifts, scour, and deposition through 2000 (Figure 13.2c). Seeps and springs developed luxuriant vegetation in response to a combination of stable substrates and abundant moisture (Figure 13.2d).

13.2.3 Sites with Residual Amphibians

In the blowdown and scorch zones, the forest overstory was killed; forest understory was largely eliminated; and forest soil, duff, and cover were buried beneath volcanic deposits. The forest was leveled in the blowdown zone, and the forest was killed but remained standing in the scorch zone (see Figure 13.2e). Streams and lakes, although disturbed, were neither created nor eliminated. Lake basins were altered to varying degrees. For example, high-elevation lakes were shielded from the eruptive blast by a protective cover of ice and snow. As the snow and ice melted, volcanic debris, nutrients, material from the pyrolized forest, and coarse wood entered the lakes, altering their physical and chemical characteristics (Dahm et al., Chapter 18, this volume; Figure 13.2f). Streams in the blowdown and scorch zones received large quantities of volcanic ejecta and coarse woody debris, and the aboveground parts of all riparian vegetation were killed (Hawkins and Sedell 1990; Figure 13.2g). Seeps in the blowdown and scorch zones lost their overstory vegetation and had woody debris deposited across or within them. Substantial quantities of blast and tephra-fall deposits were rapidly removed by water flowing over steep topography (Figure 13.2h).

13.3 Posteruption Microclimates and Amphibian Physiology

Although specific data are lacking, the eruption presumably changed moisture and temperature regimes throughout the disturbed area (*sensu* Chen et al. 1995). Loss of forest canopy would have increased surface temperatures and wind velocity near the ground, resulting in decreased relative humidity, soil moisture, and snowpack. Collectively, these changes probably resulted in warmer and drier environments. Additionally, volcanic ejecta sealed or eliminated duff, litter, and cover and restricted access to cool, moist subterranean retreats. Pumice has low water-holding capacity and high infiltration rates, creating dry surface conditions.

Body-water regulation and respiration are two key physiological factors limiting the distribution of amphibians. Amphibians have specialized skin that is highly permeable to water and from which gas exchange occurs (Stebbins and Cohen 1995). Generally, amphibians lose water at a rapid rate unless the surrounding air and substrate are saturated. Woodland salamanders (*Plethodon*) are among the most sensitive amphibians to drying conditions because they have high surface-area-to-mass ratios for cutaneous respiration (Spotila 1972). Amphibians are also responsive to temperature because they have relatively low thermal maxima (Brattstorm 1963a). The loss of forest canopy and deposition of abrasive pumice and ash over moist forest litter following volcanic disturbance would affect the ability of amphibians to survive and persist or to colonize a disturbed site.

13.4 Methods

13.4.1 Study Sites and Years of Surveys

We surveyed amphibians in four general habitat types within the pyroclastic-flow, debris-avalanche, blowdown, and scorch zones: lake and pond, stream, seep, and terrestrial (former forest; Table 13.2). Our surveys began in 1980 and continued through 2000. During these 20 years, we surveyed 248 sites within the four disturbance zones. Here we present results for two time periods: 1980 to 1985 (except streams, 1985 to 1987) and 1995 to 2000. Depending on the habitat type, the specific years within these two time frames vary (see below). The 1980 to 1985 data are used as evidence of species surviving

FIGURE 13.2. Photographs of representative study sites: (a) terrestrial habitat on the Pumice Plain (photo taken in 1983); (b) pond on the debris-avalanche deposit (photo taken in 1995); (c) stream on the debris-avalanche deposit (photo taken in 2000); (d) seep on the Pumice Plain (photo taken in 1989); (e) terrestrial habitat in the blowdown zone (photo taken in 1990); (f) lake in the blowdown zone (photo taken in 1996); (g) stream in the blowdown zone (photo taken in 1982); and (h) seep in the blowdown zone (photo taken in 1993). (Photo credits: C.M. Crisafulli.)

TABLE 13.2. Number of sites surveyed between 1980 and 1985 and between 1995 and 2000 for amphibians within the blowdown, scorch (referred to as BDSC zone in text), debris-avalanche and pyroclastic-flow zones (referred to as DAPF zone in text).

	Blowdown and scorch zones		Debris-avalanche and pyroclastic-flow zones	
Habitat type	1980 to 1985	1995 to 2000	1980 to 1985	1995 to 2000
Lake and pond	10	33	4[a]	117
Stream	10	28	3	13
Seep	0	47	5	10
Terrestrial	10	82	8	117

Locations of these sites are presented in Figure 13.1.
[a]Maratta wetland complex.

the eruption and early colonization events, and the 1995 to 2000 data are used to illustrate persistence and colonization of amphibian species during later years. Additionally, we classified the four disturbance zones into two groups based on amphibian species survival. Blowdown and scorch zones (hereafter BDSC) are pooled as sites where amphibians survived, and the debris-avalanche and pyroclastic-flow zones (hereafter DAPF) represent areas where amphibians were not known to have survived. We follow the taxonomy of Crother et al. (2000, *http://www.herplit.com/SSAR/circulars/HC29/Crother.html*).

For a subset of the lake and pond sites in the DAPF zone, referred to as the Maratta wetland complex, we present a more complete data set that includes 14 sampling years between 1983 and 2000. These data are used to illustrate a particular sequence of species invasions into newly created habitats. We then use data gathered from 117 ponds on the DAPF zone to describe broad-scale patterns of species presence and reproductive status from 1999 to 2000.

13.4.2 Sampling Methods

We surveyed amphibians with several methods. Visual-encounter surveys (Crump and Scott 1994; Crisafulli 1997), aquatic funnel trapping (Adams et al. 1997), dip netting, and mark-recapture were used to survey lake and pond systems. Visual-encounter surveys were conducted in the littoral zone (defined here as the area where the water depth is less than 1.3 m). We carefully searched the water column, benthos, aquatic vegetation, and woody debris for amphibian eggs, larvae, neotenes, and adults. To minimize disturbance to bottom sediment and to maximize detection of amphibians, snorkeling or float tubes were used in the outer littoral zone. Aquatic funnel traps were used in the littoral zones. These traps are particularly appropriate for sampling sites with dense vegetation, large amounts of woody debris, or murky water. After completing visual inspections, we swept dip nets through the benthic substrates and vegetation of the littoral zone.

We used two methods to survey streams, depending on habitat type. Pool habitats were sampled with electroshockers and dip nets. We sampled fast-water habitats (riffles and cascades) by placing a seine in the stream channel and removing all upstream substrate (gravel, cobbles, and boulders) from the channel within a defined area. We sampled a minimum of six fast-water–slow-water habitat pairs within each stream each survey year.

We used visual-encounter surveys to sample seeps. We searched the ground surface and beneath woody debris, rock rubble, bedrock faces, and rock fractures of seeps and a 15-m band of terrestrial habitat surrounding these habitats for amphibians.

Terrestrial surveys were conducted at lake and pond, seep, and stream survey locations with a visual-encounter method. We searched for amphibians on ground surfaces, in vegetation, and beneath rocks and wood within a 5-m-wide belt transect placed parallel to stream sample reaches and around seeps and lake and pond habitats. Note that terrestrial surveys were adjacent to streams and lake and pond systems rather than in areas far up hill slopes. In our study area, these habitats offer the most suitable conditions in the landscape for terrestrial amphibian species because they maintain equable moisture and temperature conditions (Spotila 1972). Studies conducted in undisturbed areas of the Pacific Northwest found that terrestrial species (e.g., ensatina and western red-backed salamander) use these habitats (McComb et al. 1993; Wilson 1993; Olson et al. 2000; Sheridan and Olson 2003).

13.4.3 Data Summary, Interpretation, and Limitations of Survey Results

Data from our surveys include compilations of species presence or absence (i.e., not detected) and the breeding status of species across sites and through time in each of the four habitat types within each of the two disturbance zone pairs (BDSC and DAPF). A species was listed as present at a site when at least one individual was observed during a survey. Breeding was confirmed if either eggs or larvae were observed.

Several issues need to be considered when interpreting the results of presence/absence surveys for amphibians.

- Amphibians are relatively small animals; are often nocturnal; and frequently reside beneath the ground, under logs and rocks, or in lake-bottom sediment.
- Many amphibians are only surface active or found in large congregations for a brief time (e.g., for 3 to 14 days).
- Each species has a unique probability of detection within a given habitat because species differ in their behaviors.
- The probability of observing a species is directly related to its abundance.

At Mount St. Helens, we wanted to document the species that survived the eruption and also those that colonized sites during the study period. We therefore timed our surveys to correspond with peak animal-activity patterns and optimal environmental conditions. Lake and pond surveys occurred when species were breeding and more than one life stage was present (i.e., eggs, larvae, and adults). This time period was April through July, depending on elevation and annual variation in snowpack, rainfall, and temperature. Large breeding aggregations, vocalization, and conspicuous eggs or larvae made species detection likely. Seep and terrestrial surveys were conducted during May and June and in October and November when animals were surface active, air temperatures were cool (typically below 16°C), and surface conditions were wet (Crisafulli 1999). We surveyed streams in July and August during summer base flow when animals were concentrated.

Although the authors are very familiar with the amphibian fauna, used numerous survey methods, and timed surveys to maximize detection of species, we cannot be certain that some species were not missed during surveys. Such sampling errors would most likely have occurred when a species was represented by few individuals. The net effect of recording a species as "absent" when it was actually present would be a reduced species-richness value, an altered species-composition list, and in the case of new colonization events an inaccurate measure of species colonization rates.

13.5 Amphibian Survival in the BDSC Zones

13.5.1 Presence and Breeding Status: 1980 to 1985

Surveys conducted during the first 5 years after the eruption allowed us to assess amphibian survival. In the BDSC zones, 9 species were sampled. Zalisko and Sites (1989) reported 2 additional species. Thus, 11 of 15 species are known to have survived the eruption (Table 13.3). Species with aquatic life histories survived more often than terrestrial forms, with 10 of the 11 surviving species having an aquatic egg and larval stage. The other surviving species is closely associated with seeps and streams. The 4 species not found included the red-legged frog and the terrestrial Larch Mountain, western red-backed, and ensatina salamanders.

TABLE 13.4. Amphibian species presence (P) and breeding (B) status at 10 blowdown-zone lakes from 1980 to 1983.

Lake	Species					
	AMGR	AMMA	TAGR	BUBO	RACA	PSRE
Elk	B	B	B	B	P	B
Fawn	B		B	B	P	B
Ghost	B			B	P	B
Hanaford	B		B	B	P	B
Meta	B		B	B	P	B
Ryan	B		B	B	P	B
Spirit (northeast bay)			P	P	P	B
St. Charles	B	B	P	P	P	P
St. Helens			B	P		
Curtis	B		P	B	P	P
Number of sites	8 (8)	2 (2)	9 (6)	10 (7)	9 (0)	9 (7)

Numbers of sites where species bred appear in parentheses. Locations of study lakes are shown in Figure 13.1.

Species codes: northwestern salamander (*Ambystoma gracile*), AMGR; long-toed salamander (*A. macrodactylum*), AMMA; rough-skinned newt (*Taricha granulosa*), TAGR; western toad (*Bufo boreas*), BUBO; Cascades frog (*Rana cascadae*), RACA; Pacific tree frog (*Pseudacris regilla*), PSRE.

13.5.1.1 Lake and Pond Species

Surveys at 10 lake and pond study sites during the first 3 years following the eruption (1980 to 1983) detected adults of six of the seven lake and pond species. Five of these species reproduced by 1981 (eggs or tadpoles present; Table 13.4). The red-legged frog was not detected, but this species is typically found at elevations lower (below 850 m) than our study sites (Nussbaum et al. 1983). For lake and pond species, the most important factor for survival appeared to be the timing of the disturbance. The eruption occurred during late spring when the majority of lakes and high-elevation forests and meadows were covered with snow or ice, which protected amphibians from the thermal and abrasive forces of the eruption. Adult frogs and aquatic forms (larvae and neotenes) of the northwestern salamander were in lake-bottom sediment. Long-toed salamanders and rough-skinned newts do not have overwintering aquatic forms in our study area and must have survived in terrestrial refugia. Similarly, the western toad likely survived in subterranean terrestrial hibernacula. Although the landscape

TABLE 13.3. Number of amphibian species present within four habitat types surveyed between 1980 and 1985 and 1995 and 2000 within the blowdown, scorch, debris-avalanche, and pyroclastic-flow zones.

Habitat	Number of species possible	Blowdown and scorch zones		Debris-avalanche and pyroclastic-flow zones	
		1980 to 1985	1995 to 2000	1980 to 1985	1995 to 2000
Lake and pond	7	6	7	2	7
Stream	3	2	3	0	0
Seep	2	2[a]	2	0	0
Terrestrial	3	0	0	0	0

[a]Data from Zalisko and Sites (1989).

TABLE 13.5. Amphibian species breeding (B) status at 10 blowdown-zone streams from 1985 to 1987.

Stream	1985 ($n = 5$)		1986 ($n = 10$)		1987 ($n = 9$)	
	ASTR	DICAMP	ASTR	DICAMP	ASTR	DICAMP
NF1	B		B		B	
NF2	B					
NF3	ns	ns	B		B	
HWF1	B		B		B	
HWF2	B		B		B	
HWF3	B		B	B	B	
Ghost	ns	ns				
Grizzly	ns	ns			B	
Meta	ns	ns			B	
Clearwater	ns	ns	B		ns	ns
Number of sites	5	0	6	1	7	0

Locations of study streams are shown in Figure 13.1.
Species codes: coastal tailed frog (*Ascaphus truei*), ASTR; giant salamanders (*Dicamptodon tenebrosus* and *D. copei*), DICAMP.
ns, stream was not sampled that year.

had been dramatically altered, successful reproduction during the first few years following the eruption indicated that conditions (e.g., water quality and food resources) remained adequate for species to persist during the first 3 years after the eruption.

13.5.1.2 Stream Species

We found coastal tailed frogs and giant salamanders during surveys of 10 streams in the blowdown zone 5 to 7 years (1985 to 1987) after the eruption (Table 13.5). Tailed frogs were present and breeding in the majority of streams sampled. In contrast, we detected giant salamanders in just 1 of the 10 study streams. Cool water temperatures, high-gradient channels, and the flushing of sediment by flowing water were likely the primary factors associated with survival. Nonetheless, amphibian mortality was widespread and severe in most basins (Hawkins and Sedell 1990; Crisafulli and Hawkins 1998). Amphibians survived in microhabitats that were protected from chronic scour, abrasion, or embedded substrates. Cascades, chutes, waterfalls, and valley-wall seeps may have served as refugia for these species. Aquatic giant salamanders (*Dicamptodon* spp.) were essentially eliminated from streams but probably survived in some ice-protected lakes. These lakes are headwaters for streams, and surviving salamanders could have been the source of colonists for highly disturbed and defaunated streams. No terrestrial adults of the two *Dicamptodon* species was found, and they probably did not survive the eruption. However, there is strong evidence that at least some adult coastal tailed frogs survived in the blowdown zone. An adult male (total length, 37 mm) was captured in a pitfall trap within the upper Smith Creek drainage on September 21, 1982. The trap was located in an area of barren open pumice with very low (~5%) plant cover and was about 100 m from a small first-order stream. Coastal tailed frog densities in the blowdown-zone streams were variable across the study sites and through time, but average densities 4 to 5 years after the eruption were remarkably high (Hawkins et al. 1988; Crisafulli and Hawkins 1998). Tadpole densities in these streams are among the highest reported in the species range (Hawkins et al. 1988).

Removal of the riparian vegetation results in increased sunlight to streams and increased water temperature, promoting increased periphyton production (Hawkins et al. 1983). Periphyton is the primary food of herbivorous coastal tailed frog tadpoles. Large diel fluctuations in water temperature were documented, but maximum temperatures remained below lethal thresholds for these species (Claussen 1973; Hawkins et al. 1988; Hawkins and Sedell 1990). The extent and duration of habitat alteration varied by landscape position and channel slope. Small, high-gradient streams flushed the fine and highly erodable volcanic material from the channel during the first two winters, exposing the original boulder and cobble substrates. Larger, lower-gradient streams tended to store more of these fines, became shallower and wider, and had fewer pools (Hawkins and Sedell 1990).

After about 8 years, deciduous shrubs and trees (*Alnus*, *Salix*, and *Populus*) colonized many stream corridors, and by 20 years posteruption, these riparian areas were typically lined with dense, continuous woody vegetation. Large amounts of coarse woody debris remain in channels and will likely persist for decades to centuries (see Figure 13.2g).

13.5.1.3 Seep Species

We did not survey seeps from 1980 to 1985. However, Zalisko and Sites (1989) found Van Dyke's and Cascade torrent salamanders in a tributary to Bean Creek during 1982. They concluded that some individuals were more than 2 years of age and had survived the eruption. The juveniles they collected must have resulted from successful posteruption reproduction. Because only a few sites were sampled between 1980 and 1983, we cannot assess the geographical extent of survival for these species.

13.5.1.4 Terrestrial Species

Surveys conducted at 10 terrestrial sites in the BDSC zone failed to detect western red-backed, Larch Mountain, and ensatina salamanders. With so few sites sampled, we cannot conclude that these species did not survive in other locations. However, we suspect that these species likely perished either during or within several months of the eruption because these salamanders require cool, moist microhabitats (Spotila 1972; Feder 1983). Animals would have been buried beneath deep deposits of dry, abrasive volcanic ejecta. Animals remaining below the tephra would have experienced depleted prey resources and low oxygen levels, eventually leading to death. Had animals reached the surface, they would have desiccated.

There is a substantial literature on the response of plethodontid salamanders to disturbances related to timber harvesting (e.g., Bury 1983; Petranka et al. 1993; Ash and Bruce 1994; deMaynadier and Hunter 1995; Welsh and Lind 1995). From these studies, we conclude that removal of forest canopy causes rapid and marked population declines or local extinctions of some plethodontids (Dupuis et al. 1995; Ash 1997; Herbeck and Larsen 1999). Most authors point to the change in microclimatic conditions as the reason for population declines. Removal of forest canopy creates warmer and drier ground surfaces, conditions that can be physiologically stressful or beyond the tolerance limits of salamanders.

The fact that 11 amphibian species survived the eruption in a variety of refugia meant that long-distance dispersal from distant source populations was not needed to reestablish populations. In the absence of these residuals, it may have taken decades to reestablish populations of these species in the BDSC zone. Moreover, the confirmed reproduction of 9 of the surviving species suggested that local sources provided colonists to adjacent defaunated or newly created habitats.

13.5.2 Presence and Breeding Status: 1995 to 2000

13.5.2.1 Lake and Pond Species

We detected breeding populations of seven lake and pond species in surveys at 33 lake and pond study sites between 1995 and 1997 within the BDSC zone (see Table 13.3). Thus, 15 to 17 years after the eruption, the entire possible lake and pond assemblage was present (Table 13.3). Species frequency of occurrence varied across the study sites; four of seven species were present at more than 76% of the sites; the remaining three species were observed less frequently (Figure 13.3a). Similarly, the number of sites where breeding occurred for the seven species was variable. For some sites, species were present but not breeding (Figure 13.3a). The average amphibian species richness of the 33 sites was 4.1 (SD = 1.1) and ranged from two to seven species (Figure 13.3, inset).

We also captured species associated with streams and seeps in our lake and pond sampling: adult tailed frogs at four sites and neotenic coastal giant salamanders at five. Coastal tailed frogs were captured under cover objects in the riparian zone, and giant salamanders were found beneath bark in the littoral zone. Streams flowing into and out of these lakes may have been the source of these animals. However, neotenic giant salamanders occur in montane lakes (Nussbaum et al. 1983), and these animals may have survived in these lakes. Van Dyke's salamanders, typically associated with seeps, occurred under rocks and wood in the riparian zone of four lakes. Headwall seeps occurred in cirques within 5 to 75 m of lake capture locations and were the likely source of these animals.

The eruption and subsequent erosion and deposition of volcanic materials led to changes in lake and pond amphibian habitat that may have been beneficial to some species. Volcanic inputs increased the total area of littoral zone as waterborne sediment was transported down streams, forming deltas and alluvial fans. Colluvium from slopes entered the lakes, creating shoals. The resulting shallow areas were colonized by aquatic plants and served as locations for amphibian ovipositioning and larval foraging. Trees surrounding lakes were

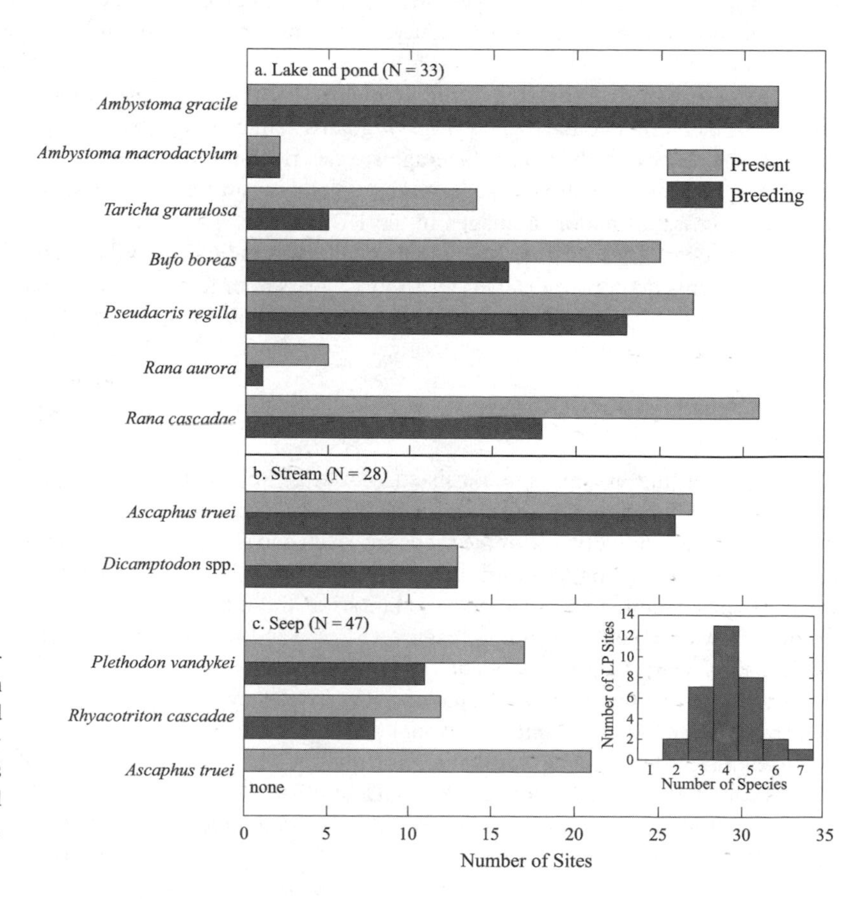

FIGURE 13.3. Presence and breeding status of amphibians at 33 lakes surveyed between 1995 and 1997 in the blowdown and scorch zones: (a) lake and pond species, (b) stream species, and (c) seep species. *Inset:* The number of amphibian species present at 33 lakes surveyed between 1995 and 1997 in the blowdown and scorch zones. LP on y-axis refers to lakes and ponds.

toppled by the blast and entered the riparian and littoral zones, creating unusually high wood levels (see Figure 13.2f). This material provided substrates for amphibian larval food (both algae and insects), places to hide, and ovipositioning sites for salamanders. In addition, the absence of overhead canopy would have increased water temperature and primary productivity in littoral zones, which may have hastened larval growth and metamorphosis.

13.5.2.2 Stream Species

The stream-breeding amphibian assemblage consisted of three species (see Table 13.1). Two of these, the coastal giant and Cope's giant salamanders, are difficult to distinguish as small larvae. Consequently, these two species were pooled as "giant salamanders" for treatment of occurrence at our sites. Thus, the potential stream species list included giant salamander and the coastal tailed frog.

Coastal tailed frog adults and tadpoles occurred in the majority of streams sampled from 1995 to 1998 (see Figure 13.3b). Three egg masses were found in BDSC-zone streams, each attached to the underside of a boulder located at the downstream end of a pool. Except for a single pitfall-trap capture, all adult frogs were observed within a few meters of water. Giant salamanders were observed in 13 (46%) of the 28 streams (Figure 13.3b). All giant salamanders captured were either larvae or neotenes; no metamorphosed individuals were seen. Metamorphosed adults were often found in undisturbed areas adjacent to the volcanically altered sites. Perhaps neoteny enables giant salamanders to survive and reproduce in the volcanically altered region.

Of the 28 streams surveyed, 27 supported amphibian populations; 14 had either coastal tailed frogs or giant salamanders present, and 13 had both species (average species richness = 1.4; SD = 0.57). The 28 streams surveyed for amphibians were dispersed across all major drainages in the BDSC zone. The presence of coastal tailed frogs and giant salamanders in 96% of these streams demonstrates that, by 15 to 17 years after the eruption, most streams in the BDSC zone supported populations of stream amphibians.

13.5.2.3 Seep Species

The seep-dwelling assemblage consisted of Van Dyke's and Cascade torrent salamanders (Table 13.1). These two species were present at relatively few of the seeps surveyed and shared sites with the tailed frog, a stream-breeder (see Figure 13.3c). Each species established breeding populations at the majority of sites where they were found (Figure 13.3c). Van Dyke's salamanders were locally abundant (about 20 animals) at a few sites but uncommon at most locations. The frequency and abundance observed at our sites at Mount St. Helens are similar to that reported by other workers. For example, McIntyre (2003) found Van Dyke's salamanders at 15 of 40 seeps, and the mean number of captures at these sites was two individuals (SE = 0.59).

Torrent salamander abundances were low at all locations, generally fewer than three animals. Because of their apparently poor dispersal capability, it is likely that Van Dyke's and Cascade torrent salamanders survived the eruption within the BDSC zone, which is consistent with the view of Zalisko and Sites (1989). Nijhuis and Kaplan (1998) found that, in the Columbia River Gorge, 27% of the animals marked in the spring were recaptured in the fall, and the mean distance moved over a 3-month period was 2.4 m (SE = 0.40; $n = 24$). McIntyre (2003) found that the mean distance moved by individuals in two populations of Van Dyke's salamanders over 4 months was 1.7 m (SE = 0.93; $n = 20$) and 0.9 m (SE = 0.22; $n = 8$), respectively.

We believe that topographic setting was likely an important factor that contributed to the survival of these species in the BDSC zone. Seeps with salamanders were typically in areas that provided some protection from the eruptive forces, such as sites in deep gorges, with north aspects (facing away from the blast), or protected behind a ridge. In addition, these sites had steep slopes (greater than 70%), cliffs, and cool groundwater, which would have ameliorated the severity of disturbance. Herbs and shrubs typically survived in areas immediately adjacent to seeps, and water temperature remained cool (less than 10°C) despite loss of canopy.

Van Dyke's and Cascade torrent salamanders are stenothermic, have low fecundity, and possess poor dispersal capabilities, which may limit their spread to defaunated habitats in the BDSC zone. Of the 47 seeps surveyed, 23 had no amphibians present. The large (49%) number of sites without amphibians is probably related to the small size and patchy distribution of seep habitats that are embedded within a nonforested landscape. Once a population in a small habitat patch undergoes extinction, the likelihood of recolonization is low in this landscape.

Coastal tailed frog adults were observed in 21 of the seeps surveyed, making it the most common species encountered in the seeps. Coastal tailed frogs were not observed breeding at these seeps. During the summer, seeps provide cool, moist conditions that sharply contrast with the hot and dry terrestrial landscape. We do not know if coastal tailed frogs are primary residents of these seeps or are using them as refugia while dispersing.

13.5.2.4 Terrestrial Species

No western red-backed, Larch Mountain, or ensatina salamanders were observed at the 82 terrestrial survey locations in the BDSC zone (see Table 13.3). Known populations of these salamanders existed north, south, and east of the BDSC zone at distances of 9 to 12 km. By 1995, small patches of surviving vegetation expanded, and initially unoccupied areas were colonized by early-successional plants. Woody debris remained abundant, but development of forest conditions was decades away. Plethodontids are well known for their sedentary habitats, and movements of numerous species are generally of the order of 3 to 50 m over periods of months or even years (Stebbins and Cohen 1995 and references therein). Lack of suitable habitat and limited dispersal capabilities are likely responsible for the absence of these species from the BDSC zones.

13.6 Amphibian Survival in the DAPF Zones

It is unlikely that amphibians survived the extreme physical effects of heat (300° to 600°C), burial (10 to 195 m), and scour that created the DAPF zones (Swanson and Major, Chapter 3, this volume). We therefore assume that any amphibian observed in the DAPF zone after the eruption was a colonist.

13.6.1 Colonization: 1980 to 1985 and 1995 to 2000

Amphibians associated with stream, seep, and terrestrial habitats were not detected during the1983 to 1985 or 1995 to 2000 samplings (see Table 13.3). The absence of these species was not surprising because streams, seeps, and terrestrial habitats remained substantially altered during the first 20 years following the eruption (see Figure 13.2a,c,d). It took several months to a few years for drainage networks to reestablish, and streams experienced significant channel shifts, high concentrations of sediment, scour, and deposition from 1980 to 2000 (Simon 1999; Swanson and Major, Chapter 3, this volume). Beginning about 1995, portions of a few stream channels were showing signs of stabilization as evidenced by cobble substrates and riparian vegetation.

The absence of coastal tailed frogs and giant salamanders from these systems was probably related to poor habitat condition, particularly the predominance of sand substrate that lined channels or filled the interstices of larger substrates. The larvae of these species require coarse, unembedded substrates for securing their prey (diatoms or invertebrates) and for cover from predators and high stream flows (Nussbaum et al. 1983; Parker 1991). Several studies in the Pacific Northwest showed that fines and substrate embeddedness can negatively affect coastal tailed frog and giant salamander larvae (Hawkins et al. 1983; Corn and Bury 1989; Welsh and Ollivier 1998). An alternate hypothesis for the absence of tailed frogs and giant salamanders from these systems is limited dispersal capability. Coastal tailed frog adults have low mobility (Daughtery and Sheldon 1982). The closest known source population was about 6 km away within the BDSC zone, a distance that may have prevented colonization. Adult giant salamanders are capable of long terrestrial movements (Corn and Bury 1989), and there were giant salamanders about 8 km away in the BDSC zone. However, only aquatic individuals (neotenes) were observed. The closest source of terrestrial adult giant salamanders was probably about 15 km away.

Suitable habitat did not exist for Van Dyke's and Cascade torrent salamanders from 1980 to 2000 in the DAPF zone. Seeps that developed in the DAPF zone exhibited little resemblance to the seeps where these salamanders are usually found (compare Figure 13.2h and 13.2d). Seeps in the DAPF zones were low gradient with water percolating through gravel, were fully exposed to sunlight, and became very warm during the summer months. Seeps that support these salamanders typically occur in steep bedrock cliffs and have cool (less than 10°C), rapidly flowing water that creates splash zones or wet areas on rubble at the base of these cliffs (Jones 1999; McIntyre 2003).

Habitat for the three forest species (Larch Mountain, western red-backed, and ensatina salamanders) did not exist in the DAPF zone from 1980 to 2000. Terrestrial vegetation cover in the DAPF zone was absent immediately after the eruption and remained sparse through 2000 (see Dale et al., Chapter 5, this volume; del Morel et al., Chapter 7, this volume). Surface substrates consisted of rocky material with scant, patchily distributed litter. Surface conditions were severe with temperatures exceeding 50°C. Coarse woody material was lacking. The forest habitat and associated cool, moist microclimates that these species require will probably require decades, or possibly centuries, to develop.

We did not sample the Maratta site until 3 years after the eruption (1983). At that time, two species were present, and species colonized incrementally for 10 years, by which time six species had arrived (Figure 13.4; Table 13.3). Anurans were the first group to establish. Within 3 years, the western toad and Pacific tree frog established, followed by the red-legged frog (year 5) and the Cascades frog (year 7). The northwestern salamander established after 9 years, and the rough-skinned newt

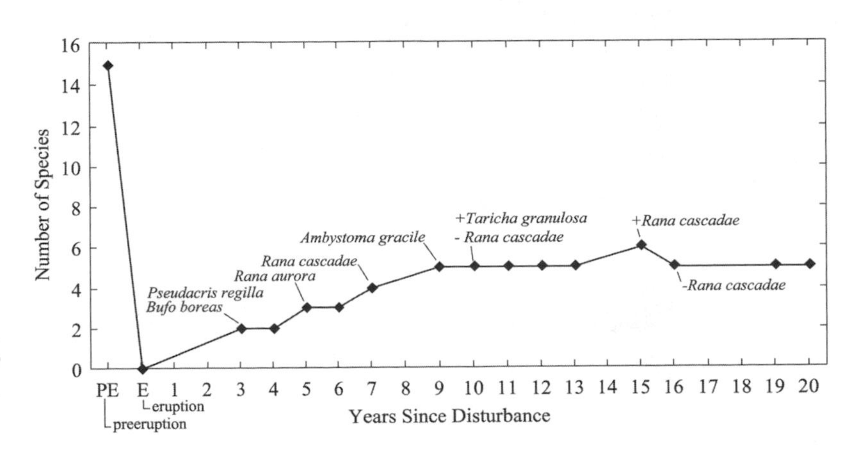

FIGURE 13.4. The sequence of amphibian-species colonization at the Maratta wetland complex (DAPF zone) based on surveys during 14 survey years between 1983 and 2000.

was first observed 10 years posteruption. No additional species colonized the Maratta study site during the next 10 years. All species except the Cascades frog were observed every year from 1989 to 2000 (Figure 13.4).

In the DAPF, species dispersal capacity and ability to use terrestrial habitats appear to be the primary factors governing colonization. Anurans are generally more mobile than salamanders and were the first to colonize. Among the four anurans, the western toad and Pacific tree frog are the most terrestrial and invaded the study site at least 2 and 3 years, respectively, before the more aquatic ranid frogs. The lower mobility of salamanders was likely responsible for their later colonization.

Observations by others show that, 1 to 2 years after the eruption, frogs and toads had colonized ponds within the DAPF zone. Karlstrom (1986) reported that the Pacific tree frog, western toad, and Cascades frog had colonized ponds in the Toutle River valley by 1985. The western toad was present on the debris-avalanche deposit during the summer of 1981 (Doug Larson, personal communication), and the Pacific tree frog was present in the area during the summer of 1982 (Karlstrom 1986).

13.6.2 Presence and Breeding Status: 1995 to 2000

The frequency of occurrence and breeding status of lake and pond species was variable among the 117 study sites (Figure 13.5). The pattern of occurrence differed between frogs and salamanders, with frogs more frequently encountered. Five of the six species established breeding populations at the study ponds. However, there were a number of sites where adults were present but breeding was not confirmed (Figure 13.5). The mean amphibian species richness of the DAPF zone ponds surveyed during 1999 and 2000 was 3.0 (SD = 1.6; $n = 117$) and ranged from zero to six species (Figure 13.5, inset).

Emergent and submergent vegetation began colonizing the littoral zones of these ponds within a few years of the eruption, and by 1995 many ponds had extensive macrophyte communities. Amphibians use vegetation for egg laying, cover, and foraging substrates. The duration of larval development is related to temperature and food levels (Duellman and Trueb 1986), and the shallow littoral zones (average maximum depth, 1.1 m; SD = 0.96; $n = 90$) of these ponds resulted in warm, highly productive waters during the summer, which led to rapid metamorphosis. A pond's hydroperiod is important to amphibians because larvae may die if a pond dries before metamorphosis occurs (Pechmann et al. 1989). Most ponds in the DAPF held water through early August of each year, enabling the majority of anuran larvae to complete metamorphosis.

The risk of metamorphosis is evident on the Pumice Plain each fall when thousands of recently metamorphosed individuals depart their moist retreats in the riparian zone of Spirit Lake. The onset of autumn rain triggers a mass dispersal of these animals in search of mesic forest environments. Instead, they find the barren Pumice Plain, and once the rains cease, these animals desiccate and die. We do not know why the relative frequency of neoteny varies so much between the Pumice Plain and the adjacent debris-avalanche zone.

13.7 Amphibian Dispersal

Dispersal of organisms into defaunated areas is a primary factor influencing colonization. Distance to source populations, species vagility, and landscape permeability are among the factors that govern the rate and pattern of dispersal. Although quantifying dispersal is challenging, we had the opportunity to document dispersal distances at Mount St. Helens for a number

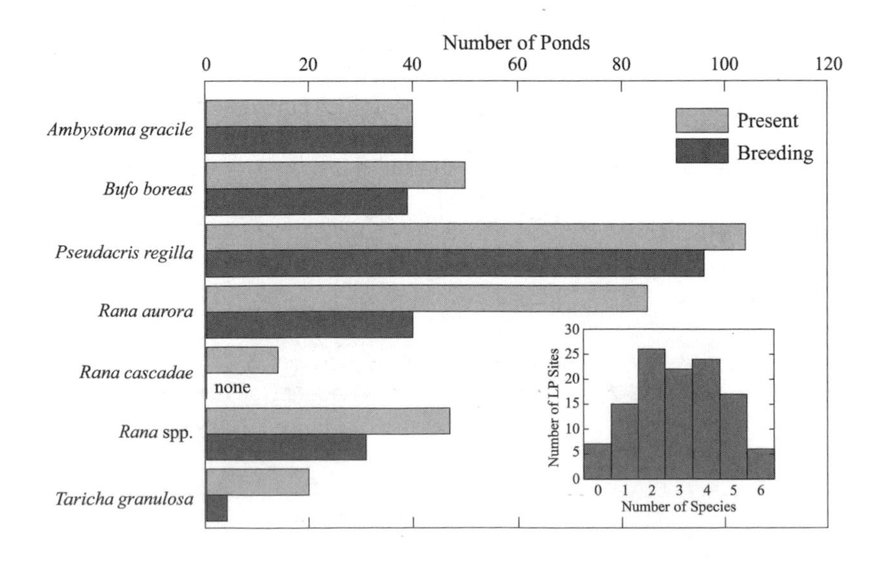

FIGURE 13.5. Presence and breeding status of lake and pond amphibians at 117 ponds surveyed between 1999 and 2000 in the debris-avalanche zone. *Inset:* The number of amphibian species present at 117 ponds surveyed between 1999 and 2000 in the debris-avalanche zone. LP on y-axis refers to lakes and ponds.

of amphibians. Some dispersal distance estimates were based on movements of marked animals, and others were based on distances to the nearest known source populations. Within the BDSC zone, survival of amphibians in numerous refugia obviated the need for dispersal from distant source populations (except for some plethodontids). These residual animals or their progeny likely colonized new habitats (e.g., ponds) that were created during the eruption and existing ponds that had lost their populations. Within the BDSC zone, several marked amphibians moved considerable distances. Two western toads moved at least 4.5 and 1.6 km from their original places of capture. Two Cascades frogs moved 1.2 and 0.75 km from their original places of capture. Dozens of recently metamorphosed northwestern salamanders dispersed more than 2 km, and a few dispersed at least 3.2 km. Individual salamanders traveled up a steep (about 70% slope) valley wall with a 245-m elevation gain through blowdown forest; and in some cases they ventured over a ridge into the next drainage, a distance of 2.5 km. On the barren Pumice Plain, two northwestern salamanders dispersing from Spirit Lake traveled 1.9 and 3.0 km.

In the debris-avalanche deposit zone, the closest known source populations were lakes within the BDSC zone, where amphibians had survived the eruption. The closest known populations of the northwestern salamander, Pacific tree frog, red-legged frog, and Cascades frog were 3.7 km away, and the closest populations for the western toad and rough-skinned newt were 5.7 km away. We observed Pacific tree frogs in the volcano's crater, which was at least 10 km from the nearest known surviving population. All distances reported here were based on the shortest map distances between the original and subsequent points of capture. The actual distances that the animals traveled would have been far greater, given microtopography and the need to go around or over obstacles.

Landscape permeability is important when considering colonization processes. The Mount St. Helens landscape appeared to be harsh and prohibitive to amphibian movement, yet thousands of successful long-distance dispersal events occurred. For much of the year, the landscape is largely impermeable because of 6 months (November to April) of snow cover and 3 months (July to September) of hot drought. However, two times each year, much of the landscape may become permeable. The first period occurs for about 6 weeks in the spring as the snowpack melts and rain and cool temperatures prevail. The second period occurs in the fall when the rains and cool temperatures return.

In the spring, all surface substrates are saturated, and it is at this time that the majority of successful dispersals likely takes place. As fall approaches, the substrates are dry, and if the precipitation is delayed, it occurs as snow, thus limiting dispersal. We have observed several "false starts" in the fall. Such episodes occur when short-term rains trigger dispersal from wet microsites only to have clearing weather trap animals in conditions that cause thermal stress or desiccation. We observed northwestern salamanders dying in this way each year.

13.8 Biotic Interactions

Since the 1980 eruption, amphibian populations were influenced by interactions with other species. Beaver (*Castor canadensis*) had profound effects on amphibian breeding by altering water levels and creating and destroying habitat. Dams constructed on stream channels in the BDSC zone caused changes in habitat and substrate that led to the loss of coastal tailed frog larvae and promoted the colonization of northwestern salamanders and Cascades frogs. Ovipositioning sites for western toads and Cascades frogs were created when dams on lake outlet channels caused lake levels to increase more than 1 m and thus lake area to increase. Beavers had their most pronounced effects in the debris-avalanche zone, where they influenced 25 of the 90 ponds surveyed. Dams caused elevated water levels, which inundated the riparian zone. Inundated plant stems were used as ovipositioning sites by northwestern salamanders. Cut willow shrubs and alder trees in riparian zones fell into the littoral zone, and their branches and twigs were also used as ovipositioning sites for northwestern salamanders. Before beaver activity, oviposition substrates were sparse and likely limited salamander reproduction.

Biotic amelioration of the landscape probably facilitated the expansion of amphibians into otherwise inhospitable terrestrial environments. Pocket gophers (*Thomomys talpoides*) are fossorial rodents that create extensive tunnel networks. Gophers were especially successful in the posteruption landscape and have spread throughout the blowdown zone, where they created numerous tunnels (Crisafulli et al., Chapter 14, this volume). Gophers had colonized the Pumice Plain 12 years after the eruption. Their tunnels were used by four amphibians (northwestern salamander, western toad, Pacific tree frog, and Cascades frog) as refugia from hot, dry surface conditions during the summer months. Summer temperatures within tunnels are 5° to 15°C cooler than is the surface and have lower vapor-pressure deficits. These tunnels are presumably "safe sites" for amphibians that are in areas that experience desiccation during periods of surface drought and may be used as travel routes. An adult male Pacific tree frog spent at least 27 days in a gopher tunnel on the Pumice Plain during August 1994. In August 1999, an adult, male northwestern salamander was in a burrow for at least 10 days during a period of very hot and dry weather. Western toads were observed on a number of occasions peering out of burrow entrances or holes created in tunnels caused by elk (*Cervus elaphus*) trampling.

Predators also likely influenced the persistence and growth of amphibian populations. During their annual breeding season, western toads congregate in great numbers. We observed common ravens (*Corvus corax*) flipping toads onto their backs, striking them with their beaks, and then separating the flesh from the poisonous skin. Only the inside-out skin, skull, and ova were left behind. On other occasions, ravens were seen flying off with toads in their beaks.

Common garter snakes (*Thamnophis sirtalis*) specialize on amphibian prey (Nussbaum et al. 1983). This species was uncommon during the first 10 years after the eruption but thereafter became increasingly abundant. Common garter snakes are good swimmers, and we observed them taking tailed frog larvae from streams in the BDSC zone. We also saw them swallowing western toad and Cascades frog adults, and many northwestern salamander terrestrial adults were found in snakes captured from the riparian zones of lakes in the BDSC zone. We also observed snakes foraging on Pacific tree frog, Cascades frog, and both tadpoles and transforming individuals of western toads.

The brook trout (*Salvelinus fontinalis*) is a nonnative species that was stocked in most of the lakes in the vicinity of Mount St. Helens to provide recreational fishing (Bisson et al., Chapter 12, this volume). Stocking began in 1913 and continued through 1979. Brook trout survived the eruption in several high-elevation lakes within the blowdown and scorch zones. Negative effects of fish on amphibians are well documented (Pilliod and Peterson 2001; Knapp et al. 2001). At Mount St. Helens, the abundance of the northwestern salamander was 10 times greater in lakes without fish ($x = 2.6$/trap; $n = 4$ lakes) than in lakes with fish ($x = 0.2$/trap; $n = 4$ lakes). In some systems, the presence of the brook trout has had a greater negative influence on the northwestern salamander than did the eruption of 1980.

13.9 Amphibian Responses to Other Large-Scale Disturbances

The lack of long-term monitoring following other large, infrequent disturbances (e.g., volcanism, wildfire, and hurricanes) precludes our ability to make strong comparisons with our data. The dearth of information is surprising, given that there are diverse amphibian assemblages found in numerous regions of the world that are subjected to these types of disturbances. The information that does exist is often largely anecdotal. Burt (1961) found a frog (*Tomodactylus angustidigitorum*) 1 year after the February 20, 1943 eruption of volcano Paricutin in Mexico in an area that received about 45 cm of ash. Two years after the eruption, the frog was more abundant, and he concluded that the species had successfully bred. The Soufriere Hills volcano erupted on the Caribbean island of Montserrat in July 1995. Shortly after the eruption, recently metamorphosed frogs (*Leptodactylus fallax*) succumbed to ashfall deposits. Daltry and Gray (1999) report that at least 10% of the species preeruption habitat had been lost to pyroclastic flows and all other habitat was altered by ashfall and volcanically derived acid rain. Despite these habitat changes, the frog was patchily abundant in 1998, and the numbers of frogs observed at some locations were comparable to those seen before the eruption. Frogs were frequently observed coated with ash but appeared to be in excellent health with no evidence of illness or physical damage.

Fire is a pervasive agent of natural disturbance; yet, as with volcanism, there has been little work evaluating its effects on amphibians. Following the severe wildfires in the Greater Yellowstone Ecosystem in the Rocky Mountains of Wyoming and Montana in 1988, C. Peterson (Idaho State University, Pocatello, Idaho, personal communication) found no significant differences in the frequency of occurrence of either Columbia spotted frogs (*Rana luteiventris*) or striped chorus frogs (*Pseudacris triseriata*) between 23 burned and 21 unburned sites. He concluded the fires did not adversely affect these species.

Hurricane Hugo caused extensive damage to the Luquillo Experimental Forest, Puerto Rico, in September 1989. Woolbright (1991, 1996, 1997) reported that the population of the frog *Eleutherodactylus coqui* increased 600% following the hurricane. He concluded that the population increases were related to increased habitat complexity of the forest floor, which was caused both by the toppling and deposition of the forest canopy and by a reduction in invertebrate predators.

These examples provide evidence that some amphibians, particularly frogs, are able to persist in areas that have been disturbed by a variety of large, infrequent disturbances. Our understanding of amphibian responses to these disturbances could be improved through future studies that describe the type and intensity (e.g., burial depth and temperature) of disturbance and which assess the initial patterns of survival, the location of source populations, and the subsequent mechanisms and rates of colonization.

13.10 Summary and the Future

Amphibian responses to the eruption varied considerably by species but even more importantly by the general habitat-type associated with each species. The coarsest level of distinction was between aquatic and terrestrial species, where the former experienced high survival and the latter appeared to have perished. Lake and pond species were clearly the most successful of the aquatic forms and were both resistant and resilient to the events of the 1980 eruption. Stream-breeders, which suffered high initial mortality, rebounded within 5 to 10 years once stream channels stabilized. Seep-dwelling species were resistant to the disturbance, yet many habitats remained unoccupied, and substantial population growth was not detected.

The timing of the eruption was a critical factor governing the extent and pattern of survival. The presence of snow and ice thermally buffered amphibians from the otherwise lethal temperatures and abrasive forces, allowing some survival. Had the eruption occurred later in the year, even by just a few weeks, the consequences could have been very different.

Steep topography and cold groundwater facilitated the survival of thermally sensitive seep species. There was not a match between the capacities of these species to survive an enormous disturbance and their ability to move in and live in its aftermath.

Neoteny was an important life-history characteristic for three salamander species that appeared to allow these species to persist (or flourish) in the posteruption landscape. Neoteny was presumably favored because metamorphosing animals perished under harsh terrestrial conditions. As forests return to the Mount St. Helens landscape, the importance of neoteny should diminish, and metamorphosis may become a more adaptive trait.

Disturbances change habitat conditions, causing the distribution and abundance of amphibians to shift. Some habitats and species increase, whereas others decrease. Forest habitat was eliminated from 600 km^2, whereas lake and pond habitat increased severalfold. Forest salamanders appeared to have been extirpated, whereas certain pond-breeders, such as the western toad and northwestern salamander, may be more abundant than before the eruption.

Invasion of ponds by lake and pond amphibians was surprisingly rapid, demonstrating an impressive ability of these amphibians to disperse several kilometers across barren terrain to colonize newly created habitats.

Amphibians associated with streams and seeps have yet to colonize the DAPF zone. There are several streams in the DAPF zone that appear capable of supporting stream species. We suspect that dispersal is currently limiting their establishment. We expect that, within the next decade, tailed frogs will colonize these systems. In contrast, we predict that giant salamanders will not colonize these streams until forest conditions allow metamorphosed individuals to survive in the terrestrial environment of the adjacent blowdown zone and then disperse to these new streams. On the other hand, there is the possibility that neotenes residing in adjacent blowdown-zone streams that are tributaries to streams in the DAPF zone could colonize through the drainage network. The seeps that were created in the pyroclastic-flow and debris-avalanche deposit zones bear little resemblance to the seeps which are typical of the region. These new seeps lack the requisite habitat features (i.e., water less than 10°C, bedrock, and spray zones) that would facilitate establishment by seep species. Moreover, these species are believed to be relatively sedentary and, therefore, unlikely to reach these small and isolated systems.

Three plethodontid salamanders (western red-backed, Larch Mountain, and ensatina) may have been extirpated from the 600-km^2 area. Each of these species occurred at several locations surrounding the areas heavily disturbed by the 1980 eruption, which could serve as source populations. However, these species have relatively low vagility. Further, these species require cool, moist microclimatic conditions that are typical of older forests. Colonization by these species will be determined by dispersal rate and the development of forest conditions. Moreover, future eruptions could reduce the colonization rate. Mount St. Helens has erupted more than 20 times in the past 4000 years, and amphibian surveys conducted within the prevailing pumice deposition zone to the northeast revealed patchy and disjunct distributions for each of these species, which contrasts with the more uniform distributions observed in other areas. We predict that ensatina and western red-backed salamanders will colonize from the margins of the scorch zone and then slowly move inward toward the core of the Pumice Plain. The Larch Mountain salamander should colonize at a much slower rate, if at all.

Acknowledgments. We are grateful for the fieldwork provided by S. Butts, T. Chestnut, D. DeGross, J. Dhundale, C. Eggleston, B. Hammersley, J. Kling, S. Lewis, E. Lund, N. Maggiulli, A. McIntyre, R. Nauman, J. Ostermiller, H. Purdom, C. Remmerde, D. Rundio, J. Tagliabue, E. Tomer, and A. Yung. K. Ronnenberg and T. Valentine assisted with figures and tables. The manuscript benefited from the comments provided by K. Aubry, E. Brodie, V. Dale, D. Skelly, and F. Swanson. J. Sedell was instrumental in getting us to sites shortly after the eruption and provided financial support over the years. We thank the USDA Forest Service, Mount St. Helens and Randle ranger districts, and the Weyerhaeuser Company for providing access to their lands. Funding for this research was provided by grants from the National Science Foundation (DEB 81-16914 and BSR 84-07213 to J.A.M. and BSR-8416127 to C.P.H.); USDA Forest Service, Pacific Northwest Research Station; and Washington Department of Fish and Wildlife.

14
Small-Mammal Survival and Colonization on the Mount St. Helens Volcano: 1980–2002

Charles M. Crisafulli, James A. MacMahon, and Robert R. Parmenter

14.1 Introduction

The eruption of Mount St. Helens on May 18, 1980, and the subsequent activity of the volcano dramatically and significantly influenced the flora and fauna of the mountain and the surrounding area in a variety of ways. Many observers expected that no organism would survive the event anywhere close to the mountain and that it would take several decades for plants and animals to reestablish.

As we know, appearances can be deceiving. In scattered refugia across the volcanically impacted area, individuals of many species survived. These "residuals" (Andersen 1982; Andersen and MacMahon 1985a), along with a host of migrant species, rapidly initiated the biotic response (succession), sometimes to a surprising degree.

We have been observing the initial effects of the 1980 eruption on mammals (Andersen 1982; Andersen and MacMahon 1985a) and their subsequent long-term responses (Andersen and MacMahon 1985b; MacMahon et al. 1989). Mammals are good candidates for study for a variety of reasons: (1) they are sensitive to microenvironmental factors and, thus, to successional changes; (2) they act as architects of the environment through their trampling, digging, and burrowing activities (Butler 1995); (3) they may alter the course of succession by changing soil characteristics and, in turn, plant establishment (Andersen et al. 1980; Andersen and MacMahon 1985b; Barnes and Dibble 1988); (4) they are consumers of animals, seeds, foliage, and even wood while acting as prey for other species; and (5) they influence ecosystem processes, such as dispersal of plants and fungi and the cycling of nutrients (Lidicker 1989; Majer 1989) (Figure 14.1).

The major objects of our study have been small mammals, those species the size of squirrels or smaller. This emphasis was occasioned by several factors: (1) small mammals can be trapped in standard, live-capture traps (Sherman traps) in numbers sufficient to address their community and population responses; (2) in the area of Mount St. Helens, the preeruption small-mammal species and their habitat needs are well documented (West 1991; Carey 1995; Carey and Johnson 1995; Gitzen and West 2002); (3) small mammals are sensitive to specific habitat variables, whereas large mammals often integrate a variety of habitats because they have very large home ranges; (4) the moderate vagility of small mammals allows us to separate residual species from migrants and, thus, to measure colonization rates; (5) small mammals are more tractable to handle for study than large mammals, which often require special techniques and equipment; (6) few large species had belowground or other (e.g., logs) refuges, so they were probably extirpated over large areas by the eruption; and (7) the mosaic of disturbance types across the landscape was at a spatial scale more suitable for the study of small species with lower vagility.

The problems of studying small mammals, however, are not insignificant. For example, all species do not enter Sherman traps equally, and species captured by these traps are not necessarily recaptured with equal frequency. Perhaps more importantly, small mammals show one of the highest temporal population variabilities of any vertebrate (Diffendorfer et al. 1996).

Throughout this study, our goals have been fairly simple: (1) to compare predisturbance community composition to the evolving, postdisturbance communities; (2) to document differences, if any, in community assembly in different portions of the complex disturbance mosaic created by the eruption; (3) to study a variety of sites long enough to avoid the common problem of successional studies in which a space-for-time substitution (chronosequence approach) is used to interpret the pattern of succession (Scrivner and Smith 1984; Fox 1996); (4) to document species-specific responses to the altered and reconstituting environment, especially in the context of colonization success; and (5) to discuss our observations in relation to general successional theory (see Dale et al., Chapter 1, this volume) and to the findings of plant ecologists studying the volcano (see Antos and Zobel, Chapter 4; Dale et al., Chapter 5; del Moral et al., Chapter 7; and Frenzen et al., Chapter 6; this volume).

FIGURE 14.1. Examples of the influences of mammals in ecological systems at Mount St. Helens following the 1980 eruption. (a) Seeds of numerous grass and herb species were carried several kilometers in the digestive tract of elk (*Cervus elaphus*) and deposited in feces in the pyroclastic-flow zone. The feces were gathered from the site and placed in a greenhouse to evaluate the presence and germination potential of propagules. (Photograph taken in 1999). (b) A decaying carcass of an elk provides food for scavengers and supplies nutrients to vegetation on and adjacent to the carcass. (Photograph taken in 1999.) (c) Herbivorous rodents forage extensively on the stems of willow shrubs (*Salix* spp.) during the winter in the blowdown zone. (Photograph taken in 1989.) (d) Prairie lupine (*Lupinis lepidus*) seeds germinate from unclaimed caches made by rodents during the previous fall in the pyroclastic-flow zone. (Photograph taken in 1990.) (e) The burrowing activity of the northern pocket gopher brings organically rich soil, seeds, and fungal spores to the surface of pumice deposits in the tephra-fall zone, creating microsites favorable for plant establishment. (Photograph taken in 1986.) (Photo credits: C.M. Crisafulli.)

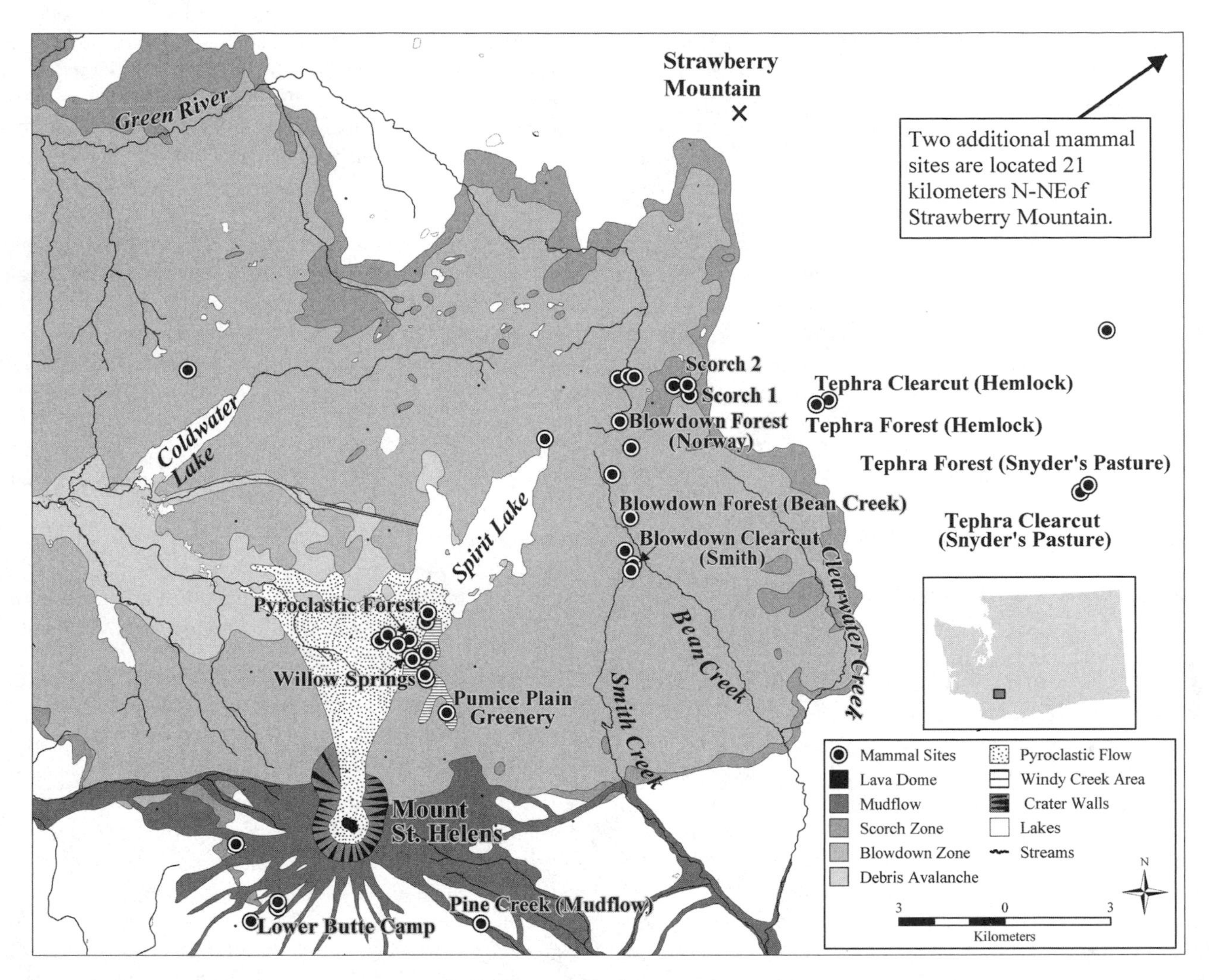

FIGURE 14.2. The Mount St. Helens Volcano region showing small-mammal sampling locations within different disturbance zones. The reference forest and clear-cut sites (not shown on map) are located on Lonetree Mountain, 21 km north-northeast of Strawberry Mountain (*top right* of figure).

14.2 Study Area and Field Methods

14.2.1 The Scene

Before the 1980 eruption, the area surrounding Mount St. Helens provided diverse habitat for small mammals (see Swanson et al., Chapter 2, this volume). Expansive, mixed-aged, coniferous forests were punctuated with numerous plantations created by clear-cut logging during the previous four decades. Natural meadows and rock outcrops scattered throughout the forest landscape were common at higher elevations. Numerous high-gradient streams created a complex drainage network and a highly dissected landscape.

The May 18, 1980 eruption caused sudden and dramatic change to Mount St. Helens and the surrounding landscape. The eruption consisted of a complex set of geophysical events acting on a highly variable landscape. Forces such as heat, wind, abrasion, scour, and burial acted singularly or in combination during the eruption to create a complex disturbed landscape. The resulting landscape consists of several structurally distinct zones reflecting the type and intensity of disturbance forces (Figure 14.2) (Swanson and Major, Chapter 3, this volume).

In 1980, it was clear that the zones created by the eruption would serve as interesting "natural experimental treatments" for assessing mammal responses to the disturbance types created during the eruption. We assumed that, by establishing study sites in undisturbed areas outside the influence of the 1980 eruption (reference areas) and in several of the disturbance zones, we could assess how disturbance intensity and habitat architecture influenced mammal species composition and relative abundances shortly after the eruption and during succession. This assumption was based on the premise that the gradient of volcanic disturbance represented a reduction in habitat complexity from late-seral, reference sites (structurally complex) to the pyroclastic-flow zone (structurally simple). The blowdown and tephra-fall zones were viewed as intermediate between these two extremes. (See Figures 10.1 to 10.4 in Parmenter et al., Chapter 10, this volume.) The disturbance zones are most distinct and homogeneous at broad spatial scales (i.e., several square kilometers). Under closer inspection, each disturbance zone is actually a mosaic of habitat patches embedded within the larger disturbance zone. The mosaics are the result of variation in topography, preeruption community composition, and site conditions. The disturbance gradient, with each of its disturbance zones, provides an opportunity to assess the influence of broad-scale habitat features on mammal composition. Within-zone sampling of habitat patches allows the opportunity to assess the influence of individual patch types on the mammal assemblage of each zone. To capture the range of variation of habitat features among and within disturbance zones required sampling both the landscape matrix and a variety of patch types (e.g., riparian areas, seeps, and refugia).

The most conspicuous posteruption changes in the first 22 years were in surface topography and vegetation. Substrate conditions and surface topography underwent substantial evolution from 1980 through 2002 (Swanson and Major, Chapter 3, this volume). Plant cover increased substantially across the landscape, particularly 10 to 22 years following disturbance, as surviving plants expanded and new plants colonized (Antos et al., Chapter 4, this volume; Dale et al., Chapter 5, this volume; Frenzen et al., Chapter 6, this volume; del Moral et al., Chapter 7, this volume; Halpern et al. 1990). In areas of high plant cover, litter accumulated and soils developed, further diversifying the surface environment. During this study, annual precipitation varied substantially (see Swanson and Major, Chapter 3, this volume). Variation in precipitation influences mammals directly by exposure and indirectly through its effects on food resources (plant production and insect biomass) and cover.

14.2.2 Field Studies

Our studies of mammals at Mount St. Helens began in 1980 and continued through 2002. We established 40 sites in several habitat types (old-growth forests, clear-cuts, meadows, and riparian and seep habitats) within reference areas (volcanically undisturbed) and several volcanic-disturbance zones (tephra-fall, scorch, blowdown, mudflow, tree-removal, and pyroclastic-flow) (see Figure 14.2). Clear-cut sites were harvested between 1961 and 1967 and supported 12- to 17-year-old noble fir (*Abies procera*) plantations at the time of the eruption. All sites were within the Pacific silver fir (*Abies amabilis*) and mountain hemlock (*Tsuga mertensiana*) vegetation zones (Franklin and Dyrness 1973). We established both "extensive" ($n = 15$) and "spot-sampling" ($n = 25$) plots. Extensive survey plots were placed in large habitat blocks representative of each disturbance zone and sampled with trapping grids or webs (Anderson et al. 1983; Parmenter et al. 2003). During 1980 and 1981, grids were established, consisting of 49 traps in a 7 × 7 matrix (at 15-m intervals). In 1982, grids were enlarged to 169 traps in a 13 × 13 matrix (at 10-m intervals). Webs replaced grids in 1987. Webs consisted of twelve 100-m radial lines, with 12 traps per line (at 5-m intervals for stations 1 through 4 near the center and at 7-m intervals for stations 5 through 12). There were 148 traps per web (144 traps on the 12 radial lines and 4 traps placed at the web center). "Spot" sampling plots were installed in "oases," small patches of surviving or colonizing plants and riparian zones, and were sampled with transects consisting of 25 traps at 5-m spacing.

Most of the extensive sites were established to monitor long-term trends of mammal succession and population dynamics and were trapped at regular intervals (e.g., 1981 to 1984, 1987, 1990, 1995, and 2000). "Spot" sampling sites were used to compare mammal composition and species abundances between matrix sites and "oases" and between riparian and upland habitats and were generally trapped for 1 or 2 years, but some were trapped much longer (e.g., 17 of 23 years). Over the years, some plots were lost and replaced because of timber harvest or salvage logging (MacMahon et al. 1989).

Small mammals were trapped with Sherman live traps baited with rolled oats and peanut butter. Cotton batting was placed in each trap as bedding material and to prevent hypothermia. Trapping sessions were 3 to 5 consecutive days for grids and webs and from 1 to 4 days for transects. Sites were sampled between one and four times annually (from June to October). A session always occurred during midsummer (late July or August). Traps remained open for the entire trapping session and were checked each morning. Northern pocket gophers (*Thomomys talpoides*) were sampled with traps constructed from PVC (polyvinyl chloride) pipe that were placed in their burrow systems (Baker and Williams 1972). Captured animals were identified to species, uniquely marked (ear tag, toe-clip, or colored nail polish on fur), weighed, examined to determine sex and reproductive condition, and released. Mammal voucher specimens from trap mortality were deposited in the Museum of Southwestern Biology, University of New Mexico, Albuquerque, New Mexico.

While implementing other projects at Mount St. Helens from 1980 to 2002, we compiled incidental observations on mammals and recorded detailed accounts of species, location, date, habitat type, behavior, and activity. These data were of particular importance for species not effectively captured in Sherman traps (e.g., mustelids and mid- to large-size rodents).

Our taxonomy follows the Integrated Taxonomic Information System online database, http://www.itis.usda.gov. Data were retrieved January 5, 2004. The taxonomy of *Peromyscus* spp. in the Pacific Northwest has received considerable attention, resulting in taxonomic revisions during the past two decades (Hogan et al. 1993). From 1980 through 1990, we did not distinguish between *P. maniculatus* and *P. oreas*, and these species were collectively treated as *Peromyscus* spp. Jones et al. (1997) dropped *P. oreas* and accepted *P. keeni* following Hogan

TABLE 14.1. Mean cover values (%) for several habitat variables measured during 1995 within each trapping web for old-growth forest and clear-cut sites in reference sites and in tephra-fall, blowdown, and pyroclastic-flow disturbance zones.

	Habitat variable						
Disturbance zone	Bryophyte	Herb	Shrub	Tree	Litter	Wood	Rock/ground
Reference:							
Clear-cut	14.2	50.2	26.1	6.8	29.9	7.3	1.8
Forest	23.3	1.5	12.8	34.1	76.3	14.1	0.0
Tephra-fall:							
Clear-cut	9.2	3.2	4.5	27.0	44.1	10.5	32.5
Forest	1.5	0.0	12.6	15.7	91.1	11.7	4.5
Blowdown:							
Clear-cut	5.7	8.9	10.0	0.0	18.5	4.1	64.1
Forest	11.5	26.9	1.9	0.0	12.7	28.6	12.4
Pyroclastic forest:	5.9	7.9	0.0	0.0	3.1	0.0	83.2

Values are based on ten 1-m^2 quadrats.

et al. (1993). In our 1995 to 2002 sampling, we recognized both *P. maniculatus* and *P. keeni*. Following Allard et al. (1987), adult mice with tails more than 96 mm long were classified as *P. keeni* and those with tails less than 96 mm were considered *P. maniculatus*. All subadult and juvenile mice were recorded as *Peromyscus* species.

In August 1995, we performed ocular estimates of percent coverage for vegetation within ten 1-m^2 quadrats by life form (bryophyte, herb, shrub, or tree), litter, wood, and bare ground/rock at each of the seven core trapping locations (Table 14.1). Quadrats were systematically placed at 5-m intervals along one of the trapping lines. Because these measurements were not repeated across trapping years and did not adequately cover the trapping web, they cannot be used as a rigorous assessment of habitat among the sites. Rather, they provide measures for several habitat features important to small mammals that aid in the interpretation of our results.

14.2.3 Data Analysis

We used data from trap sites and incidental sighting records to compile a master mammal species list to assess the rate and pattern of mammal community reassembly for each disturbance zone. We used trapping data (composition and abundance) from a subset of the "spot" sampling sites at the pyroclastic-flow and blowdown zones to assess the importance of patch types (habitat mosaic) on species relative abundance and assemblage structure. We used mammal presence/absence data at five long-term trapping sites to describe the colonization, turnover, and persistence of mammal species. We used a core set of permanent plots where intensive and repeated sampling occurred to compare the species composition and relative abundances of small mammals among disturbance zones, across habitat types, and through time. Recorded individuals for each species were plotted as a percentage of the total number of recorded individuals at each of seven sites for 8 survey years between 1981 and 2000. We performed intersite comparisons of small-mammal composition for our 2000 trapping data using Jaccard's (1908) similarity index (J), calculated as

$$J_{1,2} = a/(a + b + c)$$

where a = the number of species common to both sites 1 and 2, b = the number of species unique to site 1, and c = the number of species unique to site 2.

This formula was used to create resemblance matrices by comparing pairs of values for species composition for seven long-term monitoring sites representing the disturbance gradient and for "spot" sampling sites in the pyroclastic-flow and blowdown zones; trapping data from 2000 and 2002 were used. To evaluate the hypotheses that the Keen's deer mouse is associated with high habitat complexity (forest habitats) and that the deer mouse is a generalist associated with many habitat types (including those that are highly disturbed), we plotted the percentage of individual captures for each species across the disturbance gradient. To assess if our data conform to the expectations of the intermediate-disturbance hypothesis (Connell 1978), we plotted mammal species richness for sites across the volcanic-disturbance gradient.

14.3 Rcsults and Discussion

14.3.1 Comparison of the Mount St. Helens Small-Mammal Assemblage with the Regional Fauna

There are 35 small- and medium-size mammal species (excluding bats) indigenous to the Southern Washington Cascade Range (Dalquest 1948; Ingles 1965; Wilson and Ruff 1999). The presence and areal extent (both total coverage and juxtaposition) of habitat types in the landscape dictate, to a large degree, the distribution of these small mammals. Food and cover availability, parasites, predators, and weather influence their demography. Species-specific habitat preferences vary;

TABLE 14.2. Small- and medium-sized mammal species observed or trapped from 1980 through 2002 in areas disturbed by the May 18, 1980, eruption of Mount St. Helens and for reference sites.

			Disturbance zone (year of first observation)			
Order/family	Species (common name)	Habitat	Reference	Tephra-fall	Blowdown	Pyroclastic
Didelphimorphia:						
Didelphidae	*Didelphis virginiana* (Virginia opossum)	for			1998	
Insectivora:						
Soricidae	*Sorex monticolus* (montane shrew)	for	1981	1980	1982	1986
	Sorex palustris (northern water shrew)	rip, wat	2000		2000	1994
	Sorex trowbridgii (Trowbridge's shrew)	for	1983	1981[a]	1981[a]	1995
	Sorex vagrans (vagrant shrew)	for	1995	1984	1984	
Talpidae	*Neurotrichus gibbsii* (shrew-mole)	for	1982	1983	1983	
	Scapanus orarius (coast mole)	med			1982	
Lagomorpha:						
Leporidae	*Lepus americanus* (snowshoe hare)	for	1995	1980	1987	1983
Ochotonidae	*Ochotona princeps* (American pika)	scr		1992	1990	1994
Rodentia:						
Aplodontiidae	*Aplodontia rufa* (mountain beaver)	for	1982	1982	1999	
Castoridae	*Castor canadensis* (American beaver)	rip, wat	1981	1982	1999	
Erethizontidae	*Erethizon dorsatum* (common porcupine)	for			1997	
Geomyidae	*Thomomys talpoides* (northern pocket gopher)	med	1981	1980	1980	1992
Muridae	*Cletherionomys gapperi* (southern red-backed vole)	for	1982	1980		
	Microtus longicaudus (long-tailed vole)	for		1982	1984[b]	1983
	Microtus oregoni (creeping vole)	for	1982	1990	1983	1995
	Microtus richardsoni (water vole)	rip				1988
	Neotoma cinerea (bushy-tailed woodrat)	scr		1992		
	Ondatra zibethicus (muskrat)	rip, wat				1999
	Peromyscus keeni (Keen's mouse)	for	2000	2000	2000	1994
	Peromyscus maniculatus (deer mouse)	gen	1981	1980	1981	1982
	Phenacomys intermedius (heather vole)	med	1982	1983	1983	
Sciuridae	*Glaucomys sabrinus* (northern flying squirrel)	for	1984	1984		
	Marmota caligata (hoary marmot)	med			1983	
	Spermophilus saturatus (Cascade golden-mantled ground squirrel)	for		1980	1983	1986
	Tamius amoenus (yellow-pine chipmunk)	for	1982	1980	1982	1988
	Tamius townsendii (Townsend's chipmunk)	for	1982	1980		
	Tamiasciurus douglasii (Douglas's squirrel)	for	1982	1980	1991	1986
Dipodidae	*Zapus trinotatus* (Pacific jumping mouse)	for	1982	1982	1982	1991
Carnivora:						
Mustelidae	*Lontra canadensis* (northern river otter)	rip			1989	
	Martes americana (American marten)	for		1989		
	Mustela erminea (ermine)	for	1981	2000	1982	
	Mustela frenata (long-tailed weasel)	for	1982			1994
	Mustela vison (American mink)	rip		1980	1982	
Total number of species:		21	25	26	17	

Disturbance zones are reference (no volcanic impact), tephra-fall, blowdown, and pyroclastic-flow. Habitat codes are forest (for), generalist (gen), meadow (med), riparian (rip), scree (scr), and water (wat).

[a] Source: Steve West, University of Washington, Seattle, Washington; personal communication.

[b] Source: Adams, A.B., K.E. Hinckley, C. Hinzman, and S. R. Leffler. 1986a. Recovery of small mammals in three habitats in the northwest sector of the Mount St. Helens National Volcanic Monument. Pages 345–358 *in* S.A.C. Keller, editor. Mount St. Helens: Five Years Later. Eastern Washington University Press, Cheney, Washington, USA.

and although most species attain highest densities in certain habitat types, few are obligate habitat specialists. Broadly classified, about 22 species are associated with variously aged and structured forests, 7 are associated with riparian and open-water areas, 4 are found primarily in meadows, and 2 occur in rocky areas (scree, talus, or outcrops) (Table 14.2). Species associated with water, meadows, and rock have narrower habitat breadths than most forest-dwelling species and tend to be patchily distributed.

Given the extent and diversity of forest, meadow, and aquatic habitats that were present in the preeruption landscape, it is likely that most, perhaps all, of the 35 mammal species of the region occurred in the 600-km^2 area that received severe volcanic impact.

In what was perhaps the most comprehensive survey of small mammals in the region, West (1991) sampled 45 naturally regenerating forest stands of different ages and 8 clear-cuts in the Washington Cascade Range using a variety of sampling techniques, including visual surveys for squirrels and other diurnal species and snap-trap and pitfall-trap grids. West (1991) listed 20 rodent, lagomorph, and insectivore species, with all but one having the potential to occur in the vicinity of Mount St. Helens. The exception, the marsh shrew *Sorex bendirii*, occurred only at elevations lower than our study area at Mount St. Helens. A comparison of the species list reported by West (1991) with the species list compiled from our studies (see Table 14.2) revealed that all 19 of West's small mammals from the elevation ranges of Mount St. Helens were observed or collected between 1981 and 2000. We also recorded one additional lagomorph, the snowshoe hare (*Lepus americanus*), and 7 rodent species not listed by West (1991). The rodents included the mountain beaver (*Aplodontia rufa*), hoary marmot (*Marmota caligata*), Douglas's squirrel (*Tamiasciurus douglasii*), American beaver (*Castor canadensis*), bushy-tailed woodrat (*Neotoma cinerea*), muskrat (*Ondatra zibethicus*), and common porcupine (*Erethizon dorsatum*). In addition, we included in our study the occurrences of some taxa not surveyed by West (1991) [e.g., the Virginia opossum (*Didelphis virginiana*) and several species of small carnivores (Mustelidae; see Table 14.1)]. The 34 species recorded during our studies (see Table 14.2) appeared to comprehensively represent the small-mammal fauna of the region.

14.3.2 Small-Mammal Species Relative Abundance

We captured 5069 mammals, comprising 3301 individuals of 23 species, at 40 survey sites between 1981 and 2002. Five of the 23 species trapped accounted for 81% of all captures. In order of numerical abundance, these were the deer mouse (includes *Peromyscus maniculatus* and *P. keeni*), southern red-backed vole (*Clethrionomys gapperi*), yellow-pine chipmunk (*Tamias amoenus*), and montane shrew (*Sorex monticolus*). If 3 additional species [Cascade golden-mantled ground squirrel (*Spermophilous saturatus*), Pacific jumping mouse (*Zapus trinotatus*), and Townsend's chipmunk (*Tamias townsendii*)] are included, 92% of all recorded individuals are represented. The numerical dominance by a handful of species is typical of small-mammal communities of the region (West 1991; Carey and Johnson 1995; Gitzen and West 2002).

14.3.3 Cast of Characters

14.3.3.1 Survivors

The presence of survivors following disturbance greatly influences the rate and pattern of biological succession. Given the extreme but variable nature of the 1980 eruption, it was anticipated there would be areas of high survival as well as zones of complete mortality. Our initial studies (1980 to 1983) were conducted to document patterns of mammal survivorship across the disturbance gradient and in localized habitat patches of surviving and newly developed plant communities. We predicted mammal survival would be inversely related to disturbance intensity and influenced by species life-history traits. Survival was hypothesized to be high in the tephra-fall zone, limited to fossorial species in blowdown and tree-removal zones, and absent from the pyroclastic-flow zone.

Early trapping efforts indicated that 14 species of small mammals survived the eruption (Andersen and MacMahon 1985b). Survival was inversely related to disturbance intensity within the four volcanic disturbance zones studied: tephra-fall zone (11 species), blowdown zone (8 species), tree-removal zone (2 species), and pyroclastic flow (0 species). Thermal extremes, burial, scour, and abrasion precluded survival of any mammal in the pyroclastic-flow zone. Survival in the tephra-fall zone was widespread, and mammal species composition did not substantially differ from reference sites.

In the blowdown zone, species strongly associated with well-developed forest structure were not trapped. This group included two canopy-dwelling species [northern flying squirrel (*Glaucomys sabrinus*) and Douglas's squirrel] and the southern red-backed vole, a forest-floor species. Mammals surviving in the blowdown zone were generalists (deer mouse, yellow-pine chipmunk, and montane shrew), fossorial species [northern pocket gopher and coast mole (*Scapanus oreas*)], or species associated with early seral stages [Pacific jumping mouse, long-tailed vole (*Microtus longicaudus*), and the shrew-mole (*Neurotrichus gibbsii*), a species associated with forest soil and duff]. In the tree-removal zone, the only surviving species were the deer mouse and the northern pocket gopher. Andersen (1982) reported widespread survival of the northern pocket gopher in the blowdown and tephra-fall disturbance zones. He attributed the success of this species to its fossorial habitats, which buffered it from the volcanic forces. It also appeared that fossorial traits led to the survival of the coast mole and the shrew-mole.

We speculate that seasonal and diel timing of the eruption influenced survival in three important ways:

- Remnants of the winter snowpack were still present on north-facing slopes and in areas of wind deposition, which buffered subnivean-dwelling species from the volcanic forces. These areas of late-lying snow later became "oases" of vegetation in a matrix of relatively barren landscape.
- Some mammals (Pacific jumping mouse and chipmunks) were likely still within their subterranean winter hibernacula.
- Nocturnal animals had likely returned to their burrows or belowground retreats by the time the first eruption occurred early in the morning.

In addition, topography influenced survival as volcanic forces were impeded or deflected by ridgelines, enhancing survival on the lee of ridges. Collectively, these factors resulted in more species surviving in the blowdown and tree-removal zones than we had hypothesized. Our predicted patterns of survival in the tephra-fall and pyroclastic-flow zones were validated.

In the blowdown zone, large mammals, including elk (*Cervus elaphus*), black-tailed deer (*Odocoileus hemionus columbianus*), mountain goat (*Oreamnos americanus*), American black bear (*Ursus americanus*), and mountain lion (*Puma concolor*), sustained complete mortality. Midsized animals, such as the snowshoe hare, mountain beaver, common porcupine, hoary marmot, American beaver, American marten (*Martes americana*), American mink (*Mustela vison*), weasels (*M. frenata* and *M. erminea*), coyote (*Canis latrans*), and bobcat (*Lynx rufus*), likely perished during the eruption or departed shortly after because of the absence of requisite food and cover.

14.3.3.2 Colonizers

Reestablishment of mammals in defaunated areas required immigration from source populations. Sources could be individuals (or their progeny) that survived within the disturbed matrix or individuals which emigrated from undisturbed areas located beyond the volcanic impact zones. Dispersal ability, distance to the source population, and the habitat preferences of the species are factors most important in determining the pattern and rate of colonization. We predicted that generalist species of open habitats with herbaceous and shrubby vegetation would be the primary early colonists and that species associated with forest canopy or other habitat specialists would not colonize until habitat genesis occurred, a process that could take several decades or longer. In addition, we predicted that larger, more-mobile species would colonize first and that fossorial or small species (i.e., shrews and American shrew-mole) would be slow to colonize defaunated areas, such as the pyroclastic-flow zone.

Consistent with our predictions were the following:

- Generalist species of open and shrubby terrain, such as the deer mouse, yellow-pine chipmunk, Cascade golden-mantled ground squirrel, and the Pacific jumping mouse, were trapped at most or all sites in blowdown, tree-removal, and pyroclastic-flow zones.
- Establishment of northern pocket gophers in the pyroclastic-flow zone was a relatively prolonged process, taking 12 years to occur.
- Forest-canopy species (northern flying squirrel, Townsend's chipmunk, and Douglas's squirrel) and understory specialists (southern red-backed vole) have not colonized during the 22 years after the eruption.

Other predictions were not supported, including the colonization of forest insectivores [shrew-mole and Trowbridge's shrew (*Sorex trowbridgii*)] and two riparian specialists [northern water shrew (*Sorex palustris*) and Richardson's vole (*Microtus richardsoni*)] in some or all of the scorch, blowdown, and pyroclastic-flow zones (see Table 14.2).

From 1980 through 2002, nine species were observed in the blowdown zone and five species in the pyroclastic zone for which the establishment status is unknown. These species include the Virginia opossum, pika (*Ochotona princeps*), snowshoe hare, mountain beaver, hoary marmot, Douglas's squirrel, muskrat, common porcupine, long-tailed weasel (*M. frenata*), American mink, and northern river otter (*Lontra canadensis*) (Table 14.2). In some cases, species were represented by a single or few observations and probably were dispersing individuals. However, repeated observations of pika and the presence of several adjacent burrows of mountain beaver suggested that these species had colonized certain locations within the blowdown zone.

14.3.3.3 Species Accounts

Because of space constraints, it is not possible to chronicle the response of all mammal species that survived or colonized the Mount St. Helens landscape. Instead, listed below are the accounts for the six numerically dominant species that accounted for 90% of all captures from 1981 to 2002.

Deer mouse (includes *Peromyscus maniculatus* and *P. keeni*). The deer mouse was captured in all disturbance zones, occupied 88% (35/40) of all sites, and accounted for 52% (1730/3301) of all captures. The success and numerical dominance of the deer mouse in the posteruptive landscape were attributed to its habitat breadth, generalized diet, high fecundity, and vagility.

Southern red-backed vole (*Clethrionomys gapperi*). Red-backed voles were the second most abundant species trapped, representing 9.8% (323/3301) of all captures. Despite their abundance, voles were captured at just six sites in the reference areas, tephra-fall zone, and scorch zone. The single vole captured in the scorch zone in 2000 was likely a dispersing individual. The lack of southern red-backed voles in our surveys of the blowdown, tree-removal, mudflow, and pyroclastic-flow zones likely reflects the absence of requisite forest habitat rather than dispersal distance. As vegetation complexity and litter increase during the next decades, southern red-backed voles will likely establish.

Yellow-pine chipmunk (*Tamias amoenus*). We captured yellow-pine chipmunks in all disturbance zones and at 50% (20/40) of the sites trapped. They accounted for 9.4% (309/3301) of captures. Data from one tephra-fall site (Butte Camp) indicated that about 50% of the adults captured in 1984 persisted to late summer of 1987, suggesting that much of the population is capable of lasting 4 to 5 years.

Large quantities of downed wood, along with surviving and colonizing plants in the scorch and blowdown zones, provided food resources and cover that enabled the yellow-pine chipmunk to proliferate. The removal of the forest in the tree-removal and pyroclastic-flow zones created a barren landscape, and, within 15 years, small patches of shrubs developed. In

2002, yellow-pine chipmunks were trapped in 100% (5/5) of these habitat patches.

Montane shrew (*Sorex monticolus*). Montane shrews were captured at 45% (18/40) of the sites, represented 9.2% (305/3301) of captures, and occurred at reference sites and at tephra-fall, scorch, blowdown, tree-removal, and pyroclastic-flow zones but not at any mudflow site. The absence of montane shrews on mudflow sites probably relates to a lack of dense patches of vegetation and downed wood rather than to dispersal limitations. In the blowdown and scorch zones, montane shrews were captured in areas with high levels of woody debris or high herb coverage. In the pyroclastic-flow zone, montane shrews were captured only within riparian habitats. The presence of this small 4- to 8-g animal at several sites in the pyroclastic-flow zone indicates a remarkable dispersal capacity. Known or assumed source populations indicate a dispersal distance of several kilometers.

Cascade golden-mantled ground squirrel (*Spermophilus saturatus*). This ground squirrel was captured at 25% (10/40) of our trapping sites, including sites in the tephra-fall, mudflow, blowdown, tree-removal, and pyroclastic-flow zones, and represented 4.6% (152/3301) of captures. The numerous blowndown trees and stumps, together with surviving and colonizing vegetation in the blowdown zone, created suitable habitat that was colonized by them in 1987. By 1988 and 1996, respectively, they had colonized sites in the tree-removal and pyroclastic-flow zones where shrub communities developed, indicating that dispersal distance was not a limiting factor. This species will become increasingly abundant in response to the expansion of shrub communities.

Pacific jumping mouse (*Zapus trinotatus*). Pacific jumping mice were captured at 47.5% (19/40) of our sites and were 4.0% (133/3301) of the mammals trapped. Jumping mice were captured in every disturbance zone except the mudflow but were most frequent and abundant in riparian habitats. During 2000 sampling in the blowdown zone, jumping mice were captured at 100% (4/4) of the riparian sites and in 25% (1/4) of the upland sites. During 2002 sampling in the pyroclastic-flow zone, they were captured at 100% (5/5) of the riparian sites and 20% (1/5) of the upland sites. Jumping mice dispersed several kilometers to colonize sites once suitable riparian habitat developed. This species will become more abundant during the next decades as shrub cover increases along watercourses and in the uplands.

14.3.4 Changes in Small-Mammal Species Composition Through Time

Ecological theory of disturbance impacts and postdisturbance successional processes led us to predict that differences in the small-mammal assemblages across the volcanic disturbance gradient would reflect the quality, quantity, and distribution of suitable habitats that would provide required food and shelter. We also predicted that temporal changes in posteruption habitat characteristics would be accompanied by concomitant changes in small-mammal species composition and abundance. For example, as soil, vegetation, and invertebrate communities in the highly disturbed areas developed through succession from open meadows to closed-canopy forests, small-mammal assemblages also would change, from meadow-dwelling to forest-dwelling species. To test these predictions, we compared the temporal dynamics of species composition and relative dominance of each species at each of the study sites that represented four levels of volcanic disturbance (Figures 14.3, 14.4). The presentation of these results is organized first by forest or clear-cut habitats (preeruption conditions) and then by disturbance level (undisturbed "reference" zone, tephra-fall zone, blowdown zone, and finally the pyroclastic-flow zone). The former comparison highlights the effects of predisturbance habitat characteristics on postdisturbance community development. The latter comparison highlights the effect of disturbance intensity on posteruption mammal community assembly.

14.3.4.1 Small-Mammal Assemblages in Forest Sites

The reference forest site, first sampled in 1982, was dominated throughout the posteruption period by red-backed voles, deer mice, Townsend's chipmunks, shrew-moles, and shrew species (probably Trowbridge's shrew), although relative abundances fluctuated from year to year (see Figure 14.3). Additional species were recorded on the site (Figure 14.3). Total relative abundances of these small mammals were highest during the early 1980s (perhaps associated with periods of high precipitation; see MacMahon et al. 1989 and Swanson and Major, Chapter 3, this volume), followed by a precipitous decline by 1987. Relative abundances then increased in 1990 through 2000.

As in the reference forest, the small-mammal assemblage on the tephra-fall forest site was dominated by red-backed voles, deer mice, Townsend's chipmunks, and shrew species but differed in the early years following the eruption (see Figure 14.3). The shrew-mole, represented by a single capture (1983) at the tephra-fall site, was far less common than at the reference forest site. Only two species (five individuals) were captured in 1981, but species numbers and relative abundances increased steadily through 1983. As with the reference forest, the sample from 1987 was depauperate. By 1990 and continuing through 2000, small-mammal numbers and species were fairly comparable to those of the reference forest (Figure 14.3).

The blowdown-zone forest site supported only one species of rodent, the deer mouse, in 1981 (see Figure 14.3). The deer mouse continued to be one of the dominants through 1984. From 1987 to 2000, diurnal rodents (Cascade golden-mantled ground squirrel and yellow-pine chipmunk) became important components of the assemblage and during certain years (e.g., 1987, 1995, and 2000) were numerically dominant. As shrub and tree cover increased over the years, some forest-dwelling species began to appear, and, by 2000, a population of Trowbridge's shrew and northwestern mouse were flourishing on the

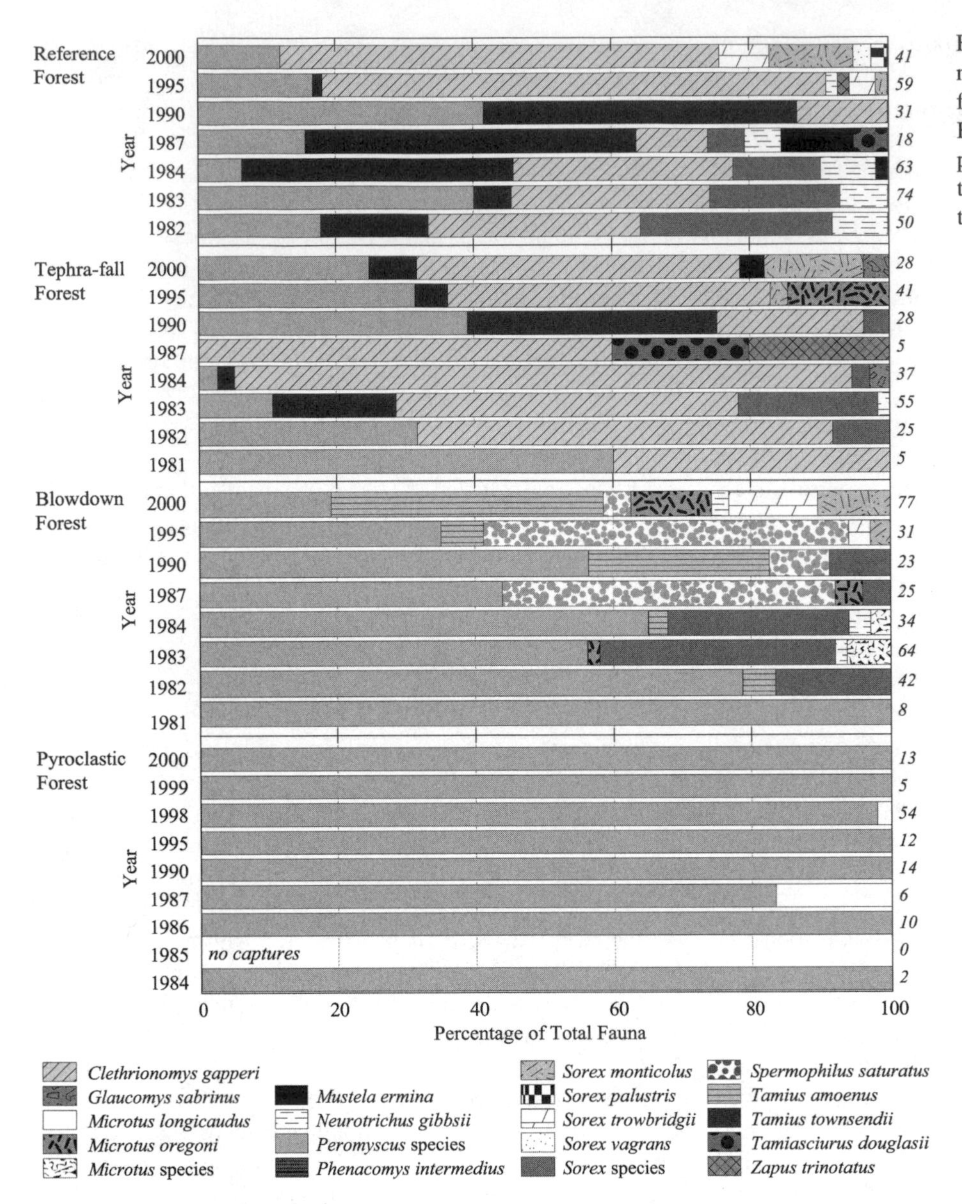

FIGURE 14.3. Relative abundances of small-mammal species trapped on representative forested sites in the vicinity of Mount St. Helens during late summer of each sampling year. *Values to the right of bars* are the total number of unique individuals captured. Sites are depicted in Figure 14.2.

site along with the creeping vole (*Microtus oregoni*) and the montane shrew (Figure 14.3). However, all the species characteristic of well-developed forests (i.e., Townsend's chipmunk, southern red-backed vole, and northern flying squirrel) failed to colonize this area by 2000.

The deer mouse dominated the small-mammal assemblage on the highly disturbed pyroclastic-flow site, albeit in low numbers in most years (see Figure 14.3). The 1998 sample exhibited a large increase, but numbers fell dramatically the following year. In addition to the deer mouse, two long-tailed voles were recorded (one each in 1987 and 1998); these individuals were apparently dispersing through the study site because no other members of the species have been collected since then (although other individuals have been recorded in nearby isolated patches of vegetation within the pyroclastic flow; see discussion following). During the 2002 sampling, three additional species were present for the first time. One individual each of yellow-pine chipmunk and Pacific jumping mouse was trapped, and numerous fresh mounds of the northern pocket gopher were present.

14.3.4.2 Small-Mammal Assemblages in Clear-Cut Sites

Although many of the same species recorded in the forested sites also occurred in the clear-cut sites, the dominant species in clear-cuts were those preferring meadow-type habitats. However, during the first 20 posteruption years, vegetation on the clear-cuts continued to develop, and, by 2000, portions of the reference and tephra-fall clear-cut sites had patches of dense tree canopy. Small areas of habitat were thus provided that were very similar to older forest stands, and these patches were eventually occupied by typical forest-dwelling small-mammal species. For example, in the reference clear-cut site,

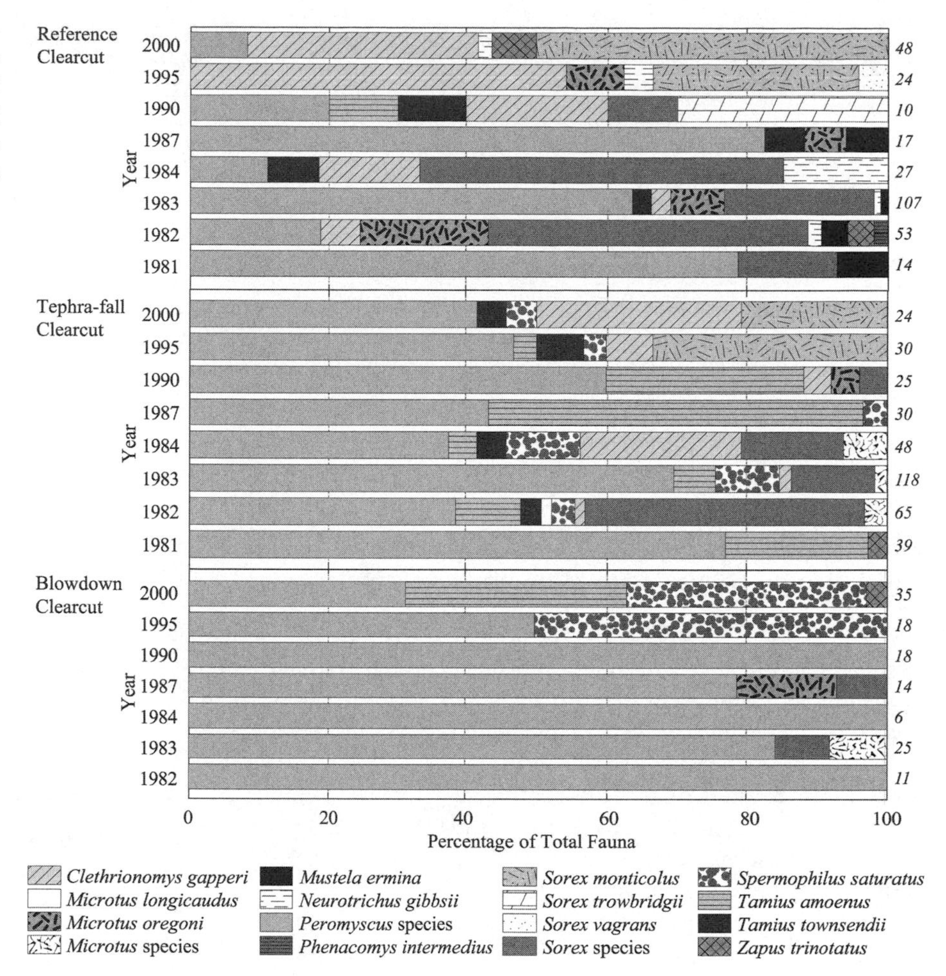

FIGURE 14.4. Relative abundances of small-mammal species trapped on representative preeruption clear-cut sites in the vicinity of Mount St. Helens during late summer of each year. *Values to the right of bars* are the total number of recorded individuals. Sites are depicted in Figure 14.2.

the small-mammal assemblage in the early 1980s was dominated by deer mice and shrew species (presumably the montane shrew) (see Figure 14.4); however, by the 1990s and 2000, forest-dwelling species, such as red-backed voles, became abundant, and the Keen's mouse and Trowbridge's shrew were present. Thus, the reference clear-cut site displayed a pattern of change consistent with expected successional responses from clear-cutting and could be used as a baseline comparison site for concomitant changes on clear-cuts that were affected by the eruption of Mount St. Helens.

On the tephra-fall clear-cut, both species composition and temporal change in the mammal assemblage were similar to those of the reference clear-cut except that the yellow-pine chipmunk was more abundant on the tephra-fall clear-cut (see Figure 14.4). The reduction in vegetation cover induced by the deposition of tephra provided excellent habitat for the yellow-pine chipmunk, which was a conspicuous component of the site from 1981 through 1990. However, as shrub and tree-canopy cover increased between 1990 and 2000, yellow-pine chipmunks dwindled in number while red-backed voles and montane shrews increased. Another forest-dwelling species, Townsend's chipmunk, was recorded occasionally in the early 1980s, and again in 1995 and 2000, but did not appear to have established populations within the clear-cut (Figure 14.4). Deer mice remained a substantial component of the assemblage throughout the study period. Thus, the impact of tephra deposition appeared to be limited to the initial vegetation alteration that favored the yellow-pine chipmunk; the remaining small-mammal species exhibited changes that paralleled the reference clear-cut assemblage.

In contrast to the reference and tephra-fall clear-cut study sites, the blowdown clear-cut site differed markedly from the paired forest site. Although both forest and clear-cut sites were dominated by deer mice shortly after the eruption, recruitment to the clear-cut was slower, and overall relative abundances of individuals were lower throughout the 20-year study period (compare Figure 14.4 with Figure 14.3). The deer mouse was the most abundant species from 1982 to 1990. By 1995, diurnal squirrels, including the Cascade golden-mantled ground squirrel and the yellow-pine chipmunk, colonized the site and gained numerical dominance. As such, the small-mammal assemblage on the blowdown clear-cut in 2000 appeared to be most similar to the assemblage found on the tephra-fall clear-cut in the early 1980s. In contrast, the blowdown clear-cut site supported both more species and individuals than did the pyroclastic-flow site (compare Figures 14.3 and 14.4).

TABLE 14.3. Similarity coefficients and numbers of shared small-mammal species in 2000 among study sites across the Mount. St. Helens disturbance gradient.

	Reference		Tephra-fall		Blowdown		
	Forest	Clear-cut	Forest	Clear-cut	Forest	Clear-cut	Pyroclastic flow
Reference:							
Forest	—	33	33	30	25	9	0
Clear-cut	3	—	33	44	36	33	14
Tephra-fall:							
Forest	3	3	—	44	15	9	0
Clear-cut	3	4	4	—	45	44	29
Blowdown:							
Forest	3	4	2	5	—	50	22
Clear-cut	1	3	1	4	5	—	33
Pyroclastic flow:	0	1	0	2	2	2	—

Upper-right portion of matrix shows percentage species similarity values; lower-left portion shows numbers of shared species.

14.3.5 Influence of Preeruption Conditions and Volcanic Disturbance Type on Developing Mammalian Communities

Comparisons of paired study plots indicate the extent to which the mammal assemblages are similar across the study plots (Table 14.3). We present these results with two caveats: (1) a difference of one species shared between two sites makes a great difference in the similarity coefficient (SC) because no pair had more than five common species, and (2) SC values are based on data from a single site from each disturbance zone and habitat type (clear-cut or forest).

The most similar pair is the blowdown forest/blowdown clear-cut (SC = 50) (see Table 14.3). One might expect that reference forest/clear-cut (SC = 33) would be the most similar because, even if the site had been clear-cut in the 1960s, a young fir plantation with a complex shrub and herb layer had developed since that time. This comparison suggests that the initial intensity of volcanic disturbance, or the lingering effects of such disturbance (e.g., tephra layer), is more important in influencing the emerging communities than the preeruption condition of a site. To examine this possibility, we looked at the similarities of all forest types and all clear-cuts across the gradient of disturbance. For forests, the SCs are, respectively, 33, 25, and 0, and for clear-cuts 44, 33, and 14. Similarity decreases as intensity of disturbance increases in both instances. This trend is further suggested by the fact that the pyroclastic flow (most severe disturbance) has the lowest overall similarity (average SC = 14.7) to other sites than any other study area, and the tephra-fall forest (least severe disturbance) has a higher overall SC (average = 22.3).

The apparent anomaly of the high similarity of blowdown forest and clear-cut might be explained as follows. A clear-cut of any sort is different from the forest because of a loss of vertical structure; mammals respond to such structural changes in various ways. Some species are lost, some increase, and some are unaffected in clear-cuts as compared to forests (Medin 1986). Toppling of trees by the volcanic blast changed the vertical structure of a site to more nearly resemble a clear-cut, except that there is more downed wood than in a clear-cut. Thus, two very different influences cause a convergence in mammalian community composition. Similar structural consequences of the blast and clear-cutting (great reduction of vertical structure) together may have caused the convergence of mammalian communities in this particular posteruption landscape.

14.3.6 Landscape Mosaic

The posteruption landscape can be viewed as a mosaic of habitat patches. At the largest spatial scale (tens to hundreds of square kilometers), disturbance zones emerged as distinct landscape features. In turn, each of these zones formed a matrix containing a number of smaller (tens to hundreds of square meters) patches of relatively favorable habitat. The most conspicuous patches were associated with small areas of surviving vegetation and the riparian zones of lakes and streams. Collectively, these were termed habitat "oases." The composition and structure of oases were complex and starkly contrasted with the relatively barren matrix. Oases comprised less than 1% of the total area of each disturbance zone (Figure 14.5). Because oases were structurally complex, we hypothesized that they would support more mammal species and higher relative abundances compared to their associated matrix. If true, oases would be biological hot spots that could be important sources for the reestablishment of mammals across the volcanic landscape. Moreover, it was expected that oases would harbor riparian specialists (northern water shrew and Richardson's vole) that would be absent from matrix habitats. Finally, we anticipated that the ecological importance of oases would be high relative to their areal extent in the landscape.

14.3.7 Species Richness, Relative Abundances, and Similarity Between Oases and Matrix Habitats

To test these hypotheses, we established paired transects in oases and matrix habitats in the pyroclastic-flow and blowdown zones. In the pyroclastic-flow zone, small mammals were

FIGURE 14.5. Aerial photograph (1:6000) of the pyroclastic-flow zone taken in August 1996. *Inset* shows close-up of verdant and structurally complex oasis habitat (*circled in white*) embedded within the relatively barren pyroclastic-flow zone.

trapped on four pairs of transects during 2000 and 2002. Results on the pyroclastic-flow zone were similar for the sampling periods of 2000 and 2002. In both years, total and mean species richness were substantially greater in oases than in the matrix (Table 14.4). Relative abundances (mean number of individuals/100 trap-nights) were also greater in the oases than in matrix habitats (Table 14.4).

Oases on the pyroclastic-flow zone were verdant, willow/herb communities that established around springs and streams. The ample moisture and dense foliage presumably provided food, cover, and nesting sites suitable for nine species of small mammals. The mammal assemblage consisted of a diverse group of riparian specialists (e.g., northern water shrew and Richardson's vole), rodents (e.g., Pacific jumping mouse, yellow-pine chipmunk, and deer mouse), and insectivores (shrews) associated with early seral habitats. In contrast, the matrix habitat was largely dry and barren and supported three species, the ubiquitous deer mouse (96% of captures), a single Pacific jumping mouse, and the northern pocket gopher. The mean similarity (Jaccard's index) among oasis sites and among matrix sites was high, indicating that there was a similar core set of species associated with each habitat (see Table 14.4). However, matrix and oasis habitats were dissimilar (Table 14.4). The average mean percentage of the total fauna that

TABLE 14.4. The total and mean number, catch per unit effort, mean percent similarity, and average mean percent similarity of small mammals from matrix and oasis habitats on the Pumice Plain at Mount St. Helens during July 2000 and September 2002 and in the blowdown zone during July 2000.

Disturbance zone	Habitat	Year	Number of sites	Total number of species	Mean number of species	CPU (catch per unit effort)	Mean percent similarity	Average mean percent of total fauna	Oasis-matrix percent similarity
Pyroclastic flow	Oases	2000	4	9	5.3 (1.5)	32.0 (8.5)	54.7 (9.4)	58.3 (17.1)	11
	Matrix	2000	4	1	1.0 (0.0)	8.5 (0.7)	100.0 (0.0)	11.0 (0.0)	
	Oases	2002	4	9	6.0 (0.8)	23.0 (10.1)	54.2 (14.4)	60.0 (8.2)	20
	Matrix	2002	4	3	1.8 (0.9)	6.0 (2.5)	55.5 (25.3)	17.5 (9.6)	
Blowdown	Oases	2000	4	10	6.0 (1.4)	19.5 (13.1)	43.5 (8.6)	60.0 (18.3)	50
	Matrix	2000	2	5	4.0 (1.8)	19.0 (12.7)	60.0 ([a])	40.0 (14.1)	

CPU, number of individual captures per 100 trap-nights; standard deviations are in parentheses.
Data for 2000 are shaded; data for 2002 were taken only on the pyroclastic flow.
[a] One comparison; no SD.

occurred at any oasis site was high compared to any matrix site (Table 14.4).

In 2000, we trapped mammals in four oases and two matrix habitats in the blowdown zone. Total and mean species richness values were greater in oases than in the matrix habitats (see Table 14.4). However, relative abundances were about the same (Table 14.4).

Oasis habitats sampled in the blowdown zone were stream riparian zones dominated by willows and herbs. The habitat structure and presence of water provided the resources to support a diverse assemblage of six rodents and four insectivores. Of these 10 species, 5 also occurred in the matrix habitat. The species unique to the oasis habitats included three insectivores (shrew-mole, northern water shrew, and Trowbridge's shrew) and the Pacific jumping mouse. No species was unique to the matrix habitat. The five mammals shared between oasis and matrix habitats were all species associated with early-successional habitats and included four rodents (northwestern mouse, yellow-pine chipmunk, Cascade golden-mantled ground squirrel, and northern pocket gopher) and the montane shrew. Blowdown-zone matrix habitat had a complex network of downed trees and abundant shrub and herb cover that provided the food, cover, and nesting requirements for these 5 species. As on the pyroclastic-flow zone, the mean similarity of the blowdown zone was fairly high for both oasis sites and matrix sites, indicating that there was a core set of species associated with each habitat. Matrix and oasis habitats had a 50.0% similarity to one another. The average mean percentage of the total fauna ($n = 10$ species) that occurred at any oasis site was higher than for any matrix site (see Table 14.4).

To assess the importance of oases across larger spatial scales, we combined all oases and matrix sites in both the pyroclastic and blowdown zones. The total species richness values for oases and matrix habitats were 15 versus 7 species, and the percent similarity between these habitat types was 46.7%.

Our results support the hypotheses that oases were habitats of greater mammal diversity, supported habitat specialists, and, at least in the pyroclastic-flow zone, had greater mammal relative abundances. Several studies addressing patterns of mammal-occurrence, species-richness, and relative-abundance values between riparian and upland sites have been conducted in the Cascade Mountains of Oregon and Washington. As was found with our results, Anthony et al. (1987), Doyle (1990), and West (2000) reported higher species-richness values in riparian areas compared to adjacent uplands. We found greater mammal abundances in riparian versus matrix habitat in the pyroclastic-flow zone, which is consistent with the findings of Doyle (1990) and Anthony et al. (1987). In the blowdown zone, the mammal abundances were largely equal between the riparian and oasis sites, which is similar to the findings of West (2000). Also consistent with our results is the affinity of certain species for riparian habitats. Each of the studies just cited detected species that had significantly higher abundances in riparian habitats and included a few shrew species, one or two voles, and the Pacific jumping mouse.

14.3.8 Spatiotemporal Variability in Small-Mammal Assemblages

Taken as a whole, this long-term study of small-mammal assemblages has revealed considerable variability in both species composition and abundances in space and time (see Figures 14.3, 14.4), a finding consistent with those of other workers (Diffendorfer et al. 1996). Two major factors contribute to this variability. First, a nonsystematic variation component results from annual fluctuations in abundance of each species. Second, an inherent systematic variation component is caused by successional changes in habitat characteristics and accumulations of recolonizing species. Both sources of variability must be considered when interpreting the 22 years of posteruption reassembly of the small-mammal assemblages.

To illustrate the variability in the sample record, species presence/absence data for five study sites have been arranged in chronological order (Figure 14.6). Two of these sites (Willow Springs in the pyroclastic-flow zone and Pumice Plain

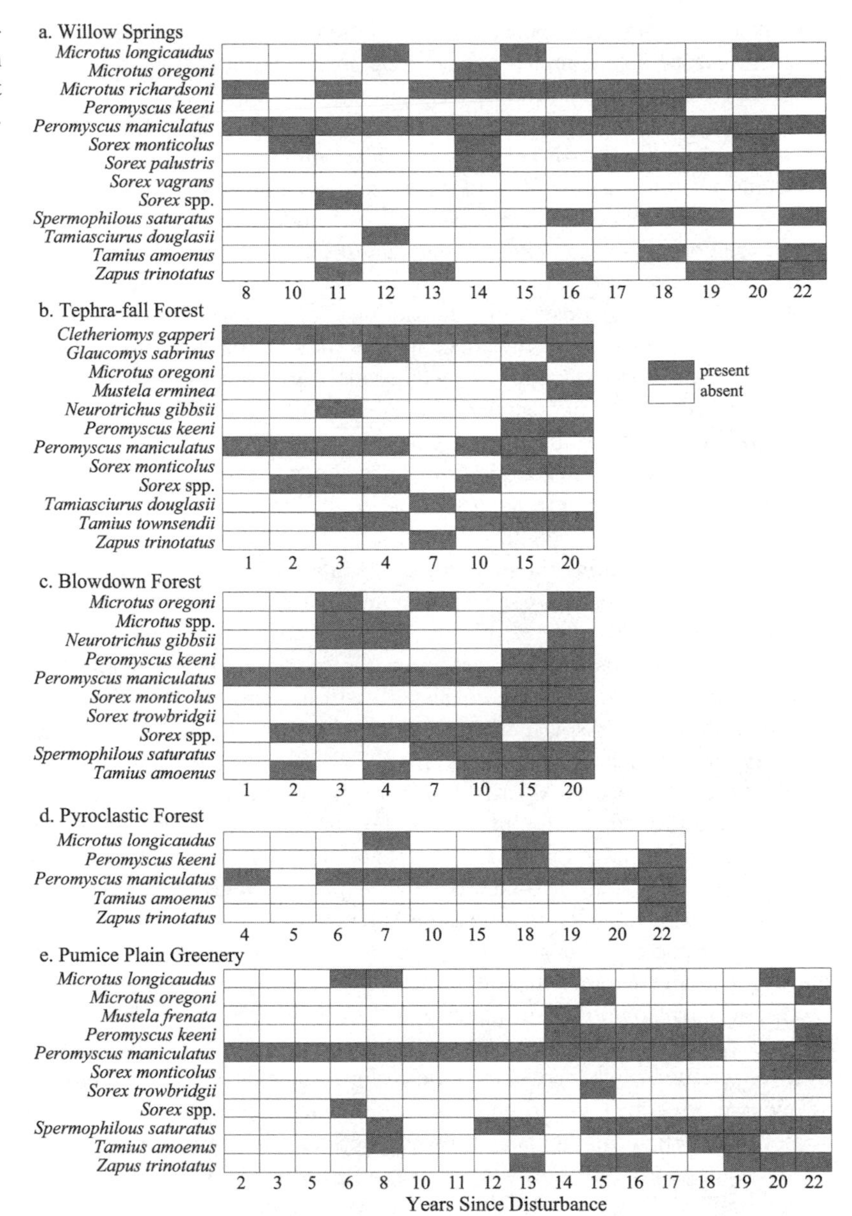

FIGURE 14.6. Temporal presence (*shaded*) and absence (*not shaded*) of small-mammal species from selected trapping sites at Mount St. Helens. Note that the species plotted and years trapped vary across sites. Sites are depicted in Figure 14.2.

Greenery in the tree-removal zone) represent relatively small, localized foci of dense vegetation in an extensive barren landscape. The other three sites are larger sampling areas (~5 ha of effective trapping area each) arbitrarily selected from a landscape matrix of similar habitat. Several species appeared to "come and go" between 1980 and 2000 while others were more consistent in their residency (see Figure 14.6). In an effort to understand these patterns, it is instructive to delve briefly into the details and assumptions of the field sampling methods used to collect these data.

Small mammals are generally secretive in habits, small in body size, and nocturnal. It is difficult, if not impossible, to enumerate every individual. Therefore, they must be "sampled" to determine their presence and relative abundance. The sampling process records only some of the animals present in a study area. Because of differences in behaviors, each species will have its own "detection probability" [i.e., the likelihood that an individual of that species would be successfully observed (trapped) during the sampling].

In our study, each study site (see Figure 14.6) was sampled with live-traps for several days during each sampling year. Given such sampling intensity, we undoubtedly missed some animals. Species present in low numbers increased the probability that our sample missed them. Even those species that are permanent, long-term residents of particular sites undergo large fluctuations in population size from year to year (e.g., Figures 14.3, 14.4) and may be missed when densities are low. This phenomenon, consisting of intermittent appearances and disappearances of species because of population levels so low that they escape detection, has been termed "pseudoturnover" (e.g., den Boer 1985). Pseudoturnover likely contributed to some of the nonsystematic variation in the patterns of species presence–absence observed in this study. The failure to detect a species when it was actually present at a site would result in inaccurate measures of species persistence and underestimate species richness. We do not know the extent of this inaccuracy in our study.

Another source of nonsystematic variation is dispersing individuals. Individuals of nonresident species may be trapped at a site as they disperse across the landscape. Upon release, they continue on in their search for suitable habitat, leaving the study area permanently. This phenomenon has the opposite statistical effect from "pseudoturnover" in that it artificially increases the estimates of resident species numbers occupying a particular site. Thus, given the nature of sampling limitations and logistical constraints in studying small mammals, some fluctuations in presence/absence data among the rarer species are expected.

In contrast to nonsystematic variation, the major changes observed (see Figure 14.6) are consistent with the predicted systematic variation associated with succession theory. In the severely disturbed environments at Mount St. Helens, one would expect to observe a general increase in species numbers through time, up to the limit of the regional pool of potential species available to colonize each habitat type. Populations of species that are particularly well adapted for a given habitat would be expected to have higher abundances in those habitats, and therefore those species would have a high probability of being recorded in each year's sample, yielding a more consistent record of occupancy within their preferred habitats. Ultimately, as habitat characteristics change through time (e.g., as tree canopy cover increases), dominance by other species would occur as the original residents become more constrained by their shrinking habitat resources. Such patterns of "relay succession" (see following discussion) have been observed only on the tephra-fall clear-cut site (see Figure 14.4), as illustrated by the decline of the yellow-pine chipmunk and rise of red-backed voles in concert with the increase in forest canopy cover.

14.4 Succession

14.4.1 Mammals and Successional Processes

In the presentation of our data, we have directly or obliquely involved a successional perspective. Here we step back and view the results of our fieldwork in the context of several succession paradigms. Our goals are these:

- To briefly consider how well changes in small-mammal community assembly following the 1980 Mount St. Helens eruption coincide with some ideas about succession derived from previous studies, usually based on plant data, and
- To compare our observations to those of ecologists who have studied plant response to these same disturbances at Mount St. Helens (del Moral and Bliss 1993).

Studies of mammalian community assembly are numerous, have occurred over a significant period (Smith-Davidson 1924), and often have a habitat-selection emphasis (M'Closkey 1975), where plant community or soil variables (Sly 1976) are the independent axes of analyses. Study sites have ranged from floodplains (Wetzel 1958) and swamps (Aldrich 1943) to forests (Andersen et al. 1980) and vary in duration from a year (Hirth 1959) to a few decades. Agents of disturbance include fire (Krefting and Ahlgren 1974; Fox 1982), mining (Verts 1957), agricultural cropping (Hirth 1959), and forest clear-cutting (Kirkland 1977; Ramirez and Hornocker 1981) among others.

Here, we use an old scheme (Clements 1916) to initiate discussion of mammalian succession at Mount St. Helens. Despite some shortcomings (McIntosh 1999), the Clementian perspective highlights specific processes that cause the phenomenon of succession and, thus, has heuristic value.

Nudation, the disturbance phase, is clearly of paramount importance in succession. The nature, timing, and extent of a disturbance determine what, if anything, remains of the original community. In some disturbances, such as the 1980 volcanic eruption at Mount St. Helens, some sites may be set back to a primary succession (e.g., those in the paths of pyroclastic flows) with regeneration depending entirely on migrants, whereas other sites regenerate based primarily on the surviving organisms (residuals). This process is especially obvious in the case of plants (del Moral and Bliss 1993) but is generally true for small mammals (see the section on Small-Mammal Assemblages in Forest Sites above). A surprise of our studies is the magnitude of residuals that occurred and their significance in creating the biotic mosaic of regeneration scenarios on the entire landscape (Andersen and MacMahon 1985a).

Migration from outside the volcanic landscape and from the epicenters of survivors in various protected sites was surprisingly rapid (MacMahon et al. 1989). Mammals were observed in the first year following the eruption, and even some of the most severely disturbed sites, distant from source populations, were occupied 6 years after the eruption (see Figure 14.6).

Ecesis, the phase of establishment, was highly variable. This phase refers to organisms not just surviving the volcanic events or being vagile enough to reach a site but also being able to breed and persist. Species such as shrews, Townsend's chipmunks, and Douglas's squirrels in the tephra-fall zone and deer mice in the tree-removal zone survived and were observed in 1980 but were not observed on the same sites in 1981 (MacMahon et al. 1989). In other areas, residuals like the northern pocket gophers did establish and begin colonization (see Figure 14.3) (Andersen 1982; MacMahon et al 1989). Yet other species (e.g., Douglas's squirrel) migrated to places like the pyroclastic-flow zone in years after 1985 but did not establish (see Table 14.2).

Reaction, the changes in the environment caused by organisms that create conditions that facilitate establishment of a different group of migrants or even residuals (those that have endured over time by some mechanism), was important at Mount St. Helens. Certainly, the presence of plants was a necessary precursor for the establishment of many small mammals. Plants provide both food and shelter. However, there are cases where the actions of residual herbivorous animals (e.g., northern pocket gophers) enhanced the establishment of plants even while the pocket gophers used belowground residual plant parts to sustain themselves (Andersen and MacMahon 1985b).

Biotic interactions, the entire suite of species–species interactions, can be positive, negative, or neutral for species in a sere. The development of mycorrhizal associations between fungi and plants undoubtedly enhanced establishment of some plants (del Moral and Bliss 1993; Allen et al., Chapter 15, this volume). Mammals we studied are herbivores or omnivores; thus, interactions such as competition, predator–prey, parasitism, or protocooperation (each mammal involved in the interaction benefits more or less equally) are likely to occur. Obviously, the interactions between the mammals and plants can be partitioned, with some imagination, into these categories. We did not observe competition with our study techniques. In other successional studies of small mammals, some workers see competition (Diffendorfer et al. 1996), and some do not (Pearson 1959). We did not note animal–animal protocooperation. Predator–prey relationships undoubtedly occurred between some of our insectivorous species and invertebrates and carnivores and other mammals. Nonetheless, our data do not show effects of these interactions on the direction of mammalian succession.

An interesting biotic interaction that probably occurred involved small mammals that eat arthropods, such as insects and spiders. We thought that plants would be required to attract arthropods for these species, thus allowing mammals to establish. An unexpected food source may have been important, at least initially, and may explain our observations of shrews establishing in areas of few plants. Airborne arthropod fallout, collected on the ground, consisted of up to 812 mg m^{-2} biomass over one summer (Crawford et al. 1995). Daily rates of arthropod deposition varied from 1.37 to 10.83 mg m^{-2} day^{-1} for two sites in the pyroclastic-flow zone over 3 years (Edwards and Sugg 1993). This potential food is probably a significant resource for populations of insectivorous small mammals.

14.4.2 Mount St. Helens Compared to Other Studies of Mammalian Succession

The literature is replete with studies of mammalian succession. We cannot summarize all these studies here, so we pick representative studies from different environments, provide some of their important conclusions, and comment on our experience.

Several studies suggest that age of successional sites is not important to mammals; rather, it is the vegetation, especially its physiognomy, assuming food is available, that drives the composition of the mammalian community (Huntley and Inouye 1987; Diffendorfer et al. 1996; Fox 1996; Swihart and Slade 1990). This trend is consistent with our observations and predictions (MacMahon 1981) and is evidenced by some species turnovers through time. In Scandinavia, *Microtus agrestis* replaces *Clethrionomys glareolous*, perhaps by competition (Hansson 1983). Studies in eastern North America often find the deer mouse (*Peromyscus maniculatus*) (Beckwith 1954) or meadow vole (*Microtus pennsylvanicus*) (Pearson 1959) in early old-field succession being replaced by the white-footed mouse (*Peromyscus leucopus*) as trees become established (Swihart and Slade 1990). In our case (Figure 14.7), the deer mouse is replaced by Keen's mouse. We also noted other replacements. Early-successional sites contain voles, montane shrew, and yellow-pine chipmunk. These species are replaced in later succession by red-backed vole, Trowbridge's shrew, and Townsend's chipmunk, respectively. Interestingly, all these replacements seem to relate to a physiognomic change in vegetation, although it is difficult to separate the importance of a plant's physiognomy and its value as a food item when plant species composition changes. For example, the northern pocket gopher occurs in early succession at Mount St. Helens and disappears as forests begin to develop. In this case, the physiognomy change is dramatic; however, the agent of change is the availability of belowground consumable plant parts (Andersen and MacMahon 1981).

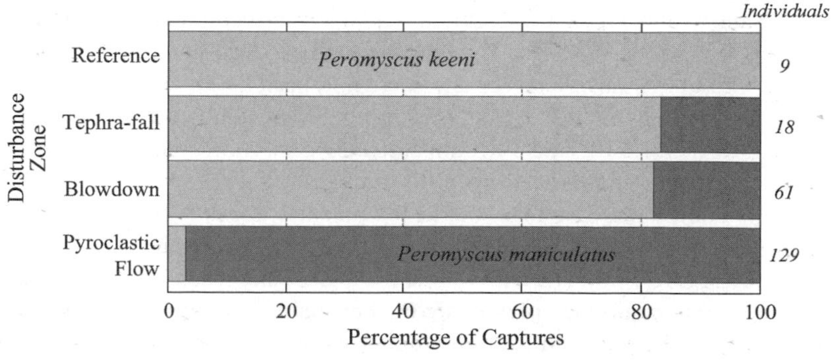

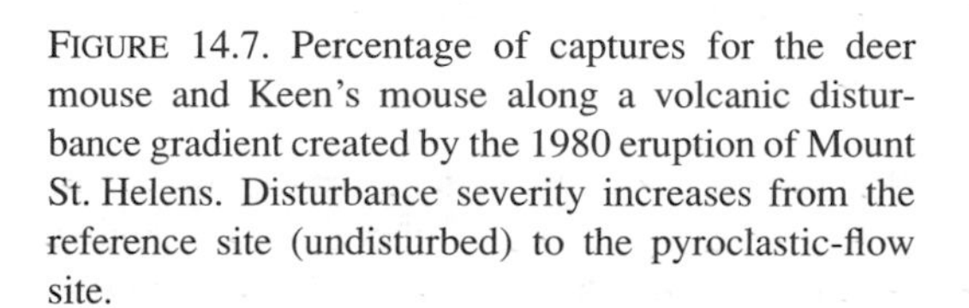
FIGURE 14.7. Percentage of captures for the deer mouse and Keen's mouse along a volcanic disturbance gradient created by the 1980 eruption of Mount St. Helens. Disturbance severity increases from the reference site (undisturbed) to the pyroclastic-flow site.

A recent manipulative study (Fox et al. 2003) conducted in Australia caused retrogression in a mammalian successional sequence by altering vegetation structure, especially density. That study along with studies of mammalian succession during the past 40 years and our long-term data set strongly suggest that vegetation structure is the driving variable for determining mammalian-assemblage composition during succession, our original broad hypothesis for these Mount St. Helens studies (MacMahon 1981).

14.4.3 Mount St. Helens Mammal Succession Compared to Some Successional Hypotheses

The theory of initial floristics differs from that of relay floristics (as hypothesized by Egler in 1954). Relay resembles a race where pioneer species are replaced sequentially and predictably by other species that are subsequently replaced by others and so on. Initial floristics suggests that any species can establish early in succession and that differences in life span explain the apparent sequencing of species over time. In the case of mammals in the Mount St. Helens landscape, our observations of some species replacements through time suggest something akin to relay floristics, whereas the complicated list of species that co-occur at some sites (see Table 14.2; Figure 14.3) suggests a process more similar to initial floristics. Life spans do not drive the sequence of mammals we observed, and Egler's hypotheses, as many others, are probably too plant oriented to apply directly to mammals.

The inhibition, facilitation, and tolerance concepts of successional pathway (Connell and Slatyer 1977) may apply better than Egler's concepts. Certainly, northern pocket gophers facilitate the establishment of other mammals through any facilitative influence they have on plant establishment. Similarly, the movement of seeds by a variety of small mammals aids dispersal and facilitates plant establishment and, in turn, animal establishment. In most cases, mammals are probably examples of tolerance (i.e., species that neither hinder nor help colonization by other mammals). No cases of inhibition are obvious, although plant consumption by a mammal could inhibit some plants, and certainly predators may have influenced mammal colonization. Species replacements of mammals are not caused by inhibition but by mammalian responses to vegetation change (Atkeson and Johnson 1979; Fox 1996).

Eugene Odum erected 24 hypotheses about successional trends (Odum 1969). These hypotheses have seldom been tested with quantitative data for any system, especially mammals. In one study of the 8 hypotheses tested, 5 were validated (Andersen et al. 1980). One of Odum's trends can be addressed with our data and those of others. Odum suggests that species diversity should increase from the developmental stages of a community to maturity. This trend is generally true for our sites (see Figure 14.6). (However, note the significant year-to-year variation, mentioned previously.) Other studies provide a myriad of conflicting results. Some studies provide support for Odum (Andersen et al. 1980), whereas others show declining species diversity (Pearson 1959) or more complicated patterns (Atkeson and Johnson 1979).

There is a general postulate in ecology that more species occur under conditions of moderate disturbance than at higher or lower intensities or frequencies of disturbance. Although this phenomenon has been noted by many authors, it was formalized by Connell (1978), reviewed by Huston (1994), and carefully scrutinized by Mackey and Currie (2001), who showed that the proposed relationship is hardly universal. Species-richness–disturbance curves studied showed no significant relationship (41%), positive monotonic relationships (29%), peaked relationships (19%), negative monotonic relationships (24%), and "U"-shaped relationships (3%). Our data (Figure 14.8) show positive monotonic curves for nearly every year of study and for all years summed.

The picture of mammal succession at Mount St. Helens is generated by what appears to be Gleasonian assembly (Owen 2001). Gleason's individualistic hypothesis (Gleason 1917, 1926, 1939), in essence, states that if an organism can get to a site and survive, it will be a component of that community. This pattern implies that, in general, groups of species do not move through time and space in lock-step with one another as a community. Rather, the components of the community respond individualistically to their various needs. The result is a varied mixture of species from place to place and time to time.

Because of their vagility, small mammals at Mount St. Helens appear to be capable of dispersing to any spot in the mosaic of habitats available. When they arrive, if conditions are conducive to their establishment and reproduction, they occur in various mixes based on their individualistic responses. With the caveat that small mammals must have food, the physical structure of the vegetation and the environment seem to be the most important determinants of their occupation of a site and the subsequent "successional" changes in the mammalian faunal composition of that site (MacMahon 1981; Fox 1982, 1990, 1996; Fox et al. 2003). Interestingly, this pattern seems consistent with a hypothesis of Platt and Connell (2003), who suggest that combinations of low and high spatial and temporal variation in disturbance effects have different consequences for successional patterns. The combination of high spatial and temporal variability of disturbance effects seems to fit the Mount St. Helens scenario. This combination predicts highly variable successional trajectories and responses to subsequent disturbances, much like those observed for small mammals at Mount St. Helens.

14.5 Comparisons with Studies at Other Volcanoes

To our knowledge, few direct parallels to our studies of Mount St. Helens exist for other volcanoes. Generally, information on mammal responses to volcanism consist of a few general

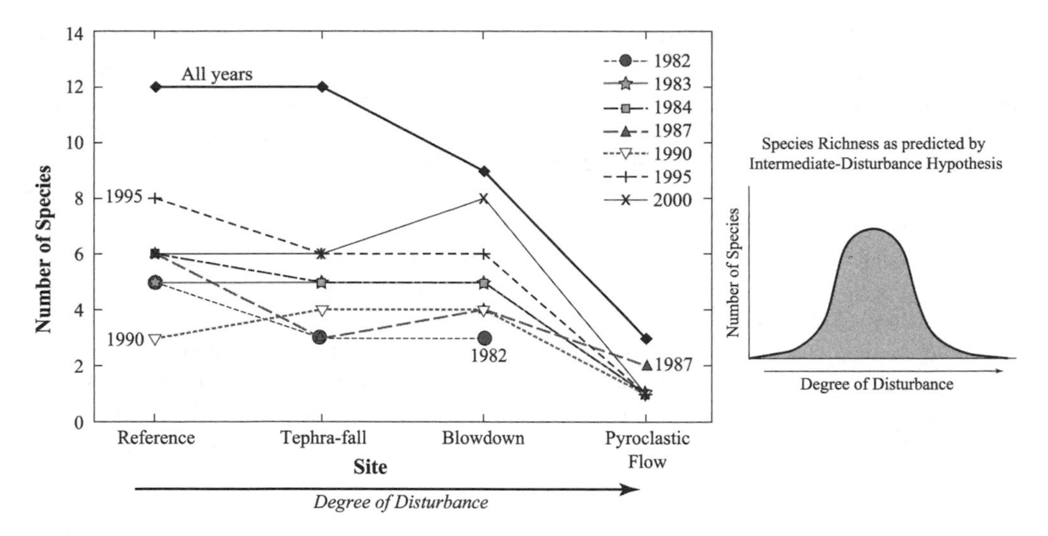

FIGURE 14.8. Small-mammal species richness for reference sites and three volcanic disturbance zones in sampled years following the 1980 eruption and for all years combined (cumulative species richness). Disturbance severity increases from the reference site (undisturbed) to the pyroclastic-flow site. Figure on the right shows the bell-shaped line that is the species-richness pattern predicted by the intermediate-disturbance hypothesis.

surveys sometime after the volcanic event to ascertain residual and colonizing species compared to a hypothetical list of preeruption inhabitants. Some volcanoes, such as Paricutin in Mexico that erupted in 1943, have received study of their mammals (Burt 1961). Other well-studied volcanoes simply had no mammalian species present [e.g., Barcena in Mexico (Brattstrom 1963b)]. Studies of most volcanic sites, worldwide, have not included observations on the reestablishment of small mammals.

Even well-studied sites where mammals have been observed, such as the Krakatau islands in Indonesia, do not provide very comparable situations. The Krakataus are islands surrounded by ocean, albeit with nearby (13 to 44 km distant) sources of potentially migrant mammals. Despite the proximity of sources, ocean water creates a formidable barrier to the migration of nonflying land mammals (Gorman 1979). This barrier is evident by the impoverishment of the current mammalian fauna of the islands. Mammals that have established on the Krakataus since the eruption in 1883 include two rats (*Rattus rattus* and *R. tiomanicus*) (Rawlinson et al. 1990), apparently brought in by humans; a pig species (*Sus* sp.), which might have been able to swim from a source island; and possibly an otter (*Lutra* sp.), known only from its tracks (Thornton 1996). Of course, highly vagile bat species have established in reasonable numbers (about 25 species) (Tidemann et al. 1990). These taxa even showed an interesting, but not unexpected, pattern whereby fruit bats (Pteropodidae) established soon after the eruption and probably acted as agents of plant-propagule dispersal. Insect-eating bats (Microchiroptera) did not establish until 50 to 70 years after the eruption. Unfortunately, we did not collect data on bat establishment. Had we done so, this region's lack of herbivorous bats would not have permitted us to draw contrasts with this sequence in any event.

In sum, our studies seem not to have parallels to other volcanoes with regard to residual mammals or details of colonizers.

14.6 Summary

Small mammals are a species-rich and trophically diverse group that effect important ecological processes in Pacific Northwest forest ecosystems. The 1980 eruption of Mount St. Helens created an important opportunity to document the initial responses and subsequent long-term successional patterns of small mammals following large, catastrophic disturbance. The complex gradient created during the eruption provided a template to evaluate the importance of disturbance severity, habitat complexity, biological residuals, oases, and dispersal during the reassembly and evolving successional landscape.

Initially surprising was the discovery that numerous small-mammal species had survived in many locations within the disturbed landscape. Survival was inversely related to disturbance intensity. In the tephra-fall zone, survival was widespread, and the mammal assemblage was typical of undisturbed areas. In the blowdown and tree-removal zones, survival was patchily distributed and related to species life-history traits (e.g., fossorial or generalists), topography, and stochastic factors, particularly the timing of the eruption and the presence of refugia created by late-lying snow. Douglas's squirrel, Townsend's chipmunk, and the southern red-backed vole are forest species that did not appear to survive the eruption in the blowdown zone. In the pyroclastic-flow zone, all mammals perished. The presence of residual small mammals within the core of the highly disturbed landscape was important because it negated the need for long-distance dispersal of individuals from distant source populations.

The primary colonists were generalist species associated with early-successional habitats (e.g., deer mouse, yellow-pine chipmunk, and Cascade golden-mantled ground squirrel), forest insectivores (e.g., Trowbridge's shrew and shrew-mole), and riparian-habitat specialists (northern water shrew and Richardson's vole). At the scale of the Mount St. Helens disturbance, dispersal ability of species was not a barrier to colonization; it just took longer for some species to arrive, and their arrival did not always coincide with their apparent vagility. By 2000, the Townsend's chipmunk, Douglas's squirrel, and southern red-backed vole had not colonized the blowdown or pyroclastic-flow zones. These species are associated with well-developed forest canopy or understory and are not expected to colonize until forest development has occurred, a process that could take several decades or longer.

Similarity coefficients of mammal assemblages at sites along the disturbance gradient decreased with increasing disturbance intensity and were generally less than 40%. Although substantial colonization and growth of vegetation occurred since the eruption, by 2000, the habitat structure remained dramatically different among the study sites. Habitat structure strongly influences mammal community composition, and we expect that the mammal assemblages will converge as plant communities, soil, and litter become more similar.

Oasis habitats played an important role in influencing the mammal community structure within the volcanic landscape. They increased the overall mammal species richness in the disturbed areas and served as biological hot spots. The faunal composition of oasis sites consists of a core group of five species (northern water shrew, Richardson's vole, Pacific jumping mouse, deer mouse, and yellow-pine chipmunk) and several other species that are present less consistently. We suggested that oases are important source populations for riparian specialists as well as for species commonly associated with developing matrix habitats. Accordingly, as oases expand and coalesce, the animals associated with these habitats will likely extend into areas formerly unoccupied.

Patterns of mammal colonization and assemblage structure did not fit any one existing model of succession. Mammals commonly associated with early- and late-successional habitat co-occurred because they happened to immigrate to a site and could survive there, even if the habitat was suboptimal. In this sense, Gleason's individualistic hypothesis (Gleason 1917, 1926, 1939) appears to provide the best explanation of mammal colonization and succession during the first two decades following the 1980 eruption of Mount St. Helens.

Acknowledgments. This work could not have been accomplished without the efforts of numerous individuals. For fieldwork, we thank D. Andersen, K. Johnson, J. Kling, E. Lund, A. McIntyre, M. Mesch, C. Parmenter, H. Purdom, L. Trippe, S. Wilson, and the many others who helped on a shorter-term basis. L. Carraway aided in the identification of shrews and voles. S. Wilson meticulously and painstakingly worked on data management. K. Ronnenberg and T. Valentine assisted with figures and tables. The manuscript benefited from the comments provided by F. Swanson, V. Dale, and two anonymous reviewers. We thank the USDA Forest Service, Mount St. Helens and Randle ranger districts, and the Weyerhaeuser Company for providing access to their lands. Funding for this research was provided by grants from the National Science Foundation (DEB 81-16914 and BSR 84-07213 to J.A.M.) and the USDA Forest Service, Pacific Northwest Research Station.

Part IV
Responses of Ecosystem Processes

15
Mycorrhizae and Mount St. Helens: Story of a Symbiosis

Michael F. Allen, Charles M. Crisafulli, Sherri J. Morris, Louise M. Egerton-Warburton, James A. MacMahon, and James M. Trappe

Mycorrhizae are symbioses between plants and fungi localized in the roots. These mutualisms represent important components in the recovery of vegetation because most plants depend on their fungal symbiont for a large portion of their soil resources, such as water, nutrients, and sometimes carbon. Although these organisms are generally rather cryptic and often unnoticed, they regulate many processes in ecosystems. The fungal hyphae, consisting of microscopic threads 2 to 10 μm in diameter, form the body of the fungus and ramify through the roots, forming a large surface area for exchanging nutrients and carbon. The hyphae then extend outward into the soil to provide nutrients and water to the host, become a sink for carbon, bind soil particles into soil aggregates, and produce sporocarps that are food for animals (Allen 1991; Smith and Read 1997; van der Heijden and Sanders 2002). External hyphae can be several meters to more than a kilometer per gram of soil. Thus, hyphae magnify the surface area of soil available for nutrient uptake and for soil-particle binding manifold compared with the roots alone.

The eruption of Mount St. Helens destroyed or buried fungi, just as it did plants and animals. The recovery of mycorrhizae is an integral part of the recovery of the vegetation. Mount St. Helens provided a natural disturbance, and, because much of the area was set aside for research, processes regulating succession could be studied over long periods. Human disturbance follows a similar parallel in that both plants and associated symbionts, such as mycorrhizal fungi, are often lost. Understanding the natural processes of succession at places such as Mount St. Helens has direct relevance to restoring human-disturbed ecological systems. The more we learn about how organisms invade and establish, the more we can subsidize those processes to hasten recovery of lands that we intentionally disturb.

15.1 Before the Eruption

Mount St. Helens sits in one of the most diverse regions in the world for mycorrhizal fungi, although the coniferous-tree diversity there is relatively low when contrasted with that of the eastern deciduous forest or tropical forest (Allen et al. 1995). By existing over many different substrates, this limited flora provides for the persistence of many belowground partners (Allen et al. 1995). The two predominant mycorrhizal types are ectomycorrhizae (EM) and arbuscular mycorrhizae (AM) [Figure 15.1 (upper)]. These two types have very different morphological and physiological structures. EM fungi tend to be very diverse. Trappe (1977) postulated that upward of 2000 species of mycorrhizal fungi formed associations with Douglas-fir (*Pseudotsuga menziesii*) alone. EM form on a number of trees, such as those in the orders Pinales (conifers) and Fagales (beeches, oaks, birches, walnuts, and bayberries). The fungi forming these mycorrhizae include some in the orders of Zygomycetes (Endogonales), Ascomycetes, and Basidiomycetes. Because EM fungi have evolved independently numerous times (Cairney 2000), they perform a suite of basic mycorrhizal functions (e.g., nutrient uptake and as carbon sinks) in many different ways. A comparatively high diversity of plants form AM, which may be linked to a lower diversity and specialization of AM fungi (Allen et al. 1995). We do not know the entire complement of fungi forming AM in this region, but 30 species were described in Gerdemann and Trappe (1974) and only about 175 are currently described worldwide (Bever et al. 2001). On Mount St. Helens, these fungi can associate with all herbs and grasses and with many trees. These fungi penetrate the root cortical cells, and external hyphae extend a few centimeters out from the root. At Mount St. Helens, this mycorrhizal group predominated in the meadows, above tree line, in riparian habitats, and in clear-cuts with a high proportion of herbaceous plants and shrubs. We found the AM genera *Acaulospora*, *Entrophospora*, *Gigaspora*, *Glomus*, and *Scutellospora* in the surrounding meadows and tephra-fall zones within the first 2 years after the eruption, indicating that they were present across the region before the eruption.

Another functional consideration of mycorrhizal fungi is that many of these fungi also form an important part of the diet for many animals. Across forest ecosystems, including Mount St. Helens, the feces of many animals are known to contain spores of EM and AM fungi (Maser et al. 1978). In forests of

FIGURE 15.1. *Upper:* Mycorrhizae from Mount St. Helens showing (a) *Cenococcum* ectomycorrhizae (EM) from noble fir (*Abies procera*) on the Pumice Plain in 2000; (b) alder (*Alnus* spp.) roots from Smith Creek showing arbuscular mycorrhizae (AM) hyphae, arbuscules, and vesicles in 1984; and (c) a sporocarp of the AM fungus *Sclerocystis coremioides* from feces of a northern pocket gopher (*Thomomys talpoides*) from 1982. *Lower:* Vertical distribution of the 1980 ash, the buried soil, and the 1800s ash from above Coldwater Lake in 1982.

the Pacific Northwest, up to 80% of the diet of the northern flying squirrel (*Glaucomys sabrinus*) is composed of truffle fungi known to form EM (McKeever 1960). Many squirrels collect EM fungi and carry them up into the canopy, where they feed on the cap and stipe, simultaneously dropping spores into turbulent air (Allen 1991). Spores are also ingested, transported, and later deposited in the feces. In meadows and early-successional clear-cuts, northern pocket gophers (*Thomomys*

TABLE 15.1. Nutrient status of soils from Mount St. Helens contrasting (1) 1980 and 2000 and (2) ash and soil.

Parameter	Value	1980[a]		2000	
		Tephra	Soil	Tephra	Soil
Organic carbon	g/kg	2.2	39.6	1.46	17.5
Nitrogen, total				0.01	0.39
Nitrogen as NO_3	mg/kg	0.9	10.9		
pH		6.6	4.9	5.4	5.1
Phosphorus, bicarbonate extractable		0.5	26	95	177
Potassium	mEq/100 g	0.011	0.12		
	mg/kg			50	40

In 1980, the sample was from Bear Meadow, and the soil consisted of that brought to the surface by gophers, with an estimated mixing ratio of 1:1, tephra to old soil. In 2000, the samples were from Meta Lake, (see Figure 15.2), and the soils were newly forming. In both cases, the tephra is from the 1980 eruption.

[a] The 1980 data are from MacMahon and Warner (1984); the 2000 data are from Morris and Allen.

talpoides) feed on bulbs, corms, and roots as well as truffles and AM sporocarps (Maser et al. 1978). Ant mounds also contain high densities of spores and mycorrhizal root fragments (Allen et al. 1984), adding to their dispersal.

Mount St. Helens has erupted many times in the geological history of the area (see Swanson et al., Chapter 2, this volume), creating a vertical legacy in soil structure. In the mid-1800s, Mount St. Helens erupted on several occasions, and several layers of tephra separated by buried forest floors can be seen below the soil layer of the pre-1980 forest [Figure 15.1 (lower)]. The buried forest floors are rich in organic matter and nitrogen, whereas the tephra layers of the 1800s are largely silica and are deficient in organic matter and nitrogen. These layers provided vastly different resources for newly developing roots following the 1980 eruption.

15.2 The Eruption

The 1980 eruption itself and the preeruption geography created a myriad of initial conditions in many areas severely affected by the eruption (Figure 15.2). The eruption consisted of a devastating mixture of tephra, 300°C (surface-sterilizing) heat, toxic gases, and strong winds that destroyed most aboveground life for many kilometers outward. Beyond the blast area, sterile tephra was deposited on the soil surface. Because the eruption occurred in May, snow still covered much of the volcano, ridgelines, and north-facing slopes at higher elevations. Both the snow and the soil provide insulation for soil organisms in these areas. As the eruption continued, pyroclastic flows emerged from the crater, spilling onto the area between the mountain and Spirit Lake and forming the Pumice Plain. This material produced a new sterile substrate where no organism, either aboveground or belowground, survived. Along the edge of the pyroclastic-flow zone, steep slopes existed that were only lightly covered by the sterile material.

We initially sampled a wide range of sites scattered across the disturbed area following the eruption (Allen et al. 1984). The resulting data led us to focus on three distinct conditions as representing the range of disturbance to soil, from only new sterile material (tephra) layered on top of existing vegetation to entirely new landscapes created by the pyroclastic flows (see Figure 15.2). The least disturbed sample area was the tephra-fall zone, such as Pinto Basin, where several centimeters of tephra was deposited on the soil and on plant leaves and branches (Seymour et al. 1983). However, below the tephra, all plant roots and organisms remained intact. The second area is the blowdown zone. In this area, trees were toppled, surface soils were ripped up as roots were torn from the soil, and a thick layer of tephra (up to a meter) was deposited. We focused our work in Bean Creek and Smith Creek. The final area that we studied was the Pumice Plain, an area covered by sterile pyroclastic-flow deposits.

15.3 Survival from the 1980 Volcanic Eruption

Surprisingly, many organisms survived the eruption, even in many unlikely places. This survival became critical to the recovery process. Three factors allowed the survival of mycorrhizal fungi within the heavily impacted blast area of Mount St. Helens:

- Insulation of soil-dwelling organisms by snow, rotten logs, and soil;
- Erosion of tephra down the steep slopes; and
- Topographic features [such as the lee side of the slope (away from the blast)] protecting organisms from hot gases.

15.3.1 Tephra-Fall Zone

Despite the locally severe effects on humans and trees, tephra fall had relatively minor direct influence on belowground organisms or processes, especially in meadows and logged areas. Most of these organisms were inactive below the soil surface and were protected by the snow and insulating properties of the soil itself (Andersen 1982; Andersen and MacMahon 1985b). In meadows, gophers, ants, and other animals burrowed from

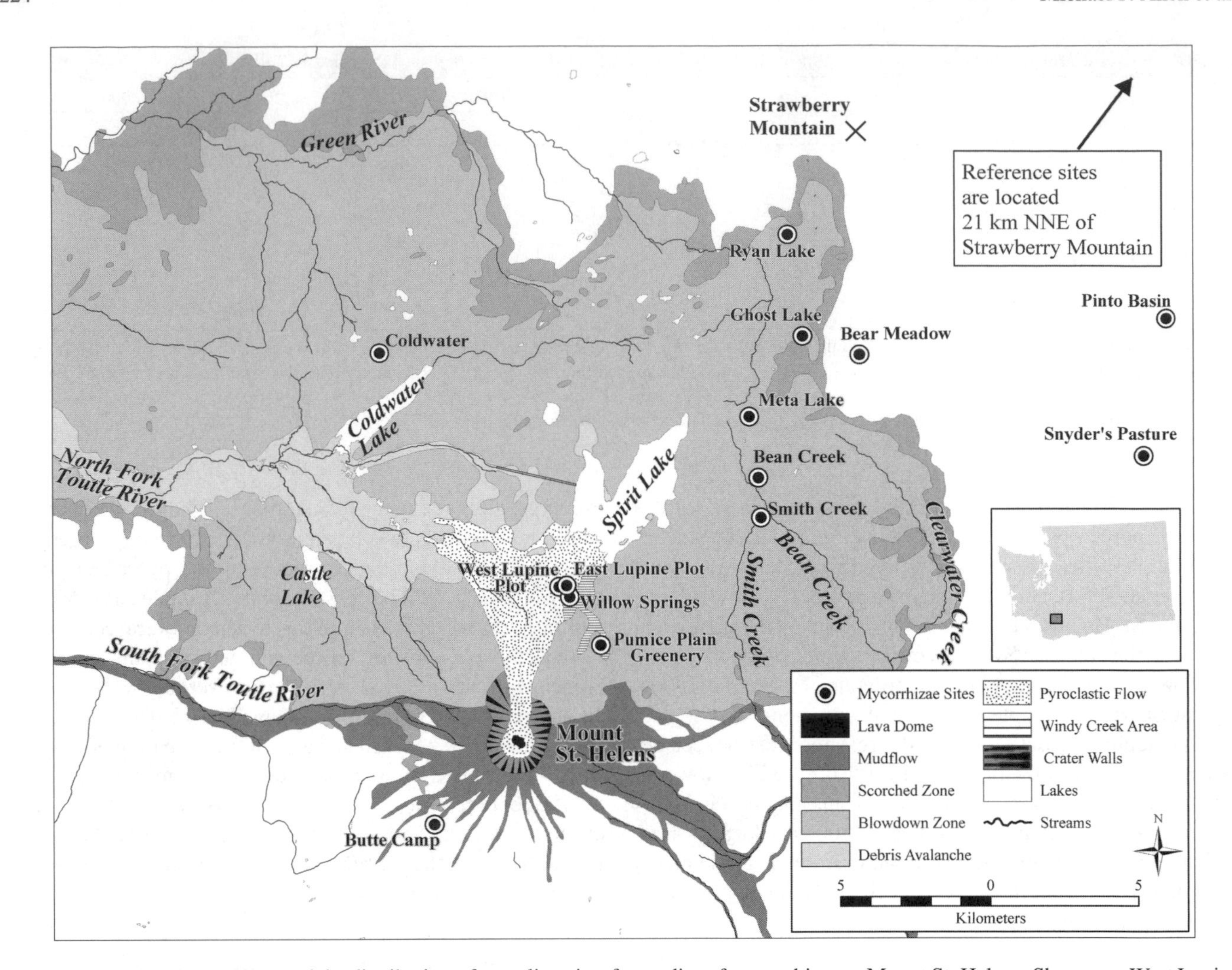

FIGURE 15.2. Disturbance types and the distribution of sampling sites for studies of mycorrhizae at Mount St. Helens. Shown are West Lupine Plot, East Lupine Plot, Willow Springs, Pumice Plain Greenery, Coldwater, Smith Creek, Bean Creek, Meta Lake, Ghost Lake, Ryan Lake, Bear Meadow, Snyder's Pasture, Pinto Basin, and Butte Camp.

the old soil up through the new tephra, mixing the buried layers with the freshly deposited tephra (Figure 15.3a). The resulting mix contained mycorrhizal fungi, roots, corms, seeds, nematodes, mites, collembola, and all other biotic elements critical for reestablishing a complex, functioning soil community. Further, by mixing the organic elements of the buried soils with the large tephra particles, the physicochemical properties of a well-aerated, almost perfect plant-growth medium resulted. The plants emerging from this soil mix were heavily mycorrhizal and healthy, whereas seedlings emerging in the neighboring, undisturbed tephra were not mycorrhizal and generally chlorotic (Allen 1987).

In some locations with shallow tephra deposits dormant and buried vegetation pushed up through the tephra, establishing vegetation on the surface. Also, large mammals, such as American elk (*Cervus elephus*), trampled and broke tephra crusts, mixing the buried soil with the surface tephra. In these areas, mycorrhizae were rapidly reestablished, and by 1985, the fungal infection in roots was no different from control areas.

In tephra-fall areas that were forested before the eruption, trees generally survived. The intact soils remained biologically active, although buried by a layer of tephra. We found little mixing of soils, unlike that mixing which occurred in clear-cuts and meadows, because pocket gophers do not inhabit forests (or occur only in very low densities there) (Andersen and MacMahon 1981). Certainly, some animals (rodents, insectivores, and ungulates) survived in tephra-fall forests and created disturbance, which probably played an important role in local recovery.

15.3.2 Blowdown Zone

Mycorrhizae are exhibiting a remarkably rapid rate of ecological recovery in the blowdown zone, largely caused by the survival of patches of plant propagules and soil organisms scattered around the area. Pocket gophers played a major role in the rate of vegetative development in this zone, and aspect and slope are also important. On lee slopes from the crater, topographic shielding protected organisms from the full brunt of

FIGURE 15.3. (a) Soil brought to the surface by northern pocket gophers that survived the eruption at Bear Meadow. The *dark-colored* soil is the buried soil mixed by gophers with the Pumiceous tephra (*light-colored* material). Notice the plant growth in the gopher-mixed soil. (Photograph from 1981.) (b) Deep-tephra-deposition area above Smith Creek. In nearby shallow areas, gopher-mixed and erosion rills brought old soil to the surface, facilitating plant and mycorrhizal recovery. Where the tephra was deep, no signs of gophers or mycorrhizae appeared, and only a few plants were present. (c) Rill erosion channels and emerging mycorrhizal plants in the slopes above Coldwater Lake in 1982. (d) Elk feces on the surface and buried just under the tephra. Notice the growth of plants from the feces. Microscopic examination showed that the roots of the established plants were mycorrhizal and that the feces themselves contained roots, hyphae, and spores of mycorrhizal fungi. (e) The northern pocket gopher within the lupine enclosure on the Pumice Plain of Mount St. Helens. This individual was brought in from Bear Meadow, remained for 24 hours, and then was returned. This prairie lupine individual was the only mycorrhizal plant on the Pumice Plain in 1983.

the blast. Also, some north-facing slopes were snow covered, providing thermal insulation from the heat. Where the slope was steep, the tephra immediately slid rapidly down slope. The combination of these factors protected soil organisms and reduced the amount of tephra through which buried plant parts had to grow and animals had to burrow to reach the surface. Above Bean Creek and Smith Creek, the tephra was rapidly eroded down slope, and gophers reemerged at several locations.

By preferentially foraging on lupine, gophers maintain ruderal plant species, such as fireweed (*Chamerion angustifolium*) and dwarf knotweed (*Polygonum minimum*). Further, they created mounds 15 to 30 cm in height with a mix of tephra and soils that was ideal for plant establishment and growth (Mahler and Fosberg 1983). At the toe of a long slope above Smith Creek, tephra accumulated to such a depth (greater than 1 m) that the gophers could not reemerge (see Figure 15.3b). In the 2000 sampling, this area remained without animal excavations and with few plants or mycorrhizae.

In the blast area north and east of Mount St. Helens, the layering of tephra deposits and buried soils influenced vegetation and mycorrhizae response to the 1980 eruption. Few roots can be found in the tephra layer of 1980 or in the layers deposited in the 1800s. Taproots penetrated these layers and radiated outward only upon reaching the buried soil layers, where roots proliferated [see Figure 15.1 (lower)]. In areas with very thick tephra layers, however, little vegetation appears, even by the year 2000 (see also Antos and Zobel, Chapter 4, this volume). As a new layer of organic matter accumulates on the surface of the tephra, a new soil with intertwined roots and fungal hyphae can be seen. Erosion initiated recovery in the blowdown zone, where rills and gullies cut down through the 1980 deposits to the old buried soil. Plants reemerging along erosion channels were associated with AM fungi and decomposer fungi (see Figure 15.3c).

At several locations above an elevation of 1200 m, patches of Pacific silver fir (*Abies amabilis*) and mountain hemlock (*Tsuga mertensiana*) survived. By 1997, the mountain hemlock trees were producing seed. These trees were all mycorrhizal, providing surviving inoculum that could be dispersed across the site. At Ryan Lake, Carpenter and colleagues (1987) found sporocarps of *Endogone pisiformis* in the fall of 1981, and at Meta Lake they reported *Laccaria laccata* in 1982. By 1990, at Ghost Lake, we found sporocarps of several mycorrhizal fungi, including species of *Boletus*.

Importantly, these isolated refuges probably served as source areas. Patches of young trees protected under snow in intact soils were essential to initiating plant and fungal recovery in the more severely disturbed surrounding areas. We found *Laccaria laccata*, often an early-seral mycorrhizal fungus, in the late 1990s in the Pumice Plain. We did not observe sporocarps of *Thelephora terrestris*, a common early-seral EM fungus, in the area. However, spores were trapped on the Pumice Plain with gel-coated surfaces (Allen 1987), and we speculate that this rather cryptic species was probably present in the area.

15.3.3 Pyroclastic-Flow Zone

Surprisingly, adjacent to the pyroclastic-flow zone, organisms survived in a few patches (Pumice Plain Greenery site), where slope, aspect, and animals favored survival and reestablishment of mycorrhizae. Near the entrance to the current Pumice Plain trail, the steep slopes facing away from the crater served as a source population. The pumice material rapidly eroded down slope, exposing old soil. Resprouting residual plants grew in these rills, and the site was vegetated by 1982. Deer mice (*Peromyscus maniculatus*) were found within those vegetated patches, and all plants observed had already formed AM symbiosis.

15.4 Invasion

Plants rapidly established across the volcanically disturbed area at Mount St. Helens, but especially in erosion gullies and in places where animals exposed buried soils with plant and microbial propagules. From these sources, immigration became important for local as well as long-distance transport of plants and fungi in the tephra-fall and blowdown zones.

15.4.1 Tephra-Fall Zone

In this zone, there was little need for invasion by organisms from the outside. At sites such as Snyder's Pasture and Pinto Basin (see Figure 15.2), existing plants retained their mycorrhizae. The soil emerging via erosion and animal disturbances contained inoculum at the same points as plant propagules initiating new vegetative patches within the first 2 years (Allen et al. 1984). Carpenter et al. (1987) listed a wide array of fungi on the soil surface, demonstrating that, within the first few years, many of the microbes had recovered.

Some likely important editing (*sensu* Franklin et al. 1985; Franklin and MacMahon 2000) affected both plants and fungi. Most of the mycorrhizal fungi reported were probably animal dispersed. A few phoenicoid (fire-following, heat-loving) fungi could be found on the tephra surface, and their dispersal mechanisms remain largely unknown but are likely windblown. Meanwhile, few aboveground fruiting mushrooms were found for several years in areas with high tephra fall.

15.4.2 Blowdown Zone

In the blowdown area, most dispersal was local, radiating out from patches of initial establishment (Allen 1987). Plants rapidly reestablished, and the species of animals present were similar to the inferred predisturbance community within 5 years (Franklin et al. 1985). Wood-decomposing and saprobic basidiomycetes and ascomycetes were commonly found by 1981 in the erosion rills within the tephra, where buried soils were exposed following the eruption. Importantly, many phoenicoid fungi were found across the blast area (Carpenter et al. 1982, 1987). Most of these fungi are likely to be saprobes. However, some related species, such as *Sphaerosporella brunnea*, are known to form EM (Danielson 1984; Byrd et al. 2000).

Riparian vegetation reemerged almost immediately, largely where stream erosion exposed preeruption soils. By 1983, alders (*Alnus* spp.) and willows (*Salix* spp.) were present and were AM. A large variety of saprobic fungi were present, such as *Hericium abietis*, *Lycoperdon*, *Gyromitra*, and *Peziza*. By

2000, *Hericium abietis* was found across many areas of the blowdown zone. Some mycorrhizal fungi, such as *Laccaria laccata* and *Russula amarissima*, may have either immigrated or survived.

Immigration by mycorrhizal fungi in the uplands of the east and west ends of the blowdown zone was caused primarily by planted seedlings of conifers. All nursery-grown seedlings were EM by the time we observed them, a few weeks after planting. We presume that these had been inoculated, perhaps intentionally, with EM fungi before planting. Thus, even though we observed fruiting of *Pisolithus tinctorius*, an EM fungus, we postulate that this fungus was brought with the planted conifers. We discontinued work on invasion of EM fungi in this area at that time.

15.4.3 Pyroclastic-Flow Zone

By 1982, animals were beginning to disperse across the Pumice Plain. Seedlings of prairie lupine (*Lupinus lepidus*) were found in two patches in the middle of the Pumice Plain. The seeds of these plants were likely dispersed by birds. In each patch, a single plant existed in 1981. By 1982, there were 12 individuals.

Elk moved across the pyroclastic-flow deposits soon after the 1980 eruption. Importantly, as they grazed in meadows in the tephra-fall and blowdown zones, they not only clipped the forbs, grasses, and shrubs but also pulled plants up, as well. It takes approximately a day for herbage to travel through the gut tracts of large herbivores, allowing for transport of inoculum a long distance. Roots with mycorrhizal fungi, corms, bulbs, and other propagules were found within the feces. These feces were deposited on the pyroclastic flow, initiating new patches of mycorrhizal plants (see Figure 15.3d). Our observations suggested that animals focused on patches of lupines and other associated plants, not scattered individual plants (Allen 1988). Hence, recovery occurs preferentially around patches and not randomly scattered across a landscape.

In 1982, we placed a pocket gopher (live-trapped at Pinto Basin) in an enclosure with an individual lupine for 24 hours at several locations on the pyroclastic-flow material (Allen et al. 1984) (see Figure 15.3e). The plants associated with this lupine patch were intensively studied. Feces of the trapped animals showed that they carried the mycorrhizal fungus *Glomus macrocarpum*. In 1983, this plant was the only mycorrhizal individual plant and the only mycorrhizal fungal species on the pyroclastic landscape (Allen and MacMahon 1988). In 1986, a long-tailed vole (*Microtus longicaudus*) was trapped at a nearby prairie lupine patch (MacMahon et al. 1989). The following year (1987), this patch was also found to have AM plants. As the lupine patches expanded, elk were attracted to them as places to forage. During these forays, they would defecate, dispersing spores and infected root fragments along with a variety of plant propagules. Importantly, although herbs such as fireweed and pearly everlasting (*Anaphalis margaritacea*) continuously invaded the Pumice Plain as windblown seed throughout the 1980s, none was ever found to be AM.

Through the 1980s, both animals and plants continued to immigrate onto the Pumice Plain. In 1988, the water vole (*Microtus richardsoni*), Cascade golden-mantled ground squirrel (*Spermophilus saturatus*), and yellow-pine chipmunk (*Tamias amoenus*) were captured. All feed on mycorrhizal fungi as well as plant materials. Also by the late 1980s, a second mycorrhizal fungus, *Sclerocystis coremioides*, was found in feces of some of the animals and in soils. During this time, AM also established in willows at Willow Springs on the Pumice Plain. Rodents were common in Willow Springs by the late 1980s. By 1990, AM fungi were found in every patch examined, albeit at very low densities (Allen et al. 1992). By 2000, 14 species of rodents and insectivores had been captured (see Crisafulli et al., Chapter 14, this volume). At the Pumice Plain lupine patch that had been initially inoculated by a pocket gopher (Allen and MacMahon 1988), about 75 to 200 elk resided during the autumn of 2000. By 2000, AM fungi were found at all sampling sites, representing four genera and 10 species (*Acaulospora morrowiae*, *Acaulospora* sp., *Glomus (Gl.) aggregatum*, *Gl. claroideum*, *Gl. geosporum*, *Gl. leptotichum*, *Gl. spurcum*, *Gl. tenue*, *Sclerocystis coremioides*, and *Scutellospora calospora*).

Wind is often postulated to be important in the dispersal of organisms onto disturbed areas. Microfungi and discomycetes, commonly wind dispersed, were found as early as July 1980 and were widespread by fall 1981 (Carpenter et al. 1982). However, wind dispersal is not as all encompassing as often perceived. Entrainment is limited to areas where wind is turbulent. Forests such as those surrounding the volcanically disturbed area are dense and tall. Most flow at the soil surface and in the surface soils where mycorrhizal fungi are located is laminar, and entrainment is infrequent (Allen et al. 1993). Thus, wind was not likely to pick up and disperse most soil-borne fungi in forests. Allen (1987) found that AM fungi probably dispersed only locally and, even then, likely only from gopher mounds in the tephra-fall or blowdown zones. This observation is supported in that, on the Pumice Plain, both Allen and MacMahon (1988) and Titus and del Moral (1998b) reported only *Glomus macrocarpum* (a large-spored AM fungus likely inoculated by the introduced pocket gopher; see invasion section). During our sampling in 2000, large spores of *Glomus*, *Scutellospora*, *Acaulospora*, and sporocarps of *Sclerocystis* were found. These fungi are not generally wind dispersed (Allen et al. 1993). Small-spored, wind-dispersed AM fungi such as *Gl. etunicatum*, *Gl. fasciculatum*, *Gl. intraradices*, *Gl. microcarpum*, *Gl. occultum*, and *Gl. tenue* were found only in the tephra-fall and blowdown zones, not on the Pumice Plain. *Glomus aggregatum*, which forms spores that group together into sporocarps, was the only small-spored species present. These observations strongly suggest that animals are responsible for most of the immigration of AM fungi onto the Pumice Plain.

Insects could be important vectors, but we were unable to demonstrate an important role for them in the Pumice Plain. For example, we collected grasshoppers that were present on plants in the lupine patches. Although spores of AM fungi were found in the gut tracts, none could be germinated.

However, for EM fungi, a different pattern may have operated. Spores of *Thelephora* spp. were trapped in wind samplers by the mid-1980s. The first sporocarps of EM fungi found were at Willow Springs and consisted of *Laccaria laccata* and *Inocybe agardhii*. We do not know the sources of the propagules, but they were at least several kilometers away (the nearest known population was Meta Lake, more than 6 km distant). These fungi are probably largely wind dispersed. They commonly appear in sterilized soil within fenced nurseries. They are also the predominant fungi found in glacial forefronts of the Kenai Fjords, Alaska, at sites restricted to wind dispersal. The Alaska sites were land areas opened by glacial retreat from the ocean; animals were restricted from invading because the patches remain surrounded by ice fields. By 1990 on the pyroclastic flow deposits, still only a few EM were present in scattered, small conifer seedlings and with some of the willows. However, by 2000, most willows and alders at Willow Springs were EM and relatively large (greater than 3 m tall). EM conifers were found scattered all across the site. *Cenococcum* sp., *Cortinarius* sp., *Inocybe maculata* (group), *Inocybe agardhii*, *Laccaria laccata*, a *Sarcoscypha* sp., and pieces of *Rhizopogon* sp. were found on the Pumice Plain. Both *Cenococcum* and *Sarcoscypha* are commonly found following fire, and both are probably wind dispersed. *Rhizopogon* is animal dispersed. *Inocybe* spp., *Laccaria* spp., and *Cortinarius* sp. spores are likely wind dispersed. Thus, by 1990, it appears that dispersal of EM fungi on the Pumice Plain was largely by wind but that animals probably were also dispersing ingested fungi throughout the first two decades.

15.5 Establishment, Persistence, and the Roles of Mycorrhizae

Once established, mycorrhizal fungi persisted at all sites. In the tephra-fall and blowdown zones, animals and erosion exposed the buried AM fungi, providing the initial inoculum. The deeper inoculum gradually declined as plants spread across the surface. By 1985, spore counts indicated an equilibration between the mounds and surrounding developing soils (Figure 15.4). On the Pumice Plain, the pattern was different: no deep inoculum survived, but immigration is still proceeding.

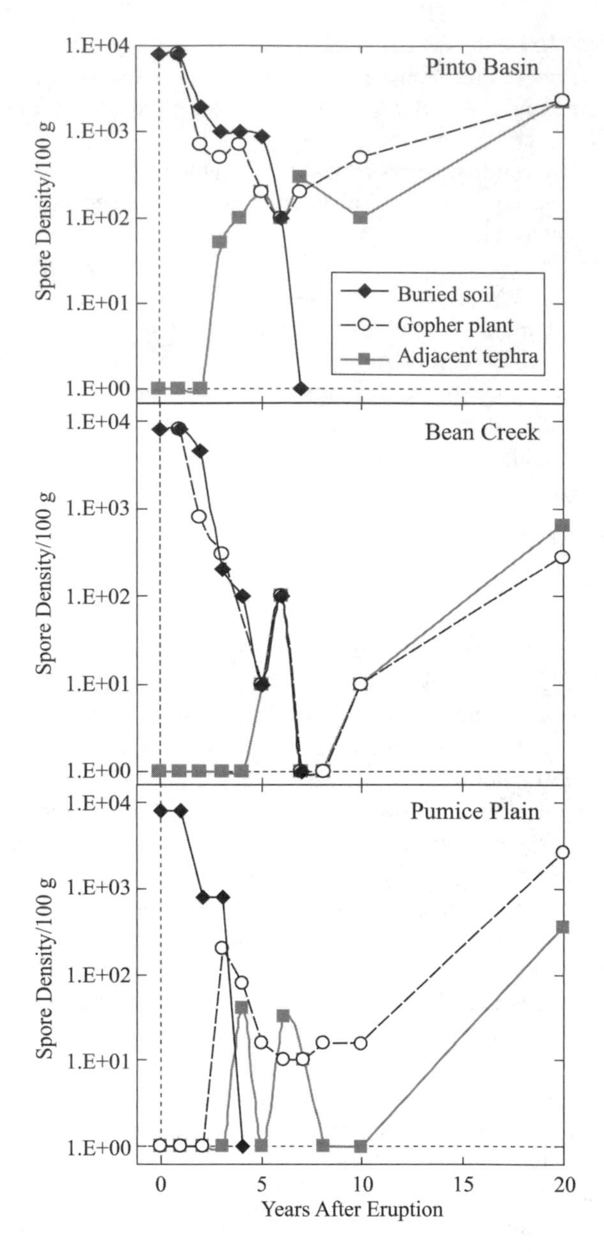

FIGURE 15.4. Recovery of arbuscular mycorrhizal fungi on Mount St. Helens based on spore numbers. Although none of the locations have recovered to preeruption soil levels, the inoculum is adequate to cause infection. Buried soils served as effective inoculum for about 3 to 5 years, after which the inoculating capability rapidly declined to zero. The inoculum rapidly migrated from the original mound in the tephra-fall zone and blast area, but only migrated out to the Pumice Plain during the past decade.

15.5.1 Tephra-Fall Zone

By the mid-1980s, plants were distributed across the tephra-fall zone, as were mycorrhizae and saprobic fungi. The tephra itself contributed a small amount of cations, phosphorus, and sulfur for plant growth (Fruchter et al. 1980). More importantly, the tephra and soils were mixed by a variety of fossorial animals, primarily pocket gophers and ants. This mixing provided increased aeration and resulted in improved plant growth and nutrient uptake in lupines and fireweed (Allen et al. 1992). The intentional mixing of pumice and soil by cultivation, observed in agricultural areas, in the region also was found to enhance plant growth (Mahler and Fosberg 1983).

Within the coniferous-forest stands, the tephra formed a persistent layer with much less mixing (Antos and Zobel, Chapter 4, this volume). Although fertility improved with the mixing, cations, phosphorus, nitrogen, and organic matter remained low (see Table 15.1). This pattern was common. In tephra from the 1800s eruptions, nitrogen and organic matter remained very low. Few fine roots or mycorrhizal fungi were found in these layers, even in the 2000 collections. Fine roots, organic matter, and mycorrhizal fungi were abundant in the forest-floor layers between the tephra layers.

Together, these observations indicate that, in forests, the coarse-grained tephra provided aeration and some cations that leached into the buried forest floor. Vertical roots grew through the tephra layers into the buried forest floor, where fine roots and mycorrhizae spread horizontally. In meadows and clear-cuts, fossorial rodents mixed the tephra and old buried soil. Vegetative regrowth was more rapid in areas of former meadows and preeruption clear-cuts than in areas previously forested, and succession to a new forest was rapid. In the forest, the existing trees largely remained, but their roots remained deeper, buried in the old soil horizons.

15.5.2 Blowdown Zone

The mechanisms of plant-community development in the blowdown zone were virtually identical to those in the high-tephra-fall zone, but the rates were slower. Small trees survived in protected patches and were mycorrhizal with a diversity of fungi. Compared with the forest survival and coverage in the tephra-fall zone, the forest in the blowdown zone is much shorter in stature and sparser. However, small- to medium-sized trees exist across the zone. Mycorrhizae were found in every area sampled by 1990. Tree growth was rapid. Importantly, in the new forests, a mature forest floor overlaying the tephra is rapidly being created.

The slowest vegetation development is occurring at sites of thick tephra accumulation such as the edge of Smith Creek (see Figure 15.3a), where tephra material eroded off hillsides, creating a deep alluvium of tephra. Few plants were found in these patches in the 2000 sampling. Lupines [mostly broadleaf lupine (*Lupinus latifolius*)] remain dominant. Tephra only, without old soil material, does not provide a good growth medium. In these sites, the mechanisms of mycorrhizal formation and recovery mimic those found on the Pumice Plain.

The dynamics of ericaceous plants probably remains one of the most interesting uncaptured stories of the ecological responses to the 1980 eruption. Ericaceous shrubs, such as huckleberries (*Vaccinium* spp.), are obligately mycorrhizal plants. Near Meta Lake, Carpenter et al. (1987) found several fungal genera, such as *Oidiodendron, Hymenoscyphus*, and *Peziza* known to form a third type of mycorrhiza, the ericoid mycorrhiza (Molina et al. 1992). These fungi may very well have survived in high-organic soils under the snow, along with the host plants. Today, huckleberry has spread and is scattered around the blowdown zone. We postulate that, in a manner similar to that of the plant family Epacridaceae of the semiarid sand plains in Australia, some individuals are forming ericoid mycorrhizae and using organic matter, such as the decomposing litter found on the surface of and immediately under the tephra, directly for nutrients (Lamont 1984).

15.5.3 Pumice Plain

The Pumice Plain of Mount St. Helens was created by sterile new substrate during the eruption in 1980. In addition to dispersal, two factors affected the establishment of vegetation: texture and nutrient status. The pumice itself was rather uniform popcorn- to golf-ball-sized particles. By 1982, the first patch of individuals of prairie lupine appeared, initiating the lupine patch. Nitrogen fixation in the lupines was already occurring in 1982; the first plants had large, red nodules, indicating leghemoglobin activity. Leghemoglobin is a useful indicator of nitrogen fixation because the red coloring indicates that oxygen is being scavenged away from the nitrogen-fixing symbionts. Soil phosphorus and cations were evenly distributed in low concentrations but were adequate for plant growth. The pumice, initially found as large stones, rapidly weathered into smaller fragments. By the late 1990s, the pumice was mostly in the sand- to silt-size particles. The lupine patch was itself changing both physically and chemically. In 1981, there was a single individual prairie lupine, but by 1983 more than a thousand individuals populated the patch, and in 1988 more than 30,000 plants were growing in the lupine patch. New patches were emerging annually across the Pumice Plain. By 1990, a line of lupine patches extended more than 3 km to Spirit Lake (Bishop et al., Chapter 11, this volume). These plants were all contributing nitrogen and organic matter and hence creating soil. The combination of organic-matter patches and breakdown of the larger pumice stones into sand- and silt-sized particles increased water-holding capacity and nitrogen concentrations. This process allowed for the invasion of a greater diversity of other plants.

In the original lupine patch, plant richness and density increased through the 1980s. In 1990, plant-species richness peaked at 29. By 1991, more than 56,000 individual plants were found. However, in 1990, we still found only two species of AM fungus, *Glomus macrocarpum* and *Sclerocystis coremioides*. In 1992, Titus and del Moral (1998b) initiated plots on the Pumice Plain using inoculum from Bear Meadows (the same location where the 1982 inoculum originated). They also reported the presence of only a single AM fungal species, *Gl. macrocarpum*.

The importance of mycorrhizae to plant establishment appears to be rather mixed. Conifers failed to establish until EM were simultaneously established (Allen 1987). Although many conifer seedlings were found across the Pumice Plain throughout the early 1980s, none was found to survive into a second growing season without mycorrhizae. By summer 2000, at least seven conifer species had established on the Pumice Plain. Stem densities were highly variable among the species present, but no species was abundant. Pacific silver fir, Douglas-fir, and western hemlock (*Tsuga heterophylla*) were the most common species, and noble fir (*Abies procera*), lodgepole pine (*Pinus contorta*), western white pine (*P. monticola*), and western red cedar (*Thuja plicata*) were uncommon or rare. The growth rate of conifers following establishment has been slow, and few trees are more than a meter tall. Many conifers were chlorotic, yet individuals have persisted for more than 10 years. Dave Wood (Chico State University, Chico, California; personal communication) monitored a 16-ha plot on the Pumice Plain from 1986 through 2000 and found annual survival rates of conifers to be about 80%. Wood also found conifers to be most abundant in washes and rills as compared to "intact" pyroclastic-flow-deposit surfaces. These recessed erosional features are where wind-dispersed inoculum (e.g., *Thelephora*) presumably would have concentrated. At these locations, substrates tend to be wetter than in adjacent intact pyroclastic-flow deposits, where water rapidly flowed down through the substrates to depths exceeding the rooting zone. AM fungal hyphae and vesicles were observed in some roots of the first surviving seedlings similar to observations from the Lyman Glacier in the North Cascades (Cazares and Trappe 1993). A mix of AM and EM fungi have been observed simultaneously invading a "normally" EM plant (Cazares and Smith 1996; Horton et al. 1998; Smith et al. 1998), which could influence plant production (Egerton-Warburton and Allen 2001). Clearly, more work on the effects of multiple mycorrhizal fungal partners is needed.

The functioning of AM is much more difficult to assess. Lupines, fireweed, and pearly everlasting clearly established without mycorrhizae. By the mid-1980s, AM fungi could be found in all sites with thick vegetation, largely surviving in patches exposed by erosion of tephra (Allen 1988; Titus et al. 1998a). Allen et al. (1992) found that AM within gopher-turned soil improved the plant phosphorus and nitrogen concentrations, and increased drought tolerance in lupines. The increased nitrogen and stomatal conductivity indicated that CO_2 fixation was enhanced, providing more rapid accumulation of both carbon and nitrogen. This increase in soil carbon and nitrogen was probably the limiting factor in plant growth on this site. Titus and del Moral (1998b) found that AM had no demonstrable effect on survival or growth on some of the early colonizing plants through the 1993 growing season, but the AM fungi themselves continued to expand.

Through the 1990s, plant composition continued to change on the Pumice Plain. Cat's-ear (*Hypochoeris radicata*) underwent rapid increases beginning about 1995. This plant is an invasive lawn weed from Europe but, as a member of the Asteraceae family, is responsive to AM. Hawkweed (*Hieracium albiflorum*), also in the Asteraceae, although present, has shown little increase through time. In the early 1990s, other associations began to emerge. Bellflower (*Campanula* sp.) was found in patches and was AM (Titus et al. 1998a). Bistort roots were found in wet areas and did not appear to be mycorrhizal. However, some species can form both AM and EM if the site is appropriate and inoculum is present.

In the summer of 2000, we reexamined the original plots and surrounding sites. The richness of plant species and the plant densities had changed little compared with 1990, but plants had spread across the landscape, including most of the originally sterile area. Prairie lupine, a good nitrogen fixer, remained among the dominant plants, suggesting that nitrogen was still limiting to plant colonization. The AM fungi have rapidly spread through these areas, and 12 taxa of AM fungi were found in the area containing only 1 or 2 species in 1983. The species of plants dominating the Pumice Plain in 2000 remained largely facultatively AM. Strawberry (*Fragaria* sp.) has been shown to be responsive to AM in growth studies (Harley and Harley 1987), and we would postulate that cat's-ear and hawkweed may also be responsive (Harley and Harley 1987). These species did not emerge on the site until after AM were present on the facultative plants. Cat's-ear, in particular, may be able to increase because of the increasing diversity and densities of AM fungi.

One of the more interesting stories may surround the family Scrophulariaceae. Beard-tongue (*Penstemon* sp.) was found to be both nonmycorrhizal and AM, depending on the location (Titus et al. 1998a). This plant was first found during the 1990s. Paintbrush (*Castilleja* sp.) was widespread by the 2000 sample. This plant is a hemiparasite on shrubs, pearly everlasting, and lupines. We did not find mycorrhizae, but that may be because of the late-August sampling. Lesica and Antibus (1986) reported that paintbrush formed AM in the early-seedling stage, before parasitizing a neighbor. At Mount St. Helens, this plant was found only after a decade following the eruption and in the presence of mycorrhizal hosts. This pattern presents an interesting opportunity for study of a very different set of response dynamics.

15.6 Surprises, Summary, and the Future

We have learned much about the invasion processes of both AM and EM in the Mount St. Helens landscape. Invasion of AM and, later, EM was largely catalyzed by animals and focused on patches of vegetation sought by large grazers. Animal behavior is essential to predicting the recovery of mycorrhizae, and succession must be studied at the patch and landscape scales. For EM, wind was a dominant vector initially, but its dispersal ability was dependent on patches of plants or physical features to create turbulence and to concentrate deposition of spores.

Establishment of mycorrhizae was much faster than we had originally anticipated. Animals rapidly moved across the harsh landscape, focusing on patches of new and invading plants. Within a decade following the eruption, a critical type of symbiosis, mycorrhizae, was well established in the tephra-fall and blowdown zones. Within two decades, a high diversity of plants and mycorrhizal fungi has gained a foothold across the sterilized material of the pyroclastic-flow deposits. At the ecosystem scale, all critical elements are now present for forest production. Only the processes of infilling species and accumulation of biomass remain.

Succession on our research plots has not been a linear progression. However, in terms of percent cover on the small plot itself, the vegetation in 2000 resembles what it looked like in 1987 or 1988 rather than 1990. In 1988 and in 2000, lupine dominated the plot, whereas it was sparse and small in 1990. The individual conifers present in 1990 had all disappeared by 2000, and new ones could be found at the patch margins.

What do these observations mean for land management and restoration? First, all elements of ecosystems are important. Persistence of patches of surviving plants and animals serve as critical initial sources of propagules. Plants spread from these patches and serve as foci for animals migrating from source areas. Large-scale source areas surrounding severe disturbances allow for the rapid invasion of fungi and plants both by wind and animals. Without surviving patches and the extensive surrounding source matrix, recovery would not have occurred within the time frame or at the scale observed. Hence, conserving source areas of both animals and microbes within close proximity to disturbance may be absolutely essential to ecosystem restoration.

Initial conditions are an important determinant of survival and successional patterns. In this case, timing was a critical feature. Without insulating snow cover, mortality would have been much greater. If lands are to be disturbed for human needs, understanding the seasonal timing of disturbance and recovery may be just as crucial to eventual restoration.

A diversity of organisms and functional groups of organisms is important to the recovery of a site, and each organism has an important timing in the successional process. Animals such as pocket gophers, often considered pests, were key elements in the recovery of plant propagules and in recovery of mycorrhizal fungi that promote plant establishment and succession. Lupines, mycorrhizal fungi, and nitrogen-fixing rhizobia formed the microsites that facilitated the reinvasion of other plant species.

Finally, although space-for-time substitution is useful as a research tool for studying successional processes, it does not substitute for long-term research on the recovery of a site with the spatial and temporal heterogeneity of the Mount St. Helens landscape or of any large, severe disturbance.

Two issues stand out for the future. First, we project that there will be an increasing diversity of mycorrhizal types and diversity of mycorrhizal fungi with increasing time. Second, in the context of successional seres, the pattern will be one of fits and starts, reversals, and advances in patches scattered across ever-larger areas. That is, until the next eruption.

Acknowledgments. We thank the USDA Forest Service for support for the 1990 and 2000 sampling and analyses. Analyses were also supported by the National Science Foundation Biocomplexity Program DEB-9981548 and support from the Center for Conservation Biology and the University of California Agricultural Experiment Station. We also thank Wendy Hodges, Edith Allen, and four reviewers for their helpful comments.

16 Patterns of Decomposition and Nutrient Cycling Across a Volcanic Disturbance Gradient: A Case Study Using Rodent Carcasses

Robert R. Parmenter

16.1 Introduction

The processes of decomposition and nutrient cycling of organic substances (plant leaf litter, woody stems, roots, and animal carcasses) are critical to the functioning of ecosystems (Swift et al. 1979; Cadisch and Giller 1997). The breakdown of dead plants and animals by scavenging animals, fungi, and bacteria is the first step in the recycling of important nutrients and is necessary for maintaining the productivity potential of soil (Lal et al. 1998; Stevenson and Cole 1999). In the case of the extremely disturbed forest ecosystem of Mount St. Helens, decomposition processes, acting on the plants and animals killed in the eruption, contributed to the early stages of soil building in landscapes covered in tephra. As plant and animal populations became reestablished after the eruption, individuals of these surviving and colonizing populations eventually died and decomposed, thereby contributing additional organic and nutrient resources to the soils. The development of soil organic fractions through decomposition in an otherwise mineral substrate (pumice and tephra) has undoubtedly facilitated the continued successional development of floral and faunal communities in the disturbed zones of the volcano and will certainly continue to do so in the future.

16.1.1 Decomposition and Nutrient-Cycling Processes

The processes of decomposition and nutrient cycling are governed by three general classes of regulating factors (Heal et al. 1997 and references therein):

- Substrate quality
- Physical and chemical environmental conditions
- Composition of the decomposer community

Substrate quality refers to the chemical and physical composition of plant litter or animal carcasses, with high-quality substrate typically characterized by a high nitrogen content (usually measured relative to the carbon content in a carbon-to-nitrogen ratio, C:N). High-quality substrates are more easily decomposed by invertebrates and microorganisms than are substrates with low nitrogen content and a greater proportion of carbon. Low-quality plant litter (particularly woody materials) has high C:N ratios, with most of the carbon in the form of cellulose, hemicellulose, and lignin, all of which are resistant to decomposer activities. Leaves of plants have more favorable C:N ratios and decompose more quickly, although species-specific differences among C:N ratios influence decomposition rates: plant species with leaves high in nitrogen decompose more quickly than leaves of species with low nitrogen concentrations (Heal et al. 1997). Nitrogen content is much higher in animal bodies than in plants, and most carbon occurs in easily decomposable forms (sugars and proteins). Not surprisingly, decomposition of animal carcasses generally occurs much more quickly than that of plant litter.

Physical and chemical environmental conditions also influence the rates of decomposition processes. Temperature and moisture levels are the two major limiting factors, with decomposition rates increasing with higher temperatures and moisture availability. Temperature governs the rates at which chemical reactions proceed and therefore controls both abiotic weathering processes as well as biotic activities of invertebrates and microorganisms. Decay rates generally increase with temperature, although mortality of some arthropod decomposers may occur at higher temperatures (greater than 30°C; e.g., Richardson and Goff 2001). Similarly, moisture levels influence decomposition processes. Dry conditions produce slow decomposition rates as desiccation reduces animal and microbial activities. Moist conditions provide water for the metabolism, growth, and reproduction of fungi, bacteria, and invertebrates (see Wachendorf et al. 1997 and references therein).

Finally, the composition of the decomposer community plays an important role in determining the decomposition rates for various litter and carcass types (Heal et al. 1997). Scavenging animals that feed upon dead animals and plant parts can quickly alter the physical and chemical substrate composition, allowing accelerated activity by fungi and bacteria. The presence or absence of these decomposer taxa is often determined by the physical and chemical environmental conditions described above; however, if conditions are favorable for the activities of decomposer animals, fungi, and bacteria, then the roles of these organisms in decomposition and nutrient cycling can be far more important than that of physical weathering of litter or carcass substrates.

16.1.2 Decomposition Processes in the Posteruption Landscape

The 1980 eruption of Mount St. Helens transformed the landscape from a mosaic of old-growth forests, clear-cuts, and nonforested areas into a series of zones of varying disturbance intensity radiating out from the crater. These zones ranged from a totally vegetation-free area with mineral substrates (rock, pumice, and ash) near the volcano, through a tree-blowdown zone with former organically rich soils buried under deposits of tephra that sometimes exceeded 50 cm, to a tree-scorch zone (standing dead trees) with thinner tephra layers, to a large zone of live forest that sustained varying amounts of tephra fall (see Swanson et al., Chapter 3, this volume). Compared to undisturbed, old-growth forest, this posteruption landscape presented a gradient of surface and subsurface temperature and moisture conditions [much like the differences observed between old-growth forests and clear-cuts (Chen et al. 1995)] because of varying amounts of shade from standing trees (live or dead) and moisture-retaining, organic soil components. In the exposed areas of pyroclastic deposits (e.g., the Pumice Plain near Spirit Lake) and the blowdown zone, surface and subsurface summer temperatures in the coarse, tephra-dominated "soil" would have been higher, and moisture levels lower, than in the scorch and tephra-fall zones, where standing trees, understory vegetation, and rich organic soils promoted cooler and moister conditions (Reynolds and Bliss 1986). Indeed, soil-surface temperatures on the Pumice Plain in August 1986 were recorded with an infrared telethermometer to be in excess of 50°C (Charles Crisafulli, USDA Forest Service, Pacific Northwest Research Station, Amboy, Washington; personal communication).

These differences in the physical environment across the volcanic disturbance gradient would be predicted to influence the rates of decomposition and nutrient cycling by altering both abiotic weathering processes and, concomitantly, the decomposer community. Although surface temperatures would increase with disturbance intensity (because of greater exposure to solar radiation), moisture levels would decrease because of increased evaporation and the desiccating effects of winds across the open landscape (Reynolds and Bliss 1986; Chen et al. 1995). As a result, the combination of increasing temperatures (potentially above the tolerances of some decomposer taxa) and decreasing moisture (reducing decomposer activities) should have caused a general decline in the rates of decomposition across the disturbance gradient.

This hypothesis, however, assumes that substrate type and quality remained constant in all the disturbance zones, an assumption that clearly was not correct. In terms of substrate quality, the initial recolonizing plants in the highly disturbed zones were herbaceous species [e.g., pearly everlasting (*Anaphalis margaritacea*), fireweed (*Chamerion angustifolium*), Merten's sedge (*Carex mertensii*), and prairie lupine (*Lupinus lepidus*); del Moral 1999a; Antos and Zobel, Chapter 4, this volume] that typically occurred in open habitats, such as clear-cuts, road cuts, burned areas, or high-elevation meadows. Such plants were rarely found in undisturbed, old-growth forests. In contrast, the litter in the reference forests was dominated by conifer litter. Hence, attempting to compare decomposition processes between leafy forbs and conifer wood/foliage (having vastly different C:N ratios) would not distinguish the effects of the disturbance gradient. However, several species of small mammals did occur naturally in the highly disturbed zones, either having survived the initial eruption in underground burrow systems or having recolonized the disturbed areas shortly after the eruption (MacMahon et al. 1989; Crisafulli et al., Chapter 14, this volume). These species, mostly rodents in the family Muridae and the genera *Peromyscus* and *Microtus*, were very similar in "substrate quality" (i.e., the chemical composition of their bodies) and occurred across the entire disturbance gradient both before and after the 1980 eruption. In addition, rodents episodically attained high densities during some years following 1980 (Crisafulli et al., Chapter 14, this volume). These population excursions, coupled with these species relatively short life spans (typically less than 1 year), provided a continual and widely available supply of natural carcasses for the decomposer community. Thus, the use of rodent carcasses seemed appropriate for testing disturbance-gradient effects on decomposition and nutrient-cycling processes at Mount St. Helens.

The purposes of this study were to assess the rates of decomposition across the disturbance gradient created by the 1980 eruption of Mount St. Helens and to test the prediction that these decomposition rates would decrease with increasing intensity of volcanic disturbance. The study used rodent carcasses as the "test standard" organic substrate because these rodent species were a natural component of the regional ecosystem, occurred commonly in all disturbance zones, and were chemically very similar in carcass composition (substrate quality). Microhabitat locations (on the ground's surface or belowground in artificial burrows) were compared within each disturbance zone. In addition to testing for disturbance-gradient influences, the study also examined the sequences of nutrient losses from the carcasses and related them to particular phenomena within the decomposition process. Finally, the results of this study were compared to results of previous decomposition studies in other ecosystems, with particular attention to differences between animal-carcass and plant-litter decomposition and nutrient-cycling processes.

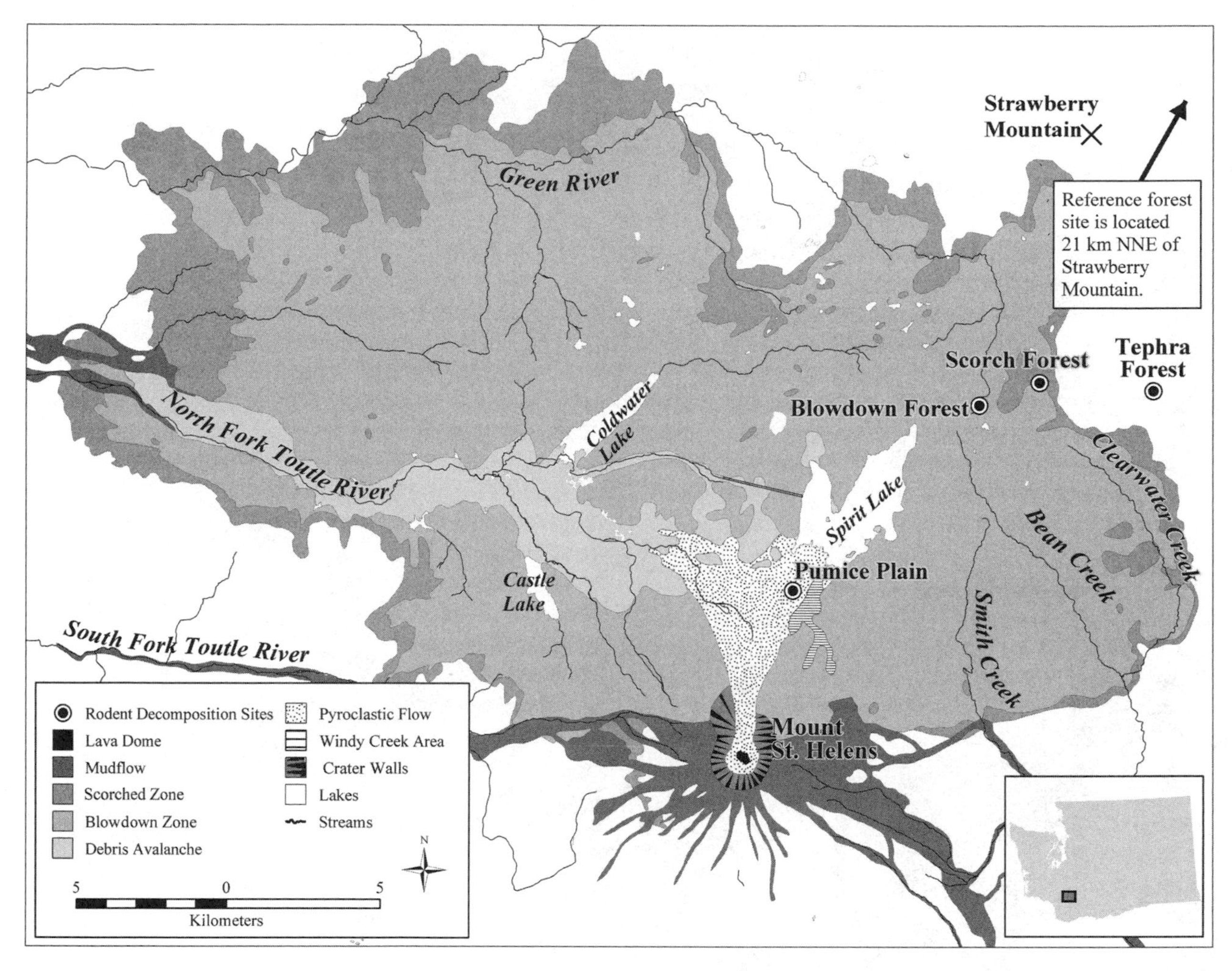

FIGURE 16.1. Locations of the study areas.

16.2 Methods

16.2.1 Study Sites

The study was conducted on five forested (or formerly forested before 1980) sites (see map, Figure 16.1) exhibiting different levels of volcanic disturbance (Figure 16.2) and forming a gradient of ambient temperature and aridity:

- The Pumice Plain in the pyroclastic-flow zone, located between the pediment slopes of the volcano and the newly formed shore of Spirit Lake.
- The blowdown zone (a previously forested site near Norway Pass).
- The scorch zone, a forested area in which standing trees were killed but not blown over in the 1980 eruption.
- The tephra-fall zone (a forested site) that received approximately 30 cm of tephra during the 1980 eruption.
- A reference site, a forest unaffected by the 1980 volcanic eruption, located on Lonetree Mountain, about 40 km north-northeast of Mount St. Helens.

These five sites provided a gradient of temperature and moisture conditions during the summer ranging from cool, moist conditions in the undisturbed "reference" forest to hot, dry conditions within the pyroclastic zone on the Pumice Plain.

16.2.2 Decomposition Study

Within each of the five study sites, carcasses of deer mice (*Peromyscus maniculatus*) and long-tailed voles (*Microtus longicaudus*) were placed in wire cages located either on the ground surface ($n = 5$/site) or belowground in artificial burrows ($n = 5$/site). Cages for surface deployment were placed along a transect line at 20-m intervals, with cage types alternating between surface and belowground microsites. The surface cages measured $25 \times 25 \times 20$ cm (L $\times$ W $\times$ H) and were made with 1.25-cm-mesh hardware cloth (see Figure 16.2). Cages were partially buried in the soil, to provide a soil–carcass contact within each cage; cages also were staked into the ground with steel rebar posts to prevent movement or dislodgement by wind, water, or animals. Artificial-burrow cages also were

FIGURE 16.2. Study sites on the Mount St. Helens volcano. (a) Pyroclastic-flow zone on the Pumice Plain with volcano crater in background. The wire-mesh cage in the foreground was used for excluding vertebrate scavengers. (b) Blowdown zone near Norway Pass. (c) Scorch zone. (d) Tephra-fall zone. (e) Reference site on Lonetree Mountain. (f) Example of naturally occurring rodent skeletal remains found on the Pumice Plain.

constructed from hardware cloth; the "burrows" were tubular, 60 cm long and 5 cm in diameter, and were dug into the ground at an approximately 45° angle; maximum carcass depth in the soil ranged from 40 to 45 cm. Paper towels were placed on top of the tube to prevent soil from filling the tube when buried. The entrance to the "burrow" was open at the ground's surface, but a hardware-cloth door prevented access by other small vertebrates. As with surface cages, the burrow tubes were staked in place with steel rebar posts.

Frozen rodent carcasses were obtained from previous studies on Mount. St. Helens. The carcasses were thawed and weighed immediately before placement in the cages. Additional "control" rodent carcasses ($n = 6$) also were weighed and then dried in a laboratory drying oven at 50°C to constant mass for the purpose of estimating the initial dry mass of carcasses placed in the field. All carcasses were put in the field during July 15–17, 1986. Carcass samples were collected at intervals of 7, 16, 37, 46, and 108 days following initial placement, with the final collection being made during October 31–November 1, 1986.

Upon collection, carcasses were weighed in the field (wet mass), and notes were taken on carcass condition and the presence and numbers of carrion-feeding arthropods. Carcasses were placed in plastic bags and put into ice-packed coolers for transport back to the laboratory, where tephra, ash, soil, and insect parts were cleaned off the carcass before further analysis. Carcasses were then dried in ovens at 50°C to constant mass and reweighed for dry mass. Dried carcasses were ground in a Wiley mill to a fine powder for nutrient analyses and calorimetry (energy content) measurements.

Nutrient concentration analyses of the "control" and field rodent carcasses were performed by the Soil Testing Laboratory at Utah State University, Logan, Utah. Nitrogen was measured using the Kjeldahl method, and other nutrients [sodium (Na), potassium (K), sulfur (S), phosphorus (P), calcium (Ca), and magnesium (Mg)] were measured with an inductively coupled argon plasma (ICAP) instrument with standard digestions and analytical procedures. Energy content (Kcal/g dry mass) was measured in the Range Science Nutritional Analysis Laboratory at Colorado State University, Fort Collins, Colorado, with an adiabatic oxygen bomb calorimeter with standard procedures.

Carcass-dry-mass data for each sample occasion were converted to percentage dry mass remaining, on the basis of the initial estimated dry mass of the carcasses. Percentage-dry-mass loss rates in each study area were calculated with one of two models, depending on how well the data from a particular study area or microsite fit each model. The first model tested was the double-negative-exponential model (Wieder and Lang 1982), which has the form:

$$PMR = Ae^{-k_{1d}t} + (1 - A)e^{-k_{2d}t}$$

where *PMR* is the percentage mass remaining (%), *A* is a mass composition proportion (%), *k* values are decomposition constants (relative mass loss per day), and *t* is time (days). This model recognizes two component fractions of materials comprising the rodent carcass, with each having a different decomposition rate (*k* value): a labile fraction (*A*) made up of easily decomposable materials (muscle and other soft tissues) characterized by a high decomposition rate (k_{1d}) and a remaining recalcitrant fraction ($1 - A$) made up of bones and hair and characterized by a much slower decomposition rate (k_{2d}). The sum of these two terms represents the total loss of mass from the carcass.

If the double-negative-exponential model did not fit the data well, a second, more simple model was used to describe the patterns of mass loss. This model was the single-negative-exponential model (Wieder and Lang 1982), of the form:

$$PMR = e^{-kt}$$

where *PMR* is the percentage mass remaining (%), *k* is the decomposition constant (relative mass loss rate per day), and *t* is time (number of days) from the start of the experiment. This model is most appropriate in cases where decomposition during the course of the study (here, 108 days) is limited to the loss of the labile fraction (soft tissues), with little observed loss of the recalcitrant fraction (bone and hair). Once calculated, the decomposition constants were then compared among sites along the disturbance gradient and between microsites (surface versus burrow).

Nutrient and energy (caloric) losses were evaluated by comparing the observed dry mass of each nutrient or energy content to the calculated initial amount for each carcass at the beginning of the study. The initial amounts were derived from the control carcasses, which provided dry:wet mass ratios and predecomposition nutrient concentrations. From each experimental carcass's wet-mass value, the estimated dry mass was obtained by multiplying the wet mass by the dry:wet mass proportion for its species (*Peromyscus* or *Microtus*). The initial amounts of each carcass's nutrient and energy content were then computed by multiplying the control-carcass concentrations by the calculated dry mass of each experimental carcass. The observed nutrient and energy amounts from the experimental carcasses at the time of their collection were then divided by the calculated initial amounts for each carcass to derive the percentage amounts remaining after *t* days of decomposition. From these values, comparisons were made to evaluate the relative rates of nutrient and energy loss from the carcasses during decomposition.

16.3 Results and Discussion

16.3.1 Composition of Rodent Carcasses

The body mass characteristics, energy values, and nutrient compositions of the control rodent carcasses are listed in Table 16.1. Although the mean carcass mass of a long-tailed vole was greater than that of a deer mouse, all other characteristics exhibited similar values between the two species.

The values for energy content and nutrient composition of the rodent carcasses were typical of small-mammal species (Gentry et al. 1975; Grodzinski and Wunder 1975) and differed markedly from plant materials. Rodent tissue energy content ranged from 5.0 to 5.5 kcal g^{-1} dry mass compared to 4.2 to 4.7 kcal g^{-1} dry mass in plant leaves, stems, and roots (Golley 1961). The rodent carcasses also proved to be a protein-rich substrate, with nitrogen concentrations in the rodents averaging 10.5% (see Table 16.1). This value was nearly 10 times higher than the nitrogen concentration in plant leaves and shoots (0.51%–1.38%) and more than 100 times higher than the nitrogen in woody-stem tissues (0.1%–0.9%; see table 4.10 in Swift et al. 1979, p. 140). The concentrations of some of the other major nutrients (phosphorus, potassium, and calcium) in the rodent carcasses also exceeded those of plant

TABLE 16.1. Composition characteristics of rodent carcasses used in the study.

Characteristic	*Peromyscus maniculatus*	*Microtus longicaudus*
Wet mass, g (mean, SE)	17.9 ± 1.2	32.0 ± 0.9
Dry mass, g (mean, SE)	5.0 ± 0.3	9.5 ± 0.2
Dry:wet mass ratio (mean, SE)	0.280 ± 0.003	0.297 ± 0.004
Total organic matter, % of dry mass	86.40	88.96
Total ash, % of dry mass	13.60	11.04
Energy content, kcal/g dry mass	5.11	5.45
Nitrogen concentration, % of dry mass	11.49	9.35
Calcium concentration, % of dry mass	3.37	3.17
Phosphorus concentration, % of dry mass	1.98	1.79
Sulfur concentration, % of dry mass	1.11	0.91
Potassium concentration, % of dry mass	1.06	1.06
Sodium concentration, % of dry mass	0.46	0.41
Magnesium concentration, % of dry mass	0.14	0.13

materials by large amounts (300%–600%; Swift et al. 1979). As such, the rodent carcasses provided a high-energy, nutrient-rich resource for decomposing organisms to use.

16.3.2 Decomposition Patterns Through Time

At the start of the experiment (July 15–17, 1986), the weather conditions were cool and rainy when the carcasses were placed in the field but became clear, hot, and dry on July 18 and remained so for the following week. During the first carcass collections (1 week into the experiment), few obvious changes in carcass condition had occurred, except that carcasses at all sites had been colonized by fly larvae (mostly flesh flies of the family Sarcophagidae). In addition, in the reference and tephra-fall sites, the surface carcasses were being eaten by ants (Formicidae: *Camponotus* sp.). At the end of the second week (July 31), the carcasses had undergone considerable change, and differences among disturbance zones and microsites had become apparent. The surface carcasses at the pyroclastic-flow, blowdown , and scorch zones appeared completely desiccated and contained no living insects (flies nor beetles), whereas the underground burrow carcasses at these sites were moist and contained many fly larvae. In the reference and tephra-fall sites, both surface and burrow carcasses were moist and contained fly larvae, although the surface carcasses supported fewer fly larvae (~50 per carcass) than did burrow carcasses (more than 500 per carcass).

By week 5 (August 21–22), all surface carcasses still appeared intact, but were dry and hollow, with most internal organs gone and no living insects present. The dried skin of the carcasses was riddled with holes from insects, and patches of fur were falling off the carcasses. Empty fly pupal cases found beneath the remaining carcasses indicated that some flies had successfully completed development on the carcass and metamorphosed into the adult stage. In contrast, the carcasses in the burrows had been reduced to a collection of cleaned bones, and only small amounts of skin and fur remained. At week 6 (August 29–30), the carcasses appeared similar to the previous week's sample, although beetle larvae (Dermestidae) had appeared at some of the surface carcasses and appeared to be feeding on the carcass skin remains. By the end of the study (week 15, October 31–November 1), the surface carcasses were still partially intact with bones and dried skin, while the burrow carcasses had been reduced to only a few of the larger bones (skull, pelvis, and femurs) and small amounts of matted fur.

16.3.3 Decomposition Patterns Across the Disturbance Gradient

The decomposition (percentage mass loss) rates in all the burrow microsites were successfully fit to the double-negative-exponential model (regression p values ranged from 0.016 to 0.066). However, this model did not significantly fit most of the surface mass-loss data sets, presumably because of the lack of observable decomposition of the bone and hair fraction. As a result, the surface data were analyzed with the single-negative-exponential model.

The decomposition rates of the rodent carcasses on the ground surface displayed an inverse relationship with disturbance intensity, in that decomposition constants (k values) decreased with increasing levels of volcanic disturbance (Figure 16.3a). The slowest rate of decomposition was observed on the pyroclastic-flow site by Spirit Lake ($k = 0.0024$), followed in increasing order by the blowdown site ($k = 0.0065$), the scorch-zone site ($k = 0.0086$), the tephra-fall site ($k = 0.0108$), and finally the reference site ($k = 0.0145$).

This sequence of decomposition constants was consistent with the hypothesis that decomposition rates would be slower in the more-exposed, high-disturbance areas than in the less-disturbed, old-growth forests; indeed, the reference-site k value was six times greater than that of the pyroclastic-flow site. If the overall hypothesis addressing the inverse effect of site disturbance intensity on k values was correct, then the k values of the five study sites would decline in a predictable order: reference > tephra fall > scorch > blowdown > pyroclastic flow.

The statistical probability that the k values of the five study sites would fall exactly in the predicted series was $p = 1/(5!) = 1/120 = 0.0083$. As shown in Figure 16.3a, the

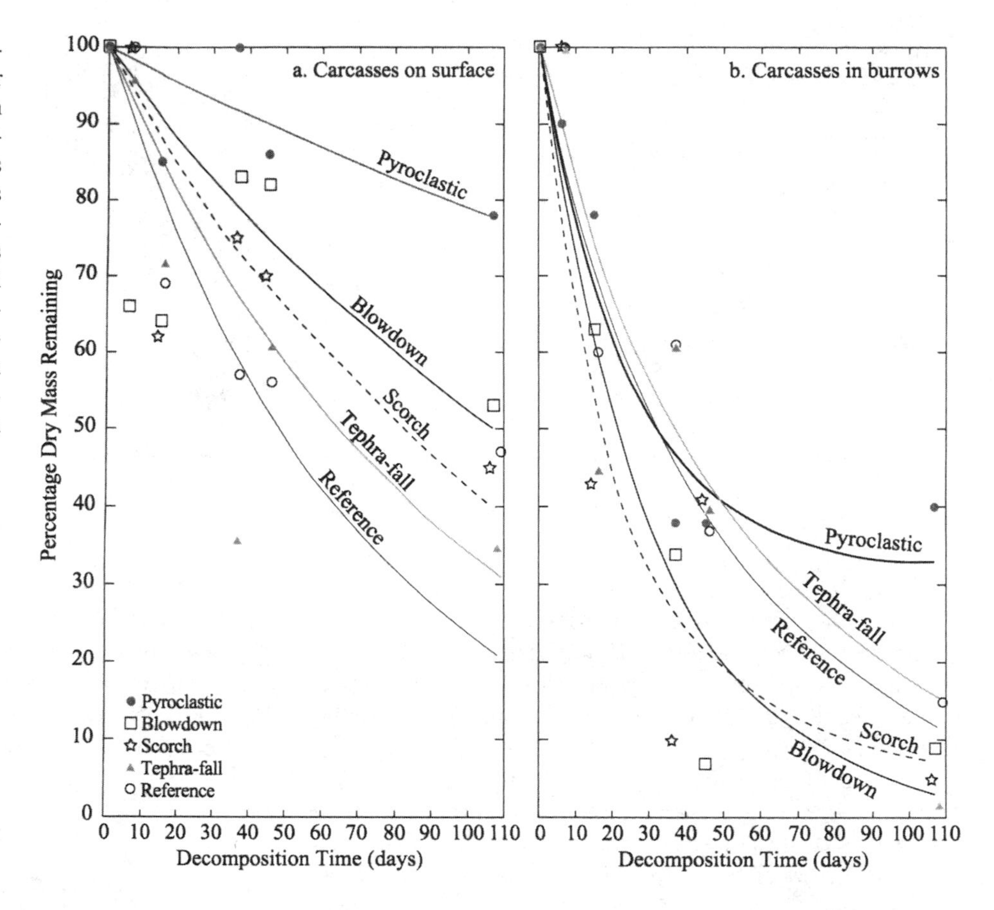

FIGURE 16.3. (a) Comparisons of single-negative-exponential-model results for rodent-carcass decomposition rates on the ground's surface across the disturbance gradient on the Mount St. Helens volcano. Note that decomposition rates (*slopes of lines*) decrease with increasing disturbance level. (b) Comparisons of double-negative-exponential-model results for rodent-carcass decomposition rates in artificial belowground burrows across the disturbance gradient on the Mount St. Helens volcano. •, pyroclastic flow; □, blowdown; ■, scorch zone; ○, tephra fall; ▲, reference.

decomposition curves for the surface carcasses followed the predicted pattern exactly, thereby supporting the hypothesis ($p < 0.01$).

In contrast to the results of the surface experiment, the results from the belowground burrow experiments showed a more consistent pattern among the study areas (Figure 16.3b). Four of the five sites exhibited similar rates of decomposition, with decomposition constants for the carcass labile fraction of $k_{1d} = 0.0419$ in the pyroclastic-flow site, $k_{1d} = 0.0328$ in the blowdown, $k_{1d} = 0.0530$ in the scorch zone, $k_{1d} = 0.0633$ in the tephra-fall, and $k_{1d} = 0.1208$ in the reference site. The decomposition constants for the recalcitrant fraction (k_{2d}) were 0.0110 in the scorch zone, 0.0161 in the tephra-fall site, 0.0191 in the reference site, and 0.0326 in the blowdown site. The pyroclastic-flow burrow treatment exhibited a k_{2d} value of 0, indicating little decomposition of bone and hair (a result corroborated by the field observations; see Figure 16.3b). With the exception of the pyroclastic-flow site, the lack of pattern across the volcanic disturbance gradient suggested that the belowground environment was generally similar across the study sites and provided comparable abiotic conditions for decomposition processes. The pyroclastic-flow site exhibited an "arrested" state of decomposition of the carcass recalcitrant fraction; it appeared that even underground, the environmental conditions on the Pumice Plain were too severe to allow for the continued breakdown of bones and hair components.

When the decomposition rates were compared between the surface and burrow experiments, it was clear that belowground decomposition processes were faster than surface processes. Rodent carcasses in four of the five burrow sites decomposed faster than those of any of the surface sites. Even on the pyroclastic flow, carcasses in burrows decomposed more quickly than those on the surface, although the absolute decomposition rates in this site's burrows were slower than the surface decomposition rates at some of the other sites. These results were consistent with the prediction that belowground conditions of moisture and temperature were more favorable to decomposer organisms (fungi, bacteria, and insects) than those on the surface, thereby allowing accelerated decomposition processes.

16.3.4 Nutrient-Loss Sequences During Carcass Decomposition

The decomposition of a rodent carcass during the first few weeks was primarily accomplished by arthropods, with physical or chemical weathering and the activities of bacteria and fungi of secondary importance. Arthropods, particularly flies and beetles, used the high-energy, protein-rich carcass as a food resource for rearing young, and as such, the soft tissues composing internal organs and musculature were the first to be devoured. Bones, skin, and hair were the most recalcitrant components of the carcass, requiring much longer periods for

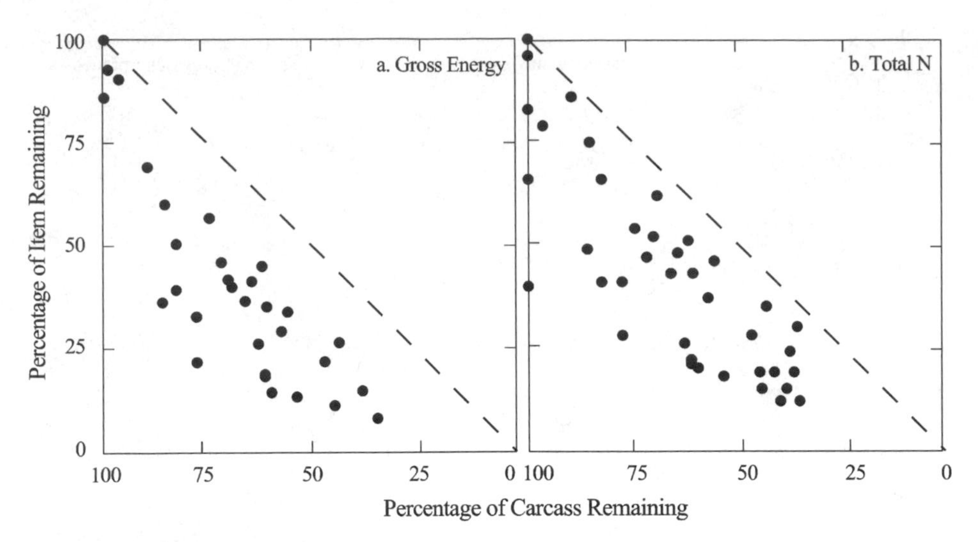

FIGURE 16.4. Scatter plots showing the relationship between carcass energy caloric content (a) and total nitrogen (b) with total remaining carcass mass. *Diagonal dashed line* represents the 1:1 loss rate (see text for details). Both caloric content and nitrogen are lost from rodent carcasses at a disproportionately high rate compared to total mass loss (i.e., all points fall below the 1:1 loss rate).

decomposition. Each of these tissues was composed of different amounts of organic and inorganic elements, and thus the sequential loss of nutrients from the carcass could be analyzed in the context of the tissue preferences of the carcass-feeding arthropods.

The main components of the carcass sought by arthropods were energy (caloric content) and protein (nitrogen content). The results of these analyses for the rodent carcasses were combined across all study areas and microsites (Figure 16.4). In the graphs, the percentage amount of energy and nitrogen remaining in the carcass is scaled against the percentage of total carcass mass remaining. A dashed diagonal line depicts the 1:1 ratio and indicates where the observed values would fall if energy content or nitrogen were lost at the same rate as total carcass mass. However, both energy content and nitrogen values fall well below the 1:1 ratio, demonstrating that both are being lost from the carcass at disproportionately high rates relative to the rate of total mass loss. This pattern is caused by the rapid removal of high-energy and nitrogen-rich tissues (internal organs and muscle) by the arthropod larvae and by the residual enrichment/accumulation of recalcitrant components.

Water-soluble salts of potassium and sodium typically occur in the rodent's blood and interstitial fluids and are subject to leaching under extremely wet conditions or to consumptive removal from the carcass by carrion-feeding arthropods. As such, these elements were lost from the carcass at rates comparable to or faster than that of nitrogen (Figures 16.4, 16.5). Sulfur, a constituent of proteins but also widely distributed in other tissues, was lost at a somewhat slower rate than was nitrogen. Calcium and magnesium, the major elements in bones, were very slow to leave the carcass (see Figure 16.5). Phosphorus is a component of bone, but also occurs in muscles in compounds such as adenosine triphosphate (ATP) and adenosine diphosphate (ADP); the portion of phosphorus in muscle tissue was lost quickly to carrion-feeding arthropods whereas the fraction in bone remained in the carcass. As a result, the net loss of phosphorus displayed a somewhat faster rate than did other bone elements (calcium and magnesium), but the loss was slower than that of the major protein constituents, nitrogen and sulfur (Figure 16.5). Arranging the results from Figures 16.4 and 16.5 in sequential order, the overall carcass loss-rate sequence was

$$K > N > Na > S > P > Mg \approx Ca$$

16.3.5 Comparisons to Plant-Litter Decomposition

In general, the decomposition of rodent carcasses observed in this study was much faster than the decomposition rates measured for a variety of plant materials (leaves and woody stems). For example, Swift et al. (1979) summarized decomposition constants (k values) for plant litter in a number of ecosystems around the world and reported that the boreal evergreen forest ecosystem had an overall annual $k = 0.21$, or a daily $k = 0.0006$, whereas the temperate deciduous forest ecosystem had an annual $k = 0.77$, or a daily $k = 0.0021$ (Swift et al. 1979, table 1.1, p. 9). In contrast, the daily decomposition constants measured during this rodent-decomposition study ranged from 0.0024 (surface carcasses on the pyroclastic flow) to 0.1208 (reference-site-burrow carcasses). This pattern is consistent with the observation from plant-decomposition studies that decomposition constants are related to the quality of the litter resource (nitrogen content and C:N ratio). That is, as litter nitrogen content increases, rates of decomposition increase (Swift et al. 1979; Cadisch and Giller 1997). Food resources richer in nitrogen attract a wider range of taxa of decomposing organisms (invertebrates, fungi, and bacteria) and make the litter easier, and more beneficial, for them to utilize. When the decomposable resource is a rodent carcass, the higher nitrogen content (greater than 10% N in rodents versus ~1% in plants) elicits a much faster response from the decomposer organisms (principally arthropods), resulting in higher decomposition rates.

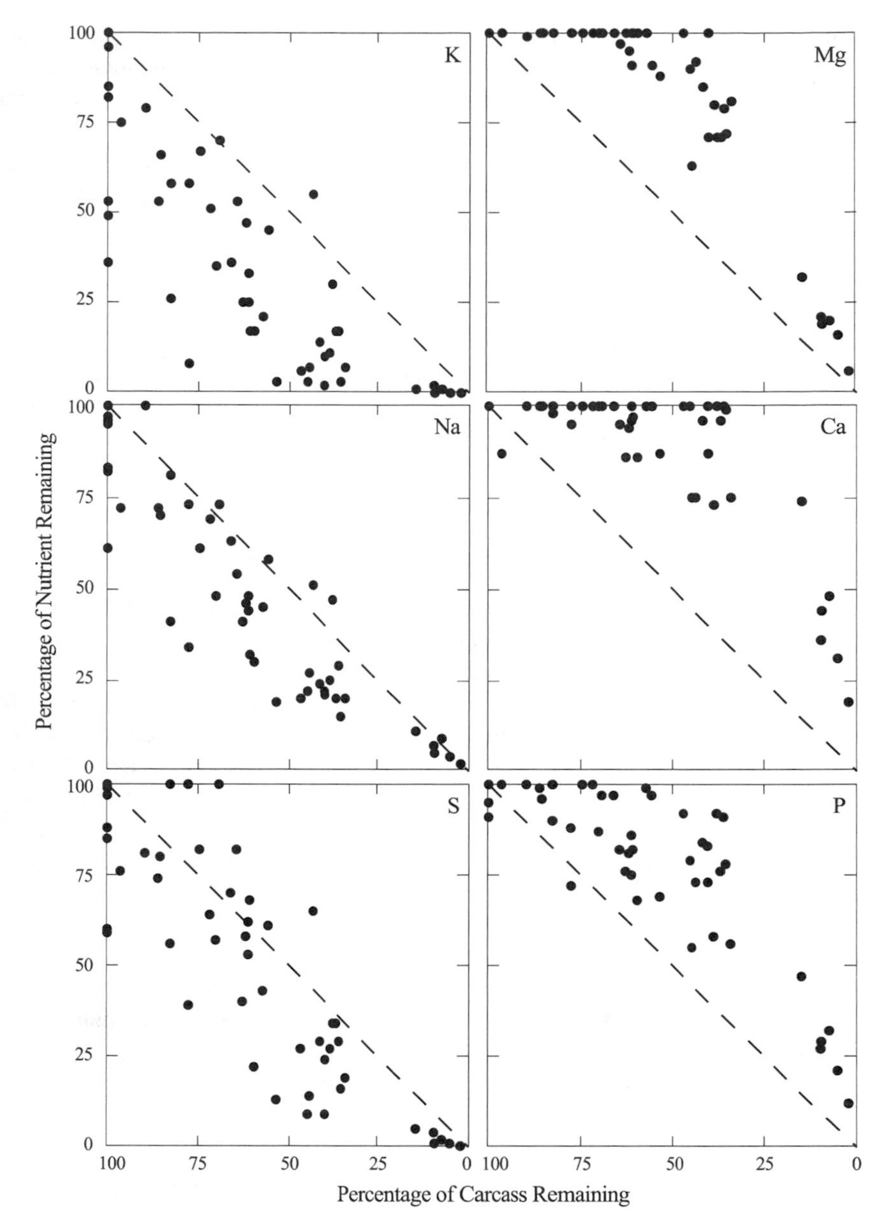

FIGURE 16.5. Scatter plots showing the relationship between various chemical elements and total remaining carcass mass. *Diagonal dashed line* represents the 1:1 loss rate (see text for details). Potassium, sodium, and sulfur are lost from rodent carcasses at a disproportionately high rate compared to total mass loss (i.e., all points fall below the 1:1 loss rate line), whereas magnesium, calcium, and phosphorus remain behind in bone materials (i.e., the majority of data points are above the 1:1 loss rate line).

Nutrient-loss sequences of rodent carcasses and plant litter exhibit both similarities and dissimilarities (Table 16.2). In both rodent and plant substrates, potassium and, in some cases, sodium are lost rapidly through leaching or tissue consumption by invertebrates, while phosphorus and calcium are often retained until later stages of decomposition. In contrast, nitrogen is lost quickly from rodent carcasses, whereas it remains in plant litter much longer. This disparity results from

TABLE 16.2. Nutrient-loss sequences from rodent carcasses compared to other substrate types.

Substrate type	Nutrient loss sequence	Source
Rodent carcass	K > N > Na > S > P > Mg ≈ Ca	This study
Fish and duck carcasses	K > Na > N > S > P > Mg ≈ Ca	Parmenter and Lamarra (1991) (in freshwater marsh)
Douglas-fir	K > Ca > N > P	Cole et al. (1967)
Oak	K > Mg ≈ P > Ca > N	Rochow (1975)
Aspen and balsam leaves	K > Na > P > Mg > Ca > N	Lousier and Parkinson (1978)
Eucalyptus leaves	Na > K > Mg > S > Ca > N > P	O'Connell (1988)
Hardwood branches (birch, maple, and beech)	K > Mg > P > S ≈ Ca > N	Gosz et al. (1973)

differences in the accessibility of nitrogen (in protein form) to the decomposer community. In rodents, the proteins in muscle and other soft tissues are readily available for consumption; but in plant litter, proteins are encapsulated within protective cell walls composed of cellulose, hemicellulose, and lignin. Decomposers must first penetrate these cell walls to gain access to the nutrients within, a process that takes considerable time.

16.4 Application of Study Results to Landscape-Level Decomposition Processes

This study produced insights into the processes of decomposition and nutrient cycling across the disturbance gradient created by the May 1980 eruption of Mount St. Helens. With a naturally occurring, "standard" organic substrate (rodent carcasses) in all disturbance zones, patterned differences in decomposition constants of carcasses on the ground's surface were observed that were consistent with prestudy predictions; however, belowground decomposition constants proved remarkably similar in all but the most disturbed environment (i.e., the pyroclastic-flow zone). In comparison with plant litter, the overall decomposition rates were much faster in rodent carcasses, and nitrogen was lost more quickly than from vegetative substrates; other nutrient-loss sequences (salt ions and minerals) appeared to be similar between plant litter and rodent carcasses.

Can the lessons learned from this rodent-based study be applied to the decomposition processes acting on the vastly larger pools of tree, forb, and grass litter across the Mount St. Helens landscape? Although studies of plant-litter decomposition have not been conducted as yet, one would predict that the more-exposed, drier environments in the pyroclastic-flow and blowdown zones would inhibit decomposition rates by reducing or preventing altogether the activities of many decomposer organisms (e.g., microarthropods, fungi, and bacteria). As vegetation continues to reoccupy the disturbed regions near the volcano, leaves and conifer needles accumulate on the ground beneath the plants or in foci of redistribution from wind and water. Through the decomposition process, these litter fragments will be recycled at a rate dependent on the local composition of the decomposer community and physicochemical environmental conditions. One would hypothesize that the decomposition rates of any particular plant species (e.g., the prairie lupine, a common inhabitant of the Pumice Plain) would be slower on the exposed, highly disturbed pyroclastic flow than in an undisturbed, moist mountain meadow because the species of decomposing organisms that would normally process this litter type may not function as well in the dry, tephra-derived soils of a pyroclastic-flow deposit. Of course, future studies are needed to determine if this is indeed the case, but the results of the rodent-carcass study would suggest that, given a common substrate, decomposition rates will vary predictably across the disturbance gradient.

The major limitation of the rodent carcass decomposition study is that it cannot provide quantitative landscape-level estimates of decomposition mass transfers and nutrient pools. Small-mammal biomass contributes only a small fraction of the total ecosystem biomass in any of the disturbance zones, and the absolute decomposition rates and nutrient-loss sequences reported here cannot be extrapolated to plant-litter decomposition. Future studies are needed that quantify the amounts and temporal dynamics of standing and down woody debris, the annual production of litter from recolonizing plants, and the accumulation of organic matter in the soils to estimate the absolute rates of decomposition, nutrient cycling, and soil carbon storage in the disturbed ecosystems on Mount St. Helens.

17 Lupine Effects on Soil Development and Function During Early Primary Succession at Mount St. Helens

Jonathan J. Halvorson, Jeffrey L. Smith, and Ann C. Kennedy

17.1 Introduction

The pyroclastic flows of Mount St. Helens remain important to scientists seeking to understand the mechanisms of early succession. As in other primary-succession systems, biotic and abiotic development on these sites has been strongly influenced by legume colonists. Legumes are postulated to be critical contributors to nutrient pools during early succession, especially in infertile volcanic substrates, and are thought to facilitate colonization and growth of subsequent species that are limited by soil organic matter and by availability of critical nutrients such as nitrogen (Chapin et al. 1986; Franz 1986; Mooney et al. 1987; Vitousek et al. 1987; Chapin et al. 1994; Ritchie and Tilman 1995).

Two species of lupines, broadleaf lupine (*Lupinus latifolius*), an upright deciduous species, and prairie lupine (*Lupinus lepidus*), a prostrate wintergreen species (Braatne and Chapin 1986; Braatne and Bliss 1999), were among the initial colonists of cool mudflows (lahars) and pyroclastic deposits at Mount St. Helens. Photosynthesis, litter input, and symbiotic nitrogen fixation by lupines contributed to the formation of highly localized soil-resource islands, characterized by higher concentrations of carbon and nitrogen than in the surrounding soil, and supported larger and more diverse active populations of heterotrophic soil microorganisms in the soil (Halvorson et al. 1991b; Halvorson and Smith 1995).

Lupines growing on pyroclastic deposits have influenced revegetation of the site by other plants (Morris and Wood 1989; del Moral and Bliss 1993; del Moral and Wood 1993a,b; del Moral 1998), insect distributions (Fagan and Bishop 2000; Bishop 2002), soil resources, and microorganisms (Allen et al. 1984; Allen 1987, Carpenter et al. 1987; Halvorson et al. 1991b; Halvorson and Smith 1995; Titus and del Moral 1998b).

This chapter summarizes research about the impacts of lupines on soil that in turn influenced plant and soil microbial-community development of Mount St. Helens pyroclastic deposits. The following sections provide conceptual frameworks for viewing succession as a belowground process, describe the rationale and analytical approach of lupine-soil research at the pyroclastic study site during the first 20 years after the eruption, summarize research findings, and suggest topics for future research.

17.2 Conceptual Approaches to Soil Succession

17.2.1 Succession as a Belowground Process

Studies of aboveground succession often emphasize development of ecosystem form by focusing on predictable changes to the species structure of communities over time. Alternatively, studies of belowground succession of ecosystems more often focus on development of soil function because succession implies a series of changes to infertile substrates that result in increased aboveground productivity; ecosystem services; and development of traits, such as resistance and resilience to perturbations (Parr et al. 1992). This fundamental link between soil attributes and processes and ecosystem development, function, and integrity is increasingly recognized (Thompson et al. 2001). Although often viewed as an aboveground process, early succession may instead be viewed as primarily a belowground phenomenon fueled by the substantial proportion of net fixed carbon transferred through roots to heterotrophic microorganisms. Fresh photosynthates are needed by symbiotic bacteria for dinitrogen (N_2) fixation, whereas litter inputs, carbon flow through roots (termed rhizodeposition), root sloughage, and live root material provide the carbon substrates whose supply controls microbial activity and thus nutrient cycling (Lynch and Whipps 1991). In addition to nutrient cycling, microorganisms found in the zone of soil immediately adjacent to plant roots (the rhizosphere) affect other plant–environment relationships (e.g., soil stabilization and water-holding capacity), plant–plant relationships (e.g., allelopathy and biocontrol), and plant–microbial interactions (e.g., competition, phytotoxicity, and disease) (Kennedy 1997).

Effects of disturbance and mechanisms of ecosystem response to the eruption at Mount St. Helens are discussed in

Chapter 1 in this book and elsewhere (del Moral and Bliss 1993; Dale et al. 2005) and reflect, in general, an aboveground, plant-centered perspective (see Lindahl et al. 2002). However, other approaches usually associated with soil genesis are also conceptually useful for studies of soil ecosystem development of Mount St. Helens pyroclastic-flow substrates. The state-factor approach, popularized by Jenny (1941), can be paraphrased to describe soil succession as a function of parent material being acted upon by climate and biotic activity over relief through time (Nuhn 1987). Soils undergoing primary succession are initially very similar to their parent material and are thus strongly influenced by the type and magnitude of the initial site-formation processes (Dale et al. 1998; Turner and Dale 1998). As succession proceeds, soil ecosystems are increasingly influenced by vegetation and the cumulative effects of climate. In 1986, soil textural properties on the Mount St. Helens pyroclastic-flow deposits were still largely unchanged from the initial deposit, but lupines affected the formation of an incipient A horizon and accumulations of windborne sands and other material (Nuhn 1987).

With another conceptual approach borrowed from pedology, soil-ecosystem succession can be viewed as resulting from changes in the net balance of soil additions, losses, translocations, and transformations (Simonson 1959). Physical additions and losses, resulting from mechanical processes such as deposition or erosion, and changes in soil chemistry, influenced by precipitation and leaching, may be particularly important during early succession. Important transformations or translocations during succession are often mediated by biotic activities that conduct energy through or recycle matter within system boundaries and are characterized by feedback loops and the exchange of information. For example, microbially mediated decomposition and transformation of soil organic matter are the primary driving forces in nutrient cycling, which plays a significant role in ecosystem development and functioning (Smith 1994).

17.2.2 Studying Succession in a Soil Ecosystem: A Problem of Integration

Measuring accumulations of individual soil properties does not equate to measuring soil succession. Although conceptually clear, the terms ecosystem and succession can be operationally difficult to characterize or monitor. No single operational definition of succession in a soil ecosystem seems entirely satisfactory, in part because the definition is context dependent, not simply related to a set of abiotic or biotic standards, such as the accumulation of nutrients, rates of energy and mass cycling, or presence of a particular species. For example, even though nitrogen fixers such as lupines are associated with primary succession, there is no clear correlation between their presence and nitrogen accumulation in soil and no consensus about their role in succession (Walker 1993). Although availability of soil nutrients seems to affect characteristics such as the growth rate of dominant species or community richness, patterns of plant development do not appear to be closely linked to heterogeneity of nutrients in the soil (Collins and Wein 1998).

Because soil ecosystems are complex, scientists generally agree that several kinds of data, including physical, chemical, and biological properties, must be integrated to evaluate the whole (Gregorich et al. 1994; Doran and Parkin 1996). However, further work is needed to link above- and belowground processes for studies of succession. Data from plot-scale, relatively short duration studies must be scaled appropriately to be useful for understanding long-term, landscape-scale processes. The idea of studying succession from a soil perspective also implies the emergence of collective soil properties that are not predictable as a simple function of individual soil components and whose spatial and temporal patterns need not be correlated to those of any individual soil variable (Halvorson et al. 1995a, 1997). Various approaches for integration of individual soil properties have been advocated and applied to agricultural land at field scales (Halvorson et al. 1995b); to regional scales (Brejda et al. 2000a,b); and to less intensively managed landscapes, including rangeland and forestland (Herrick and Whitford 1995; Karlen et al. 1998; Burger and Kelting 1999; Herrick et al. 2002).

17.3 Rationale and Methods of Research

Much of the research described here was conducted on the Pumice Plain pyroclastic-flow deposits (46°12′ N 122°11′ W), located at an elevation of about 1160 m near Spirit Lake, where the substratum is rocky volcanic material deposited by numerous pyroclastic flows that occurred on the north face of Mount St. Helens (see Swanson et al., Chapter 3, this volume). This volcanic parent material was initially dominated by coarse (30- to 200-mm) particles, possessed little structure or organic material, and was very low in nutrients (del Moral and Clampitt 1985; Nuhn 1987; del Moral and Bliss 1993). Prairie lupine (*Lupinus lepidus*) was the predominant lupine colonist, appearing on pyroclastic-flow deposits soon after the eruption as isolated individuals and later comprising nearly monospecific patches of irregular size and distribution. More recently, individual patches have converged, and other plant species have become established (see del Moral, Chapter 7, this volume).

Early studies of the impact of lupines on Mount St. Helens soil succession determined the timing and magnitude of nitrogen fixation by lupine colonists to correlate nitrogen fixation with lupine physiology and to provide data needed for constructing a nitrogen budget (Halvorson et al. 1991a, 1992). Both daily and seasonal patterns of nitrogen fixation were investigated in 1986 and 1987 by determining the activity of nitrogenase, the enzyme that fixes nitrogen, by measuring the reduction of acetylene to ethylene by nitrogen-fixing organisms found in nodules on lupine roots. Nitrogenase activity was related to nitrogen fixation in lupines more directly by using stable-isotope (in this case, nonradioactive ^{15}N) dilution techniques (Hauch and Weaver 1986). With this approach, the amount of nitrogen fixed is determined by comparing the proportion of ^{15}N in lupines to that in nonfixing plants grown in field soil with ^{15}N-containing fertilizer added to it.

Studies soon shifted to understanding more about the relationship between lupines and changes in properties especially patterns of soil carbon and nitrogen and soil microbial communities. Immediately following the May 18, 1980, eruption, the pyroclastic-flow deposits seemed to be a clean slate of nearly sterile substrate with low fertility, little spatial variability or stratification of soil properties, and a high degree of homogeneity. Lupines were expected to strongly influence patterns of ecosystem development by affecting soil genesis, soil microorganisms, and subsequent plant establishment (Engle 1983; Franz 1986; Nuhn 1987; Rossi 1989; Fagan and Bishop 2000).

Lupine influence on soil pools of carbon and nitrogen and on microbial activity was evaluated by comparing lupine root-zone soil to uncolonized soil collected in 1987 from the pyroclastic-flow deposits and several other locations along a gradient of carbon and nitrogen levels related to volcanic disturbance by the 1980 eruption (Halvorson et al. 1991b). Lupine effects, it was hypothesized, would be most clearly observed in sites with low total soil carbon and nitrogen and would subsidize soil microorganisms, resulting in a larger soil microbial biomass and more rapid nutrient cycling under lupines than in uncolonized soil. Soil was again sampled in 1990 from under living and dead prairie lupine, live broadleaf lupine, and bare uncolonized soil to determine the vertical distribution of microorganisms and nutrient pools, observe the effects of additions of lupine biomass on soil microbial activity, and relate any differences in soil microbial activity to differences in lupine tissue composition (Halvorson and Smith 1995). For these studies, sieved ($\leq$2 mm) soils were analyzed for pH with a 1:1 deionized soil-water paste. Fresh soil was extracted with 2 N KCl for determination of inorganic nitrogen (NH_4 nitrogen and NO_3 nitrogen) with an autoflow colorimetric procedure. Total organic carbon was measured titrometrically after wet oxidation, and total soil nitrogen was determined colorimetrically after Kjeldahl digestion. Microbial biomass carbon was determined with the substrate-induced respiration method (SIR-C) and product-formation equations. Soil respiration and nitrogen mineralization rates were determined from aerobic incubations.

Studies of the effects of lupines on soil development were initially limited to the analysis of relatively small numbers of samples collected beneath plants versus samples collected "away" from plants. This binary approach was sufficient to detect small-scale enriched zones in the soil under lupine plants, called resource islands and characterized by greater concentrations of soil nutrients, and larger, more active populations of soil microorganisms. This approach, however, was unable to provide detailed knowledge of the size and internal dynamics of resource islands, important for understanding energy-flux, mass-transport, and nutrient-cycling processes at a scale beyond the individual plant. To gain more information about the spatial relationships of resources in pyroclastic soil, geostatistics were employed with samples collected in 1991.

The use of geostatistics to model the spatial variability of ecologically important soil characteristics has been well documented (Warrick et al. 1986; Robertson 1987; Rossi et al. 1992; Goovaerts 1998). In particular, geostatistics have been used to model changes in soil patterns during succession (Robertson et al. 1988; Gross et al. 1995) and relationships between individual plants and resource-island patterns in the surrounding soil (Jackson and Caldwell 1993; Halvorson et al. 1994, 1995a). In general, geostatistics characterize the similarity of samples as a function of distance or direction, termed spatial autocorrelation. The model of this relationship is then used to estimate values at unsampled locations, often applying some variation of an estimation procedure called kriging (Isaaks and Srivastava 1989).

Recently collected data have been used to place the rates of carbon and nitrogen accumulation in pyroclastic-flow deposits in historical context and to relate patterns of organic matter in pyroclastic-flow soil to the diversity and function of soil microbial communities (Frohne et al. 2001; Halvorson and Smith 2001). Samples collected in 2000 from the pyroclastic flow and other sites around Mount St. Helens were used to determine how soil properties had changed over time, the rates of carbon and nitrogen accumulation in soil, and relationships between populations of soil microorganisms and soil-carbon quantity and quality. These studies used many of the same analytical procedures as earlier work except that total soil carbon and nitrogen were determined by dry combustion, and soil enzymes, phosphatase (pNP), and dehydrogenase (TPF) were assayed with standard methods.

Recent studies have also investigated patterns of fatty-acid methyl esters (FAME) and of carbon-substrate use in pyroclastic soil (Frohne et al. 1998, 2001). FAME are derived from important membrane constituents of plants and microorganisms, called phospholipids, found in the soil and can be used to characterize both the current soil microbial community (i.e., cellular lipids) and the paleogeochemistry (i.e., the extracellular lipids) (Kennedy 1994; Buyer and Drinkwater 1997; Zelles 1999; Hill et al. 2000; Pinkart et al. 2002). Patterns of FAME have been used to distinguish soil ecosystems from one another and to measure the adaptation of soil microorganisms to changing soil conditions (Ibekwe and Kennedy 1999; Drijber et al. 2000; Mummey et al. 2002).

We extracted FAME from archived pyroclastic-flow-deposit samples from previous studies and fresh samples collected in 1997 and in 2000 to compare FAME near lupine plants to that in uncolonized soil. FAME profiles, composite information about the amount and identity of individual fatty acids in a soil sample, were summarized and compared with principal-component analysis (PCA), a statistical approach that can be used to study the correlations of large numbers of variables by grouping them into "factors" so that the variables within each factor are more highly correlated with variables in that factor than with variables in other factors. The spatial variation of FAME profiles was also characterized with transect or gridded samples and geostatistics.

Patterns of carbon-substrate use, determined by exposing microorganisms to different carbon substrates, have also been suggested as a means for assessing the potential function of soil microbial communities or the quality of substrates in the soil available to heterotrophic soil microorganisms (Sicilianoa and Germidaa 1998; Schutter and Dick 2001; Widmer et al. 2001). Pyroclastic soil collected in 2000 was used to distinguish the metabolic characteristics of soil microorganisms in lupine-influenced soil from those in bare soil by measuring the patterns of use of 31 carbon sources with Biolog Ecoplates (Biolog, Hayward, CA, USA).

17.4 Summary of Results and Discussion

17.4.1 Nitrogen Fixation by Lupines

Nitrogenase activity in prairie lupine growing at a pyroclastic site exhibited significant diurnal trends (with lowest rates at night) apparently related to plant photosynthesis. Nitrogen-fixing activity also followed seasonal trends with high rates in June, very low levels in August, the dry and warm part of the summer season, and a partial recovery of nitrogenase activity in September after precipitation resumed. Adult lupine carbon and nitrogen composition also varied during the growing season, with trends correlated to seasonal patterns of nitrogenase activity.

The ^{15}N isotope-dilution technique showed nitrogenase activity to be related to seasonal nitrogen fixation in both prairie and broadleaf lupine (Halvorson et al. 1992). During one season of growth, both species fixed about 60% of their nitrogen with some evidence of preferential allocation to aboveground biomass. Prairie lupine fixed about 18.1 mg N g^{-1} biomass or an average of 15.4 mg N per plant, whereas broadleaf lupine fixed an average of 16.3 mg N g^{-1} biomass, equivalent to 22.9 mg per plant. Average net carbon fixation during the same period was 355 and 589 mg per plant for prairie and broadleaf lupine, respectively.

The potential rates of nitrogen fixation by lupines growing on the pyroclastic-flow deposits were similar to values reported for other volcanic sites (Kerle 1985). In 1986 and 1987, these nitrogen inputs were most important at the local scale with total annual fixation rates estimated to be less than 0.05 kg N ha^{-1} because prairie lupines were distributed into small patches covering less than 1% of the pyroclastic surface area. By 2000, even though lupines covered more than 50% of some parts of the pyroclastic-flow area (Bishop et al., Chapter 11, this volume), total annual symbiotic nitrogen additions may not have been the predominant source of nitrogen input (Engle 1983; Halvorson et al. 1992), a conclusion reached in other studies (e.g., Wojciechowski and Heimbrook 1984; Holtzmann and Haselwandter 1988; Walker 1993). By comparison, atmospheric nitrogen inputs at the pyroclastic-flow site are about 2 kg N ha^{-1} $year^{-1}$ if similar to Oregon Cascade Range forest sites (Sollins et al. 1980; Vanderbilt et al. 2003).

17.4.2 The Importance of Nitrogen Fixation by Seedlings During Early Succession: Age-Specific Nutrient Cycling

The relationship between age cohort and nutrient cycling in early succession systems was postulated by Halvorson et al. (1991a), who examined the possible importance of lupine seedlings as agents of carbon and nitrogen inputs into Mount St. Helens pyroclastic deposits. Greenhouse-grown prairie and broadleaf lupine had different patterns of germination, growth, and allocation of carbon and nitrogen that were apparently related to phenology. Prairie lupine germinated rapidly and directed the majority of its resources to aboveground growth. Broadleaf lupine germinated slowly, allocating more carbon and nitrogen to belowground growth. Significant rates of nitrogenase activity were observed in both species within 2 weeks after planting, suggesting an early reliance on atmospherically derived nitrogen by young seedlings, at least those growing in infertile soil. The greatest amount of nitrogenase activity per gram of nodule occurred about 6 weeks after planting and then declined in a pattern similar to that observed for adult lupines in the field. The greatest total plant nitrogenase activity occurred after peak nodule activity because of increasing nodule biomass.

These patterns, together with estimates of seedling population density and mortality, suggest dense populations of seedlings might be significant contributors to soil nitrogen pools because they allocate a large proportion of their photosynthate to nitrogen fixation and move carbon and nitrogen into soil relatively rapidly. Seedling mortality could be particularly important during early succession because the availability of the carbon and nitrogen to other organisms would depend upon lupine decomposition. Although seedling populations and environmental conditions vary from year to year, many seedlings do not survive summer drought, insect predation, or winter conditions during the first year (Wood and del Moral 1987; Braatne 1989). Lower C:N ratios of seedling biomass compared to adult biomass (Halvorson et al. 1991a) would probably result in relatively rapid mineralization by soil microorganisms (Waksman and Tenney 1927).

Established adult lupines are more likely to withstand the prolonged periods of heat and low precipitation at Mount St. Helens that can kill up to 60% to 100% of seedlings (Braatne and Chapin 1986; Braatne 1989). Thus, adult lupines contribute proportionally smaller annual amounts of carbon and nitrogen to the ecological system over several seasons through leaf senescence, rhizodeposition, and mortality, which are influenced by phenological characteristics (e.g., whether the plant is wintergreen or deciduous). In perennial lupine species, adult lupine carbon and nitrogen may be sequestered for several years in living material or in more decomposition-resistant forms after death. However, carbon and nitrogen sequestered in adult lupines may be released in an episodic "pulse" if mass mortality of adult lupines occurs because of factors such as insects (Bishop et al., Chapter 11, this volume).

17.4.3 Lupine Impacts on Soil Properties

Lupines began to influence soil development soon after colonization, and their effects are clearly evident more than 20 years after the eruption. In 1987, pyroclastic-flow deposits contained only small amounts of carbon and nitrogen, less than 5% of a minimally disturbed forest soil (Halvorson et al. 1991b). However, significantly higher pH and concentrations of carbon and nitrogen and a significantly lower average soil C:N ratio were recorded in lupine root-zone soil than in uncolonized soil (Table 17.1, 1987 data). Total inorganic-nitrogen concentrations in pyroclastic-flow deposits were small but comprised a significant proportion of the total amount of soil nitrogen, about 5% of lupine root-zone soil and 15% of uncolonized soil. In general, lupine root-zone soil contained similar amounts of NO_3^+ nitrogen but significantly more NH_4^+ nitrogen than did uncolonized soil. Potential nitrogen mineralization rates measured in the laboratory were of the same order of magnitude as total inorganic-nitrogen pools in the field, suggesting the transformation of nitrogen from organic material into the inorganic forms used by plants is rapid and important in early succession within low-carbon and -nitrogen sites.

Soil pH in pyroclastic-flow deposits, first reported to be around 6.0 in 1980 and 1981 (Engle 1983; del Moral and Clampitt 1985), decreased to about 5.2 under lupines and to 5.0 in uncolonized soil by 1987. A significant drop in pH through time was thought to be the effect of acid precipitation (Nuhn 1987). Pyroclastic-flow deposits had pH values comparable to undisturbed forest sites in 1987 but had almost no detectable buffering capacity (Halvorson 1989).

Stratification of soil organic carbon is a feature in the top meter of soil under forest vegetation growing in cool, humid climates (Jobbagy and Jackson 2000), a pattern also developing under lupines. By 1990, soil under lupines contained significantly higher concentrations of total Kjeldahl nitrogen (TKN), total organic carbon (TOC), and microbial-biomass carbon near the surface (0 to 5 cm) that decreased significantly with depth. Less evidence for stratification was observed in uncolonized soil, which also contained much lower concentrations of carbon and nitrogen. Soil under the deciduous broadleaf lupine contained significantly more TKN, TOC, and water-soluble carbon than that under dead or live prairie lupine, a wintergreen species, or in uncolonized soil, showing the influence of nutrient inputs from leaf litter. Carbon in surface soil was strongly and positively linearly correlated with nitrogen (Halvorson and Smith 1995).

In 2000, the soil under prairie lupine exhibited greater electrical conductivity (EC) and higher concentrations of carbon and nitrogen than uncolonized soil at the pyroclastic-flow site (Table 17.1). When compared to other Mount St. Helens sites

TABLE 17.1. Average chemical and biological soil-quality indicators of the Mount St. Helens pyroclastic flow.

Year	Surface	pH[a]	EC[b]	Total soil carbon[c]	Total soil nitrogen[d]	C:N ratio[e]	Soil respiration[f]	Net nitrogen mineralization[g]	Dehydrogenase[h]	Phosphatase[h]	Microbial-biomass carbon[i]
2000											
0–5 cm	Never colonized	5.05^{A}	76^{B}	1284^{B}	114^{B}	13.9^{A}	4.1^{A}	0.11^{A}	4^{B}	10^{B}	93^{B}
0–5 cm	Under live *Lupinus lepidus*	5.21^{A}	138^{A}	4604^{A}	384^{A}	12.8^{A}	13.6^{A}	0.21^{A}	24^{A}	134^{A}	377^{A}
1987											
0–20 cm	Never colonized	5.04^{X}	NM	590^{X}	17^{X}	33.5^{X}	1.5^{X}	0.32^{X}	NM	NM	3
0–20 cm	Under live *L. lepidus*	5.19^{Y}	NM	840^{Y}	53^{Y}	14.7^{Y}	5.9^{Y}	0.37^{X}	NM	NM	7

Data are from Halvorson (1989), Halvorson et al. (1991b), and Halvorson and Smith (2001). Significant differences between the two types of soil samples, collected under live *Lupinus lepidus* and in uncolonized areas, are indicated by different letters. For 2000, differences are based on Bonferroni adjusted tests (denoted A or B); $p \leq 0.05$. For 1987, differences are based on Duncan's multiple range test (denoted by X or Y).
NM, not measured.
[a] Determined 1:1 (soil:water).
[b] $\mu S\ cm^{-1}$.
[c] mg kg^{-1} soil, assayed by wet combustion 1987 and by dry combustion in 2000.
[d] mg kg^{-1} soil, assayed as Kjeldahl nitrogen in 1987 and by dry combustion in 2000.
[e] Test of significance for C:N done on log_{10}-transformed data.
[f] mg CO_2 carbon kg^{-1} soil day^{-1} after 10 (1987) or 11 (2000) days ($n = 3$).
[g] mg N kg^{-1} soil day^{-1} after 10 or 11 days ($n = 3$).
[h] mg TPF (dehyrogenase) or pNP (phosphatase) kg^{-1} soil.
[i] Derived from substrate-induced respiration; mg kg^{-1} soil.

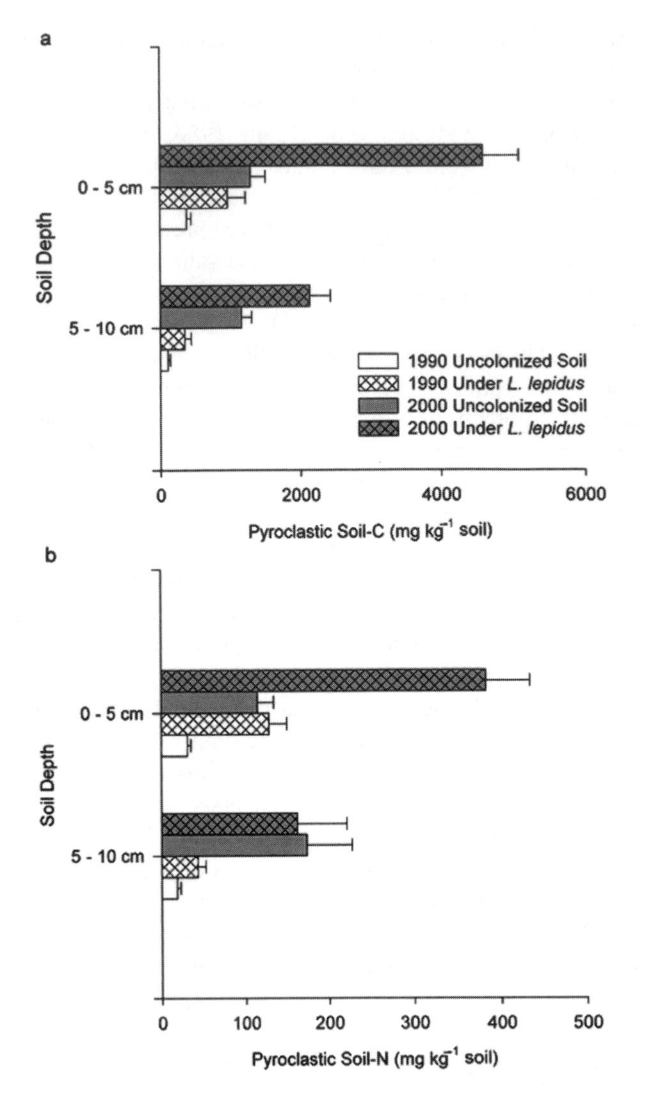

FIGURE 17.1. Vertical distribution of mean (a) soil carbon and (b) soil nitrogen in pyroclastic substrate in 1990 and 2000. *Error bar* is the standard error.

that received only air-fall deposits of tephra, the pyroclastic-flow site had significantly higher EC, less carbon and nitrogen, and lower C:N ratios. Samples collected in 2000 also showed the greatest concentrations of both carbon and nitrogen at the surface under prairie lupine, decreasing significantly with depth (Figure 17.1). However, no significant stratification of carbon and nitrogen was observed in uncolonized soil. Similar patterns were observed for soil enzymes, suggesting lupines are increasing both the quantity and the spatial hetcrogeneity of chemical and biological variables related to soil quality.

17.4.4 Rates of Carbon and Nitrogen Accumulation in Soil

Soil carbon and nitrogen are increasing at the pyroclastic-flow site at rates comparable to those of other North American volcanic soils. Initial concentrations were indistinguishable from zero soon after the eruption (Engle 1983), but by 1987 total organic carbon and nitrogen had accumulated in uncolonized soil at rates of about 84 and 2.4 mg kg^{-1} $year^{-1}$ and under lupines at about 120 and 7.6 mg kg^{-1} $year^{-1}$ (Halvorson et al. 1991b). In comparison, Griggs (1933) observed nitrogen concentrations equivalent to accumulation rates of 3.3 mg kg^{-1} $year^{-1}$ 13 years after the initial measurements of Alaskan Katmai ash deposits by Shipley (1919). Similarly, Dickson and Crocker (1953) reported nitrogen concentrations equivalent to accumulation rates of 3.5 and 8.7 mg kg^{-1} $year^{-1}$ in the 0- to 12.7-cm depth of 27- and 205-year-old deposits from Mount Shasta, California. In these same deposits, they measured organic carbon concentrations equivalent to 44 mg kg^{-1} $year^{-1}$ and 188 mg kg^{-1} $year^{-1}$, respectively.

By 2000, carbon and nitrogen had increased in uncolonized soil at an average annual rate of about 64 and 5.7 mg kg^{-1} soil, respectively, compared to 230 and 19.2 mg kg^{-1} soil in lupine-colonized soil (Figure 17.2). These concentrations are equivalent to net increases of about 42 kg C ha^{-1} $year^{-1}$ and 3.7 kg N ha^{-1} $year^{-1}$ for uncolonized soil and about 150 kg C ha^{-1} $year^{-1}$ and 12.5 kg N ha^{-1} $year^{-1}$ for lupine-colonized soil in the surface 5 cm, assuming a bulk density of 1.3. Significant ($p > 0.05$) quadratic orthogonal contrasts indicate that rates of accumulation are increasing. Between 1997 and 2000, carbon and nitrogen increased in uncolonized soil by an average of 128 and 168% (241 and 24 mg kg^{-1} soil $year^{-1}$) and under lupines by 42% and 68% (452 and 52 mg kg^{-1} soil $year^{-1}$). Increasing rates of accumulation after an initial period of quiescence have also been reported for other locations undergoing primary (Chapin et al. 1994) and secondary (Zak et al. 1990) succession.

17.4.5 Lupine Effects on Soil Microbial Carbon and Activity

Within 18 months after the 1980 eruption, pyroclastic-flow substrates were dominated by bacteria but exhibited low respiration rates (Engle 1983). Similarly, Halvorson et al. (1991b) reported very low amounts of active soil-microbial-biomass carbon and cumulative respiration in pyroclastic-flow soil in 1987. However, overall microbial activity was correlated with soil carbon and nitrogen and was significantly higher in prairie lupine root zones than in uncolonized soil. The average soil-microbial-biomass carbon to soil carbon ratio was less than 1% in uncolonized soil and under lupines, indicating very low substrate availability. However, this result may also be an artifact of analytical difficulties encountered in measuring very low values for microbial and total soil carbon. Adding glucose, a carbon substrate easily used by soil microorganisms, to soil collected from several sites along a volcanic disturbance gradient increased respiration in all soils tested. The greatest relative responses were observed in pyroclastic-flow soil at the low end of the carbon and nitrogen gradient. Respiration responses in pyroclastic soil occurred only after a relatively long lag time compared to other soils, indicating a small soil

FIGURE 17.2. Change in concentration of (a) soil carbon, and (b) soil nitrogen in Mount St. Helens pyroclastic deposits after the May 18, 1980 eruption.

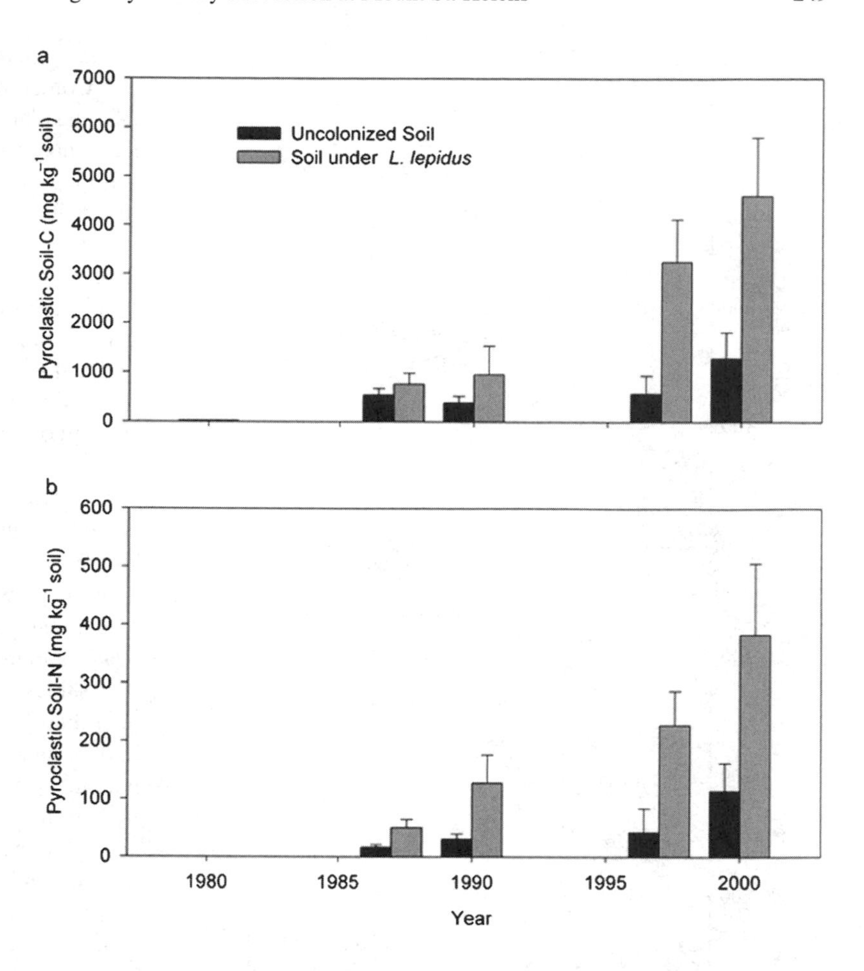

microbial population size and low in situ resource availability to microorganisms.

Samples of pyroclastic soil collected in 1990 were also used to detect patterns of soil-microbial-biomass carbon and carbon and nitrogen mineralization from lupine biomass-amended and nonamended soils (Halvorson and Smith 1995). Similar to TOC and TKN, surface soil under broadleaf lupine contained the most microbial carbon, followed by dead and live prairie lupine and uncolonized soil, respectively. The average soil-microbial-biomass carbon to soil carbon ratio in the 0- to 5-cm depth under lupines was about 24%, and in uncolonized soil 10%, higher than in samples collected only 3 years earlier. These ratios, also higher than the 2% to 3% common to equilibrium agricultural systems reported by Anderson and Domsch (1989), infer a carbon-limited soil ecosystem, but one in which much of the carbon input from lupines or other sources is available for microbial use (i.e., it is not recalcitrant or occluded). Rather than accumulating as soil organic matter, this carbon is incorporated into soil microbial biomass. In addition, carbon inputs might be rapidly metabolized by soil microorganisms, and the apparently high soil-microbial-biomass carbon to soil carbon ratios might be the result of a priming effect (Kuzyakov et al. 2000), where soil respiratory activity (used to calculate the microbial biomass carbon values) is disproportionately stimulated by very small inputs of carbon or nitrogen in "trigger solutions" (De Nobili et al. 2001). Evidence for priming effects associated with carbon inputs by Mount St. Helens lupines was also observed in the soil of relatively undisturbed sites (Halvorson et al. 1991b) and in pyroclastic-flow soil, where the relationship between SIR-C and both TKN and TOC was curvilinear (Figure 17.3) (Halvorson and Smith 1995).

The rates at which microbial processes can proceed are governed by the characteristics of microbial populations, by the environment, and by the quantity and bioavailability of carbon substrates and nutrients. Carbon mineralization potentials (the initial amount of decomposable carbon substrate in the soil, C_0) and carbon mineralization rate constants (the speed at which it decomposes, k) were estimated for surface soil (0 to 5 cm), collected in 1990, with a nonlinear least-squares approach described by Smith et al. (1980) with 20-day aerobic incubation data. Bare, uncolonized soil contained significantly smaller pools of readily mineralizable carbon and thus had a lower average mineralization potential ($C_0 = 10$ ppm) than did soil under live prairie lupine ($C_0 = 67$ ppm), which, in turn, had mineralization potentials significantly lower than those of soils under dead prairie lupine ($C_0 = 114$ ppm). Conversely, mineralization rate constants were significantly higher for bare

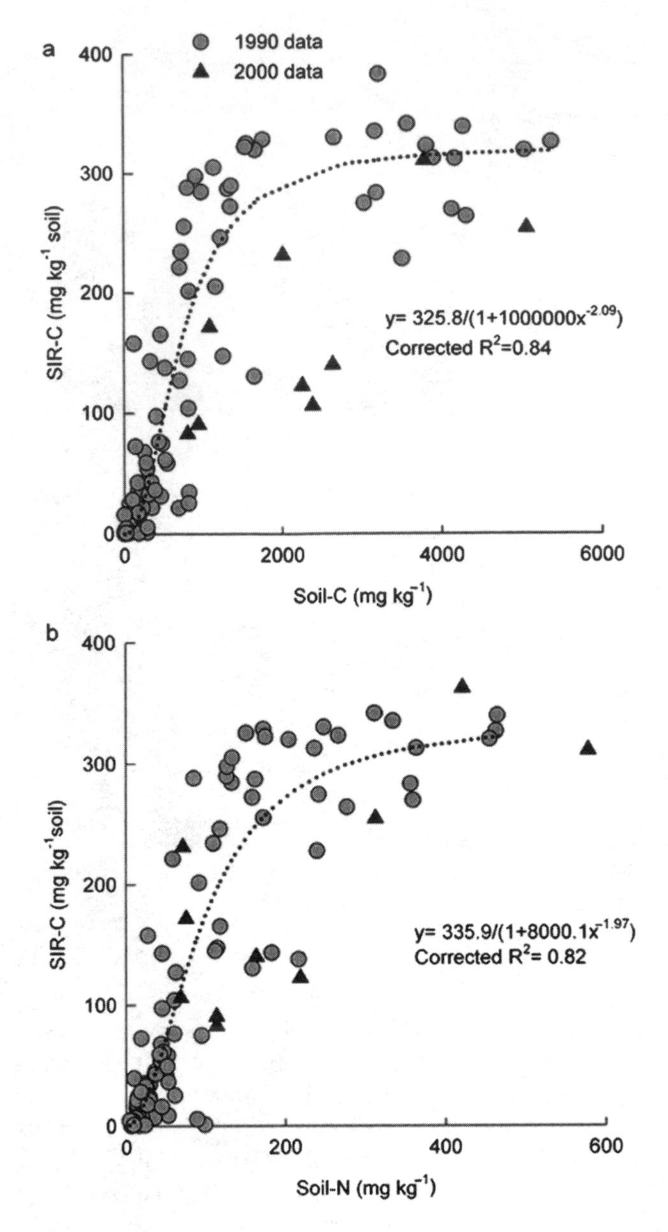

FIGURE 17.3. Relationship between microbial-biomass carbon (substrate-induced respiration method, SIR-C) and (a) soil carbon or (b) soil nitrogen observed in pyroclastic deposits (0–5 cm) for 1990 and 2000.

soil ($k = 0.16$ per day) than either live or dead prairie lupine, $k = 0.10$ or 0.09 per day, respectively. Assuming that the carbon mineralization rate is proportional to the amount of carbon substrate remaining in the soil (i.e., first-order kinetics), these data indicate the average residence time for soil carbon in pyroclastic-flow soil was extremely short, less than 2 weeks, and thus biologically active (Paul and Clark 1989; Herrick and Wander 1997). This analysis, together with observations of low amounts of TOC, TKN, and soluble carbon in soil and zero net nitrogen mineralization in some lupine-biomass-amended soils, indicates that much of the carbon in early Mount St. Helens pyroclastic-flow substrates was contained in microbial biomass and also that microbial populations were nitrogen limited. Competition for inorganic nitrogen between microorganisms and plants may be an important mechanism for controlling early succession.

Lupines continued to augment biological indicators of soil quality in 2000 (Table 17.1). Soil collected under lupines contained significantly more microbial-biomass carbon than did uncolonized soil, a value about two orders of magnitude greater than in 1987 and about a 67% increase from 1990. Compared to 1990, ratios of soil-microbial-biomass carbon to soil carbon declined under lupines to about 8% but remained nearly unchanged in uncolonized soil, about 7%. This pattern indicates that soil organic matter is accumulating or that microbial populations are being limited in part by some other nutrient, such as nitrogen (see Figure 17.3 and following discussion). Development of larger or more active microbial populations under lupines was also reflected in significantly more soil dehydrogenase and phosphatase, important enzymes that affect biological oxidation of soil organic matter and plant nutrition (Tabatabai 1994), and in a higher metabolic quotient (qCO_2), the amount of basal respiration per unit of soil microbial biomass. Greater microbial-biomass-carbon and enzyme activity under lupines implies an increased biological potential to mineralize soil carbon, nitrogen, and phosphorus.

Patterns of qCO_2 reflect how soil microorganisms and the average quality of carbon and nitrogen substrates are closely related (Smith 1994) and may indicate a shift from a carbon-limited to a nitrogen-limited soil ecosystem (Smith 2002). Comparatively high values of qCO_2 have been attributed to low microbial efficiency in disturbed, stressed, or developing ecosystems or in systems with relatively poor carbon-substrate quality. Several studies have shown that average qCO_2 decreases with the age of the ecosystem and is higher in soils without vegetation or in monocropped agroecosystems (Smith 2002). However, other studies suggest qCO_2 may be insensitive to disturbance and ecosystem development and unable to distinguish between the effects of disturbance and stress (Wardle and Ghani 1995). Various estimates of qCO_2 derived from 1987 data previously reported by Halvorson et al. (1991b) show little difference between uncolonized and lupine soil but indicate that qCO_2 for pyroclastic-flow substrates was higher than that for other, less-disturbed, Mount St. Helens sites (Wardle and Ghani 1995; Smith 2002).

More recent comparisons suggest average qCO_2 in the 0- to 5-cm soil under prairie lupines is higher than in uncolonized soil. Metabolic quotients under lupines doubled between 1990 and 2000 from about 30 to 58 μg CO_2 carbon per μg SIR-C g^{-1} soil $day^{-1} \times 10^3$. In contrast, qCO_2in uncolonized soil declined from about 18 to 9 μg CO_2 carbon per μg SIR-C g^{-1} soil $day^{-1} \times 10^3$ (Halvorson and Smith 1995). The increasing qCO_2 observed under lupines suggests that the activity, number, or type of soil microorganisms is increased by the quantity or quality of carbon and nitrogen inputs by plants.

In 2000, average rates of soil respiration from aerobic laboratory incubations were more than 2 times higher than in

1987 (see Table 17.1). In contrast, net nitrogen mineralization rates in 2000 were lower than in 1987 and accounted for a smaller proportion of total soil nitrogen, a pattern of succession suggested by Vitousek et al. (1989). Soil respiration and net nitrogen mineralization were higher under lupines than in uncolonized soil in 1987 and 2000.

Soil microbial biomass remains significantly and positively correlated to concentrations of soil carbon and nitrogen ($p < 0.05$) (see Figure 17.3). Data for samples collected in 2000 compare well with models developed in 1990 by Halvorson and Smith (1995) that predict a saturation or carrying capacity of about 330 mg biomass C kg^{-1} soil or about 430 kg microbial biomass C ha^{-1} in the surface 10 cm. However, soil-microbial-biomass-carbon data observed in 2000 were somewhat below model predictions in relation to soil carbon. This relationship could occur if average carbon-substrate availability is decreasing as the recalcitrant forms of soil carbon continue to accumulate in the soil or if soil microbial populations are increasingly limited by the supply of nitrogen in the soil because of greater competition for nitrogen by plants.

17.4.6 Fatty-Acid Methyl Ester Profiles

Lupines have influenced the quantity and quality of fatty acids in pyroclastic-flow soil over time in comparison to uncolonized sites (Frohne et al. 1998, 2001). Soil collected in 1988 under live and dead lupines and stored air dried contained significantly greater proportions of those fatty acids associated with bacteria (15:0), gram-negative bacteria (16:1ω7c), and fungi or plants (18:1ω9c; 18:2ω6c) than did bare soil. Similarly, samples collected in 1997 and analyzed while still fresh showed higher proportions of fatty acids associated with gram-positive bacteria (16:0 iso); fungi or plants (18:1ω9c; 18:2ω6c); eukaryotes, mosses, and higher plants (23:0); and protozoa (20:4ω6c) under live and dead prairie lupine than in uncolonized bare soil (Vestal and White 1989; Findlay 1996; Sundh et al. 1997; Frohne et al. 1998; Zelles 1999; Pinkart et al. 2002). In comparison, very little fatty-acid material was recovered from either stored or fresh uncolonized soil, confirming the extremely low fertility of pyroclastic-flow material and the continuing direct link between carbon inputs by plants and development of more active and diverse soil ecosystems during succession.

Principal-component analyses of fatty-acid profiles extracted from archived samples of uncolonized soil collected from 1981 to 1988 showed that samples collected in the 1980s were less variable and distinct than were samples collected in 1997. The latter samples were likely influenced by the presence of fatty acids often found in higher plants, mosses, and eukaryotes. The presence of these fatty acids in samples of uncolonized soil collected in 1997 might be attributed to cryptogamic crust growing on the pyroclastic surface and present on about 50% of the samples, changes in the archived samples related to storage (Petersen and Klug 1994), or the influence of plant roots from nearby lupines.

Spatial analysis fatty-acid profiles collected in 1997 showed that the influence of lupine plants extends beyond the plant surface cover (Frohne et al. 1998). Although fatty-acid profiles for soil collected from directly under lupine plants were distinct and more variable than those from uncolonized soil, profiles from uncolonized locations within 0.5 m of a plant were more similar to those under lupine plants than were profiles from farther away.

Samples collected in 2000 revealed that lupines are associated with greater quantities and a more diverse array of fatty acids in relation to their inputs of carbon and nitrogen into pyroclastic soil. As in previous studies, Frohne et al. (2001) reported significantly higher concentrations of soil carbon inside lupine patches (about 4000 mg kg^{-1}), more than five and seven times greater, respectively, than at the patch edge (about 800 mg kg^{-1}) or in uncolonized soil (550 mg kg^{-1}). In correlation to soil carbon, they identified at least 61 fatty acids in the soil under prairie lupine, significantly more than the 21 found in soil at the edge of the lupine patch or the 16 identified in uncolonized soil. Both the amount of soil carbon and the number of fatty acids were observed to change dramatically over short distances, increasing significantly within 50 cm along a transect extending from areas with little plant growth into zones of relatively dense plant cover.

The number of identified fatty acids was significantly, but nonlinearly, related to concentrations of soil carbon (Figure 17.4). At low concentrations of soil carbon (0% to 0.25%), the number of fatty acids was strongly related to soil carbon, but at higher concentrations the incremental increase in the number of fatty acids per unit of soil carbon decreased.

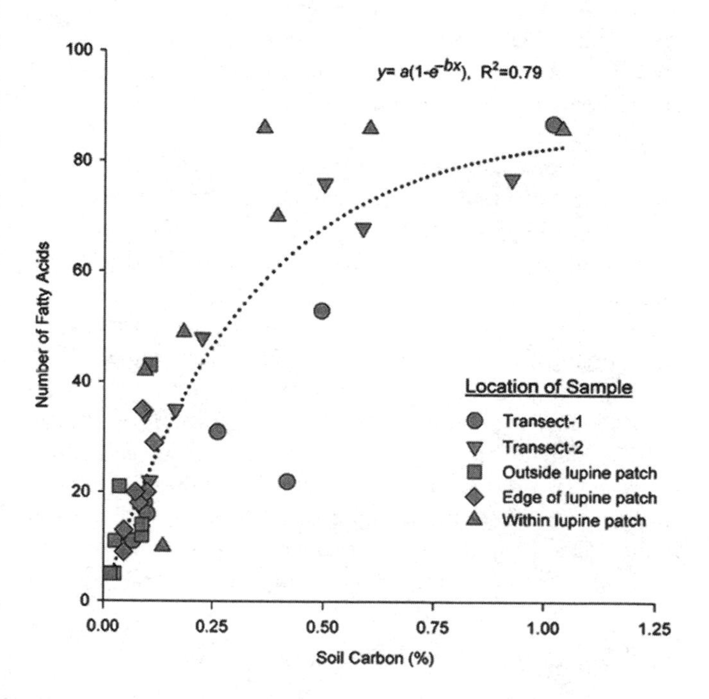

FIGURE 17.4. Relationship between pyroclastic soil carbon (0–5 cm) and the number of identified fatty acids. Samples were collected in July 2000.

Low concentrations of soil carbon were most often associated with uncolonized locations, where carbon and nitrogen accumulate more slowly in the soil from a relatively diverse variety of sources, transported by mechanisms such as animal feces, carcasses, seed rain, water, or wind deposition. There, establishment and composition of soil microbial communities would be conditioned by chance encounters between opportunistic microorganisms and unpredictable supplies of resources. Hence, fatty-acid profiles in uncolonized pyroclastic soils, influenced by both substrate and microbial diversity, would be controlled by microscale versions of low-probability stochastic events thought to be important for determining early development of devastated landscapes (del Moral and Bliss 1993; Turner et al. 1998). The number of fatty acids (i.e., substrate diversity) is thus strongly linked to the quantity of soil carbon in uncolonized locations. Conversely, at locations with more soil carbon, relatively large inputs of carbon from a few predominant plant sources (e.g., lupines) may disproportionately increase the amount of the soil-carbon pool size relative to the variety of fatty acids.

Uncoupling of the relationship between the quantity of inputs and substrate diversity may favor the development of larger but relatively less diverse microbial communities because of microbial competition and specialization. However, in carbon- and nitrogen-limited pyroclastic soil, lupine inputs appear to increase the complexity of fatty-acid profiles and support larger, more active microbial populations. Principal components of the fatty-acid profiles under lupines were distinguishable and more variable than those in uncolonized soil or at the edge of a lupine patch in 2000 (Figure 17.5a). Similar to total soil carbon and the number of fatty acids, principal-component data varied over distances of less than 1 m, as revealed by changes in the first principal component observed at the boundary between uncolonized soil and lupine patch (Figure 17.5b).

More diverse numbers of fatty acids observed under lupines were reflected in substrate-utilization patterns. Microorganisms from samples of uncolonized and patch-edge soil were able to utilize only a few of the substrates offered to them, and not every sample replicate responded the same way, suggesting highly variable and very small population numbers. Microorganisms from samples collected beneath lupine plants were able to utilize more than half of the substrate types offered to them and did so comparatively rapidly (Frohne et al. 2001). Relatively rapid response to a greater variety of substrate types and less variable patterns of substrate utilization provide further evidence for the presence of larger or more active soil microbial communities with the capability to metabolize many kinds of substrates in the soil under lupines.

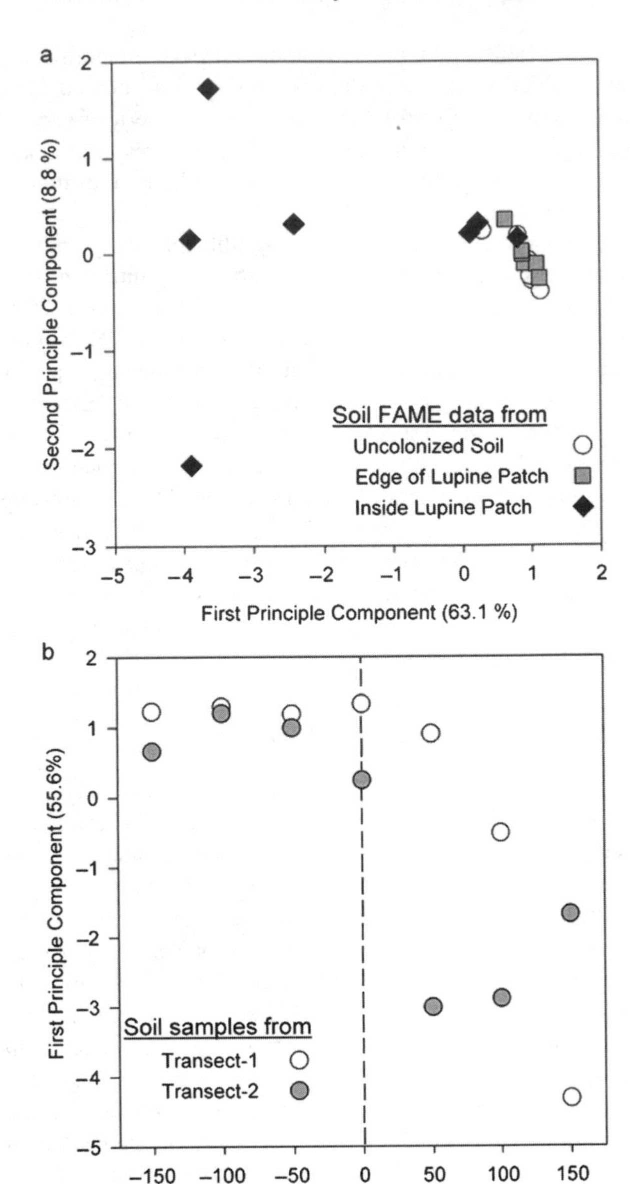

FIGURE 17.5. Plots of (a) the first two principal components calculated for soil fatty-acid methyl ester (FAME) data from samples collected in 2000 and (b) the first principal component for surface soil samples (0–5 cm) collected at 50-cm spacing along transects across lupine patch boundaries.

17.4.7 Lupine Effects on Spatial Patterns

The first studies of the Mount St. Helens pyroclastic-flow deposits using geostatistics detected spatial autocorrelation for surface characteristics such as soil moisture, temperature, and pH at scales of less than a meter, probably influenced most by microtopography and microclimate (Rossi 1989). Concentrations of soil carbon and nitrogen in uncolonized pyroclastic deposits were too low for reliable spatial comparisons whereas plant distribution, probably also influenced by microtopography and microclimate, exhibited only slight, small-scale spatial autocorrelation and evidence of some spatial aggregation.

Small-scale spatial patterns were again observed at the pyroclastic-flow site in June 1991 during a study to determine the relationship between the pattern of soil variables and plant locations. Samples ($n = 97$) of surface soil (0 to 5 cm) were collected in a small, modified grid pattern (3 × 3 m) and analyzed for several soil properties, including moisture content, pH, electrical conductivity, inorganic nitrogen, net nitrogen mineralization, microbial-biomass carbon, and carbon mineralization (respiration) rate. Spatial autocorrelation was characterized with variography (nonergodic correlograms) and spatially interpolated with kriging.

Spatial patterns were most apparent for soil variables associated with carbon but not for variables related to nitrogen because concentrations of nitrogen in pyroclastic-flow substrates were so low that they affected the reliability of measurements. Concentrations of inorganic nitrogen were less than 2 mg kg^{-1} with only slight autocorrelation and little apparent relationship to lupine plant location. Similarly, during a 10-day aerobic incubation, net nitrogen mineralization rates revealed little spatial autocorrelation, and maps produced by kriging indicated a random distribution across the sampling grid. Conversely, those soil variables influenced by carbon inputs from lupines, such as microbial-biomass carbon and soil respiration, showed significant spatial autocorrelation made even more evident after accounting for the confounding effects of spatial outliers. Kriged maps of the distribution of the carbon-related variables in the sample grid showed several discrete biological hotspots associated with plant location. These locations supported up to 40 times more microbial-biomass carbon and respiration rates more than an order of magnitude greater than the surrounding uncolonized substrate.

Another modified grid design (4 × 4 m) was again employed in 1997 to examine the spatial distribution of surface-soil (0 to 5 cm) properties in an area of transition between uncolonized soil and a lupine patch (Frohne et al. 1998). Lupines appeared to influence the pattern of soil moisture, with spatial autocorrelation accounting for about 75% of the total variability. Geostatistics indicated less spatial autocorrelation for pH in samples separated by very small distances (about 25 cm) than for samples separated by about 50 cm, indicative of a patchy distribution of pH, perhaps related to the location of lupine plants. Geostatistics explained more than 70% of the overall plot variation for electrical conductivity but suggested patterns larger than the extent of the sampling plot. Kriged maps indicated that concentrations of soil variables (such as soil carbon and nitrogen, pH, and electrical conductivity) and biological properties (such as microbial-biomass carbon and fatty-acid profiles) were not randomly distributed. Instead, they were often higher inside the lupine patch than they were in nearby uncolonized soil, with relatively sharp transition boundaries.

When contrasted against the background of uncolonized deposits, soil patterns related to lupines were distinct and easily measured at the pyroclastic-flow site shortly after the eruption. Patterns were often observed over short distances for those variables influenced by randomly distributed carbon and nitrogen inputs, such as single-plant colonists or animal droppings, or for those influenced by microclimate or microtopography. Soil variables exhibiting larger-scale patterns were associated with patterns in the deposition of the pyroclastic material, topography, and the flow of water.

17.4.8 Important Patterns for Succession: Abundance or Availability?

The importance of the carbon and nitrogen inputs by lupines, especially during the earliest stages of succession, may lie not in landscape-scale estimates of the rate of nitrogen inputs but rather in the temporal and spatial context of those inputs. The first lupines colonized extremely infertile pyroclastic-flow substrates, and under such conditions the ability to fix nitrogen proved to be an important advantage that allowed lupines to establish and persist. Individual lupines were an early creator of small-scale soil patterns and a direct linkage between above- and belowground processes, serving as a conduit for inputs of carbon and nitrogen into the soil. These early inputs were disproportionately important for the development of soil organic matter and the microbial communities responsible for nutrient retention and cycling and for increasing the availability of nitrogen pools in the soil, which is needed for the subsequent success of other plant species.

Patches of lupines can be thought of as resource islands that are composed of ensembles of covarying soil properties correlated spatially within the landscape. They are associated with living plants but also reinforce patterns of predecessors (i.e., areas of previous plant mortality), creating a historical context of nutrient availability. Understanding the importance of resource islands to other organisms, especially in studies of nutrient-limited ecosystems like Mount St. Helens, requires information about both abundance and availability of resources. Emphasizing the likelihood of encountering "enough" resources, at the "right" time, rather than focusing only on the locations of highest concentrations or rates of accumulation, may change the way we define resource islands or evaluate ecosystem development.

Lupines influence the average probability of subsequent plant colonists or soil microorganisms encountering resources in nonlimiting quantities and/or quality (Halvorson et al. 1995a). Aboveground, lupine plants ameliorate the local microclimate, provide cover and food for invertebrates and vertebrates, promote aeolian deposition of fine soil particles, and trap windborne seeds. Belowground, the nutrient inputs from lupine colonists create distinct zones of relative nutrient abundance. The likelihood of a particular seed or microorganism encountering those resources is affected by its proximity to the lupine and by dispersal patterns, a function of time, distance, propagule abundance, transport mechanisms, and random events. Locally dispersed lupine seeds and microorganisms in the lupine rhizosphere have the highest chance for encountering nutrients, followed by seeds that are preferentially transported to the lupine vicinity by biotic vectors, such as small vertebrates or windborne seeds trapped by lupine plants themselves. These patterns indicate a reinforcing feedback between lupine colonists and other resource-island-promoting mechanisms, such as the growth of other plants or heterotrophic microorganisms. Propagules separated from lupine resource islands by greater distance or those affected mostly by random dispersal processes have the lowest spatial probability of encountering resource islands.

In addition to spatial uncertainty, low substrate availability to heterotrophic soil microorganisms and competition for resources between lupines and microorganisms may decrease the temporal probability of encountering resources by other plant species. On extremely infertile sites, both plants and heterotrophic microorganisms are limited by the quantity of soil nitrogen. Generally, carbon inputs from plants drive microbially mediated activities, such as nitrogen fixation and decomposition, thereby increasing the availability of nitrogen to plants in a cooperative or facilitatory manner (see Connell and Slatyer 1977). However, data from pyroclastic deposits suggest soil microorganisms may actively compete with plants for nitrogen during early succession, when quantities are limiting. During such times, plant–microbial interactions might be competitive or antagonistic (Smith 1994; Lindahl et al. 2002). Competition for nitrogen by soil microorganisms, fueled with carbon substrates supplied by lupines, would limit its availability to nonlegume plants and reinforce the advantage of nitrogen fixation for lupine colonists. Mineralization would become a significant source of nitrogen to plant colonists only at times when the amount of nitrogen available in the soil exceeded the needs of soil microorganisms (Halvorson and Smith 1995). However, as soil nutrient pools increase, so does the likelihood of encountering them. Consequently, the importance of incremental carbon and nitrogen inputs by lupines will diminish together with the spatial and temporal uncertainty of encountering them. In addition to providing a simple source of carbon substrates for heterotrophic microorganisms, other plant inputs, such as polyphenolic compounds and organic acids, may increase in importance for influencing the activity of soil microorganisms, nutrient nitrogen and phosphorus availability, and tolerance of low-pH soil conditions (Jones 1998; Hättenschwiler and Vitousek 2000; Hocking 2001).

17.5 Future Research Needs

Pyroclastic deposits at Mount St. Helens provide a unique opportunity to investigate the impacts of legumes (lupines) on ecosystem development and to come to a more complete understanding of how the aboveground processes associated with succession relate to and are affected by changes in belowground resource pools and processes. However, more information is needed about how the quantity and quality of plant inputs, soil microorganisms, and nutrient cycling change as soils develop. Important temporal and spatial patterns of belowground succession need to be identified together with appropriate scales of measurement. Appropriate experimental designs are needed that can translate small-scale, short-term spatial and temporal patterns of microbial processes, such as nitrogen mineralization, nitrification, and denitrification, into longer-term landscape-scale patterns of carbon and nitrogen cycling. This achievement will require determining which soil and microbial variables are most important at different spatial and temporal scales, what measurements should be taken and when, and what complex interactions among soil variables should be taken into consideration (Halvorson et al. 1997).

18 Response and Recovery of Lakes

Clifford N. Dahm, Douglas W. Larson, Richard R. Petersen, and Robert C. Wissmar

18.1 Introduction

The 1980 eruption of Mount St. Helens devastated vast forestlands, triggered massive landslides and mudflows, and emplaced timber and volcanic debris in nearby lakes. Tephra and pyrolyzed forest debris rained down on dozens of subalpine, oligotrophic lakes scattered across a fan-shaped area affected by the lateral blast of the May eruption. This blast area, which encompasses the debris-avalanche, pyroclastic-flow, blowdown, and scorch zones, covers roughly 570 km^2 north of the volcano. An additional, extensive zone northeast of the blast area received tephra fall.

Runoff from melting glaciers and snowfields on the flanks of the volcano generated enormous mudflows. Quickly gathering momentum, the debris surged into river valleys radiating out from the volcano's base. The massive debris avalanche that swept down the North Fork of the Toutle River blocked tributary streams on either side of the river with an estimated 2.5 km^3 of water-saturated sediment. Dammed at their confluence with the main stem of the river, tributary valleys began to fill with water and formed new lakes. Some of these were small ponds that eventually disappeared; others persisted. But two of the lakes, named Coldwater and Castle, became prominent, established bodies of water in less than a year.

At Spirit Lake, the debris-avalanche deposit blocked the lake's only outlet, the North Fork Toutle River. The lake was thereafter impounded in a closed, hydrologically unstable basin by a debris dam 150 to nearly 200 m thick. To stabilize the lake and to prevent it from either breaching or overtopping the dam, water was artificially withdrawn and discharged downstream, first by pumping and later through a newly constructed tunnel 2600 m long (U.S. Army Corps of Engineers 1987).

Four major factors determined the extent of limnological disturbance:

- Location of the lakes relative to the trajectory of blast materials;
- Abundance of terrestrial organic matter deposited in lakes;
- Emplacement and alteration of mineral deposits; and
- Subsequent in-lake biogeochemical processes

(Wissmar et al. 1982a). Although most lakes inside the blast area were abruptly and substantially altered, some were affected much less, especially those that were shielded from the brunt of the eruption by ice cover or intervening ridges. The few lakes located outside the blast area along the south and west slopes of the volcano all received tephra fall of 5 to 25 cm.

The eruption provided a rare opportunity to study lake response and recovery in the wake of volcanic disturbance. Scientists predicted that recovery (broadly defined as the return to preeruption chemical and biological conditions) would require up to 20 years (Wissmar et al. 1982a,b). Information about lake response and recovery in other volcanically impacted regions was scarce, consisting largely of posteruption studies of Lake Asabatchye on Russia's Kamchatka Peninsula (Kurenkov 1966) and of several lakes in southwestern Alaska (Eicher and Rounsefell 1957). But these studies, limited in scope and duration, were of little help in reliably predicting the recovery of lakes at Mount St. Helens.

Our objectives were threefold:

- To characterize the initial posteruption limnology of lakes within and outside the blast zone;
- From this baseline, to track lake response and recovery over several years; and
- To document the limnological development of the two newly created lakes, Coldwater and Castle, with emphasis on biological colonization.

This chapter reports our findings and addresses three important questions: (1) What was learned about the response of lakes to a large-scale, infrequent disturbance? (2) What factor(s) contributed most to lake recovery? And (3) what was learned about colonization of the new lakes?

We distinguish among four categories of lakes, as shown in Figure 18.1: (1) Spirit Lake, which received a variety of volcanic impacts, including lateral blast, debris avalanche, pyroclastic flows, and tephra fall; (2) the newly created lakes,

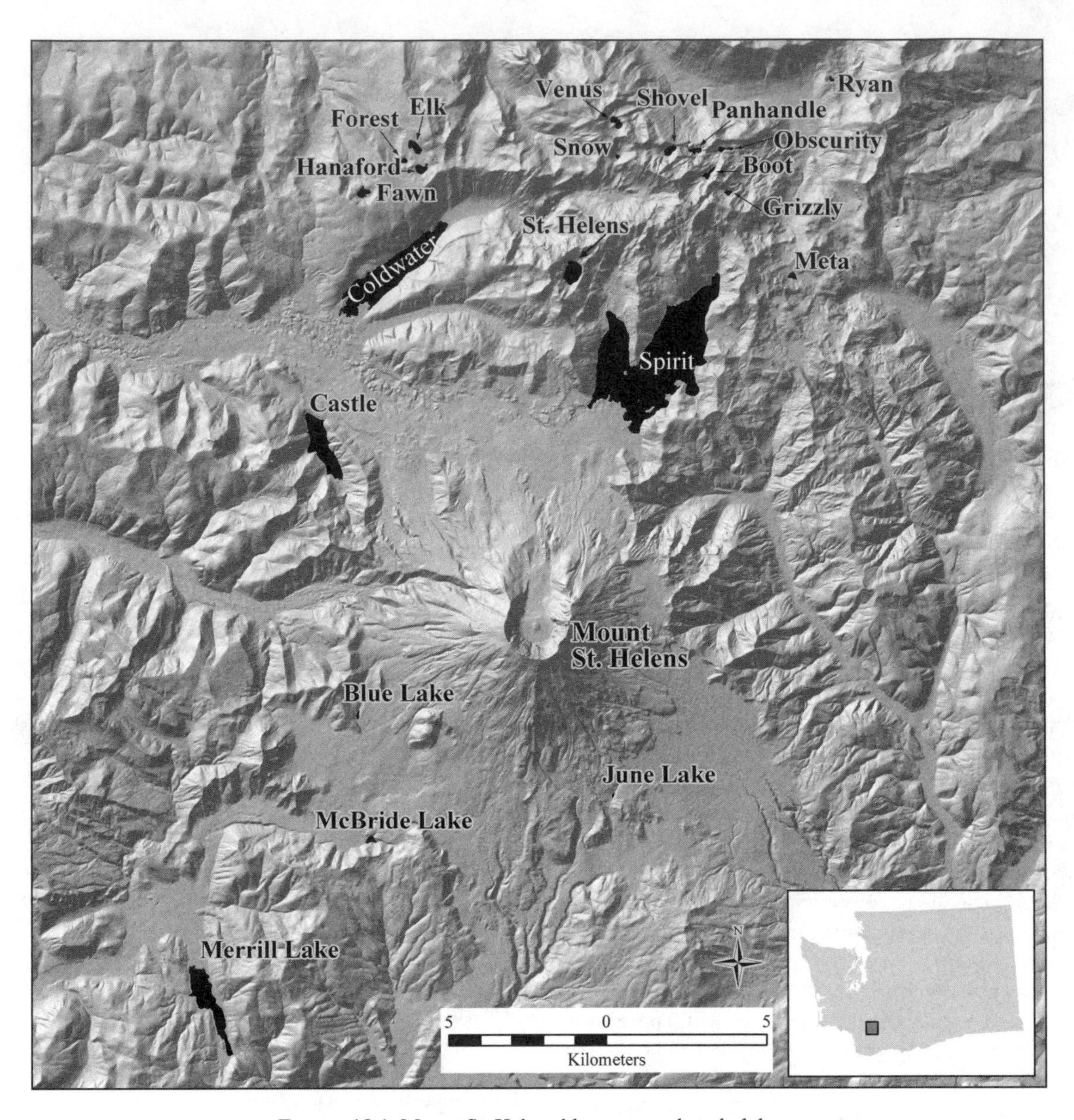

FIGURE 18.1. Mount St. Helens blast area and study lakes.

Coldwater and Castle; (3) lakes in the tree-blowdown zone and along the periphery of the scorch zone; and (4) lakes outside the blowdown and scorch zones that were barely affected morphologically, but periodically received tephra fall. Lakes in Category 3 (in the blowdown and scorch zones) include Boot, Fawn, Grizzly, Hanaford, Meta, Obscurity, Panhandle, Ryan, Shovel, Snow, St. Helens, and Venus. Category 4 lakes (Blue, June, McBride, and Merrill) were used as reference lakes because they received minimal volcanic impact.

18.2 Preeruption Limnology of Mount St. Helens Lakes

In 1980, Mount St. Helens was surrounded by 35 to 40 subalpine lakes of glacial, landslide, and volcanic origin. Lake-surface elevations ranged from 475 to 1500 m (above mean sea level, msl). Lakes at the higher elevations were ice covered 5 to 8 months of the year.

Preeruption limnological information for Mount St. Helens lakes is scarce (but see Bortleson et al. 1976; Dethier et al. 1980; Wissmar et al. 1982a; Crawford 1986). The data consist mostly of a few physical and chemical measurements. Information about phytoplankton, periphyton, other aquatic plants, zooplankton, and other aquatic invertebrates is almost nonexistent. Planktonic and benthic organisms in Spirit Lake in 1937 were described as "sparse"; dissolved oxygen concentrations at the surface and at 53 m were 6.7 and 7.4 mg l^{-1}, respectively; Secchi-disk transparency was 9.2 m; and surface pH was 6.8 (Crawford 1986). Size of lakes examined varied considerably, ranging from Elk Lake (area = 12 ha; maximum depth = 14 m) to Spirit Lake with an area of 530 ha and a maximum depth of 58 m when it was measured by Wolcott (1973).

Generally, the lakes were oligotrophic, although late-summer hypolimnetic oxygen depletion was reported for Fawn and Spirit lakes (Bortleson et al. 1976). Concentrations of dissolved and particulate organic and inorganic substances were low. Nutrient concentrations (total phosphorus and total inorganic nitrogen) were usually less than 10 $\mu g\ l^{-1}$. Metal concentrations were so low that most were below minimum detection limits. Secchi-disk visibility ranged from 7 m at Fawn Lake to 14 m at Spirit Lake. Planktonic chlorophyll *a* was present in "trace" quantities; chlorophyll *a* concentrations of 0.3 $\mu g\ l^{-1}$ were reported for Spirit Lake (Wissmar et al. 1982b). The zooplankton community in Spirit Lake, described as being "diverse," consisted of the water fleas *Daphnia* sp. and *Bosmina* sp., the calanoid copepod *Diaptomus* sp., and unidentified cyclopoids (Wissmar et al. 1982b). Crawford (1986) reported that the potential for fish production in Spirit Lake was "relatively small," amounting to roughly 1100 g ha^{-1} $year^{-1}$.

18.3 Scope of Posteruption Lake Research

Posteruption lake research got under way in early June 1980, about 2 weeks after the eruption, and continued intermittently until September 1994. Lake researchers were faced with formidable logistical and safety problems. Volcanic events had obliterated roads and hiking trails, making most of the lakes inaccessible by land. This problem, coupled with numerous potential hazards, such as renewed eruptions, toxic chemicals, Legionnaires' disease, and unstable terrain (Baross et al. 1981; Campbell et al. 1984; Larson and Glass 1987), necessitated the use of helicopters to rapidly and safely airlift researchers to and from lakes. Additionally, entry was greatly restricted because of concerns for public health (because of the threat of Legionnaires' disease) and safety. Consequently, lake-research efforts were sporadic and short term, resulting in a fragmented limnological database punctuated by a 10-year data gap for most lakes. The exception is Spirit Lake, where investigators conducted extensive limnological studies between 1980 and 1986. Castle and Coldwater lakes also received considerable attention, especially in 1989 and 1990.

Posteruption lake research began with a survey of four lakes (Spirit, Fawn, St. Helens, and Venus) by the U.S. Geological Survey on June 6–8, 1980. Investigators analyzed surface-water samples for 25 to 30 ionic constituents and solids (Dethier et al. 1980; Dion and Embrey 1981; Embrey and Dion 1988). This work was followed on June 30 by a second team of investigators (Wissmar et al. 1982a,b), who collected surface-water samples at 14 lakes: Spirit, Castle, Coldwater, Boot, Fawn, Hanaford, Panhandle, Ryan, St. Helens, Venus, Blue, June, Merrill, and McBride. Although most of these lakes were revisited in 1980 and 1981, research was centered on the chemistry and microbial processes in Spirit, Castle, Coldwater, and Ryan lakes (Dahm et al. 1981, 1982, 1983; Baross et al. 1982; Staley et al. 1982; Ward et al. 1983; McKnight et al. 1984; Lilley et al. 1988).

During fish surveys between July 1980 and October 1984, the Washington Department of Game obtained limnological data for several lakes, including Boot (June 1981), Grizzly (October 1984), Hanaford (September 1980), Meta (July 1980, September 1980, and June 1983), Obscurity (August 1984), Panhandle (June 1981), St. Helens (August 1984), Shovel (August 1984), and Snow (August 1984). Investigators recorded surface temperature and Secchi-disk transparency and collected surface-water samples for dissolved oxygen, alkalinity, hardness, and pH analyses (Crawford 1986).

In 1983, as the U.S. Army Corps of Engineers proceeded to discharge water from Spirit Lake, various state and federal agencies cooperated in an intensive study of the lake to determine how and to what extent these discharges might threaten domestic water supplies, fish, and other valuable water resources downstream in the Toutle River system. This effort was terminated in October 1986 (Larson and Glass 1987; Larson 1993). In 1989 and 1990, limnological studies resumed at Castle and Coldwater lakes, with emphasis on biological colonization of newly created lakes (Kelly 1992; Petersen 1993; Vogel et al. 2000). Between 1991 and 1994, these investigators and others revisited lakes that had not been surveyed for at least 10 years (Menting 1995; Scharnberg 1995; Carpenter 1995; Baker 1995). The most recent limnological survey of Spirit Lake was conducted in July 1994.

Sampling and analytical techniques varied considerably among investigators, whose research objectives and capabilities were often different. Fieldwork was done almost entirely from rubber inflatable rafts. Generally, water samples were collected from the surface and from various depths with messenger-activated, polyvinyl chloride (PVC) Van Dorn tubes, or Niskin bottles. Samples were dispensed into various containers, including 300-ml glass BOD (biological oxygen demand) bottles for dissolved-oxygen analyses (Winkler method) and acid-washed, polyethylene, screw-cap bottles for chemical and biological analyses. Samples for dissolved oxygen were preserved with $MnSO_4$ and alkali-iodide-azide reagents. Samples for metal and phytoplankton analyses were preserved with HNO_3 and Lugol's solution, respectively. Conductivity and pH were measured onsite with portable instruments. Temperature was measured with a YSI thermistor probe (model 43TD, scale calibrated in 0.1°C units). Vertical light penetration and lake-water transparency were determined with a protomatic underwater photometer (unfiltered) and a 20-cm-diameter Secchi disk, respectively. Zooplankton samples were collected by towing a plankton net (mesh aperture = 64 μm, intake diameter = 30 cm) vertically over distances ranging between 5 and 25 m at a rate of about 1 m s^{-1}. The collected samples were preserved with formalin. Other field and laboratory analytical procedures are published elsewhere (Dahm et al. 1981, 1982, 1983; Baross et al. 1982; Larson and Geiger 1982; Staley et al. 1982; Wissmar et al. 1982a,b, 1988; Ward et al. 1983; Sedell and Dahm 1984; Larson and Glass 1987; Lilley et al. 1988).

18.4 Posteruption Limnological Effects

18.4.1 Spirit Lake

18.4.1.1 Basin Morphometry

The influx of volcanic and forest debris greatly altered the morphometric features of Spirit Lake. The quantity of incoming debris was sufficient to displace the lake-surface elevation upward by about 60 m over its preeruption level of 975 m (above msl). The deposition of these materials produced a shallower, expanded basin, with its storage capacity reduced by 10% or more. Because the lake's outlet was blocked, however, the lake continued to rise during the next 3 years, reaching a surface elevation of 1056 m in April 1983. At that elevation, the lake's volume stood at 0.349 km^3, or nearly double its preeruption volume of 0.197 km^3. By September 1984, as the lake was being artificially drawn down, volume had fallen to 0.333 km^3. By October 1986, elevation and volume were down to 1048 m and 0.255 km^3, respectively, at which point the debris-avalanche deposits were deemed to have little probability of being breached (Larson and Glass 1987).

The posteruption depth of Spirit Lake was not precisely determined. Between March and May 1981, the U.S. Geological Survey mapped the lake's bathymetry, reporting that several of the map's contours were "uncertain" (Meyer and Carpenter 1983). This map indicates that the lake's "maximum" depth (34 m) was found in the west bay. Before the eruption, the lake's deepest point (58 m) was in the east bay, which was shoaled to a depth of 29 m by the eruption. As the lake continued to rise, the west bay depth reached 45 m. By October 1986, however, in response to lake drawdown, "maximum" depth was back to 34 m (Larson and Glass 1987).

Shoaling from blast deposits more than doubled the lake's preeruption surface area of 5.3 km^2. In November 1982, just before lake drawdown, the surface area had expanded to 12.50 km^2. As drawdown continued, surface area decreased to 12.08 km^2 in September 1983, 11.29 km^2 in September 1984, 11.30 km^2 in September 1985, and 11.01 km^2 in October 1986 (Larson and Glass 1987).

Following the eruption, thousands of shattered trees floated on the lake surface, forming an immense log raft that covered roughly 40% of the lake's surface. Winds regularly pushed the raft around the lake, causing logs to roll and abrade one another. This action produced detrital organic matter, much of which remained in suspension or precipitated onto the lake bottom. The limnological effects of the log raft are unknown, but investigators have suggested several effects, including (1) shade, thus inhibiting light availability and, hence, photosynthetic activity; (2) nutrient source, particularly carbon; (3) interference with whole-lake mixing; and (4) a biological substrate influencing nutrient cycling and productivity.

18.4.1.2 Optical Properties

An unmistakable limnological attribute of Spirit Lake before the eruption was exceptional water clarity. Bortleson et al. (1976) recorded a maximum Secchi-disk measurement of 14 m in September 1974. Following the eruption, lake waters were soon blackened by extremely high concentrations of organic compounds leached from forest debris in the watershed and in the lake. Sulfides and other chemically reduced metals also contributed to this blackish appearance. On June 30, 1980, the Secchi disk was visible to a depth of only a few centimeters. The average light-extinction coefficient then was 6.05 m^{-1}, which was considerably higher than the coefficient (0.17 m^{-1}) obtained before the eruption on April 4 (Wissmar et al. 1982b).

Lake optical properties improved considerably by 1983. Average Secchi depths doubled between 1983 and 1986, from 3 to 6 m. The maximum recorded posteruption Secchi depth of 9 m, obtained twice during summer and fall 1986, was similar to Secchi depths recorded in 1937 and 1974 (Larson and Glass 1987). This improvement was attributed, in part, to the steady decline of dissolved organic carbon (DOC), which had contributed to the lake's extreme discoloration. Posteruption color of Spirit Lake waters fell from 400 Pt units (platinum units; see Wetzel 2001) during summer 1980 to an average of 29 Pt units during summer 1985. Secchi depths of 9 and 7 m were recorded in September 1989 and July 1994, respectively.

18.4.1.3 Water Chemistry and Bacterial Metabolism

Concentrations of most inorganic chemical constituents rose dramatically, along with concentrations of dissolved and particulate organic matter (Table 18.1). The lake's posteruption concentrations of iron and manganese were, respectively, more than 5000 and 1000 times higher than preeruption levels. The lake contained 33 times more phosphorus (mostly as phosphate) after the eruption than before. Dissolved organic acids became increasingly important as buffering compounds, accounting for most of the lake's surplus alkalinity (McKnight et al. 1988; Wissmar et al. 1990). Total organic carbon concentrations rose from 1.3 mg C l^{-1} (preeruption) to 41 mg C l^{-1} after the eruption. DOC concentrations, rising to more than 60 times the preeruption level, were comparable to or greater than those reported for swamps, bogs, or blackwater rivers (Baross et al. 1982; Thurman 1985).

Posteruption increases in inorganic nitrogen concentrations (ammonium, nitrite, and nitrate) were considerably less than increases of other major ions (see Table 18.1). Low nitrogen concentrations resulted from the low nitrogen content of volcanic and forest-debris deposits in lakes. NH_4 accounted for most of the posteruption nitrogen enrichment, increasing preeruption levels by a factor of 5. NH_4 concentrations averaged 24 $\mu g\ l^{-1}$ for lakes in the debris-avalanche areas and 19 $\mu g\ l^{-1}$ for lakes in blowdown areas. Spirit Lake and the other sampled blast-area lakes had NH_4 concentrations ranging from 2 to 76 $\mu g\ l^{-1}$, with five lakes having concentrations at or near the minimum detection limit of 2 $\mu g\ l^{-1}$.

In Spirit Lake, dissolved organic nitrogen (DON) concentrations were low in relation to DOC. The C:N ratio (the ratio of dissolved organic carbon to dissolved organic nitrogen) was 240. During July and August 1980, DOC concentrations

TABLE 18.1. Water chemistry, Spirit Lake, shortly before and after the eruption, 1980.

	April 4	June 30
pH	7.35	6.21
Total alkalinity, mg l^{-1} as $CaCO_3$	6.95	150.5
Carbon, organic, dissolved, mg l^{-1}	0.83	39.9
Carbon, organic, particulate, mg l^{-1}	0.435	0.570
Ca, dissolved, mg l^{-1}	2.15	66.9
Mg, dissolved, mg l^{-1}	0.48	13.2
Na, dissolved, mg l^{-1}	2.00	67.2
K, dissolved, mg l^{-1}	0.40	16.0
Mn, dissolved, μg l^{-1}	Below mdl[a]	4480.
Fe, dissolved, μg l^{-1}	Below mdl	1080.
Cu, dissolved, μg l^{-1}	Below mdl	109.3
Pb, dissolved, μg l^{-1}	Below mdl	25.9
Zn, dissolved, μg l^{-1}	Below mdl	24.2
As, dissolved, μg l^{-1}	Below mdl	5.2
Cr, dissolved, μg l^{-1}	0.624	3.1
Sb, dissolved, μg l^{-1}	Below mdl	19.5
Al, dissolved, μg l^{-1}	9.98	301.
P, dissolved, μg l^{-1}	7.21	236.
PO_4, μg l^{-1}	2.85	707.
NO_3-N, μg l^{-1}	3.72	9.0
NO_2-N, μg l^{-1}	1.84	12.0
NH_3-N, μg l^{-1}	1.19	16.0
N, organic, particulate, μg l^{-1}	47.0	70.0
Si, dissolved, mg l^{-1}	4.82	21.9
SiO_4, dissolved, mg l^{-1}	5.06	72.1
Cl, dissolved, μg l^{-1}	1.04	88.7
SO_4, dissolved, μg l^{-1}	0.795	124.9

Source: Wissmar et al. (1982a).
[a]Minimum detection limits (mdl), unknown.

increased because of continued leaching of organic debris and because of microbial decomposition of organic matter buried in lake-bottom sediments. DOC increased from 39.9 mg C l^{-1} on June 30 to 51.1 mg C l^{-1} on August 10. Geothermal springs emanating from debris-avalanche deposits accounted, in part, for the high DOC concentrations in Spirit Lake waters (Baross et al. 1982).

As the summer of 1980 progressed, lake bacteria proliferated rapidly in response to massive loadings of organic and inorganic material, especially DOC, iron, manganese, and sulfur (Baross et al. 1982; Staley et al. 1982). Within a month of the eruption, total bacteria in surface waters numbered more than 4×10^8 cells ml^{-1}, a density thought to be unprecedented in natural waters and one that exceeded the range (6.4×10^6 to 3×10^8 cells ml^{-1}) for other greatly impacted lakes (Baross et al. 1982).

The elimination of green plants, including phytoplankton and other algae, effectively halted photosynthetic activity. Consequently, the lake ecosystem was no longer fueled by solar energy fixed by aquatic plant life. Instead, heterotrophic and chemosynthetic bacterial activity became the preeminent, if not the only, source of organic production. Chemosynthetic bacteria derived energy by oxidizing reduced metals (such as iron and manganese), reduced sulfur, and reduced gases.

The bacteria proceeded to decompose and oxidize organic matter found abundantly in lake water and lake-bottom sediment. This process soon depleted all dissolved oxygen except in the uppermost 1 or 2 m, where wind-driven mixing maintained dissolved-oxygen concentrations at 2 to 3 mg l^{-1}. By August, however, no oxygen was detected at 0.2 m, a condition that persisted throughout the late summer and fall of 1980.

Intense anaerobic microbial activity in lake sediment and water during summer and fall 1980 had other effects as well, including (1) reduction of transition metals, such as manganese and iron; (2) high production of carbon dioxide, hydrogen sulfide, methane, and other biogenic gases; and (3) uptake and denitrification of almost all available inorganic nitrogen. Sulfur- and manganese-oxidizing bacteria also proliferated, eventually comprising up to 30% of total bacteria in some samples. In the presence of high concentrations of metals, sulfates, and biogenic gases, metal oxidation reactions and nitrogen-cycle processes (nitrogen fixation, nitrification, and denitrification) became key in regulating the overall rate of bacterial activity (Baross et al. 1982; Ward et al. 1983).

Because of low nitrogen concentrations, pioneering organisms capable of nitrogen fixation became predominant. An anaerobic, dark-nitrogen-fixation rate of 0.27 mg N fixed l^{-1} day^{-1} (1-m depth) was determined. This value is possibly the highest rate of anaerobic dark nitrogen fixation ever reported for lakes (Baross et al. 1982). Because nitrogen fixation does not require light, it is likely that anaerobic nitrogen fixation was occurring throughout the water column and in the sediment, although this case was not established because of limited depth-profile sampling during 1980. Bacterial nitrogen fixation was the principal source of essential nitrogen that sustained microbial metabolism during summer and fall 1980 (Baross et al. 1982; Dahm et al. 1983; Ward et al. 1983).

Denitrifying bacteria were also abundant. The predominant denitrifiers were the sulfur- and manganese-oxidizing bacteria, which were active throughout the water column, except in oxygenated surface waters.

Included among the mix of abundant bacteria were *Klebsiella* sp. and *Legionella* sp., both potentially pathogenic. Two new strains of *Legionella* were discovered and named *Legionella spiritensis* and *L. sainthelensi* (Campbell et al. 1984; Larson and Glass 1987). One or more of the *Legionella* strains were suggested as the cause of a mild form of Legionnaires' disease contracted by several scientists and others working in the vicinity of Spirit Lake in the months following the eruption (Baross et al. 1981; Tison et al. 1983; Campbell et al. 1984; Larson and Glass 1987).

Winter storms typical of the Pacific Northwest brought high precipitation to the Mount St. Helens area beginning in late October 1980. Between November 1, 1980, and April 1, 1981, precipitation runoff and input from numerous small streams increased the lake's volume by nearly 30%, from 0.153 to 0.194 km^3 (Meyer and Carpenter 1983). Precipitation runoff had a dilution effect, which altered lake-water chemistry. By April 1981, concentrations of many ionic constituents, including sodium, potassium, magnesium, calcium, iron, manganese, silica, and especially sulfate, had decreased considerably from levels found during summer and fall of 1980 (Dahm et al. 1981). Dissolved iron and manganese concentrations in Spirit Lake had each fallen by nearly 50%, averaging 540 and 3050 $\mu g\ l^{-1}$, respectively, throughout the water column. By September 1981, iron was barely present in surface waters, and the surface-water concentration of manganese stood at 2570 $\mu g\ l^{-1}$.

Lake-water dilution may have also reduced bacterial density overall. Dahm et al. (1981) reported that the aggregate density of all bacteria in Spirit Lake had diminished during the preceding winter by two orders of magnitude. Nevertheless, bacterial densities remained high during 1981, and the unusual microbial assemblage observed in 1980 was still present. Manganese and sulfur-oxidizing bacteria were abundant in cultured lake-water samples collected on April 30 and June 29, 1981. Viable organisms numbered between 1.5×10^3 and 2.3×10^5 cells ml^{-1} (Dahm et al. 1981).

The oxygen content had also increased by spring 1981. Oxygen concentration, measured in late April 1981, ranged from 8.8 mg l^{-1} (74% saturation) at the lake's surface to 5.8 mg l^{-1} (45% saturation) near the lake bottom at 20 m. Oxygen restoration was attributable largely to autumnal lake turnover and subsequent continuous vertical mixing of the lake during winter. Reduced water temperatures, lake-water dilution, and diminished microbial activity also contributed to reoxygenation.

Oxygen use by various microbial processes continued at high rates, although atmospheric inputs of oxygen maintained measurable daytime oxygen levels in surface waters. However, substantial quantities of dissolved inorganic and organic materials were still present. DOC concentration, which had reached 50 mg C l^{-1} in August 1980, remained at 16 to 18 mg C l^{-1} during spring 1981. DOC provided a rich energy source for oxygen-consuming bacteria, which proliferated. Consequently, oxygen concentration began to diminish rapidly between April and June 1981.

By midsummer 1981, after the lake had become thermally well stratified, bacterial decomposition of in-lake organic deposits depleted hypolimnetic dissolved oxygen. Consequently, anaerobic microbial processes in lake sediment (fermentative carbon and nitrogen decomposition, methanogenesis, and metal and sulfate reduction) supplied the lake with various energy sources, including DOC, CH_4, DON, NH_4, H_2S, Fe^{+2}, and Mn^{+2}. These sources were metabolized by aerobic bacteria through DOC and DON mineralization; methane oxidation; nitrification; and oxidation of sulfur, manganese, and iron. These processes also contributed to hypolimnetic oxygen depletion, allowing anaerobic conditions to build during summer. By August 1981, the lake was once more anoxic except in surface waters (Figure 18.2).

Microbial nitrogen-cycling processes changed in 1981. Nitrogen fixation was no longer detected at every depth, as it had been during 1980. Denitrification rates were below the detection limit on April 30, when dissolved oxygen was present throughout the water column. In June, however, denitrification was detected as oxygen concentrations fell. Nitrification was also detected, however, suggesting that the major nitrogen source for microbial metabolism was shifting from anaerobic nitrogen fixation to ammonium and nitrate (see Figure 18.2). At this time, however, the relative importance of nitrification was small because oxygen conditions were still marginal and other sources of nitrogen (particulate organic nitrogen, dissolved organic nitrogen, and fixed nitrogen) were still available for microbial use. Eventually, as oxygen conditions improved, nitrification would become increasingly important.

Methane was a major energy source during 1981, its concentration greatly exceeding atmospheric saturation levels (see Figure 18.2). In April 1981, surface-water samples contained more CH_4 (0.8 mg l^{-1}) than did samples collected in August 1980 (0.25 to 0.49 mg l^{-1}). CH_4 emanated from sediment, which provided highly active sites of methanogenesis.

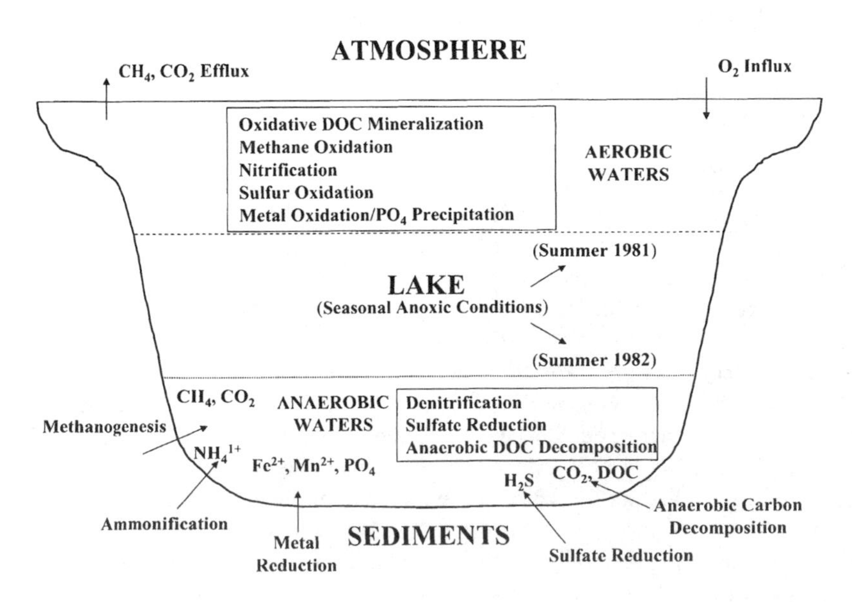

FIGURE 18.2. Conceptual model of biogeochemical processes in the most heavily impacted lakes during transitional years, 1981 and 1982.

Methanogens, which obtain energy from either hydrogen or acetate produced by anaerobic heterotrophs, were strongly linked to anaerobic reactions in the interstitial waters of sediment. During summertime thermal stratification, epilimnetic waters were slightly supersaturated with CH_4, whereas hypolimnetic waters contained ever-increasing CH_4 concentration during summer and fall because of hypolimnetic oxygen depletion and stagnation. These findings indicated the start of aerobic CH_4 oxidation in response to gradual improvement in oxygen availability. During the early posteruption years, CH_4 oxidation became an important pathway for microbial biomass production (Lilley et al. 1988).

During the winter and spring of 1981–1982, lake volume continued to increase, reaching 0.326 km^3 in August 1982 (Meyer and Carpenter 1983). This buildup resulted in further dilution, as evidenced by diminishing values for alkalinity, DOC, and various metallic and inorganic nonmetallic constituents. The iron concentration was barely detectable, and the manganese concentration was below 1000 $\mu g\ l^{-1}$. Microbial oxygen demand had also slowed considerably. The lake's upper 10 m remained well oxygenated (about 10 mg l^{-1}) throughout the summer of 1982 (Lilley et al. 1988). This level was a significant improvement in lake oxygenation, suggesting that lake recovery [defined as "the return to oxygenated water year-round at all depths" (Dahm et al. 1981)] was under way.

By summer 1982, DOC concentration had fallen sharply. This drop resulted from (1) intensive DOC metabolism by heterotrophic bacteria and (2) dilution of lake waters by precipitation runoff water that contained little DOC. Between September 1980 and April 1981, bacterial metabolic activity accounted for at least 56% of the DOC reduction: average water-column DOC concentration fell from 47.1 to 17.0 mg C l^{-1}. Between September 1981 and July 1982, however, 77% of DOC reduction (from 16.8 to 10.5 mg C l^{-1}) was attributed to dilution (Dahm et al. 1981).

The concentration of dissolved organic acids was also lowered as the result of dilution and microbial mineralization. Consequently, organic acid alkalinity decreased whereas carbonate alkalinity increased in response to bicarbonate ions derived from chemical weathering, microbially mediated reactions, and possible anionic exchange (Wissmar et al. 1990). Since 1982, the lake has become well buffered, and pH readings have stabilized around 7.5 (Wissmar 1990).

Direct measurements of CH_4 oxidation rates were first made during the summer of 1982. In September, the CH_4 concentration was near saturation in the upper one-third of the lake. The CH_4 concentration in anoxic hypolimnetic waters reached supersaturation levels, with the maximum concentration (0.2 mg l^{-1}) recorded near the lake bottom at 35 m. The CH_4 oxidation rate was high throughout the hypolimnion, with the maximum rate (0.001 mg l^{-1} day^{-1}) occurring at 35 m (Lilley et al. 1988). This rate is equivalent to the formation of 2.7 mg bacterial biomass m^{-2} day^{-1}, based on a conversion factor of 50% (Rudd and Hamilton 1978).

As the lake became more chemically dilute and oxygen increased, new groups of microorganisms capable of exploiting these changing conditions began to predominate (Baross et al. 1982). Obligate aerobic organisms, for example, could now compete for resources in the lake. The large in-lake deposits of volcanic and forest debris continued to be a substantial source of detrital energy for the microbial community. The lake also received large quantities of sediment transported from surrounding watersheds by overland surface runoff and stream inflows during winter and spring (Collins and Dunne 1988). The log raft supplied additional organic matter.

Nutrient conditions shifted between 1980 and 1982, from initial conditions represented by scarcity of inorganic nitrogen, nitrogen limitation, abundant phosphorus, and low N:P ratios to conditions observed in 1982 (more inorganic nitrogen, less phosphorus, and higher N:P ratios). N:P ratios during 1982 ranged between 8 and 38, which exceeded the 1980 ratios when microbial activity was nitrogen limited. Many N:P ratios in 1982 exceeded the standard published molar ratio of 16:1, indicating that phytoplankton primary production was phosphorus limited (Rhee 1978). The transition from nitrogen limitation to either phosphorus limitation or combined nitrogen and phosphorus limitation occurred within 2 years after the eruption. Large reductions in PO_4 concentrations coincided with similarly large reductions of dissolved iron, suggesting that PO_4/Fe^{3+} coprecipitation may have been an important mechanism for phosphorus removal.

By 1983, the recovery of Spirit Lake was well under way. As the lake proceeded to increase in volume, reaching 0.349 km^3 in April 1983, dilution continued to play an important role in the recovery process. Bacterial densities, which had reached unprecedented levels in 1980, had diminished to levels described by Wetzel (2001) as typical of moderately productive lakes. Various microbial metabolic processes, including nitrogen fixation, denitrification, manganese and sulfur oxidation, and DOC metabolism, were now proceeding at moderate or even undetectable rates. Concentrations of inorganic nitrogen, inorganic phosphorus, iron, and manganese were approaching preeruption levels. Only a small fraction of the 1980–1981 supply of labile DOC remained. By 1986, DOC concentration was only 10% of levels measured during summer 1980. However, the continuing influx of geothermal seepwaters (Table 18.2) from or through the debris-avalanche deposit possibly contributed to the persistence of relatively high ion concentrations (Larson and Glass 1987).

Various factors (photochemical reactions, oxygenated water column, and microbial transformations) shifted the remaining DOC into a more heterogeneous mixture of refractory organic compounds (McKnight et al. 1982, 1984, 1988). But CH_4, also an important energy source, continued to fuel lake bacteria. Methane oxidation rates throughout the water column increased from 7.8 mg C m^{-2} day^{-1} in 1982 to 32.2 mg C m^{-2} day^{-1} in 1986 (Lilley et al. 1988). Assuming that 50% of this carbon was converted to biomass (Rudd and Hamilton 1978), bacterial production from CH_4 oxidation ranged from

TABLE 18.2. Water chemistry, Spirit Lake and seepwaters, 1986 and 1994.

	September 1986[a]		July 1994	
	Spirit Lake[b]	Seep	Spirit Lake[b]	Seep
Total alkalinity, mg l^{-1} as $CaCO_3$	96	181	NA[c]	NA
Conductivity, μmhos cm^{-1}	470	3600	374	1970
Total dissolved solids, mg l^{-1}	451	4250	NA	NA
Ca, mg l^{-1}	31.7	NA	27.1	135
Mg, mg l^{-1}	16.2	NA	5.5	20.2
Na, mg l^{-1}	158	NA	48.4	276
K, mg L l^{-1}	14.5	NA	3.2	15.2
Cl, mg l^{-1}	60[d]	NA	26.5	103
SO_4, mg l^{-1}	150	900	106	580
HCO_3, mg l^{-1}	NA	NA	57.2	100
Fe, dissolved, μg l^{-1}	35	200	NA	NA
Mn, dissolved, μg l^{-1}	410	1200	NA	NA
SiO_2, mg l^{-1}	25	115	NA	NA

[a]Larson and Glass (1987).
[b]Surface sample.
[c]Not available.
[d]Mean value, $n = 66$.

3.9 to 16.1 mg C m^{-2} day^{-1}, which was comparable to phytoplankton production up until 1985. Methane oxidation rates were lowest in surface waters and highest in the hypolimnion, where oxygen concentration was extremely low. Lilley et al. (1988) reported that maximum methane oxidation rates were measured at depths where the oxygen concentration was less than 2.5 mg l^{-1}. At these depths and in sediment interstitial waters, bacteria capable of oxidizing CH_4 reached maximum numbers because of the substantial amounts of CH_4 produced by microbial methanogenesis.

Although oxygen levels had increased overall by 1983, persistent microbial oxygen consumption created hypolimnetic oxygen deficits in Spirit Lake during summertime thermal stratification. By mid-August 1983, for example, the near-bottom oxygen concentration had diminished to 1.0 mg l^{-1} or less. Within a month, the entire hypolimnion had become anoxic. During 1984, however, hypolimnetic dissolved oxygen was not entirely depleted, even to the time of thermal destratification in late October. In 1985, however, hypolimnetic oxygen concentration again fell to zero. In 1986, the lake's hypolimnetic oxygen supply increased over 1985 conditions by lasting 2 weeks longer during thermal stratification (Figure 18.3; Larson and Glass 1987).

In September 1989, dissolved-oxygen concentration ranged from 10.8 mg l^{-1} at the surface to 3.8 mg l^{-1} near the bottom. In July 1994, surface and near-bottom concentrations were 8.9 and 5.6 mg l^{-1}, respectively. The lake may never fully "recover" if the criterion for recovery is a well-oxygenated hypolimnion year round. Even before the eruption, near-bottom waters (50 to 52 m) became considerably undersaturated in late summer (Bortleson et al. 1976).

18.4.1.4 Phytoplankton

Phytoplankton were scarcely evident during summer and fall 1980 (Larson and Geiger 1982; Ward et al. 1983). Fewer than 15 taxa were found, and densities ranged between less than 1 and 100 algal units ml^{-1} (an algal unit is defined as an individual cell, colony, or filament; Sweet 1986). Phytoplankton cells were mostly nonliving frustules of the diatoms *Cyclotella* sp., *Melosira* sp., *Cocconeis* sp., *Nitzschia* sp., and *Meridion* sp. (Larson and Geiger 1982). These cells were probably the remnants of populations that had existed before the eruption. A few unicellular green algae, blue-green algae, and diatoms were isolated and grown on culture media inoculated with Spirit Lake water. Spirit Lake isolates included *Chlorella* sp., *Nitzschia* sp., and an unidentified small unicellular alga (Ward et al. 1983).

Phytoplankton continued to be sparse during 1981–1982 (Dahm et al. 1981; Ward et al. 1983). Total phytoplankton density in June 1981 was only 0.7 algal units ml^{-1}, with *Pseudanabaena* sp. and *Navicula* sp. comprising roughly 60% of the phytoplankton assemblage (Ward et al. 1983). Total density was less than 1.0 algal units ml^{-1} in August 1982, with 82% of the total assemblage composed of *Cryptomonas* sp. and unidentified flagellates (Smith and White 1985).

Both species diversity and abundance of phytoplankton increased dramatically between 1983 and 1986. During this period, 135 phytoplankton species were identified, with diatoms comprising 84% of the phytoplankton community (Larson and Glass 1987). In August 1983, blooms of *Asterionella formosa* and *Diatoma tenue elongatum* produced a combined density of 2588 algal units ml^{-1}. Occasionally during 1985 and 1986, up to 96% of the community consisted of *Rhodomonas minuta*, a motile cryptophyte. In April 1986 *Asterionella formosa* reached a density of 3645 algal units ml^{-1}, representing 97% of the community. By 1985, phytoplankton densities and species composition were comparable to those found in subalpine, oligotrophic/mesotrophic lakes in the Washington–Oregon Cascades (Johnson et al. 1985; Sweet 1986; Larson et al. 1987).

The phytoplankton assemblage in September 1989 included diatoms and cryptophytes. The diatom *Cyclotella comta* was dominant, averaging 168 algal units ml^{-1}, or 67% of the assemblage. The dominance of *C. comta* is highly unusual in subalpine lakes of the Pacific Northwest and is typically found in lakes that are mesotrophic or eutrophic (Sweet 1986). Small flagellates (*Rhodomonas minuta* and *Cryptomonas erosa*) were common (mean = 107 algal units ml^{-1}), representing 27% of the assemblage. Replicate surface-water samples collected from Spirit Lake during summer 1993 yielded 17 taxa, including 11 diatoms, 2 chrysophytes, 2 chlorophytes, 1 cryptophyte, and 1 pyrrophyte (Baker 1995). In July 1994, total phytoplankton densities ranged from 122 to 242 algal units ml^{-1}. *Cyclotella comta* was still dominant, comprising 92% and 57% of the assemblage at depths of 7 and 14 m, respectively. Near bottom, at 26 m, *Rhodomonas minuta* and *C. comta* were predominant at 68% and 29%, respectively. *Fragilaria*

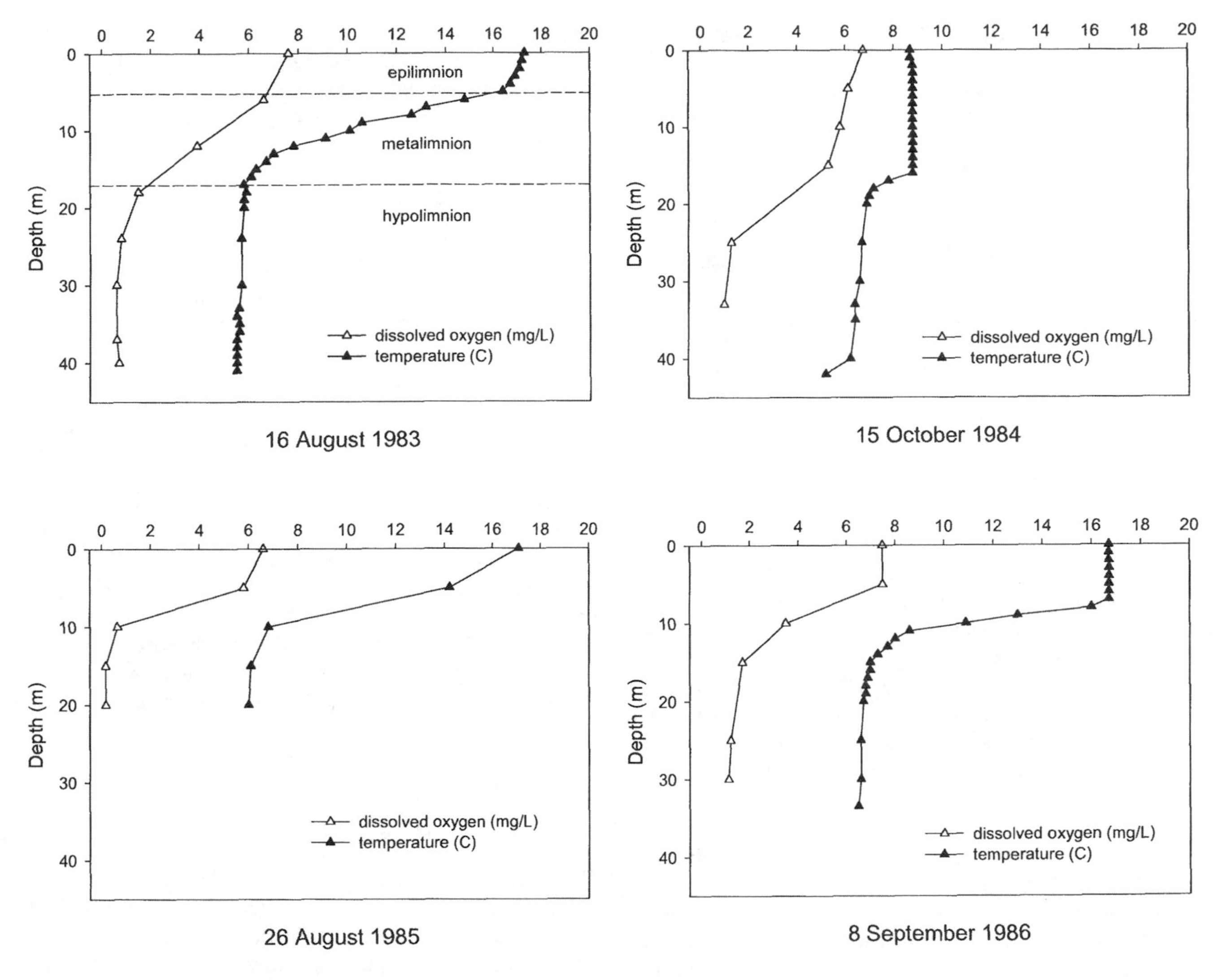

FIGURE 18.3. Temperature–dissolved oxygen profiles, Spirit Lake, 1983 to 1986.

crotonensis, a pennate diatom described as "salt-indifferent" and typical of eutrophic lakes (Sweet 1986), was also present.

After 1982, phytoplankton primary production increased rapidly as light availability and other environmental conditions (e.g., inorganic nitrogen) favorable to algae improved. Between 1984 and 1986, summertime areal production rates ranged from 12.6 to 41.7 mg C m^{-2} h^{-1}. These rates fell within the range for low to moderately productive lakes worldwide (Wetzel 2001) and were close to average summer production rates for Crater Lake, Oregon (Larson 1972). By 1986, algal photosynthesis had largely replaced microbial chemosynthesis as the means of production (Figure 18.4).

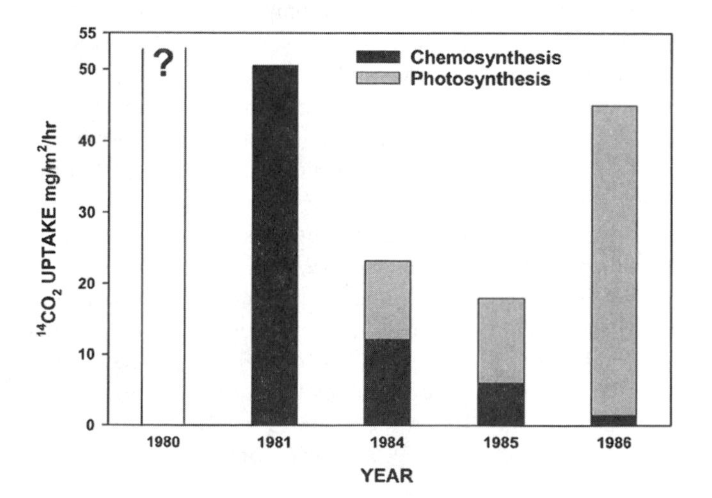

FIGURE 18.4. Limnological recovery of Spirit Lake as indicated by shift from chemosynthetic to photosynthetic means of organic production.

18.4.1.5 Zooplankton and Other Invertebrates

The fate of invertebrates is largely unknown, although Crawford (1986) reported that macroinvertebrates were not found. Biological communities were simply buried under the

TABLE 18.3. Zooplankton taxa, Spirit Lake, 1982 to 1994.

July 1982 to October 1983[a]
Cladocerans: *Bosmina* sp., *Ceriodaphnia* sp., *Daphnia* sp.
Copepods: *Diaptomus* sp., *Macrocyclops albidus*, *Cyclops* sp., unidentified cyclopoids, unidentified calanoids, nauplii
Rotifers: *Asplanchna* sp., *Filinia* sp., *Keratella quadrata*, *Keratella* sp., *Kellicottia longispina*

September 1989
Cladocerans: *Chydorus sphaericus*, *Daphnia pulex*
Copepods: *Leptodiaptomus tyrrelli*, *Ectocyclops phaleratus*, nauplii

July 1994
Cladocerans: *Alona* sp., *Bosmina longirostris*, *Ceriodaphnia pulchella*, *Chydorus sphaericus*, *Daphnia schødleri*
Copepods: *Cyclops varicans rubellus*, *Diaptomus tyrelli*, diaptomid copepodids, cyclopoid copepodids, harpacticoid copepodids, nauplii
Rotifers: *Filinia terminalis*, *Keratella quadrata*, *Lecane* sp., *Polyarthra dolichoptera*, *Trichotria* sp., *Trichocerca* sp.

[a] Larson and Glass (1987).

overwhelming deluge of volcanic and forest debris. Most organisms that happened to survive the initial impact were probably killed shortly afterward by the abrupt loss of oxygen and possibly by the buildup of toxic compounds in the lake.

Zooplankton were first observed in Spirit Lake in July 1982. Taxa included *Ceriodaphnia* sp., *Macrocyclops albidus*, and copepod nauplii. Zooplankton densities ranged between 250 and 17.8×10^3 organisms m^{-3} (Table 18.3; Larson and Glass 1987).

By 1983, the lake contained relatively large numbers of zooplankton and aquatic insects. Rotifers were extremely abundant during July 1983, with *Keratella* sp. densities reaching 124×10^3 organisms m^{-3} on July 17. Cladocerans and copepods were also present but were considerably less abundant. Nine vertical net tows taken between July and October 1983 yielded an average of 1546 cladocerans and 391 copepods m^{-3} (see Table 18.3; Larson and Glass 1987). In October 1984, swarms of aquatic insects (corixids) and amphipods were observed and photographed in near-shore waters by SCUBA divers (Larson and Glass 1987). The large predatory zooplankter *Chaoborus* sp. first appeared in samples collected during summer 1986. The absence of fish undoubtedly influenced both zooplankton density and species composition, although by 1986 the lake's zooplankton community began to resemble communities typical of subalpine, oligotrophic/mesotrophic lakes in the Washington–Oregon Cascades.

In September 1989, vertical net tows from depths of 9 and 22 m yielded cladocerans, copepods, and *Chaoborus* sp. Total density averaged 1104 organisms m^{-3}. *Daphnia pulex* was dominant, comprising 92% of the assemblage in the 9-m tow and 81% in the 22-m tow. *Chaoborus* sp. was scarce (8 individuals m^{-3}), appearing only in the 22-m haul (Table 18.3).

Vertical net tows on July 28, 1994, yielded cladocerans, copepods, rotifers, and *Chaoborus* larvae (see Table 18.3). Zooplankton were most abundant in the 0- to 5-m stratum, led by copepods (1491 organisms m^{-3}) and followed by cladocerans (1030 organisms m^{-3}) and rotifers (197 organisms m^{-3}). *Chaoborus* (17 organisms m^{-3} in a vertical tow between near bottom at 20 m to the surface) was least abundant in the upper 5 m of the lake.

18.4.1.6 Benthic Algae and Macrophytes

During 1990–1992, profuse growths of Eurasian watermilfoil (*Myriophyllum spicatum*) were first observed in the littoral zone along the south shore of Spirit Lake. Milfoil had never been reported in the lake before the 1980 eruption and was probably introduced after the eruption and proliferated in the now shallower, more-fertile lake waters. Because public access to Spirit Lake is still largely restricted and is off limits to recreation of any sort, milfoil was probably introduced by waterfowl flying in from milfoil-infested lakes and ponds located outside the blast area. An abundance of benthic algae (*Rhizoclonium hieroglyphicum* and *Chara vulgaris*) was associated with the milfoil (Larson 1991). By July 1994, other plants (*Ceriophyllum demersum*, *Nitella* sp., and *Potamogeton* sp.) had become established (personal communication, C. Crisafulli, USPA). In 2000, aside from observations indicating that milfoil was not abundant, little was known about the distribution, identification, and abundance of macrophytes in Spirit Lake.

18.4.2 Coldwater and Castle Lakes

18.4.2.1 Basin Morphometry

Coldwater and Castle lakes, lacking outlets, continued to rise behind the debris-avalanche deposit blocking tributaries to the North Fork of the Toutle River. Both lakes were partly covered with log rafts, but the logs were eventually extracted for lumber. As the lakes increased in volume, they threatened to breach or overtop the debris-avalanche deposit and surge destructively downstream. In 1981, the U.S. Army Corps of Engineers proceeded to construct outlets, or overflow channels, to draw the lakes down and stabilize surface elevations. Release flows from Coldwater Lake commenced in July 1981, followed by Castle Lake in November 1981. In a matter of weeks, the lakes were drawn down to their current surface elevations and volumes (Table 18.4).

Depth soundings and other morphometric measurements were first obtained during summer 1990. These data (Table 18.4) were used to construct bathymetric charts (Kelly 1992).

18.4.2.2 Optical Properties

The waters of both lakes during 1980 and 1981 appeared blackish and were essentially opaque. In June 1980, Wissmar et al. (1982b) reported light-extinction coefficients of 7.21 m^{-1} for Coldwater Lake and 12.43 m^{-1} for Castle Lake. In June 1981, the rate of vertical light extinction in Coldwater Lake was still high, with only 1% of surface light intensity remaining at a depth of 1.3 m (Dahm et al. 1981). Secchi-disk transparency

TABLE 18.4. Basin morphometry, Coldwater and Castle lakes, 1990 (Kelly 1992).

	Coldwater	Castle
Surface elevation, m	762	796
Maximum length, km	5.11	2.37
Maximum breadth, km	0.985	0.744
Area, km^2	3.10	1.07
Volume, km^3	0.0838	0.0216
Maximum depth, m	62	32
Mean depth, m	27	20
Relative depth, %[a]	3.1	2.8
Shoal area, % less than 3 m	9.5	5.8
Shoreline, km	14.1	6.24
Shoreline development[b]	2.26	1.70
Volume development[c]	1.31	1.88
Watershed area, km^2	49.45	8.67
Retention time, years	1.7	2.5

[a] Relative depth is the ratio of the maximum depth in meters to the square root of the area in hectares (Hutchinson 1957).
[b] Shoreline development is the ratio of the length of the shoreline to the length of the circumference of a circle of area equal to that of the lake (Hutchinson 1957).
[c] Volume development is the ratio of the volume of the lake to that of a cone of basal area equal to the surface area with its height equal to the maximum depth (Hutchinson 1957).

in Coldwater Lake was 1.0 m in October 1981; Secchi-disk readings in Castle Lake during September averaged 2.7 m.

During 1989 and 1990 (May through October), Secchi-disk readings in Coldwater and Castle lakes averaged 7.5 m (n = 17) and 8.0 m (n = 12), respectively. Extinction coefficients were 0.30 m^{-1} in Coldwater Lake and 0.34 m^{-1} in Castle (Kelly 1992). At Coldwater Lake in 1993, light extinction averaged 0.27 m^{-1}, and Secchi-disk readings ranged between 5.7 and 8.0 m. At Castle Lake, the average light-extinction coefficient was 0.36 m^{-1} and the Secchi-disk range was 7.3–7.9 m (Carpenter 1995).

18.4.2.3 Water Chemistry and Bacterial Metabolism

Chemical and microbial conditions in Coldwater and Castle lakes during summer 1980 were similar to those found in Spirit Lake. By late summer 1980, both lakes were completely anoxic because of high microbial oxygen demand, although aeration may have oxygenated surface layers to some extent. Microbial proliferation produced large concentrations of hydrogen sulfide, methane, ammonia, and phenolic compounds. Pathogenic bacteria were also present, including *Legionella* sp. and *Klebsiella pneumoniae* (Dahm et al. 1981).

Microbial activity rose in response to increased DOC concentrations, reaching 149 mg l^{-1} in Castle Lake by June 30 (Wissmar et al. 1982b). This concentration was one of the highest ever reported for a lake. The DOC concentration in Coldwater Lake on June 30 was 23.1 mg C l^{-1} but rose to 114 mg C l^{-1} by late summer (Baross et al. 1982). High DOC concentrations in Coldwater Lake possibly resulted, in part, from geothermal springs entering the lake. Whether these springs originated from the hot secondary pyroclastic flows at the north end of the lake basin or from the cooler debris avalanche to the south is unknown. Geothermal waters entering Coldwater Lake had temperatures as high as 80°C and DOC concentrations of more than 4000 mg C l^{-1} (Baross et al. 1982). Conversely, DON concentrations were very low (0.690 mg N l^{-1} in Coldwater Lake), resulting in high DOC:DON ratios of 165 in Coldwater Lake and 1019 in a geothermal seep entering the lake. Anaerobic, dark-nitrogen-fixation values of up to 0.16 mg N fixed l^{-1} day^{-1} were reported for Coldwater Lake (Baross et al. 1982).

During the winter and spring of 1980–1981, as lake volumes continued to increase, lake-water dilution resulted in lower values for alkalinity, DOC, and various metallic and inorganic nonmetallic constituents. By April 1981, iron concentrations were barely detectable, and manganese concentrations were below 1000 μg l^{-1}. Microbial oxygen demand also slowed as bacterial densities diminished. By midsummer, however, bacterial densities had risen to nearly 1980 levels. Maximum bacterial densities approached 10^8 organisms ml^{-1}, with samples yielding an abundance of manganese- and sulfur-oxidizing bacteria. Although denitrifying and nitrogen-fixing bacteria were still abundant, their importance as major nitrogen sources was gradually being supplanted by nitrifying bacteria as the oxygen supply increased. During summer 1981, nitrification was detected in both lakes. Methane emanated from lake-bottom sediment, where methanogenesis was active (Dahm et al. 1981).

By 1989, the oxygen supply in Coldwater and Castle lakes had increased substantially. However, hypolimnetic microbial oxygen demand in Coldwater Lake was still sufficiently strong to lower oxygen saturation in near-bottom waters to below 40% by late summer and fall. In Castle Lake, near-bottom waters (depth, 30 m) were essentially anoxic during September and October 1990 (Kelly 1992). Summertime mean areal oxygen deficits during 1989 and 1990 averaged 455 mg O_2 m^{-2} day^{-1} in Coldwater Lake and 450 mg O_2 m^{-2} day^{-1} in Castle Lake (Kelly 1992). Wetzel (2001) referred to these values as typical of mesotrophic or eutrophic lakes.

Ionic concentrations in Coldwater Lake continued to decline until 1984–1985 when chemically rich water from Spirit Lake was discharged into South Coldwater Creek via the outlet tunnel designed to stabilize the rising volume of Spirit Lake (Larson and Glass 1987). South Coldwater Creek entered the south end of Coldwater Lake near the lake's outlet. These inflows were apparently large enough to alter the lake's chemistry, as evidenced by an increase in ionic concentrations (Menting 1995). The creek also transported large sediment loads, especially during storms, creating a large delta consisting of material eroded from the 1980 deposits. Growth of the delta was slow but ongoing. As the delta grew, it filled much of the lake's south end. Consequently, South Coldwater Creek began to shift away from the lake toward the lake's outflow channel, eventually bypassing the lake. Thus, water from Spirit Lake was largely diverted, resulting in lower ionic concentrations in Coldwater Lake (Menting 1995).

TABLE 18.5. Water chemistry, Coldwater and Castle lakes, 1980[a] versus 1993.[b]

	Coldwater		Castle	
	1980	1993	1980	1993
Total alkalinity, $\mu Eq\ l^{-1}$	1520	313	4500	451
Ca, $mg\ l^{-1}$	91.00	8.79	94.30	7.66
Mg, $mg\ l^{-1}$	14.70	1.92	21.21	1.27
Na, $mg\ l^{-1}$	88.01	5.06	112.01	2.18
K, $mg\ l^{-1}$	17.00	0.75	39.99	0.29
SO_4, $mg\ l^{-1}$	312.01	21.66	120.01	5.91
Cl, $mg\ l^{-1}$	115.03	2.62	142.03	0.94
PO_4-P, $\mu g\ l^{-1}$	2.2	0.9	13.9	3.1
NO_2-N + NO_3-N, $\mu g\ l^{-1}$	3.9	46.9	67.7	2.4
NH_3-N, $\mu g\ l^{-1}$	10.8	3.8	48.0	3.6
SiO_4, $mg\ l^{-1}$	22.21	13.49	44.64	15.02

[a] 1980 data from Wissmar et al. (1982a).
[b] 1993 data from Menting (1995) and Carpenter (1995).

Ionic and nutrient concentrations in both Coldwater and Castle lakes had greatly diminished by 1993 (Table 18.5). In Coldwater Lake, the importance of nitrification as a nitrogen source was reflected by the factor-of-10 increase in inorganic nitrogen. In Castle Lake, however, the reverse was true, suggesting that the lake's oxygen supply was remaining low enough overall to limit the role of nitrification. In addition, particulate organic nitrogen (PON) concentration in Castle Lake during June 1980 was 1.31 $mg\ l^{-1}$, or more than 4 times greater than the PON concentration in Coldwater Lake and nearly 20 times greater than the concentration in Spirit Lake (Wissmar et al. 1982a). This elevated value may be related to the relatively high inorganic nitrogen concentrations reported by Wissmar et al. (1982a) for Castle Lake in 1980.

18.4.2.4 Phytoplankton and Benthic Algae

During August and September 1980, Ward et al. (1983) investigated algal activity in Coldwater and Castle lakes. At Castle Lake, a few unicellular green algae, blue-green algae, and diatoms were isolated and grown on culture media inoculated with water obtained from the lake. In August, total algal density was 0.5 algal units ml^{-1}, with 60% of the sample consisting of diatom fragments. In June 1981, total density had increased to 280 algal units ml^{-1}, with flagellates (cryptomonads) comprising 67% of the sample and pennate diatoms 33% (Ward et al. 1983).

Maximum phytoplankton densities during September and October 1981 were 0.053 algal units ml^{-1} in Coldwater Lake and 2.0 algal units ml^{-1} in Castle Lake. The phytoplankton community in Coldwater Lake consisted of only *Gonyaulax* sp. and *Cryptomonas* sp. The Castle Lake community was far more diverse, featuring 17 taxa (Table 18.6) and several unidentified flagellates and coccoids.

TABLE 18.6. Phytoplankton taxa, Coldwater and Castle lakes, 1981 to 1989.

Castle Lake, September–October 1981
Diatoms: *Achnanthes* sp., *Cyclotella* sp., *Melosira* sp., *Navicula* sp., *Nitzschia* sp., *Surirella* sp.
Cryptophytes: *Cryptomonas* sp.
Chlorophytes: *Carteria* sp., *Chlamydomonas* sp., *Scenedesmus* sp.
Chrysophytes: *Dinobryon* sp., *Mallomonas* sp.
Euglenophytes: *Euglena* sp.
Cyanophytes: *Gleocapsa* sp.
Pyrrophytes: *Anabaena* sp., *Gonyaulax* sp., *Glenodinium* sp.

Castle Lake, July–August 1986
Cryptophytes: *Cryptomonas* sp., *Rhodomonas* sp.
Chrysophytes: *Chrysochromulina* sp., *Dinobryon bavaricum*

Castle Lake, 1989[a]
Diatoms: *Achnanthes* sp.
Cryptophytes: *Cryptomonas erosa*, *Cryptomonas ovata*, *Rhodomonas minuta*, *Rhodomonas* sp.
Chlorophytes: *Chlamydomonas* sp., *Chlorella vulgaris*, *Gloeocystis* sp., *Oocystis pusilla*, *Oocystisparva* sp.
Chrysophytes: *Ochromonas* sp., *Chromulina* sp., *Chromulina* sp., *Chrysochromulina* sp., *Dinobryon* sp.
Cyanophytes:*Anabaena* sp., *Chroococcus varius*, *Chroococcus rufescens*, *Chroococcus dispersus*, *Chroococcus* sp., *Synechococcus elongatus*
Pyrrophytes: *Glenodinium pulviscula*, *Glenodinium pulvis*, *Gymnodinium* sp.

Coldwater Lake, July–August 1986
Diatoms: *Cyclotella stelligera*
Cryptophytes: *Cryptomonas* sp., *Rhodomonas* sp.
Chrysophytes: *Chrysochromulina* sp., *Dinobryon bavaricum*

Coldwater Lake, 1989[a]
Diatoms: *Asterionella formosa*, *Achnanthes minutissima*, *Anomoeoneis vitrea*, *Cymbella minuta*, *Cyclotella meneghiniana*, *Diatoma tenue*, *Fragilaria crotonensis*,*Hannea arcus*, *Nitzschia* sp., *Navicula* sp., *Synedra rumpens*
Cryptophytes: *Cryptomonas erosa*, *Cryptomonas ovata*, *Rhodomonas minuta*
Chlorophytes: *Ankistrodesmus* sp., *Chlamydomonas* type, *Elaktothrix gelatinosa*, *Gloeocystis* sp., *Oocystis* sp., *Quadrigula closteroides*, *Sphaerocystis schroterii* Chrysophytes: *Dinobryon* sp., *Kephyrion* sp.
Cyanophytes: *Chroococcus* sp.

[a] Kelly (1992).

Total phytoplankton densities in Coldwater and Castle lakes during July and August 1986 averaged 255 and 224 algal units ml^{-1}, respectively. Chrysophytes (*Dinobryon bavaricum* and *Chrysochromulina* sp.) and small, motile cryptophytes (*Rhodomonas* sp. and *Cryptomonas* sp.) were the predominant forms, comprising 74% of the phytoplankton community in Coldwater Lake and 90% in Castle Lake. The diatom *Cyclotella stelligera* was the most common, but it was present only in Coldwater Lake at densities averaging 165 algal units ml^{-1} (see Table 18.6).

Kelly (1992) reported that total phytoplankton densities in Coldwater and Castle lakes during 1989 (May to October) averaged 317 and 1824 algal units ml^{-1}, respectively. Coldwater taxa included diatoms, cryptophytes, chlorophytes, chrysophytes, and cyanophytes. Diatoms, especially *Cyclotella meneghiniana*, were dominant. In Castle Lake, flagellates and the dinoflagellate *Glenodinium pulvis* dominated the phytoplankton community. Other taxa included chlorophytes, chrysophytes, cyanophytes, pyrrophytes, and a single diatom, *Achnanthes* sp. (see Table 18.6).

18.4.2.5 Zooplankton

In October 1981, zooplankton densities at the surface and at 12 m in Coldwater Lake were essentially identical (56 and 60 organisms m^{-3}, respectively). Collections yielded a cladoceran (*Ceriodaphnia reticulata*), a rotifer (*Keratella quadrata*), and a cyclopoid copepodite. Zooplankton densities in Castle Lake were higher (117 and 130 organisms m^{-3} at the surface and at 6 m, respectively), with only rotifers found (Table 18.7).

TABLE 18.7. Zooplankton taxa, Coldwater and Castle lakes, 1981 to 1994.

Coldwater Lake, October 1981
- Cladocerans: *Ceriodaphnia reticulata*
- Copepods: cyclopoid copepodids
- Rotifers: *Keratella quadrata*

Coldwater Lake, August–September 1982
- Cladocerans: *Alona* sp., *Ceriodaphnia* sp., *Daphnia* sp.
- Copepods: calanoid copepodids, nauplii
- Rotifers: *Asplanchna* sp., *Keratella* sp.

Coldwater Lake, June–September 1983
- Cladocerans: *Daphnia* sp.
- Copepods: calanoid copepodids, nauplii
- Rotifers: *Asplanchna* sp., *Keratella* sp.

Coldwater Lake, 1989[a]
- Cladocerans: *Daphnia pulicaria*, *Daphnia rosea*, *Bosmina longirostris*, *Chydorus sphaericus*, *Alona rectangula*, *Alonella nana*
- Copepods: *Cyclops vernalis*, *Diaptomus tyrelli*, *Paracyclops fimbriatus*
- Rotifers: *Asplanchna priodonta*, *Asplanchna* sp., *Brachionus* sp., *Conochilus unicornis*, *Conochilus* sp., *Filinia terminalis*, *Kellicottia longispina*, *Keratella quadrata*, *Notholca squamula*, *Polyarthra dolichoptera*, *Synchaeta oblonga*, *Squatinella* sp., *Trichocerca* sp.

Coldwater Lake, 1992–1994[b]
- Cladocerans: *Alonella nana*, *Bosmina longirostris*, *Chydorus sphaericus*, *Daphnia pulex*, *Daphnia pulicaria*, *Daphnia rosea*, *Holopedium gibberum*
- Copepods: *Diaptomus oregonensis*, *Diaptomus tyrelli*, calanoid copepodids, cyclopoid copepodids, nauplii
- Rotifers: *Asplanchna priodonta*, *Euchlanis dilatata*, *Filinia terminalis*, *Harringia* sp., *Kellicottia longispina*, *Keratella cochlearis*, *Keratella quadrata*, *Lecane* sp., *Lepadella* sp., *Polyarthra* sp., *Synchaeta* sp., *Trichocerca* sp.

Castle Lake, August–September 1982
- Cladocerans: *Alona* sp., *Ceriodaphnia* sp., *Daphnia* sp.
- Copepods: calanoid copepodids, nauplii
- Rotifers: *Asplanchna* sp., *Keratella* sp.

Castle Lake, June–September 1983
- Cladocerans: *Alona* sp., *Daphnia* sp.
- Copepods: calanoid copepodids, nauplii
- Rotifers: *Asplanchna* sp., *Keratella* sp.

Castle Lake, 1989[a]
- Cladocerans: *Alona* sp., *Bosmina longirostris*, *Ceriodaphnia pulchella*, *Daphnia pulicaria*
- Copepods: *Tropocyclops prasinus*, *Cyclops* copepodite, copepod nauplii
- Rotifers: *Brachionus* sp., *Conochilus* sp., *Filinia terminalis*, *Kellicottia longispina*, *Keratella quadrata*, *Keratella cochlearis*, *Polyarthra dolichoptera*, *Synchaeta lakowitziana*

Castle Lake, 1992–1994[b]
- Cladocerans: *Alonella costata*, *Bosmina longirostris*, *Chydorus sphaericus*, *Daphnia pulex*
- Copepods: *Orthocyclops modestus*, cyclopoid copepodids, nauplii
- Rotifers: *Asplanchna priodonta*, *Branchionis calyciflorus*, *Branchionus angularis*, *Conochilus unicornis*, *Euchlanis dilatata*, *Filinia terminalis*, *Keratella cochlearis*, *Keratella quadrata*, *Lecane* sp., *Polyarthra doliochoptera*, *Pleosoma lenticulare*

[a] Kelly (1992).
[b] Scharnberg (1995).

Samples collected during August and September 1982 in Coldwater Lake yielded cladocerans (*Ceriodaphnia* sp., *Daphnia* sp., and *Alona* sp.), rotifers (*Keratella* sp., *Asplanchna* sp., and unidentified species), calanoid copepods, and cyclopoid nauplii. Total density was 8850 individuals m^{-3}, with 80% consisting of *Daphnia* sp. and 18% *Asplanchna* sp. The zooplankton community in Castle Lake during July 1982 was similar, although total density was less (5710 individuals m^{-3}), and the predominant forms were *Alona* sp. and unidentified rotifers (68% and 23%, respectively; see Table 18.7).

During summer 1983, total zooplankton densities in Coldwater and Castle lakes were almost identical, 4380 and 4300 organisms m^{-3}, respectively. *Daphnia* sp. constituted 73% of the zooplankton assemblage in Coldwater Lake and 70% in Castle Lake. Cyclopoid nauplii, calanoids, and rotifers (*Asplanchna* sp., *Keratella* sp., and unidentified species) were also present in both lakes, but the cladoceran *Alona* sp. was found only in Castle Lake (Table 18.7).

During 1989 (May to October), total zooplankton densities averaged 3088 organisms m^{-3} ($n = 33$ net tows) in Coldwater Lake and 2552 organisms m^{-3} ($n = 16$ net tows) in Castle Lake (Kelly 1992). In Coldwater Lake, *Daphnia pulicaria* was the predominant zooplankter, especially in surface waters. Other cladocerans were present, and *Diaptomus tyrelli* was the major copepod. Rotifers were occasionally abundant: total rotifer density in the lake's upper 5 m on July 1, for example, was 9600 organisms m^{-3}, or 63% of the zooplankton assemblage (Kelly 1992). The zooplankton community in Castle Lake was dominated by rotifers during summer and by cladocerans during fall. Copepods were rarely found (see Table 18.7; see also Kelly 1992).

Vogel et al. (2000) studied zooplankton colonization in Coldwater and Castle lakes by comparing their zooplankton assemblages with a model of expected community composition (Dodson 1992). The two lakes differed principally by the fact that Castle Lake lacked calanoids, which was attributed to a tendency for calanoid eggs to sink in the water column and, thus, be unavailable for dispersal from other nearby lakes.

18.4.3 Blowdown- and Scorch-Zone Lakes

18.4.3.1 Basin Morphometry

The morphometric features of blowdown- and scorch-zone lakes were altered far less than those of Spirit Lake. The material deposited in these lakes was mostly forest debris blown in from surrounding watersheds along with blast material and tephra. At St. Helens Lake, for example, logs and smaller woody debris covered roughly one-third of the lake surface. The lakes also received varying amounts of tephra fallout and material from overland surface runoff and inflowing streams. Crawford (1986) reported that large or extensive "deltas of ash" occupied the mouths of streams entering Grizzly and Obscurity lakes. Tephra deposits in Ryan Lake were described as "heavy," and deposits around St. Helens Lake were more than 200 cm thick, second only to those at Spirit Lake (Crawford 1986). The influx and deposition of these materials, both forest debris and volcanic ejecta, altered basin shapes, areas, depths, and volumes to some extent. The exact measure of these alterations is largely unknown, however, because pre- and posteruption morphometric data are apparently unavailable.

18.4.3.2 Optical Properties

By 1990, lake-water transparency in most blowdown- and scorch-zone lakes had generally improved to preeruption levels (Table 18.8; Carpenter 1995). During the early 1990s, light penetration was greatest in St. Helens Lake (Secchi-disk depth = 18 m; light extinction coefficient = 0.12 m^{-1}) and least in Ryan Lake (Secchi-disk depth = 4 m; light-extinction coefficient = 0.71 m^{-1}). Both Ryan Lake and Meta Lake (Secchi-disk depth = 4.8 m; extinction coefficient = 0.50 m^{-1}) remained stained, however, possibly because of runoff from watersheds still deeply covered with woody debris.

TABLE 18.8. Optical properties, blowdown- and scorch-zone lakes, 1980[a] versus 1993[b].

	Hanaford		Fawn		St. Helens			
	1980	1993	1980	1993	1980	1993		
Light extinction, m^{-1}	4.28	0.23	3.04	0.38	nd	0.12		
Secchi-disk depth, m	nd	9.0–13.5	7.0	6.5–7.7	9.1[c]	18.3–23		
	Panhandle		Boot		Venus		Ryan	
	1980	1993	1980	1993	1980	1993	1980	1993
Light extinction, m^{-1}	16.38	0.21	14.3	0.39	3.75	0.13	7.82	0.71
Secchi-disk depth, m	nd	nd	nd	nd	6.2[d]	16.5	nd	4.0–5.5

nd, no data.
[a] 1980 light-extinction data from Wissmar et al. (1982a); Secchi-disk data from Crawford (1986).
[b] 1993 data from Menting (1995) and Carpenter (1995).
[c] Reading taken August 29, 1984.
[d] Reading taken September 14, 1974.

TABLE 18.9. Water chemistry, blowdown- and scorch-zone lakes, 1980[a] versus 1993.[b]

	Hanaford		Fawn		St. Helens	
	1980	1993	1980	1993	1980	1993
Total alkalinity, $\mu Eq\ l^{-1}$	461	187	367	350	110	150
Ca, $mg\ l^{-1}$	57.80	7.64	62.70	10.23	13.00	8.15
Mg, $mg\ l^{-1}$	5.59	2.12	6.25	2.11	1.32	1.40
Na, $mg\ l^{-1}$	58.01	2.65	43.50	2.78	11.00	3.04
K, $mg\ l^{-1}$	23.99	0.39	11.00	0.35	2.50	0.83
SO_4, $mg\ l^{-1}$	195.01	24.74	165.01	20.96	37.90	22.50
Cl, $mg\ l^{-1}$	86.02	1.26	64.01	1.55	14.50	3.51
PO_4-P, $\mu g\ l^{-1}$	2.2	0.9	0.9	0.9	2.8	3.7
NO_2-N + NO_3-N, $\mu g\ l^{-1}$	2.0	11.9	9.4	10.6	1.7	nd
NH_3-N, $\mu g\ l^{-1}$	3.8	3.1	42.6	2.8	2.5	5.0
SiO_4, $mg\ l^{-1}$	13.96	9.20	17.33	10.27	12.47	8.74

	Panhandle		Boot		Venus		Ryan	
	1980	1993	1980	1993	1980	1993	1980	1993
Total alkalinity, $\mu Eq\ l^{-1}$	99	76	195	91	24	70	409	271
Ca, $mg\ l^{-1}$	20.00	2.39	24.40	3.10	5.87	1.39	27.80	6.39
Mg, $mg\ l^{-1}$	2.46	0.61	2.98	0.62	0.93	0.56	2.85	1.61
Na, $mg\ l^{-1}$	15.30	1.10	16.90	1.17	5.90	0.94	17.60	2.10
K, $mg\ l^{-1}$	5.30	0.33	3.80	0.31	1.10	0.31	5.80	1.13
SO_4, $mg\ l^{-1}$	58.70	8.24	69.70	9.59	15.95	6.45	79.80	13.83
Cl, $mg\ l^{-1}$	30.01	0.55	32.01	0.48	7.10	0.58	27.01	0.75
PO_4-P, $\mu g\ l^{-1}$	3.7	0.30	4.0	0.90	2.5	0.0	5.0	2.5
NO_2-N + NO_3-N, $\mu g\ l^{-1}$	2.7	47.6	3.2	43.8	2.7	24.8	3.6	4.6
NH_3-N, $\mu g\ l^{-1}$	2.2	1.5	3.8	1.0	2.9	1.7	7.6	4.8
SiO_4, $mg\ l^{-1}$	13.00	9.20	14.31	10.12	9.98	7.05	19.24	12.57

nd, no data.
[a] 1980 data from Wissmar et al. (1982a).
[b] 1993 data from Menting (1995) and Carpenter (1995).

18.4.3.3 Water Chemistry and Bacterial Metabolism

The posteruption water chemistry of blowdown- and scorch-zone lakes varied considerably. The chemistry of Ryan and Hanaford lakes was altered far more than that reported for Venus and St. Helens (Wissmar et al. 1982a). In either case, blowdown- and scorch-zone lakes were impacted far less than lakes located adjacent to debris-avalanche deposits (Coldwater, Castle, and Spirit). The DOC concentration in Fawn Lake (16 km from the crater), for example, was 1.3 mg $C\ l^{-1}$, or less than 1% of the concentrations found in Castle Lake (149 to 221 mg $C\ l^{-1}$). The TN:TP ratio (molar ratio of dissolved inorganic nitrogen to soluble reactive phosphorus) for blowdown- and scorch-zone lakes ranged between 6 and 16, suggesting that nitrogen was the limiting nutrient affecting microbial activity.

At Ryan Lake in 1980, near-surface concentrations of dissolved oxygen were low, with saturation values ranging from 11% to 72%. Thus, with some oxygen present, the C:N ratio at the surface was 47. In the lake's anoxic hypolimnion, however, the ratio was 377. During July and August 1980, DOC concentrations increased because of continued leaching of organic debris and because of microbial decomposition of organic matter buried in lake-bottom sediment (Dahm et al. 1983). Maximum denitrification was measured near the lake bottom at a depth of 6 m (Dahm et al. 1983). Nitrification was detectable only near the lake surface where some oxygen was available. Ryan Lake appeared to have been affected much more than were the other lakes in the blowdown and scorch zones for several reasons: (1) the lake is very small, shallow, and low in elevation; (2) the lake was not covered with snow or ice at the time of the eruption; (3) the lake has a large ratio of littoral zone to deep water; and (4) the lake is located in a fairly steep-sided bowl at the edge of the blowdown zone, where large amounts of organic matter were deposited.

Ionic concentrations continued to decline after 1986 (Table 18.9). In deeper lakes, such as Venus and St. Helens, concentrations declined at slower rates, possibly because of their greater volumes (Petersen 1993; Menting 1995). In some lakes (Hanaford, Fawn, and Ryan), ionic concentrations in epilimnetic waters decreased during summer. Conversely, concentrations in hypolimnetic waters increased because of an influx of groundwater or leachate from tephra and other volcanic deposits on the lake bottom. These estimates were made using a simple dilution model (Welch 1979; Menting 1995) using lake morphology and flows based on watershed area, precipitation gauges, and rainfall chemistry (Berner and Berner 1987). Initial posteruption concentrations in blowdown- and scorch-zone lakes varied by the inverse of lake volume; that is, lakes with larger volumes, such as St. Helens and Venus, had lower initial ion concentrations because of a dilution effect. Although distance from the crater would be expected to play some role in determining initial concentrations, no correlation between distance and concentration was detected, likely because of the effects of the extreme variation in topography on the dispersal of tephra.

Although the simple dilution model generally predicted the decline in concentration of individual ions, concentrations were significantly higher than those predicted by simple dilution alone, suggesting weathering of tephra and other volcanic deposits in lake watersheds. Additionally, differences among individual ionic concentrations were clearly evident. Menting (1995) related these differences to variations in weathering rates among individual ions, both in the watershed and on the lake bottom. Sodium and potassium exhibited the greatest declines, possibly because of their lower rates of release from the weathering of the tephra and other volcanic deposits. Calcium and magnesium, which appear to be more easily released from the tephra, had higher relative concentrations a decade after the eruption. Fruchter et al. (1980), Dethier et al. (1980), and McKnight et al. (1981a,b) reported that concentrations of water-soluble constituents were highly variable from location to location but that calcium, magnesium, chloride, and sulfate generally showed the highest release rates from tephra. Thus, it appears that the unusually high concentrations of ions observed in the lakes immediately after the eruption declined as a consequence of dilution modified by rapid weathering of the tephra. Nevertheless, the scant preeruption chemical data imply that ion concentrations in blast-area lakes were still above preeruption levels in 1991. For example, the conductivity of Fawn Lake during 1974 [30 μmhos cm^{-1} (Dion and Embrey 1981)] had increased by nearly a factor of four (109 μmhos cm^{-1}) by 1991. At Venus Lake, the increase was similar, that is, from 9 μmhos cm^{-1} in 1974 (Dion and Embrey 1981) to 37 μmhos cm^{-1} in 1991. Petersen (1993) and Menting (1995) have noted, however, that the concentrations of common ions in lakes around Mount St. Helens (including the reference lakes outside the blowdown and scorch zones) are generally higher than similar subalpine lakes located nearby in the southern Washington Cascade Range (Landers et al. 1987).

Carpenter (1995) reported that nitrogen, phosphorus, and silica concentrations during 1992 and 1993 were similar to those reported for undisturbed oligotrophic/mesotrophic subalpine lakes in the Oregon Cascade Range (Johnson et al. 1985). Carpenter (1995) applied various indices, including N:P ratios, alkaline phosphatase, and the depletion of nitrogen and phosphorus pools during the growing season, to determine if either nitrogen or phosphorus was limiting. This work led to a ranking of the lakes on the basis of whether they were nitrogen limited or phosphorus limited (or both) 12 to 14 years after the 1980 eruption. Although some lakes (Meta, Castle, Ryan, Fawn, Merrill, McBride, and Blue) were ranked as slightly nitrogen limited and others (Hanaford, Coldwater, St. Helens, Venus, and Panhandle) as slightly phosphorus limited, Carpenter (1995) believed that all the lakes were limited by both nitrogen and phosphorus. (Note that Carpenter's ranking included lakes throughout the Mount St. Helens region, the results of which are reported here even though Section 18.4.3 refers only to lakes in the blowdown and scorch areas.) These results tend to agree with observations made shortly after the eruption, indicating that nitrogen cycling in several lakes appeared to control other important processes (Dahm et al. 1983; Wissmar et al. 1988). The results are also consistent with a general review of nutrient-enrichment experiments (Elser et al. 1990), suggesting that colimitation by nitrogen and phosphorus is the most common case.

Nutrient bioassay experiments directly confirmed Carpenter's results. These experiments, designed after Dodds and Priscu (1990) and conducted during summer 1993, indicated that Fawn and Hanaford lakes were limited by both nitrogen and phosphorus and that Castle and Venus lakes were primarily limited by phosphorus (Petersen and Carpenter 1997). At Ryan Lake, the results were unusual: nitrogen additions, either alone or combined with phosphorus, caused an increase in carbon fixation, suggesting nitrogen limitation. In contrast, chlorophyll *a* concentrations were decreased by nitrogen additions but increased by phosphorus additions, suggesting phosphorus limitation. Possibly, the nitrogen additions stimulated heterotrophic carbon uptake, but at the expense of autotrophs that were also competing for nitrogen. Because the woody-debris load in Ryan Lake was still substantial in 1993, heterotrophic activity continued to predominate over autotrophic primary production, which may be unusual for blowdown- and scorch-zone lakes. Baker (1995) reported that protozoan biomass in Ryan Lake during 1993–1994 was much greater than phytoplankton biomass.

18.4.3.4 Phytoplankton

The posteruption condition of phytoplankton communities in blowdown- and scorch-zone lakes is scarcely known. Wissmar et al. (1982b) obtained chlorophyll *a* data for Hanaford, Fawn, St. Helens, Panhandle, Boot, Venus, and Ryan, reporting that values ranged from "0.0" μg l^{-1} (in four lakes) to 0.7 μg l^{-1} in Venus Lake.

At Ryan Lake, a mat of *Anabaena* sp. and other algae covered the littoral area during summer and fall 1980. Photosynthetically active phytoplankton (*Anabaena* sp., *Microcystis* sp., *Tribonema* sp., *Navicula* sp., *Nitzschia* sp., and unidentified cryptomonads) was present in the epilimnion (Ward et al. 1983). Below 2 m, the lake was anaerobic and devoid of viable algae, with only heterotrophic and chemosynthetic bacteria present.

Smith and White (1985) found that the phytoplankton in surface-water samples collected from Meta Lake during August 1982 consisted almost entirely (90.5%) of the chrysophyte *Dinobryon cylindricum* var. *palustris*. The diatoms *Synedra delicatissima* var. *angustissima* and *Nitzschia acicularis* were also present (0.8% and 0.3% of the total community, respectively), as were chrysophyte statocysts (3.9%). Total phytoplankton density was 8230 algal units ml^{-1}.

During 1992–1993, similarities were observed among lakes in the blowdown and scorch zones regarding phytoplankton assemblages (Baker 1995). In Fawn and Hanaford lakes, similarities included dominance by the diatom *Cyclotella comta*. In Panhandle and Boot lakes, *Glenodinium* sp., *Dinobryon* sp., and *Fragilaria crotonensis* were the dominant taxa (Baker 1995).

Use of the indicator species analysis method (Dufrene and Legendre 1997) revealed that 5 phytoplankton species (*Cryptomonas erosa*, *Glenodinium* sp., *Kephyrion* sp., *Ochromonas* sp., and *Planktosphaeria gelatinosa*) were associated with blowdown- and scorch-zone lakes, while 17 other species were associated with lakes outside the blowdown and scorch zones (Cynthia Baker, Portland State University, personal communication). Some phytoplankton species in blowdown- and scorch-zone lakes have been identified as mixotrophic species (organisms that are both heterotrophic and autotrophic, such as phagotrophic phytoflagellates; Sanders and Porter 1988), whose success may be caused by the relatively large amounts of decaying organic material remaining in lakes and watersheds.

Because blowdown- and scorch-zone lakes were highly enriched with organic matter, a condition inducing lake eutrophication, Baker (1995) applied a trophic state index (TSI) based on phytoplankton biovolume to ascertain the current trophic states of the lakes:

$$\mathrm{TSI} = [\log \text{ base } 2(B + 1) \times 5]$$

where B is the phytoplankton biovolume in $\mu m^3\ ml^{-1}$ divided by 1000; TSI ranges from 0 to 100, with lower values representing oligotrophy and higher values eutrophic lakes (Sweet 1986). Accordingly, eight lakes (Blue, Castle, Coldwater, Hanaford, June, Merrill, St. Helens, and Venus) were categorized as "oligotrophic," and seven others (Boot, Fawn, McBride, Meta, Panhandle, Ryan, and Spirit) as "mesotrophic." (Note that Baker's analysis included lakes throughout the Mount St. Helens region, the results of which are reported here even though Section 18.4.3 refers only to lakes in the blowdown/scorch areas.)

18.4.3.5 Zooplankton and Other Invertebrates

Following a preliminary survey of blowdown- and scorch-zone lakes during summer and fall 1980, Crawford (1986) reported that relatively small numbers of benthic macroinvertebrates (crayfish, leeches, insects, and amphipods) survived in several lakes, including Fawn, Hanaford, Meta, and Ryan. Stomach contents of fish caught in Meta and Hanaford lakes in September 1980 included beetles, dragonflies, chironomids (80% to 90%) and amphipods (6% to 15%). Benthic invertebrates (leeches and chironomid larvae) and zooplankton (calanoid copepods) were observed in Fawn Lake. Ryan Lake contained "remnant" invertebrate populations, mostly chironomids, mayflies, and dragonflies. An "abundance" of *Daphnia* sp. and calanoid copepods was found in St. Helens and Venus lakes (Crawford 1986).

During May and August 1981, trout-stomach analyses of remnant-population fish indicated that chironomids and beetles were the principal food organisms for fish in Panhandle, Fawn, and Elk lakes, with chironomids comprising 91% to 100% of the food organisms in Panhandle and Fawn lakes and beetles representing the predominant food source (91%) for Elk Lake trout. Minor food organisms included phantom midges, stoneflies, black flies, and copepods. *Daphnia* sp. and calanoid copepod densities in St. Helens and Venus lakes were considerably larger than they had been during 1980, notably in St. Helens Lake, where *Daphnia* sp. numbered 2260 organisms m^{-3}. During August 1982, crayfish, amphipods (*Gammarus*), beetles, corixids, and mayflies were observed in Fawn Lake.

Trout captured at Venus Lake in July 1983 foraged largely on copepods (*Diaptomus* sp.), chironomids, and salamanders. Trout-stomach analyses during August 1984 indicated that the major food organisms were mayflies (100% in St. Helens fish), caddis flies (80% in Shovel Lake fish), and chironomids (94% in Obscurity Lake fish) (Crawford 1986).

Scharnberg (1995) used cluster analysis to identify three distinct clusters of lakes based on their zooplankton community structure. (Note that Scharnberg's analysis included lakes throughout the Mount St. Helens region, the results of which are reported here even though Section 18.4.3 refers only to lakes in the blowdown/scorch areas.) He attributed clusters to predation and lake trophic status. Predation effects were identified according to palatability of prey, which was estimated from species biovolumes (Wetzel and Likens 1979) and length–frequency analysis (Edmondson and Winberg 1971).

Zooplankton assemblages in lakes around Mount St. Helens (both within and outside the blowdown and scorch zones) were similar to assemblages reported for subalpine lakes near Mount Rainier in Washington and Mount Hood in Oregon (Scharnberg 1995). This similarity indicated that zooplankton communities in blast-area lakes had largely recovered. Some minor differences were noted, however, including the following:

- Mount St. Helens lakes contained far greater numbers of *Chaoborus* sp.
- *Ceriodaphnia* and *Diaphanosoma* species were found only in Mount St. Helens lakes.
- *Keratella* sp., *Kellicottia* sp., *Polyarthra* sp., and *Conochilus* sp. dominated the rotifer communities in Mount St. Helens lakes and in Mount Rainier lakes, but not in lakes around Mount Hood.
- *Conochilus* sp. was frequently dominant in Mount Rainier lakes, whereas *Keratella* sp. was dominant in Mount St. Helens lakes, possibly because of substantially greater numbers of *Chaoborus* sp.

18.4.4 Tephra-Fall-Zone Lakes

Watersheds surrounding the tephra-fall lakes received airborne deposits of pumice and other volcanic fallout. These deposits,

TABLE 18.10. Optical properties, tephra-fall lakes, 1980[a] versus 1993.[b]

	Merrill		McBride		June		Blue	
	1980	1993	1980	1993	1980	1993	1980	1993
Light extinction, m^{-1}	0.56	0.29	0.59	0.94	nd	0.24	0.71	nd
Secchi-disk depth, m	nd	7.3–8.2	nd	1.6[c]	nd	nd	nd	12.8

nd, no data.
[a] 1980 light-extinction data from Wissmar et al. (1982a); Secchi-disk data from Crawford (1986).
[b] 1993 data from Menting (1995) and Carpenter (1995).
[c] Questionable reading; obtained in October under low-light conditions.

up to 25 cm thick in places, were subsequently eroded and washed into the lakes, forming alluvial fans where streams and overland runoff entered. Tephra-fall deposits had little if any effect on lake-water optical properties, except perhaps at Merrill Lake (Table 18.10). Wissmar et al. (1982b) reported that the physical properties of these lakes, including their optical properties presumably, were "unaffected" by the eruption. The chemistry of tephra-fall-zone lakes was also "unaffected," according to Wissmar et al. (1982b). DOC concentrations in Merrill, McBride, and Blue lakes were lowest, averaging 1.0 mg C l^{-1}. The DOC concentration in June Lake was 2.28 mg C l^{-1}, which fell within the DOC range for blowdown- and scorch-zone lakes (i.e., 1.34 to 10.97 mg C l^{-1}; Wissmar et al. 1982b). Water-chemistry data are compared between 1980 and 1993 in Table 18.11.

Bacterial densities were markedly smaller, ranging between 5×10^5 and 8×10^5 cells ml^{-1}. Thirty-one water samples were collected from Merrill Lake between 1983 and 1985 and analyzed for total coliforms, fecal coliforms, standard plate count, *Klebsiella pneumoniae*, *Pseudomonas aeruginosa*, and *Legionella* sp. by the Washington Department of Social and Health Services. These analyses indicated that bacterial densities and composition in Merrill Lake were similar to those occurring in subalpine lakes in the Washington Cascades that were not affected by the volcanic eruption (Larson and Glass 1987).

18.5 Summary and Conclusions

Following the 1980 eruption of Mount St. Helens, scientists had the rare opportunity to track geochemical responses and biological succession patterns of microbes, plants, and animals leading to the recolonization and often nearly complete recovery of these ecosystems. These processes occurred more rapidly than predicted: Within 2 years after the eruption, the lakes had recovered substantially from their initial posteruption conditions.

Nowhere else was this recovery more remarkable than at Spirit Lake, where limnological deterioration was most severe. Early investigators encountered the following conditions: (1) substantial changes to lake-basin morphometry and blockage of the lake's outlet; (2) massive organic loading, causing proliferation of heterotrophic bacteria and subsequent oxygen depletion; (3) elimination of oxygen-requiring organisms; (4) high concentrations of reduced elements, particularly iron, manganese, and sulfur, because the redox potential was lowered by anoxic conditions; (5) rapid shift from photosynthetic to chemosynthetic production as chemosynthetic bacteria proliferated in a chemically reduced environment; (6) intensive microbial activity yielding high concentrations of carbon dioxide, hydrogen sulfide, and methane; (7) bacterial denitrification, which reduced nitrate and nitrite concentrations while generating molecular nitrogen; (8) predominance by organisms

TABLE 18.11. Water chemistry, tephra-fall lakes, 1980[a] versus 1993[b].

	Merrill		McBride		June	
	1980	1993	1980	1993	1980	1993
Total alkalinity, μEq l^{-1}	126	216	146	413	563	683
Ca, mg l^{-1}	2.17	2.54	3.62	4.28	4.06	5.04
Mg, mg l^{-1}	0.42	0.50	0.83	1.09	1.52	2.13
Na, mg l^{-1}	3.70	1.64	2.80	3.32	8.80	11.04
K, mg l^{-1}	0.30	0.09	0.75	0.23	1.57	1.38
SO_4, mg l^{-1}	0.69	0.57	1.50	0.35	nd	8.82
Cl, mg l^{-1}	1.20	0.77	4.48	1.93	nd	4.65
PO_4-P, μg l^{-1}	1.9	1.9	1.2	4.0	31.0	51.1
NO_2-N + NO_3-N, μg l^{-1}	1.5	0.6	2.7	12.6	37.3	46.1
NH_3-N, μg l^{-1}	2.2	0.0	2.4	1.0	22.6	1.8
SiO_4, mg l^{-1}	10.99	9.81	15.82	13.49	50.45	36.78

nd, no data.
[a] 1980 data from Wissmar et al. (1982a).
[b] 1993 data from Menting (1995) and Carpenter (1995).

capable of fixing nitrogen because of nitrogen limitation; (9) lake waters greatly discolored by high concentrations of humic substances and sulfur, which, combined with high concentrations of suspended particulate matter, resulted in near-zero lake-water transparency; (10) abundant presence of pathogenic bacteria, including *Legionella* and *Klebsiella*; and (11) greatly elevated water temperatures.

When the lakes on Mount St. Helens were first visited in June 1980, scientists described some lakes (Spirit, Castle, Coldwater, and Ryan, for example) as "enrichment cultures," in which extremely high concentrations of both dissolved and particulate organic matter, metals, phosphorus, and other energy sources had prompted an explosion of microbial activity. Heterotrophic oxidation of organic carbon soon depleted dissolved-oxygen concentrations, creating anaerobic lakes in some instances. As nitrogen became limiting, pioneering microbes capable of nitrogen fixation predominated. Intense microbial activity mobilized various metals, particularly iron and manganese; reduced sulfate to sulfides; and evolved gases, such as methane, carbon dioxide, carbon monoxide, and hydrogen. Under anaerobic conditions, bacteria used nitrate and sulfate as alternative sources of oxygen. Fermentation of organic matter and methanogenesis occurred throughout the water column and in sediment.

By September 1980, biotic communities in the most severely impacted lakes differed greatly from those in lakes that were either less disturbed or outside the blast area. Metazoans were eliminated, and algal photosynthesis was essentially halted. Anaerobic heterotrophic and chemosynthetic bacteria prevailed. Anaerobic nitrogen-fixing bacteria supplied reduced nitrogen to bacterial populations. The lakes remained entirely anoxic, or nearly so, and extremely high concentrations of DOC, methane, iron, and manganese persisted. Under anoxic conditions, denitrifying and sulfate-reducing bacteria accounted for most microbial decomposition of organic detritus. These processes occurred mostly in sediment, although they greatly influenced lake-water chemistry.

During winter and spring 1980–1981, high runoff greatly diluted and flushed chemically enriched lake waters and dramatically altered the course of recovery. By spring 1981, bacterial densities had decreased by two orders of magnitude.

During the next two years, high concentrations of DOC in blast-area lakes continued to influence microbial activity. As DOC concentrations decreased, however, microbial oxygen consumption rapidly diminished. By 1983, oxygen concentrations in blast-area lakes had increased markedly, with epilimnetic waters remaining well oxygenated year round and hypolimnetic waters becoming progressively less oxygen depleted each year. By 1986, recovery of blast-area lakes was evident:

- Bacterial levels were back to near preeruption levels.
- Concentrations of most ionic constituents, especially iron and manganese, were significantly lower.
- Nitrogen, phosphorus, and silica concentrations were similar to levels reported for undisturbed subalpine oligotrophic/mesotrophic lakes.
- Lake-water transparency was considerably greater, allowing sufficient light penetration for benthic- and pelagic-zone photosynthesis.
- Phytoplankton and zooplankton communities were beginning to resemble those found in undisturbed oligotrophic/

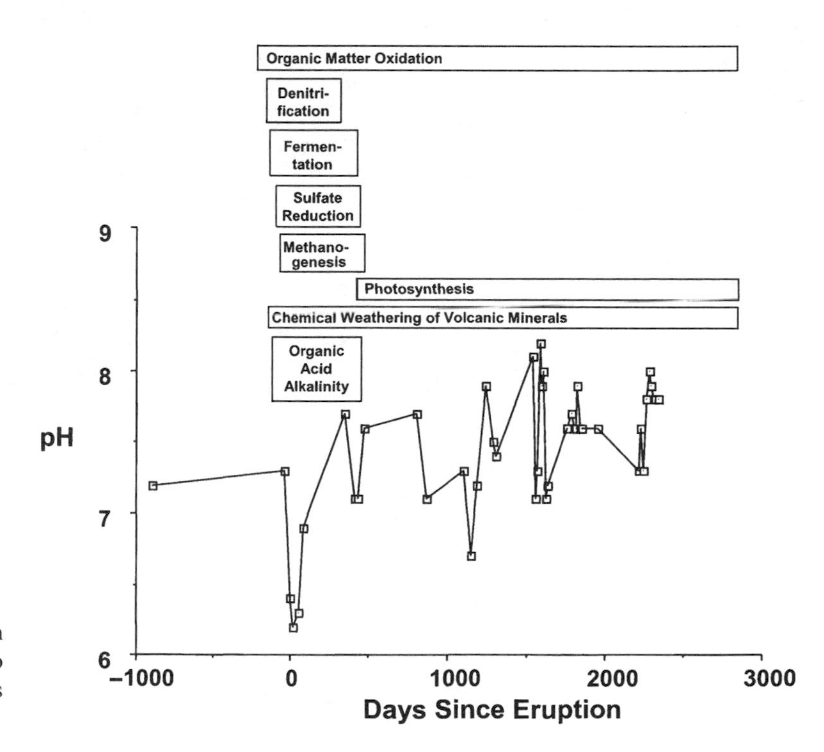

FIGURE 18.5. Summary of biogeochemical processes in blast-zone lakes, 1980 to 1990, as related to changes to pH. The length of each process rectangle indicates its period of major influence.

mesotrophic lakes in the Washington and Oregon Cascades.
- Algal photosynthesis, not microbial chemosynthesis, was now the principal means of energy fixation.
- Benthic and pelagic macroinvertebrates were abundant.
- Habitat conditions in most lakes were favorable for restocked fish populations (Figure 18.5).

The rapid recovery of blast-zone lakes demonstrated the vigor and resiliency of lake ecosystems. Nevertheless, despite substantial improvements in lake water quality and the reestablishment of more typical lake biological communities, the blast-zone lakes are today very different from what they were before the eruption of Mount St. Helens. Spirit Lake, in particular, will probably never return to its preeruption condition. In 1990, vast beds of rooted vegetation choked the south end of the lake, an area shoaled by organically laden, chemically reduced muds and fed by warm, chemically enriched geothermal seepwaters originating in the debris avalanche materials at the base of Mount St. Helens (Larson 1991, 1993). Concentrations of some inorganic chemicals remain well above their preeruption levels, probably because of continued geothermal inflows and runoff from the lake's denuded, erosion-prone, geochemically rich watershed. Indeed, centuries will pass before the lake's watershed is restored to its preeruption status as an old-growth forest.

Moreover, some lakes continue to exhibit hypolimnetic oxygen deficits during summertime thermal stratification, although this may have been a typical condition even before the eruption, as evidenced by the 1974 oxygen profiles for Fawn and Spirit lakes (Bortleson et al. 1976). Because of the continuing influx of dissolved organic matter from sediment and watershed sources, the rate of hypolimnetic oxygen consumption by heterotrophic microbial activity has been sufficiently high during summer and fall to deplete dissolved oxygen at lower depths in some lakes.

The tremendous influx of organic matter and nutrients to blast-zone lakes created a complex enrichment culture for opportunistic microbes to flourish and quickly become dominant, as exemplified by the great abundance of nitrogen-fixing and manganese- and sulfur-oxidizing bacteria. The rise and proliferation of these organisms suggest that vestigial microbial communities can survive in refugia, such as hot springs, hydrothermally altered soils, anaerobic sediment, and geothermal water. There, confined to specific microhabitats, the microbes remain dormant while waiting for conditions that trigger their resurgence. Eventually, their time will come again when the volcano reawakens, perhaps abruptly, and once more bursts destructively across the lands and waters around its base.

Acknowledgments. This work was funded, in part, by National Science Foundation grants DEB-811307 and BSR-8407429; USDA Forest Service contracts PNW 80-178, 80-277, and 80-289; and U.S. Army Corps of Engineers (Portland District) contract DACW57-81-C-0084. Funds were also provided to Richard Petersen and his graduate students by the U.S. Forest Service (1988 to 1993) and by the Aldo Leopold Wilderness Research Institute (1993 to 1995). We thank Virginia Dale of Oak Ridge National Laboratory and Charles Crisafulli, Peter Frenzen, Fred Swanson, and James Sedell of the USDA Forest Service for their support and encouragement. We also thank Charles Crisafulli, Fred Swanson, and three unnamed reviewers for their comments on the initial draft of the manuscript. We are deeply grateful to Dianne McDonnell and James Thibault, Department of Biology, University of New Mexico, who produced the illustrations. Zooplankton identifications were provided by Judy Li, Oregon State University, for samples collected between 1980 and 1986.

Part V
Lessons Learned

19 Ecological Perspectives on Management of the Mount St. Helens Landscape

Virginia H. Dale, Frederick J. Swanson, and Charles M. Crisafulli

19.1 Introduction

The dramatic change and dynamic nature of recently disturbed landscapes often create major challenges for management of public safety and natural resources. This was certainly the case at Mount St. Helens following the 1980 eruption. The eruption triggered an immediate response that entailed search and rescue of missing people and protection of human health and property. Monitoring geological hazards and further volcanic activity was a key tool for providing warnings to the public and aided the State of Washington, USDA Forest Service, and other agencies in decisions regarding access, pending and current dangers, and area closures. As volcanic activity quieted and biotic and geomorphic change commenced, the perspectives of environmental scientists became pertinent to land- and water-management issues.

The sequence of management issues at Mount St. Helens forms a framework for considering the perspectives and roles of environmental scientists in management of the area. Before March 20, 1980, 123 years had elapsed since the previous eruption of Mount St. Helens, and management of the area focused on recreation and forestry for wood production. Between March 20 and May 18, 1980, the mountain underwent a period of mild volcanic activity, and concern focused on the hazards it posed to people and property. Immediately after the massive eruption, search and rescue efforts became all consuming. Thereafter, concerns gradually shifted to long-term management of the hazards and the commercial, educational, recreational, and research opportunities of the area. During this period, environmental scientists became strongly engaged. During the first few years after the initial eruption, their engagement was highly energetic. The pace of their engagement decreased to quite modest in the mid-1980s and to very little in the 1990s and 2000s. These shifting roles were influenced by changes in the eruptive behavior of the volcano; the progress of construction projects to minimize hazards; and the rate of geomorphic and ecological change in uplands, rivers, and lakes.

In this chapter, we present a brief synopsis of the roles of environmental scientists in management issues related to the 1980 eruption of Mount St. Helens. First, the chapter describes the geographic and temporal contexts of the eruption and surrounding landscape. Then, it reviews selected examples of hazard- and land-management issues involving environmental sciences and scientists. Finally, it summarizes the advisory role of environmental scientists in protecting natural processes and features, human life, property, and commercial development. The discussion focuses on the posteruption period when environmental scientists were most active and when the ecological changes considered in this book were taking place.

19.2 Broad-Scale Context: Resource Management Before and After the Eruption

Before 1980, patterns of resource use and ecological conditions in the Mount St. Helens area strongly reflected landownership patterns and the related management objectives and approaches. Much of the land within about 30 km of the volcano was a mosaic of federal, state, and private ownership, managed primarily for timber production and recreation (Figure 19.1). The Gifford Pinchot National Forest managed most of the mountain and lands to the north and east. Some of the federal forest lands to the east had a distinctive pattern of dispersed-patch clear-cuts with plantations established in the previous 30 years, whereas lands on the volcano below tree line and to the north primarily supported older, natural forests. Mount St. Helens, Spirit Lake, and the Mount Margaret area north of the volcano were intensively used for recreation by hikers, hunters, fishers, campers, miners, and transient and permanent residents of private cabins and lodges.

State-owned land was concentrated on the south and west side of Mount St. Helens and managed by the Washington Department of Natural Resources (DNR) largely for forest products. The DNR implemented clear-cutting and intensive

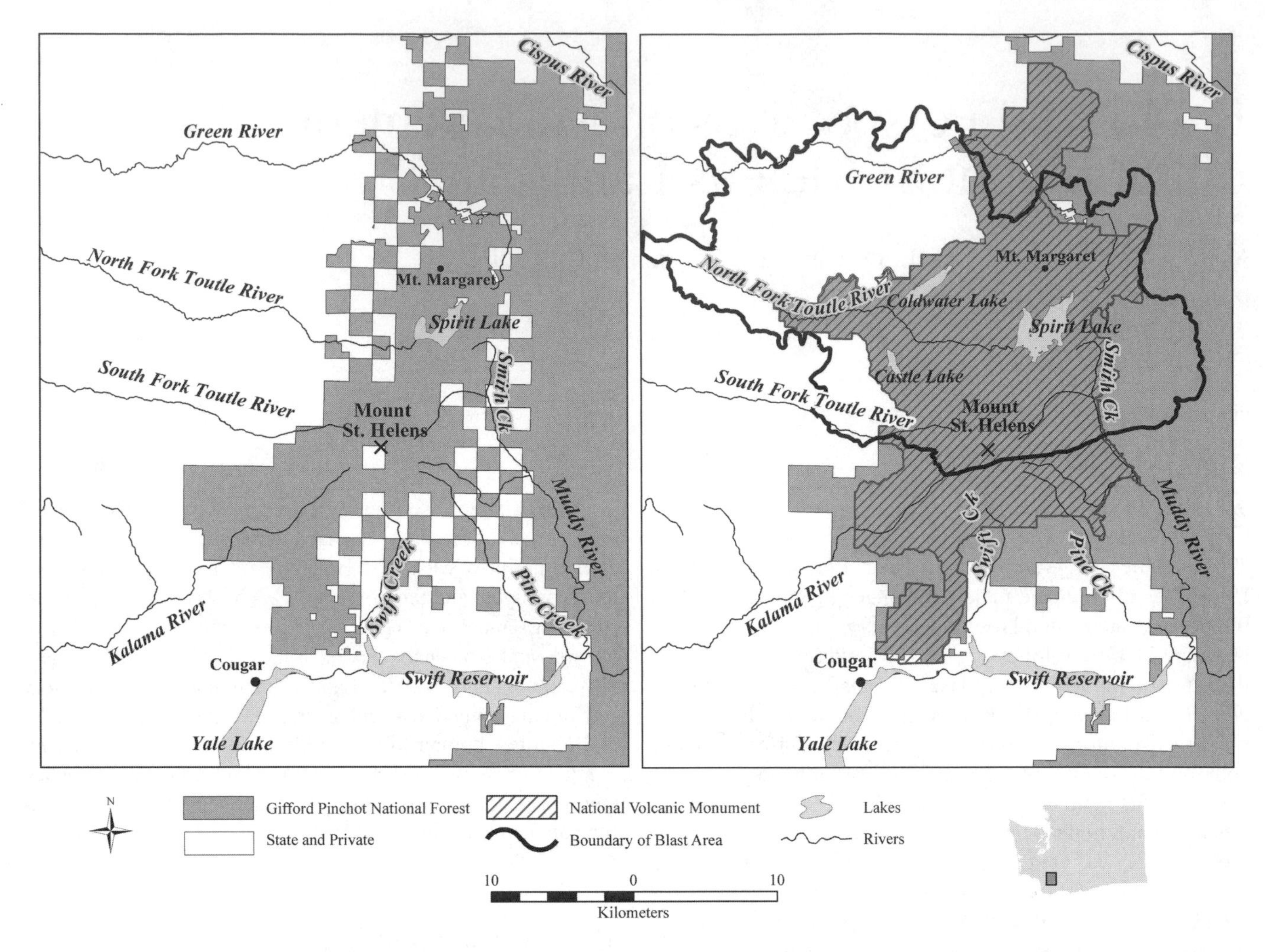

FIGURE 19.1. Map of ownership pattern of the Mount St. Helens area before (left) and after (right) establishment of the National Volcanic Monument.

forestry practices to establish plantations dominated by Douglas-fir (*Pseudotsuga menziesii*).

In the decades before 1980, the extensive private industrial forest lands west of the volcano were used largely for resource extraction, principally by the major landowner, the Weyerhaeuser Company, and to a lesser extent for recreation. Several decades of timber extraction and road construction had removed much of the native forest from private lands, and young conifer plantations grew in their place.

The May 18, 1980, eruption profoundly altered the landscape. Fifty-seven people died, many millions of dollars worth of timber was blown down, and the forest road network was either obliterated or made impassable by tephra deposits and toppled trees. Private cabins and lodges were swept away or buried by deep volcanic material. A logging camp and 18 bridges along the Toutle River and several bridges on the Muddy River were washed away by the mudflows. Great concern was expressed about future mudflows and floods.

The long-term management of the vast area impacted by the volcano was immediately under discussion. It was clear that some of the federal lands would be the focus of geological studies and tourism and would likely be dedicated to such activities. However, the future of land not in the major impact zones was less certain. After much debate and some movement in alternative management directions, Congress passed the Mount St. Helens National Volcanic Monument Act in 1982, which established the 43,300-ha Monument, set policy for management of the area, and changed some of the land-ownership patterns (see Figure 19.1). A critical feature of the legislation was direction for "allowing geologic forces and ecological succession to continue substantially unimpeded" (Public Law 97–243, Sect 4.b.1). This charge set up a tension between protection of natural processes and protection of people, property, and commercial development. There was substantial involvement by environmental scientists in the decisions leading up to the designation of the Monument and in its proscribed management. Scientists also had roles in implementing management practices through advisory panels and other activities. These roles are considered next in terms of four time periods.

19.3 Temporal Context of Management Issues

19.3.1 Before March 21, 1980

In the decades before the eruptive events of 1980 began, resource-management concerns at Mount St. Helens focused on timber, wildlife, fish, and recreation. The Forest Service, DNR, and private companies extracted timber as a renewable resource and provided access to lands for fishing, hunting, hiking, and camping. The Washington Department of Fish and Game managed fishing and hunting. Very few people lived close to the mountain year round. Visitors to the region appreciated the beauty of the symmetrical, snow-capped mountain, but did not consider the volcano to be a hazard.

In a prescient scientific analysis during the 1970s, United States Geological Service (USGS) geologists Dwight ("Rocky") Crandell and Donal Mullineaux surveyed potential volcanic hazards of Mount St. Helens and other Cascade Range volcanoes. They found that Mount St. Helens had been distinctively active during the Holocene (the past 10,000 years), leading them to predict that Mount St. Helens would erupt soon (Crandell and Mullineaux 1978): "Mount St. Helens has had a long history of spasmodic explosive activity, and we believe it to be an especially dangerous volcano. In the future Mount St. Helens probably will erupt violently and intermittently just as it has in the recent geologic past, and these future eruptions will affect human life, property, agriculture, and general economic welfare over a broad area."

Environmental scientists had little direct role in management of the area before 1980, in part because concepts about ecological effects of human activities and natural-disturbance processes in the region were just developing. Ecologists had described plant species (St. John 1976) and vegetation communities in the area (Franklin 1966) and the relationships of those communities to soil and topography (Franklin and Dyrness 1973). The Forest Service's Area Ecology Program had conducted vegetation inventories in the late 1970s at several locations in the vicinity of Mount St. Helens as part of a regional effort to characterize plant associations. Herpetologists and fish biologists had conducted numerous surveys in the area as well (Crawford 1986; Crisafulli et al., Chapter 13, this volume). The 1980 eruption occurred just as the field of ecology was recognizing the contributions it could make to management of forest, stream, and lake systems and to outdoor recreation (e.g., Lubchenco 1991; Swanson and Franklin 1992).

19.3.2 March 20 to May 18, 1980

During the initial eruptive period, USGS and university geologists and geophysicists began monitoring the deforming volcano, gas emissions, and earthquakes (Thompson 2000). This information provided a basis for making short-term predictions of impending volcanic events. Local officials and the public grew more wary of a potential major eruption and its impacts, and officials established zones of restricted entry to protect the inquisitive. An interagency center for tracking volcano developments was established at the Gifford Pinchot National Forest to provide information to the USDA Forest Service, emergency personnel, and the public. Washington Governor Dixie Lee Ray declared a state of emergency, assumed control of the zone of restricted entry, and called in the National Guard to provide security. Logging continued in restricted-entry zones during regular weekday hours, but recreational activity, including use of personal cabins, was prohibited.

Ecologists immediately became interested in studying ecological effects of the eruption. The Mount St. Helens Research and Education Coordinating Committee was established in Vancouver, Washington, at the request of the USGS to coordinate requests for access to the mountain for scientific and educational purposes (Allen 1982). It was composed of representatives from the Oregon and Washington State geology departments and universities in the region. When it was possible to obtain the requisite permits from the Committee to enter the restricted area, some environmental scientists made observations and collected samples. These incipient studies were interrupted by the massive eruption on May 18.

19.3.3 May 18, 1980, and Immediately After

The violent eruption focused attention of federal, state, and local authorities on search and rescue for the dozens of people feared lost. These efforts were complicated by poor weather, continued minor eruptions of tephra, and the large number of agencies attempting to help with this natural disaster. The Federal Emergency Management Agency (FEMA) quickly assumed the role of coordinating communication regarding search and rescue and established an information center (Brown 1982). By May 21, when the state was declared a national disaster area, attention turned to protecting lives and property. USGS geologists and University of Washington geophysicists were responsible for assessing the potential of future volcanic activity and communicating associated hazards to government officials who were responsible for communicating with the public. Environmental scientists could not gain access to the central portion of the volcanically disturbed area, so they focused their late-May sampling on the margins of the volcanically disturbed area.

There was immediate concern about the health hazards of the huge volume of ash spread from the newly formed crater to most of eastern Washington and the states beyond. The Washington Department of Social and Health Services (DSHS), the Centers for Disease Control and Prevention (CDC), and the National Institute for Occupational Health and Safety (NIOSH) worked together to determine that the ash did not cause cancer, but it did induce short-term acute respiratory irritation. Furthermore, the ash could cause long-term chronic respiration illness for individuals who had high and long-term exposure, a group that included scientists on the mountain.

19.3.4 Management Issues After May 1980

For the next few years, Mount St. Helens produced a series of minor eruptions and mudflows of diminishing magnitude. High levels of water and sediment yields persisted, particularly in the Toutle River, where the debris-avalanche deposit provided a nearly limitless sediment source. Also during this period, forest, lake, and stream systems began to display their ecological responses to the disturbances: surviving plants sprouted; surviving animals emerged from their burrows and protective snow and ice cover; dispersal of fungi, plants, and animals proceeded; ecological communities developed; and physical and biogeochemical change proceeded in terrestrial and aquatic systems.

Management issues involving ecological concerns after the 1980 eruption had two immediate goals: responding to hazards and meeting resource and recreation objectives of government and private landowners. The establishment of the Mount St. Helens Monument in 1982 set scientific research and education as major goals. The strength and relative importance of these goals evolved over time as volcanic, hydrologic, ecological, and management circumstances changed. Geographically, both the goals and the actions differed according to whether the lands were managed by the Forest Service, state, private forestry companies, or local residents and corporations.

The May 18 eruption created a host of new environmental hazards. The St. Helens Forest Lands Research Cooperative (1980) was formed of members drawn from the forestry research staff of DNR, the Forest Service, and the Weyerhaeuser Company to identify immediate major environmental threats. The Cooperative recognized that floods or mudflows could occur from unstable deposits, tephra-covered hill slopes, breaches of dams blocking new lakes, and sediment-filled channels. Also, both the blast-toppled and standing dead trees across large expanses were considered potentially susceptible to fires and insects. Catastrophic wildfire or insect outbreaks could affect living forests and the potential to salvage standing and downed dead trees. The diversity and complexity of the dangers was a major challenge to resource managers. Ecologists, foresters, geologists, engineers, and hydrologists were called upon to assess these risks by various agencies, including the U.S. Army Corps of Engineers (the Corps), FEMA, Forest Service, USGS, Washington Department of Game, and the Soil Conservation Service (now named the Natural Resources Conservation Service). For flood and sediment hazards downstream from the volcano, the issues and responses were partially determined by the needs of the cities of Kelso and Longview and the Columbia River shipping channel.

19.3.4.1 Warning of Impending Hazards

After the eruption, the need to establish and maintain a long-term monitoring and warning system was paramount. USGS scientists provided information on volcanic activity and potential hydrologic hazards. FEMA, state, and county officials delivered information about hazards and evacuation procedures to the public. Environmental scientists had little role in the hazard protection system, but did benefit by having a safer work environment.

19.3.4.2 Managing Hill-Slope Erosion

There was great concern by local citizens and government officials that storm-water runoff and sediment shed from the tephra-covered hill slopes would cause extreme flooding and channel change in downstream areas. The St. Helens Forest Lands Research Cooperative (1980) identified erosion and its impacts as major threats, which led to two management questions:

- Would salvage logging increase or decrease water and sediment runoff?
- Could aerial application of plant seeds lead to vegetation cover that would suppress water and sediment movement?

Experimental studies showed that salvage logging reduced erosion. Activities that mixed tephra layers, especially with older soil, greatly increased water-infiltration capacity, thereby reducing water and sediment movement from hill slopes (Swanson and Major, Chapter 3, this volume). Furthermore, field studies led to the unanticipated conclusion that ground disturbance reduced rather than increased erosion in the blast area. In some parts of the blast area, small gullies cut down to the old soil surface, released plants buried in the preeruption soil, and allowed buried seed to germinate. Hence, an active salvage and tree-planting program was initiated by the Forest Service, DNR, and timber industry, in which tephra was mixed with underlying soils as tree seedlings were planted, which planting trials and erosion studies had shown to be important for seedling survival and establishment (Winjum et al. 1986).

Emergency funds were available for a few months after the May eruption to protect life and property. In an effort to control erosion, the Soil Conservation Service proposed a $16.5 million program of seeding a nonnative-grass/legume/fertilizer mix in fall 1980 over much of the blast area and debris-avalanche zone (USDA Soil Conservation Service 1980). An initial proposal to spread nonnative seed over the entire volcanic area drew a sharp, negative response from ecologists, as evidenced by a resolution from the Second International Congress of Systematic and Evolutionary Biology in July 1980, which stated "this Congress vigorously opposes any proposal for mass seeding of grasses or any other species on the newly created substrate." Subsequently, the U.S. Congress reduced the allocation to a $2 million program of aerial application by the SCS over about 8660 ha of Weyerhaeuser, state, and national forest land in the fall of 1980 (Carlson et al. 1982). The expenditures of emergency funds for seeding had to occur before the emergency funding expired and before the fall rains, even though no native seeds were available.

Distributing nonnative plant seeds was largely ineffective for erosion control. The timing of the seed distribution was not appropriate for successful germination (Stroh and Oyler 1981; Klock 1982). Erosion was not reduced (Swanson and Major, Chapter 3, this volume). Where nonnative seeds did establish plants, they were associated with greater mortality of native conifers (Dale 1991; Dale and Adams 2003). Building upon these observations from Mount St. Helens and studies elsewhere, the practice of broadcast seeding of nonnative plant species onto newly disturbed sites has now been supplanted by a commitment by the Natural Resource Conservation Service and others to use native species, where possible (e.g., see Brindle 2003).

19.3.4.3 Managing the Flow of Water from Lakes Blocked by the Debris Avalanche

The massive debris avalanche of May 18 raised the elevation of the blockage to Spirit Lake and obstructed Coldwater and Castle creeks and many small tributaries of the Toutle River, forming new lakes with potentially unstable outlets. As impoundments gradually filled with water, it was anticipated that they would overtop or erode through their blockages or that the dams would fail, as had occurred elsewhere (Costa 1991; Parkes et al. 1992). Therefore, the Corps created stable drainage outlets for Castle and Coldwater lakes to discharge over bedrock and through armored channels and intentionally breached some of the small impoundments.

Controlling the level of Spirit Lake posed a complex problem that required both short- and long-term management strategies. Communities and logging operations downstream were greatly concerned about the potential for catastrophic damage from a breach of the lake. Hence, the Corps designed a drainage system that would maintain the lake's surface elevation at a level that would reduce threats to downstream areas. From November 1982 until spring 1985, the Corps controlled the lake level by pumping water through a 1113-m-long pipeline over the western Pumice Plain and debris-avalanche deposit while a 2450-m-long tunnel was constructed from the western shore of Spirit Lake through bedrock to upper South Coldwater Creek. Since the tunnel opened in May 1985, Spirit Lake water has followed this course to the North Fork Toutle River. Hence, the short-term measure provided time to determine and implement long-term solutions.

Development and operation of the pumping and tunnel systems for Spirit Lake and the construction of outlet channels on the dammed lakes involved trade-offs between protecting human life and property and maintaining natural processes and features. Although the threat of major mudflows from breaching of lakes was vastly reduced, the western Pumice Plain and debris-avalanche area had a massive canyon cut through the 1980 deposits, lake levels were artificially regulated, south Coldwater Creek's hydrologic regime was dramatically altered, and a road was built across an ecologically sensitive portion of the landscape to facilitate the construction and maintenance of the pumping station. Moreover, human-created barriers prevented anadromous fish from reaching Spirit Lake and north Coldwater Creek. Additionally, the Spirit Lake tunnel requires periodic closure for maintenance; during this time, lake level can rise substantially, grossly reconfiguring the ecology of the littoral and riparian zones. These compromises were readily accepted because the threat to human life and property was so severe. However, proposals to remove the large mat of wood floating on Spirit Lake were resisted by ecologists interested in letting this unusual natural feature and its ecological role to persist (see Dahm et al., Chapter 18, this volume). Furthermore, water-quality issues were considered in the decision of whether to keep Spirit Lake water in the Toutle River system or to divert it to the Lewis River. Initially, water quality in Spirit Lake was poor (Dahm et al., Chapter 18, this volume), and microbiologists discovered the pathogenic bacteria *Legionella* and *Klebsiella* in Spirit Lake (Dahm et al., Chapter 18, this volume), which contributed to the decision to keep the water drained from Spirit Lake in its natural watershed. Natural processes vastly improved water quality before flow from Spirit Lake to the Toutle River was reestablished.

These water-flow-management efforts illustrate the benefits of cooperation among interested parties in resolving hazardous situations. The Corps worked with local communities and federal and state agencies to resolve potential crises that could have resulted from an impoundment being breached. Similar combinations of short- and long-term strategies might be appropriate when consequences of no action are so dire.

19.3.4.4 Managing Downstream Sedimentation and Floods

Because of the immediate threats to human life and property downstream from large deposits, sedimentation concerns began to be addressed in the first weeks and months after the eruption and are still a problem. The Corps undertook a series of short- and long-term measures to control sediment delivery and channel capacity. Immediate steps included dredging segments of river channel where sediment accumulated and where disposal sites were nearby. Two low sediment-retention structures constructed by the Corps in 1980 and 1981 on the Toutle River were quickly filled with sediment and breached. Subsequently, the Corps completed a 55-m-high, 504-m-long sediment-retention structure on the North Fork of the Toutle River in 1988 with an anticipated lifetime of five decades (see Figure 3.5), but by April 1998 it had filled above all discharge pipes and was flowing over the spillway. The large discharge over the spillway has caused more sediment than expected to accumulate downstream of the structure. During summer 2004, the Corps, USGS, Forest Service, and Washington Department of Fish and Wildlife began meeting to discuss possible options for sediment control in the basin.

The sediment-retention structure created a new environmental problem. The structure impeded upstream movement of fish, including several stocks of salmon and steelhead trout (Bisson et al., Chapter 12, this volume). To mitigate these barrier effects, a fish-collection facility was constructed about 1 km downstream of the structure, in which migrating fish are captured and trucked above the structure and placed in tributaries to spawn. Young fish are able to negotiate the structure as they migrate downstream to the ocean, but they frequently experience mortality or injury during high flows (Bisson et al., Chapter 12, this volume). This fish-collection facility causes adult salmonids returning to spawn in the upper reaches of the North Fork of the Toutle River to be dependent on humans to transport them upstream. The sediment-retention structure has altered natural sediment transport and storage, hydrology, and river habitat and has hindered the pace and altered the pattern of geomorphic and ecological response. The effectiveness of the retention structure in terms of the environmental problems it created can best be gauged over several more decades by its long-term success in reducing downstream hazards from erosion as well as the ability of the fish-collection system to maintain the upstream movement of fish stocks.

19.3.4.5 Salvaging Downed Wood and Establishing Timber Operation

Salvaging standing dead and downed wood became a point of disagreement between those people promoting timber salvage for its wood products value and those promoting natural processes. Immediately after the eruption, the timber industry, some elected officials, and others were alarmed about the loss to the Pacific Northwest economy of more than 650 million board feet (or 1.5 million m^3) of wood (USDA Forest Service 1981a). The Forest Service, DNR, and private timber companies called for rapid salvaging of the downed wood to avoid loss due to decay, wildfires, or outbreaks of the Douglas-fir beetle (*Dendroctonous pseudotsugae*) and silver fir beetles (*Pseudohylesinus granulatus* and *P. grandis*) starting in dead trees and spreading to live trees. Yet, local ecologists did not agree with this dire prediction. In retrospect, it is apparent that no management action was necessary to avert insect outbreaks because the Mount St. Helens ash was a potent natural insecticide that caused rapid insect mortality by disrupting their ability to maintain water balance (Edwards and Schwartz 1980; Shanks and Chase 1981). Similarly, fire was not a likely scenario because the nonflammable tephra coated woody debris and limited the potential for fire to spread.

In the 1980s, management strategies were shifting from an emphasis on removal of wood from streams for the sake of maximizing extraction of wood products and fish passage to retention or even placement of wood in streams to maintain or create complex habitat. As was common practice at that time, the Forest Service, DNR, and private forest companies removed wood from streams in the first few years after the May 18 eruption (Franklin et al. 1988). However, the Forest Service retained wood in a few stream reaches so that stream ecologists could observe its effect on recovery of aquatic ecosystems, including fish (Franklin et al. 1988). These experiments showed that wood left in large alluvial valley streams increased habitat complexity, which supported more trout but also retained sediment in the channel (Baker 1989).

Objections notwithstanding, plans to salvage the standing and downed wood were promptly put in place by the private timber companies and the Forest Service. The Weyerhaeuser Company quickly harvested much of the downed wood on their lands (Snyder 1999). The Forest Service performed an environmental assessment (EA) as required by the National Environmental Policy Act (NEPA) to determine the appropriate way to handle damaged timber on 5600 ha in the Gifford Pinchot National Forest. The stated goal was to avoid loss of wood to insects, decay, or fire and to remove from streams timber that "threatens downstream improvement" (USDA Forest Service 1981a). The management scheme selected from the set of alternatives put forth in the EA would have salvaged more than 1 million m^3 of damaged timber. However, there was a strong protest about this decision from local stakeholders and some environmental scientists who wanted to preserve much of the disturbed land for research, recreation, and education. Subsequently, the EA decision was overridden by legislation designating 43,300 ha as a national monument. Neither broad-scale insect outbreaks nor fire occurred in the vicinity of Mount St. Helens. Even so, more than two decades later, the ecological consequences of salvage logging after natural disturbances are still debated (Lindenmayer et al. 2004).

The issue of salvage logging leads to the question of how managers should respond when confronted by different opinions about proposed management plans. These differences occurred as a result of the developing state of ecological knowledge and conflicting values. Some groups supported the economic gain to the communities that would occur with salvage logging, while others preferred preservation of natural processes, encouragement of tourism, and focus on the potential for scientific investigations. Furthermore, consequences of actions are sometimes not known for decades or even centuries. The apparent inaccuracies of science-based judgment regarding salvaging of downed trees in the blast area raised intriguing management issues concerning decision making in the face of uncertain or conflicting scientific evidence. Where possible, the decision makers chose a strategy of quick response for economic gain versus waiting and seeing the long-term effects. The issue is really about risk. What and how much risk are resource managers willing to accept, given various levels of uncertainty? In some situations at Mount St. Helens where scientific opinions differed, an outside review helped clarify the differences (Brown 1982). In the case of salvage logging, land managers proceeded (as they saw appropriate) with the NEPA process, and it was a flood of concern to Congress that led to expansion of the area set aside as the Mount St. Helens National Volcanic Monument and therefore protected from salvage logging.

19.3.4.6 Protecting Natural and Scientific Values

The large scientific and public outcry about lands being set aside from active management concerned not only salvage logging but also the extent of lands to be protected. In October

1981, the Forest Service designated an Interpretive Area of 33,350 ha, including 10,270 ha of state and private lands, with the primary objective of protecting "significant geological features in the impact area of the volcano . . . " (USDA Forest Service 1981b). Protests were led by citizens and scientists over the absence of several key ecological features and the amount of timber planned for salvage on federal land. Petitions signed by more than 112 scientists were sent to Congressional delegates, who called for hearings, which gave scientists and others the opportunity to voice their concerns publicly (U.S. Congress 1982). Subsequently, Congress passed the Mount St. Helens National Volcanic Monument Act in August 1982 (Public Law 97-243). The boundaries of the compromise plan were drawn with input from the political, geological, ecological, environmental activist, and local communities. Involvement of ecological scientists meant that the boundaries included sites that had been affected by all types of volcanic disturbances as well as unaffected areas that could serve as controls in scientific studies. Land exchanges among the diverse owners through trades and purchases were necessary to develop the contiguous block of land that became the National Volcanic Monument.

A major goal in establishing the Mount St. Helens Volcanic Monument was preserving the land for research, recreation, and education. The Monument Act designated that a comprehensive management plan be developed to guide science, education, and recreation in the Monument. In 2005, the selected plan is still being implemented by the Forest Service to balance (1) protection of sensitive sites and processes for scientific research; (2) development and operation of access and support facilities for public interpretation; and (3) needs for safety, public access, and tourism (USDA Forest Service 1985). The Plan retained certain aspects of the management of the Mount St. Helens landscape before 1980 and gave fresh direction for the future. Initially, some expressed concern about the ability of the Forest Service to manage the land for this mix of values. However, during the past 25 years the Forest Service has balanced protection of habitats and research with creating opportunities for education and recreation while attending to public safety.

An important component of the Act was establishment of a Scientific Advisory Board to advise and recommend to the Secretary of Agriculture ways to manage the science being conducted at the Monument and to "retain the natural and ecologic and geologic processes and integrity of the Monument" [Public Law 97-243, Sec. 7 (a) (1), (2), Aug. 26, 1982]. Board members were appointed by the secretaries of Agriculture and Interior, Governor of the State of Washington, and Director of the National Science Foundation. By establishing the Scientific Advisory Board, the Act positioned scientists in an unprecedented role as part of the management process. The Board advised the Forest Service on such issues as fish stocking in lakes, a cross-Monument highway, the sediment-retention structure, and the preferred alternative for the Comprehensive Management Plan for the Monument. The existence of the Board provided a formal mechanism for scientists to have input to the evolving management questions. This perspective was especially important for balancing the goal of sustaining naturalness with providing for human needs and wants, desired commodities, and services from the area. However, this Board was disbanded 10 years after its first meeting, as directed by the Monument Act. Termination of the Board left the Monument without a recognized scientific voice to provide recommendations on management and other issues requiring scientific input. Input from the science community on new and recurring issues has been handled on an ad hoc basis since 1992.

A notable feature of the Monument is its acceptance of dynamism, both in the wording of the Act to let natural processes proceed substantially unimpeded and in management of the public-access system. This recognition of change is unusual for natural areas and contrasts with the admonition from the Leopold Committee on Wildlife Management in the National Parks (Leopold et al. 1963) that such areas be managed to "represent a vignette of primitive America." The concept of freezing in time the condition of a place is unrealistic in the face of active volcanism, ongoing ecological and geomorphic change, and policies evolving to meet the changing needs of the people.

Predictably, tension developed between preserving natural processes and features, as stipulated in the Monument Act, while also protecting opportunities for economic development. Differences in these objectives often put geologists and ecologists, who wished to preserve "naturalness," in conflict with groups supporting commercial development, recreation, and other developments. Although most disagreements over land use and development of the volcanic area were settled by the establishment of the Monument in 1982, some issues linger after more than two decades, as reflected in persistent proposals for a cross-Monument highway. The management implications of effectively working with contentious issues are that the scientist must stay informed and involved in the management process to protect the interests of science, which also serves to preserve education and interpretation opportunities.

19.3.4.7 Restoring Fish, Wildlife, and Ecological Resources

Various practices were established to benefit fish and wildlife at Mount St. Helens, and each approach was influenced by management objectives of the landowners and managers (Franklin et al. 1988). Within the Monument, the Forest Service policy is not to enhance habitat. Beyond the borders of the Monument, several types of enhancement measures have been used. For example, snags (i.e., standing dead trees) were left in place to serve as nesting sites, perches, food resources, and roosts (see Figure 3.1), and artificial structures, such as nesting boxes, raptor roosts, and perches, were also created for use by birds.

The Washington Department of Fish and Game worked with Weyerhaeuser Company and the Rocky Mountain Elk Foundation to restore populations of elk by developing winter-range foraging areas. The resulting pastures of grasses and legumes were used extensively by elk during the winter months when high-elevation forage was covered by snow. However, subsequent studies showed that in areas without the seeding, abundant natural forage developed from resprouting plants that survived the eruption and from plants that had established from windblown seeds (Merrill et al. 1987). Similarly, the absence of forest for thermal cover was initially thought by scientists to restrict elk colonization, but this turned out not to be a constraint. Elk avoided thermal stress by spending the heat of the day in wallows or lying on the ground where heat was lost through conduction. Hunting was limited or prohibited, depending on location, so the populations were not strongly influenced by harvesting. A combination of rapidly developed, high-quality forage, a string of relatively mild winters, and the near absence of hunting resulted in a remarkable rebound of elk populations in the Mount St. Helens area within 5 years of the eruption.

Unique spatial patterns of elk herbivory were created by the pattern of landowners with differing management objectives. Weyerhaeuser's extensive (120,000 ha) tree farm provided much forage during the first decade following the eruption when trees were still small, but palatable forage declined markedly as the forest matured. The reduced carrying capacity of the maturing tree farm shifted elk foraging to the Monument, where studies have demonstrated that elk herbivory altered plant community composition and structure (see Dale et al., Chapter 5, this volume). During the autumn hunting season, hundreds of elk temporarily reside on the Pumice Plain and debris-avalanche deposit, where they seek refuge from hunters on adjacent lands. The management of large roaming mammals is complex when adjacent lands are managed according to different objectives, with potentially severe consequences for some of the land uses and areas.

Fish stocking of streams and, particularly, lakes raised conflicts among fishermen, scientists, and resource managers (Bisson et al., Chapter 12, this volume). Ecologists argued for retaining natural organisms and processes, and the Washington Department of Fish and Game and anglers wanted to restock lakes, which had been popular fishing sites before 1980. Generally, states regulate fish and wildlife, and landowners regulate access and manage habitat, so this dispute about fish stocking in the Monument was a states rights issue. The Scientific Advisory Board made a recommendation to the Monument that allowed for some lakes to be stocked and others not. The Board's decision, which was accepted by the Forest Service, environmental scientists, and Washington Department of Fish and Wildlife, was based on maintaining the lakes with the greatest scientific value in a fishless condition and not stocking lakes that currently did not support fish. Stocking was not a large issue because fish had survived in most lakes and were sustained by self-perpetuating populations. Since 1980, fish have been stocked by Washington Department of Fish and Wildlife in 6 of the 35 lakes in the Monument and additional lakes in lands within the blast area owned by Weyerhaeuser Company. However, several of the lakes that were fishless in 1980 and were intended to be managed in a fishless condition have since developed fish populations; apparently through a combination of natural colonization and illegal stocking.

The scientific community viewed Spirit Lake as the most important body of water to keep in a fishless state to study patterns of recovery, but it was apparently clandestinely stocked by anglers. By 2000, with many large fish present in Spirit Lake, the Washington Department of Fish and Wildlife and sportsman groups wanted access to the lake for angling purposes, but according to state policy, fishing is not permitted in the lake. This situation has resulted in contention among the agencies and the public.

The next broad-scale consideration of fishing regulations for lakes in the Mount St. Helens area will occur when the management plan for the Gifford Pinchot National Forest is revised in 2009, as legislatively required. Monument officials decided to postpone case-by-case requests for open fishing until that time, when many aspects of the Monument's Comprehensive Management Plan will be evaluated to see if change in current policy is warranted. In 2005, only a few lakes capable of supporting fish remained fishless.

19.3.4.8 Providing Access and Recreation for the Public

Public access to, through, and within the Mount St. Helens area has involved long-standing traditions and arguments. Before the 1980 eruption, people could drive to the north side of the mountain, where they hiked, swam, boated, camped, and fished. The extent of roadless area and the possible Congressional designation of some areas as wilderness were being debated. After the 1980 eruption, reestablishment of road access and new forms of access, such as proposals for a tram to transport visitors to a high-elevation vista or an east–west public road through the Monument, took on new significance. The recurring theme of access within the Monument has been the balance among protecting natural resources for preservation and research, providing for recreation, and maintaining public safety and access for those with limited mobility. The area outside the Monument has emphasized economic development.

Since the eruption, access into the disturbed area has varied. Backcountry recreation has been a major theme of the Monument, particularly hikes along the 320 km of trails that were installed since the eruption and climbs to the summit to view the surrounding landscape from the crater rim. However, most visitors to the Monument view the landscape from their cars via one of the three main roads that traverse the volcanic area. Mountain bike and snowmobile use is popular in the Monument, but fishing has not been a popular activity, despite substantial investments in infrastructure (e.g., a boat-launching area and a fish-cleaning facility at Coldwater Lake). A few years after the 1980 eruption, many trails were reopened or constructed, but new hazards exist in the volcanic landscape, and some of the trails are unstable. Heavy visitor traffic on trails, including horses and other pack animals that often disperse seeds of nonnative species, could interfere with natural processes of succession, particularly in sensitive areas. For these reasons, planners devised trail routes and travel restrictions to balance visitor access with protection of research opportunities. Access to volcanically disturbed lands is often limited by gated roads or area closures.

A recurring proposal for a cross-Monument highway passing in front of the volcano's crater is an example of conflicts over access (Adams 2002). The visitor centers have only west-side access from Interstate 5 via State Route 504, which was reconstructed after the eruption. Opportunities for viewing the mountain from the east are distinctive and spectacular, yet difficult and time consuming to access. Construction of a cross-Monument highway has been a strong desire of Lewis, Cowlitz, and Skamania counties and their municipalities, which all seek financial benefits from a direct route through the Monument. However, to date, these proposals have been countered by concerns about safety and protecting Spirit Lake and the Pumice Plain ecological systems. A 2004 study of the economic impact of the proposed Monument highway was considered by the Washington State legislature but not approved by the Governor after loud public outcry. Clearly, trade-offs among access, recreation, scientific research, safety, and economic development will continue to be discussed.

19.3.4.9 Providing Interpretive Programs and Other Educational Opportunities

Interpretative programs play a vital role in conveying information to the public about geological events, restoration, and ecological change at Mount St. Helens. Interpretation is presented to more than 1 million visitors each year to Mount St. Helens through visitor center displays and exhibits, road and trail kiosks and signs, and onsite interpretive talks. The Forest Service, Weyerhaeuser Company, Cowlitz County, and Washington State Parks and Recreation Commission each operate visitor centers; all five centers are on the west side of the volcano. Each center has a specific focal theme; these themes include an overview of the mountain, salvage logging and reforestation, ecological recovery, and geological events. These visitor centers are the primary Mount St. Helens destination for most visitors (based on road tallying devices and Forest Service surveys). As access improved, new visitor centers were constructed progressively closer to the volcano. Visitor center displays, films, and onsite interpretive programs draw strongly from science findings, allowing the latest discoveries to be incorporated into interpretive material long before this information is published in conventional scientific outlets.

A mutually beneficial relationship quickly developed between the scientific and educational communities after the 1980 eruption. In the first years, scientists often shared their expertise while being provided access to sampling sites on the volcano by news helicopters. Also, some scientific research received financial support from organizations that allowed the public to pay to assist field research (e.g., Earthwatch and the School for Field Studies). Scientists have transferred information learned at Mount St. Helens to a great variety of audiences through annual training workshops for interpretive program staff; onsite review of interpretive talks; preparation of Monument trail signage; popular articles for the Monument's newspaper; development of visitor center exhibits; media interviews; presentations to organizations, groups, and schools; and scientific symposia presentations, field trips, and publications. Managers, elected officials, and scientists recognize the important role that scientific research has at Mount St. Helens in terms of producing exciting, new information about the dynamic volcanic landscape and biota that is so important in developing interpretive themes for visitors to the Monument and educational curricula for the classroom.

The Mount St. Helens landscape and its story have earned an important place in public education throughout the world. Major anniversaries of the May 18 eruption have been marked with copious press attention. Many ecology textbooks present examples from Mount St. Helens. Scientists have been important commentators in media presentations concerning geological and ecological developments at the volcano. Informative web sites have been developed, as well.

19.4 Conclusions

Management concerns at Mount St. Helens emerged in a sequence of overlapping stages beginning with broad-scale assessment of volcanic hazards on a geological time scale (Crandell and Mullineaux 1978). Next, protection of human life and property in the posteruption landscape was addressed. Some weeks and months later, concerns about resource protection and extraction, such as salvage logging, grew in importance. The importance of different concerns was determined, in part, by the rates at which the volcano quieted, hydrologic hazards abated, control measures were put in place, and the long-term land-use policy was determined. As the physical environment began to stabilize, focus shifted to ecological restoration of forests, fish and wildlife, recreation, interpretation, and revitalization of local economies. Thus, the contributions of environmental scientists were more important in the later stages.

The experiences at Mount St. Helens in the quarter century since the eruption reveal that the role of environmental scientists has been largely advisory. They provided information to the decision makers, but decisions were not always made solely on environmental science perspectives. In several cases, hazards to life or property overrode environmental concerns; and in some situations, environmental science was not relevant to management issues. When it was germane, environmental science provided information and perspectives on management at a variety of spatial and temporal scales. These perspectives influenced management decisions made at Mount St. Helens.

Since the major eruption of Mount St. Helens, environmental scientists have contributed to the decision-making process by communicating information, building consensus, maintaining credibility, and discovering options for new management and research directions (Dale 2002). Communication of information occurred via field tours, coverage by the press, scientific papers, and many other venues. Some nonfederal scientists became actively involved in the preservation of natural processes by lobbying for the creation of the Monument, writing letters to members of Congress, developing and endorsing petitions, and participating in Congressional hearings. Scientists helped build consensus within the scientific community about the management plan for the Monument by sharing information, teaching, developing analyses, and taking part in scientific advisory groups. Consensus among scientists about the area to be protected as a national monument and how its management would be implemented arose as scientists shared information with each other and with policy makers. Credibility of the science was maintained by publication of many peer-reviewed articles from Mount St. Helens studies. In a review of papers on vegetation response to volcanic events, more than half of the papers were based on the Mount St. Helens work. Finally, scientists have been effective in presenting options for resource management and protection. For example, scientists were involved in discussions of proposed areas for inclusion in the Monument, and they participated intensively in planning a management structure for the Monument to meet policy objectives.

Finally, scientists took advantage of the natural experimental conditions created by the eruption to explore the ecological implications of varying management strategies. By making observations along gradients of disturbance severity, the diversity of ecological responses to disturbance was documented (e.g., Dale et al. 1998). Developing understanding of the multifaceted nature of ecological responses helps avoid overly simple management solutions to complex ecological and natural resource problems. It also helps match broad-scale goals with solutions. Alternative management practices were evaluated when scientists and managers used experiments to assess and monitor effects of different treatments. Thus, the advice of scientists in questions dealing with management at Mount St. Helens led to changes in direction, improvements in management actions, and better understanding of the implications of ecological system responses to large disturbances.

The Mount St. Helens experience has provided some general lessons about the role of environmental scientists in coping with the aftermath of large-scale disturbances:

- Risks to human life or welfare sometimes outweigh potential harm to ecological systems.
- Input from environmental scientists is usually advisory.
- Environmental decision making is enhanced by advice from ecologists who have a persistent involvement with the concern.
- A formal, institutional context of providing advice makes the information more accessible and timely.
- The early involvement of ecologists in societal responses to large-scale disturbances can help avoid some problems.
- Ecologists tend to provide a long-term and broad-scale perspective on how actions can affect environmental systems, a necessary but not always integral component of decision making.

Acknowledgments. We thank three anonymous reviewers for helpful comments on drafts of the manuscript. Discussions with Wendy Adams were quite helpful. This chapter represents the interpretations and views of the authors and not the agencies for which they work. Oak Ridge National Laboratory is managed by UT-Battelle, LLC, for the U.S. Department of Energy under contract DE-AC05-00OR22725. We thank the Forest Service, Pacific Northwest Research Station, and Mount St. Helens National Volcanic Monument for continuing financial and logistical support.

20 Overview of Ecological Responses to the Eruption of Mount St. Helens: 1980–2005

Charles M. Crisafulli, Frederick J. Swanson, and Virginia H. Dale

20.1 Introduction

The sensational 1980 eruption of Mount St. Helens and the subsequent ecological responses are the most thoroughly studied volcanic eruption in the world. The posteruption landscape was remarkable, and nearly a quarter century of study has provided a wealth of information and insight on a broad spectrum of ecological and physical responses to disturbance. The eruption and its effects on ecological and geophysical systems have many dimensions: a complex eruption affected an intricate landscape containing forests, meadows, lakes, and streams populated by diverse fauna and flora. This complexity created a rich environment and an exemplary living laboratory for study. Because the volcano is in close proximity to major metropolitan areas, scientists were able to perform reconnaissance trips and establish a network of permanent plots within days to months of the eruption. These early observations enabled scientists to assess the initial impacts of the eruption, which was important in understanding the subsequent quarter century of invasion and succession.

Suddenly, and almost beyond comprehension, at 8:32 a.m. on May 18, 1980, and lasting for little more than 12 hours, the eruption of Mount St. Helens transformed more than 600 km^2 of lush, green forest and meadows and clear, cold lakes and streams to a stark gray, ash- and pumice-covered landscape (see Figure 1.1; Swanson et al., Chapter 2, this volume; Swanson and Major, Chapter 3, this volume). The area influenced by the eruptive events will respond to them for hundreds or even thousands of years. However, even within the 24 years since the eruption, substantial change took place as hill slopes gradually turned from gray to green, opaque lakes cleared, and streams flushed sediment from their channels. Some of the initial ecological responses are well advanced; others have been set back by secondary disturbances; and yet others, such as soil development, will respond to the eruption over millennia.

The major 1980 eruption created distinctive disturbance zones that differed in the types and magnitudes of impacts on terrestrial and aquatic systems, including the types and amounts of surviving organisms and other legacies of the preeruption ecological systems (Figure 20.1). Thereafter, the natural system consisting of surviving and colonizing plants, fungi, animals, and microbes began responding to the new conditions. During the subsequent decades, species diversity, plant cover, and vegetation structure (the size and shape of plants) developed rapidly. Vegetation in 2005 ranged from herbs and scattered shrub cover in the severely disturbed pyroclastic-flow zone to the continuous canopy of young forest in tree plantations around the perimeter of the blast area. The story of this collective ecological response to the 1980 eruption of Mount St. Helens involves both successional change over time at individual sites and development of landscape patterns.

The 1980 eruption provided a special opportunity for scientists from a variety of disciplines to study ecological survival and establishment after a large disturbance, but several caveats challenge efforts to integrate this information. Developing a synthesis of ecological responses to the eruption is complicated by three factors:

- No ongoing, coordinated program integrated research efforts across taxa, research themes, disturbance zones, or time. Consequently, individuals and small groups of investigators operated relatively independently and focused narrowly on specific components of the overall story.
- Many studies initiated shortly after the eruption were terminated after a few years; others began many years after the eruption; and only a handful were maintained through the first 25 years after the eruption. Of the long-term studies, several have maintained regular measurements, producing outstanding long-term data sets.
- Most ecological studies at Mount St. Helens have sampled relatively small spatial scales, using different sampling designs, which further limit comparative analyses.

Even so, the collective work at Mount St. Helens provides a wealth of information and new insights on the responses of ecological composition, structure, and function to a large, catastrophic disturbance.

Relevant fields of science that might provide theoretical guidance in unraveling and communicating the complex

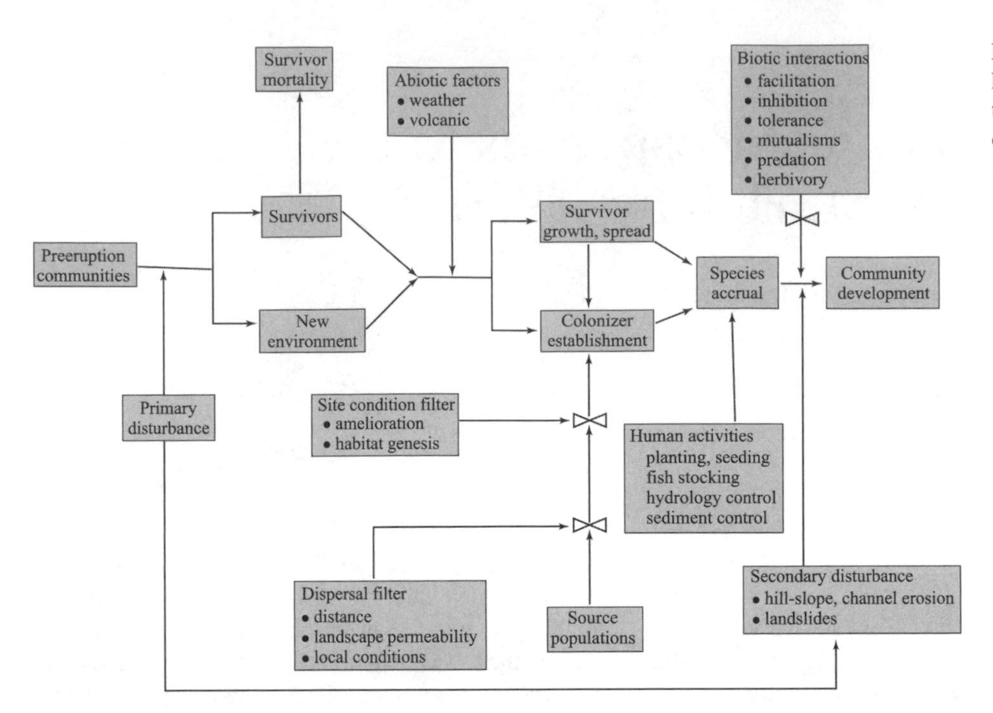

FIGURE 20.1. Generalized depiction of key biological, human, and physical factors and their flow paths influencing succession at Mount St. Helens after the 1980 eruption.

ecological responses to the 1980 eruption have themselves been seeking generality with only limited success [McIntosh (1999) reviewed succession, and White and Jentsch (2001) considered disturbance ecology]. Consequently, there was no unified theory from which to evaluate the ecological responses to the 1980 eruption. Studies at Mount St. Helens can provide information and insight for interpreting effects and responses to disturbances that may lead to the refinement of extant theory or the development of new constructs. Some of the lessons from Mount St. Helens have been consistent with findings from other disturbance studies, while others appear to be unique to the 1980 eruption of the volcano.

This closing chapter gives an overview and reflects on the physical and biological factors involved in survival and succession. First, we discuss the patterns and processes that were of prime importance for several taxonomic groups across the various disturbance zones. Perspectives on the development of landscape patterns are also discussed. Next, we describe the landscape of 2005 and the factors that have shaped its development. Then, we briefly place the Mount St. Helens experience of 1980 to 2005 in the context of the longer-term disturbance history of the area. The chapter concludes with key summary points from the Mount St. Helens experience.

20.2 Factors Influencing Survival

Surprisingly, numerous species survived the 1980 eruption in many locations throughout most of the disturbance zones (Table 20.1). Organisms survived by means of a wide variety of factors that were related either to characteristics of the disturbance and the species or to chance. The type, number, and extent of organisms that survived the eruption varied across the disturbed landscape, creating a complex mosaic of survival. We illustrate mechanisms of survival with the following examples.

20.2.1 Effects of Physical Disturbance and Topography on Survival

The specific disturbance types (e.g., directed blast), mechanisms (e.g., heating), and intensities (e.g., degree of heating) during the 1980 eruption strongly influenced survival of organisms (Swanson and Major, Chapter 3, this volume). The suite of physical processes operating during the eruption created a full spectrum of survival from nearly complete retention of biota, biotic structures, and abiotic features (such as soil, cliffs, and streams) to complete extirpation of life and loss (or gross reconfiguration) of landforms and drainage networks (see Table 20.1). Disturbance mechanisms of heating, burial, impact force, and abrasion damaged plants, fungi, and animals to varying degrees among the different volcanic processes and along the flow path of each process. The number of surviving species was inversely related to disturbance intensity (Table 20.1). In the case of the most extreme disturbance at Mount St. Helen, no multicellular organisms survived in the pyroclastic-flow zone (Means et al. 1982), where a new landscape was created by thick (greater than 10 m), 800°C pumice deposits (Table 20.1) in an area that was also profoundly altered by the debris avalanche and blast (Swanson and Major, Chapter 3, this volume). The much cooler debris avalanche transported a few living plant fragments that could establish (Adams et al. 1987), but the great thickness and high impact force of the debris-avalanche deposit precluded any in situ

TABLE 20.1. Survival of plant and animal taxa in several volcanic-disturbance zones following the May 18, 1980, eruption of Mount St. Helens.

Disturbance zone	Mean % vegetation cover	Average number of herbaceous species per square meter	Species of small mammals	Species of large mammals	Species of birds	Species of lake fish	Species of amphibians	Species of reptiles	References[a]
Pyroclastic-flow zone	0.0	0.0	0	0	0	0	0	0	1,2,3,6,7,9
Debris-avalanche deposit	0.0	0.0	0	0	0	0	0	0	1,2
Mudflow central flow path	0.0	0.0	0	0	0	na	0	0	5
Blowdown zone			8	0	0	4	11	1	3,7,8,9
Preeruption clear-cut	3.8	0.0050							1
Forest without snow	0.06	0.0021							1
Forest with snow	3.3	0.0064							1
Scorch zone	0.4	0.0038		0	0	2	12	1	1,7,8,9
Tephra-fall zone outside of blast			11	4	21	3	12	3	3,7,8,9
Tephra depth:									
23 mm	27.0	3.1							4
45 mm	28.2	2.0							4
75 mm	34.3	2.4							4
150 mm	10.5	0.2							4

Data are based on surveys within the first few years after the eruption.
[a]See the references listed below for details.
1. Means et al. (1982).
2. Adams and Adams (1982).
3. Andersen and MacMahon (1985b).
4. Antos and Zobel (1985b).
5. Frenzen et al., Chapter 6, this volume.
6. Crisafulli et al., Chapter 13, this volume.
7. Crisafulli et al., Chapter 14, this volume.
8. Bisson et al., Chapter 12, this volume.
9. Crisafulli and MacMahon, unpublished data.

survival of plants or animals (Table 20.1). At the other extreme of disturbance intensity, survival was exceedingly high for many taxonomic groups in the tephra-fall zone because deposits were emplaced with low impact force and temperature and thickness was generally quite limited (e.g., less than 20 cm) (Table 20.1). In zones of intermediate disturbance intensity, where deposit thickness was typically less than 1 m, many organisms survived in a great variety of refuges, such as under snow and within the soil and decaying logs. In these disturbance zones where biological legacies were very important, local topography created fine-scale heterogeneity in deposit thickness, which affected survival of plants and animals. Some topographic settings (e.g., rock outcroppings, cliffs, and ridges) protected sites from the brunt of the blast, while elsewhere landforms caused deposition of thick deposits (e.g., valley floors and benches), which smothered the life beneath.

20.2.2 Effects of Biological Attributes on Survival

Life history, habitat associations, life form, and organism size, among other biological attributes, influenced the survival of species during the 1980 eruption. Of primary importance were species characteristics that provided ways for organisms to avoid the brunt of disturbance events by being either temporarily away from the area or in protected habitats. Life-history traits, for example, aided the survival of anadromous fish that had a portion of their populations at sea during the period of greatest disturbance impacts (Bisson et al., Chapter 12, this volume) and the survival of many migratory birds that had yet to return to their summer breeding grounds when the spring eruption occurred. Zooplankton have eggs and resting stages that settle to the bottom of lakes and can remain viable for decades; those eggs provided a source of stored propagules (Vogel et al. 2000).

Cryptic habits of organisms, habitat preferences, and life forms contributed to the survival of some organisms but doomed others. Animals that live beneath the ground, such as the northern pocket gopher (*Thomomys talpoides*) and numerous invertebrate species, survived in soil or in large, decaying logs. Likewise, entire aquatic communities survived in many lakes, where they were buffered from the eruption by cold water and a protective layer of ice and snow. In contrast, species living in exposed habitats, such as resident birds and arboreal rodents, perished in the blast area. Amphibians associated with water had higher survival rates than did entirely terrestrial species (Crisafulli et al., Chapter 13, this volume). The most common plant survivors in the blast area were plants with buds located belowground, where soil provided protection (Adams et al. 1987). Mosses and low-stature herbs sustained heavy mortality in the tephra-fall zone, whereas erect shrubs and tree saplings experienced widespread survival (Antos and Zobel, Chapter 4, this volume).

Organism size was critical to survival, and small organisms experienced higher survival rates than did large organisms.

Small organisms were more likely to have been in protected locations, such as belowground or in decaying logs, whereas large organisms were typically exposed to volcanic forces. However, the effect of size had to be considered in relation to the scale of the disturbance process. In the area affected by the towering, hot cloud of the lateral blast, large-stature individuals experienced greater mortality than did smaller ones. Tall, mature conifers, for example, were toppled, while small saplings and shrubs survived under snow cover. In contrast, on the margins of mudflows, commonly only a few meters thick, tall trees survived because they were able to resist the force of the flow and hold their unaffected crowns above the flow (Frenzen et al., Chapter 6, this volume). Similarly, large mammals, such as elk (*Cervus elaphus*) and black bear (*Ursus americanus*), experienced complete mortality in the blast area, but many small mammalian species survived in subsurface refuges (Crisafulli et al., Chapter 14, this volume).

20.2.3 Effects of Timing of the Eruption on Survival

Effects of the 1980 eruption were strongly influenced by timing of the event at scales ranging from the time of day to the successional status of the ecological systems affected by the disturbance. The 8:32 a.m. beginning of the eruption, for example, occurred when many nocturnal mammals, such as mice and voles, had returned to their subterranean daytime retreats, where they were protected from the blast and tephra fall. The season of the eruption was significant in terms of the status of snow and ice in terrestrial and aquatic systems, the migratory status of seasonally transient visitors to the area (e.g., migratory birds and salmon), and the phenology of plants. Bud break, for example, had not yet occurred in many upper-elevation areas, so plants avoided damage to foliage that might have occurred if the eruption had been just a few weeks later. The status of succession across the Mount St. Helens landscape before the eruption influenced the ability of species present to survive. In the first few years after 1980, for example, preeruption clear-cuts supported more surviving species because early-successional communities were dominated by herb species with the ability to sprout vigorously from belowground, perennating structures, unlike most understory species of mature forests (Means et al. 1982). Had the eruption occurred at another time, the ecological consequences would have been markedly different. These sensitivities of ecological response to the timing of disturbance highlight the importance of chance in determining effects of disturbance, both in the near term and long term.

20.3 Factors Influencing Succession

Ecological succession at Mount St. Helens was influenced by

- Survivors of the eruption;
- Physical conditions, including secondary disturbances;
- New colonists that dispersed into the disturbed area, established, grew, spread, and interacted with other species and their environment; and
- Human activities.

These elements of succession provide a conceptual framework for discussing succession at Mount St. Helens (see Figure 20.1). Briefly, following the eruption, a subset of the organisms and structures survived, some of which later perished because of their inability to tolerate the new conditions, and the remaining survivors grew and spread, and were joined by colonizing species. The colonizing species were determined by source populations, species vagility, and a series of dispersal and site-condition filters. Species accrued to the point where biological communities developed and a myriad of biotic interactions commenced. Climate, human activities, and secondary disturbances strongly influenced the rate and pattern of succession.

The many processes of succession are highly complex and inextricably intertwined, which makes a discussion of each of the elements of succession in isolation somewhat artificial. Nonetheless, to discuss key aspects of succession, each is presented separately next. During the first quarter century after the 1980 eruption, all mechanisms of succession, for example, facilitation, inhibition, and nearly all forms of species–species interactions, have occurred in the Mount St. Helens landscape. Frequently, these successional processes and species interactions operate at the same time and place, and alternative sequences are common.

20.3.1 Influence of Survivors on Succession

Surviving fungi, plants, and animals played numerous important roles in many of the disturbance zones following the 1980 eruption. Survivors were source populations for adjacent areas, ameliorated site conditions, established important ecological linkages among biota, and served as habitat and food resources, permitting or promoting the persistence of other survivors and the colonization of new species. Survivors produced seeds, spores, and offspring, which served as source populations for adjacent areas where those species did not survive. The in situ reproduction of nearly all the flora and fauna of the southern Washington Cascade Range from thousands of epicenters of survival was paramount. This process greatly reduced the potentially lengthy process of dispersal from distant sources and strongly influenced the rate and spatial pattern of succession.

Survivors developed important ecological linkages among organisms and facilitated the establishment and spread of other species. Within days of the eruption, for example, surviving gophers mixed underlying mineral soil with tephra. These gopher-modified soils contained fungal spores and buried seeds and were suitable sites for seedling establishment and plant growth (Andersen and MacMahon 1985b; Allen et al., Chapter 15, this volume). Surviving plants and dead wood provided food resources and habitat for colonizing plants and animals.

Large conifer snags provided the nesting and foraging sites necessary for several birds to colonize the blast area (Crisafulli and Hawkins 1998). Small mammals transported seeds from surviving patches of vegetation to barren areas where they later germinated. Shading from surviving plants ameliorated hot surface conditions and created safe sites for wind-dispersed seeds. In this way, surviving plants and other structures were nuclei of establishment that promoted the colonization and spread of plants, which in turn attracted animals. Through time, these areas of surviving plants developed into large patches of vegetation supporting complex communities.

Surviving animals also became important consumers in posteruption communities (i.e., predators herbivores, scavengers, and decomposers), in some cases impeding colonization by other species. Survivors did not always persist in the new landscape because of desiccation, heat, lack of nutrients, and other causes of mortality (see Figure 20.1).

20.3.2 Dispersal

With the elimination or reduction in species and the creation of open terrain, dispersal became a very important initial process of succession within the most severely disturbed areas at Mount St. Helens. The pattern and rate of dispersal was influenced by the distance to source populations, local wind patterns, landscape permeability, and species and propagule mobility. Distance and landscape permeability were filters restricting the number of species capable of reaching a location within the disturbed landscape (see Figure 20.1). The distance to source populations was determined by the size and shape of the disturbance zones for the most severely disturbed, legacy-free areas, such as the pyroclastic-flow and debris-avalanche zones. Wind flows along the river valley may best explain the high number of wind-dispersed seeds in the center of the debris-avalanche deposit (Dale 1989). In zones with high survival, such as parts of the blast area, local sources of colonists of most species were often nearby in the numerous scattered epicenters of survival. These in situ sources of propagules greatly accelerated the pace of succession by reducing the importance of species vagility, dispersal distance, and landscape permeability. In areas lacking survivors or in areas where a species was extirpated during the eruption, highly mobile species such as birds, ballooning spiders, windblown seeds, and large mammals were the first to arrive. Within 10 years after the eruption, most of the plant species with long-distance dispersal capabilities and that were present elsewhere in the southern Washington Cascade Range had reached most disturbed areas. During the second decade, many animal populations became widely distributed, and animals disseminated plant species that have limited dispersal capacity, such as species with berries and fleshy fruits.

For less-mobile organisms, distance appeared to be a barrier to colonization. For example, tailed frogs (*Ascaphus truei*) survived in many watersheds within the blowdown zone, but by 2003 they had yet to colonize streams in the pyroclastic-flow or debris-avalanche zones despite the presence of suitable habitat (Crisafulli et al., Chapter 13, this volume). Many individual organisms and propagules of diverse taxonomic groups exhibited surprising ability to traverse substantial distances (e.g., several kilometers), sometimes over harsh terrain, to colonize more-favorable habitats (Dahm et al., Chapter 18, this volume; Crisafulli et al., Chapter 13, this volume; Crisafulli et al., Chapter 14, this volume). Many of these movements appeared to be random dispersal events, suggesting that chance was an important aspect of dispersal (del Moral et al., Chapter 7, this volume). Overall, a remarkably diverse assemblage of fungal, plant, and animal species or their propagules had successfully dispersed through each disturbance zone within the first quarter century after the eruption.

20.3.3 Secondary Disturbance and Succession of Geophysical Processes

Secondary disturbance by erosion, scour, and deposition processes, such as lateral channel migration and small landslides, is a direct result of the primary disturbance processes and influenced succession in many areas. The type and magnitude of these processes varied in time across the volcanic landscape, creating a geophysical succession of prominent processes. Because the new deposits were easily eroded and the terrain was steep, remobilization of tephra was common during the first few years after the eruption. Small gullies that developed in fall and winter cut down to preeruption soil, facilitating plant survival and growth by releasing perennial rootstocks, exposing soil, and trapping windblown seeds. However, other physical processes, such as shifting river channels and landslides, retarded succession by scouring or burying plants and animals, thereby resetting succession in substantial parts of the landscape. Landslides and sediment that traveled down lake-inlet streams created shoal habitat, alluvial fans, and deltas in many lakes, which promoted establishment of emergent wetland plant communities. Many streams on the Pumice Plain and debris-avalanche deposit were subjected to chronic secondary disturbance from 1980 and continued through the following decades, which set back development of riparian vegetation and associated animal communities. With the exceptions of continued river-channel instability in the debris-avalanche deposits and precipitation-triggered landslides during 1996 and 1997, secondary geophysical disturbances lessened dramatically after the early 1980s.

20.3.4 Site Amelioration

Site amelioration occurs when the conditions on a site improve, typically with respect to soil, microclimate, and other physical conditions, and enable a species to establish, grow, and reproduce. The harsh physical conditions of recently disturbed sites greatly influence the process and pace of succession. During the dry, hot summers, the fresh geological material deposited over much of the area disturbed by Mount St. Helens was more like a desert than the lush Pacific Northwest forest. Site amelioration was of great importance during the first few years after the eruption when the new deposits blanketed the ground, and survivors, where present, were in their initial stages of response. The fresh substrates had low nutrient status, little moisture-retention capability, and limited shade in the blast area and pyroclastic-flow and debris-avalanche zones, which hindered establishment and growth. Thus, materials deposited during the 1980 eruption were far from optimal for plant growth and required weathering and nutrient input from windblown material, precipitation, and (where present) plants and animals for biotic development to proceed. The volcanic events also produced high concentrations of sediment, especially fine material, in the benthic layer and water column of lakes and streams. These conditions impeded establishment and growth of immigrating fungi, plants, and animals. Amelioration was most important where biological legacies were few or absent.

A variety of physical processes, such as erosion that created favorable microsites, ameliorated site conditions and promoted establishment and growth. Physical weathering by freeze-thaw and wet-dry cycles fragmented surface tephra particles. The resulting production of fine-textured, inorganic material, combined with accumulation of organic matter, hastened the development of soil and increased water-holding capability and nutrient content. Continued flow through stream systems over the years has removed a great deal of fine sediment from the channels and exposed cobble streambeds characteristic of Cascade Range streams. Suspended sediment settled to the bottom of lakes, resulting in increased light transparency, which was necessary for photosynthesis.

Animals also played a role in site amelioration. Northern pocket gophers created vast tunnel systems, which were used by amphibians and other animals to escape from the hot, dry surface conditions during the summer (Crisafulli et al., Chapter 14, this volume). Wind-dispersed insects landed in surprisingly large numbers and mass on nutrient-poor pyroclastic-flow deposits, where they died and their bodies contributed nitrogen and other nutrients and trace elements important for plant growth (Edwards and Sugg, Chapter 9, this volume).

As time elapsed, substrates were modified, species colonized, and conditions generally improved so that site amelioration presumably became less critical in many locations. Further amelioration will undoubtedly be required for species requiring the development of duff or other characteristics to establish.

20.3.5 Establishment

Establishment of individuals and populations involves several steps; organisms must either survive in or disperse to the disturbed area and then find suitable conditions under which they can colonize, grow to maturity, and reproduce. Critical factors limiting establishment included the ability of seeds and spores to germinate and grow in the fresh volcanic substrates and, in the case of animals, the presence of appropriate habitat and food resources. Substrate quality influenced which species were able to establish because seed germination and plant growth were initially less than optimum on the new volcanic material. Microtopography created slope, moisture, and shading conditions that were important to plant establishment and growth. Vertebrates with high vagility established in lockstep with the development of suitable habitat, with little apparent lag time between their establishment and the presence of suitable habitat. Thus, habitat structure was a reliable predictor of vertebrate establishment.

Specific characteristics of species either promoted or limited those species establishment. For example, neoteny, the attainment of sexual maturity while maintaining larval characteristics, enabled three species of salamanders to establish populations in some disturbance zones. For these species, suitable habitat was available for the aquatic neotenic forms, but the forest habitat required by terrestrial life-history stages was absent. Thus, the portion of the population that metamorphosed into terrestrial adults often perished.

Failure of some organisms to establish may reflect inadequate habitat or dispersal limitations or both. Many of the species that have not colonized may have been limited by their habitat requirements rather than by dispersal limitation, such as species associated with forest canopy. The diversity of habitats that had developed on the landscape by 2005 suggests that even many late-seral-forest understory species and woodland salamanders could exist in certain microsites (i.e., seeps, upper reaches of streams, and ravines) in the highly disturbed zones and that it is dispersal distance that has precluded their establishment. For example, many of the plant species that reached the Pumice Plain lacked the requirements for establishment, and those that could readily establish did not arrive because of low dispersal capability (del Moral et al., Chapter 7, this volume).

A tremendous number of organisms and propagules dispersed into disturbed areas and established; however, several hundred known and perhaps many more unknown species arrived but failed to establish (Edwards and Sugg, Chapter 9, this volume). As succession proceeds and appropriate habitat develops, it is anticipated that many of these species will become established. The simultaneous establishment of both early- and late-successional species has important long-term consequences for succession. In particular, several late-successional conifer tree species established on the debris-avalanche and pyroclastic-flow deposits in the early 1980s.

Many of the establishment events that occurred at Mount St. Helens have gone against conventional wisdom and suggest that we lack a clear understanding of the conditions under which many species may establish. Chance also plays a large role in the establishment of many species, particularly as it relates to the dispersal of propagules and organisms by animals and to the settling of wind-dispersed seeds onto favorable safe sites.

20.3.6 Mechanisms of Succession and Biotic Interactions

Interactions among species are very important components of succession because they influence the rate and pattern of colonization, the accumulation of biomass and plant cover, population dynamics, and trophic interactions. Biotic interactions in the Mount St. Helens landscape have been incredibly diverse, and all of these phenomena are integral to succession. Connell and Slatyer (1977) proposed three broad models of mechanisms of succession: facilitation, tolerance, and inhibition. Each of these has been observed at Mount St. Helens, often at the same place and time, demonstrating that multiple mechanisms of succession are likely to occur simultaneously, and any one model should not be expected to have universal explanatory power. Facilitation was the most conspicuous mechanism of succession at Mount St. Helens and involved virtually all taxonomic groups. Facilitation is exemplified by patches of lupines on the Pumice Plain that trapped seeds and detritus and helped contribute nitrogen to the soil, thereby facilitating establishment of additional plant and animal species (del Moral et al., Chapter 7, this volume; Halvorson et al., Chapter 17, this volume). Although Connell and Slatyer's (1977) facilitation model was specific to plant–plant interaction, we note that microbes and animals also facilitated the establishment of other species. For example, microbes in Spirit Lake swiftly altered nutrient and oxygen concentrations and ushered in suites of new bacteria (Dahm et al., Chapter 18, this volume). Beavers (*Castor canadensis*) created impoundments on all major streams in the blowdown zone, thereby establishing habitat that supported colonization by several pond-breeding amphibians (Crisafulli et al., Chapter 13 this volume).

Inhibition, where one species has a negative effect on another, was conspicuous in the Mount St. Helens landscape in the cases of herbivory on lupines by larvae of several lepidopteran species and where woody vegetation shaded the ground, resulting in the decline of sun-requiring herbs. Tolerance was evident on the Pumice Plain and debris-avalanche deposit, where late-successional tree species colonized alongside early-successional herbs. Numerous boom-and-bust cycles occurred as colonizing species encountered favorable conditions upon arrival at Mount St. Helens and their populations flourished, essentially unregulated. However, these burgeoning populations were eventually found by herbivores, predators, or parasites, which in turn grew rapidly and decimated their resource base, only to have their own populations eventually collapse.

Interactions such as predation, herbivory, parasitism, and mutualism that permeate complex, species-rich ecosystems were particularly notable in the sparse ecological systems characteristic of the early stages of response to the disturbances at Mount St. Helens. As communities grew in complexity, virtually all forms of species interactions could be found in the area.

20.3.7 Species Accrual and Community Structure

The 1980 eruption greatly reduced or eliminated vegetative cover and structure and the species richness and abundances of virtually all taxonomic groups. However, over the subsequent years, a remarkable reassembly process unfolded as survivors grew and spread and were joined by migrants that colonized and filled in much of the landscape and interacted with one another and their environment. Over time, all disturbance zones experienced increases in the number of species, percent vegetation cover, complexity of vegetation structure, abundance of animals, and community complexity. For example, plant-species richness was very low (commonly no species was present) in the most severely disturbed terrestrial sites following the eruption. In 13 years, however, at least 150 species of vascular plants had colonized these areas (Titus et al. 1998b) [as compared to the almost 300 species that St. John (1976) reported from the slopes of the mountain before the eruption]. Terrestrial animals also underwent rapid colonization, and by 2000, even the most heavily disturbed zone supported nearly all small mammals indigenous to the southern Washington Cascade Mountain Range (Crisafulli et al., Chapter 14, this volume). Colonization by aquatic taxa was also rapid; for example, 135 species of phytoplankton established in Spirit Lake between 1983 and 1986 (Dahm et al., Chapter 18, this volume). During the first quarter century since the eruption, a steady accrual of species has occurred, with few examples of species replacements (Parmenter et al., Chapter 10, this volume; Dahm et al., Chapter 18, this volume). During the accrual process, species from all taxonomic groups of the regional pool, spanning all trophic levels had colonized the Mount St. Helens landscape. Many of these species flourished in the posteruption conditions.

Vegetation cover (percent of land area covered by live plant material) increased dramatically in all disturbance zones from 1980 to 2005. The rate of cover increase was greatest in areas with high survival, which corresponded with lowest disturbance intensity. In areas undergoing primary succession (the debris-avalanche and pyroclastic-flow zones), cover went from zero in 1980 to as high as 70% by 2000, but typically ranged from 10% to 40%. Cover increases were commonly related to the growth form of the species present. Many plants spread by vegetative growth from stolons or rhizomes. In the case of prostrate plants, the rooting of stems made them dominant players in many of the disturbance zones. Cover increased dramatically in the ponds created by the eruption, and by 2000 the percent cover of emergent and submergent species frequently exceeded 80%.

Herbivory and secondary disturbances have strongly shaped percent cover of many plant communities. Herbivory by several lepidopteran larvae reduced the abundance and cover of their target species on the Pumice Plain (Bishop, Chapter 11, this volume). Elk herbivory on the debris-avalanche deposit resulted in an increase in plant cover and richness as compared to ungrazed areas (Dale et al., Chapter 5, this volume).

Cover of nonnative species experienced an especially high increase with elk grazing, yet understory species typical of the Pacific Northwest forests declined in cover. Secondary geophysical disturbances also dramatically affected cover and in some cases eliminated the plant cover in large areas.

After the eruption, vegetation structure (physiognomy) became increasingly complex within each disturbance zone; this had important consequences for succession. As composition or dominance of plant species changed and plants grew, vegetation structure increased, which created habitat for a host of new species. This change was particularly true of riparian zones that developed a continuous growth of willow during the second posteruption decade and was followed by the colonization of numerous animals. Differences in physiognomy across the disturbance zones were related to the dominant life forms, stem density, and growth rates.

20.3.8 Human Influences on Succession

Human actions following the 1980 eruption modified the composition and structure of both aquatic and terrestrial systems in ways that both accelerated and retarded succession (Dale et al., Chapter 19, this volume) (see Figure 20.1). These human activities were undertaken to reduce hazardous conditions; to salvage economic value; and to enhance restoration, tourism, and recreation. Some management actions, such as broadcast seeding of nonnative plant species in an effort to control erosion, had minimal success over much of the treated area. In other locations, it had deleterious effects on the native flora (Dale et al., Chapter 5, this volume). In many cases, multiple objectives involving tradeoffs with protection of ecological values were at stake. For example, extensive salvage of blown-down timber eliminated snags for their wood value, for reduction of future fire hazard, and for the safety of loggers removing valuable toppled trees. But this activity also removed habitat for cavity-nesting birds. By 2005, the forest that was planted after salvage logging had grown into dense plantations with animal assemblages that differed substantially from areas undergoing natural succession. For example, field inventories of the avifauna revealed much higher species abundance and more numerous foraging guilds in the naturally regenerating blast area than in dense, young plantation forests that had been salvage logged and planted with tree seedlings. Some lakes were stocked with fish, which altered the lake ecosystems (Bisson et al., Chapter 12, this volume). Some of these management actions also required ongoing human intervention, such as management of the sediment-retention structure on the Toutle River and the associated transportation of fish moving upstream.

20.4 The Mount St. Helens Landscape of 2005

The Mount St. Helens landscape shortly after the eruption in 1980 appeared stark, simple, and seemingly sterile (see Figure 1.1b). However, within a few months, it became clear that life persisted in much of the landscape. During the next two decades, the biology of lakes, streams, and terrestrial systems responded dramatically (Figure 20.2). In the areas nearly devoid of biological legacies (the debris-avalanche and pyroclastic-flow zones), much of the land was covered with herbaceous vegetation by 2005. That vegetation included lupines and numerous other herbs and grasses, with scattered shrubs and low densities of conifer and deciduous tree saplings (Figure 20.2a,b). Animal communities in these habitats tended to be depauperate and to include species characteristic of open terrain. Dense thickets of willow (*Salix* spp.) and alder (*Alnus* spp.) grew at most seeps and along streams with stable flow regimes. These deciduous woody plant communities provided habitat for several diverse animal assemblages, most notably mammals, birds, and invertebrates. The large number of animals that established in these structurally complex habitats greatly increased the overall diversity of the pyroclastic-flow and debris-avalanche zones. Streams with flashy flow regimes and unstable bed and bank material remained quite dynamic environments, which impeded establishment of riparian and aquatic communities. Accordingly, these streams had a low diversity of aquatic insect species. Lakes stabilized physically and biologically and quickly attained characteristics similar to undisturbed lakes in the southern Washington Cascade Range. The newly formed ponds on the debris-avalanche deposit included a broad array of wetland types, ranging in hydrologic regime from ephemeral to perennial. By 2005, these highly productive ponds supported complex, species-rich plant communities and were used extensively by a diverse assemblage of aquatic and semiaquatic animals.

In the blast area, where natural succession was substantially influenced by living and dead biological legacies, vegetation patterns were quite heterogeneous by 2005 (see Figure 20.2c). Herbaceous vegetation and large shrubs provided extensive cover, shrub-dominated riparian and lakeshore vegetation was well developed, and numerous individual and patches of conifer trees were present. These patches of conifers were legacies that were shielded as small saplings under snowpack at the time of the eruption and have since grown several meters tall and developed the characteristics of a young natural forest. Many of the large standing, dead trees and downed logs were decomposing and fragmenting. In the scorch zone in 2005, the variety of community types was based largely on the extensive biological legacies that remained after the eruption. In areas where snow was present during the eruption, vigorous stands of conifers and understory species were growing among the boles of standing dead trees. In areas where snow was not present during the eruption, shrubs and early successional herbs were dominant. Areas of standing, dead forest had high bird-species diversity. In sharp contrast to naturally regenerating portions of the blowdown and scorch zones, areas that were salvaged logged and planted with conifer seedlings in the first few years after the eruption became dense plantations with trees commonly exceeding 10 m in height. Animal assemblages in parts of the blast area were very diverse, and for vertebrates included the majority of species indigenous to the southern Washington

FIGURE 20.2. Photographs of several disturbance zones at Mount St. Helens: (a) pyroclastic-flow zone with invader hotspot (2002); (b) debris-avalanche deposit (2003); (c) blowdown forest in upper Green River (2004); (d) Clearwater Valley showing plantations punctuated with landslides.

Cascade Range. In contrast, the plantation forests had fewer vertebrate species, but those present tended to be at high densities, reflecting the simpler and extensive habitat structure and composition.

By 2005, most streams in the blast area had flushed away the fine volcanic sediment, exposing coarse substrates typical of the region, and their riparian zones were lined with shrubs that provided shade and stabilized daily temperature. Their biota was composed of a suite of vertebrates and invertebrates typical of the region. Blast-area lakes had regained physical and biological conditions similar to undisturbed lakes, despite the lack of coniferous forest along their shores and in-flowing streams (Dahm et al., Chapter 18, this volume).

Vegetation characteristics of the tephra-fall zone in 2005 were substantially different between sites that had been clear-cut before the eruption and those which had been forested. Vegetation and animals in preeruption clear-cuts were typical of clear-cuts of the same age that were not disturbed by the eruption and supported a very dense growth of grasses, herbs, and shrubs. In contrast, tephra-fall forest sites have been slow to regenerate an understory component, which remains sparse and depauperate. Understory legacy species, primarily conifer saplings and ericaceous shrubs, dominate most stands, with numerous understory herbs associated with logs and tree buttresses.

Mudflow-impacted streams and riparian areas were located mainly at lower elevation and bordered by forested systems. In many cases, they have tributary streams little affected by the eruption. These and other factors have led to speedy ecological responses and recovery approximating preeruption conditions, except for the North Fork Toutle River, which continues to carry exceptionally high sediment loads from the debris-avalanche deposit. In most areas of mudflow impact, the river position has stabilized, and dense stands of alder and willow have developed on the floodplain, although gravel bars within the floodway are turned over annually. Fish populations have established through natural colonization and stocking, but the upstream movement of anadromous species is blocked by the sediment-retention structure on the North Fork of the Toutle River and requires human intervention.

The diverse patterns of terrestrial vegetation can be viewed in a simplified framework of being composed of sites of slow or rapid development situated within a matrix of intermediate conditions (see Figure 20.2a,b,c). Sites with substantial legacies of preeruption plants developed relatively rapidly, forming *survivor hotspots* for organisms that lived through the eruption. Survivor hotspots occurred where the smothering effects of the new deposits were ameliorated by erosion, snowpack, bioturbation, or other processes and where deposits were thin enough that organisms could penetrate them from either above or below. In contrast, areas with thick deposits that were dominated by primary succession had some sites with physical conditions favorable to ecological establishment where high densities of immigrant species and biomass accumulated. These *invader hotspots*, sites with concentrations of invading plants (Figure 20.2a,b), were located on substrates with adequate physical stability, nutrients, and water, such as the stable flows of groundwater seeps and spring-fed streams. Sites with very slow establishment of plant and animal communities, termed *coldspots*, occurred on areas of severe nutrient and water limitation, very poor microclimate conditions, and persistent or pulsed secondary disturbance (see Figure 20.2d). These hotspots and coldspots occur in a matrix of intermediate site conditions, where the pace of biotic development was determined mainly by the initial disturbance processes, by local factors, and by subsequent in situ soil development.

Biotic landscape patterns at Mount St. Helens have changed with time, and in some cases earlier patterns have been overprinted by more recent ones. Some small, initial survivor hotspots have coalesced, creating much large patches. Invader hotspots began developing within the first few years after the eruption, but typically were represented by a few, small individuals that took 6 to 10 years to develop into compositionally and structurally distinct communities covering many square meters or even hectares. Hotspots became most conspicuous along stream courses, which by the 1990s were lined with verdant growth of deciduous shrubs and trees in sharp contrast with the matrix areas of the pyroclastic-flow and blowdown zones. In some cases, survivor hotspots served as stepping stones for immigrants that then colonized the matrix. In other cases, surviving species achieved negligible colonization of neighboring matrix areas, probably because of unsuitable habitat conditions (Fuller and del Moral 2003). Planted forests in the outer parts of the blast area developed homogeneous, closed-canopy forest cover, overprinting the earlier, more complex patterns of natural vegetation, which had been dominated by survivor hotspots. Some parts of the blast area switched from a gray landscape in the summer of 1980 to scattered survivor hotspots of green within the gray matrix, followed by development of the continuous green canopy of plantation forest punctuated by gray, deforested patches created by secondary disturbance (e.g., small landslides and channel shifting) (see Figure 20.2d).

Lake, stream, and terrestrial systems were strikingly different in their rates of ecological response following the 1980 eruption (Figure 20.3). These contrasting rates were regulated by

- The degree to which the various systems became nutrient enriched or nutrient limited,
- The pace of adjustment to those changes,

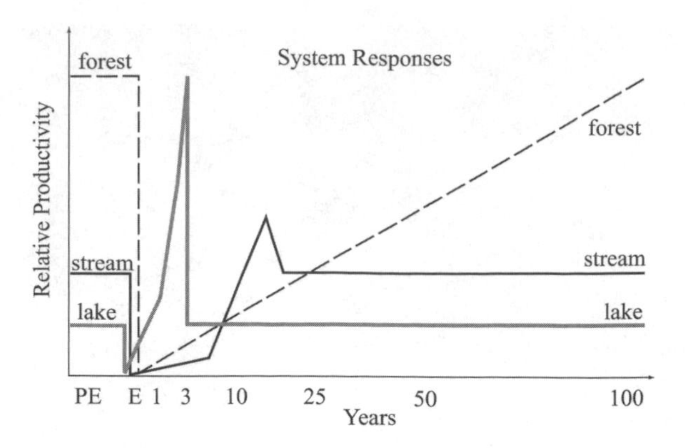

FIGURE 20.3. Conceptual response curves depicting temporal changes in productivity in terrestrial, lake, and stream systems after the 1980 eruption of Mount St. Helens. *PE*, preeruption values; *E*, values at time of eruption.

- Physical factors that promoted or inhibited biotic response, and
- The speed of species maturation and longevity.

Systems that became nutrient enriched underwent a rapid rate of change relative to systems that became nutrient impoverished. The most disturbed lakes, for instance, were greatly enriched in nutrients by the eruption and underwent rapid biogeochemical cycling that returned water quality and other aspects of habitat to conditions approaching undisturbed lakes of the southern Washington Cascade Range within several years (Dahm et al., Chapter 18, this volume). Succession in streams was more strongly influenced by the rates of change of sediment levels, habitat structure, and condition of riparian vegetation (Hawkins and Sedell 1990). For example, in the blowdown zone, 10 to 15 years of physical habitat and vegetation development led to the establishment of faunal communities typical of forested streams in the region. In contrast, by 2005 streams in the pyroclastic-flow and debris-avalanche zones remained in the early stages of ecological response because of persistently high sediment levels and failure of riparian vegetation development. Succession of terrestrial plant and animal communities was slower than lakes or streams, and in all but the tephra fall zone, sites of natural forest development were in the early stages of development by 2005. It will likely be a century or more before these forests become mature conifer-dominated communities typical of the Pacific Northwest. This slow pace of response was driven in part by the slow pace of establishment, the slow growth and great longevity of individual trees, and the protracted development of structural heterogeneity of forest stands (Franklin et al. 2002).

20.5 Ecological Response at Mount St. Helens: Broader Contexts

This book has focused on ecological change in the Mount St. Helens landscape from 1980 to 2005, but it is important to view this period in the context of longer-term ecological and volcanic change in the vicinity of Mount St. Helens and of volcanic and other major disturbance in the Pacific Northwest and elsewhere.

20.5.1 Mount St. Helens Events from 1980 to 2005 in Relation to the Longer-Term, Local Disturbance History

Geological and ecological events associated with the 1980 eruption of Mount St. Helens had attributes both characteristic of and quite different from the disturbance regime of the landscape over the preceding millennia (see Table 2.1). Earlier disturbances by volcanic flows were more restricted to the lower flanks of the volcano, except in the cases of extensive mudflows down rivers draining the mountain. Tephra fall from the 1980 eruption was more limited in thickness and extent than several of the previous eruptions but was consistent with past patterns of principal deposition being northeast of the vent. The most distinctive features of the 1980 eruption were the debris avalanche and the lateral blast, which are rare events that had a relatively large areal extent. Legacies of earlier eruptive episodes in the forms of tephra deposits blanketing hill slopes and landforms, such as the blockage forming Spirit Lake, affected the ecological response to the 1980 eruption.

Nonvolcanic disturbance processes have also been integral components of the southern Washington Cascade landscape, and these processes interact with volcanic events and their aftermath. Wildfire has been a prominent process in this landscape, typically functioning as stand-replacement disturbance, leading to establishment of the extensive conifer forests so characteristic of the west slope of the Cascade Mountains (Agee 1993; Yamaguchi 1993). The frequency of wildfire 10 to 20 km northeast of the volcano was similar to the frequency of substantial tephra deposition, and both occur more frequently that the maximum lifetimes of individuals of dominant tree species (Yamaguchi 1993).

20.5.2 Mount St. Helens 1980 Eruption in a Pacific Northwest Regional Context

Large-scale, infrequent disturbances have been important forces shaping ecological systems of the western Cascade Range for millennia. Extensive and severe disturbances, such as catastrophic stand-replacement wildfire and volcanism, create large areas where succession proceeds from early seral stages. These early stages of succession are very important in providing habitat for a wide array of the region's biota and may be integral to maintaining regional biodiversity. Distinctive, large areas of early-successional habitat and their associated large source pool of early seral species are becoming increasingly rare in the Pacific Northwest because of recent and current management direction. Fire suppression has reduced the number of large fires; and when they have occurred, managers have been quick to undertake salvage logging and planting of conifers, which removes legacy structures (e.g., snags and downed wood) and suppresses herb and shrub vegetation (which provides vital, diverse habitat and food resources for many species). Also, clear-cut logging and subsequent intensive silvicultural practices have created vast areas of conifer plantations in the Pacific Northwest landscape, greatly limiting the extent of the early seral (herb and shrub) successional stage. The net result of these activities is the reduction in area of early-successional herb- and shrub-dominated communities and a highly heterogeneous successional process in which plant communities develop at different rates and with different species compositions. The resulting habitat diversity is important to a wide suite of organisms (e.g., lepidopterans, birds, and amphibians). The 1980 eruption of Mount St. Helens and establishment of the National Volcanic Monument created conditions that allowed a natural succession to occur and will likely play a vital function for the larger Pacific Northwest region.

20.5.3 Mount St. Helens in Relation to Major Disturbances Elsewhere

Many other disturbance events, both large and small, have been studied around the globe. Among well-studied volcanic eruptions, the 1980 Mount St. Helens events stand out in terms of complexity, real-time observations, and diversity and intensity of detailed posteruption analyses by geologists and ecologists. However, the event was far from the grandest volcanic eruption in terms of magnitude of energy release, size of affected areas, or severity of ecological disturbance. Flood basalt flow and massive pyroclastic flows, for instance, have affected much larger areas (Harris 1988) with more profound and protracted ecological impacts. The 1980 Mount St. Helens events lack lava flows, which are common in many volcanic eruptions and geological settings.

Similarly, Mount St. Helens events of 1980 were dwarfed in extent and impact by some nonvolcanic events, such as bolide impacts on the earth that may have caused major, worldwide extinctions observed in the geological record. Yet, the 1980 eruption of Mount St. Helens has provided a useful example of volcanic-disturbance processes and associated ecological responses in the context of analysis of large, infrequent disturbances (Turner et al. 1997; Turner and Dale 1998). In comparisons with wildfire and windstorms, the Mount St. Helens eruption had notable distinctions, including widespread deposition of fresh rock debris. These tephra deposits have gradients of thickness over kilometers, localized sites of greater or lesser thickness affecting survival and colonization, and spatial superposition of processes and deposits, all of which complicate interpretation of disturbance effects.

20.6 Conclusions

Mount St. Helens after the 1980 eruption is a landscape of strong impressions. The initial disturbances were immense and highly varied in the types of processes involved and ecological systems influenced. The affected ecological systems exhibited remarkable resilience in many ways. Yet, a quarter century after the 1980 eruption, we have witnessed only the initial stages of the ecological responses. Most areas have undergone rapid, dramatic change in substrate condition, species diversity, and vegetative cover and physiognomy since the eruption. On the other hand, biological responses in some parts of the landscape have been quite slow, particularly where secondary disturbances have repeatedly reset biological development. Forests of this region undergo succession spanning a millennium or more, so we have viewed only a tiny fraction of the potential duration of successional sequences.

Prior analysis of succession and disturbance has been useful to varying degrees for interpreting events at Mount St. Helens to date. During more than a century of debate, many concepts of succession have been put forth, yet none of these appears to be comprehensive enough to adequately capture the dynamic and highly variable processes that occur during succession. The Mount St. Helens landscape provides examples of most, if not all, extant models of succession and illustrates that many mechanisms of succession can operate concurrently in one location. Hence, the Mount St. Helens experience provides a clear example of how complex succession may be and demonstrates that most models are far too simplistic to account for all the variation that occurs as systems respond to disturbance. Some of this complexity appears to be related to random processes, chance, and contingencies that unfold during the initial disturbance and subsequently as succession proceeds. These properties of succession are difficult to incorporate into mechanistic models and have likely limited the extent to which succession can be predicted precisely. Disturbance ecology, another scientific point of view relevant to interpreting the Mount St. Helens story, took shape as a discipline after the 1980 eruption, so the studies at Mount St. Helens have contributed to the development of the field. However, both succession concepts and the field of disturbance ecology have struggled to find generality (McIntosh 1999; White and Jentsch 2001), and neither field has sufficient scope to give a comprehensive framework for interpreting events at Mount St. Helens. Therefore, we summarize the story of Mount St. Helens in terms of 11 lessons that have been affirmed or have expanded thinking concerning ecological response to major physical disturbance.

1. **Mechanisms of disturbance** varied substantially among the types of disturbance processes that operated at Mount St. Helens in 1980. Consideration of the specific mechanisms of disturbance rather than disturbance types provides greater ability to interpret and predict ecological responses to disturbance. Pursuing an understanding of effects of specific mechanisms and their intensity on organisms is an ongoing theme of disturbance ecology.
2. **Gradients of disturbance intensity** occurred systematically across the different disturbance processes involved in the 1980 eruption of Mount St. Helens. Variation in impact force and depth of new deposits, for example, created gradients of disturbance severity reflecting the varying physical forces that occur with distance from the disturbance source. This pattern contributes to both coarse- and fine-scale spatial heterogeneity of disturbance effects across the impacted landscape, making it difficult to disentangle effects of disturbance gradients from effects of dispersal from distant seed sources.
3. **Biotic and abiotic legacies** were exceedingly abundant in much of the disturbed landscape at Mount St. Helens, and they played many vital roles governing the patterns and rates of succession. Areas with surviving organisms and dead biotic structures were among the most complex habitats in the posteruption landscape and promoted establishment of many additional species. Chemical legacies strongly influenced the response of aquatic systems, and legacy soils were important for supporting biological legacies and colonizers. The single most important characteristic influencing the pace and pattern of succession following disturbance may indeed be the type, amount, and distribution of biotic and abiotic legacies.
4. **Rapid ecological response** was perhaps the most striking impression of the posteruption landscape at Mount St. Helens. Several factors contributed to the fast pace at which biota spread, colonized, and grew throughout the disturbed area. The most important contributing factors were the presence of numerous, widely distributed survivors; important dead biological legacies; a diverse source population in the surrounding landscape; unconsolidated volcanic deposits in which plants could establish roots and animals could burrow; and a moist, maritime climate. During the initial decades following the eruption, species from all taxonomic groups of the regional pool of species, represented by all trophic levels, had colonized the Mount St. Helens landscape.
5. **Variability of system response trajectories** was profound among lakes, streams, and forests. This difference appeared to be driven primarily by nutrient levels and physical conditions. Lakes became nutrient enriched and responded rapidly and dramatically, attaining uncharacteristically high levels of primary productivity within days of the eruption. In contrast, terrestrial systems had greatly reduced productivity because vegetation was stripped away and the ground was covered with nutrient-poor volcanic deposits. By 2005, lakes had returned to preeruption productivity levels, whereas productivity of terrestrial systems had increased but remained far below that of a mature forest. This observation illustrates the complexity in evaluating the time required for a landscape to recover following disturbance and has important implications for management of areas following both natural disturbances and those caused by humans.
6. **Spatial heterogeneity** is a common hallmark of large, disturbed landscapes because of the complexities of both the primary disturbance processes themselves and the responses to them. One important consequence of this

heterogeneity is that biological legacies, which serve as sources of propagules for recolonization, may be widely dispersed across the affected landscape. This pattern may greatly accelerate ecological response at the landscape scale and influence alternative successional pathways.

7. **Secondary disturbances** and **succession of geophysical processes** are important consequences of some large, intense disturbances and are integral components of ecological responses to the primary disturbance events. It is important to recognize these phenomena as succession proceeds and to distinguish effects of biological and physical factors in environmental change.
8. **Successional processes** were multifaceted. Numerous processes of succession occurred in the area disturbed by the 1980 eruption, including facilitation, tolerance, inhibition, and relay succession. In many cases, these processes occurred concomitantly at one location and illustrate the complexity of succession. This observation underscores the limits of any one existing mechanistic model to explain succession and our inability to predict when and where certain processes will occur. In addition, nearly all forms of biotic interactions, such as herbivory, predation, and mutualism, were established among organisms shortly after the eruption and played an important role in succession.
9. **Human activities** influenced patterns and rates of ecological response. Even though a large area was designated for natural processes to proceed, many human concerns led to diverse and persistent ecological changes. The overriding concern for human life and welfare, for example, caused parts of the area to be manipulated to reduce downstream sediment movement. There are likely to be long-lasting effects of these interventions, such as the designed program to truck anadromous fish upstream and the continuing, unintended spread of nonnative plant species. Modification of fish and game species, such as stocking and harvest, affects these populations and the ecosystems in which they play critical roles, as does the salvage of downed trees to avoid economic loss.
10. **Chance and contingency** were important factors in ecological responses to the eruption. Chance was paramount in how the timing of the eruption influenced survival at multiple scales. Furthermore, the great size and complexity of the disturbed areas created many opportunities for surprising developments, such as side-by-side occurrence of early- and late-seral species. The importance of chance contributes to the difficulty of developing a general theory for predicting ecological responses to such events. Contingencies were important in determining how sequences of events, such as the order of species colonization, can affect succession. Contingencies are particularly important in succession because it is often the concomitant occurrence of several low-probability circumstances that greatly influences survival and colonization.
11. **Large, disturbed areas experiencing slow succession** may become integral components of regional biodiversity, especially in areas where fire suppression and intensive culture of forests aim to maximize the dominance of conifer forest canopy. Early-seral patches provide complex habitat and diversity of herb and shrub species that are hosts to very diverse communities and food webs involving birds, invertebrates, and other taxa.

Change will be the primary theme in the dynamic Mount St. Helens landscape during the coming decades. The extent and pace of this change will be influenced by the highly variable and complex processes of ecological succession and influenced by landscape position, topography, climate, and further biotic, human, and geophysical forces. We anticipate that the landscape will continue to develop toward a complex mosaic of vegetation patch types, both among and within the various disturbance zones, but that woody species, particularly shrubs, will become increasingly abundant, eventually cloaking most of the landscape. A variety of scattered coniferous and deciduous trees will likely emerge from the shrubs, giving the landscape an open parkland-forest appearance. Eventually, shade-tolerant conifers will become more abundant, and the landscape will appear more uniformly forested but still will express substantial variation because of physical and biological legacies of the 1980 eruption and subsequent successional processes. During this succession, streams and lakes will become more heavily shaded and receive increasing inputs from the terrestrial system. Given the eruptive history of Mount St. Helens, a major volcanic disturbance is likely to occur within the life span of long-lived species in the landscape, such as Douglas-fir (Yamaguchi 1993). Yet, we expect that many biotic, landform, and soil legacies of the 1980 eruption will be evident several centuries in the future.

Bibliography

Note: Because the bibliographic citations in the text refer to the first (and sometimes the second) author's last name and the date of publication, the entries in this bibliography are arranged by first author's last name and then by date of publication (including the a, b, or c used to differentiate among ambiguous citations). Within a year, entries having the same last name are ordered by the first author's initials and then by the second author's last name.

Aarssen, L.W., and G.A. Epp. 1990. Neighbor manipulations in natural vegetation: A review. Journal of Vegetation Science **1**:13–30.

Adams, V.D., and A.B. Adams. 1982. Initial recovery of the vegetation on Mount St. Helens. Pages 105–113 *in* S.A.C. Keller, editor. Mount St. Helens: One Year Later. Eastern Washington University Press, Cheney, Washington, USA.

Adams, A.B., K.E. Hinkley, C. Hinzman, and S.R. Leffler. 1986a. Recovery of small mammals in three habitats in the northwest sector of the Mount St. Helens National Volcanic Monument. Pages 345–358 *in* S.A.C. Keller, editor. Mount St. Helens: Five Years Later. Eastern Washington University Press, Cheney, Washington, USA.

Adams, A.B., J.R. Wallace, J.T. Jones, and W.K. McElroy. 1986b. Plant ecosystem resilience following the 1980 eruptions of Mount St. Helens, Washington. Pages 182–207 *in* S.A.C. Keller, editor. Mount St. Helens: Five Years Later. Eastern Washington University Press, Cheney, Washington, USA.

Adams, A.B., V.H. Dale, E.P. Smith, and A.R. Kruckeberg. 1987. Plant survival, growth form and regeneration following the 18 May 1980 eruption of Mount St. Helens, Washington. Northwest Science **61**:160–170.

Adams, A.B., and V.H. Dale. 1987. Vegetative succession following glacial and volcanic disturbances in the Cascade Mountain Range of Washington, U.S.A. Pages 70–147 *in* D.E. Bilderback, editor. Mount St. Helens 1980: Botanical Consequences of the Explosive Eruptions. University of California Press, Berkeley, California, USA.

Adams, M.J. 1993. Summer nests of the tailed frog (*Ascaphus truei*) from the Oregon Coast Range. Northwestern Naturalist **74**:15–18.

Adams, M.J., K.O. Richter, and W.P. Leonard. 1997. Surveying and monitoring amphibians using aquatic funnel traps. Pages 47–54 *in* D.H. Olson, W.P. Leonard, and R.B. Bury, editors. Sampling Amphibians in Lentic Habitats: Methods and Approaches for the Pacific Northwest. Northwest Fauna No. 4. Society for Northwestern Vertebrate Biology, Olympia, Washington, USA.

Adams, W. 2002. Policy implications of the 1980 eruptions of Mount St. Helens. Bachelor's thesis. Davidson College, Davidson, North Carolina, USA.

Adler, P.B., C.M. D'Antonio, and J.T. Tunison. 1998. Understory succession following a dieback of *Myrica faya* in Hawai'i Volcanoes National Park. Pacific Science **52**:69–78.

Agee, J.K. 1993. Fire Ecology of Pacific Northwest Forests. Island Press, Washington, District of Columbia, USA.

Aldrich, J.W. 1943. Biological survey of the bogs and swamps in northeastern Ohio. American Midland Naturalist **30**:346–402.

Allard, M.W., S.J. Gunn, and I.F. Greenbaum. 1987. Mensural discrimination of chromosomally characterized *Peromyscus oreas* and *P. maniculatus*. Journal of Mammalogy **68**:402–406.

Allen, J.E. 1982. Scientific access to Mount St. Helens, 1980. Page 235 *in* S.A.C. Keller, editor. Mount St. Helens One Year Later. Eastern Washington University Press, Cheney, Washington, USA.

Allen, M.F., J.A. MacMahon, and D.C. Andersen. 1984. Reestablishment of Endogonaceae on Mount St. Helens: Survival of residuals. Mycologia **76**:1031–1038.

Allen, M.F. 1987. Re-establishment of mycorrhizas on Mount St. Helens: Migration vectors. Transactions of the British Mycological Society **88**:413–417.

Allen, M.F. 1988. Re-establishment of VA mycorrhizae following severe disturbance: Comparative patch dynamics of a shrub desert and subalpine volcano. Proceedings of the Royal Society of Edinburgh **94B**:63–71.

Allen, M.F., and J.A. MacMahon. 1988. Direct VA mycorrhizal inoculation of colonizing plants by pocket gophers (*Thomomys talpoides*) on Mount St. Helens. Mycologia **80**:754–756.

Allen, R.B., and R.K. Peet. 1990. Gradient analysis of forests of the Sangre de Cristo Range, Colorado. Canadian Journal of Botany **68**:193–201.

Allen, M.F. 1991. The Ecology of Mycorrhizae. Cambridge University Press, New York, New York, USA.

Allen, M.F., C. Crisafulli, C.F. Friese, and S. Jeakins. 1992. Reformation of mycorrhizal symbioses on Mount St. Helens, 1980–1990: Interactions of rodents and mycorrhizal fungi. Mycological Research **96**:447–453.

Allen, M.F., E.B. Allen, C.N. Dahm, and F.S. Edwards. 1993. Preservation of biological diversity in mycorrhizal fungi: Importance and human impacts. Pages 81–108 *in* G. Sundnes, editor. International

Symposium on Human Impacts on Self-Recruiting Populations. The Royal Norwegian Academy of Sciences, Trondheim, Norway.

Allen, E.B., M.F. Allen, D.J. Helm, J.M. Trappe, R. Molina, and E. Rincon. 1995. Patterns and regulation of arbuscular and ectomycorrhizal plant and fungal diversity. Plant and Soil **170**:47–62.

Ambers, R.K.R. 2001. Using the sediment record in a western Oregon flood-control reservoir to assess the influence of storm history and logging on sediment yield. Journal of Hydrology **244**: 181–200.

Analytical Software. 2000. *Statistix7®for Windows*. Analytical Software, Tallahassee, Florida, USA.

Andersen, D.C., J.A. MacMahon, and M.L. Wolfe. 1980. Herbivorous mammals along a montane sere: Community structure and energetics. Journal of Mammalogy **61**:500–519.

Andersen, D.C., and J.A. MacMahon. 1981. Population dynamics and bioenergetics of a fossorial herbivore, *Thomomys talpoides* (Rodentia: Geomyidae), in a spruce-fir sere. Ecological Monographs **51**:179–202.

Andersen, D.C. 1982. Observations on *Thomomys talpoides* in the region affected by the eruption of Mount St. Helens. Journal of Mammalogy **63**:652–655.

Andersen, D.C., and J.A. MacMahon. 1985a. Plant succession following the Mount St. Helens volcanic eruption: Facilitation by a burrowing rodent, *Thomomys talpoides*. American Midland Naturalist **114**:62–69.

Andersen, D.C., and J.A. MacMahon. 1985b. The effects of catastrophic ecosystem disturbance: The residual mammals at Mount St. Helens. Journal of Mammalogy **66**:581–589.

Andersen, D.C., and J.A. MacMahon. 1986. An assessment of ground-nest depredation in a catastrophically disturbed region, Mount St. Helens, Washington. Auk **108**:622–626.

Anderson, D.R., K.P. Burnham, G.C. White, and D.L. Otis. 1983. Density estimation of small-mammal populations using a trapping web and distance sampling methods. Ecology **64**:674–680.

Anderson, J. 1983. The habitat distribution of species of the tribe Bembidiini (Coleoptera, Carabidae) on banks and shores in northern Norway. Notulae Entomologicae **63**:131–142.

Anderson, T.H., and K.H. Domsch. 1989. Ratios of microbial biomass carbon to total carbon in arable soils. Soil Biology and Biochemistry **21**:471–479.

Anderson, G.L., J.D. Hanson, and R.H. Haas. 1993. Evaluating Landsat Thematic Mapper derived vegetation indices for estimating above-ground biomass on semiarid rangelands. Remote Sensing of Environment **45**:165–175.

Anderson, C.H., and M.R. Vining. 1999. Observations of glacial, geomorphic, biologic, and mineralogic development in the crater of Mount St. Helens, Washington. Washington Geology **27**(2):9–19.

Anthony, R.G., E.D. Forsman, G.A. Green, G. Witmer, and S.K. Nelson. 1987. Small mammal populations in riparian zones of different-aged coniferous forest. The Murrelet **68**:94–102.

Antos, J.A., and D.B. Zobel. 1982. Snowpack modification of volcanic tephra effects on forest understory plants near Mount St. Helens. Ecology **63**:1969–1972.

Antos, J.A., and D.B. Zobel. 1984. Ecological implications of belowground morphology on nine coniferous forest herbs. Botanical Gazette **145**:508–517.

Antos, J.A., and D.B. Zobel. 1985a. Plant form, developmental plasticity and survival following burial by volcanic tephra. Canadian Journal of Botany **63**:2083–2090.

Antos, J.A., and D.B. Zobel. 1985b. Recovery of forest understories buried by tephra from Mount St. Helens. Vegetatio **64**:105–114.

Antos, J.A., and D.B. Zobel. 1985c. Upward movement of underground plant parts into deposits of tephra from Mount St. Helens. Canadian Journal of Botany **63**:2091–2096.

Antos, J.A., and D.B. Zobel. 1986. Seedling establishment in forests affected by tephra from Mount St. Helens. American Journal of Botany **73**:495–499.

Antos, J.A., and D.B. Zobel. 1987. How plants survive burial: A review and initial responses to tephra from Mount St. Helens. Pages 246–261 *in* D.E. Bilderback, editor. Mount St. Helens 1980: Botanical Consequences of the Explosive Eruptions. University of California Press, Berkeley, California, USA.

Antos, J.A., and R. Parish. 2002. Structure and dynamics of a nearly steady-state subalpine forest in south-central British Columbia. Oecologia (Berlin) **130**:126–135.

Aplet, G.H., R.F. Hughes, and P.M. Vitousek. 1998. Ecosystem development on Hawaiian lava flows: Biomass and species composition. Journal of Vegetation Science **9**:17–26.

Arnett, R.H., Jr., and M.C. Thomas. 2001. American Beetles. Volume 1. Archostemata, Myxophaga, Adephaga, Polyphaga: Staphyliniformia. CRC Press, Boca Raton, Florida, USA.

Arnett, R.H., Jr., M.C. Thomas, P.E. Skelley, and J.H. Frank. 2002. American Beetles. Volume 2. Polyphaga: Scarabaeoidea through Curculionoidea. CRC Press, Boca Raton, Florida, USA.

Ash, A.N., and R.C. Bruce. 1994. Impacts of timber harvesting on salamanders. Conservation Biology **8**:300–301.

Ash, A.N. 1997. Disappearance and return of plethodontid salamanders to clearcut plots in the southern Blue Ridge Mountains. Conservation Biology **11**:983–989.

Ashlock, P.D., and W.C. Gagne. 1983. A remarkable new micropterous *Nysius* species from the aeolian zone of Mauna Kea, Hawai'i (Hemiptera: Lygaeidae). International Journal of Entomology **25**:47–55.

Ashmole, N.P., and M.J. Ashmole. 1988. Arthropod communities supported by biological fallout on recent lava flows on the Canary Islands. Entomologica Scandinavica (Suppl.) **332**:67–88.

Ashmole, N.P., P. Oromi, M.J. Ashmole, and J.L. Martin. 1996. The invertebrate fauna of early successional volcanic habitats in the Azores. Boletim do Museu Municipal do Funchal **48**:5–39.

Ashmole, N.P., and M.J. Ashmole. 1997. The land fauna of Ascension Island: New data from caves and lava flows, and a reconstruction of the prehistoric ecosystem. Journal of Biogeography **24**:549–589.

Atkeson, T.D., and A.S. Johnson. 1979. Succession of small mammals on pine plantations in the Georgia Piedmont. American Midland Naturalist **101**:385–392.

Atkinson, I.A.E. 1970. Successional trends in the coastal and lowland forest of Mauna Loa and Kilauea volcanoes, Hawaii. Pacific Science **24**:387–400.

Aubry, K.B., and P.A. Hall. 1991. Terrestrial amphibian communities in the southern Washington Cascade Range. Pages 327–338 *in* L.F. Ruggiero, K.B. Aubry, A.B. Carey, and M.H. Huff, editors. Wildlife and Vegetation of Unmanaged Douglas-Fir Forests. General Technical Report PNW-GTR-285. USDA Forest Service, Pacific Northwest Research Station, Portland, Oregon, USA.

Bach, C.E. 1990. Plant successional stage and insect herbivory: Flea beetles on sand-dune willow. Ecology **71**:598–609.

Bach, C.E. 1994. Effects of a specialist herbivore (*Altica subplicata*) on *Salix cordata* and sand dune succession. Ecological Monographs **64**:423–445.

Bach, C.E. 2001a. Long-term effects of insect herbivory and sand accretion on plant succession on sand dunes. Ecology **82**:1401–1416.

Bach, C.E. 2001b. Long-term effects of insect herbivory on responses by *Salix cordata* to sand accretion. Ecology **82**:397–409.

Baker, R.J., and S.L. Williams. 1972. A live trap for pocket gophers. Journal of Wildlife Management **36**:1320–1322.

Baker, A.D. 1989. Variation in Trout Abundance in Relation to Habitat Quality in a Recently Disturbed Stream: Patterns at Two Spatial Scales. Master's thesis. Utah State University, Logan, Utah, USA.

Baker, C. 1995. Phytoplankton Communities in Mount St. Helens Lakes. Master's thesis. Portland State University, Portland, Oregon, USA.

Baker, W.L., and G.M. Walford. 1995. Multiple stable states and models of riparian vegetation succession on the Animas River, Colorado. Annals of the Association of American Geographers **85**:320–338.

Banks, N.G., and R.P. Hoblitt. 1981. Summary of temperature studies of 1980 deposits. Pages 295–313 *in* P.W. Lipman and D.R. Mullineaux, editors. The 1980 Eruptions of Mount St. Helens, Washington. Professional Paper 1250. U.S. Geological Survey, Washington, District of Columbia, USA.

Barnes, W.J., and E. Dibble. 1988. The effects of beaver in riverbank forest succession. Canadian Journal of Botany **66**:40–44.

Baross, J.A., J.R. Sedell, and C.N. Dahm. 1981. Letter to the editor and editor's reply. National Geographic **160**:3.

Baross, J.A., C.N. Dahm, A.K. Ward, M.D. Lilley, and J.R. Sedell. 1982. Initial microbiological response in lakes to the Mt. St. Helens eruption. Nature (London) **296**:49–52.

Baross, J.A., and S.E. Hoffman. 1985. Submarine hydrothermal vents and associated gradient environments as sites for the origin and evolution of life. Origins of Life and Evolution of the Biosphere **15**:327–345.

Bart, M.L. 1999. Sedimentary and Geomorphic Evolution of a Portion of the North Fork Toutle River, Mount St. Helens National Volcanic Monument, Washington. Master's thesis. Washington State University, Pullman, Washington, USA.

Baruch, Z., and G. Goldstein. 1999. Leaf construction cost, nutrient concentration, and net CO_2 assimilation of native and invasive species in Hawaii. Oecologia (Berlin) **121**:183–192.

Bazzaz, F.A. 1979. The physiological ecology of plant succession. Annual Review of Ecology and Systematics **10**:351–372.

Beaman, J.H. 1962. The timberlines of Iztaccihuatl and Popocatepetl, Mexico. Ecology **43**:377–385.

Beard, J.S. 1976. The progress of plant succession on the Soufriere of St. Vincent: Observations in 1972. Vegetatio **31**:69–77.

Beardsley, G.F., and W.A. Cannon. 1930. Note on the effects of a mud-flow at Mt. Shasta on the vegetation. Ecology **11**:326–336.

Beckwith, S.L. 1954. Ecological succession on abandoned farm lands and its relationship to wildlife management. Ecological Monographs **24**:349–376.

Behnke, R.J. 2002. Trout and Salmon of North America. The Free Press, New York, New York, USA.

Belyea, L.R., and J. Lancaster. 1999. Assembly rules within a contingent ecology. Oikos **86**:402–416.

Benda, L., and T. Dunne. 1997. Stochastic forcing of sediment supply to channel networks from landsliding and debris flow. Water Resources Research **33**:2849–2863.

Bendix, J. 1997. Flood disturbance and the distribution of riparian species diversity. The Geographical Review **87**:468–483.

Bennett, R.N., and R.M. Wallsgrove. 1994. Tansley review no. 72: Secondary metabolites in plant defence mechanisms. New Phytologist **127**:617–633.

Bentley, B.L., and N.D. Johnson. 1994. The impact of defoliation by a tussock moth, *Orygia vetusta*, on a nitrogen-fixing legume, *Lupinus arboreus*. Pages 169–178 *in* J.I. Sprent and D. McKey, editors. Advances in Legume Systematics 5: The Nitrogen Factor. Royal Botanic Gardens, Kew, UK.

Bernays, E.A., and R.F. Chapman. 1994. Host–Plant Selection by Phytophagous Insects. Chapman & Hall, London, UK.

Berner, E.K., and R.A. Berner. 1987. The Global Water Cycle: Geochemistry and the Environment. Prentice-Hall, Englewood Cliffs, New Jersey, USA.

Beschta, R.L. 1978. Long-term patterns of sediment production following road construction and logging in the Oregon Coast Range. Water Resources Research **14**:1011–1016.

Bever, J.D., P.A. Schultz, A. Pringle, and J.B. Morton. 2001. Arbuscular mycorrhizal fungi: More diverse than meets the eye, and the ecological tale of why. Bioscience **51**:923–931.

Bierzychudek, P. 1999. Looking backwards: Assessing the projections of a transition matrix model. Ecological Applications **9**:1278–1287.

Bilby, R.E. 1984. Characteristics and frequency of cool-water areas in a western Washington stream. Journal of Freshwater Ecology **2**:593–602.

Bilby, R.E., and P.A. Bisson. 1987. Emigration and production of hatchery coho salmon (*Oncorhynchus kisutch*) stocked in streams draining an old-growth and a clear-cut watershed. Canadian Journal of Fisheries and Aquatic Sciences **44**:1397–1407.

Bilby, R.E., B.R. Fransen, and P.A. Bisson. 1996. Incorporation of nitrogen and carbon from spawning coho salmon into the trophic system of small streams: Evidence from stable isotopes. Canadian Journal of Fisheries and Aquatic Science **53**:164–173.

Bilderback, D.E., editor. 1987. Mount St. Helens 1980: Botanical Consequences of the Explosive Eruptions. University of California Press, Berkeley, California, USA.

Bilderback, D.E., and C.E. Carlson. 1987. Effects of Persistent Volcanic Ash on Douglas-Fir in Northern Idaho. USDA Forest Service Intermountain Research Station, Missoula, Montana, USA.

Bilderback, D.E., and J.H. Slone. 1987. Persistence of Mount St. Helens ash on conifer foliage. Pages 282–299 *in* D.E. Bilderback, editor. Mount St. Helens 1980. Botanical Consequences of the Explosive Eruptions. University of California Press, Berkeley, California, USA.

Biondi, F., A. Gershunov, and D.R. Cayan. 2001. North Pacific decadal climate variability since 1661. Journal of Climate **14**:5–10.

Bishop, J.G., and D.W. Schemske. 1998. Variation in flowering phenology and its consequences for lupines colonizing Mount St. Helens. Ecology **79**:534–546.

Bishop, J.G. 2002. Early primary succession on Mount St. Helens: The impact of insect herbivores on colonizing lupines. Ecology **83**:191–202.

Bisson, P.A., J.L. Nielsen, and J.W. Ward. 1988. Summer production of coho salmon stocked in Mount St. Helens streams 3–6 years after the 1980 eruption. Transactions of the American Fisheries Society **117**:322–335.

Bjornn, T.C., and D.W. Reiser. 1991. Habitat requirements of salmonids in streams. American Fisheries Society Special Publication **19**:83–138.

Black, R.A., and R.N. Mack. 1984. Aseasonal leaf abscission in *Populus* induced by volcanic ash. Oecologia (Berlin) **64**:295–299.

Black, R.A., and R.N. Mack. 1986. Mount St. Helens ash: Recreating its effects on the steppe environment and ecophysiology. Ecology **67**:1289–1302.

Blanchette, C.A. 1997. Size and survival of intertidal plants in response to wave action: A case study with *Fucus gardneri*. Ecology **78**:1563–1578.

Blaustein A.R., and D.B. Wake. 1990. Declining amphibian populations: A global phenomena? Trends in Ecology and Evolution **5**:203–204.

Blaustein, A.R., and D.B. Wake. 1995. The puzzle of declining amphibian populations. Scientific American **272**(4):52–57.

Blessing, B.J., E.P Phenix, L.C. Jones, and M.G. Raphael. 1999. Nests of Van Dyke's salamander (*Plethodon vandykei*) from the Olympic Peninsula, Washington. Northwestern Naturalists **80**:77–81.

Boerner, R.E.J., B.G. DeMars, and P.N. Leicht. 1996. Spatial patterns of mycorrhizal infectiveness of soils along a successional chronosequence. Mycorrhizae **6**:79–90.

Bond, C.E. 1963. Distribution and Ecology of Freshwater Sculpins, Genus *Cottus*, in Oregon. Doctoral dissertation. University of Michigan, Ann Arbor, Michigan, USA.

Bortleson, G.C., N.P. Dion, J.B. McConnell, and L.M. Nelson. 1976. Reconnaissance data on lakes in Washington. Volume 4. Water-Supply Bulletin 43. Washington Department of Ecology, Olympia, Washington, USA.

Bottema, S., and A. Sarpaki. 2003. Environmental change in Crete: A 9000-year record of Holocene vegetation history and the effect of the Santorini eruption. Holocene **13**:733–749.

Braatne, J.H., and D.M. Chapin. 1986. Comparative water relations of four subalpine plants at Mount St. Helens. Pages 163–172 *in* S.A.C. Keller, editor. Mount St. Helens: Five Years Later. Eastern Washington University Press, Cheney, Washington, USA.

Braatne, J.H. 1989. Comparative Physiological and Population Ecology of *Lupinus lepidus* and *Lupinus latifolius* Colonizing Early Successional Habitats on Mount St. Helens. Doctoral dissertation. University of Washington, Seattle, Washington, USA.

Braatne, J.H., and L.C. Bliss. 1999. Comparative physiological ecology of lupines colonizing early successional habitats on Mount St. Helens. Ecology **80**:891–907.

Bradshaw, A.D. 1993. Introduction: Understanding the fundamentals of succession. Pages 1–4 *in* J. Miles and D.W.H. Walton, editors. Primary Succession on Land. Blackwell Scientific, Oxford, UK.

Bråkenhielm, S., and L. Qinghong. 1995. Comparison of field methods in vegetation monitoring. Water, Air and Soil Pollution **79**:75–87.

Brantley, S.R., and R.B. Waitt. 1988. Interrelations among pyroclastic surge, pyroclastic flow, and lahars in Smith Creek valley during first minutes of 18 May 1980 eruption of Mount St. Helens, USA. Bulletin of Volcanology **50**:304–326.

Brantley, S.R., and B. Myers. 2000. Mount St. Helens: From the 1980 Eruption to 2000. Fact Sheet 036-00. U.S. Geological Survey, Vancouver, Washington, USA.

Brattstrom, B.H. 1963a. A preliminary review of the thermal requirements of amphibians. Ecology **44**:238–255.

Brattstrom, B.H. 1963b. Barcena volcano, 1952: Its effect on the fauna and flora of San Benedicto Island, Mexico. Pages 499–524 *in* J.L. Gressitt, editor. Pacific Basin Biogeography. Bishop Museum Press, Honolulu, Hawaii, USA.

Bray, J.R., and J.T. Curtis. 1957. An ordination of the upland forest communities of southern Wisconsin. Ecological Monographs **27**:325–349.

Breiman, L., J.H. Friedman, R.A. Olshen, and C.J. Stone. 1984. Classification and Regression Trees. Wadsworth International Group, Belmont, California, USA.

Brejda, J.J., D.L. Karlen, J.L. Smith, and D.L. Allen. 2000a. Identification of regional soil quality factors and indicators: II. Northern Mississippi loess hills and Palouse prairie. Soil Science Society of America Journal **64**:2125–2135.

Brejda, J.J., T.B. Moorman, D.L. Karlen, and T.H. Dao. 2000b. Identification of regional soil quality factors and indicators: I. Central and southern high plains. Soil Science Society of America Journal **64**:2115–2124.

Brindle, F.A. 2003. Use of native vegetation and biostimulants for controlling soil erosion on steep terrain. Transportation Research Record **1819**:203–209.

Brown, W.L. 1982. Areas of controversy. Pages 165–166 *in* S.A.C. Keller, editor. Mount St. Helens One Year Later. Eastern Washington University Press, Cheney, Washington, USA.

Brugman, M.M., and A. Post. 1981. Effects of Volcanism on the Glaciers of Mount St. Helens, Washington. Circular 850-D. U.S. Geological Survey, Reston, Virginia, USA.

Brunet, J., G. von Oheimb, and M. Diekmann. 2000. Factors influencing vegetation gradients across ancient-recent woodland borderlines in southern Sweden. Journal of Vegetation Science **11**:515–524.

Bruno, J.F., J.J. Stachowicz, and M.D. Bertness. 2003. Inclusion of facilitation into ecological theory. Trends in Ecology and Evolution **18**:119–125.

Bryant, J.P., F.S. Chapin, III, and D.R. Klein. 1983. Carbon/nutrient balance of boreal plants in relation to vertebrate herbivory. Oikos **40**:357–368.

Bryant, J.P., and F.S. Chapin, III. 1986. Browsing-woody plant interactions during boreal forest plant succession. Pages 213–225 *in* K. Van Cleve, F.S. Chapin, III, P.W. Flanagan, L.A. Viereck, and C.T. Dyrness, editors. Forest Ecosystems in the Alaskan Taiga. Springer-Verlag, New York, New York, USA.

Bryant, J.P. 1987. Feltleaf willow-snowshoe hare interactions: Plant carbon/nutrient balance and floodplain succession. Ecology **68**:1319–1327.

Burger, J.A., and D.L. Kelting. 1999. Using soil quality indicators to assess forest stand management. Forest Ecology and Management **122**:155–166.

Burnham, R. 1994. Plant deposition in modern volcanic environments. Transactions of the Royal Society, Edinburgh: Earth Sciences **84**:275–281.

Burrows, F.M. 1973. Calculation of the primary trajectories of plumed seeds in steady winds with variable convection. New Phytologist **72**:647–664.

Burt, W.H. 1961. Some effects of Volcan Paricutin on vertebrates. Occasional Paper No. 620. Museum of Zoology, University of Michigan, Ann Arbor, Michigan, USA.

Bury, R.R. 1983. Differences in amphibian populations in logged and old growth redwood forest. Northwest Science **57**:167–178.

Bush, M.B., R.J. Whittaker, and T. Partomihardjo. 1992. Forest development on Rakata, Panjang and Sertung: Contemporary dynamics (1979–1989). GeoJournal **28**:185–199.

Butler, D.R. 1995. Zoogeomorphology: Animals as Geomorphic Agents. Cambridge University Press, New York, New York, USA.

Buyer, J.S., and L.E. Drinkwater. 1997. Comparison of substrate utilization and fatty acid analysis of soil microbial communities. Journal of Microbiological Methods **30**:3–11.

Byrd, K.B., T.V. Parker, D.R. Vogler, and K.W. Cullings. 2000. The influence of clear-cutting on ectomycorrhizal fungus diversity in a lodgepole pine (*Pinus contorta*) stand, Yellowstone National Park, Wyoming, and Gallatin National Forest, Montana. Canadian Journal of Botany **78**:149–156.

Cadish, G., and K.E. Giller, editors. 1997. Driven by Nature: Plant Litter Quality and Decomposition. CAB International, Oxford, UK.

Cain, S.A. 1950. Life-forms and phytoclimate. Botanical Review **16**:1–32.

Cairney, J.W.G. 2000. Evolution of mycorrhiza systems. Naturwissenschaften **87**:467–475.

Callaway, R.M., and L.R. Walker. 1997. Competition and facilitation: A synthetic approach to interactions in plant communities. Ecology **78**:1958–1965.

Cameron, K.A., and P.T. Pringle. 1990. Avalanche-generated debris flow of 9 May 1986, at Mount St. Helens, Washington. Northwest Science **64**:159–164.

Campbell, J., W.F. Bibb, M.A. Lambert, S. Eng, A.G. Steigerwalt, J. Allard, C.W. Moss, and D.J. Brenner. 1984. *Legionella sainthelensi*: A new species of *Legionella* isolated from water near Mount St. Helens. Applied and Environmental Microbiology **47**: 369–373.

Campbell, D.R. 2001. Keystone Herbivores and Their Impact on Vegetation and Successional Dynamics Within the Debris Avalanche Deposit at Mount St. Helens National Volcanic Monument. Master's thesis. University of Wisconsin at Stevens Point, Stevens Point, Wisconsin, USA.

Carey, D.B., and M. Wink. 1994. Elevational variation of quinolizidine alkaloid contents in a lupine (*Lupinus argenteus*) of the Rocky Mountains. Journal of Chemical Ecology **20**:849–857.

Carey, A.B. 1995. Sciurids in Pacific Northwest managed and old-growth forests. Ecological Applications **5**:648–661.

Carey, A.B., and M.L. Johnson. 1995. Small mammals in managed, naturally young, and old-growth forests. Ecological Applications **5**:336–352.

Carlson, J.R., J.R. Stroh, J.A. Oyler, and F. Reckendorf. 1982. Erosion control revegetation on land damaged by the 1980 Mt. St. Helens eruption. Pages 174–198 *in* R.I. Cudny and J. Etra, editors. Proceedings of the Fifth High Altitude Revegetation Workshop. Information Series No. 48. Colorado Water Resources Research Institute, Colorado State University, Fort Collins, Colorado, USA.

Carpenter, S.E., J.M. Trappe, and G.A. Hunt. 1982. Observations on fungal succession on recent volcanic deposits of Mount St. Helens. Pages 36–44 *in* Proceedings, 39th Meeting. Oregon Academy of Sciences, Newberg, Oregon, USA.

Carpenter, S.E., J.M. Trappe, and J.F. Ammirati. 1987. Observations of fungal succession in the Mount St. Helens devastation zone, 1980–1983. Canadian Journal of Botany **65**:716–728.

Carpenter, K.D. 1995. Indicators of Nutrient Limited Plankton Growth in Lakes Near Mount St. Helens, Washington. Master's thesis. Portland State University, Portland, Oregon, USA.

Case, R., and J.B. Kauffman. 1997. Wild ungulate influences on the recovery of willows, black cottonwood, and thin-leaf alder following cessation of cattle grazing in northeastern Oregon. Northwest Science **71**:115–126.

Cazares, E., and J.M. Trappe. 1993. Vesicular endophytes in roots of Pinaceae. Mycorrhiza **2**:153–156.

Cazares, E., and J.E. Smith. 1996. Occurrence of vesicular-arbuscular mycorrhizae in *Pseudotsuga menziezieii* and *Tsuga heterophylla* seedlings grown in Oregon Coast Range soils. Mycorrhiza **6**: 65–67.

Cederholm, C.J., D.H. Johnson, R.E. Bilby, L.G. Dominguez, A.M. Garrett, W.H. Graeber, E.L. Greda, M.D. Kunze, B.G. Marcot, J.F. Palmisano, R.W. Plotnikoff, W.G. Pearcy, C.A. Simenstad, and P.C. Trotter. 2001. Pacific salmon and wildlife: Ecological contexts, relationships, and implications for management. Pages 628–684 *in* D.H. Johnson and T.A. O'Neil, editors. Wildlife-Habitat Relationships in Oregon and Washington. Oregon State University Press, Corvallis, Oregon, USA.

Chapin, F.S., III, P.M. Vitousek, and K. Van Cleve. 1986. The nature of nutrient limitation in plant communities. American Naturalist **127**:48–58.

Chapin, D.M., and L.C. Bliss. 1988. Soil–plant water relations of two subalpine herbs from Mount St. Helens. Canadian Journal of Botany **66**:809–818.

Chapin, D.M., and L.C. Bliss. 1989. Seedling growth, physiology, and survivorship in a subalpine, volcanic environment. Ecology **70**:1325–1334.

Chapin, F.S., III, L.R. Walker, C.L. Fastie, and L. Sharman. 1994. Mechanisms of primary succession following deglaciation at Glacier Bay, Alaska. Ecological Monographs **62**:149–175.

Chapin, D.M. 1995. Physiological and morphological attributes of two colonizing plant species on Mount St. Helens. American Midland Naturalist **133**:76–87.

Chavez, P.S. 1996. Image-based atmospheric corrections—revisited and improved. Photogrammetric Engineering and Remote Sensing **9**:1025–1036.

Chen, J., J.F. Franklin, and T.A. Spies. 1995. Growing-season microclimatic gradients from clearcut edges into old-growth Douglas-fir forests. Ecological Applications **5**:74–86.

Christiansen, R.L., and D.W. Peterson. 1981. Chronology of the 1980 eruptive activity. Pages 17–30 *in* P.W. Lipman and D.R. Mullineaux, editors. The 1980 Eruptions of Mount St. Helens, Washington. Professional Paper 1250. U.S. Geological Survey, Washington, District of Columbia, USA.

Christy, J.A. 2000. Classification and Catalogue of Native Plant Communities in Oregon. Oregon Natural Heritage Program, Portland, Oregon, USA.

Clark, B.P. 1986. New look-up tables. Landsat Technical Notes **1**:1–8.

Clarke, K.R. 1993. Non-parametric multivariate analyses of changes in community structure. Australian Journal of Ecology **18**: 117–143.

Clarkson, B.R., and B.D. Clarkson. 1983. Mt. Tarawera: 2. Rates of change in the vegetation and flora of the high domes. New Zealand Journal of Ecology **6**:107–119.

Clarkson, B.D. 1990. A review of vegetation development following recent (less than 450 years) volcanic disturbance in North Island, New Zealand. New Zealand Journal of Ecology **14**:59–71.

Clarkson, B.R., L.R. Walker, B.D. Clarkson, and W.B. Silvester. 2002. Effect of *Coriaria arborea* on seed banks during primary succession on Mt. Tarawera, New Zealand. New Zealand Journal of Botany **40**:629–638.

Claussen, D.L. 1973. The thermal relations of the tailed frog, *Ascaphus truei*, and the Pacific tree frog, *Hyla regilla*. Comparative Biochemistry and Physiology **44**:137–153.

Clements, F.E. 1916. Plant Succession: An Analysis of the Development of Vegetation. Publication No. 242. Carnegie Institution, Washington, District of Columbia, USA.

Cochran, V.L., D.F. Bezdicek, L.F. Elliott, and R.I. Papendick. 1983. The effect of Mount St. Helens' volcanic ash on plant growth and mineral uptake. Journal of Environmental Quality **12**:415–417.

Cole, D.W., S.P. Gessel, and S.F. Dice. 1967. Distribution and cycling of nitrogen, phosphorus, potassium, and calcium in a second growth Douglas-fir ecosystem. Pages 197–232 *in* H.E. Young, editor. Symposium on Primary Productivity and Mineral Cycling in Natural Ecosystems. University of Maine Press, Orono, Maine, USA.

Coley, P.D., J.P. Bryant, and F.S. Chapin, III. 1985. Resource availability and plant antiherbivore defense. Science **230**:285–289.

Collins, B.D., and T. Dunne. 1986. Erosion of tephra from the 1980 eruption of Mount St. Helens. Geological Survey of America Bulletin **97**:896–905.

Collins, B.D., and T. Dunne. 1988. Effects of forest land management on erosion and revegetation after the eruption of Mount St. Helens. Earth Surface Processes and Landforms **13**:193–205.

Collins, B., and G. Wein. 1998. Soil heterogeneity effects on canopy structure and composition during early succession. Plant Ecology **138**:217–230.

Connell, J.H., and R.O. Slatyer. 1977. Mechanisms of succession in natural communities and their role in community stability and organizations. American Naturalist **111**:1119–1144.

Connell, J.H. 1978. Diversity in tropical rain forests and coral reefs. Science **199**:1302–1310.

Cook, R.J., J.C. Barron, R.I. Papendick, and G.J. Williams, III. 1981. Impact on agriculture of the Mount St. Helens eruptions. Science **211**:16–18.

Corn, S.P., and R.R. Bury. 1989. Logging in western Oregon: Responses of headwater habitats and stream amphibians. Forest Ecology and Management **29**:39–57.

Corn, S.P. 2000. Amphibian declines: Review of some current hypotheses. Pages 663–696 *in* D.W. Sparling, G. Linder, and C. Bishop, editors. Ecotoxicology of Amphibians and Reptiles. Society of Environmental Toxicology and Chemistry, Pensacola, Florida, USA.

Costa, J.E. 1991. Nature, mechanics, and mitigation of the Val Pola landslide, Valtellina, Italy, 1987–1988. Zeitschrift für Geomorphologie **35**:15–38.

Courtney, S.P., and S. Courtney. 1982. The "edge effect" in butterfly oviposition: Causality in *Anthocaris cardamines* and related species. Ecological Entomology **7**:131–137.

Courtney, S.P. 1986. The ecology of pierid butterflies: Dynamics and interactions. Advances in Ecological Research **15**:51–131.

Cousins, S.A., and O. Eriksson. 2001. Plant species occurrences in a rural hemiboreal landscape: Effects of remnant habitats, site history, topography and soil. Ecography **24**:461–469.

Covich, A.P., and T.A. Crowl. 1990. Effects of hurricane storm flow on transport of woody debris in a rain forest stream (Luquillo Experimental Forest, Puerto Rico). Pages 197–205 *in* J.H. Krishna, V.Q. Aponte, and F. Gomez-Gomez, editors. Tropical Hydrology and Caribbean Water Resources. American Water Resources Association, San Juan, Puerto Rico, USA.

Covich, A.P., T.A. Crowl, S.L. Johnson, D. Varza, and D.L. Certain. 1991. Post-Hurricane Hugo increases in Atyid shrimp abundance in a Puerto Rican montane stream. Biotropica **23**:448–454.

Covich, A.P., and W.H. McDowell. 1996. The stream community. Pages 434–459 *in* D.P. Reagan and R.B. Waide, editors. The Food Web of a Tropical Rain Forest. The University of Chicago Press, Chicago, Illinois, USA.

Cowles, H.C. 1899. The ecological relations of vegetation on the sand dunes of Lake Michigan. Botanical Gazette **27**:95–117, 167–202, 281–308, 361–391.

Crandell, D.R., and D.R. Mullineaux. 1978. Potential hazards from future eruptions of Mount St. Helens volcano, Washington. Bulletin 1383-C. U.S. Geological Survey, Washington, District of Columbia, USA.

Crandell, D.R., and R.P. Hoblitt. 1986. Lateral blasts at Mount St. Helens and hazard zonation. Bulletin of Volcanology **48**:27–37.

Crandell, D.R. 1987. Deposits of pre-1980 pyroclastic flows and lahars from Mount St. Helens, Washington. Professional Paper 1444. U.S. Geological Survey, Washington, District of Columbia, USA.

Crawford, B.A. 1986. Recovery of Game Fish Populations Impacted by the May 18, 1980, Eruption of Mount St. Helens. Part II. Recovery of Surviving Fish Populations Within the Lakes in the Mount St. Helens National Volcanic Monument and Adjacent Areas. Fishery Management Report 85-9B. Washington Department of Game, Vancouver, Washington, USA.

Crawford, R.L., P.M. Sugg, and J.S. Edwards. 1995. Spider arrival and primary establishment on terrain depopulated by volcanic eruption at Mount St. Helens. American Midland Naturalist **133**:60–75.

Crisafulli, C.M. 1997. A habitat-based method for monitoring pond-breeding amphibians. Pages 83–112 *in* D.H. Olson, W.P. Leonard, and R.B. Bury, editors. Sampling Amphibians in Lentic Habitats: Methods and Approaches for the Pacific Northwest. Northwest Fauna No. 4. Society for Northwestern Vertebrate Biology, Olympia, Washington, USA.

Crisafulli, C.M., and C.P. Hawkins. 1998. Ecosystem recovery following a catastrophic disturbance: Lessons learned from Mount St. Helens. Pages 23–26 *in* M.J. Mac, P.A. Opler, C.E. Puckett Haecker, and P.D. Doran, editors. Status and Trends of the Nation's Biological Resources. Volumes 1, 2. U.S. Geological Survey, Reston, Virginia, USA.

Crisafulli, C.M. 1999. Survey protocol for Larch Mountain salamander (*Plethodon larselli*). Pages 253–310 *in* D.H. Olson, editor. Standardized Survey Protocols for Amphibians Under the Survey and Manage and Protection Buffer Provisions. USDA Forest Service, Pacific Northwest Research Station, Olympia, Washington, USA.

Criswell, C.W. 1987. Chronology and pyroclastic stratigraphy of the May 18, 1980, eruption of Mount St. Helens, Washington. Journal of Geophysical Research **92**(B10):10237–10266.

Crocker, R.L., and J. Major. 1955. Soil development in relation to vegetation and surface age at Glacier Bay, Alaska. Journal of Ecology **43**:427–448.

Crother, B.I. 2000. Scientific and Standard English Names of Amphibians and Reptiles of North America North of Mexico, with Comments Regarding Confidence in Our Understanding. SSAR Herpetological Circular 29. Society for the Study of Amphibians and Reptiles, Marceline, Missouri, USA.

Crump, M.L., and N.J. Scott, Jr. 1994. Visual encounter surveys. Pages 84–91 *in* W.R. Heyer, M.A. Donnelly, R.W. McDiarmid, L.C. Hayek, and M.S. Foster, editors. Measuring and Monitoring Biological Diversity: Standard Methods for Amphibians. Smithsonian Institution Press, Washington, District of Columbia, USA.

Csecserits, A., and T. Redei. 2001. Secondary succession on sandy old-fields in Hungary. Applied Vegetation Science **4**:63–74.

Cummans, J. 1981. Mudflows Resulting from the May 18, 1980, Eruption of Mount St. Helens, Washington. Circular 850-B. U.S. Geological Survey, Reston, Virginia, USA.

Curtis, J.T. 1959. The Vegetation of Wisconsin. University of Wisconsin Press, Madison, Wisconsin, USA.

Dahm, C.N., J.A. Baross, A.K. Ward, M.D. Lilley, R.C. Wissmar, and A. Devol. 1981. North Coldwater Lake and Vicinity: Limnology, Chemistry, and Microbiology. U.S. Army Corps of Engineers, Portland, Oregon, USA.

Dahm, C.N., J.A. Baross, M.D. Lilley, A.K. Ward, and J.R. Sedell. 1982. Lakes in the blast zone of Mount St. Helens: Chemical and microbiological responses following the May 18, 1980, eruption. Pages 98–137 *in* W.H. Funk, editor. Conference on Mount St. Helens: Effects on Water Resources. Report No. 41. Washington State University, Pullman, Washington, USA.

Dahm, C.N., J.A. Baross, A.K. Ward, M.D. Lilley, and J.R. Sedell. 1983. Initial effects of the Mount St. Helens eruption on nitrogen cycle and related chemical processes in Ryan Lake. Applied and Environmental Microbiology **45**:1633–645.

Dale, V.H. 1986. Plant recovery on the debris avalanche at Mount St. Helens. Pages 208–214 *in* S.A.C. Keller, editor. Mount St. Helens: Five Years Later. Eastern Washington University Press, Cheney, Washington, USA.

Dale, V.H. 1989. Wind dispersed seeds and plant recovery on the Mount St. Helens debris avalanche. Canadian Journal of Botany **67**:1434–1441.

Dale, V.H. 1991. Mount St. Helens: Revegetation of Mount St. Helens debris avalanche 10 years post-eruption. National Geographic Research & Exploration **7**:328–341.

Dale, V.H., A. Lugo, J. MacMahon, and S.T.A. Pickett. 1998. Ecosystem management in the context of large, infrequent disturbances. Ecosystems **1**:546–557.

Dale, V.H., S. Brown, R.A. Haeuber, N.T. Hobbs, N. Huntly, R.J. Naiman, W.E. Riebsame, M.G. Turner, and T.J. Valone. 2000. Ecological principles and guidelines for managing the use of land. Ecological Applications **10**:639–670.

Dale, V.H., L.A. Joyce, S. McNulty, R.P. Neilson, M.P. Ayres, M.D. Flannigan, P.J. Hanson, L.C. Irland, A.E. Lugo, C.J. Peterson, D. Simberloff, F.J. Swanson, B.J. Stocks, and B.M. Wotton. 2001. Climate change and forest disturbances. BioScience **51**:723–734.

Dale, V.H. 2002. Science and decision making. Pages 139–152 *in* R. Costanza and S.E. Jorgensen, editors. Understanding and Solving Environmental Problems in the 21st Century: Toward a New, Integrated Hard Problem Science. Elsevier, The Hague, The Netherlands.

Dale, V.H., and W.M. Adams. 2003. Plant reestablishment 15 years after the debris avalanche at Mount St. Helens, Washington. Science of the Total Environment **313**(1-3):101–113.

Dale, V.H., J. Delgado-Acevedo, and J. MacMahon. 2005. Effects of modern volcanic impacts on vegetation. Pages 227–249 *in* J. Marti and G.G.J. Ernst, editors. Volcanoes and the Environment. Cambridge University Press, Cambridge, UK.

Dalquest, W.W. 1948. Mammals of Washington. Museum of Natural History, University of Kansas, Lawrence, Kansas, USA.

Daltry, J.C., and G. Gray. 1999. Effects of volcanic activity on the endangered mountain chicken frog (*Leptodactylus fallax*). Froglog **32**:1–2.

Dammerman, K.W. 1948. The fauna of Krakatau, 1883–1933. Verhandelingen Koniklijke Nederlansche Akademie van Wetenschappen, Afdeling Natuurkunde II **44**:1–594.

Danielson, R.M. 1984. Ectomycorrhizal formation by the operculate discomycete *Sphaerosporella brunnea* (Pezizales). Mycologia **76**:454–461.

Danin, A. 1999. Sandstone outcrops—A major refugium of Mediterranean flora in the xeric part of Jordan. Israel Journal of Plant Sciences **47**:179–187.

Daubenmire, R.F. 1959. Canopy cover method of vegetation analysis. Northwest Science **33**:43–64.

Daugherty, C.H., and A.L. Sheldon. 1982. Age-specific movement patterns of the tailed frog, *Ascaphus truei*. Herpetologica **38**:468–74.

Day, T.A., and R.G. Wright. 1989. Positive plant spatial association with *Eriogonum ovalifolium* in primary succession on cinder cones: Seed trapping nurse plants. Vegetatio **80**:37–45.

De Nobili, M., M. Contin, C. Mondini, and P.C. Brookes. 2001. Soil microbial biomass is triggered into activity by trace amounts of substrate. Soil Biology and Biochemistry **33**:1163–1170.

Decae, A.E. 1987. Dispersal: Ballooning and other mechanisms. Pages 348–356 *in* W. Nentwig, editor. Ecophysiology of Spiders.Springer-Verlag, Berlin, Germany.

del Moral, R. 1983. Initial recovery of subalpine vegetation on Mount St. Helens. American Midland Naturalist **109**:72–80.

del Moral, R., and C.A. Clampitt. 1985. Growth of native plant species on recent volcanic substrates from Mount St. Helens. American Midland Naturalist **114**:374–383.

del Moral, R., and D.M. Wood. 1988a. Dynamics of herbaceous vegetation recovery on Mount St. Helens, Washington, USA, after a volcanic eruption. Vegetatio **74**:11–27.

del Moral, R., and D.M. Wood. 1988b. The high elevation flora of Mount St. Helens. Madrona **35**:309–319.

del Moral, R. 1993. Mechanisms of primary succession on volcanoes: A view from Mount St. Helens. Pages 79–100 *in* J. Miles and D.H. Walton, editors. Primary Succession on Land. Blackwell Scientific, London, UK.

del Moral, R., and L.C. Bliss. 1993. Mechanisms of primary succession: Insights resulting from the eruption of Mount St. Helens. Pages 1–66 *in* M. Began and A. Fitter, editors. Advances in Ecological Research. Volume 24. Academic Press, London, UK.

del Moral, R., and D.M. Wood. 1993a. Early primary succession on a barren volcanic plain at Mount St. Helens, Washington. American Journal of Botany **80**:981–991.

del Moral, R., and D.M. Wood. 1993b. Early primary succession on the volcano Mount St. Helens. Journal of Vegetation Science **4**:223–234.

del Moral, R., J.H. Titus, and A.M. Cook. 1995. Early primary succession on Mount St. Helens, Washington, USA. Journal of Vegetation Science **6**:107–120.

del Moral, R. 1998. Early succession on lahars spawned by Mount St Helens. American Journal of Botany **85**:820–828.

del Moral, R., and S.Y. Grishin. 1999. Volcanic disturbances and ecosystem recovery. Pages 137–160 *in* L.R. Walker, editor.

Ecosystems of Disturbed Ground. Ecosystems of the World 16. Elsevier, New York, New York, USA.

del Moral, R. 1999a. Plant succession on pumice at Mount St. Helens, Washington. American Midland Naturalist **141**:101–114.

del Moral, R. 1999b. Predictability of primary successional wetlands on pumice, Mount St. Helens. Madroño **46**:177–186.

del Moral, R., and C. Jones. 2002. Vegetation development on pumice at Mount St. Helens, USA. Plant Ecology **162**:9–22.

del Moral, R. 2000a. Local species turnover on Mount St. Helens. Pages 195–197 *in* P. White, editor. Proceedings of the 41st Symposium of the International Association for Vegetation Science. Opulus Press, Uppsala, Sweden.

del Moral, R. 2000b. Succession and species turnover on Mount St. Helens, Washington. Acta Phytogeographica Suecica **85**:53–62.

Delong, D.M. 1966. Insects. Pages 97–120 *in* A. Mirsky, editor. Soil Development and Ecological Succession in a Deglaciated Area of Muir Inlet, Southeast Alaska. Report 20. Ohio State University Institute of Polar Studies, Columbus, Ohio, USA.

deMaynadier, P.G., and M.L. Hunter, Jr. 1995. The relationship between forest management and amphibian ecology: A review of the North American literature. Environmental Review **3**:230–261.

den Boer, P.J. 1985. Fluctuations of density and survival of carabid populations. Oecologia (Berlin) **67**:322–330.

Dethier, D.P., D. Frank, and D.R. Pevear. 1980. Chemistry of Thermal waters and Mineralogy of the New Deposits at Mount St. Helens—A Preliminary Report. Open-File Report 81-80. U.S. Geological Survey, Vancouver, Washington, USA.

Dickson, B.A., and R.L. Crocker. 1953. A chronosequence of soils and vegetation near Mt. Shasta, California. II. The development of the forest floor and the carbon and nitrogen profiles of the soils. Journal of Soil Science **4**:142–156.

Diffendorfer, J.E., R.D. Holt, N.A. Slade, and M.S. Gaines. 1996. Small-mammal community patterns in old fields: Distinguishing site-specific from regional processes. Pages 421–466 *in* M.L. Cody and J.A. Smallwood, editors. Long-Term Studies of Vertebrate Communities. Academic Press, San Diego, California, USA.

Dinehart, R.L., J.R. Ritter, and J.M. Knott. 1981. Sediment data for streams near Mount St. Helens, Washington. Volume 100. 1980 Water Year Data. Open-File Report 81-822. U.S. Geological Survey, Vancouver, Washington, USA.

Dion, N.P., and S.S. Embrey. 1981. Effects of Mount St. Helens Eruption on Selected Lakes in Washington. Circular 850-G. U.S. Geological Survey, Reston, Virginia, USA.

Dlugosch, K., and R. del Moral. 1999. Vegetational heterogeneity along environmental gradients. Northwest Science **43**:12–18.

Dobran, F., A. Neri, and M. Todesco. 1994. Assessing the pyroclastic flow hazard of Vesuvius. Nature (London) **367**:551–554.

Dodds, W.K., and J.C. Priscu. 1990. A comparison of methods for the assessment of nutrient deficiency in phytoplankton in a large oligotrophic lake. Canadian Journal of Fisheries and Aquatic Sciences **47**:2328–2338.

Dodds, W.K., K. Gido, M.R. Whiles, K.M. Fritz, and W.J. Matthews. 2004. Life on the edge: The ecology of great plains prairie streams. BioScience **54**:205–216.

Dodson, S. 1992. Predicting crustacean zooplankton species richness. Limnology and Oceanography **37**:848–856.

Donohue, K., D.R. Foster, and G. Motzkin. 2000. Effects of the past and the present on species distributions: Land-use history and demography of wintergreen. Journal of Ecology **88**:303–316.

Doran, J.W., and T.B. Parkin. 1996. Quantitative indicators of soil quality: A minimum data set. Pages 25–37 *in* J.W. Doran and A.J. Jones, editors. Methods for Assessing Soil Quality. Soil Science Society of America, Madison, Wisconsin, USA.

Doyle, A.T. 1990. Use of riparian and upland habitats by small mammals. Journal of Mammalogy **71**:14–23.

Drake, D. 1991. Communities as assembled structures: Do rules govern pattern? Trends in Ecology and Evolution **5**:159–164.

Drake, V.A., and A.G. Gatehouse. 1995. Insect Migration: Tracking Resources Through Space and Time. Cambridge University Press, Cambridge, UK.

Drijber, R.A., J.W. Doran, A.M. Parkhurst, and D.J. Lyond. 2000. Changes in soil microbial community structure with tillage under long-term wheat-fallow management. Soil Biology and Biochemistry **32**:1419–1430.

Drury, W.H., and I.C.T. Nisbet. 1973. Succession. Journal of the Arnold Arboretum, Harvard University **54**:331–368.

Duellman, W.E., and L. Trueb. 1986. Biology of Amphibians. The Johns Hopkins University Press, Baltimore, Maryland, USA.

Dufrene, M., and P. Legendre. 1997. Species assemblages and indicator species: The need for a flexible asymmetrical approach. Ecological Monographs **67**:345–366.

Dunne, T., and L.B. Leopold. 1981. Flood and Sedimentation Hazards in the Toutle and Cowlitz Valleys as a Result of the Mount St. Helens Eruption. Federal Emergency Management Agency, Seattle, Washington, USA.

Dupuis, L.A., J.N. Smith, and F. Bunnell. 1995. Relation of terrestrial-breeding amphibian abundance to tree-stand age. Conservation Biology **9**:645–653.

Easterling, D.R., T.R. Karl, E.H. Mason, P.Y. Hughes, D.P. Bowman, R.C. Daniels, and T.A. Boden, editors. 1996. United States Historical Climatology Network (U.S. HCN) Monthly Temperature and Precipitation Data. ORNL/CDIAC-87, NDP-019/R3. Carbon Dioxide Information Analysis Center, Oak Ridge National Laboratory, Oak Ridge, Tennessee, USA.

Eastman, J.R., and M. Fulk. 1993. Long sequence time series evaluation using standardized principal components. Photogammetric Engineering & Remote Sensing **59**:1307–1312.

Edmonds, R.L., and H.E. Erickson. 1994. Influence of Mount St. Helens ash on litter decomposition. I. Pacific silver fir needle decomposition in the ash-fall zone. Canadian Journal of Forest Research **24**:826–831.

Edmondson, W.T., and G.G. Winberg, editors. 1971. IBP Handbook No. 17: A Manual of Methods for the Assessment of Secondary Productivity in Fresh Waters. Blackwell Scientific, Oxford, UK.

Edwards, J.S., and L.M. Schwartz. 1980. Mount St. Helens ash: A natural insecticide. Canadian Journal of Zoology **59**:714–715.

Edwards, J.S., R.L. Crawford, R.M. Sugg, and M. Peterson. 1986. Arthropod colonization in the blast zone of Mount St. Helens: Five years of progress. Pages 329–333 *in* S.A.C. Keller, editor. Mount St. Helens: Five Years Later. Eastern Washington University Press, Cheney, Washington, USA.

Edwards, J.S. 1986a. Arthropods as pioneers: Recolonization of the blast zone on Mount St. Helens. Northwest Environmental Journal **2**:263–273.

Edwards, J.S. 1986b. Derelicts of dispersal: Arthropod fallout on Pacific Northwest volcanoes. Pages 196–203 *in* W. Danthanarayana, editor. Insect Flight: Dispersal and Migration. Springer-Verlag, New York, New York, USA.

Edwards, J.S. 1987. Arthropods of aeolian ecosystems. Annual Review of Entomology **32**:163–179.

Edwards, J.S., and P. Sugg. 1993. Arthropod fallout as a resource in the recolonization of Mount St. Helens. Ecology **74**:954–958.

Edwards, J.S. 2005. Animals and volcanoes: Survival and revival. Pages 250–272 *in* J. Marti and G.G.J. Ernst, editors. Volcanoes and the Environment, Cambridge University Press, Cambridge, UK.

Egerton-Warburton, L.M., and M.F. Allen. 2001. Endo- and ectomycorrhizae in *Quercus agrifolia* Nee. (Fagaceae): Patterns of root colonization and effects on seedling growth. Mycorrhiza **11**:283–290.

Eggler, W.A. 1941. Primary succession on volcanic deposits in southern Idaho. Ecological Monographs **11**:277–298.

Eggler, W.A. 1948. Plant communities in the vicinity of the volcano El Paracutin, Mexico, after two and a half years of eruption. Ecology **29**:415–436.

Eggler, W.A. 1959. Manner of invasion of volcanic deposits by plants with further evidence from Paracutin and Jurullo. Ecological Monographs **29**:267–284.

Eggler, W.A. 1963. Plant life of Paracutin volcano, Mexico, eight years after activity ceased. American Midland Naturalist **69**:38–68.

Eggler, W.A. 1971. Quantitative studies of vegetation on sixteen young lava flows on the island of Hawaii. Tropical Ecology **12**:66–100.

Egler, F.E. 1954. Vegetation science concepts. I. Initial floristic composition. A factor in old-field vegetation development. Vegetatio **4**:412–417.

Eglitis, A. 1984. The spruce beetle in Glacier Bay National Park and Preserve. Pages 30–31 *in* J.D. Wood, I. Gladziszewski, A. Worley, and C. Veqvist, editors. Proceedings of the 1st Glacier Bay Science Symposium. National Park Service, Atlanta, Georgia, USA.

Eicher, G.J., and G.A. Rounsefell. 1957. Effects of lake fertilization by volcanic activity on abundance of salmon. Limnology and Oceanography **2**:70–76.

Elser, J.J., E.R. Marzolf, and C.R. Goldman. 1990. Phosphorus and nitrogen limitation of phytoplankton growth in the freshwaters of North America: A review and critique of experimental enrichments. Canadian Journal of Fisheries and Aquatic Science **47**:1468–1477.

Elser, J.J., W.F. Fagan, R.F. Denno, D.R. Dobberfuhl, A. Folarin, A. Huberty, S. Interlandi, S.S. Kilham, E. McCauley, K.L. Schulz, E.H. Siemann, and R.W. Sterner. 2000. Nutritional constraints on terrestrial and freshwater foodwebs. Nature (London) **408**:578–580.

Embrey, S.S., and N.P. Dion. 1988. Effects of the 1980 Eruption of Mount St. Helens on the Limnological Characteristics of Selected Lakes in Western Washington. Report 87-4263. U.S. Geological Survey, Vancouver, Washington, USA.

Endo, E.T., S.D. Malone, L.L. Noson, and C.S. Weaver. 1981. Locations, magnitudes, and statistics of the March 20–May 18 earthquake sequence. Pages 93–107 *in* P.W. Lipman and D.R. Mullineaux, editors. The 1980 Eruptions of Mount St. Helens, Washington. Professional Paper 1250. U.S. Geological Survey, Washington, District of Columbia, USA.

Engle, M.S. 1983. Carbon, Nitrogen and Microbial Colonization of Volcanic Debris on Mt. St. Helens. Master's thesis. Washington State University, Pullman, Washington, USA.

Eriksson, O., and A. Eriksson. 1998. Effects of arrival order and seed size on germination of grassland plants: Are there assembly rules during recruitment? Ecological Research **13**:229–239.

Faegri, K. 1986. Plant succession at Nigardsbreen. Pages 99–103 *in* H.J.B. Birks, editor. The Cultural Landscape: Past, Present and Future, Excursion Guide. Botanical Institute, University of Bergen, Bergen, Norway.

Fagan, W.F., and J.G. Bishop. 2000. Trophic interaction during primary succession: Herbivores slow a plant reinvasion at Mount St. Helens. American Naturalist **155**:238–251.

Fagan, W.F., J.G. Bishop, and J.D. Shade. 2004. Spatially structured herbivory and primary succession at Mount St. Helens: Field surveys and experimental growth studies suggest a role for nutrients. Ecological Entomology **29**:398–409.

Fairchild, L.H., and M. Wigmosta. 1983. Dynamic and volumetric characteristics of the 18 May 1980 lahars on the Toutle River, Washington. Pages 131–153 *in* Proceedings of the Symposium on Erosion Control in Volcanic Areas, July 6–9, 1982, Seattle and Vancouver, Washington. Technical Memorandum No. 1908. Public Works Research Institute, Tsukuba, Japan.

Fairchild, L.H. 1985. Lahars at Mount St. Helens, Washington. Master's thesis. University of Washington, Seattle, Washington, USA.

Fairchild, L.H. 1987. The importance of lahar initiation processes. Pages 51–62 *in* J.E. Costa and G.F. Wieczorek, editors. Debris Flows/Avalanches: Process, Recognition, and Mitigation. Reviews in Engineering Geology. Geological Society of America, Boulder, Colorado, USA.

Farrell, F.M. 1991. Models and mechanisms of succession: An example from a rocky intertidal community. Ecological Monographs **61**:95–113.

Feder, M.E. 1983. Integrating the ecology and physiology of plethodontid salamanders. Herpetologica **39**:291–310.

Fiksdal, A.J. 1981. Infiltration rates of undisturbed and disturbed Mount St. Helens ash. Pages 34–35 *in* Abstracts of the 24th Annual Meeting, Association of Engineering Geologists, Denver, Colorado, USA.

Findlay, R.H. 1996. The use of phospholipid fatty acids to determine microbial community structure. Pages 4.1.4/1–17 *in* A.D.L. Akkermans, J.D. van Elsas, and F.J. de Bruijn, editors. Molecular Microbial Ecology Manual. Kluwer Academic, Dordrecht, The Netherlands.

Fink, J.H., M.C. Malin, R.E. D'Alli, and R. Greeley. 1981. Rheological properties of mudflows associated with the spring 1980 eruptions of Mount St. Helens Volcano, Washington. Geophysical Research Letters **8**:43–46.

Fisher, R.V., and H.-U. Schmincke. 1984. Pyroclastic Rocks. Springer-Verlag, New York, New York, USA.

Fitter, A.H. 1986. Effect of benomyl on leaf phosphorus concentration in alpine grasslands: A test of mycorrhizal benefit. New Phytologist **103**:767–776.

Fosberg, R.F. 1959. Upper limits of vegetation on Mauna Loa, Hawaii. Ecology **40**:144–146.

Foster, D.R., D.H. Knight, and J.F. Franklin. 1998. Landscape patterns and legacies resulting from large, infrequent forest disturbances. Ecosystems **1**:497–510.

Fox, B.J. 1982. Fire and mammalian secondary succession in an Australian coastal heath. Ecology **63**:1332–1341.

Fox, B.J. 1990. Changes in the structure of mammal communities over successional time scales. Oikos **59**:321–329.

Fox, B.J. 1996. Long-term studies of small-mammal communities from disturbed habitats in eastern Australia. Pages 467–501 *in* M.L. Cody and J.A. Smallwood, editors. Long-Term Studies of Vertebrate Communities. Academic Press, San Diego, California, USA.

Fox, B.J., J.E. Taylor, and P.T. Thompson. 2003. Experimental manipulation of habitat structure: A retrogression of the small mammal succession. Journal of Animal Ecology **72**:927–940.

Foxworthy, B.L., and M. Hill. 1982. Volcanic Eruptions of 1980 at Mount St. Helens: The First 100 Days. Professional Paper 1249. U.S. Geological Survey, Washington, District of Columbia, USA.

Franklin, J.F. 1966. Vegetation and Soils in the Subalpine Forest of the Southern Washington Cascade Range. Doctoral dissertation. Washington State University, Pullman, Washington, USA.

Franklin, J.F. 1972. Cedar Flats Research Natural Area. Supplement to Federal Research Natural Areas in Oregon and Washington: A Guidebook for Scientists and Educators. USDA Forest Service, Pacific Northwest Forest and Range Experiment Station, Portland, Oregon, USA.

Franklin, J.F., and C.T. Dyrness. 1973. Natural vegetation of Oregon and Washington. General Technical Report PNW-8. Pacific Northwest Forest and Range Experiment Station, Portland, Oregon, USA. (Republished by Oregon State University Press in 1988.)

Franklin, J.F., and C. Wiberg. 1979. Goat Marsh Research Natural Area. Supplement No. 10 to Federal Research Natural Areas in Oregon and Washington: A Guidebook for Scientists and Educators (1972). USDA Forest Service, Pacific Northwest Forest and Range Experiment Station, Portland, Oregon, USA.

Franklin, J.F., and M.A. Hemstrom. 1981. Aspects of succession in the coniferous forests of the Pacific Northwest. Pages 212–229 *in* D.C. West, H.H. Shugart, and D.B. Botkin, editors. Forest Succession: Concepts and Application. Springer-Verlag, New York, New York, USA.

Franklin, J.F., J.A. MacMahon, F.J. Swanson, and J.R. Sedel. 1985. Ecosystem responses to the eruption of Mount St. Helens. National Geographic Research **1**:198–216.

Franklin, J.F., P.M. Frenzen, and F.J. Swanson. 1988. Re-creation of ecosystems at Mount St. Helens: Contrasts in artificial and natural approaches. Pages 1–37 *in* J. Cairnes, editor. Rehabilitating Damaged Ecosystems. Volume 2. CRC Press, Boca Raton, Florida, USA.

Franklin, J.F., and J.A. MacMahon. 2000. Messages from a Mountain. Science **288**:1183–1185.

Franklin, J.F., T.A. Spies, R. Van Pelt, A.B. Carey, D.A. Thornburgh, D.R. Berg, D.B. Lindenmayer, M.E. Harmon, W.S. Keeton, D.C. Shaw, K. Bible, and J.Q. Chen. 2002. Disturbances and structural development of natural forest ecosystems with silvicultural implications, using Douglas-fir forests as an example. Forest Ecology and Management **155**:399–423.

Franz, E.H. 1986. A dynamic model of the volcanic landscape: Rationale and relationships to the theory of disturbance. Pages 143–146 *in* S.A.C. Keller, editor. Mount St. Helens: Five Years Later. Eastern Washington University Press, Cheney, Washington, USA.

Frehner, H.F. 1957. Development of Soil and Vegetation on Kautz Creek Flood Deposit in Mount Rainier National Park. Master's thesis. University of Washington, Seattle, Washington, USA.

Frenzen, P.M., and J.F. Franklin. 1985. Establishment of conifers from seed on tephra deposited by the 1980 eruptions of Mount St. Helens, Washington. American Midland Naturalist **114**:84–97.

Frenzen, P.M., M.E. Krasney, and L.P. Rigney. 1988. Thirty-three years of plant succession on the Kautz Creek mudflow, Mount Rainier National Park, Washington. Canadian Journal of Botany **66**:130–137.

Frenzen, P.M., and C.M. Crisafulli. 1990. Mount St. Helens 10 years later: Past lessons and future promise. Northwest Science **64**:263–267.

Frenzen, P.M. 1992. Mount St. Helens: A laboratory for research and education. Journal of Forestry **90**:14–18, 37.

Frenzen, P.M., A.M. Delano, and C. M. Crisafulli. 1994. Mount St. Helens: Biological Responses Following the 1980 Eruptions—An Indexed Bibliography and Research Abstracts (1980–93). PNW-GTR-342. USDA Forest Service, Pacific Northwest Research Station, Portland, Oregon, USA.

Fridriksson, S. 1987. Plant colonization of a volcanic island, Surtsey, Iceland. Arctic and Alpine Research **19**:425–431.

Fridriksson, S., and B. Magnusson. 1992. Development of the ecosystem on Surtsey with reference to Anak Krakatau. GeoJournal **28**:287–291.

Frohne, P., J.J. Halvorson, M. Ibekwe, and A.C. Kennedy. 1998. Use of FAME analysis to characterize microbial communities found in Mount St. Helens pyroclastic substrate. Poster presented October 1998 at the 90th Annual Meeting of the American Society of Agronomy, Crop Science Society of America, and Soil Science Society of America, Baltimore, Maryland, USA.

Frohne, P., J.J. Halvorson, and A.C. Kennedy. 2001. Microbial community structure and function in Mount St. Helens pyroclastic material 20 years after the eruption. Poster presented August 2001 at the 86th Annual Meeting of the Ecological Society of America, Madison, Wisconsin, USA.

Fruchter, J.S., D.E. Robertson, J.C. Evans, K.B. Olsen, E.A. Lepel, J.C. Laul, K.H. Abel, R.W. Sanders, P.O. Jackson, N.S. Wogman, R.W. Perkins, H.H. Van Tuyl, R.H. Beauchamp, J.W. Shade, J.L. Daniel, R.L. Erickson, G.A. Sehmel, R.N. Lee, A.V. Robinson, O.R. Moss, J.K. Briant, and W.C. Cannon. 1980. Mount St. Helens ash from the 18 May 1980 eruption: Chemical, physical, mineralogical and biological properties. Science **209**:1116–1125.

Fuller, R.N., and R. del Moral. 2003. The role of refugia and dispersal in primary succession on Mount St. Helens, Washington. Journal of Vegetation Science **14**:637–644.

Garwood, N.C., D.P. Janos, and N. Brokaw. 1979. Earthquake-caused landslides: A major disturbance to tropical forests. Science **205**:997–999.

Gentry, J.B., L.A. Briese, D.W. Kaufman, M.H. Smith, and J.G. Wiener. 1975. Elemental flow and standing crops for small mammal populations. Pages 205–221 *in* F.B. Golley, K. Petrusewicz, and L. Ryszkowski, editors. Small Mammals: Their Productivity and Population Dynamics. Cambridge University Press, Cambridge, UK.

Gerdemann, J.W., and J.M. Trappe. 1974. The Endogonaceae in the Pacific Northwest. Mycologia Memoir **5**:1–76.

Gignoux, J., J. Clobert, and J.C. Menaut. 1997. Alternative fire resistance strategies in savanna trees. Oecologia (Berlin) **110**:576–583.

Gitzen, R.A., and S.D. West. 2002. Small mammal response to experimental gaps in the southern Washington Cascades. Forest Ecology and Management **168**:187–199.

Gleason, H.A. 1917. The structure and development of the plant association. Bulletin of the Torrey Botanical Club **43**:463–481.

Gleason, H.A. 1926. The individualistic concept of the plant association. Bulletin of the Torrey Botanical Club **53**:1–20.

Gleason, H.A. 1939. The individualistic concept of the plant association. American Midland Naturalist **21**:92–110.

Glenn-Lewin, D.C. 1980. The individualistic nature of plant community development. Vegetatio **43**:141–146.

Glenn-Lewin, D.C., R.K. Peet, and T.T. Veblin. 1992. Plant Succession Theory and Prediction. Chapman & Hall, London, UK.

Glicken, H., W. Meyer, and M. Sabol. 1989. Geology and Ground-Water Hydrology of Spirit Lake Blockage, Mount St. Helens, Washington, with Implications for Lake Retention. Bulletin 1789. U.S. Geological Survey, Washington, District of Columbia, and Denver, Colorado, USA.

Glicken, H. 1998. Rockslide-debris avalanche of May 18, 1980, Mount St. Helens volcano, Washington. Bulletin of the Geological Survey of Japan **49**(2-3):55–106.

Golley, F.B. 1961. Energy values of ecological materials. Ecology **42**:581–584.

Goovaerts, P. 1998. Geostatistical tools for characterizing the spatial variability of microbiological and physico-chemical soil properties. Biology and Fertility of Soils **27**:315–334.

Gorman, M. 1979. Island Ecology. Chapman & Hall, London, UK.

Gosz, J.R., G.E. Likens, and F.H. Bormann. 1973. Nutrient release from decomposing leaf and branch litter in the Hubbard Brook Forest, New Hampshire. Ecological Monographs **47**:173–191.

Grant, G.E., and A.L. Wolff. 1991. Long-term patterns of sediment transport after timber harvest, western Cascades Mountains, Oregon, USA. Pages 31–40 *in* N.E. Peters and D.E. Walling, editors. Sediment and Stream Water Quality in a Changing Environment: Trends and Explanations. International Association of Hydrological Sciences Press, Wallingford, Oxfordshire, UK.

Green, D.M. 1997. Perspectives on amphibian population declines: Defining the problem and searching for answers. Pages 291–308 *in* D.M. Green, editor. Herpetological Conservation. Volume 1. Amphibians in Decline. Canadian Studies of a Global Problem. Society for Study of Amphibians and Reptiles, St. Louis, Missouri, USA.

Gregorich, E.G., M.R. Carter, D.A. Angers, C.M. Monreal, and B.H. Ellert. 1994. Towards a minimum data set to assess soil organic matter quality in agricultural soils. Canadian Journal of Soil Science **74**:367–385.

Griggs, R.F. 1917. The Valley of Ten Thousand Smokes. The National Geographic Magazine **31**:13–68.

Griggs, R.F. 1918a. The recovery of vegetation at Kodiak. Ohio Journal of Science **19**:1–57.

Griggs, R.F. 1918b. The great hot mudflow of the Valley of 10,000 Smokes. Ohio Journal of Science **19**:117–142.

Griggs, R.F. 1918c. The Valley of Ten Thousands Smokes: An account of the discovery and exploration of the most wonderful volcanic region in the world. The National Geographic Magazine **33**:10–68.

Griggs, R.F. 1919. The beginnings of revegetation in Katmai Valley. Ohio Journal of Science **19**:318–342.

Griggs, R.F. 1922. The Valley of Ten Thousand Smokes. National Geographic Society, Washington, District of Columbia, USA.

Griggs, R.F. 1933. The colonization of Katmai ash, a new and "inorganic" soil. American Journal of Botany **20**:92–113.

Grime, J.P. 1977. Evidence for the existence of three primary strategies in plants and its relevance to ecological and evolutionary theory. American Naturalist **111**:1169–1194.

Grime, J.P. 1979. Plant Strategies and Vegetation Processes. John Wiley & Sons, New York, New York, USA.

Grishin, S.Y. 1994. Role of *Pinus pumila* in primary succession on the lava flows of volcanoes of Kamchatka. Pages 240–250 *in* W.C. Schmidt and F.K. Holtmeier, editors. Proceedings: International Workshop on Subalpine Stone Pines and Their Environment: The Status of Our Knowledge. INT-GTR-309. USDA Forest Service, Intermountain Research Station, Ogden, Utah, USA.

Grishin, S.Y., R. del Moral, P. Krestov, and V.P. Verkholat. 1996. Succession following the catastrophic eruption of Ksudach volcano (Kamchatka, 1907). Vegetatio **127**:129–153.

Grodzinski, W., and B.A. Wunder. 1975. Ecological energetics of small mammals. Pages 173–204 *in* F.B. Golley, K. Petrusewicz, and L. Ryszkowski, editors. Small Mammals: Their Productivity and Population Dynamics. Cambridge University Press, Cambridge, UK.

Gross, K.L., K.S. Pregitzer, and A. J. Burton. 1995. Changing scales of nitrogen availability in early and late successional plant communities. Journal of Ecology **83**:357–367.

Gu, L.H., D.D. Baldocchi, S.C. Wofsy, J.W. Munger, J.J. Michalsky, S.P. Urbanski, and T.A. Boden. 2003. Response of a deciduous forest to the Mount Pinatubo eruption: Enhanced photosynthesis. Science **299**:2035–2038.

Haagen, E. 1990. Soil Survey of Skamania County Area, Washington. USDA Soil Conservation Service, Spokane, Washington, USA. Available online at www.fs.fed.us/gpnf/forest-research/ei/workshop/docs/skamania-soil-tephra.doc.

Haberle, S.G., J.M. Szeicz, and K.D. Bennett. 2000. Late Holocene vegetation dynamics and lake geochemistry at Laguna Miranda, XI Region, Chile. Revista Chilena de Historia Natural **73**:655–669.

Halpern, C.B., and M.E. Harmon. 1983. Early plant succession on the Muddy River mudflow, Mount St. Helens, Washington. American Midland Naturalist **110**:97–106.

Halpern, C.B., P.M. Frenzen, J.E. Means, and J.F. Franklin. 1990. Plant succession in areas of scorched and blown-down forest after the 1980 eruption of Mount St. Helens, Washington. Journal of Vegetation Science **1**:181–194.

Halvorson, J.J. 1989. Carbon and Nitrogen Contributions to Mount St. Helens Volcanic Sites by Lupines. Doctoral dissertation. Washington State University, Pullman, Washington, USA.

Halvorson, J.J., R.A. Black, J.L. Smith, and E.H. Franz. 1991a. Nitrogenase activity, growth, and carbon and nitrogen allocation in wintergreen and deciduous lupine seedlings. Functional Ecology **5**:554–561.

Halvorson, J.J., J.L. Smith, and E.H. Franz. 1991b. Lupine influence on soil carbon, nitrogen and microbial activity in developing ecosystems at Mount St. Helens. Oecologia (Berlin) **87**:162–170.

Halvorson, J.J., E.H. Franz, J.L. Smith, and R.A. Black. 1992. Nitrogenase activity, nitrogen fixation, and nitrogen inputs by lupines at Mount St. Helens. Ecology **73**:87–98.

Halvorson, J.J., H. Bolton, Jr., J.L. Smith, and R.E. Rossi. 1994. Measuring resource islands using geostatistics. Great Basins Naturalist **54**:313–328.

Halvorson, J.J., and J.L. Smith. 1995. Decomposition of lupine biomass by soil microorganisms in developing Mount St. Helens pyroclastic soils. Soil Biology & Biochemistry **27**:983–992.

Halvorson, J.J., J.L. Smith, H. Bolton, Jr., and R.E. Rossi. 1995a. Defining resource islands using multiple variables and geostatistics. Soil Science Society of America Journal **59**:1476–1487.

Halvorson, J.J., J.L. Smith, and R.I. Papendick. 1995b. Integration of multiple parameters to evaluate soil quality: A field example. Biology and Fertility of Soils **21**:2207–2214.

Halvorson, J.J., J.L. Smith, and R.I. Papendick. 1997. Issues of scale for evaluating soil quality. Journal of Soil and Water Conservation **52**:26–30.

Halvorson, J.J., and J.L. Smith. 2001. Increasing carbon and nitrogen in Mount St. Helens pyroclastic soil: Measuring soil quality and function after 20 years. Poster presented August 2001 at the 86th Annual Meeting of the Ecological Society of America, Madison, Wisconsin, USA.

Hanley, T.A., and R.D. Taber. 1980. Selective plant species inhibition by elk and deer in three conifer communities in western Washington. Forest Science **26**:97–107.

Hanley, T.A. 1984. Habitat patches and their selection by wapiti and black-tailed deer in a coastal montane coniferous forest. Journal of Applied Ecology **21**:423–436.

Hansen, E.M., and K.J. Lewis. 1997. Compendium of Conifer Diseases. American Phytopathological Society (APS) Press, St. Paul, Minnesota, USA.

Hanski, I. 1999. Metapopulation Ecology. Oxford University Press, Oxford, UK.

Hanson, H.C. 1962. Dictionary of Ecology. Philosophical Library, New York, New York, USA. Second edition: M. Allaby, editor. 1998. A Dictionary of Ecology. Oxford University Press, New York, New York, USA.

Hansson, L. 1983. Competition between rodents in successional stages in taiga forests: *Microtus agrestis* vs. *Clethrionomys glareolus*. Oikos **40**:258–266.

Hardison, J.H. 2000. Post-Lahar Channel Adjustment, Muddy River, Mount St. Helens, Washington. Master's thesis. Colorado State University, Colorado, USA.

Harley, J.L., and E.L. Harley. 1987. A check-list of mycorrhiza in the British flora. New Phytologist (Suppl.) **105**:1–102.

Harper, J.L. 1977. Population Biology of Plants. Academic Press, New York, New York, USA.

Harr, R.D. 1981. Some characteristics and consequences of snowmelt during rainfall in western Oregon. Journal of Hydrology **53**:277–304.

Harris, E., R.N. Mack, and M.S.B. Ku. 1987. Death of steppe cryptogams under the ash from Mount St. Helens. American Journal of Botany **74**:1249–1253.

Harris, S.L. 1988. Fire Mountains of the West: The Cascade and Mono Lake Volcanoes. Mountain Press, Missoula, Montana, USA.

Haruki, M., and S. Tsuyuzaki. 2001. Woody plant establishment during the early stages of volcanic succession on Mount Usu, northern Japan. Ecological Research **16**:451–457.

Hatch, M.H. 1965. The Coleoptera of the Pacific Northwest. Part iv. University of Washington Press, Seattle, Washington, USA.

Hättenschwiler, S., and P.M. Vitousek. 2000. The role of polyphenols in terrestrial ecosystem nutrient cycling. Trends in Ecology and Evolution **15**:238–243.

Hauch, R.D., and R.W. Weaver, editors. 1986. Field Measurement of Dinitrogen Fixation and Denitrification. Soil Science Society of America, Madison, Wisconsin, USA.

Hawkins, C.P., M.L. Murphy, N.H. Anderson, and M.A. Wilzbach. 1983. Density of fish and salamanders in relation to riparian canopy and physical habitat in streams of the northwestern United States. Canadian Journal of Fisheries and Aquatic Sciences **40**:1173–1185.

Hawkins, C.P, L.J. Gottschalk, and S.S. Brown. 1988. Densities and habitat of tailed frog tadpoles in small streams near mount St. Helens following the 1980 eruption. Journal of the North American Benthological Society **7**:246–252.

Hawkins, C.P., and J.R. Sedell. 1990. The role of refugia in the recolonization of streams devastated by the 1980 eruption of Mount St. Helens. Northwest Science **64**:271–274.

Heal, O.W., J.M. Anderson, and M.J. Swift. 1997. Plant litter quality and decomposition: An historical overview. Pages 3–30 *in* G. Cadish and K.E. Giller, editors. Driven by Nature: Plant Litter Quality and Decomposition. CAB International, Oxford, UK.

Heath, J.P. 1967. Primary conifer succession, Lassen Volcanic National Park. Ecology **48**:270–275.

Heiniger, P.H. 1989. Arthropoden auf Schneefelden und in schneefrieen Habitaten im Jungfraugebiet (Berner Oberland, Schweiz). Mitteilungen der Schweizerischen Entomologischen Gesellschaft **62**:375–386.

Hemstrom, M.A., and J.F. Franklin. 1982. Fire and other disturbances of the forests of Mount Rainier National Park. Quaternary Research **18**:32–51.

Hemstrom, M.A., and W.H. Emmingham. 1987. Vegetation changes induced by Mount St. Helens eruptions on previously established forest plots. Pages 188–209 *in* D.E. Bilderback, editor. Mount St. Helens 1980: Botanical Consequences of the Explosive Eruptions. University of California Press, Berkeley, California, USA.

Hendrix, L.B. 1981. Post-eruption succession on Isla Fernandina, Galapagos. Madroño **28**:242–254.

Herbeck, L.A., and D.R. Larsen. 1999. Plethodontid salamander response to silvicultural practices in Missouri Ozark forests. Conservation Biology **13**:623–632.

Herrick, J.E., and W.G. Whitford. 1995. Assessing the quality of rangeland soils: Challenges and opportunities. Journal of Soil and Water Conservation **50**:247–242.

Herrick, J.E., and M.M. Wander. 1997. Relationships between soil organic carbon and soil quality in cropped and rangeland soils: The importance of distribution, composition, and soil biological activity. Pages 405–425 *in* R. Lal, J.M. Kimble, R. Follett, and B.A. Stewart, editors. Soil Processes and the Carbon Cycle. CRC Press, Boca Raton, Florida, USA.

Herrick, J.E., J.R. Brown, A.J. Tugel, P.L. Shaver, and K.M. Havstada. 2002. Application of soil quality to monitoring and management: Paradigms from rangeland ecology. Agronomy Journal **94**: 3–11.

Hill, M., and H. Gauch. 1980. Detrended correspondence analysis, an improved ordination technique. Vegetatio **42**:47–58.

Hill, G.T., N.A. Mitkowskia, L. Aldrich-Wolfeb, L.R. Emelea, D.D. Jurkoniea, A. Fickea, S. Maldonado-Ramireza, S.T. Lyncha, and E.B. Nelsona. 2000. Methods for assessing the composition and diversity of soil microbial communities. Applied Soil Ecology **15**:25–36.

Hinckley, T.M., H. Imoto, K. Lee, S. Lacker, Y. Morikawa, K.A. Vogt, C.C. Grier, M.R. Keyes, R.O. Teskey, and V. Seymour. 1984. Impact of tephra deposition on growth in conifers: The year of the eruption. Canadian Journal of Forest Research **14**:731–739.

Hinkley, T.K., editor. 1987. Chemistry of Ash and Leachates from the May 18, 1980, Eruption of Mount St. Helens, Washington. Professional Paper 1397. U.S. Geological Survey, Reston, Virginia, USA.

Hirose, T., and M. Tateno. 1984. Soil nitrogen patterns induced by colonization of *Polygonu mcuspidatum* on Mt. Fuji. Oecologia (Berlin) **61**:218–223.

Hirth, H.F. 1959. Small mammals in old field succession. Ecology **40**:417–425.

Hixon, M.A., and W.N. Brostoff. 1996. Succession and herbivory: Effects of differential fish grazing on Hawaiian coral-reef algae. Ecological Monographs **66**:67–90.

Hobbs, N.T. 1996. Modification of ecosystems by ungulates. Journal of Wildlife Management **60**:695–713.

Hoblitt, R.P., C.D. Miller, and J.W. Vallance. 1981. Origin and stratigraphy of the deposit produced by the May 18 directed blast. Pages 401–419 *in* P.W. Lipman and D.R. Mullineaux, editors. The 1980 Eruptions of Mount St. Helens, Washington. Professional Paper 1250. U.S. Geological Survey, Washington, District of Columbia, USA.

Hocking, P.J. 2001. Organic acids exuded from roots in phosphorus uptake and aluminum tolerance of plants in acid soils. Advances in Agronomy **74**:63–97.

Hogan, K.M., C. Hedein, H.S. Koh, S.K. Davis, and I.F. Greenbaum. 1993. Systematic and taxonomic implications of karyoptic, electrophoretic, and mitochondrial-DNA variation in *Peromyscus* from the Pacific Northwest. Journal of Mammalogy **74**:819–831.

Holtzmann, H.P., and K. Haselwandter. 1988. Contributions of nitrogen fixation to nitrogen nutrition in an alpine community (*Caricetum curvulae*). Oecologia (Berlin) **76**:298–302.

Horn, E.M. 1968. Ecology of the pumice desert, Crater Lake National Park. Northwest Science **42**:141–149.

Horton, T.R., E. Cazares, and T.D. Bruns. 1998. Ectomycorrhizal, vesicular arbuscular and dark septate fungal colonization of bishop pine (*Pinus muricata*) seedlings in the first 5 months of growth after wildfire. Mycorrhiza **8**:11–18.

Howarth, F.G. 1979. Neogeoaeolian habitats on new lava flows on Hawaii Island: An ecosystem supported by windblown debris. Pacific Insects **20**:133–144.

Howarth, F.G. 1987. Evolutionary ecology of aeolian and subterranean habitats in Hawaii. Trends in Ecology and Evolution **2**:220–223.

Huebert, B., P. Vitousek, J. Sutton, T. Elias, J. Heath, S. Coeppicus, S. Howell, and B. Blomquist. 1999. Volcano fixes nitrogen into plant-available forms. Biogeochemistry **47**:111–118.

Hunter, M.D., and P.W. Price. 1992. Playing Chutes and Ladders: Heterogeneity and the relative roles of bottom-up and top-down forces in natural communities. Ecology **73**:724–732.

Hunter, M.D., and J.N. McNeil. 1997. Host-plant quality influences diapause and voltinism in a polyphagous insect herbivore. Ecology **78**:977–986.

Huntley, N., and R.S. Inouye. 1987. Small mammal populations of an old-field chronosequence: Successional patterns and associations with vegetation. Journal of Mammalogy **68**:739–745.

Huston, M.A. 1994. Biological Diversity: The Coexistence of Species on Changing Landscapes. Cambridge University Press, New York, New York, USA.

Hutchinson, G.E. 1957. A Treatise on Limnology. I. Geography, Physics, and Chemistry. John Wiley, New York, New York, USA.

Ibekwe, A.M., and A.C. Kennedy. 1999. Fatty acid methyl ester (FAME) profiles as a tool to investigate community structure. Plant and Soil **206**:151–161.

Inbar, M., J.L. Hubp, and L.V. Ruiz. 1994. The geomorphological evolution of the Paricutin cone and lava flows, Mexico, 1943–1990. Geomorphology **9**:57–76.

Inbar, M., H.A. Ostera, C.A. Parica, M.B. Pemesaland, and F.M. Salani. 1995. Environmental assessment of 1991 Hudson volcano eruption ashfall effects on southern Patagonia region, Argentina. Environmental Geology **25**:119–125.

Independent Scientific Advisory Board: R.E. Bilby, P.A. Bisson, C.C. Coutant, D. Goodman, R. Gramling, S. Hanna, E. Loudenslager, L. McDonald, D. Philipp, and B. Riddell. 2003. Review of Salmon and Steelhead Supplementation. ISAB Report 2003-3. Northwest Power Planning Council, Portland, Oregon, USA.

Ingles, L.G. 1965. Mammals of the Pacific States: California, Oregon, and Washington. Stanford University Press, Stanford, California, USA.

Irland, L.C. 1998. Ice storms and forest impacts. The Science of the Total Environment **262**:231–242.

Isaaks, E.H., and R.M. Srivastava. 1989. An Introduction to Applied Geostatistics. Oxford University Press, New York, New York, USA.

Jaccard, P. 1908. Nouvelles recherches sur la distribution florale. Bulletin de la Societé Vandoise de Science Naturelle **44**:223–270.

Jackson, M.T., and A. Faller. 1973. Structural analysis and dynamics of the plant communities of Wizard Island, Crater Lake National Park. Ecological Monographs **43**:441–461.

Jackson, R.B., and M.M. Caldwell. 1993. The scale of nutrient heterogeneity around individual plants and its quantification with geostatistics. Ecology **74**:612–614.

Janda, R.J., K.M. Scott, K.M. Nolan, and H.A. Martinson. 1981. Lahar movement, effects, and deposits. Pages 461–478 *in* P.W. Lipman and D.R. Mullineaux, editors. The 1980 Eruptions of Mount St. Helens, Washington. Professional Paper 1250. U.S. Geological Survey, Washington, District of Columbia, USA.

Janda, R.J., D.F. Meyer, and D. Childers. 1984. Sedimentation and geomorphic changes during and following the 1980–1983 eruptions of Mount St. Helens, Washington (in two parts). Shin-Sabo, Journal of the Erosion Control Engineering Society of Japan **37**(2):10–21, **37**(3):5–19.

Jenny, H. 1941. Factors of Soil Formation. McGraw-Hill, New York, New York, USA.

Jenny, H., J. Vlamis, and W.E. Martin. 1950. Greenhouse assay of fertility of California soils. Hilgardia **20**:1–8.

Jobbagy, E.G., and R.B. Jackson. 2000. The vertical distribution of soil organic carbon and its relation to climate and vegetation. Ecological Applications **10**:423–436.

Johnson, M.G., and R.L. Beschta. 1980. Logging, infiltration capacity, and surface erodibility in western Oregon. Journal of Forestry **78**:334–337.

Johnson, D.M., R.R. Petersen, D.R. Lycan, J.W. Sweet, M.E. Neuhaus, and A.L. Schaedel. 1985. Atlas of Oregon Lakes. Oregon State University Press, Corvallis, Oregon, USA.

Johnson, K.A. 1986. Effects of a Volcanic Disturbance on Small Mammals, with Special Reference to Deer Mouse (*Peromyscus maniculatus*) Populations. Doctoral dissertation. Utah State University, Logan, Utah, USA.

Johnson, N.D., B. Liu, and B.L. Bentley. 1987. The effects of nitrogen fixation, soil nitrate, and defoliation on the growth, alkaloids, and nitrogen levels of *Lupinus succulentus*. Oecologia (Berlin) **74**:425–431.

Johnson, N.D., and B.L. Bentley. 1988. Effects of dietary protein and lupine alkaloids of growth and survivorship of *Spodoptera eridania*. Journal of Chemical Ecology **14**:1391–1403.

Johnson, N.D., L.P. Rigney, and B.L. Bentley. 1989. Short-term induction of alkaloid production in lupines: Differences between N_2 fixing and nitrogen-limited plants. Journal of Chemical Ecology **15**:2425–2434.

Jones, J.K., Jr., R.S. Hoffman, D.W. Rice, C. Jones, R.J. Baker, and M.D. Engstrom. 1997. Revised Checklist of North American Mammals North of Mexico. Occasional Paper of the Museum 146. Texas Tech University, Lubbock, Texas, USA.

Jones, D.L. 1998. Organic acids in the rhizosphere: A critical review. Plant and Soil **205**:25–44.

Jones, L.C. 1999. Survey protocol for the Van Dyke's salamander (*Plethodon vandykei*). Pages 201–252 *in* D.H. Olson, editor. Standardized Survey Protocols for Amphibians Under the Survey and Manage and Protection Buffer Provisions of the Northwest Forest Plan. Version 3.0. USDA Forest Service, Portland, Oregon, USA.

Kamijo, T., K. Kitayama, A. Sugawara, S. Urushimichi, and K. Sasai. 2002. Primary succession of the warm-temperate broad-leaved forest on a volcanic island, Miyake-jima, Japan. Folia Geobotanica **37**:71–91.

Kareiva, P. 1994. Space: The final frontier for ecological theory. Ecology **75**:1.

Karlen, D.L., J.C. Gerdner, and M.J. Rosek. 1998. A soil quality framework for evaluating the impact of CRP. Journal of Production Agriculture **11**:56–60.

Karlstrom, E.L. 1986. Amphibian recovery in the north fork Toutle River debris avalanche area of Mount St. Helens. Pages 334–344 *in* S.A.C. Keller, editor. Mount St. Helens: Five Years Later. Eastern Washington State University Press, Cheney, Washington, USA.

Kaufmann, R. 2001. Invertebrate succession on an alpine glacier foreland. Ecology **82**:2261–2278.

Keddy, P.A. 1989. Competition. Chapman & Hall, London, UK.

Keith, T.E., T.J. Casadevall, and D.A. Johnston. 1981. Fumarole encrustations: Occurrence, mineralogy, and chemistry. Pages 239–250 *in* P.W. Lipman and D.R. Mullineaux, editors. The 1980 Eruptions of Mount St. Helens. Washington Professional Paper 1250. U.S. Geological Survey, Washington, District of Columbia, USA.

Keller, S.A.C., editor. 1982. Mount St. Helens: One Year Later. Eastern Washington University Press, Cheney, Washington, USA.

Keller, S.A.C., editor. 1986. Mount St. Helens: Five Years Later. Eastern Washington University Press, Cheney, Washington, USA.

Kelly, V.J. 1992. Limnology of Two New Lakes, Mount St. Helens, Washington. Master of Science thesis. Portland State University, Portland, Oregon, USA.

Kennedy, A.C. 1994. Carbon utilization and fatty acid profiles for characterization of bacteria. Pages 543–556 *in* R.W. Weaver, S. Angle, P. Bottomley, D. Bezdiecek, S. Smith, A. Tabatabai, and A. Wollum, editors. Methods of Soil Analysis. Part 2. Microbiological and Biochemical Properties. Soil Science Society of America, Madison, Wisconsin, USA.

Kennedy, A.C. 1997. The rhizosphere and spermosphere. Pages 389–407 *in* D.M. Sylvia, J.J. Furhman, P.G. Hartel, and D.A. Zuberer, editors. Principles and Applications of Soil Microbiology. Prentice Hall, Upper Saddle River, New Jersey, USA.

Kent, M., N.W. Owen, P. Dale, R.M. Newnham, and T.M. Giles. 2001. Studies of vegetation burial: A focus for biogeography and biogeomorphology. Progress in Physical Geography **25**:455–482.

Kerle, E.A. 1985. The Ecology of Lupines in Crater Lake National Park, Oregon. Master's thesis. Oregon State University, Corvallis, Oregon, USA.

Kershaw, K.A., and J.H. Looney. 1985. Quantitative and Dynamic Plant Ecology. Third edition. Arnold, London, UK.

Kielland, K., and J.P. Bryant. 1998. Moose herbivory in taiga: Effects on biogeochemistry and vegetation dynamics in primary succession. Oikos **82**:377–383.

Kirkland, G.L.J. 1977. The responses of small mammals to the clear cutting of northern Appalachian forests. Journal of Mammalogy **58**:600–609.

Kitayama, K., D. Mueller-Dombois, and P.M. Vitousek. 1995. Primary succession of Hawaiian montane rain forest on a chronosequence of eight lava flows. Journal of Vegetation Science **6**:211–222.

Klock, G.O. 1982. Stabilizing ash-covered timberlands with erosion control seeding and fertilization. Pages 164–190 *in* Proceedings from the Conference: Mt. St. Helens: Effects on Water Resources. Washington State University Press, Pullman, Washington, USA.

Knapp, R.A., K.R. Matthews, and O. Sarnelle. 2001. Resistance and resilience of alpine lake fauna to fish introductions. Ecological Monographs **71**:401–421.

Knops, J.M.H., D. Wedin, and D. Tilman. 2001. Biodiversity and decomposition in experimental grassland ecosystems. Oecologia (Berlin) **126**:429–433.

Koch, C. 1960. The tenebrionid beetles of South West Africa. Bulletin of the South African Museum Association **7**:73–85.

Kochy, M., and H. Rydin. 1997. Biogeography of vascular plants on habitat islands, peninsulas, and main lands in an east-central Swedish agricultural landscape. Nordic Journal of Botany **17**:215–223.

Korosec, M.A., J.G. Rigby, and K.L. Stoffel. 1980. The 1980 Eruption of Mount St. Helens, Washington. Part I: March 20–May 19, 1980. Information Circular No. 71. Washington Department of Natural Resources, Olympia, Washington, USA.

Koss, J.E., F.G. Ethridge, and S.A. Schumm. 1994. An experimental study of the effects of base-level change on fluvial, coastal plain, and shelf systems. Journal of Sediment Research B **64**:90–98.

Kovanen, D., D. Easterbrook, and P. Thomas. 2001. Holocene eruptive history of Mount Baker, Washington. Canadian Journal of Earth Sciences **38**:1355–1366.

Krefting, L.W., and C.E. Ahlgren. 1974. Small mammals and vegetation changes after fire in mixed conifer-hardwood forest. Ecology **55**:1391–1398.

Kroh, G.C., J.D. White, S.K. Heath, and J.E. Pinder, III. 2000. Colonization of a volcanic mudflow by an upper montane coniferous forest at Lassen Volcanic National Park, California. American Midland Naturalist **143**:126–140.

Kruckeberg, A.R. 1987. Plant life on Mount St. Helens before 1980. Pages 3–23 *in* D.E. Bilderback, editor. Mount St. Helens 1980: Botanical Consequences of the Explosive Eruptions. University of California Press, Berkeley, California, USA.

Kunin, W.E. 1999. Patterns of herbivore incidence on experimental arrays and field populations of ragwort, *Senecio jacobea*. Oikos **84**:515–525.

Kuntz, M.A, P.D. Rowley, N.S. MacLeod, R.L. Reynolds, L.A. McBroome, A.M. Kaplan, and D.J. Lidke. 1981. Petrography and particle-size distribution of pyroclastic-flow, ash-cloud, and surge deposits. Pages 525–539 *in* P.W. Lipman and D.R. Mullineaux, editors. The 1980 Eruptions of Mount St. Helens. Washington Professional Paper 1250. U.S. Geological Survey, Washington, District of Columbia, USA.

Kurenkov, I.I. 1966. The influence of volcanic ashfall on biological processes in a lake. Limnology and Oceanography **11**:426–429.

Kuwayama, S. 1929. Eruption of Mt. Komagatake and insects (in Japanese). Kontyu **3**:271–273.

Kuzyakov, Y., J.K. Friedel, and K. Stahr. 2000. Review of mechanisms and quantification of priming effects. Soil Biology and Biochemistry **32**:1485–1498.

Labus, P., R.L. Whitman, and M.B. Nevers. 1999. Picking up the pieces: Conserving remnant natural areas in the post-industrial landscape of the Calumet Region. Natural Area Journal **19**:180–187.

Lal, R., J.M. Kimble, R.F. Follett, and B.A. Stewart. 1998. Soil Processes and the Carbon Cycle. CRC Press, Boca Raton, Florida, USA.

Lamberti, G.A., S.V. Gregory, C.P. Hawkins, R.C. Wildman, L.R. Ashkenas, and D.M. Denicola. 1992. Plant-herbivore interactions in streams near Mount St. Helens. Freshwater Biology **27**:237–247.

Lamont, B.B. 1984. Specialised modes of nutrition. Pages 236–245 *in* J.S. Pate and J.S. Beard, editors. Kwongan, Plant Life of the Sandplain. University of Western Australia Press, Nedlands, Western Australia, Australia.

Lamont, B.B., E.T.F. Witkowski, and N.J. Enright. 1993. Post-fire litter microsites: Safe for seeds, unsafe for seedlings. Ecology **74**:501–512.

Landers, D.H., J.M. Eilers, D.F. Brakke, W.S. Overton, P.E. Kellar, M.E. Silverstein, R.D. Schonbrod, R.E. Crow, R.A. Linthurst, J.M. Omernik, S.A. Teague, and S.P. Meier. 1987. Western Lake Survey, Phase 1. U.S. Environmental Protection Agency, Washington, District of Columbia, USA.

Larson, D.W. 1972. Temperature, transparency, and phytoplankton productivity in Crater Lake, Oregon. Limnology and Oceanography **17**:410–417.

Larson, D.W., and N.S. Geiger. 1982. Existence of phytoplankton in Spirit Lake near active volcano Mount St. Helens, Washington, U.S.A. Archives of Hydrobiology **93**:375–380.

Larson, D.W., C.N. Dahm, and N.S. Geiger. 1987. Vertical partitioning of the phytoplankton assemblage in ultraoligotrophic Crater Lake, Oregon, U.S.A. Freshwater Biology **18**:429–442.

Larson, D.W., and M.W. Glass. 1987. Spirit Lake, Mount St. Helens, Washington: Limnological and Bacteriological Investigations. Volumes 1, 2. Final Report. March 1987. U.S. Army Corps of Engineers, Portland, Oregon, USA.

Larson, D.W. 1991. Aquatic vegetation in volcanically impacted Spirit Lake near Mount St. Helens, Washington. Aquaphyte **11**(2):1–6.

Larson, D.W. 1993. The recovery of Spirit Lake. American Scientist **81**:166–176.

Lawrence, D.B. 1939. Continuing research on the flora of Mount St. Helens. Mazama **21**(12):49–54.

Lawrence, D.B. 1941. The "floating island" lava flow of Mount St. Helens. Mazama **23**(12):56–60.

Lawrence, D.B. 1954. Diagrammatic history of the northeast slope of Mount St. Helens, Washington. Mazama **36**(13):41–44.

Lawrence, R.L., and W.J. Ripple. 1998. Comparisons among vegetation indices and bandwise regression in a highly disturbed, heterogeneous landscape: Mount St. Helens, Washington. Remote Sensing of Environment **64**:91–102.

Lawrence, R.L., and W.J. Ripple. 2000. Fifteen years of revegetation of Mount St. Helens: A landscape-scale analysis. Ecology **81**:2742–2752.

Leavesley, G.H., G.C. Lusby, and R.W. Lichty. 1989. Infiltration and erosion characteristics of selected tephra deposits from the 1980 eruption of Mount St. Helens, Washington. Hydrological Sciences Journal **34**:339–353.

Lehre, A.K., B.D. Collins, and T. Dunne. 1983. Post-eruption sediment budget for the North Fork Toutle River drainage, June 1980–June 1981. Zeitschrift für Geomorphologie (Suppl.) **46**:43–165.

Leopold, A.S., S.A. Cain, C.M. Cottam, I.M. Gabrielson, and T.L. Kimball. 1963. Wildlife Management in the National Parks. Advisory Board on Wildlife Management, Washington District of Columbia, USA. [The Leopold Report was reprinted in several places, including Transactions of the North American Wildlife and Natural Resources Conference **28**:28–45.]

Lesica, P., and R.K. Antibus. 1986. Mycorrhizal status of hemiparasitic vascular plants in Montana, USA. Transactions of the British Mycological Society **86**:341–343.

Lettenmaier, D.P., and S.J. Burges. 1981. Estimation of Flood Frequency Changes in the Toutle and Cowlitz River Basins Following the Eruption of Mt. St. Helens. Technical Report No. 69. Department of Civil Engineering, University of Washington, Seattle, Washington, USA.

Levy, S. 1997. Pacific salmon bring it all back home. BioScience **47**:657–660.

Lichatowich, J.A. 1999. Salmon Without Rivers. Island Press, New York, New York, USA.

Lidicker, W.A.J. 1989. Impacts of non-domesticated vertebrates on California grasslands. Pages 135–150 *in* L.F. Huenneke and H. Mooney, editors. Grassland Structure and Function: California Annual Grassland. Kluwer Academic, Dordrecht, The Netherlands.

Lieder, S.A. 1989. Increased straying by adult steelhead, *Salmo gairdneri*, following the 1980 eruption of Mount St. Helens. Environmental Biology of Fishes **24**:219–229.

Lilley, M.D., J.A. Baross, and C.N. Dahm. 1988. Methane production and oxidation in lakes impacted by the May 18, 1980, eruption of Mount St. Helens. Global Biogeochemical Cycles **2**:357–370.

Lincoln, R., G. Boxshall, and P. Clark. 1998. A Dictionary of Ecology, Evolution and Systematics. Second edition. Cambridge University Press, New York, New York, USA.

Lindahl, B.O., A.F.S. Taylor, and R.D. Finlay. 2002. Defining nutritional constraints on carbon cycling in boreal forests: Towards a less 'phytocentric' perspective. Plant and Soil **242**:123–135.

Lindenmayer, D.B., D.R. Foster, J.R. Franklin, M.L. Hunter, R.F. Noss, F.A. Schmiegelow, and D. Perry. 2004. Salvage harvest policies after natural disturbance. Science **303**:13.

Lindroth, C.H. 1961. The ground beetles (Carabidae, excl. Cicindelidae) of Canada and Alaska, Part 2. Opuscula Entomologica (Suppl.) **20**:1–200.

Lindroth, C.H. 1963. The ground beetles (Carabidae, excl. Cicindelidae) of Canada and Alaska, Part 3. Opuscula Entomologica (Suppl.) **24**:201–408.

Lindroth, C.H. 1966. The ground beetles (Carabidae, excl. Cicindelidae) of Canada and Alaska, Part 4. Opuscula Entomologica (Suppl.) **29**:409–648.

Lindroth, C.H. 1968. The ground beetles (Carabidae, excl. Cicindelidae) of Canada and Alaska, Part 5. Opuscula Entomologica (Suppl.) **33**:649–944.

Lindroth, C.H., H. Anderson, H. Bödvardsson, and S.H. Richter. 1973. Surtsey Iceland. The Development of a New Fauna, 1963–1970. Terrestrial Invertebrates. Entomologica Scandinavica, Supplement 5. Zoologiska Museet, Lund, Sweden.

Lipman, P.W., and D.R. Mullineaux, editors. 1981. The 1980 Eruptions of Mount St. Helens, Washington. Professional Paper 1250. U.S. Geological Survey. U.S. Government Printing Office, Washington, District of Columbia, USA.

Lisle, T.E. 1995. Effects of coarse woody debris and its removal on a channel affected by the 1980 eruption of Mount St. Helens, Washington. Water Resources Research **31**:1797–1808.

Lousier, J.D., and D. Parkinson. 1978. Chemical element dynamics in decomposing leaf litter. Canadian Journal of Botany **56**:2795–2812.

Louw, G.N., and M.K. Seely. 1982. Ecology of Desert Organisms. Longmans, New York, New York, USA.

Lubchenco, J. 1983. *Littorina* and *Fucus*: Effects of herbivores, substratum heterogeneity, and plant escapes during succession. Ecology **64**:1116–1123.

Lubchenco, J. 1991. The Sustainable Biosphere Initiative: An ecological research agenda. A report from the Ecological Society of America. Ecology **72**:371–412.

Lucas, R.E. 1985. Recovery of Game Fish Populations Impacted by the May 18, 1980, Eruption of Mount St. Helens. Part I. Recovery of Winter-Run Steelhead in the Toutle River Watershed. Fishery Management Report 85-9A. Washington Department of Game, Olympia, Washington, USA.

Lucas, R.E., and K. Pointer. 1987. Wild Steelhead Spawning Escapement Estimates for Southwest Washington Streams—1987. Fishery Management Report 87-6. Washington Department of Wildlife, Fisheries Management Division, Olympia, Washington, USA.

Lucas, R.E., and J. Weinheimer. 2003. Recovery of Fish Populations Affected by the May 18, 1980, Eruption of Mount St. Helens. Washington Department of Fish and Wildlife, Olympia, Washington, USA.

Lucht, W., I.C. Prentice, R.B. Myneni, S. Sitch, P. Friedlingstein, W. Cramer, P. Bousquet, W. Buermann, and B. Smith. 2002. Climatic control of the high-latitude vegetation greening trend and Pinatubo effect. Science **296**:1687–1689.

Lynch, J.M., and J.M. Whipps. 1991. Substrate flow in the rhizosphere. Pages 15–24 *in* D.L. Keister and P.B. Cregan, editors. The Rhizosphere and Plant Growth. Kluwer Academic, Dordrecht, The Netherlands.

M'Closkey, R.T. 1975. Habitat succession and rodent distribution. Journal of Mammalogy **56**:950–955.

Macdonald, G.A. 1972. Volcanoes. Prentice-Hall, Englewood Cliffs, New Jersey, USA.

Mack, R.N. 1981. Initial effects of ashfall from Mount St. Helens on vegetation in eastern Washington and adjacent Idaho. Science **213**:537–539.

Mack, R.N., D. Simberloff, W.M. Lonsdale, H. Evans, M. Clout, and F.A. Bazzaz. 2000. Biotic invasions: Causes, epidemiology, global consequences, and control. Ecological Applications **10**:689–710.

Mackey, R.L., and D.J. Currie. 2001. The diversity–disturbance relationship: Is it generally strong and peaked? Ecology **82**:3479–3492.

MacMahon, J.A. 1981. Successional processes: Comparisons among biomes with special reference to probable roles of and influences on animals. Pages 277–304 *in* D.C. West, H.H. Shugart, and D.B. Botkin, editors. Forest Succession, Concepts and Application. Springer-Verlag, New York, New York, USA.

MacMahon, J.A. 1982. Mount St. Helens revisited. Natural History **91**:14, 18–22.

MacMahon, J.A., and N. Warner. 1984. Dispersal of mycorrhizal fungi: Processes and agents. Pages 28–41 *in* S.E. Williams and M.F. Allen, editors. VA Mycorrhizae and Reclamation of Arid and Semi-Arid Lands. Scientific Report Number SA1261. Wyoming Agricultural Experiment Station, Laramie, Wyoming, USA.

MacMahon, J.A., R.R. Parmenter, K.A. Johnson, and C.M. Crisafulli. 1989. Small mammal recolonization on the Mount St. Helens volcano: 1980–1987. American Midland Naturalist **122**: 365–387.

Madej, M.A., and V. Ozaki. 1996. Channel response to sediment wave propagation and movement, Redwood Creek, California, USA. Earth Surface Processes and Landforms **21**:911–927.

Magnússon, B., S.H. Magnússon, and B.D. Sigursson. 2001. Vegetation succession in areas colonized by the introduced Nootka lupin (*Lupinus nootkatensis*) in Iceland. Publication No. 207. Agricultural Research Institute, Reykjavik, Iceland (in Icelandic with English summary).

Maguire, B., Jr. 1963. The passive dispersal of small aquatic organisms and their colonization of isolated bodies of water. Ecological Monographs **33**:161–185.

Mahler, R.L., and M.A. Fosberg. 1983. The influence of Mount St. Helens volcanic ash on plant growth and nutrient uptake. Soil Science **135**:197–201.

Majer, J.D., editor. 1989. Animals in Primary Succession. Cambridge University Press, Cambridge, UK.

Major, J.J., and B. Voight. 1986. Sedimentology and clast orientations of the 18 May 1980 southwest flank lahars, Mount St. Helens, Washington. Journal of Sedimentary Petrology **56**:691–705.

Major, J.J., T.C. Pierson, R.L. Dinehart, and J.E. Costa. 2000. Sediment yield following severe volcanic disturbance: A two-decade perspective from Mount St. Helens. Geology **28**:819–822.

Major, J.J., L.E. Mark, and K.R. Spicer. 2001. Response of peak-flow discharges to the catastrophic 1980 Mount St. Helens eruption. EOS, Transactions of the American Geophysical Union **82**(47):F511.

Major, J.J. 2003. Post-eruption hydrology and sediment transport in volcanic river systems. Water Resources Impact **5**:10–5.

Major, J.J. 2004. Posteruption suspended sediment transport at Mount St. Helens: Decadal-scale relationships with landscape adjustments and river discharges. Journal of Geophysical Research (Earth Surface) **109**(1):F01002, doi:10.1029/2002JF000010.

Major, J.J., T.C. Pierson, and K.M. Scott. 2005. Debris flows at Mount St. Helens, Washington, USA. Pages 643–690 *in* M. Jakob and O. Hungr, editors. Debris Flow Hazard Assessment and Related Phenomenon. Springer Praxis Books, New York, New York, USA.

Major, J. J., and T. Yamakoshi. 2005. Decadal-scale change of infiltration characteristics of a tephra-mantled hillslope at Mount St. Helens, Washington. Hydrological Processes (in press).

Malanson, G.P., and D.R. Butler. 1991. Floristic variation among gravel bars in a subalpine river, Montana, USA. Arctic and Alpine Research **23**:273–278.

Malanson, G.P. 1993. Riparian Landscapes. Cambridge University Press, Cambridge, UK.

Mann, D.H., J.S. Edwards, and R.I. Gara. 1980. Diel activity patterns in snow field foraging invertebrates on Mount Rainier, Washington. Arctic and Alpine Research **12**:359–368.

Mantua, N.J., S.R. Hare, Y. Zhang, J.M. Wallace, and R.C. Francis. 1997. A Pacific interdecadal climate oscillation with impacts on salmon production. Bulletin of the American Meteorological Society **78**:1069–1079.

Manuwal, D.A. 1991. Spring bird communities in the southern Washington Cascade Range. Pages 177–205 *in* L.F. Ruggiero, K.B. Aubry, A.B. Carey, and M.H. Huff, editors. Wildlife and Vegetation of Unmanaged Douglas-Fir Forests. PNW-GTR-285. USDA Forest Service, Pacific Northwest Research Station, Portland, Oregon, USA.

Markow, T.A., B. Raphael, D. Dobberfuhl, C.M. Breitmeyer, J.J. Elser, and E. Pfeiler. 1999. Elemental stoichiometry of *Drosophila* and their hosts. Functional Ecology **13**:78–84.

Marks, D.J., J. Kimball, D. Tingey, and T. Link. 1998. The sensitivity of snowmelt processes to climate conditions and forest cover during rain-on-snow: A case study of the 1996 Pacific Northwest flood. Hydrological Processes **12**:1569–1587.

Maron, J.L. 1998. Insect herbivory above and belowground: Individual and joint effects on plant fitness. Ecology **79**:1281–1293.

Martin, D.J., L.J. Wasserman, R.P. Josnes, and E.O. Salo. 1982. Effects of the eruption of Mount St. Helens on salmon populations and habitat of the Toutle River. Pages 235–254 *in* Proceedings of the Mount St. Helens Effects on Water Resources Symposium. Publication 41. Washington Water Resources Center, Pullman, Washington, USA.

Martin, D.J., L.J. Wasserman, R.P. Jones, and E.O. Salo. 1984. Effects of the Mount St. Helens Eruption on Salmon Populations and Habitat in the Toutle River. FRI-UW-8412. Fisheries Research Institute, University of Washington, Seattle, Washington, USA.

Martin, D.J., L.J. Wasserman, and V.H. Dale. 1986. Influence of riparian vegetation on posteruption survival of coho salmon fingerlings on the west-side streams of Mount St. Helens, Washington. North American Journal of Fisheries Management **6**:1–8.

Martin, K.M. 2001. Wildlife in alpine and subalpine habitats. Pages 239–260 *in* D.J. Johnson and T.A. O'Neil, managing directors. Wildlife-Habitat Relationships in Oregon and Washington. Oregon State University Press, Corvallis, Oregon, USA.

Martinson, H.A., S.D. Finneran, and L.J. Topinka. 1984. Changes in Channel Geomorphology of Six Eruption-Affected Tributaries of the Lewis River, 1980–1982, Mount St. Helens, Washington. Open-File Report 84-614. U.S. Geological Survey, Portland, Oregon, USA.

Martinson, H.A., H.E. Hammond, W.W. Mast, and P.D. Mango. 1986. Channel Geometry and Hydrological Data for Six Eruption-Affected Tributaries of the Lewis River, Mount St. Helens, Washington, Water Years 1983–1984. Open-File Report 85-631. U.S. Geological Survey, Portland, Oregon, USA.

Maser, C., J.M. Trappe, and R.A. Nussbaum. 1978. Fungal–small mammal interrelationships with emphasis on Oregon coniferous forest. Ecology **59**:799–809.

Mastin, L.G. 1994. Explosive tephra emissions of Mount St. Helens, 1989–1991: The violent escape of magmatic gas following storms? Geological Society of America Bulletin **106**:175–185.

Masuzawa, T. 1985. Ecological studies on the timberline of Mount Fuji. I. Structure of plant community and soil development on the timberline. Botanical Magazine Tokyo **98**:15–28.

Matson, P. 1990. Plant-soil interactions in primary succession at Hawaii Volcanoes National Park. Oecologia (Berlin) **85**:241–246.

Matthews, J.A. 1992. The Ecology of Recently-Deglaciated Terrain: A Geoecological Approach to Glacier Forelands and Primary Succession. Cambridge University Press, Cambridge, UK.

Mattson, W.J. 1980. Herbivory in relation to plant nitrogen content. Annual Review of Ecology and Systematics **11**:119–161.

Maun, M.A. 1998. Adaptations of plants to burial in coastal sand dunes. Canadian Journal of Botany **76**:713–738.

Mazzoleni, S., and M. Ricciardi. 1993. Primary succession on the cone of Vesuvius. Pages 101–112 *in* J. Miles and D.W.H. Walton, editors. Primary Succession on Land. Blackwell Scientific, London, UK.

McCabe, G.J., and M.D. Dettinger. 1999. Decadal variations in strength of ENSO teleconnections with precipitation in the western United States. International Journal of Climatology **19**:1399–1410.

McComb, W.C., K. Margarial, and R.G. Anthony. 1993. Small mammal and amphibian abundance in streamside and upslope habitats of mature Douglas-fir stands, western Oregon. Northwest Science **67**:7–15.

McCook, L.J. 1994. Understanding ecological community succession: Causal models and theories, a review. Vegetatio **110**:115–147.

McCune, B., and M.J. Mefford. 1999. PC-ORD, multivariate analysis of ecological data, Version 4.0. MjM Software Design, Gleneden Beach, Oregon, USA.

McIntosh, R.P. 1999. The succession of succession: A lexical chronology. Bulletin of the Ecological Society of America **80**:256–265.

McIntyre, A. 2003. Ecology of Populations of Van Dyke's Salamanders in the Cascade Range of Washington State. Master's thesis. Oregon State University, Corvallis, Oregon, USA.

McKeever, S. 1960. Food of the northern flying squirrel in northeastern California. Journal of Mammalogy **41**:270–271.

McKnight, D.M., G.L. Feder, and E.A. Stiles. 1981a. Toxicity of Mount St. Helens Ash Leachate to a Blue-Green Alga. Circular 850-F. U.S. Geological Survey, Reston, Virginia, USA.

McKnight, D.M., G.L. Feder, and E.A. Stiles. 1981b. Toxicity of volcanic-ash leachate to a blue-green alga. Results of a preliminary bioassay experiment. Environmental Science and Technology **15**:362–364.

McKnight, D.M., W.E. Periera, M.L. Ceazan, and R.C. Wissmar. 1982. Characterization of dissolved organic material in surface waters within the blast zone of Mount St. Helens, Washington. Organic Geochemistry **4**:85–92.

McKnight, D.M., J.M. Klein, and R.C. Wissmar. 1984. Changes in the organic material in lakes in the blast zone of Mount St. Helens, Washington. Circular 850-L. U.S. Geological Survey, Reston, Virginia, USA.

McKnight, D.M., K.A. Thorn, R.L. Wershaw, J.M. Bracewell, and G.W. Robertson. 1988. Rapid changes in dissolved humic substances in Spirit Lake and South Fork Castle Lake, Washington. Limnology and Oceanography **33**(6, part 2):1527–1541.

McNaughton, S.J. 1985. Ecology of a grazing ecosystem: The Serengeti. Ecological Monographs **53**:259–294.

McNeill, S., and T.R.E. Southwood. 1978. The role of nitrogen in the development of insect/plant relationships. Pages 77–98 *in* J.B. Harborne, editor. Biochemical Aspects of Plant and Animal Coevolution. Academic Press, London, UK.

McPhail, J.D. 1967. Distribution of freshwater fishes in western Washington. Northwest Science **41**:1–11.

McPhail, J.D., and C.C. Lindsey. 1986. Zoogeography of the freshwater fishes of Cascadia (the Columbia system and rivers north to the Stikine). Pages 615–637 *in* C.H. Hocutt and E.O. Wiley, editors. The Zoogeography of North American Freshwater Fishes. John Wiley & Sons, New York, New York, USA.

Means, J.E., W.A. McKee, W.H. Moir, and J.F. Franklin. 1982. Natural revegetation of the northeastern portion of the devastated area. Pages 93–103 *in* S.A. Keller, editor. Mount St. Helens: One Year Later. Eastern Washington University Press, Cheney, Washington, USA.

Medin, D.E. 1986. Small Mammal Responses to Diameter-Cut Logging in an Idaho Douglas-Fir Forest. Research Note INT-362. USDA Forest Service Intermountain Research Station, Missoula, Montana, USA.

Menting, V.L. 1995. The Biogeochemistry of Lakes in the Mount St. Helens Blast Zone. Master's thesis. Portland State University, Portland, Oregon, USA.

Merrill, E., K. Raedeke, and R. Taber. 1987. The Population Dynamics and Habitat Ecology of Elk in the Mount St. Helens Blast Zone. College of Forest Resources, University of Washington, Seattle, Washington, USA.

Merritt, R.W., and K.W. Cummins. 1996. An Introduction to the Aquatic Insects of North America. Third edition. Kendall/Hunt, Dubuque, Iowa, USA.

Meyer, W., and P.J. Carpenter. 1983. Filling of Spirit Lake, Washington, May 18, 1980 to July 31, 1982. Open-File Report 82-771. U.S. Geological Survey, Vancouver, Washington, USA.

Meyer, D.F., and R.J. Janda. 1986. Sedimentation downstream from the 18 May 1980 North Fork Toutle River debris avalanche deposit, Mount St. Helens, Washington. Pages 68–86 *in* S.A.C. Keller, editor. Mount St. Helens: Five Years Later. Eastern Washington University Press, Cheney, Washington, USA.

Meyer, D.F., K.M. Nolan, and J.E. Dodge. 1986. Post-Eruption Changes in Channel Geometry of Streams in the Toutle River Drainage Basin, 1980–1982, Mount St. Helens, Washington. Open-File Report 85-412. U.S. Geological Survey, Vancouver, Washington, USA.

Meyer, D.F., and J.E. Dodge. 1988. Post-Eruption Changes in Channel Geometry of Streams in the Toutle River Drainage Basin, 1983-1985, Mount St. Helens, Washington. Open-File Report 87-549. U.S. Geological Survey, Vancouver, Washington, USA.

Meyer, D.F., and H.A. Martinson. 1989. Rates and processes of channel development and recovery following the 1980 eruption of Mount St. Helens, Washington. Hydrological Sciences Journal **34**:115–127.

Meyer, D.F. 1995. Stream-Channel Changes in Response to Volcanic Detritus Under Natural and Augmented Discharge, South Coldwater Creek, Washington. Open-File Report 94-519. U.S. Geological Survey, Vancouver, Washington, USA.

Miles, J., and D.W.H. Walton, editors. 1993. Primary Succession on Land. Blackwell Scientific, London, UK.

Minshall, G.W., J.T. Brock, and J.D. Varley. 1989.Wildfire and Yellowstone's stream ecosystems. Bioscience **39**:707–715.

Minshall, G.W. 2003. Responses of stream benthic macroinvertebrates to fire. Forest Ecology and Management **178**:155–161.

Mizuno, N., and K. Kimura. 1996. Vegetational recovery in the mud flow (lahar) area. *In* M. Nanjo, editor. Restoration of Agriculture in Pinatubo Lahar Areas. International Research of the Faculty of Agriculture, Tohoku University, Tohoku, Japan.

Molina, R., H. Massicotte, and J.M. Trappe. 1992. Specificity phenomena in mycorrhizal symbiosis: Community-ecological consequences and practical implications. Pages 357–423 *in* M.F. Allen, editor. Mycorrhizal Functioning: An Integral Plant-Fungal Process. Chapman & Hall, New York, New York, USA.

Montllor, C.B., E.A. Bernays, and R.V. Barbehenn. 1990. Importance of quinolizidine alkaloids in the relationship between larvae of *Uresiphita reversalis* (Lepidoptera: Pyralidae) and a host plant, *Genista monpessulana*. Journal of Chemical Ecology **16**:1853–1865.

Mooney, H.A., P.M. Vitousek, and P.A. Matson. 1987. Exchange of materials between terrestrial ecosystems and the atmosphere. Science **238**:926–932.

Moore, J.G., and T.W. Sisson. 1981. Deposits and effects of the May 18 pyroclastic surge. Pages 421–438 *in* P.W. Lipman and D.R. Mullineaux, editors. The 1980 Eruptions of Mount St. Helens, Washington. Professional Paper 1250. U.S. Geological Survey, Washington, District of Columbia, USA.

Moore, J.G., and D.A. Clague. 1992. Volcano growth and evolution of the island of Hawaii. Geological Society of America Bulletin **104**:1471–1484.

Morris, W.F., and D.M. Wood. 1989. The role of lupine in succession on Mount St. Helens: Facilitation or inhibition? Ecology **70**:697–703.

Moyer, T.C., and D.A. Swanson. 1987. Secondary hydroeruptions in pyroclastic flow deposits: Examples from Mount St. Helens. Journal of Volcanology and Geothermal Research **32**:299–319.

Mueller-Dombois, D., and H. Ellenberg. 1974. Aims and Methods of Vegetation Ecology. John Wiley & Sons, New York, New York, USA.

Mueller-Dombois, D., and H. Ellenberg. 2002. Aims and Methods of Vegetation Ecology. Second edition. Blackburn Press, Caldwell, New Jersey, USA.

Mullineaux, D.R., and D.R. Crandell. 1981. The eruptive history of Mount St. Helens. Pages 3–15 *in* P.W. Lipman and D.R. Mullineaux, editors. The 1980 Eruptions of Mount St. Helens, Washington. Professional Paper 1250. U.S. Geological Survey, Washington, District of Columbia, USA.

Mullineaux, D.R. 1996. Pre-1980 tephra-fall deposits erupted from Mount St. Helens. Professional Paper 1563. U.S. Geological Survey, Washington, District of Columbia, USA.

Mummey, D.L., P.D. Stahl, and J.S. Buyer. 2002. Microbial biomarkers as an indicator of ecosystem recovery following surface mine reclamation. Applied Soil Ecology **21**:251–259.

Mundie, J.H. 1974. Optimization of the salmonid nursery stream. Journal of the Fisheries Research Board of Canada **31**:1827–1837.

Munn, T., editor. 2002. Encyclopedia of Global Environmental Change. Wiley, New York, New York, USA.

Nakamura, T. 1985. Forest succession in the subalpine region of Mt. Fuji, Japan. Vegetatio **64**:15–27.

Nakashizuka, T., S. Iida, W. Suzuki, and T. Tanimoto. 1993. Seed dispersal and vegetation development on a debris avalanche on the Ontake volcano, Central Japan. Journal of Vegetation Science **4**:537–542.

NASA. 2001. Landsat 7 Science Data Users Handbook. http://ltpwww.gsfc.nasa.gov/IAS/handbook/handbook_toc.html.

Newnham, R.M., and D.J. Lowe. 1991. Holocene vegetation and volcanic activity, Auckland Isthmus, New Zealand. Journal of Quaternary Science **6**:177–93.

Nicholas, A.P., P.J. Ashworth, M.J. Kirkby, M.G. Macklin, and T. Murray. 1995. Sediment slugs: Large-scale fluctuations in fluvial sediment transport rates and storage volumes. Progress in Physical Geography **19**:500–519.

Nickelson, T.E., M.F. Solazzi, and S.L. Johnson. 1986. Use of hatchery coho salmon (*Oncorhynchus kisutch*) presmolts to rebuild wild populations in Oregon coastal streams. Canadian Journal of Fisheries and Aquatic Sciences **43**:2443–2449.

Nijhuis, M.J., and R.H. Kaplan. 1998. Movement patterns and life history characteristics in a population of the Cascade torrent salamander (*Rhyacotriton cascade*) in the Columbia River gorge, Oregon. Journal of Herpetology **32**:301–304.

Nishi, H., and S. Tsuyuzaki. 2004. Seed dispersal and seedling establishment of *Rhus trichocarpa* promoted by a crow (*Corvus macrorhynchos*) on a volcano in Japan. Ecography **27**:311–322.

North, M., and J. Franklin. 1990. Post-disturbance legacies that enhance biological diversity in a Pacific Northwest old-growth forest. Northwest Environmental Journal **6**:427–429.

Nuhn, W.W. 1987. Soil Genesis on the 1980 Pyroclastic Flow of Mount St. Helens. Master's thesis. University of Washington, Seattle, Washington, USA.

Nussbaum, R.A. 1976. Geographic Variation and Systematics of Salamanders of the Genus *Dicamptodon* Strauch (Ambystomatidae). Miscellaneous Publication No. 149. Museum of Zoology, University of Michigan, Ann Arbor, Michigan, USA.

Nussbaum, R.A., and C.K. Tait. 1977. Aspects of the life history and ecology of the Olympic salamander, *Rhyacotriton olympicus* (Gaige). The American Midland Naturalist **98**:176–199.

Nussbaum, R.A., E.D. Brodie, and R.M. Storm. 1983. Amphibians and Reptiles of the Pacific Northwest. University of Idaho Press, Moscow, Idaho, USA.

O'Connell, A.M. 1988. Nutrient dynamics in decomposing litter in karri (*Eucalyptus diversicolor* F. Muell.) forests of south-western Australia. Journal of Ecology **76**:1186–1203.

Odum, E.P. 1969. The strategy of ecosystem development. Science **164**:262–270.

Odum, E.P. 1983. Basic Ecology. Saunders, New York, New York, USA.

Ohsawa, M. 1984. Differentiation of vegetation zones and species strategies in the subalpine region of Mt. Fuji. Vegetatio **57**:15–52.

Oksanen, L. 1990. Exploitation ecosystems in heterogeneous habitat complexes. Evolutionary Ecology **4**:220–234.

Olds, C. 2002. Fisheries Studies at the Sediment Retention Structure on the North Fork Toutle River 1993, 2001, and 2002. Washington Department of Fish and Wildlife, Olympia, Washington, USA.

Olff, H., and M.E. Ritchie. 1998. Effects of herbivores on grassland plant diversity. Trends in Ecology and Evolution **13**:261–265.

Olson, J.S. 1958. Rates of succession and soil changes on southern Lake Michigan sand dunes. Botanical Gazette **119**:125–169.

Olson, D.H., S.S. Chan, G. Weaver, P. Cunningham, A. Moldenke, R. Progar, P.S. Muir, B. McCune, A. Rosso, and E.B. Peterson. 2000. Characterizing stream, riparian, upslope habitats and species in Oregon managed headwater forests. Pages 83–88 *in* J. Wigginton and R. Beschta, editors. Riparian Ecology and Management in Multi-Land Use Watersheds. Proceedings of the International Conference of the American Water Resources Association, 30 August 2000, Portland, Oregon. TPS-00-2. American Water Resources Association, Middleburg, Virginia, USA.

Oner, M., and S. Oflas. 1977. Plant succession on the Kula volcano in Turkey. Vegetatio **34**:55–62.

Orwig, C.E., and J.M. Mathison. 1981. Forecasting considerations in Mount St. Helens affected rivers. Pages 272–292 *in* Proceedings, Conference on Mount St. Helens: Effects on Water Resources. Washington Water Research Center, Washington State University, Pullman, Washington, USA.

Owen, J.G. 2001. Gleasonian Pattern of Mammalian Distribution at a Macrogeographical Scale in Texas. Occasional Paper of the Museum 207. Texas Tech University, Lubbock, Texas, USA.

Paine, A.M., D.F. Meyer, and S.A. Schumm. 1987. Incised channel and terrace formation near Mount St. Helens, Washington. Pages 389–390 *in* R.L. Beschta, T. Blinn, G.E. Grant, G.G. Ice, and F.J. Swanson, editors. Erosion and Sedimentation in the Pacific Rim. Publication 165. International Association of Hydrological Sciences, Christchurch, New Zealand.

Paine, R.T., M.J. Tegner, and E.A. Johnson. 1998. Compounded perturbations yield ecological surprises. Ecosystems **1**:535–545.

Parker, M.S. 1991. Relationship between cover availability and larval giant salamander density. Journal of Herpetology **25**:355–357.

Parker, B.R., D.W. Schindler, D.B. Donald, and R.S. Anderson. 2001. The effects of stocking and removal of a nonnative salmonid on the plankton of an alpine lake. Ecosystems **4**:322–334.

Parkes, A., J.T. Teller, and J.R. Flenley. 1992. Environmental history of the Lake Vaihiria drainage-basin, Tahiti, French-Polynesia. Journal of Biogeography **19**:431–447.

Parmenter, R.R., and V.A. LaMarra. 1991. Nutrient cycling in a freshwater marsh: The decomposition of fish and waterfowl carrion. Limnology and Oceanography **36**:976–987.

Parmenter, R.R., T.L. Yates, D.R. Anderson, K.P. Burnham, J.L. Dunnum, A.B. Franklin, M.T. Friggens, B.C. Lubow, M. Miller, G.S. Olson, C.A. Parmenter, J. Pollard, E. Rexstad, T.M. Shenk, T.R. Stanley, and G.C. White. 2003. Small mammal density estimation: A field comparison of grid-based versus web-based density estimators. Ecological Monographs **73**:1–26.

Parr, J.F., R.I. Papendick, S.B. Hornick, and R.E. Meyer. 1992. Soil quality: Attributes and relationship to alternative and sustainable agriculture. American Journal of Alternative Agriculture **7**: 5–11.

Parsons, G.L., G. Cassis, A.R. Moldenke, J.D. Lattin, N.H. Anderson, J.C. Miller, P. Hammond, and T.D. Schowalter. 1991. Invertebrates of the H.J. Andrews Experimental Forest, Western Cascade Range, Oregon. V: An Annotated List of the Insects and Other Arthropods. PNW-GTR-290. USDA Forest Service, Pacific Northwest Research Station. Portland, Oregon, USA.

Partomihardjo, T., E. Mirmanto, and R.J. Whittaker. 1992. Anak Krakatau's vegetation and flora circa 1991, with observations on a decade of development and change. GeoJournal **28**:233–248.

Paul, E.A., and F.E. Clark. 1989. Soil Microbiology and Biochemistry. Academic Press, San Diego, California, USA.

Pearson, P.G. 1959. Small mammals and old field succession on the Piedmont of New Jersey. Ecology **40**:249–255.

Pechmann, J.H.K., D.E. Scott, J.W. Gibbons, and R.D. Semlitsch. 1989. Influence of wetland hydroperiod on diversity and abundance of metamorphosing juvenile amphibians. Wetlands Ecology Management **1**:3–11.

Pechmann, J.H.K., D.E. Scott, R.D. Semlitsch, J.P. Caldwell, L.J. Vitt, and J.W. Gibbons. 1991. Declining amphibian populations: The problem of separating human impacts from natural fluctuations. Science **253**:892–895.

Pechmann, J.H.K., and H.M. Wilbur. 1994. Putting amphibian declines in perspective: Natural fluctuations and human impacts. Herpetologica **50**:65–84.

Petersen, R.R. 1993. Recovery of lakes located in the blast zone of Mount St. Helens. Internationale Vereinigung für Theoretische und Angewandte Limnologie: Verhandlungen **25**:366–369.

Petersen, R.R., and K.D. Carpenter. 1997. Nutrient limitation in five lakes near Mount St. Helens, Washington. Internationale Vereinigung für Theoretische und Angewandte Limnologie: Verhandlungen **26**:377–380.

Petersen, S.O., and M.J. Klug. 1994. Effects of sieving, storage, and incubation temperature on the phospholipid fatty acid profile of a soil microbial community. Applied and Environmental Microbiology **60**:2421–2430.

Petranka, J.W., M.E. Eldrige, and K.E. Haley. 1993. Effects of timber harvesting on southern Appalachian salamanders. Conservation Biology **7**:363–370.

Pfitsch, W.A., and L.C. Bliss. 1988. Recovery of net primary production in subalpine meadows of Mount St. Helens following the 1980 eruption. Canadian Journal of Botany **66**:989–997.

Phillips, K.N. 1941. Fumaroles of Mount St. Helens and Mt. Adams. Mazama **23**(12):37–42.

Phillips, E.L. 1964. Washington Climate for These Counties: Clark, Cowlitz, Lewis, Skamania, and Thurston. EM 2462. Agricultural Extension Service, Washington State University, Pullman, Washington, USA.

Pickett, S.T.A., and P.W. White, editors. 1985. The Ecology of Natural Disturbance and Patch Dynamics. Academic Press, New York, New York, USA.

Pierson, T.C. 1985. Initiation and flow behavior of the 1980 Pine Creek and Muddy River lahars, Mount St. Helens, Washington. Geological Society of America Bulletin **96**:1056–1069.

Pierson, T.C. 1999. Transformation of water flood to debris flow following the eruption-triggered transient lake breakout from the crater on March 19, 1982. Pages 19–36 *in* T.C. Pierson, editor. Hydrologic Consequences of Hot-Rock/Snowpack Interactions at Mount St. Helens Volcano, Washington, 1982–1984. Professional Paper 1586. U.S. Geological Survey, Reston, Virginia, and Denver, Colorado, USA.

Pilliod, D.S., and C.R. Peterson. 2001. Local and landscape effects on introduced trout on amphibians in historically fishless watersheds. Ecosystems **4**:322–334.

Pinkart, H.C., D.B. Ringelberg, Y.M. Piceno, S.J. Macnaughton, and D.C. White. 2002. Biochemical approaches to biomass measurements and community structure analysis. Pages 101–113 *in* C.J. Hurst, G.R. Knudsen, M.J. McInerney, L.D. Stetzenbach, and R.L. Crawford, editors. Manual of Environmental Microbiology. Second edition. American Society for Microbiology (ASM) Press, Washington, District of Columbia, USA.

Platt, W.J., and J.H. Connell. 2003. Natural disturbances and directional replacement of species. Ecological Monographs **73**:507–522.

Polis, G.A., and S.D. Hurd. 1996. Linking marine and terrestrial food webs: Allochthonous input from the ocean supports high secondary productivity on small islands and coastal land communities. American Naturalist **147**:396–423.

Polis, G.A., W.B. Anderson, and R.D. Holt. 1997. Toward an integration of landscape and food web ecology: The dynamics of spatially subsidized food webs. Annual Review of Ecology and Systematics **28**:289–316.

Power, M.E. 1990. Effects of fish in river food webs. Science **250**:811–814.

Prach, K., and P. Pyšek. 1994. Spontaneous establishment of woody plants in Central Europe derelict sites and their potential for reclamation. Restoration Ecology **2**:190–197.

Prach, K., and P. Pyšek. 1999. How do species dominating in succession differ from others? Journal of Vegetation Science **10**:383–392.

Price, P.W. 1992. Plant resources as the mechanistic basis for insect herbivore population dynamics. Pages 139–173 *in* M.D. Hunter, T. Ohgushi, and P.W. Price, editors. Effects of Resource Distribution and Animal-Plant Interactions. Academic Press, San Diego, California, USA.

Pringle, P.T., and K.A. Cameron. 1999. Eruption-triggered lahar on May 14, 1984. Pages 81–103 *in* T.C. Pierson, editor. Hydrologic Consequences of Hot-Rock/Snowpack Interactions at Mount St. Helens Volcano, Washington, 1982–1984. Professional Paper 1586. U.S. Geological Survey, Reston, Virginia, and Denver, Colorado, USA.

Pugnaire, F.I., C. Armas, and F. Valladares. 2004. Soil as a mediator in plant–plant interactions in a semi-arid community. Journal of Vegetation Science **15**:85–92.

Quinn, T.P., R.S. Nemeth, and D.O. McIsaac. 1991. Patterns of homing and straying by fall chinook salmon in the lower Columbia River. Transactions of the American Fisheries Society **120**:150–156.

Ramirez, P.J., and M. Hornocker. 1981. Small mammal populations in different-aged clearcuts in northwestern Montana. Journal of Mammalogy **62**:400–403.

Raunkiaer, C. 1934. The Life Forms of Plants and Statistical Plant Geography. Clarendon Press, Oxford, UK.

Rawlinson, P.A., H.T. Widjoya, M.N. Hutchinson, and G.W. Brown. 1990. The terrestrial vertebrate fauna of the Krakatau Islands, Sunda Strait, 1883–1986. Philosophical Transactions of the Royal Society, London, B **328**:3–28.

Redding, J.M., and C.B. Schreck. 1982. Mount St. Helens ash causes sublethal stress responses in steelhead trout. Pages 300–307 *in* Proceedings from the Conference, Mount St. Helens: Effects on Water Resources. Report Number 41. Washington Water Research Center, Washington State University, Pullman, Washington, USA.

Reimers, P.E., and C.E. Bond. 1967. Distribution of fishes in tributaries of the lower Columbia River. Copeia **3**:541–550.

Rejmanek, M., R. Haagerova, and J. Haager. 1982. Progress of plant succession on the Paracutin Volcano: 25 years after activity ceased. American Midland Naturalist **108**:194–198.

Renwick, J.A., and F.S. Chew. 1994. Oviposition behavior in Lepidoptera. Annual Review of Entomology **39**:377–400.

Reynolds, G.D., and L.C. Bliss. 1986. Microenvironmental investigations of tephra covered surfaces at Mount St. Helens. Pages 147–152 *in* S.A.C. Keller, editor. Mount St. Helens: Five Years Later. Eastern Washington University Press, Cheney, Washington, USA.

Rhee, G.-Y. 1978. Effects of N:P atomic ratios and nitrate limitation on algal growth, cell composition, and nitrate uptake. Limnology and Oceanography **23**:10–25.

Richardson, M.S., and M.L. Goff. 2001. Effects of temperature and intraspecific interactions on the development of *Dermestes maculatus* (Coleoptera: Demestidae). Journal of Medical Entomology **38**:347–351.

Risacher, F., and H. Alonso. 2001. Geochemistry of ash leachates from the 1993 Lascar eruption, northern Chile. Implication for recycling of ancient evaporites. Journal of Volcanology and Geothermal Research **109**:319–337.

Ritchie, M.E., and D. Tilman. 1995. Responses of legumes to herbivores and nutrients during succession on a nitrogen-poor soil. Ecology **76**:2648–2655.

Ritchie, M.E., D. Tilman, and J.M.H. Knops. 1998. Herbivore effects on plant and nitrogen dynamics in oak savanna. Ecology **79**:165–177.

Riviere, A. 1982. Plant recovery and seed invasion on a volcanic desert, the crater basin of Usu-san, Hokkaido. Seed Ecology/Shushi Seitai **13**:11–18.

Robertson, G.P. 1987. Geostatistics in ecology: Interpolating with known variance. Ecology **68**:744–748.

Robertson, G.P., M.A. Huston, F.C. Evans, and J.M. Tiedje. 1988. Spatial variability in a successional plant community: Patterns of nitrogen availability. Ecology **69**:1517–1524.

Rochow, J.J. 1975. Mineral nutrient pool and cycling in a Missouri forest. Journal of Ecology **63**:985–994.

Rosenbaum, J.G., and R.B. Waitt, Jr. 1981. Summary of eyewitness accounts of the May 18 eruption. Pages 53–67 *in* P.W. Lipman and D.R. Mullineaux, editors. The 1980 Eruptions of Mount St. Helens. Washington Professional Paper 1250. U.S. Geological Survey, Washington, District of Columbia, USA.

Rosenfeld, C.L., and G.L. Beach. 1983. Evolution of a Drainage Network: Remote Sensing Analysis of the North Fork Toutle River, Mount St. Helens, Washington. Report WRRI-88. Water Resources Institute, Oregon State University Press, Corvallis, Oregon, USA.

Rossi, R.E. 1989. The Geostatistical Interpretation of Ecological Phenomena. Doctoral dissertation. Washington State University, Pullman, Washington, USA.

Rossi, R.E., D.J. Mulla, A.G. Journel, and E.H. Franz. 1992. Geostatistical tools for modeling and interpreting ecological spatial dependence. Ecological Monographs **62**:277–314.

Rouse, J.W., R.H. Haas, J.A. Schell, D.W. Deering, and J.C. Harlan. 1973. Monitoring the Vernal Advancement and Retrogradation (Greenwave Effect) of Natural Vegetation. Final report. National Aeronautics and Space Administration, Goddard Space Flight Center, Greenbelt, Maryland, USA.

Rowley, P.D., M.A. Kuntz, and N.S. MacLeod. 1981. Pyroclastic flow deposits. Pages 489–512 *in* P.W. Lipman and D.R. Mullineaux, editors. The 1980 Eruptions of Mount St. Helens, Washington. Professional Paper 1250. U.S. Geological Survey, Washington, District of Columbia, USA.

Rudd, J.W.M., and R.D. Hamilton. 1978. Methane cycling in a eutrophic shield lake and its effects on whole lake metabolism. Limnology and Oceanography **23**:337–348.

Ruggiero, L.F., K.B. Aubry, A.B. Carey, and M.H. Huff. 1991. Wildlife and Vegetation of Unmanaged Douglas-Fir Forests. PNW-GTR-285. USDA Forest Service, Pacific Northwest Research Station, Portland, Oregon, USA.

Running, S.W., D.L. Peterson, M.A. Spanner, and K.B. Teuber. 1986. Remote sensing of coniferous forest leaf area. Ecology **67**:273–276.

Russell, K. 1986. Revegetation trials in a Mount St. Helens eruption debris flow. Pages 231–248 *in* S.A.C. Keller, editor. Mount St. Helens: Five Years Later. Eastern Washington University Press, Cheney, Washington, USA.

St. Helens Forest Lands Research Cooperative. 1980. Minutes of the Technical Needs Workshop, September 4–5, 1980. St. Helens Forest Lands Research Cooperative, Olympia, Washington, USA.

St. John, H. 1976. The flora of Mt. St. Helens, Washington. The Mountaineer **70**(7):65–77.

Sarkar, S. 1996. Ecological theory and amphibian declines. BioScience **46**:199–207.

Sanders, R.W., and K.G. Porter. 1988. Phagotrophic phytoflagellates. Advances in Microbial Ecology **10**:167–192.

Sarna-Wojcicki, A.M., S. Shipley, R.B. Waitt, D. Dzurisin, and S.H. Wood. 1981. Areal distribution, thickness, mass, volume, and grain size of air-fall ash from the six major eruptions of 1980. Pages 577–600 *in* P.W. Lipman and D.R. Mullineaux, editors. The 1980 Eruptions of Mount St. Helens, Washington. Professional Paper 1250. U.S. Geological Survey, Washington, District of Columbia, USA.

Savage, M., B. Sawhill, and M. Askenazi. 2000. Community dynamics: What happens when we rerun the tape? Journal of Theoretical Biology **205**:515–526.

Schade, J.D., M. Kyle, S.E. Hobbie, W.F. Fagan, and J.J. Elser. 2003. Stoichiometric tracking of soil nutrients by a desert insect herbivore. Ecology Letters **6**:1–6.

Scharnberg, L.D. 1995. Zooplankton Community Structure in Lakes near Mount St. Helens, Washington. Master's thesis. Portland State University, Portland, Oregon, USA.

Schilling, S.P., P.E. Carrara, R.E. Thompson, and E.Y. Iwatsubo. 2004. Posteruption glacier development within the crater of Mount St. Helens, Washington, USA. Quaternary Research **61**:325–329.

Schindler, D.E., R.A. Knapp, and P.R. Leavitt. 2001. Alteration on nutrient cycles and algal production resulting from fish introductions into mountain lakes. Ecosystems **4**:308–322.

Schmincke, H.U., C. Park, and E. Harms. 1999. Evolution and environmental impacts of the eruption of Laacher See Volcano (Germany) 12,900 a BP. Quaternary **61**:61–72.

Schopmeyer, C.S. 1974. Seeds of Woody Plants in the United States. Agricultural Handbook 450. USDA Forest Service, Washington, District of Columbia, USA.

Schutter, M., and R. Dick. 2001. Shifts in substrate utilization potential and structure of soil microbial communities in response to carbon substrates. Soil Biology and Biochemistry **33**:1481–1491.

Scott, K.M. 1988. Origins, Behavior, and Sedimentology of Lahars and Lahar-Runout Flows in the Toutle-Cowlitz River system. Professional Paper 1447-A. U.S. Geological Survey, Washington, District of Columbia, USA.

Scriber, J.M., and R.C. Lederhouse. 1992. The thermal environment as a resource dictating geographic patterns of feeding specialization of insect herbivores. Pages 429–466 *in* M.D. Hunter, T. Ohgushi, and P.W. Price, editors. Effects of Resource Distribution and Animal–Plant Interactions. Academic Press, San Diego, California, USA.

Scrivner, J.H., and H.D. Smith. 1984. Relative abundance of small mammals in four successional stages of spruce-fir forest in Idaho. Northwest Science **58**:171–176.

Sedell, J.R., and C.N. Dahm. 1984. Catastrophic disturbances to stream ecosystems: Volcanism and clear-cut logging. Pages 531–539 *in* M.J. Klug and C.A. Reddy, editors. Current Perspectives in Microbial Ecology. American Society for Microbiology, Washington, District of Columbia, USA.

Segura, G., L.B. Brubaker, J.F. Franklin, T.M. Hinckley, D.A. Maguire, and C. Wright. 1994. Recent mortality and decline in mature *Abies amabilis*: The interaction between site factors and tephra deposition from Mount St. Helens. Canadian Journal of Forest Research **24**:1112–1122.

Segura, G., T.M. Hinckley, and L.B. Brubaker. 1995a. Variations in radial growth of declining old-growth stands of *Abies amabilis* after tephra deposition from Mount St. Helens. Canadian Journal of Forest Research **25**:1484–492.

Segura, G., T.M. Hinckley, and C.D. Oliver. 1995b. Stem growth responses of declining mature *Abies amabilis* trees after tephra deposition from Mount St. Helens. Canadian Journal of Forest Research **25**:1493–1502.

Seymour, V.A., T.M. Hinckley, Y. Morikawa, and J.F. Franklin. 1983. Foliage damage in coniferous trees following volcanic ashfall from Mt. St. Helens. Oecologia (Berlin) **59**:339–343.

Shanks, C.H., and D.L. Chase. 1981. Effect of volcanic ash on adult *Otiorhynchus* (Coleoptera: Curculionidae). Melandria **37**: 63–66.

Sheridan, C.D., and D.H. Olson. 2003. Amphibian assemblages in zero-order basins in the Oregon Coast Range. Canadian Journal of Forest Research **33**:1452–477.

Shipley, J.W. 1919. The nitrogen content of volcanic ash in Katmai eruption of 1912. Ohio Journal of Science **19**:213–223.

Shipley, S., and A.M. Sarna-Wojcicki. 1983. Maps showing distribution, thickness, and mass of late Pleistocene and Holocene tephra from major volcanoes in the northwestern United States: A preliminary assessment of hazards from volcanic ejecta to nuclear reactors in the Pacific Northwest. Miscellaneous Field Studies Map MF-1435. U.S. Geological Survey, Reston, Virginia, USA.

Sicilianoa, S.D., and J.J. Germidaa. 1998. Biolog analysis and fatty acid methyl ester profiles indicate that pseudomonad inoculants that promote phytoremediation alter the root-associated microbial community of *Bromus biebersteinii*. Soil Biology and Biochemistry **30**:1717–1723.

Sidle, R.C., A.J. Pearce, and C.L. O'Loughlin. 1985. Hillslope Stability and Land Use. Water Resources Monograph 11. American Geophysical Union, Washington, District of Columbia, USA.

Simon, A. 1999. Channel and Drainage-Basin Response of the Toutle River System in the Aftermath of the 1980 Eruption of Mount St. Helens, Washington. Open-File Report 96-633. U.S. Geological Survey, Vancouver, Washington, USA.

Simonson, R.W. 1959. Outline of a generalized theory of soil genesis. Soil Science Society of America Proceedings **23**:152–156.

Singer, F.J., L.C. Zeigenfuss, R.G. Cates, and D.T. Barnett. 1998. Elk, multiple factors, and persistence of willows in national parks. Wildlife Society Bulletin **26**:419–428.

Sinton, D.S., J.A. Jones, J.L. Ohmann, and F.J. Swanson. 2000. Windthrow disturbance, forest composition, and structure in the Bull Run basin, Oregon. Ecology **81**:2539–2556.

Slater, J.R. 1934. Notes on northwestern amphibians. Copeia **3**:140–141.

Slater, J.R. 1955. Distribution of Washington Amphibians. Occasional Paper No. 16. Department of Biology, College of Puget Sound, Tacoma, Washington, USA, pp. 122–154.

Sly, G.R. 1976. Small mammal succession on strip-mined land in Vigo County, Indiana. American Midland Naturalist **95**:257–267.

Smathers, G.A., and D. Mueller-Dombois. 1974. Invasion and Recovery of Vegetation after a Volcanic Eruption in Hawaii. National Park Service Science Monograph Series, No. 5. U.S. Government Printing Office, Washington, District of Columbia, USA.

Smith, B.D. 1966. Effect of the plant alkaloid sparteine on the distribution of the aphid *Acyrthosiphon spartii* (Koch.). Nature (London) **212**:213–214.

Smith, H.W., R. Okazaki, and J. Aarstad. 1968. Recent volcanic ash in soils of Northeastern Washington and Northern Idaho. Northwest Science **42**:150–160.

Smith, J.L., R.R. Schnabel, B.L. McNeal, and G.S. Campbell. 1980. Potential errors in the first-order model for estimating soil nitrogen mineralization potentials. Soil Science Society of America Journal **44**:996–1000.

Smith, M.A., and M.J. White. 1985. Observations on lakes near Mount St. Helens: Phytoplankton. Archives Hydrobiologia **104**:345–362.

Smith, R.D., and F.J. Swanson. 1987. Sediment routing in a small drainage basin in the blast zone at Mount St. Helens, Washington, USA. Geomorphology **1**:1–13.

Smith, J.L. 1994. Cycling of nitrogen through microbial activity. Pages 91–119 *in* J.L. Hatfield and B.A. Stewart, editors. Soil Biology: Effects on Soil Quality. Lewis, Boca Raton, Florida, USA.

Smith, S., and D.J. Read. 1997. Mycorrhizal Symbiosis. Academic Press, New York, New York, USA.

Smith, J.E., K.A. Johnson, and E. Cazares. 1998. Vesicular mycorrhizal colonization of seedlings of Pinaceae and Betulaceae after spore inoculation with *Glomus intraradices*. Mycorrhiza **7**:279–285.

Smith, J.L. 2002. Soil quality: The role of microorganisms. Pages 2944–2957 *in* G. Bitton, editor. Encyclopedia of Environmental Microbiology. Volume 5. John Wiley & Sons, New York, New York, USA.

Smith-Davidson, V.G. 1924. Animal communities of a deciduous forest succession. Ecology **9**:479–500.

Snyder, M. 1999. Still growing after 100 years: Weyerhaeuser Company celebrates its centennial. Forest History Today Fall:2–8.

Söderström, G., B. Svennsson, K. Vessby, and A. Glimskär. 2001. Plants, insects and birds in semi-natural pastures in relation to local habitat and landscape factors. Biodiversity and Conservation **10**:1839–1863.

Sollins, P., C.C. Grier, F.M. McCorison, J.K. Cromack, R. Fogel, and R.L. Fredriksen. 1980. The internal element cycles of an old-growth douglas-fir ecosystem in Western Oregon. Ecological Monographs **50**:261–285.

Spencer, C.N., K.O. Gabel, and F.R. Hauer. 2003. Wildfire effects on stream food webs and nutrient dynamics in Glacier National Park, USA. Forest Ecology and Management **178**:141–153.

Spotila, J.R. 1972. Role of temperature and water in the ecology of lungless salamanders. Ecological Monographs **42**:95–125.

Sprules, W.G. 1974. The adaptive significance of paedogenesis in North American species of *Ambystoma* (Amphibia: Caudata): An hypothesis. Canadian Journal of Zoology **52**:393–400.

Staley, J.T., L.G. Lehmicke, F.E. Palmer, R.W. Peet, and R.C. Wissmar. 1982. Impact of Mount St. Helens eruption on bacteriology of lakes in the blast zone. Applied and Environmental Microbiology **43**:664–670.

Stebbins, R.C., and N.H. Cohen. 1995. A Natural History of Amphibians. Princeton University Press, Princeton, New Jersey, USA.

Stehlik, I. 2000. Nunataks and peripheral refugia for alpine plants during quaternary glaciation in the middle part of the Alps. Botanica Helvetica **110**:25–30.

Stermitz, F.R., G.N. Belofsky, D. Ng, and M.C. Singer. 1989. Quinolizidine alkaloids obtained by *Pedicularis semibarbata* (Scrophulariaceae) from *Lupinus fulcratus* (Leguminosae) fail to influence the specialist herbivore *Euphydryas editha* (Lepidoptera). Journal of Chemical Ecology **15**:2521–2530.

Stevens, R.G., J.K. Winjum, R.R. Gilchrist, and D.A. Leslie. 1987. Revegetation in the western portion of the Mount St. Helens blast zone during 1980 and 1981. Pages 210–227 *in* D.E. Bilderback, editor. Mount St. Helens 1980: Botanical Consequences of the Explosive Eruptions. University of California Press, Berkeley, California, USA.

Stevenson, F.J., and M.A. Cole. 1999. Cycles of Soils: Carbon, Nitrogen, Phosphorus, Sulfur, Micronutrients. John Wiley & Sons, New York, New York, USA.

Stober, Q.J., B.D. Ross, C.L. Melby, P.A. Dinnel, T.H. Jagielo, and E.O. Salo. 1981. Effects of Suspended Volcanic Sediment on Coho and Chinook Salmon in the Toutle and Cowlitz Rivers. Fisheries Research Institute, University of Washington, Seattle, Washington, USA.

Stroh, J.R., and J.A. Oyler. 1981. Soil Conservation Service Seeding Evaluation on Mount St. Helens: Assessment of Grass-Legume Seeding in the Mount St. Helens Blast Area and the Lower Toutle River Mudflow. USDA Soil Conservation Service, Spokane, Washington, USA.

Strong, D.R., J.L. Maron, P.G. Connors, A. Whipple, S. Harrison, and R.L. Jefferies. 1995. High mortality, fluctuation in numbers, and heavy subterranean insect herbivory in bush lupine, *Lupinus arboreus*. Oecologia (Berlin) **104**:85–92.

Stuvier, M., and G.W. Pearson. 1993. High-precision bidecadal calibration of the radiocarbon time scale, A.D. 1950–500 B.C. and 2500–6000 B.C. Radiocarbon **35**:1–23.

Sugg, P.M. 1989. Arthropod Populations at Mount St. Helens: Survival and Revival. Doctoral dissertation. University of Washington, Seattle, Washington, USA.

Sugg, P.M., L. Greve, and J.S. Edwards. 1994. Neuropteroidea from Mount St. Helens and Mount Rainier: Dispersal and immigration in volcanic landscapes. Pan-Pacific Entomologist **70**:212–221.

Sugg, P.M., and J.S. Edwards. 1998. Pioneer aeolian community development on pyroclastic flows after the eruption of Mount St. Helens, Washington, U.S.A. Arctic and Alpine Research **30**: 400–407.

Sumioka, S.S., D.L. Kresch, and K.D. Kasnick. 1998. Magnitude and Frequency of Floods in Washington. Water-Resources Investigation Report 97-4277. Washington Department of Transportation, Washington State Department of Ecology, and U.S. Geological Survey; Tacoma, Washington, and Denver, Colorado, USA.

Sundh, I., M. Nilsson, and P. Borga. 1997. Variation in microbial community structure in two boreal peatlands as determined by analysis of phospholipid fatty acid profiles. Applied Environmental Microbiology **63**:1476–1482.

Swan, L.W. 1963. Aeolian zone. Science **140**:77–78.

Swanson, F.J. 1981. Fire and geomorphic processes. Pages 401–420 *in* H.A. Mooney, T.M. Bonnicksen, N.L. Christensen, J.E. Lotan, and W.A. Reiners, editors. Proceedings of the Conference on Fire Regimes and Ecosystem Properties. General Technical Report. WO-26. U.S. Department of Agriculture, Forest Service, Washington, District of Columbia, USA.

Swanson, F.J., M.M. Swanson, and C. Woods. 1981. Analysis of debris avalanche erosion in steep forest lands: An example from Mapleton, Oregon, U.S.A. Pages 67–75 *in* Erosion and Sediment Transport in Pacific Rim Steeplands. Publication No. 132. International Association of Hydrological Sciences, Christchurch, New Zealand.

Swanson, F.J., R.L. Fredriksen, and F.M. McCorison. 1982a. Material transfer in a western Oregon forested watershed. Pages 233–266 *in* R.L. Edmonds, editor. Analysis of Coniferous Forest Ecosystems in the Western United States. US/IBP Synthesis Series 14. Hutchinson Ross, Stroudsburg, Pennsylvania, USA.

Swanson, F.J., R.J. Janda, T. Dunne, and D.N. Swanston, editors. 1982b. Sediment Budgets and Routing in Forested Drainage Basins. General Technical Report PNW-141. USDA Forest Service, Portland, Oregon, USA.

Swanson, D.A., T.J. Casadevall, D. Dzurisin, S.D. Malone, C.G. Newhall, and C.S. Weaver. 1983a. Predicting eruptions at Mount St. Helens, June 1980 through December 1982. Science **221**:1369–1376.

Swanson, F.J., B. Collins, T. Dunne, and B.P. Wicherski. 1983b. Erosion of tephra from hillslopes near Mount St. Helens and other volcanoes. Pages 183–222 *in* Proceedings of the Symposium on Erosion Control in Volcanic Areas. Technical Memorandum No. 1908. Public Works Research Institute, Ministry of Construction, Tsukuba-shi, Ibaraki-ken, Japan.

Swanson, F.J. 1986. Debris slides in the eastern blast zone. Pages 29–31 *in* Proceedings, Field Trip Guidebook and Abstracts. American Geomorphological Field Group Conference Sept. 3–6, 1986, Cispus Center, Washington. U.S. Geological Survey, Vancouver, Washington, USA.

Swanson, F.J., T.K. Kratz, N. Caine, and R.G. Woodmansee. 1988. Landform effects on ecosystem patterns and processes. BioScience **38**:92–98.

Swanson, F.J., and J.F. Franklin. 1992. New forestry principles from ecosystem analysis of Pacific Northwest forests. Ecological Applications **2**:262–274.

Swanson, F.J., S.L. Johnson, S.V. Gregory, and S.A. Acker. 1998. Flood disturbance in a forested mountain landscape. BioScience **48**:681–689.

Sweet, J.W. 1986. A Survey and Ecological Analysis of Oregon and Idaho Phytoplankton. U.S. Environmental Protection Agency, Seattle, Washington, USA.

Swift, M.J., O.W. Heal, and J.M. Anderson. 1979. Decomposition in Terrestrial Ecosystems. Studies in Ecology, Volume 5. University of California Press, Berkeley, California, USA.

Swihart, R.K., and N.A. Slade. 1990. Long-term dynamics of an early successional small mammal community. American Midland Naturalist **123**:372–382.

Tabacchi, E., A.-M. Planty-Tabacchi, and O. Décamps. 1990. Continuity and discontinuity of the riparian vegetation along a fluvial corridor. Landscape Ecology **5**:9–20.

Tabatabai, M.A. 1994. Soil enzymes. Pages 775–833 *in* R.W. Weaver, S. Angle, P. Bottomley, D. Bezdicek, S. Smith, A. Tabatabai, and A. Wollum, editors. Methods of Soil Analysis, Part 2: Microbial and Biochemical Properties. Soil Science Society of America, Madison, Wisconsin, USA.

Tacoma Power and Washington Department of Fish and Wildlife. 2004. Cowlitz River fisheries and hatchery management plan (FHMP). Public Review Draft. Tacoma Power Natural Resources and Washington Department of Fish and Wildlife; Tacoma, Washington, and Olympia, Washington, USA.

Tagawa, H., E. Suzuki, T. Partomihardjo, and A. Suriadarma. 1985. Vegetation and succession on the Krakatau Islands, Indonesia. Vegetatio **60**:131–145.

Taylor, B.W. 1957. Plant succession on recent volcanoes in Papua. Journal of Ecology **45**:233–243.

Teal, J.M. 1957. Community metabolism in a temperate cold spring. Ecological Monographs **27**:283–302.

ter Braak, C.J.F. 1986. Canonical correspondence analysis: A new eigenvector technique for multivariate direct gradient analysis. Ecology **67**:1167–1179.

Thomas, D.W., and S.D. West. 1991. Forest age associations of bats in the southern Washington Cascade and Oregon Coast Ranges. Pages 295–303 *in* L.F. Ruggiero, K.B. Aubry, A.B. Carey, and M.H. Huff, editors. Wildlife and Vegetation of Unmanaged Douglas-Fir Forests. PNW-GTR-285. USDA Forest Service, Pacific Northwest Research Station, Portland, Oregon.

Thompson, J.N., and P.W. Price. 1977. Plant plasticity, phenology, and herbivore dispersion: Wild parsnip and the parsnip webworm. Ecology **58**:1112–1119.

Thompson, R.G. 1979. Larvae of North American Carabidae with a key to the tribes. Pages 209–291 *in* T.L. Erwin, G.E. Ball, and D.R. Whitehead, editors. Carabid Beetles: Their Evolution, Natural History and Classification.Junk, The Hague, The Netherlands.

Thompson, D. 2000. Volcano Cowboys: The Rocky Evolution of a Dangerous Science. Thomas Dunne, New York, New York, USA.

Thompson, J.N., O.J. Reichman, P.J. Morin, G.A. Polis, M.E. Power, R.W. Sterner, C.A. Couch, L. Gough, R. Holt, D.U. Hooper, F. Keesing, C.R. Lovell, B.T. Milne, M.C. Molles, D.W. Roberts, and S.Y. Strauss. 2001. Frontiers of ecology. Bioscience **51**: 15–24.

Thoreau, H.D. 1993. Faith in a Seed. Island Press, Washington, District of Columbia, USA.

Thornton, I.W.B., T.R. New, R.A. Zann, and P.A. Rawlinson. 1990. Colonization of the Krakatau Islands, Indonesia, by animals: A perspective from the 1980's. Philosophical Transactions of the Royal Society, London, B **328**:131–165.

Thornton, I.W.B. 1996. Krakatau. The Destruction and Reassembly of an Island Ecosystem. Harvard University Press, Cambridge, Massachusetts, USA.

Thornton, I.W.B. 2000. The ecology of volcanoes: Recovery and assembly of living communities. Pages 1057–1081 *in* H. Sigurdson, editor in chief. Encyclopedia of Volcanoes. Academic Press, San Diego, California, USA.

Thurman, E.M. 1985. Organic Geochemistry of Natural Waters. Martinus Nijhoff/Dr. W. Junk, Dordrecht, The Netherlands.

Tidemann, C.R., D.J. Kitchener, R.A. Zann, and I.W.B. Thornton. 1990. Recolonization of the Krakatau Islands and adjacent areas of West Java, Indonesia, by bats (Chiroptera), 1883–1986. Philosophical Transactions of the Royal Society, London, B **328**:123–130.

Tilman, D., D. Wedin, and J. Knops. 1996. Productivity and sustainability influenced by biodiversity in grassland ecosystems. Nature (London) **379**:718–720.

Tison, D.L., J.A. Baross, and R.J. Seidler. 1983. *Legionella* in aquatic habitats in the Mount Saint Helens blast zone. Current Microbiology **9**:345–348.

Titus, J.H., J. Christy, D. Vander Schaaf, and J.S. Kagan. 1996. Native Wetland, Riparian and Upland Plant Communities and Their Biota in the Willamette Valley, Oregon. Oregon Natural Heritage Program, The Nature Conservancy, Portland, Oregon, USA.

Titus, J.H., and R. del Moral. 1998a. Seedling establishment in different microsites on Mount St. Helens, Washington, USA. Plant Ecology **134**:13–26.

Titus, J.H., and R. del Moral. 1998b. The role of mycorrhizal fungi and microsites in primary succession on Mount St. Helens. American Journal of Botany **85**:370–375.

Titus, J.H., and R. del Moral. 1998c. Vesicular-arbuscular mycorrhizae influence Mount St. Helens pioneer species in greenhouse experiments. Oikos **81**:495–510.

Titus, J.H., R. del Moral, and S. Gamiet. 1998a. The distribution of vesicular-arbuscular mycorrhizae on Mount St. Helens. Madroño **45**:162–170.

Titus, J.H., S. Moore, M. Arnot, and P. Titus. 1998b. Inventory of the vascular flora of the blast zone, Mount St. Helens, Washington. Madroño **45**:146–161.

Titus, J.H., P.J. Titus, and R. del Moral. 1999. Wetland development in primary and secondary successional substrates: Fourteen years after the eruption of Mount St. Helens, Washington, USA. Northwest Science **73**:186–204.

Titus, J.H., and S. Tsuyuzaki. 2003a. Distribution of plants in relation to microsites on recent volcanic substrates on Mount Koma, Hokkaido, Japan. Ecological Research **18**:91–98.

Titus, J.H., and S. Tsuyuzaki. 2003b. Influence of a non-native invasive tree on primary succession at Mt. Koma, Hokkaido, Japan. Plant Ecology **169**:307–315.

Trappe, J.M., J.F. Franklin, R.F. Tarrant, and G.M. Hansen, editors. 1968. Biology of Alder. Proceedings, Northwest Scientific Association, Fortieth Annual Meeting, Pullman, Washington, April 14–15, 1967. Pacific Northwest Forest and Range Experiment Station, USDA Forest Service, Portland, Oregon.

Trappe, J.M. 1977. Selection of fungi for ectomycorrhizal inoculation in nurseries. Annual Review of Phytopathology **15**:203–222.

Triska, F.J., J.R. Sedell, and S.V. Gregory. 1982. Coniferous forest streams. Pages 292–332 *in* R.L. Edmonds, editor. Analysis of Coniferous Forest Ecosystems in the Western United States. US/IBP Synthesis Series 14. Hutchinson Ross, Stroudsburg, Pennsylvania, USA.

Tsuyuzaki, S. 1987. Origin of plants recovering on the volcano Usu, Northern Japan, since the eruptions of 1977 and 1978. Vegetatio **73**:53–58.

Tsuyuzaki, S. 1989. Analysis of revegetation dynamics on the volcano Usu, northern Japan, deforested by 1977–1978 eruptions. American Journal of Botany **76**:1468–1477.

Tsuyuzaki, S. 1991. Species turnover and diversity during early stages of vegetation recovery on the volcano Usu, northern Japan. Journal of Vegetation Science **2**:301–306.

Tsuyuzaki, S. 1994. Fate of plants from buried seeds on Volcano Usu, Japan, after the 1977–1978 eruptions. American Journal of Botany **81**:395–399.

Tsuyuzaki, S., and R. del Moral. 1994. Canonical correspondence analysis of early volcanic succession on Mt. Usu, Japan. Ecological Research **9**:143–150.

Tsuyuzaki, S. 1995. Vegetation recovery patterns in early volcanic succession. Journal of Plant Research **108**:241–248.

Tsuyuzaki, S., and R. del Moral. 1995. Species attributes in early primary succession. Journal of Vegetation Science **6**: 517–522.

Tsuyuzaki, S. 1996. Species diversity analyzed by density and cover in an early volcanic succession. Vegetatio **122**:151–156.

Tsuyuzaki, S., and M. Haruki. 1996. Tree regeneration patterns on Mount Usu, northern Japan, since the 1977–78 eruptions. Vegetatio **126**:191–198.

Tsuyuzaki, S., and J.H. Titus. 1996. Vegetation development patterns in erosive areas on the Pumice Plains of Mount St. Helens. American Midland Naturalist **135**:172–177.

Tsuyuzaki, S. 1997. Wetland development in early stages of volcanic succession. Journal of Vegetation Science **8**:353–360.

Tsuyuzaki, S., J.H. Titus, and R. del Moral. 1997. Seedling establishment patterns on the Pumice Plain, Mount St. Helens, Washington. Journal of Vegetation Science **8**:727–734.

Tsuyuzaki, S., and M. Goto. 2001. Persistence of seed bank under thick volcanic deposits twenty years after eruptions of Mount Usu, Hokkaido Island, Japan. American Journal of Botany **88**:1813–1817.

Tsuyuzaki, S. 2002. Vegetation development patterns on skislopes in lowland Hokkaido, northern Japan. Biological Conservation **108**:239–246.

Tu, M., J.H. Titus, S. Tsuyuzaki, and R. del Moral. 1998. Composition and dynamics of the wetland seed bank on Mount St. Helens, Washington, USA. Folia Geobotanica **33**:3–16.

Turner, M.G., V.H. Dale, and E.E. Everham, III. 1997. Crown fires, hurricanes, and volcanoes: A comparison among large-scale disturbances. Bioscience **47**:758–768.

Turner, M.G., W.I. Baker, C.J. Peterson, and R.K. Peet. 1998. Factors influencing succession: Lessons from large, infrequent natural disturbances. Ecosystems **1**:511–523.

Turner, M.G., and V.H. Dale. 1998. Comparing large infrequent disturbances: What have we learned? Ecosystems **1**:493–496.

U.S. Army Corps of Engineers. 1987. Spirit Lake Outlet Tunnel, Spirit Lake, Washington. U.S. Army Corps of Engineers, Portland, Oregon, USA.

U.S. Congress, House Committee on Agriculture and House Committee on Interior and Insular Affairs. 1982. Mount St. Helens National Volcanic Areas. Joint hearings on H.R. 5281, H.R. 5773, and H.R. 5787. 97th Congress, 2nd Session. U.S. Government Printing Office, Washington, District of Columbia, USA.

USDA. No date. Integrated Taxonomic Information System (ITIS) website: http://www.itis.usda.gov.

USDA Forest Service. 1981a. Environmental Assessment: Mount St. Helens Volcano Timber Salvage. Gifford Pinchot National Forest, Vancouver, Washington, USA.

USDA Forest Service. 1981b. Mount St. Helens Land Management Plan. Final Environmental Impact Statement. Gifford Pinchot National Forest, Vancouver, Washington, USA.

USDA Forest Service. 1985. Mount St. Helens National Volcanic Monument. Final Environmental Impact Statement Comprehensive Management Plan. Record of Decision. USDA Forest Service, Portland, Oregon, USA.

USDA Soil Conservation Service. 1980. Assessment of Cowlitz River, Toutle River, and Blast Area Damage Resulting from May 18, 1980, Mt. St. Helens Eruption, Washington. USDA Soil Conservation Service, Spokane, Washington, USA.

Vallance, J.W., and K.M. Scott. 1997. The Osceola mudflow from Mount Rainier: Sedimentology and hazards implications of a huge clay-rich debris flow. Geological Society of America Bulletin **109**:143–163.

Vallance, J.W. 2000. Lahars. Pages 601–616 *in* H. Sigurdsson, editor. Encyclopedia of Volcanoes. Academic Press, San Diego, California, USA.

Van Cleve, K., L.A. Viereck, and G.M. Marion. 1993. Introduction and overview of a study dealing with the role of salt-affected soils in primary succession on the Tanana River floodplain, interior Alaska. Canadian Journal of Forest Research **23**:879–888.

van der Heijden, M.G.A., and I.R. Sanders. 2002. Mycorrhizal Ecology. Springer-Verlag, New York, New York, USA.

van der Maarel, E., and M.T. Sykes. 1997. Rates of small-scale species mobility in alvar limestone grassland. Journal of Vegetation Science **8**:199–208.

Van Dyke, E.C. 1928. The American species of *Pteroloma* (Coleoptera: Silphidae) and a new Japanese species. Bulletin of the Brooklyn Entomological Society **23**:19–27.

Vanderbilt, K.L., K. Lajtha, and F.J. Swanson. 2003. Biogeochemistry of unpolluted forested watersheds in the Oregon Cascades: Temporal patterns of precipitation and stream nitrogen flux. Biogeochemistry **62**:87–117.

Veblen, T.T., and D.H. Ashton. 1978. Catastrophic influences on the vegetation of the Valdivian Andes, Chile. Vegetatio **36**:149–167.

Verts, B.J. 1957. The population and distribution two species of *Peromyscus* on some Illinois strip-mined land. Journal of Mammalogy **38**:53–59.

Vestal, J.R., and D.C. White. 1989. Lipid analysis in microbial ecology. BioScience **39**:535–541.

Vitousek, P.M., L.R. Walker, and L.D. Whiteaker. 1987. Biological invasion by *Myrica faya* alters ecosystem development in Hawaii. Science **238**:802–804.

Vitousek, P.M., P.A. Matson, and K. Van Cleve. 1989. Nitrogen availability and nitrification during succession: Primary, secondary, and old-field seres. Plant and Soil **115**:229–239.

Vogel, A.H, R.R. Petersen, and V.J. Kelly. 2000. Zooplankton colonization of two new lakes, Mt. St. Helens, Washington, U.S.A. Internationale Vereinigung für Theoretische und Angewandte Limnologie: Verhandlungen **27**:785–790.

Voight, B. 1981. Time scale for the first moments of the May 18 eruption. Pages 69–86 *in* P.W. Lipman and D.R. Mullineaux, editors. The 1980 Eruptions of Mount St. Helens, Washington. Professional Paper 1250. U.S. Geological Survey, Washington, District of Columbia, USA.

Voight, B., H. Glicken, R.J. Janda, and P.M. Douglas. 1981. Catastrophic rockslide avalanche of May 18. Pages 347–378 *in* P.W. Lipman and D.R. Mullineaux, editors. The 1980 Eruptions of Mount St. Helens, Washington. Professional Paper 1250. U.S. Geological Survey, Washington, District of Columbia, USA.

Voight, B., R.J. Janda, H. Glicken, and P.M. Douglass. 1983. Nature and mechanics of the Mount St. Helens rockslide-avalanche of 18 May 1980. Geotechnique **33**:243–273.

Wachendorf, C., U. Irmler, and H.-P. Blume. 1997. Relationships between litter fauna and chemical changes of litter during decomposition under different moisture conditions. Pages 135–144 *in* G. Cadish and K.E. Giller, editors. Driven by Nature: Plant Litter Quality and Decomposition. CAB International, Oxford, UK.

Waitt, R.B., Jr. 1981. Devastating pyroclastic density flow and attendant airfall of May 18: Stratigraphy and sedimentology of deposits. Pages 439–458 *in* P.W. Lipman and D.R. Mullineaux, editors. The 1980 eruptions of Mount St. Helens, Washington.

Professional Paper 1250. U.S. Geological Survey, Washington, District of Columbia, USA.

Waitt, R.B., Jr., and D. Dzurisin. 1981. Proximal air-fall deposits from the May 18 eruption: Stratigraphy and field sedimentology. Pages 601–616 *in* P.W. Lipman and D.R. Mullineaux, editors. The 1980 Eruptions of Mount St. Helens, Washington. Professional Paper 1250. U.S. Geological Survey, Washington, District of Columbia, USA.

Waitt, R.B., V.L. Hansen, A.M. Sarna-Wojcicki, and S.H. Wood. 1981. Proximal air-fall deposits of eruptions between May 24 and August 1980: Stratigraphy and field sedimentology. Pages 617–628 *in* P.W. Lipman and D.R. Mullineaux, editors. The eruptions of Mount St. Helens, Washington. Professional Paper 1250. U.S. Geological Survey, Washington, District of Columbia, USA.

Waitt, R.B., Jr., T.C. Pierson, N.S. MacLeod, R.J. Janda, B. Voight, and R.T. Holcomb. 1983. Eruption-triggered avalanche, flood, and lahar at Mount St. Helens: Effects of winter snowpack. Science **221**:1394–1397.

Waitt, R.B. 1989. Swift snowmelt and floods (lahars) caused by great pyroclastic surge at Mount St. Helens volcano, Washington, 18 May 1980. Bulletin of Volcanology **52**:138–157.

Waksman, S.A., and F.G. Tenney. 1927. The composition of natural organic materials and their decomposition in the soil. II. Influence of age of plant upon the rapidity and nature of its decomposition: Rye plants. Soil Science **24**:317–333.

Wali, M.K. 1999. Ecological succession and the rehabilitation of disturbed terrestrial ecosystems. Plant and Soil **213**:195–220.

Walker, L.R., and F.S. Chapin, III. 1986. Physiological controls over seedling growth in primary succession on an Alaskan floodplain. Ecology **67**:1508–1523.

Walker, L.R., J.C. Zasada, and F.S. Chapin, III. 1986. The role of life history processes in primary succession on an Alaskan floodplain. Ecology **67**:1243–1253.

Walker, L.R., and F.S. Chapin, III. 1987. Interactions among processes controlling successional change. Oikos **50**:131–135.

Walker, L.R. 1993. Nitrogen fixers and species replacements in primary succession. Pages 249–272 *in* J. Miles and D.W.H. Walton, editors. Primary Succession on Land. Blackwell Scientific, Oxford, UK.

Walker, L.R., and R. del Moral. 2001. Primary succession. *In* Encyclopedia of Life Sciences (electronic journal). Elsevier, The Hague, The Netherlands.

Walker, L.R., B.D. Clarkson, W.B. Silvester, and B.R. Clarkson. 2003. Colonization dynamics and facilitative impacts of a nitrogen-fixing shrub in primary succession. Journal of Vegetation Science **14**:277–290.

Walker, L.R., and R. del Moral. 2003. Primary Succession and Ecosystem Rehabilitation. Cambridge University Press, Cambridge, UK.

Wall, D.H., P.V.R. Snelgrove, and A.P. Covich. 2001. Conservation priorities for soil and sediment invertebrates. Pages 99–124 *in* M.E. Soule and G.H. Orians, editors. Conservation Biology: Research Priorities for the Next Decade. Island Press, Washington, District of Columbia, USA.

Walter, H. 1973. Vegetation of the Earth in Relation to Climate and Eco-Physiological Conditions. Springer-Verlag, New York, New York, USA.

Ward, A.K., J.A. Baross, C.N. Dahm, M.D. Lilley, and J.R. Sedell. 1983. Qualitative and quantitative observations on aquatic algal communities and recolonization within the blast zone of Mount St. Helens, 1980 and 1981. Journal of Phycology **19**: 238–247.

Wardle, D.A., and A. Ghani. 1995. A critique of the microbial quotient (qCO_2), a bioindicator of disturbance and ecosystem development. Soil Biology and Biochemistry **27**:1601–1610.

Waring, R.H., and J.F. Franklin. 1979. Evergreen coniferous forests of the Pacific Northwest. Science **204**:1380–1386.

Warrick, A.W., D.E. Myers, and D.R. Nielsen. 1986. Geostatistical methods applied to soil science. Pages 53–82 *in* A. Klute, editor. Methods of Soil Analysis, Part 1. Physical and Mineralogical Methods. American Society of Agronomy, Madison, Wisconsin, USA.

Waters, T.F. 1983. Replacement of brook trout by brown trout over 15 years in a Minnesota stream: Production and abundance. Transactions of the American Fisheries Society **112**:137–146.

Weber, M.H. 2001. Patterns in Forest Succession and Mortality Following Burial by Mudflow at Cedar Flats, Mount St. Helens, Washington. Master's thesis. Portland State University, Portland, Oregon, USA.

Weiher, E., and P. Keddy. 1995. The assembly of experimental wetland plant communities. Oikos **73**:323–335.

Weins, J.A. 1978. Nongame bird communities in northwestern coniferous forests. Pages 19–31 *in* R.M. DeGraff, editor. Proceedings of the Workshop on Nongame Bird Habitat Management in Coniferous Forests of the Western United States. PNW-GTR-64. USDA Forest Service, Pacific Northwest Forest and Range Experiment Station, Portland, Oregon, USA.

Welch, E.B. 1979. Lake restoration by dilution. Pages 133–139 *in* Lake Restoration. EPA-400/5-79-001. U.S. Environmental Protection Agency, Washington, District of Columbia, USA.

Welsh, H.H., and A.J. Lind. 1995. Habitat correlates of the Del Norte salamander, *Plethodon elongatus*, in northwestern California. Journal of Herpetology **29**:198–210.

Welsh, H.H., and L.M Ollivier. 1998. Stream amphibians as indicators of ecosystem stress: A case study from California's redwoods. Ecological Applications **8**:1118–1132.

West, S.D. 1991. Small mammal communities in the southern Washington Cascade Range. Pages 268–283 *in* L.F. Ruggiero, K.B. Aubry, A.B. Carey, and M.H. Huff, technical editors. Wildlife and Vegetation of Unmanaged Douglas-Fir Forests. General Technical Report PNW-GTR-285. USDA Forest Service, Pacific Northwest Research Station, Portland, Oregon, USA.

West, S.D. 2000. Terrestrial small mammals. Chapter 10 *in* Effectiveness of Riparian Management Zones in Providing Habitat for Wildlife. Final Report. Timber, Fish and Wildlife Cooperative Monitoring Committee, Washington Department of Natural Resources, Olympia, Washington, USA.

Wetzel, R.M. 1958. Mammalian succession on midwestern floodplains. Ecology **39**:262–271.

Wetzel, R.G., and G.E. Likens. 1979. Limnological Analyses. Second edition. Saunders, Philadelphia, Pennsylvania, USA.

Wetzel, R.G. 2001. Limnology: Lake and River Ecosystems. Third edition. Academic Press, New York, New York, USA.

White, P.S., and S.T.A. Pickett. 1985. Natural disturbance and patch dynamics: An introduction. Pages 3–13 *in* S.T.A. Pickett and P.W. White, editors. The Ecology of Natural Disturbance and Patch Dynamics. Academic Press, New York, New York, USA.

White, T.C.R. 1993. The Inadequate Environment: Nitrogen and the Abundance of Animals. Springer-Verlag, New York, New York, USA.

White, P.S., and A. Jentsch. 2001. The search for generality in studies of disturbance and ecosystem dynamics. Progress in Botany **62**:399–450.

Whitman, R.P., T.P. Quinn, and E.L. Brannon. 1982. Influence of suspended volcanic ash on homing behavior of adult chinook salmon. Transactions of the American Fisheries Society **111**:63–69.

Whittaker, R.H. 1953. A consideration of climax theory: The climax as a population and pattern. Ecological Monographs **23**:41–78.

Whittaker, R.J., M.B. Bush, and K. Richards. 1989. Plant recolonization and vegetation succession on the Krakatau Islands, Indonesia. Ecological Monographs **59**:59–123.

Whittaker, R.J., M.B. Bush, T. Partomihardjo, and N.M. Asquith. 1992. Ecological aspects of plant colonization of the Krakatau Islands. GeoJournal **28**:201–210.

Whittaker, R.J., S.F. Schmitt, S.H. Jones, T. Partomihardjo, and M.B. Bush. 1998. Stand biomass and tree mortality from permanent forest plots on Krakatau, Indonesia, 1989–1995. Biotropica **30**:519–529.

Whittaker, R.J., T. Partomihardjo, and S.H. Jones. 1999. Interesting times on Krakatau: Stand dynamics in the 1990s. Philosophical Transactions of the Royal Society, London, B **354**:1857–1867.

Widmer, F., A. Fließbach, E. Laczkó, J. Schulze-Aurich, and J. Zeyer. 2001. Assessing soil biological characteristics: A comparison of bulk soil community DNA-, PLFA-, and Biolog^{TM}-analyses. Soil Biology and Biochemistry **33**:1029–1036.

Wieder, R.K., and G.E. Lang. 1982. A critique of the analytical methods used in examining decomposition data obtained from litter bags. Ecology **63**:1636–1642.

Willson, M.F., and K.C. Halupka. 1995. Anadromous fish as key-stone species in vertebrate communities. Conservation Biology **9**:489–497.

Wilson, A.G. 1993. Distribution of Van Dyke's Salamander, *Plethodon vandykei* Van Denburgh. Washington Department of Wildlife, Olympia, Washington, USA.

Wilson, J.B. 1999. Assembly rules in plant communities. Pages 251–271 *in* E. Weiher and P. Keddy, editors. Ecological Assembly Rules: Perspectives, Advances, Retreats. Cambridge University Press, Cambridge, UK.

Wilson, D.E., and S. Ruff. 1999. The Smithsonian Book of North American Mammals. Smithsonian Institution Press, Washington, District of Columbia, USA.

Winjum, J.K., J.E. Keatley, R.G. Stevens, and J.R. Gutzwiler. 1986. Regenerating the blast zone of Mount St. Helens. Journal of Forestry **84**(5):29–35.

Wink, M., T. Hartmann, L. Witte, and J. Rheinheimer. 1982. Interrelationship between quinolizidine alkaloid producing legumes and infesting insects: Exploitation of the alkaloid-containing phloem sap of *Cystisus scoparius* by the broom aphid *Aphis cystisorum*. Zeitschrift für Naturforschung C, A Journal of Biosciences **37**:1081–1086.

Wink, M., and D.B. Carey. 1994. Variability of quinolizidine alkaloid profiles of *Lupinus argenteus*. Biochemical Systematics and Ecology **22**:663–669.

Wink, M., C. Meissner, and L. Witte. 1995. Patterns of quinolizidine alkaloids in 56 species of the genus *Lupinus*. Phytochemistry **38**:139–153.

Winner, W.E., and H.A. Mooney. 1980. Responses of Hawaiian plants to volcanic sulfur dioxide: Stomatal behavior and foliar injury. Science **210**:789–791.

Winner, W.E., and T.J. Casadevall. 1981. Fir leaves as thermometers during the May 18 eruption. Pages 315–320 *in* P.W. Lipman and D.R. Mullineaux, editors. The 1980 Eruptions of Mount St. Helens, Washington. Professional Paper 1250. U.S. Geological Survey, Washington, District of Columbia, USA.

Winner, W.E., and T.J. Casadevall. 1983. The effects of the Mount St. Helens eruption cloud on fir (*Abies* sp.) needle cuticles: Analysis with scanning electron microscopy. American Journal of Botany **70**:80–87.

Wise, D.H. 1993. Spiders in Ecological Webs. Cambridge University Press, Cambridge, UK.

Wissmar, R.C., A.H. Devol, A.N. Nevissi, and J.R. Sedell. 1982a. Chemical changes of lakes within the Mount St. Helens blast zone. Science **216**:175–178.

Wissmar, R.C., A.H. Devol, J.T. Staley, and J.R. Sedell. 1982b. Biological responses of lakes in the Mount St. Helens blast zone. Science **216**:178–181.

Wissmar, R.C., J.A. Baross, M.D. Lilley, and C.N. Dahm. 1988. Nitrogen cycling in altered and newly created lakes near the Mount St. Helens volcano. Journal of Freshwater Ecology **4**:551–568.

Wissmar, R.C. 1990. Recovery of lakes in the 1980 blast zone of Mount St. Helens. Northwest Science **64**:268–270.

Wissmar, R.C., D.M. McKnight, and C.N. Dahm. 1990. Contributions of organic acids to alkalinity in lakes within the Mount St. Helens blast zone. Limnology and Oceanography **35**:535–542.

Wojciechowski, M.F., and M.E. Heimbrook. 1984. Dinitrogen fixation in alpine tundra, Niwot Ridge, Front Range, Colorado, USA. Arctic and Alpine Research **16**:1–10.

Wolcott, E.E. 1973. Lakes of Washington, Volume 1: Western Washington. Water-Supply Bulletin 14. Washington Department of Ecology, Olympia, Washington, USA.

Wood, D.M. 1987. Pattern and Process in Primary Succession in High-Elevation Habitats on Mount St. Helens. Doctoral dissertation. University of Washington, Seattle, Washington, USA.

Wood, D.M., and R. del Moral. 1987. Mechanisms of early primary succession in subalpine habitats on Mount St. Helens. Ecology **68**:780–790.

Wood, D.M., and R. del Moral. 1988. Colonizing plants on the Pumice Plains, Mount St. Helens, Washington. American Journal of Botany **75**:1228–1237.

Wood, D.M., and M.C. Anderson. 1990. The effect of predispersal seed predators on colonization of *Aster ledophyllus* on Mount St. Helens. American Midland Naturalist **123**:193–201.

Wood, D.M., and W.F. Morris. 1990. Ecological constraints to seedling establishment on the Pumice Plains, Mount St. Helens, Washington. American Journal of Botany **77**:1411–1418.

Wood, D.M., and R. del Moral. 2000. Seed rain during early primary succession on Mount St. Helens, Washington. Madroño **47**: 1–9.

Woods, K.D. 2000. Dynamics in late-successional hemlock-hardwood forests over three decades. Ecology **81**:110–126.

Woolbright, L.L. 1991. The impact of hurricane Hugo on forest frogs of Puerto Rico. Biotropica **23**:462–467.

Woolbright, L.L. 1996. Disturbance influences long-term population patterns in the Puerto Rican frog, *Eleutherodactylus coqui* (Anura: Leptodactylidae). Biotropica **28**:493–501.

Woolbright, L.L. 1997. Local extinction of anuran amphibians in the Luquillo experimental forest of northeastern Puerto Rico. Journal of Herpetology **31**:572–576.

Wydoski, R.S., and R.R. Whitney. 1979. Inland Fishes of Washington. University of Washington Press, Seattle, Washington, USA.

Wydoski, R.S., and R.R. Whitney. 2003. Inland Fishes of Washington. Second edition. University of Washington Press, Seattle, Washington, USA.

Yamaguchi, D.K. 1993. Forest history, Mount St. Helens. National Geographic Research & Exploration **9**:294–325.

Yamaguchi, D.K., and R.P. Hoblitt. 1995. Tree-ring dating of pre-1980 flowage deposits at Mt. St. Helens, Washington. Geological Society of America Bulletin **107**:1077–1093.

Yoder, B.J., and R.H. Waring. 1994. The normalized difference vegetation index of small Douglas-fir canopies with varying chlorophyll concentrations. Remote Sensing of Environment **49**:81–91.

Young, T.P., J.M. Chase, and R.T. Huddleston. 2001. Community succession and assembly. Ecological Restoration **19**:1–18.

Zak, D.R., D.F. Grigal, S. Gleeson, and D. Tilman. 1990. Carbon and nitrogen cycling during old-field succession: Constraints on plant and animal biomass. Biogeochemistry **11**:111–129.

Zalisko, E.J., and R.W. Sites. 1989. Salamander occurrence within Mt. St. Helens blast zone. Herpetological Review **20**(4): 84–85.

Zanaboni, A., and G.G. Lorenzoni. 1989. The importance of hedges and relict vegetation in agroecosystems and environment reconstitution. Agriculture Ecosystems and Environment **27**: 155–161.

Zelles, L. 1999. Fatty acid patterns of phospholipids and lipopolysaccharides in the characterization of microbial communities in soil: A review. Biology and Fertility of Soils **29**:111–129.

Zobel, D.B., and J.A. Antos. 1982. Adventitious rooting of eight conifers into a volcanic tephra deposit. Canadian Journal of Forest Research **12**:717–719.

Zobel, D.B., and J.A. Antos. 1985. Response of conifer shoot elongation to tephra from Mount St. Helens. Forest Ecology and Management **12**:83–91.

Zobel, D.B., and J.A. Antos. 1986. Survival of prolonged burial by subalpine forest understory plants. American Midland Naturalist **115**:282–287.

Zobel, D.B., and J.A. Antos. 1987a. Composition of rhizomes of forest herbaceous plants in relation to morphology, ecology and burial by tephra. Botanical Gazette **148**:490–500.

Zobel, D.B., and J.A. Antos. 1987b. Survival of *Veratrum viride*, a robust herbaceous plant, when buried by volcanic tephra. Northwest Science **61**:20–22.

Zobel, D.B., and J.A. Antos. 1991a. 1980 tephra from Mount St. Helens: Spatial and temporal variation beneath forest canopies. Biology and Fertility of Soils **12**:60–66.

Zobel, D.B., and J.A. Antos. 1991b. Growth and development of natural seedlings of *Abies* and *Tsuga* in old-growth forest. Journal of Ecology **79**:985–998.

Zobel, D.B., and J.A. Antos. 1992. Survival of plants buried for eight growing seasons by volcanic tephra. Ecology **73**:698–701.

Zobel, D.B., and J.A. Antos. 1997. A decade of recovery of understory vegetation buried by volcanic tephra from Mount St. Helens. Ecological Monographs **67**:317–344.

Glossary

Some definitions of ecological terms in this glossary are derived from the *Encyclopedia of Global Environmental Change* (Munn 2002); the *Dictionary of Ecology* (Hansen 1972); or *A Dictionary of Ecology, Evolution and Systematics* (Lincoln et al. 1998). Some definitions of geological terms are from *Volcanic Eruptions of 1980 at Mount St. Helens: The First 100 Days* (Foxworthy and Hill 1982); *The 1980 Eruptions of Mount St. Helens, Washington* (Lipman and Mullineaux 1981); *Volcanoes* (MacDonald 1972); and *The 1980 Eruption of Mount St. Helens, Washington. Part I: March 20–May 19, 1980* (Korosec et al. 1980).

Abundance: The total number of individuals of a taxon or of taxa in an area, volume, population, or community.

Adventitious roots: Roots that develop without a pattern.

Adaptation: The process of adjustment of an individual organism to environmental stress.

Aeolian zone: The habitat of communities supported by allochthonous inputs of nutrients carried by winds.

Aerobic: Growing or occurring only in the presence of molecular oxygen.

Age class: A category comprising individuals of a given age within a population; cohort.

Aggradation: The filling up of a channel by sediment deposition.

Airfall (Ashfall): Volcanic ash and coarser material that has fallen through the air from an eruption cloud. A deposit so formed is commonly well sorted and layered.

Allele: Any of the different forms of a gene occupying the same locus.

Allochthonous: Exogenous; originating outside of and transported into a given system or area.

Allogeneic succession: The replacement of one community by another as a result of extrinsic changes in the environment.

Allopatric: Used of populations, species, or taxa occupying different and disjunct geographical areas.

Alluvial deposit: Sediment deposited by flowing water.

Amphibiotic: Having aquatic larval stages and terrestrial adults.

Amplexus: Precopulation pairing; the grasping of the female by the male before copulation.

Anadromous: Migrating from saltwater to freshwater, as in the case of a fish moving from the sea into a river to spawn.

Anaerobic: Growing or occurring in the absence of molecular oxygen.

Andesite: A volcanic rock containing 53% to 63% silica (SiO_2); has moderate viscosity in the molten state.

Annual: Pertaining to a year; having an annual periodicity; living for 1 year.

Arboreal: Living in trees; adapted for life in trees.

Anuran: Frogs and toads.

Ash: Fine (less than 4-mm-diameter) particles of pulverized rock blown from an explosion vent. Measuring less than 0.1 inch in diameter, ash particles may be either solid or molten when first erupted. A common variety is vitric ash, glassy particles formed by gas bubbles bursting through liquid magma.

Ashfall (Airfall): Volcanic ash and coarser material that has fallen through the air from an eruption cloud. A deposit so formed is commonly well sorted and layered.

Ashflow (see also Pyroclastic flow): A turbulent mixture of gas and rock fragments, most of which are ash-sized particles, ejected violently from a crater or fissure. Ashflows are usually of high temperature and move rapidly over the land surface.

Assemblage: A group of species occurring together in the same geographical area.

Autochthonous: Endogenous; produced within a given habitat, community, or system.

Autogenic succession: The replacement of one community by another as a result of intrinsic changes in the environment brought about by the community itself.

Autotrophic: Capable of synthesizing complex organic substances from simple inorganic substrates.

Autotrophic food webs: The network of producing and consuming interactions among organisms based on algal production within streams; a stream in which all or most of the organic matter present is derived from within the stream and not from drainage or input from the surrounding land.

Avifauna: The bird fauna of an area or period.

Basalt: A volcanic rock containing less than 53% silica (SiO_2); has low viscosity in the molten state.

Bathymetry: The measurement of ocean or lake depth and the study of the floor topography.

Belt transect: A narrow belt of predetermined width set out across a study area and within which the occurrence or distribution of plants and/or animals is recorded.

Benthic: Pertaining to the sea bed, river bed, or lake floor.

Biodiversity: The variety of organisms considered at all levels, from genetic variants of a single species through arrays of species to arrays of genera, families, ecosystems, and still higher levels; the totality of biological diversity.

Biological legacy: Organisms, propagules, and organically derived structures and patterns that remain following disturbances; includes surviving plants, animals, seeds, spores of microorganisms, and plant or animal organic matter, such as large woody debris and above- and belowground litter (e.g., roots) that affect soil properties.

Biological reassembly: The patterns and processes that occur in species accumulation and community reorganization following an ecological disturbance.

Biomass: Any quantitative estimate of the total mass of organisms comprising all or part of a population or any other specified unit or within a given area at a given time; measured as volume, mass, or energy; standing crop; standing stock.

Biome: A biogeographical regional ecological community characterized by distinctive life forms and principal plant or animal species.

Biota: The total flora and fauna of a given area.

Biotic: Pertaining to life or living organisms; caused by, produced by, or comprising living organisms.

Bioturbation: The mixing of a sediment by the burrowing, feeding, or other activity of living organisms, forming a bioturbated sediment.

Blast, lateral: Rapid movement of a cloud of hot ash and coarser fragments across the landscape.

Blast area or zone: The area subjected to the effects of a lateral blast, including the tree-removal zone; the blowdown zone of toppled trees; and the fringe scorch zone of standing, dead vegetation, where the hot blast cloud lacked force necessary to topple trees but the hot gas cloud scorched foliage.

Blowdown: Toppling of forest vegetation by the shock wave and winds, including that accompanying explosive volcanic activity.

Blowdown zone (also **Down-tree zone** and **Tree-down zone**)**:** The area where forest is toppled (e.g., by the lateral blast of a volcanic eruption).

Boreal: Pertaining to cool- or cold-temperate regions of the northern hemisphere; the northern coniferous zone and taiga.

Boulder: Sediment particle greater than 256 mm in diameter.

Breccia: A coarse-grained rock composed of large (greater than sand sized), angular, broken rock fragments cemented in a fine-grained matrix.

Bryophyte: Mosses and liverworts.

Chemotrophic: An organism that derives its energy from endogenous, light-independent chemical reactions.

Chlorosis: A discoloration of green plants caused by disease, lack of necessary nutrients, or pollution.

Circumboreal: Distributed around the high latitudes of the northern hemisphere.

Climax: A more or less stable biotic community that is in equilibrium with existing environmental conditions and that represents the terminal stage of an ecological succession.

Cobble: Sediment particle 64 to 256 mm in diameter.

Colluvial deposit: Material transported to a site by gravity.

Colonization: The successful invasion of a new habitat by a species.

Community: Any number of organisms belonging to a number of different species that co-occur in the same habitat or area and interact through trophic and spatial relationships.

Competition: The simultaneous demand by two or more organisms or species for an essential common resource that is or potentially is in limited supply (exploitation competition); the detrimental interaction between two or more organisms or species seeking a common resource that is not limiting (interference competition).

Competitive exclusion: The exclusion of one species by another when they compete for a common resource that is in limited supply.

Connectivity: The interrelationships between different components of an ecosystem.

Conspecific: Belonging to the same species.

Contagious distribution: A distribution pattern in which values, observations, or individuals are more aggregated or clustered than in a random distribution, indicating that the presence of one individual or value increases the probability of another occurring nearby.

Corridor: A generally linear, more or less continuous connection between disjunct habitat patches of the same type.

Coverage: That part of a sampled area covered by a particular plant species or individual plant canopy; typically expressed as a percentage.

Crater: A steep-sided, usually circular or arcuate depression formed by either explosion or collapse of a volcanic vent.

Crepuscular: Active during the twilight hours; of the dusk and dawn.

Cross-pollination: Transfer of pollen from one flower to the stigma of a flower on another plant of the same species.

Cryptobiotic: Organisms that are typically hidden or concealed in crevices or under stones.

Cryptofauna: The fauna of protected or concealed microhabitats.

Debouch: To flow out.

Dacite: A volcanic rock containing 63% to 68% silica (SiO_2); has high viscosity in the molten state.

Debris avalanche: Rapid flowage of large, unsorted masses of rock material, possibly in combination with liquid water or snow and ice, that moves rapidly down slope under the influence of gravity. Collapse of a volcano's flank can trigger massive debris avalanches.

Debris-avalanche deposit: Accumulations of poorly sorted, rock material emplaced by debris avalanches. These can form extensive areas of irregular terrain with hummocks (irregular mounds) up to tens of meters high and undrained depressions, which become ponds. Deposits may be up to several hundreds of meters thick.

Defaunated: Depleted of animals or of symbiotic microorganisms.

Deme: A local interbreeding group; also used loosely to refer to any local group of individuals of a given species.

Diel: Daily; pertaining to a 24-hour period.

Disjunct: Distinctly separate; used of a discontinuous range in which one or more populations are separated from other potentially interbreeding populations by sufficient distance to preclude gene flow between them.

Dispersal: Outward spreading of organisms or propagules from their point of origin or release; one-way movement of organisms from one home site to another; the outward extension of a species range, typically by a chance event.

Disturbance: "Any relatively discrete event in time that disrupts ecosystem, community, or population structure and changes resources, substrate availability, or the physical environment" (White and Pickett 1985). Disturbances are an integral part of most ecological systems. Rather than being catastrophic agents of destruction, many disturbances are normal, perhaps even integral, parts of long-term system dynamics.

Dome: A steep-sided mass of highly viscous lava extruded from a volcanic vent, typically circular in plan view with a rounded or flat top and a spiny surface.

Down-tree zone (also **Blowdown zone** and **Tree-down zone**)**:** The area where forest is toppled (e.g., by the lateral blast of a volcanic eruption).

Ecosystem: The level in the biological hierarchy of organization that includes the biological community and the physical environment in which it occurs.

Edaphic: Pertaining to or influenced by the nature of the soil.

Ejecta: Material that is thrown out by a volcano, including pyroclastic material and lava bombs.

Endemic: Native to and restricted to a particular geographical region.

Epilimnion: The warm upper layer of circulating water above the thermocline in a lake.

Eruption cloud: A cloud of gas, ash, and larger rock fragments blown away from an eruption column by the wind.

Eruption column: The vertical column of gas, ash, and larger rock fragments rising from a crater or other vent during a volcanic eruption.

Eruption: The violent or tranquil processes by which solid, liquid, and gaseous materials are ejected into the atmosphere or onto the Earth's surface by volcanic activity.

Establishment: Growing and reproducing successfully in a given area.

Eutrophication: Overenrichment of a water body with nutrients, resulting in excessive growth of organisms and depletion of the oxygen concentration.

Extant: Already existing.

Facilitation: Enhancement of the behavior or performance of an organism or system by the presence or actions of other organisms.

Fauna: The entire animal life of a given region, habitat, or geological stratum.

Fecundity: Ability to produce offspring.

Flora: The plant life of a given region, habitat, or geological stratum.

Fossorial: An animal adapted for digging or burrowing into the substratum (soil).

Freshet: A great rise in, or sudden overflowing of, a small stream, usually caused by heavy rains or rapid snowmelt.

Fumerole: A vent or opening through which issue steam, hydrogen sulfide, and other gases. Fumeroles may be "rooted," having magma as sources of heat, or "rootless," where the heat is derived from surficial deposits, such as pyroclastic-flow deposits.

Gravel: Sediment particles greater than 2 mm diameter.

Guild: A group of species having similar ecological resource requirements and foraging strategies and therefore having similar roles in a community.

Habitat: The locality, site, and particular type of local environment occupied by an organism.

Herbivory: Feeding on plants.

Heterotrophic food webs: The network of producing and consuming interactions among organisms based on terrestrial leaf litter and other allochthonous food sources.

Hummock: A rounded, conical, steep-sided mound up to several tens of meters in height. At Mount St. Helens, hummocks formed as a result of the debris-avalanche flow.

Hydroperiod: The duration that water remains in a basin, such as a pond or lake.

Hydrophobicity: Water-repellent soil conditions resulting from chemical or physical properties of soil particles.

Hydrophyte: A perennial plant with renewal buds below the water and with submerged or floating leaves; also, any plant adapted to live in water or very wet habitats.

Hygrophilous: Thriving in moist habitats.

Indigenous: Native to a particular area.

Inhibition: Any process that acts to restrain reactions or behavior.

Isopach: A line of equal thickness of deposit (e.g., of tephra on a landscape).

Keystone species: Organisms that disproportionately affect the organization of their communities.

Lagomorph: Animals of the order Lagomorpha, which have two pairs of upper incisors, one behind the other (e.g., rabbits, hares, and pikas).

Lahar: Indonesian term for volcanic mudflows (see **Mudflow**). Lahars can have various types of origin, including hot tephra falling from an eruption column onto snow, melting, and mixing with it to form a cool slurry flowing down a volcano's flank.

Landscape: An area that includes a heterogeneity of land covers.

Lateral blast: Rapid movement of a cloud of hot ash and coarser fragments across the landscape.

Lava: Liquid magma that has been extruded in liquid form onto the surface of the Earth. Most commonly refers to streams of liquid rock that flow from a vent; also cooled, solidified rock of this origin.

Lava flow: A stream of liquid rock that flows from a vent; also cooled, solidified rock of this origin.

Legacy: Physical and both live and dead biological components of a system that persist through a disturbance.

Legacy, biological: Organisms, propagules, and organically derived structures and patterns that remain following disturbances; includes surviving plants, animals, seeds, spores of microorganisms, and plant or animal organic matter, such as large woody debris and above- and belowground litter (e.g., roots) that affect soil properties.

Life form: Categories of plants that have similar positions of their dormant buds relative to the surface (e.g., trees and shrubs).

Limnology: The study of lakes, ponds, and other standing waters and their associated biota.

Littoral zone: The shore of a lake to a depth of about 10 m.

Magma: Naturally occurring hot, mobile rock material, generated within the Earth and capable of intrusion within the Earth or extrusion onto the Earth's surface.

Magmatic: Process or feature related to magma, such as processes driven by gas released from magma.

Metamorphosis: A marked structural transformation during the development of an organism, often representing a change from larval stage to adult.

Metapopulation: A group of partially isolated populations of one species that are distributed across a landscape in such a way that the spatial arrangement of the populations can influence the long-term persistence of the species. The number of populations present at any given time is governed by the spatial structure of suitable and unsuitable habitat, as well as by a balance between local extirpation and colonization.

Microsite: Locations where physical conditions are amenable for establishment of organisms; usually small areas that occur in a matrix of relatively inhospitable habitat.

Mudflow (see also **Lahar**)**:** Flowage of water-saturated earth commonly moving rapidly and with a high degree of fluidity. Mudflow deposits are generally thin (up to 1 m in thickness) and moderately sorted sediment, covering inundated floodplain areas and vegetation. Lahar is a more specific term for mudflows of volcanic origin.

Mudflow deposits: Accumulations of sediment transported by mudflows. The deposits are generally thin (less than a few meters in thickness) and moderately sorted sediment, covering inundated floodplain areas and vegetation.

Mutualism: A symbiosis in which both organisms benefit, frequently a relationship of complete dependence; interdependent association.

Mycorrhizae: A mutualistic association between the fine roots of plants and specialized soil fungi; the fungi provide the plant with certain nutrients, such as phosphorus and water, and in return receive simple carbohydrates from the plant.

Neoteny: Produced by retardation of somatic development, such that sexual maturity is attained in an organism retaining juvenile characteristics.

Nitrogen fixation: The reduction of gaseous nitrogen to ammonia or to another inorganic or organic compound by microorganisms.

Oligotrophic: Having low primary productivity; pertaining to waters having low levels of the mineral nutrients required by green plants. Used of a lake in which the hypolimnion does not become depleted of oxygen during the summer.

Otoliths: Granules of calcium carbonate in a vertebrate's inner ear.

Oviposition: The act or process of depositing eggs.

Parasitism: An obligatory symbiosis between individuals of two different species in which the parasite is metabolically dependent on the host and the host is typically adversely affected but rarely killed.

Pebble: Sediment particle 4 to 64 mm in diameter.

Perennial: A plant that persist for several years with a period of growth each year.

Perrenation: The survival of plants from year to year with an intervening period of reduced activity.

Phoenicoid: Fire-following; heat-loving.

Phreatic eruption (explosion): A steam explosion produced when water comes in contact with hot volcanic rocks.

Physiognomy: The appearance of vegetation as determined by the life form of the dominant plants.

Phytomass: Plant biomass; any quantitative estimate of the total mass of plants in a stand, in a population, or within a given area at a given time.

Plankton: Passively floating organisms in a body of water.

Plethodontid: A salamander that is a member of the lungless salamander family, Plethodontidae.

Population: The level in the biological hierarchy that consists of co-occurring groups of individuals of one species.

Predation: The consumption of one animal (the prey) by another animal (the predator); also used to include the consumption of plants by animals and the partial consumption of a large prey organism by a smaller predator.

Primary succession: An ecological succession commencing on a substrate that does not support any organism or contain any biological remnants from previous life.

Propagule: Any part of a plant that, when separated from the parent plant, can give rise to a new individual (e.g., seeds and root or shoot fragments).

Protocooperation: An interaction between or among animals in which each participant benefits more or less equally.

Pumice: Light-colored, low-density, frothy, glassy volcanic rock formed by expansion of gas in erupting lava. Fragments range from ash to large blocks.

Pyroclastic: Pertaining to fragmented (clastic) rock formed by volcanic explosion or ejection from a volcanic vent.

Pyroclastic flow: Rapid movement of extremely hot (often more than 700°C), turbulent gases and fragmental material across a land surface from a volcanic vent. The denser, basal part of a pyroclastic flow hugs the ground and follows topography, moving with great force and speed (up to 200 km/h). A lower-density ash-cloud phase extends well above the ground surface and moves more slowly.

Pyroclastic-flow deposits: Accumulations of sediment transported by pyroclastic flows, which commonly include basal deposits of poorly sorted, gravelly deposits topped by finer-grained, layered deposits from the ash-cloud phase of transport.

Ranid frogs: Belonging to the family Ranidae, including the red-legged and Cascades frogs.

Reassembly, biological: The patterns and processes that occur in species accumulation and community reorganization following an ecological disturbance.

Redds: Sites where fish deposit eggs in sediment on a streambed.

Residuals: Surviving individual organisms, vegetative tissue that can regenerate, seeds, and components of the microbial and fungal soil community.

Rhizome: An underground stem that bears buds in axils of reduced scale-like leaves and serves as a means of perrenation and vegetative propagation.

Rhizosphere: The soil immediately surrounding plant roots that is influenced structurally or biologically by the presence of such roots; root zone.

Rhyolite: A volcanic rock containing greater than 68% silica (SiO_2); has very high viscosity in the molten state.

Richter scale: A logarithmic scale for measuring the magnitude of seismic disturbances; the range of magnitudes includes 1.5 for the smallest detectable tremor and 8.5 for a devastating earthquake.

Rill erosion: The formation of many small (several centimeters in width and depth) channels by erosion caused by surface-water movement.

Riparian zone: Pertaining to, living on, or situated on the banks of streams and lakes.

Safe sites: Places in an environmentally unfavorable landscape with physical conditions that favor survival and establishment of organisms.

Sand: Sediment particles 0.002 to 2 mm diameter.

Scorch zone (also **Seared zone** and **Standing-dead zone**)**:** The area of standing, dead trees around the perimeter of the blowdown zone, where hot gases and rock fragments in the blast cloud killed foliage and caused other injury to the trees, leading to tree death.

Seared zone: The area of standing, dead trees around the perimeter of the blowdown zone, where hot gases and rock fragments in the blast cloud killed foliage and caused other injury to the trees, leading to tree death.

Secondary succession: The gradual changes that occur over time in biological and physical conditions after a site has been disturbed, where some legacies of earlier ecological systems remain.

Seiche: A wave that oscillates in lakes, bays, or gulfs from a few minutes to a few hours as a result of seismic, atmospheric, or other disturbances.

Sere: The sequence of communities that replace each other over time during succession (adj., seral).

Silt: Sediment particles less than 0.002 mm diameter.

Sink habitat: A locale in which net immigration of a population at the site remains constant or even grows at a rate that differs from the rate of immigration. Sink habitats are associated with source areas from which the organisms migrate.

Snowmelt: Runoff produced by the melting of snow and ice.

Source habitat: A locale that supports a net increase in population both within itself and within another locale. (See **Sink habitat**.)

Species: The level in the biological organization that consists of organisms capable of interbreeding.

Standing-dead zone (also **Scorch zone** and **Seared zone**)**:** The area of standing, dead trees around the perimeter of the blowdown zone, where hot gases and rock fragments in the blast cloud killed foliage and caused other injury to the trees, leading to tree death.

Stenothermic: Tolerant of a narrow range environmental temperatures.

Succession: The process of gradual replacement of one species population by another over time and the concurrent change in ecosystem properties after a site has been disturbed. The concept can be extended to the replacement of one kind of community by another, the progressive changes in vegetation and animal life that may culminate in dominance by a community that is stable until the next disturbance. Succession refers to changes that occur over 1 to 500 years and not to seasonal changes in populations and communities.

Tephra: Fragmental rock material ejected from a volcano during an eruption and deposited by airfall. It is typically composed of ash (less than 4 mm in diameter), lapilli (4- to 32-mm particles), and blocks (angular stones larger than 32 mm).

Thermocline: A horizontal temperature-discontinuity layer in a lake in which the temperature falls by at least 1°C per meter of depth.

Thermophilic: Thriving in warm environmental conditions; used of microorganisms having an optimum growth rate above 45°C.

Tolerance: The ability of an organism to endure extreme conditions; the range of an environmental factor within which an organism or population can survive; also, the situation where organisms best able to tolerate prevailing conditions are favored.

Tree-down zone (also **Blowdown zone** and **Down-tree zone**)**:** The area where forest is toppled (e.g., by the lateral blast of a volcanic eruption).

Trophic levels: The sequence of stages in a food chain or food pyramid, from producer to primary, secondary, or tertiary consumer.

Tuff: A compacted pyroclastic deposit of volcanic ash.

Vagility: Ability to move about freely.

Vent: Opening in the Earth from which volcanic material is emitted.

Volant: Adapted for flying or gliding.

Wildlife: Animals and vegetation (but especially animals) living in a natural, undomesticated state.

Index

T